Fundamentals of Materials Science

Eric J. Mittemeijer

Fundamentals of Materials Science

The Microstructure–Property Relationship
Using Metals as Model Systems

Second Edition

 Springer

Eric J. Mittemeijer
Max Planck Institute for Intelligent Systems
Institute for Materials Science
University of Stuttgart
Heisenbergstr. 3
D-70569 Stuttgart
Germany

ISBN 978-3-030-60058-7   ISBN 978-3-030-60056-3   (eBook)
https://doi.org/10.1007/978-3-030-60056-3

1ˢᵗ edition: © Springer-Verlag Berlin Heidelberg 2010
2ⁿᵈ edition: © Springer Nature Switzerland AG 2011, 2021

This Springer imprint is published by the registered company Springer Nature Switzerland AG
The registered company address is: Gewerbestrasse 11, 6330 Cham, Switzerland

"Schrijven is het heilige in de alledag."

Willem Brakman (1922–2008) in:

De Gifmenger (2003)

"To be sure, I never expected the influence of vacancies to be so large."

Roman Smoluchowski (1910–1996) in: Trans. AIME (1947)

For Marion

Preface to the Second Edition

This second edition was long in the making. Of course I was pleased by the request of the publisher, quite some years ago now, to prepare a second edition, since the book evidently is used and liked by students, but also by researchers, as I found out. After publication of the first edition, I noticed, also on the basis of remarks by users of the book, some inaccuracies, partly printing errors. Of course, I was eager to correct these in a next edition.

However, I felt that correcting inaccuracies was not enough; much more revision appeared desirable. In the Preface to the first edition I have outlined what is strived for with this book: "to present a treatise on the basics of materials science that has a fundamental character", and that can "be used also in the beginning of a materials science and engineering study". As a consequence "the material contained in the book is much more comprehensive than what can normally be offered in an introductory course on materials science" and thus "the book will be used by the reader also at later stages of his/her study, because a fundamental background may be quickly grasped on the basis of what this book offers". To draft text under such boundary conditions is no easy task. At a number of places in the book distinct extension and some restructuring was possible and would lead to quality improvement of the book. Moreover at many places some minor reformulation was thought to be desirable. As long as I was in active duty, it appeared that I could not find the time and energy (again) to perform such extensive book-writing task. Thus, as it happened, I found myself taking on this project only after formal retirement.

More or less serious modifications have been applied to all chapters of the previous edition. Most drastic revision and extension has been applied to Chap. 8 ("Diffusion"). For example, new sections have been included which a.o. now provide comprehensive discussion of the Kirkendall effect on the basis of the Darken treatment (deliberately left out in the first edition (see the Preface), but now included, also as a response to reader comments) and the role of ("self-") stress. In Chap. 6, a new section is devoted to "supermicroscopy", in particular as recognition of recent developments of microscopy beyond the resolution limit. No doubt that these additions to the book present material normally *not* offered in a freshman course. However, as consequence of the imposed constraints, described explicitly in the above paragraph, the treatment is always on a basic, fundamental level and, I believe, easily understandable, thus providing useful preparation for reading and studying an advanced textbook on the topic concerned. At the same time, it remains possible to select straightforwardly those sections evidently corresponding to introductory courses on materials science.

New *Intermezzi* and a further *Epilogue* and a number of partly extensive footnotes have been added, fully in the spirit of the first edition (see the Preface).

I am very grateful to a number of colleagues who provided me with their comments on the first edition. Here I want to especially mention Prof. Yongchang Liu (Tianjin University) who not only hinted at a number of deficiencies in the first edition, but also presided over the translation of the book in Chinese.

My former coworker, Dr. Ralf Schacherl (University of Stuttgart), modified quite a number of the figures of the first edition and also prepared the new figures in this second edition. His assistance and loyalty throughout the years are greatly appreciated.

Unfortunately C., my companion on my writing desk while I was drafting the first edition, has died soon after the first edition was published and thus the present endeavour has been a more solitary occupation…

My wife Marion, having had no easy life in the last years and rightly expecting that time for relaxed life without professional duties would be forthcoming after both our retirements, yet accepted my embarkation on this project. As the very last words written for this book, I want to express my deepest gratitude for her unrelenting understanding and critical support, already now for a large part of my life. This book is dedicated to her.

Stuttgart, Germany   Eric J. Mittemeijer
March 2020

Preface to the First Edition

Science and Engineering; Metals and Other Materials; the Microstructure

The German and Dutch languages have single, almost identical words for the field of "Materials Science and Engineering": "Materialkunde" and "Materiaalkunde", respectively. Thereby applications of materials serving mankind and the development of the corresponding basis of knowledge and understanding of nature have been indicated in a unified way. The intertwined nature of science and engineering is a decisive characteristic of this multidisciplinary field. Yet, as its title indicates, this book is devoted to materials *science* and much less to materials *engineering*. The reason for this restriction is twofold: firstly, a theoretical background is a prerequisite for any engineer to be successful, and thus any study in this field must start with providing a scientific basis, and, secondly, including a coverage of the synthesis and treatment of materials in practical applications would have made this book either too bulky or, to keep the amount of information offered manageable, too superficial.

The implication from the above is that it is intended to present a treatise on the basics of materials science that has a fundamental character. This may seem an impossible undertaking, as at the same time the book is meant to be used also in the beginning of a materials science and engineering study. For a start it implies that one largely has to abandon usage of mathematical techniques the reader is not familiar with yet. It is my conviction that this does not impede transmitting physical and chemical understanding. Of course, then some important results of advanced theories have to be introduced and accepted without proof, but this is no serious obstacle in order to develop a sound basis of the basics of the field. On the contrary, in this way one is best prepared for later to absorb separate, advanced courses on, say, quantum mechanics and materials thermodynamics and kinetics. If this book realizes these aspirations sufficiently satisfactorily, then this book will be used by the reader also at later stages of his/her study, because a fundamental background may be quickly grasped on the basis of what this book offers. Also therefore the material contained in the book is much more comprehensive than what can normally be offered in an introductory course on materials science. Or, phrased in another way, the book should provide useful preparation for reading and studying advanced textbooks on topics as "chemical bonding", "diffusion", "lattice defects", etc., dealt with here in, only, chapters. There is no lack of such textbooks. But I do feel that there is a need for a book as the present one in the light of my experience with existing introductory texts for the field of materials science which I consider as often to be too superficial and too phenomenological of nature.

Adopting the above philosophy I have made some, sometimes difficult choices in writing this book. This can be illustrated by what has been left out. For example, I did not include detailed quantitative discussions on dislocation dynamics (Chap. 5), the derivation of phase diagrams from the dependence of the Gibbs energy on composition (Chap. 7), the Kirkendall effect and the corresponding Darken treatment (Chap. 8) or the (intrinsic) elastic anisotropy (Chap. 11). Those topics which do have been treated in this book invariably are of paramount importance to the materials scientist and have been dealt with in a fundamental way to an extent widely surpassing what can possibly be presented in a freshman's course (e.g. the chapters on "Crystallography" (Chap. 4), "Phase Transformations" (Chap. 9) and "Mechanical Strength of Materials" (Chap. 11)). This does not impede at all the use of this book in a beginner's course already, and especially makes this book useful throughout an entire undergraduate and even graduate study as an introduction and solid background against which more detailed monographs and specialized texts can be studied. Such is the task of this book.

It is claimed here to offer "fundamentals of materials science", and thus this is a book about materials phenomena rather than materials. It has to be admitted that in the text, some apparent emphasis has been laid on metals as class of materials. This should then be discussed as follows.

An obvious, not very important but not to be ignored, observation is the following. The great majority of the naturally occurring elements in the Periodic System (92) are metals; only a limited number of non-metal elements exist (about 15). It is true that a few of these non-metals are of extreme importance for life on earth (C, N, O and H). It is also true that life of man would not have the slightest resemblance with how it is now were it not for the application of metals. It may also be relevant to remark here that the category of metals is in fact even much larger than one may naively expect on the basis of the classical division of the elements given above: any substance may be made *metallic upon densification*. Thus, hydrogen can be made metallic under high pressure and silicon becomes metallic upon melting. The background of this behaviour, i.e. why this is so, is discussed in Sect. 3.5 in this book. This leaves unimpeded that other categories of materials, man-made or not, as silicon-based components in microelectronics, ceramics, polymers and biomaterials, are crucial materials as well. However, and now the cardinal argumentation follows, understanding of the fundamental properties of materials is largely independent of the type of material considered. The knowledge and science of crystallography, diffusion, the thermodynamics and kinetics of phase transformations, etc., are *not* confined to a specific class of materials.

Materials science has developed as a discipline from the time that metals were considered as the perhaps most important materials in the world (see R.W. Cahn, *The Coming of Materials Science*, Pergamon (Elsevier Science), Amsterdam, 2001). This view needs no longer be held, but it explains that our knowledge on materials behaviour has been developed with metallic materials as the type of material that was subject of investigation. Material classes, metals, ceramics, polymers, biomaterials, etc., most distinctly differ particularly in their way of synthesis (a topic not dealt with at all in this book) and applications (polymers and biomaterials serve as examples). However, their microstructure–property relationships are predominantly based on the same concepts. Historically, such research on microstructure–property relationships was done first for metals. It should be recognized that concepts developed first for metals are needed and used now to characterize and explain the behaviour of newer classes of materials, as already demonstrated for ceramics, semiconductors and also polymers. This remains true observing strikingly specific properties associated with a certain material class; rubber elasticity serves as an example (see Sect. 11.6). New and future classes of materials will be dealt with on the basis of the same body of knowledge. The complexity of the materials classes appears to increase inversely with their "age". The above leads to the conclusion that there is another reason why "metals" as a material class are of special importance to the materials scientist: *metals provide the simplest class of materials where one can best start to investigate the concepts behind material behaviour.*

Hence, in a book dedicated to materials science and less to engineering, as indicated above, it is justified and understandable that of the existing material classes the class of metals is emphasized, just as a simple consequence of most fundamental research on material behaviour having been done and still being done on metallic materials.[1] This does not at all obstruct the transfer of fundamental, general knowledge on materials properties, which is the goal of this book.

---

[1]As an anecdote, I here recall that decades ago I attended a conference in London where A. H. Cottrell, the author of a famous booklet on "Theoretical Structural Metallurgy" that I have cherished until today, presented a lecture on the role of metals in society. He showed, on the basis of sound references, that if one would have believed the predictions of those who said that polymers would take over the role of metals in the automotive industry, that one would have driven plastic cars already in the seventies (of the twentieth century, I have to add now). The message of his remark was: those who advocate the application and predict future importance, if not dominance, of a certain, new material class, extrapolate the properties of these materials into the future, but do as if the "classical" materials, against which the new materials are compared, are not the subject of on-going research and further development. In this sense new materials are chasing moving targets. Concerning the example discussed here: the emergence of HSLA (high strength low alloy) steels was ignored or not observed by the protagonists of plastics. This does not mean that plastics eventually cannot take over the role of metals in cars, but even today that has happened only partly. The point is: each time a new material emerges (quasicrystals, high T_c superconductors, carbon nanotubes, graphene, etc.) one is tempted to overexaggerate its possibilities for application. One should not forget, as a warning signal, that the, for a long time with great emphasis, much promoted idea of the development of the fully ceramic combustion engine has been buried, as it seems once and for all. A critical and yet open attitude towards any new, sensational presentation of a new material is in order.

The notion perhaps most typifying the field of materials science is: the *"microstructure of a material"*. The microstructure of a material comprises all aspects of the atomic arrangement of the material that should be known in order to understand its properties. Confining ourselves to mostly solid, crystalline materials, the microstructure not in the first place concerns the idealized crystal structure, but in particular the *imperfections*, as the compositional inhomogeneity, the amount and distribution of phases, the grain size and shape and their parameter-distribution functions, the grain(crystal)-orientation distribution (called texture or preferred orientation), the grain boundaries and surface, the concentrations and distributions of defects as vacancies, dislocations, stacking and twin faults, and, not least, distortions as due to strains/stresses, etc. (A special feature of this book is the chapter on "Analysis of the Microstructure; Analysis of Lattice Imperfections" (Chap. 6)). As may be anticipated from this still incomplete listing, the microstructure to a very large extent determines the properties of a material. The central issue of materials science may be formulated as: to develop *models that provide the relation between the microstructure and the properties*. Such an integrated and bridging the length scales (from micro to meso to macro) approach is the feature distinguishing materials science from merely solid-state physics and solid-state chemistry. If this book succeeds in conveying also this message, I, as author, can be more than satisfied.

Science is not an abstract activity performed by flawless gods. At a number of places side remarks, as footnotes or *"intermezzi"* and *"epilogues"*, have been inserted which, for example, may refer to an illuminating historical development or point at an existing controversy. This has been done in an effort to indicate what the process of science involves and that insight often is the result of a long struggle and not of unrestrictedly eternal value.

Stuttgart, Germany   Eric J. Mittemeijer
August 2010

A Word of Thanks and Homage

This book has been written in a period of time stretching over more than five years. It emerged, at least partly, as the outcome of courses that I have taught on in particular the basics of materials science (a freshman course), and also advanced courses on materials thermodynamics, on microscopy and diffraction, on surface engineering and on solid state kinetics, in Delft and in Stuttgart. Inevitably, my own research work in the field of materials science and engineering has driven me to my understanding of the field as also represented in this basic book. My dual job, as a Professor at the university and as a Director at a research institute "par excellence", has only strengthened my conviction that, in order to be an outstanding teacher for students, one has to be an excellent, active researcher; teaching forces you to think about the roots of knowledge, already only because of fundamental and sometimes brilliant questions by students. To respond to such questions, the mentality of an enthusiastic, active researcher at the frontiers of science is essential. I am greatly indebted to my undergraduate students and graduate, Ph.D. students of the past and present: by their interaction with me I have learned immensely.

My own, initiating involvement with materials science is connected with two names in particular. Professor W. G. Burgers (see also Sects. 5.2.3 and 10.2.1) taught, as an Emeritus, for the last time (it must have been in about 1970/1971) a course on the physical chemistry of the solid state that I attended. At the time, he was of course a rather old man, but, while lecturing, still had the enthusiasm and dynamic aura of a young man (which was the more remarkable in view of his physical handicap), presenting in this lecture course basic, established knowledge, interspersed with anecdotes and research results obtained by others and himself, and in particular thereby was capable to transfer to the audience his love for science. This lecture course has brought about my decision to turn to materials science, for my master's project and, later, my Ph.D. project, not realizing that it would be the playing field for my whole career. Moreover, the sketched philosophy behind Burgers' lectures and, I believe, his style, have influenced largely my own lecturing until today and thereby also this book.

Professor B. Okkerse, already for many years retired successor of Prof. Burgers, has been my Ph.D. supervisor. Although being a gifted experimentalist (see

Sect. 8.6.1), he is even more an extremely well organized man who led and cared for his team in an impeccable manner. I have learned from him that one can manage multiple jobs at the same time: the discipline required to write this book, through the years, while not neglecting my teaching and research supervising tasks, and the way I lead my research department, in many respects are derived from my experience with him.

This basic textbook unavoidably also testifies to the research I conducted with colleagues and coworkers through the years, by the inclusion of many illustrations and examples. I find these examples appropriate, but do not claim that other suitable examples could not have been chosen from the existing literature as well: the reason I largely drew from my own research work simply is that I know these examples best.

I am grateful to all those who assisted me in the course of this book-writing project. Here I want especially mention (in alphabetical order): Dr. P. van Aken, Dipl.-Ing. E. Jägle, Dr. A. Leineweber, Dr. W. Sigle and Dr. U. Welzel, who all read parts of (earlier versions of) the text and provided me with useful, critical remarks. Dr. A. Leineweber made a first draft of a major part of the chapter on crystallography and discussed with me successive versions of the chapter. Dipl.-Ing. J. Aufrecht carried out the enormous task of preparing most figures, and drafting many of them on the basis of only sparse suggestions by me. Of course, any errors remaining in this book (and there will be) are my sole responsibility.

The almost last sentences I write here are devoted to my wife (and secretary), Marion. We met and married rather late in life. The time left for life together may thus be relatively limited. Writing a book as this one, next to professional duties already requiring investment of an unusual amount of daily time, implies that this has been predominantly done during evenings, nights and weekends. This is a burden for any relationship and is especially troubling if you feel that time is running. Marion has accepted this, because she understood how important finishing this self-given assignment was to me. I cannot express in words how grateful I am for her understanding, acceptance and patience: she made this book possible. As long as I am in active duty, I will not again embark on a project as this one. Indeed, it seems that many scientific textbooks and monographs are written after retirement...

Finally, there is C. She spends many hours in the evenings and nights on my writing desk accompanying my struggling with purring as background sound and leaving her black hairs between my papers and manuscripts, later found by my students and coworkers who certainly misinterpret them. This book is also a memory to her.

Stuttgart, Germany   Eric J. Mittemeijer
August 2010

About the Author

Eric J. Mittemeijer was born in 1950 in Haarlem, The Netherlands. He studied "chemical technology", with specialization physical chemistry, at the Delft University of Technology and acquired his "ingenieur (= Ir.)" degree (comparable to a M.Sc. degree) in 1972 and his Ph.D. degree in 1978. From 1985 till 1998, he was full Professor of solid-state chemistry at the Delft University of Technology. From 1998 till 2017, he was Director at the Max Planck Institute for Metals Research (later renamed in Max Planck Institute for Intelligent Systems) in Stuttgart in conjunction with a full Professorship in materials science at the University of Stuttgart. He was Dean of the Study Course Materials Science of the University of Stuttgart and Speaker of the International Max Planck Research School on Advanced Materials. He has (co)authored about 700 scientific papers in international scientific journals, published a number of books and has received a number of honours for his scientific work. He can be contacted at: e.j.mittemeijer@is.mpg.de.

Contents

Chapter 1
Introduction

1.1 The Notion Material

"Materials" can be said to emerge by human action: a material is a *substance with a present or an expected future application* for mankind.[1] So not all substances are materials.

Examples of materials with current applications:

- wood
- steel
- nylon

Examples of materials which, at the time of writing of this book, are considered as having potential for future applications:

---

[1] This definition avoids rambling, unspecific discussions as in Materials Research Bulletin, 37 (2012), 95.

© Springer Nature Switzerland AG 2021
E. J. Mittemeijer, *Fundamentals of Materials Science*,
https://doi.org/10.1007/978-3-030-60056-3_1

- high-T_c superconductors
- graphene and fullerenes (as carbon nanotubes and buckyballs).[2]

One can distinguish between *natural* materials and *man-made* materials. Natural materials are found on earth in the state as they will be used, more or less. Wood or gold and also copper (in their elementary, native form!) serve as examples. A man-made material is the product of some process carried out by humans. It goes without saying that steel provides a classical example; of more recent times TiN-layers, used for wear protection and decorative purposes (they exhibit a gold shining), can be mentioned out of a myriad of new materials.

It has often been attempted to identify *material classes*. The most common subdivision is a historic one, which is still used today:

- metals (Fe, Cu, ...; for more information about what is a metal, see further);
- ceramics (metal–nonmetal combinations: like metal oxides, metal carbides and metal nitrides);
- polymers (consisting of mainly C and H and characterized by the enormous size of their molecules).

Sometimes semiconductors (intermediates between conductors and isolators, as follows from their name) are mentioned separately.

The revolutionary development of new materials nowadays has led to proposals for definition of separate material classes, e.g. composites, biomaterials and biomimetic materials, quasicrystals, carbon nanotubes, etc.

The list of material classes can be made endless. Every subdivision is problematic, as upon close inspection the borders between the classes as a matter of fact are diffuse. This becomes immediately clear if one tries to give sharp, exact definitions of what a metal, ceramic, polymer is. The descriptive remarks already given above are in any case insufficient.

As an illustration in the following, a tentative description of a metal is given.

---

[2] Graphene is an *one-atom thick* sheet of carbon atoms arranged, defect-free, in a two-dimensional network of hexagons; as such it is the basic building unit of graphite, which consists of a specific three-dimensional stacking of graphene layers. A material thinner than graphene is inconceivable. It is very strong (harder than diamond, another, three-dimensional arrangement of carbon atoms; cf. Chaps. 3 and 4) but it can be bent easily. Its electrical conductivity at room temperature is the largest known of all solids. The graphene sheet can be rolled up and wrapped up such that three-dimensional structures, fullerenes, are created: cylindrically shaped (carbon) nanotubes and soccer-ball-symmetry-like structures called buckyballs (the most well-known representative is the C_{60} molecule, buckminsterfullerene: a sphere composed of interlocking hexagons (20) and pentagons (12)). These materials appear to have unusual chemical and physical (mechanical, optical, thermal and electrical) properties, suggesting important applications. One reason to refer to graphene and fullerenes at this place, in the introductory chapter of this book, is the recognition that these materials are examples of "new" materials which are in fact "old" materials, which have been around us for a long time already, notably in soot (fullerenes) and pencil mark (graphene), but they have not been noticed until (very) recently: fullerene in 1985 (buckyball) and 1991 (nanotube), and graphene in 2004. Not always "new" materials are discovered by novel and possibly revolutionary synthesis routes.

1.2 The Notion Metal

Metals show good electrical and thermal conductivity (this leads to *functional* applications like electricity cables), and they posses high mechanical strength in combination with good toughness (this leads to *structural* applications like tools). Further metals often show the typical "metallic" shine or lustre. Perhaps the best definition of the type considered here for a metal is by its temperature coefficient of its electrical resistance, for metals show an electrical resistivity that typically increases distinctly with temperature. Also. the thermal conductivity of a metal decreases with increasing temperature.

It has been tacitly assumed in the above that in fact aggregates of metal atoms are considered. If an individual metal atom would be considered, the description of properties given just does not apply. Apparently, we were concerned with describing what could be called "the metallic state of matter".

Yet, metals can be defined on the basis of the individual atoms. To this end, we turn to the Periodic Table of the elements, also called Periodic System (see Fig. 2.9). From about 1860 till 1870 Mendeleev, Meyer and others discovered the Periodic Law: a periodic reoccurrence of typical properties with increasing atomic mass. The elements were arranged for increasing atomic number in periods (rows; horizontal) and groups (columns; vertical) such that the elements of a group have similar properties. On this basis, it was possible to indicate where elements were missing (i.e. were undiscovered yet) in the Periodic Table and what properties these elements would have. It took fifty more years of research before the underlying reason for the occurrence of the Periodic Table was established: the electronic structure of the atoms (say, the "arrangement" of the electrons in the atoms) has specific features leading straightforwardly to the regularities expressed by the Periodic Table (see Chap. 2). It follows that most of the naturally occurring elements (about 90) are metals; only about 15 elements can be considered as outspoken non-metals (see Fig. 2.9). Elements of intermediate character are B, Si, As, Te, ...

1.3 Models and Experiments

Models play an important role in materials science. They provide connections between the *structure* and the *properties* of a material, and this is what materials science is all about!

A model is not the reality. A model is a construct of our thinking that provides an explanation of certain observations made on a certain system. Thus the model describes what we already know ("experiences"), but did not "understand", i.e. a unifying, theoretical concept lacked. At the same time, the model allows to predict what can happen under certain conditions for the system for which the model was devised. As long as the results of such experiments are in agreement with the (predictions of the) model, the model is considered as a satisfactory description of nature.

If this is no longer the case, the model must be modified or a new model has to be developed.

In many cases models known to be inadequate to explain the results of all experiments are used yet. For example, it is well known that mass in atoms is very inhomogeneously distributed: the atom (radius of the order 10^{-10} m) is composed of a tiny nucleus (radius of the order 10^{-14} m), containing practically all mass, surrounded by a number of electrons representing a negligible amount of mass (see Chap. 2). Yet, in many cases we conceive atoms as massive (ping-pong) balls, for example in order to discuss crystal structures (see Sect. 4.2): we tend to adopt the simplest model for explaining those observations that are under discussion.

And indeed, a relatively simple model, not to be expected to expose all details of an observed phenomenon, can in many cases, if not always, be much more appropriate to exhibit and explain the principal workings of nature than detailed "ab initio" calculations departing from the interactions between elementary particles. Too much detail may conceal rather than reveal the inner structure of our material world.

The limitation of models as a means for describing nature may be demonstrated with a simple example.[3] One can observe that a coffee machine dispenses a cup of coffee after insertion of a coin. There are at least two possible explanations for this observation, i.e. two models of reality can be given. The machine could consist of system of motors, pumps, gear-wheels, etc., which operate with electricity. Or some dwarf could live inside the machine who is trained to prepare the cup of coffee once the coin is inserted. It is not possible to decide definitively if a model is true, even if it explains all observations to date, until one is able to open the machine and investigate the interior. This "opening" of the system is something we normally cannot do and hence we do develop models! Experiments are therefore crucial to determine whether one model is nearer to reality than another model. For the funny example considered one could propose to stop feeding the machine with electricity. If then no longer coffee is provided upon inserting the coin, the advocates of the "motor/pumps/gear-wheels" model would immediately say that their model has been validated by this experiment. However, the advocates of the "dwarf" model could suppose that the dwarf needs electricity to see what he/she does and just refuses to operate in the dark. This may illustrate that it is not easy to settle debates regarding the appropriateness/validity of one or the other model. Indeed, the history of science is full of such long-standing debates; as a matter of fact this is one of the most characteristic traits of scientific research activity.

The "uncertainty principle" of quantum mechanical theory (see Chap. 2) implies that we never will be able to "open and look into" an individual atom. So in any case our description of the atom will forever remain a model, that at best explains all available experimental data, but of which it never can be claimed that it is truly identical with nature.

---

[3] This example is not original; it has developed from one I read, as a student, in a book (D.K. Sebera, *Electronic Structure and Chemical Bonding*, Blaisdell Publishing Company, New York, 1964), and which has not left me ever since.

1.4 Bridging Length Scales

A cardinal assignment in materials science is to bridge the length scales, so that on the basis of knowledge on the atomic scale, the properties of macroscopic specimens can be well explained: understanding of the forces acting between the smallest building units of matter does not at all imply that macroscopic behaviour can be described.

The chemical bonding of atoms is derived from fundamental physical, electronic interactions. Aggregates of atoms, e.g. molecules and crystals, can thus develop. This can lead to condensed microstructures composed of many aggregates (crystals) which can contain many defects. At this scale the thermodynamic and kinetic approaches govern the description of material behaviour. At last the world of engineering is entered where macroscopic averages of the properties of material workpieces, as its mechanical strength, are wanted. These sentences serve to introduce the various length scales as follows:

(i)   *The atom*
      Two types of models will be considered: the electronic structure of the atom, as a nucleus with surrounding electrons (e.g. see Fig. 2.2), and the atom as a massive ball (see Sect. 1.5).

(ii) *The arrangement of atoms in space*
      Atoms can be arranged chaotically, i.e. randomly (as in (ideal) gases), or strictly regularly, exhibiting, for example translational, symmetry (as in crystals; the whole crystal is built up from unit cells in a massive arrangement; see Chap. 4) or, on the basis of an initially fully random arrangement, with a preference for unlike atoms to be neighbours leading to short range order (as in so-called amorphous solids).

(iii) *Specimens composed of crystals having defects (crystal imperfection)*
      Polycrystalline specimens composed of many crystals (grains) have a lot of grain boundaries; the individual crystals have structure defects as dislocations (see Chap. 5) and contain internal strain fields. These defects in the specimens influence their properties greatly: the mechanical strength (see Chap. 12), the diffusion and phase-transformation properties (see Chaps. 8, 9 and 10), the corrosion resistance, etc.

(iv) *Workpieces*
      Material bodies as, for example, machines, are constituted of parts, often made of various different materials. The safe design of such compounded, macroscopic workpieces requires a knowledge base of the macroscopic behaviour of the materials utilized.

The different length scales to be considered by the materials scientist may thus also be summarized crudely and tentatively as follows:

- the *nanoscale* (nm range), where an atomistic approach is needed;
- the *microstructure scale* (nm–μm–mm range); understanding material behaviour at this intermediate length scale, where the outcome of the concerted action of

individual units, as e.g. dislocations and grains, is experienced, is the core of materials science (and thereby provides its clearest distinction with fields as solid state chemistry and physics, where the importance of the microstructure for material application is not considered);

– the *macroscale* (mm–m range, and beyond) where compounded components as parts of engineering constructions are applied. Only at this length scale and for the description of certain properties the material can be considered as a "continuum".

Separate models and approaches are necessary for the different length scales, e.g. Molecular Dynamics and Monte Carlo methods can be applied for atomistic simulations on the nanoscale; microstructure models can be developed for understanding dislocation mobility and grain interaction upon external loading; and structural mechanics models, possibly expressed in finite element algorithms, can be used to describe the mechanical properties of macroscopic workpieces. The challenge for the materials scientist to traverse the transition regions between the different length scales can also be described as the endeavour to provide the transition from local (individual atom/defect) to non-local (polycrystalline, macroscopic) descriptions.

The distinction between materials science and materials engineering is gradual and not outspoken, as may be felt after having read the above. Yet, it can be said that this book concentrates on materials science rather than on materials engineering, as the focal point of our interest will be the microstructure.

1.5   Understanding of Nature, the Role of Science; Magic, Discovery and Models

At the end of this chapter, some personal notes reflecting on what has been said above may be in order.

Magic precedes knowledge. In a book about materials science this cannot be illustrated better than by a citation, translated into English, from an old German book devoted to the heat treatment of metals. As elucidation: metals are often annealed and cooled (often fast, then the cooling is called "quenching") to generate certain microstructures which are associated with favourable properties, as a high hardness (see Chaps. 9 and 12). In 1920, C. Scholz wrote in a book called "Härte-Praxis" (= the practice of hardening; published by Springer Verlag):

> During a travel through Saxony, I was led into a *mysterious* heat-treatment shop ("Härtestube"). *An old man with beard* was standing at the furnace and carried out a hardening treatment while he *murmured prayers* during the quenching. The result appeared to be rather satisfactory. After having watched the man for some time, it struck me that *depending on the weight of the workpieces he grumbled a shorter or longer prayer*. It took considerable time to convince the man that the steel was not hardened as the result of the prayer, but as the consequence of the well chosen time for the quenching.

The reason that parts of the above text have been represented deliberately in an italic fashion by the present author is self-explanatory: the atmosphere of magic and the invocation of unknown powers are connected with what is not understood.

Science strives for increase of knowledge, so that not understood realities get a rational basis. An important role is reserved for the scientific discovery. Bystanders are often impressed by such discoveries, for one reason or another. This can be illustrated by means of an example closely related to this book as well.

X-rays play a very important role to unravel the internal structure of materials (see, in particular, Chaps. 4 and 6). This is due to the strongly penetrative power of X-rays. Especially this aspect has had a great impact on the human society of more than a century ago when the X-rays were discovered in 1895 by Wilhelm Röntgen. Figure 1.1 shows a first X-ray image revealing the skeleton of a *living* human being, who was "X-rayed". This image has been reproduced in important journals of the time, deeply impressing the men and women of the epoch. Skeletons were known of course, from cemeteries or so, but here one could see, so to speak, "death" manifesting itself already in one's own living body!

After the discovery, it is the time for the explanation. A theory/model is developed that provides a logical basis for the observation. From an intellectual point of view, this is definitively the most satisfying part of science making, i.e. scientific research. Yet, at the same time, the result of such theorizing is uncertain and most subject to decay: *observations are immutable, but insight can become deeper and more comprehensive in the course of time*.

Sometimes it can take a long time before such profound understanding has been achieved. The diffraction of X-rays by crystalline material can be understood as that these solids are composed of very small particles, atoms, which are arranged in a regular, periodic manner in space (Chap. 4). This arrangement can be represented by visualizing the atoms as small (ping-pong) balls. The simplest way to stack these ping-pong balls, and thus atoms, in space appears to be the so-called "closest packing of spheres", which represents the "natural" way for the spheres to realize the most intimate contact among themselves.

Quite a number of years ago I made a stroll on the market in Kiev. Oranges and other fruits were sold. The merchants exhibited their merchandise as beautifully as possible and had piled the oranges, apples, etc. in a neat way on the counter of their stands. The photo reproduced here (Fig. 1.2) shows the result: this is a closest packing of spheres: the spheres occupy a fraction of $\pi/(3\sqrt{2})$, i.e. $\approx 74\%$, of space. The type of packing shown here images, for example, the natural packing of copper or aluminium atoms in their crystalline modifications (this is the so-called face-centred cubic crystal structure; see Sect. 4.2.1).

Already in 1609, Kepler conjectured that such an arrangement of spheres is the closest packing of spheres in three dimensional space. It then may come as a surprise that proof for this supposition was long in the making. It took almost 400 years before in 1998 Hales and Ferguson claimed to have proven Kepler's conjecture. However, there is a problem of principle with their proof. The proof requires the massive use of computers and thus software codes. The (12!) referees of the 250 pages long manuscript submitted by Hales to the journal Annals of Mathematics

Fig. 1.1 X-ray image revealing the skeleton inside the body of *living* human beings. The image (Deutsches Museum, Munich) in fact is a composite one (sections were taken from three different persons). The photographs (radiograms) were made by Ludwig Zehnder in 1896 (taken from J. Darius, Beyond Vision, Oxford University Press, Oxford, 1984)

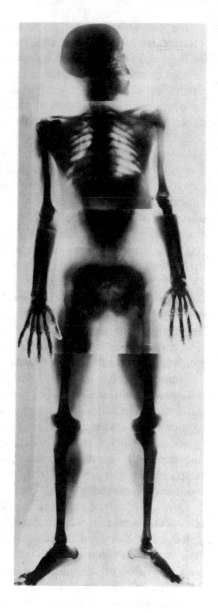

were eventually not able to confirm the correctness of the proof. The problem boils down to the question how to proof that a computer functions correctly, apart from the fact that Hales and Ferguson used also commercial software of which it is especially

Fig. 1.2 A closest packing of identical spheres, i.e. the number of contacts between the spheres is maximized, is illustrated by a stack of oranges in three dimensions (of course, the oranges are only approximately of identical, spherical shape). The arrangement shown represents a cubic close packing leading to a (crystal) structure also named "face-centred cubic" (see Sect. 4.2.1.2; in terms of the treatment in Chap. 4: the Bravais translation lattice is face-centred cubic and the motif is one orange). The packing density is $\pi/(3\sqrt{2})$, i.e. about 74% of the available space is occupied. This highest packing density is also realized by the hexagonal close-packed arrangement of identical spheres, not shown here (see Sect. 4.2.1.3)

difficult (secret machine codes) to proof that it operates failure proof.[4,5] The final paper appeared in printed form in 2005 with a cautionary, editorial remark implying that a final guarantee of validity could not be given, as not all computer calculations could be verified… Hales and coworkers later (in 2012) published a "formal proof" of Kepler's conjecture on the basis of "proof checking software", which should eliminate any remaining doubt about the correctness of the (a) proof by means of a computer.

A chaotic, random arrangement of atoms, implying that long-range translational order (cf. Chap. 4) as in crystals does not occur, can be modelled by the oranges crudely thrown on a pile and as shown on the counters of the stands on markets in (at least the countries (where I have lived) of) western Europe, which contrasts with the "civilized", regular way of stacking the oranges at the market in Kiev (see above). Also such lack of ordering in space of the atoms has relevance for real materials: amorphous solids do exist. In this case the experimental evidence for the amorphous structure requires the absence of crystalline reflections in the X-ray diffraction pattern (cf. Chap. 4). The experimental verification of the occurrence of

---

[4] Another famous mathematical problem that only could be proven (after 125 years; by Appel and Haken in 1976) on the basis of the massive involvement of computers is the so-called "four colour problem": assigning a colour to each country on a map, four different colours suffice to assure that neighbouring countries on the map are of different colour.

[5] Similar questions can also be asked (and (partly) answered) for higher dimensional spaces (see H. Cohn and N. Elkies, Annals of Mathematics, 157 (2003), 689–714).

short range ordering in amorphous solids, as due to the tendency of an atom to have unlike atoms as direct neighbours, is much more ambiguous than the experimental validation of the occurrence of the crystalline modification.

In the end, in the natural sciences, as materials science, and in contrast with pure mathematical problems as discussed above, a model cannot be proven with the guarantee of eternal "truth". *The universe of knowledge is enclosed by an outer border that moves away from us: we know more and more as time proceeds, and there is no end to it, but we appear to be unable to grasp it all*. There will always be place to pose questions which cannot be answered. This is where religious convictions start and which designates a limit of the field covered by this book.

Chapter 2
Electronic Structure of the Atom; the Periodic Table

2.1 Protons, Neutrons and Electrons

Atoms consist of a nucleus that is surrounded by a "cloud" of electrons. Protons, elementary particles carrying positive unit charge ($e = 1.602 \times 10^{-19}$ C), and neutrons, elementary particles carrying no charge, form together the nucleus of diameter of the order 10^{-14} m. Electrons, elementary particles carrying negative unit charge, have only about 1/1836 the mass of a proton, but are located within a relatively enormously large space of diameter of the order 10^{-10} m. Hence, with a view to mass distribution, the atom is largely "empty".

The number of protons in the core, Z, must be equal to the number of surrounding electrons, to assure charge neutrality. The atomic number, Z, identifies the element; the number of neutrons may vary. Atoms of the same sort, i.e. with the same number of protons, but with different numbers of neutrons, are called "isotopes".

2.2 Rutherford's Model (1911)

On the basis of scattering (bombardment) experiments with α-particles ($^4_2\text{He}^{2+}$ particles) Rutherford concluded that the atom consists of a tiny nucleus and a swarm of electrons. This contrasted strongly with the earlier model of the atom due to (J.J.) Thomson, involving a more or less uniform distribution of positively charged matter containing a dispersion of small electrons. However, according to classical electrodynamics, the Rutherford planet-like model of the atom is unstable: electrons revolving about the nucleus are accelerated particles and consequently should emit electromagnetic energy (i.e. radiation) and thus spiral towards and collapse onto the nucleus (see Fig. 2.1). This is evidently incompatible with the observation of sharp lines in atomic absorption or emission spectra, suggesting that only specific changes of energy are allowed (for the electrons in the atoms).

© Springer Nature Switzerland AG 2021
E. J. Mittemeijer, *Fundamentals of Materials Science*,
https://doi.org/10.1007/978-3-030-60056-3_2

Fig. 2.1 Rutherford's
planet-like model of the
atom. According to classical
electrodynamics, electrons
revolving about the nucleus
are accelerated particles and
consequently should emit
electromagnetic energy (i.e.
radiation) and thus spiral
towards and collapse onto
the nucleus

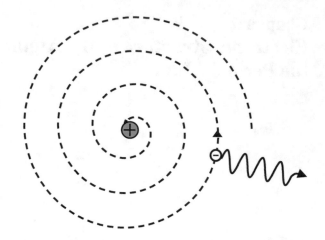

2.3 Bohr's Model (1913)

Bohr developed the first so-called quantum model of the atom. He based his model on
the postulate by Max Planck who in 1900 had concluded that energy is not continuous
but quantized, i.e. energy transfer occurs by a stream of small, not further divisible,
packets (units) of energy, called quanta.

The energy of a quantum of radiation is not a constant of nature but variable:
it depends on the frequency of the radiation considered. The energy of one such
quantum, E_{quantum}, obeys (Planck's relation):

$$E_{\text{quantum}} = h\upsilon \tag{2.1}$$

where h is Planck's constant ($= 6.626 \times 10^{-34}$ Js; a genuine constant of nature) and
υ denotes the frequency of the radiation.

Bohr conjectured that the sharply defined spectral lines in an atomic emission or
absorption spectrum are indicative of transitions of electrons in the atom from one
specific energy state to another specific energy state, in association with the emission
or absorption of (precisely) one quantum of energy per transition. Accordingly, Bohr,
still adopting a planet system of electrons circling the nucleus, allowed only specific
orbits of the electrons by quantization of the impulse (angular) momentum of the
electron in its orbit (see Eq. (2.9) further below). For the rest, classical mechanics
was used: the electrostatic (Coulomb) force between the nucleus of positive charge
Ze and the electron of negative charge $-e$, has to be balanced by the centrifugal force.
Considering a one electron system, i.e. a hydrogen-like atom with nucleus of charge
Ze and only one electron, and for a circular electron orbit (later Sommerfeld extended
Bohr's model to incorporate elliptical electron orbits as well), the following results
are obtained for the radius r of the allowed orbits:

$$r = c_1 \left(\frac{n^2}{Z} \right) \tag{2.2}$$

and for the (total = kinetic + potential) energy E of the electron in these orbits:

$$E = -c_2 \left(\frac{Z^2}{n^2} \right) \tag{2.3}$$

where c_1 and c_2 are positive constants and where $n = 1, 2, 3, \ldots$ represents in an explicit way the quantization of the energy of the electron energy in the atom and is called the principal quantum number. Note that, by convention, the (potential) energy of the electron at infinite distance of the nucleus is taken equal to nil and this causes the (potential and also total) energy of an electron near the nucleus to be negative [cf. Eq (2.3)].

It follows from Eqs. (2.2) and (2.3) that the differences between the radii of the successive orbits *increase* with n (see Figs. 2.2 and 2.4) and that the differences between the energy levels of the successive orbits *decrease* with n (see Figs. 2.3 and 2.4). If an electron "jumps" form a "higher" to a "lower" trajectory, the energy difference between these two states is emitted as energy. For example for the transition characterized by $n = 2 \rightarrow n = 1$ it follows for the energy difference = energy released:

$$\Delta E_{2 \rightarrow 1} = \frac{3}{4} c_2 Z^2 \tag{2.4}$$

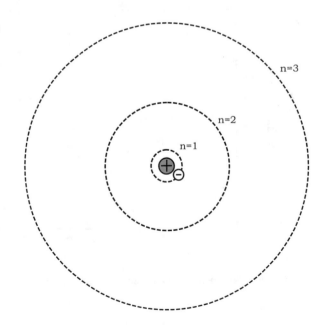

Fig. 2.2 Bohr's model of the atom (actually a hydrogen-like atom is considered: nucleus of positive charge Ze; one electron of negative charge e). The radii of the orbits for different values of the principal quantum number, n, have been scaled according to the prescription given by Eq. (2.2)

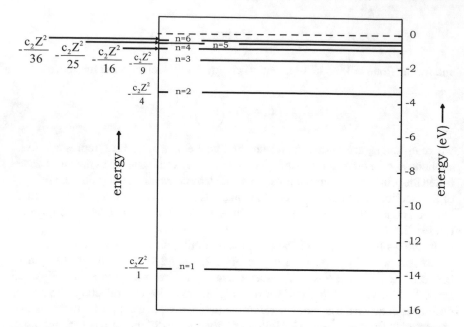

Fig. 2.3 Bohr's model of the atom (actually a hydrogen-like atom is considered: nucleus of positive charge Ze; one electron of negative charge e). The energy levels of the orbits for different values of the principal quantum number, n, are given by Eq. (2.3). The energy levels at the right ordinate pertain to the case $Z = 1$ (hydrogen atom). (1 eV $= 1.602 \times 10^{-19}$ J)

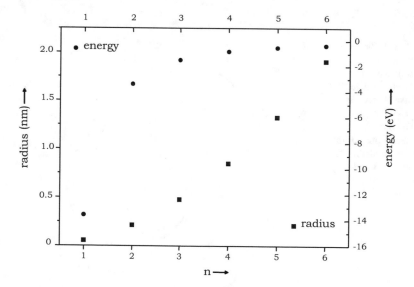

Fig. 2.4 Energies and radii of the electron orbits according to Bohr's model for the hydrogen atom

This energy is emitted as radiation. The wavelength of this radiation, λ, can be calculated using the Planck relation recognizing that $\Delta E_{2 \to 1}$ is one quantum:

$$\Delta E_{2 \to 1} = h\upsilon = \frac{hc}{\lambda} \tag{2.5}$$

with c as the constant velocity of light (2.998×10^8 m/s). The thus predicted wavelengths for hydrogen(-like) atoms agree extremely well with the experimentally observed values. However, the Bohr model could not satisfactorily explain the behaviour (spectra) of many electron systems and furthermore was considered as flawed owing to the incorporation of, as it was felt, too many assumptions/postulates. A more fundamental approach was necessary.

2.4 The Wave or Quantum Mechanical Model (Heisenberg/Schrödinger 1926); Quantum Numbers

The "dualistic" nature of light has fascinated or bothered (depending on one's personal point of view) generations of scientists:

1. *Interference phenomena* (scattering of light by a grating) are explained assuming that light is a wave phenomenon, an approach dating back to Huygens who lived in the seventeenth century, although the medium through which the "wave" would propagate (called "ether") was finally proven to be non-existent.

2. The perhaps most revealing phenomenon demonstrating the particle nature of light is the *photo-electric effect*, showing that only if the frequency of the incident light exceeds a critical value, the incident light is able to "kick out" an electron from the surface adjacent region of the irradiated solid considered. This implies that light (electro-magnetic radiation) can be conceived as a stream of "particles" (light quanta = photons), each of which has, apparently, an energy proportional to the frequency, which is compatible with Eq. (2.1). Light quanta of energy corresponding to a frequency below the critical value have not enough energy to overcome the binding energy of the electron in the solid, upon "collision" with this electron. Indeed, increasing the intensity of the incident light of under-critical frequency is of no avail: no photoelectron is produced (for further discussion of the photo-electric effect, see Sect. 3.5.1 [Eq. (3.13)].

De Broglie (1923) recognized that the introduction of an integer as the principal quantum number in the theory of Bohr, for describing the dynamics of the, particulate, electron system of an atom, was odd: in the physics of phenomena until then integers only appeared in the wave theory of interference and vibration (for identifying the occurrences of extinctions and nodes, respectively). Consequently he conjectured that not only electro-magnetic radiation but also matter in motion has a dualistic nature: both a wave aspect and a particulate aspect. Hence, because for a photon it holds (Eq. (2.1) and the famous Einstein relation):

$$E = h\upsilon = \frac{hc}{\lambda} \quad \text{and} \quad E = mc^2$$

and thus $\lambda = h/mc$, with c as the velocity of light, by analogy, de Broglie then proposed that for matter in motion it should hold:

$$\lambda = \frac{h}{m\upsilon} \tag{2.6}$$

with υ as the velocity of the material object considered. Shortly after this stipulation it was experimentally shown that a stream of electrons indeed can give rise to diffraction phenomena associating a wavelength to the electron stream according to Eq. (2.6) [experiments by Davisson and German (1927) and by Thomson (1927) (son of J.J. Thomson; see Sect. 2.2)].

The second major step was taken by Heisenberg in 1926. In classical mechanics it is presupposed that position and momentum of a moving particle can be known both exactly and simultaneously. Then, standard kinematical theory involves that the future and past position and velocity of the material object considered are completely predictable and retrievable, respectively. This is called **"determinism"**. Heisenberg recognized that measuring either the position or the momentum (velocity) of a particle implied *interaction* with this particle and thereby a certain *uncertainty* is introduced. This plays a great role for the dynamics of particles on the atomic scale. The quantization of the energy causes that the energy of the particle, that, for the measurement of position or momemtum of the object (particle), is interacting with that object (particle), has a lower limit. On this basis the so-called "uncertainty relation" was derived:

$$\Delta x \cdot \Delta p > \frac{h}{4\pi} \tag{2.7}$$

where x denotes the position and p represents the momentum of the moving particle. The uncertainties Δx and Δp actually are the standard deviations of the corresponding distributions of x and p. If Δx and Δp would have been defined as the maximal uncertainties or the mean uncertainties, the quantity h or the quantity $h/2\pi$, respectively, would have appeared at the right-hand side of the inequality.

Intermezzo: A "derivation" of the Uncertainty Relation;
Diffraction of Moving Particles at a Slit
Consider Fig. 2.5. A monochromatic pencil of parallel rays of light is incident on an opaque screen 1 containing a slit of width d defined by the boundaries A and B. If the slit width is of the same order of magnitude as the wavelength of the incident light, the propagation of the light is not restricted to the column of width d defined by the boundaries A′ and B′ on a screen 2 obtained by extrapolation of the incident light rays: on the screen 2 light is also observed at positions above

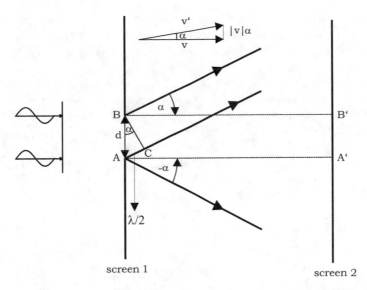

Fig. 2.5 Illustration of Heisenberg's uncertainty principle. Electrons passing the opaque screen 1 through the slit AB hit screen 2 not only on the projected slit A′B′ but also outside of it: The direction of the velocity of the electron after passage of the slit AB can diverge over an angle 2α: The velocity of the electron, a vectorial quantity, has obtained a component parallel to screen 2. The distribution of the component of the velocity parallel to screen 2 has maximal values characterized by the angles $+\alpha$ and $-\alpha$ (see further text)

B′ and below A′; see what follows. The Huygens principle involves that every point of the incident wavefront arriving at AB can be considered as the source of a new wave. In the direction defined by the angle α the waves emanating from the points A and B have a common tangent, wavefront, indicated by BC. The waves from B and C have a path difference equal to AC. If this path difference is equal to one half of the wavelength (i.e. $\sin\alpha \approx \alpha = (\lambda/2)/d$), then the waves from B and C have opposite phase. As a consequence, in the direction given by the angle α, there are no other pairs of waves from points on AB of opposite phase: there are no waves emanating from the wavefront AB which could extinguish each other; only the waves emanating from B and C are extinguished. For smaller values of α (even) none of the waves emanating of the wavefront AB are extinguished. Hence, the region of light observed on screen 2 (at distance from screen 1 very much larger than the slit width d) is more extended than the projection A′B′ of AB on screen 2. Evidently, the light rays responsible for the central light region on screen 2 are not parallel but diverge over an angle 2α, where the factor 2 recognizes that the deviation (diffraction) can occur both in the upward direction (considered above) and in the downward direction (in the two-dimensional consideration pertaining to Fig. 2.5).

Now the pencil of incident parallel rays of light is replaced by a beam of particles (e.g. electrons) of constant mass m propagating parallel to each other with constant velocity v. Each particle is associated with a wave of wavelength given by the de Broglie equation (Eq. (2.6)). Hence, according to the above discussion, the direction of the velocity of the particle[1] after passage of the slit can diverge over an angle 2α. The velocity, a vectorial quantity (indicated by **v**; the symbol v used here (a scalar) denotes the magnitude of **v**), has obtained a component parallel to screen 2: the distribution of the component of the velocity parallel to screen 2 has maximal values characterized by the angles $+\alpha$ and $-\alpha$. For small values of α the maximal values of the velocity component parallel to screen 2 are given by the products $+v\alpha$ and $-v\alpha$ (cf. Fig. 2.5). Hence the maximal spread in the component of the velocity parallel to screen 2 is [cf. Fig. 2.5 and Eq. (2.6)]:

$$\Delta v = 2v\alpha = 2v\frac{\left(\frac{\lambda}{2}\right)}{d} = \frac{h}{(md)}$$

and thus it holds for the uncertainty in the component of the impuls parallel to screen 2:

$$\Delta p = m\Delta v = \frac{h}{d}$$

The uncertainty in the position (parallel to screen 2) is of course given by the slit width d which will now be denoted by the symbol Δx. Hence, from the above formula for Δp it then immediately follows:

$$\Delta x \cdot \Delta p = h$$

which is nothing else than the uncertainty relation of Heisenberg for Δx and Δp taken as maximal uncertainties [cf. discussion below Eq. (2.7)].

Thus, the position of an electron may be measured with high accuracy (in the example discussed in the *intermezzo* above the slit width is made infinitely small), but the act of measurement transfers so much energy to the electron that its momentum (velocity) becomes undefined (in the example discussed in the *intermezzo* above the central light region on screen 2 then becomes of infinitely large lateral dimensions). This is called **"indeterminacy"**. A wave, of fixed wavelength [and thus fixed frequency and thus fixed momentum: $p = mc = h v/c$ (see above)], associated, according to Eq. (2.6), with a particle as an electron, has in principle an infinite spatial extension and thereby the position of the electron is indeterminate. Hence, the consequence of the uncertainty relation is: **loss of causality**: the non-measured

---

[1] The velocity is a vector: it has direction and magnitude.

property of a particle is uncertain to a degree implied by the degree of certainty with which the measured property is determined [cf. Eq. (2.7)]. Atomic-scale processes, as the scattering of an individual electron, can, for the individual atomic-scale particle, only be described in terms of "probabilities". Statistical averages become practical certainties only for infinitely large numbers of those atomic-scale particles, as we meet on the macroscale ("daily life/human scale").

The above indicated recognition by Heisenberg, that measurements on an elementary particle involve interaction with the elementary particle and thereby introduce intrinsic uncertainty in the parameter values sought for, can be given a broader implication in the following side note. The scientific method, as we know it, departs from observations on a system by interaction using tools, i.e. instruments. As a consequence our observations do not pertain to nature directly: our instruments provide data which are unavoidably affected by measurement errors: *precision* (reproducibility) is only realizable within a certain range and *accuracy* (i.e. the degree to which the "true" value of a certain parameter has been determined, i.e. even if the reproducibility would be perfect) leaves space for debate on the trueness of the model of nature that is based on the measured data. This consideration explains the huge efforts undertaken by scientists to develop instruments providing ever deeper, more detailed looks into the mechanisms of nature and their enormous pains to adjust and calibrate their apparatuses, leading to better accuracy in association with higher precision (cf. Chap. 6 and see footnote 3 in Chap. 7).

Now returning to the "quantum mechanical world", and having accepted that a moving particle can be associated with a wave phenomenon (de Broglie) and that the spatial extension of the corresponding wave introduces indeterminacy regarding its position (Heisenberg), the question arises how to describe the dynamics of such a particle.

2.4.1 The Probability Amplitude

Schrödinger (1926) proposed an equation that in principle allows the calculation of the amplitude, ψ, of the wave associated with the particle as a function of position (x, y, z). This time independent Schrödinger (wave) equation is used for describing the properties of systems in stationary states: the time dependency of the wave amplitude is not considered and one thereby does not study changes in the atomic state *during* a transition (e.g. electron excitation) but concentrates on the states before and after the transition. This equation is a second order partial differential equation. Exact solutions have been obtained only for one electron systems. The realm of quantum mechanics is thus mainly concerned with developing approximate methods for carrying out approximate calculations for many (more than one) particle (electron) systems.

The amplitude of the wave function has no physical meaning: there is no undulating medium through which the wave propagates. So, for example, we cannot speak of the amplitude of the wave as a measure of displacement experienced by that medium, as compared to an average level (as is possible for a wave propagating through a liquid, where the displacement of the surface of the liquid at a certain location is given by the local amplitude of the wave). The square of the wave amplitude at the position (x, y, z) does have a physical meaning: it represents the probability density for finding the particle concerned in the volume element $\Delta x \Delta y \Delta z$ at the position (x, y, z). This can be compared with the calculation of the intensity of light which is given by the square of the amplitude of the electromagnetic wave that represents the light (there is also no undulating medium ("ether") through which the light wave propagates; see at the begin of this Sect. 2.4). Against this background the amplitude of the wave function is also called "probability amplitude".

By calculation of the spatial distribution of ψ^2 for an electron, in general the distribution in space of the probability density for the electron to be in the volume element $\Delta x \Delta y \Delta z$ at a certain position (x, y, z) is obtained. This result can be conceived, for the special case of a one electron system, as an (time averaged) electron density distribution, thereby "smearing out" the electron (see Figs. 2.12, 2.13, 2.14, 2.15, 2.16 and 2.17 discussed in Sect. 2.6). Such a simple pictorial description of the electron distribution is impossible for a two or more electron system, as the probability for one electron to be at a specific location (x, y, z) depends on the coordinates of all other electrons.

2.4.2   Characterizing the Possible Energy States; the Quantum Numbers

The solutions of the time independent Schrödinger equation depend on the imposed boundary conditions. The wave phenomenon associated with a moving particle is spatially confined: the probability for finding the electron considered at infinite distance from the nucleus must be zero. Such conditions lead to the recognition that only a limited number of stationary energy states is allowed. These energies are called the *eigenvalues* for the system and the corresponding wave functions are called the *eigenfunctions*. Thus the energy of a moving particle becomes quantized. It appears that the possible energy states can be characterized, i.e. are fully determined by, a set of numbers: the quantum numbers.

A simple way to visualize the occurrence of quantum numbers is as follows. Consider Fig. 2.6. The electron moving around the atom in its orbit, as in the Bohr model (see Fig. 2.2), can be considered as a wave phenomenon (Eq. (2.6)). In general the orbit length, i.e. the circumference of the circular orbit considered in Fig. 2.6, is not equal to an integral multiple of the wavelength of the electron wave. Then at a certain location at the orbit considered, more than one value for the amplitude of the wave function occurs, i.e. destructive interference takes place, and the electron state considered is non-existent. Hence, only if the orbit length equals an integral multiple of the wavelength, the electron state, the electron orbit, is an allowed one:

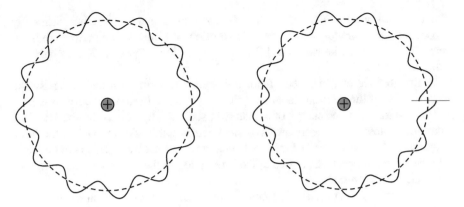

Fig. 2.6 Electron wave on orbit as in Bohr's model of the atom. Left part of the figure: constructive (self-)interference. Right part of the figure: destructive (self-)interference

$$2\pi r = n\lambda, \quad n = 1, 2, \ldots \tag{2.8}$$

Substitution of λ according to Eq. (2.6) directly leads to

$$mvr = \frac{nh}{2\pi} \tag{2.9}$$

implying the quantization of the impulse (angular) momentum of the electron circling the nucleus, as postulated by Bohr (see Sect. 2.3). In this way the electron is conceived as a stationary wave, not as a particle rotating around the nucleus, and thereby the problem due to classical electrodynamics, involving that an electron revolving around the nucleus is an accelerated particle that should emit energy and consequently collapse on the nucleus (Sect. 2.2), has been removed at last. It is usual to use the notion "orbit" for the trajectory followed by a moving electron-particle in the atomic models proposed until and including the model of Bohr; the notion "orbital" is used for designating a stationary state of an electron in the wave-mechanical model.

Recognizing the three-dimensional nature of the electron wave, it seems obvious that three (orthogonal) linear waves, as indicated in Fig. 2.6, are needed to describe the three-dimensional wave phenomenon corresponding to the electron. This reasoning suggests that three quantum numbers are required to characterize an allowed stationary electron state; i.e. one anticipates that each dimension in which an electron can move introduces one quantum number.

1.   *The principal quantum number, n*

This quantum number has effectively first been postulated by Bohr (see above and Sect. 2.3). It can take integer values 1, 2, … Values $n = 1, 2, 3, \ldots$ are also indicated by $K, L, M, \ldots$ etc.

2.   *The subsidiary, secondary quantum number, l*

This quantum number is usually called the azimuthal quantum number, originally introduced as a consequence of the introduction of ellipses as electron orbits by Sommerfeld. It can take values 0, 1, 2, ..., $n - 1$. Values $l = 0, 1, 2, 3, ...$ are also indicated by $s, p, d, f, ...$

In the absence of a magnetic field, the quantum numbers n and l specify the energy levels of the electrons in a many electron system. In a one electron system, a hydrogen-like atom, the energy of an electron state is fully defined by the principal quantum number n; the energy levels of the electron states of different l for the same n are equal: "degeneration" of the electron state specified by n. This is not exactly true, because minor effects, as due to the interaction of the electron with the spin of the nucleus, are ignored.

The energy level of the allowed electron states increases for increasing l at constant n, and also the energy level increases with n at constant l (see Fig. 2.7) Because the differences between energy levels decrease with increasing n, "l states (orbitals)" of relatively low l value and associated with the principal quantum number $n + 1$ may have a lower energy than "l states" of relatively high l value and associated with the principal quantum number n. Thus a $4s$ state may have a lower energy than a $3d$ state (see Fig. 2.8). This effect depends on the nucleus charge (Ze) and hence atomic number. For elements of high atomic number ("heavy atoms"), the order of the energy levels is governed by the principal quantum number n, i.e. the above discussed "disordering" of the energy levels does not occur: the energy of a $3d$ state is lower than that of a $4s$ state (see discussion in Sect. 2.5).

In the presence of a magnetic field two more, magnetic quantum numbers have to be specified for identifying the energy of an electron state.

3.   *The magnetic quantum number,* m_l

This quantum number can take values: $-l, -(l - 1), ..., 0.0, ...(l - 1), l$. In total there are $2l + 1$ values of m_l.

Fig. 2.7 Energy levels and occupation of atomic orbitals by electrons for the carbon atom

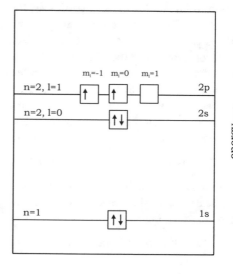

Fig. 2.8 Possible energy
levels of the atomic orbitals
for an atom in the fourth
period of the Periodic Table

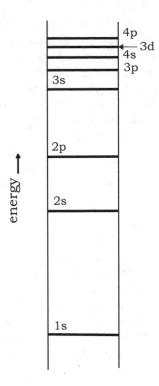

4.   *The spin-quantum number,* m_s.

Solving the Schrödinger equation leads to (only) the three quantum numbers n, l and m_l (see above discussion). The fourth (second magnetic) quantum number was introduced because spectroscopic evidence indicated that energy states calculated as single states by solving the Schrödinger equation, i.e. energy states specified by three quantum numbers, were split into two states in a magnetic field. This effect is ascribed to the existence of an intrinsic angular momentum of the electron, which can be conceived to be due to the spinning of the electron around an axis and thereby the electron has a magnetic moment. In a magnetic field only two states of electron spinning are allowed (either to the right ("spin up") or to the left ("spin down") around an axis), which are characterized by two values of the spin-quantum number m_s: $-1/2$ and $+1/2$.

2.5   The Pauli Exclusion Principle and the "Aufbau Prinzip"

Recognizing that nature strives for a state of minimal energy, one may expect that the ground state for an atom would involve that all electrons in the atom occupy the lowest allowed energy level, indicated by $n = 1$ and $l = 0$, in the absence of a magnetic field. This appears to be impossible for an atom with more than two electrons. Pauli (1925) formulated a principle, which has to be conceived as an empirical law (and thus cannot be derived from first principles), that, for electrons in an atom and in a derived, specific way, can be expressed as follows: *two electrons in an atom cannot have the same set of (four) quantum numbers.*[2] It is difficult to underestimate the importance of this recognition.

The Pauli exclusion principle leads, together with the above indicated listing of possible energy states on the basis of the allowable quantum numbers (e.g. see Figs. 2.7 and 2.8), directly to the recipe for derivation of the electron structure (i.e. the ground state) of the atoms of the elements, which, in principle, runs as follows:

- The number of electrons in an atom equals Z, the atomic number. These electrons are put on the allowable energy levels, starting with the lowest level and going upwards:
- $n = 1$, and consequently $l = 0$ with $m_l = 0$. Then two electrons (one for $m_s = -1/2$ and one for $m_s = +1/2$) can be taken up in the K shell. This is a pair of s electrons. If there is a pair of s electrons in the n scale we write for this electron configuration: ns^2;
- $n = 2$, and consequently $l = 0$ with $m_l = 0$ and $l = 1$ with $m_l = -1, 0, 1$. Hence, a total of 8 electrons can be taken up in the L shell: two s electrons ($l = 0$) and six p ($l = 1$) electrons. We write for this configuration: ns^2np^6;
- Etc., etc.

Given the availability of degenerated quantum states (cf. Sect. 2.4.2 under 2.) for the electron, as the three types of p orbitals (corresponding to $m_l = -1, 0, 1$; for the notion "orbital", see Sect. 2.4.2) in the absence of an external magnetic field, one may wonder how, for example, the $2p$ subshell becomes filled. Starting from hydrogen, upon increasing the atomic number, the carbon atom is the first atom where this "dilemma" occurs: two electrons have to be placed in the $2p$ subshells with three p orbitals available. The Pauli exclusion principle would allow a configuration with a pair of electrons in one p orbital, but then with opposite spin, or a configuration with the two electrons each in a separate p orbital, with either the same (parallel) or opposite (antiparallel) spin. Electrons repel each other, because they have the same type of charge. The state of minimal energy for the two electrons considered then is the one where they are as far as possible separate (here by occupying p orbitals of different m_l) with parallel spins (which obstructs electron pairing in one orbital according to the Pauli exclusion principle). This has become known as "Hund's rule": The ground state of an atom reflects the highest possible "multiplicity", i.e. *the electron configuration in a partly filled subshell of degenerated orbitals is the one given by the largest number of (unpaired) electrons with parallel spin.* Thereby the Coulomb repulsion of the electrons is minimized. So the electron configuration of carbon can be given, more specifically than by $1s^22s^22p^2$, by $1s^2(\uparrow\downarrow)2s^2(\uparrow\downarrow)2p(\uparrow)2p(\uparrow)$ (with

---

[2] The Pauli principle applies not only to isolated atoms but also to aggregates of bound atoms, as in molecules and crystals (see Sects. 3.4 and 3.5).

the arrow, up or down, indicating "spin up" or "spin down"; see end of Sect. 2.4): a pair of electrons of opposite spins in both the $1s$ shell and the $2s$ subshell and two unpaired electrons of parallel spin in two of the three $2p$ orbitals; the third $2p$ orbital remains empty (see Fig. 2.7). Similarly, nitrogen in its ground state has three unpaired electrons of parallel spin in the three $2p$ orbitals and oxygen in its ground state has a pair of electrons of opposite spin in one of the $2p$ orbitals and two unpaired electrons of parallel spin in the two remaining $2p$ orbitals. Etc.[3]

On this basis—it is called the "*Aufbau Prinzip*" (= "building principle")—a list of electron configurations for the ground states of the elements in the Periodic Table can be provided, by filling the orbitals successively: see Table 2.1.

As discussed under 2. in Sect. 2.4.2, the $3d$ states can have a higher energy than the $4s$ states, etc. (Fig. 2.8). Thus, proceeding in the Periodic Table from Ar (filled $3p$ subshell) to K and Ca, the $4s$ subshell is filled, instead of the $3d$ subshell. Next, now that the $4s$ subshell has been filled, starting with Sc, the $3d$ subshell becomes filled: the first transition series.

Thus, for the elements of the two *transition metal series* (first series: Sc to Zn; second series: Y to Cd) d electrons have to be added of a principal quantum number, n, smaller and a second quantum number, l, larger than those for the "outermost" electrons. Here "outermost" has to be discussed as follows. For the transition elements in period 4 (fourth row in the Periodic Table) the "outer" ($4s$) electrons are more tightly bound than the $3d$ electrons. This is due to the "penetration" of the $4s$ electrons, as expressed by the distinct probability of finding a $4s$ electron close to the nucleus (see Sect. 2.6), which is significantly larger than for a $3d$ electron. Therefore the $4s$ subshell becomes filled before the $3d$ shell is filled (see above discussion). Yet, the *most probable* location of finding a $4s$ electron is more remote from the nucleus than for a $3d$ electron. Hence, on the basis of this reasoning, the electrons of higher principal quantum number (the $4s$ electrons for the transition series considered) are therefore the "outer" electrons for the transition elements.

A similar discussion holds for the series of *rare earth* or *lanthanide metals* and of *actinide metals* (sometimes also called second series of rare earth metals), where an f subshell of lower principal quantum number is filled.

Half-filled and wholly filled subshells have a relatively high stability. Hence, although for Cr the electron configuration $1s^2 2s^2 2p^6 3s^2 3p^6 3d^4 4s^2$ is expected according to the "Aufbau Prinzip", the reality is: $1s^2 2s^2 2p^6 3s^2 3p^6 3d^5 4s^1$. Similarly, for Cu the electron configuration is expected to be $1s^2 2s^2 2p^6 3s^2 3p^6 3d^9 4s^2$, but $1s^2 2s^2 2p^6 3s^2 3p^6 3d^{10} 4s^1$ is observed: $3d^5$ and $3d^{10}$ are preferred over $3d^4$ and $3d^9$, so to speak: "at the cost of transferring one $4s$ electron to the $3d$ subshell".

---

[3] The net magnetic moment of an atom is related to the number of unpaired electrons with parallel spin, and thereby these electrons provide the key to understanding the magnetic properties of a material (see Sect. 3.5).

Table 2.1 Electron configurations of the elements (ground states; data taken from W.C. Martin, A. Musgrove, S. Kotochigova, and J.E. Sansonetti, NIST Standard Reference Database 111)

		1s	2s	2p	3s	3p	3d	4s	4p	4d	4f	5s	5p	5d	5f	6s	6p	6d	6f	7s	7p	
1	Hydrogen	1																				
2	Helium	2																				
3	Lithium	2	1																			
4	Beryllium	2	2																			
5	Boron	2	2	1																		
6	Carbon	2	2	2																		
7	Nitrogen	2	2	3																		
8	Oxygen	2	2	4																		
9	Fluorine	2	2	5																		
10	Neon	2	2	6																		
11	Sodium	2	2	6	1																	
12	Magnesium	2	2	6	2																	
13	Aluminum	2	2	6	2	1																
14	Silicon	2	2	6	2	2																
15	Phosphorus	2	2	6	2	3																
16	Sulfur	2	2	6	2	4																
17	Chlorine	2	2	6	2	5																
18	Argon	2	2	6	2	6																
19	Potassium	2	2	6	2	6		1														
20	Calcium	2	2	6	2	6		2														
21	Scandium	2	2	6	2	6	1	2														
22	Titanium	2	2	6	2	6	2	2														
23	Vanadium	2	2	6	2	6	3	2														
24	Chromium	2	2	6	2	6	5	1														
25	Manganese	2	2	6	2	6	5	2														
26	Iron	2	2	6	2	6	6	2														
27	Cobalt	2	2	6	2	6	7	2														
28	Nickel	2	2	6	2	6	8	2														
29	Copper	2	2	6	2	6	10	1														
30	Zinc	2	2	6	2	6	10	2														
31	Gallium	2	2	6	2	6	10	2	1													
32	Germanium	2	2	6	2	6	10	2	2													
33	Arsenic	2	2	6	2	6	10	2	3													
34	Selenium	2	2	6	2	6	10	2	4													
35	Bromine	2	2	6	2	6	10	2	5													
36	Krypton	2	2	6	2	6	10	2	6													
37	Rubidium	2	2	6	2	6	10	2	6			1										
38	Strontium	2	2	6	2	6	10	2	6			2										
39	Yttrium	2	2	6	2	6	10	2	6	1		2										
40	Zirconium	2	2	6	2	6	10	2	6	2		2										
41	Niobium	2	2	6	2	6	10	2	6	4		1										
42	Molybdenum	2	2	6	2	6	10	2	6	5		1										
43	Technetium	2	2	6	2	6	10	2	6	6		1										
44	Ruthenium	2	2	6	2	6	10	2	6	7		1										
45	Rhodium	2	2	6	2	6	10	2	6	8		1										
46	Palladium	2	2	6	2	6	10	2	6	10												
47	Silver	2	2	6	2	6	10	2	6	10		1										
48	Cadmium	2	2	6	2	6	10	2	6	10		2										
49	Indium	2	2	6	2	6	10	2	6	10		2	1									
50	Tin	2	2	6	2	6	10	2	6	10		2	2									
51	Antimony	2	2	6	2	6	10	2	6	10		2	3									
52	Tellurium	2	2	6	2	6	10	2	6	10		2	4									
53	Iodine	2	2	6	2	6	10	2	6	10		2	5									
54	Xenon	2	2	6	2	6	10	2	6	10		2	6									

Elements 21–30: 1st transition series

Elements 39–48: 2nd transition series

(continued)

Table 2.1 (continued)

		1s	2s	2p	3s	3p	3d	4s	4p	4d	4f	5s	5p	5d	5f	6s	6p	6d	6f	7s	7p
55	Cesium	2	2	6	2	6	10	2	6	10		2	6			1					
56	Barium	2	2	6	2	6	10	2	6	10		2	6			2					
57	Lanthanum	2	2	6	2	6	10	2	6	10		2	6	1		2					
58	Cerium	2	2	6	2	6	10	2	6	10	2	2	6			2					
59	Praseodymium	2	2	6	2	6	10	2	6	10	3	2	6			2					
60	Neodymium	2	2	6	2	6	10	2	6	10	4	2	6			2					
61	Promethium	2	2	6	2	6	10	2	6	10	5	2	6			2					
62	Samarium	2	2	6	2	6	10	2	6	10	6	2	6			2					
63	Europium	2	2	6	2	6	10	2	6	10	7	2	6			2					
64	Gadolinium	2	2	6	2	6	10	2	6	10	7	2	6	1		2					
65	Terbium	2	2	6	2	6	10	2	6	10	9	2	6			2					
66	Dysprosium	2	2	6	2	6	10	2	6	10	10	2	6			2					
67	Holmium	2	2	6	2	6	10	2	6	10	11	2	6			2					
68	Erbium	2	2	6	2	6	10	2	6	10	12	2	6			2					
69	Thulium	2	2	6	2	6	10	2	6	10	13	2	6			2					
70	Ytterbium	2	2	6	2	6	10	2	6	10	14	2	6			2					
71	Lutetium	2	2	6	2	6	10	2	6	10	14	2	6	1		2					
72	Hafnium	2	2	6	2	6	10	2	6	10	14	2	6	2		2					
73	Tantalum	2	2	6	2	6	10	2	6	10	14	2	6	3		2					
74	Tungsten	2	2	6	2	6	10	2	6	10	14	2	6	4		2					
75	Rhenium	2	2	6	2	6	10	2	6	10	14	2	6	5		2					
76	Osmium	2	2	6	2	6	10	2	6	10	14	2	6	6		2					
77	Iridium	2	2	6	2	6	10	2	6	10	14	2	6	7		2					
78	Platinum	2	2	6	2	6	10	2	6	10	14	2	6	9		1					
79	Gold	2	2	6	2	6	10	2	6	10	14	2	6	10		1					
80	Mercury	2	2	6	2	6	10	2	6	10	14	2	6	10		2					
81	Thallium	2	2	6	2	6	10	2	6	10	14	2	6	10		2	1				
82	Lead	2	2	6	2	6	10	2	6	10	14	2	6	10		2	2				
83	Bismuth	2	2	6	2	6	10	2	6	10	14	2	6	10		2	3				
84	Polonium	2	2	6	2	6	10	2	6	10	14	2	6	10		2	4				
85	Astatine	2	2	6	2	6	10	2	6	10	14	2	6	10		2	5				
86	Radon	2	2	6	2	6	10	2	6	10	14	2	6	10		2	6				
87	Francium	2	2	6	2	6	10	2	6	10	14	2	6	10		2	6			1	
88	Radium	2	2	6	2	6	10	2	6	10	14	2	6	10		2	6			2	
89	Actinium	2	2	6	2	6	10	2	6	10	14	2	6	10		2	6	1		2	
90	Thorium	2	2	6	2	6	10	2	6	10	14	2	6	10		2	6	2		2	
91	Protactinium	2	2	6	2	6	10	2	6	10	14	2	6	10	2	2	6	1		2	
92	Uranium	2	2	6	2	6	10	2	6	10	14	2	6	10	3	2	6	1		2	
93	Neptunium	2	2	6	2	6	10	2	6	10	14	2	6	10	4	2	6	1		2	
94	Plutonium	2	2	6	2	6	10	2	6	10	14	2	6	10	6	2	6			2	
95	Americium	2	2	6	2	6	10	2	6	10	14	2	6	10	7	2	6			2	
96	Curium	2	2	6	2	6	10	2	6	10	14	2	6	10	7	2	6	1		2	
97	Berkelium	2	2	6	2	6	10	2	6	10	14	2	6	10	9	2	6			2	
98	Californium	2	2	6	2	6	10	2	6	10	14	2	6	10	10	2	6			2	
99	Einsteinium	2	2	6	2	6	10	2	6	10	14	2	6	10	11	2	6			2	
100	Fermium	2	2	6	2	6	10	2	6	10	14	2	6	10	12	2	6			2	
101	Mendelevium	2	2	6	2	6	10	2	6	10	14	2	6	10	13	2	6			2	
102	Nobelium	2	2	6	2	6	10	2	6	10	14	2	6	10	14	2	6			2	
103	Lawrencium	2	2	6	2	6	10	2	6	10	14	2	6	10	14	2	6			2	1?
104	Rutherfordium	2	2	6	2	6	10	2	6	10	14	2	6	10	14	2	6	2?		2	

Bracket annotations (right margin):

- 57–71: lanthanides (rare earth metals)
- 57–80: 3rd transition series
- 89–103: actinides (2nd series of rare earth metals)
- 89–104: 4th transition series

Comparing the "Aufbau Prinzip" (cf. Table 2.1) with the Periodic Table (see Fig. 2.9a) it can be concluded that the arrangements of elements in groups apparently corresponds to a similar electron configuration in the outer shell of the atoms of the elements in a group: group I (group of the alkali elements; hydrogen is normally not considered to belong fully to this group, see discussion in Sect. 3.2): one s electron in the outer shell; group II (group of the alkaline earth elements): two s electrons in the outer shell; group VII (group of the halogen elements): two s and five p electrons in the outer shell; group VIII (group of the noble or "inert" gases): two s electrons and six p electrons in the outer shell.

The usual presentation of the Periodic Table is given in Fig. 2.9a. An alternative presentation of the Periodic Table, emphasizing the electron configuration of the atom as guiding principle for the "Periodic Law", is provided by Fig. 2.9b. The last presentation has as remarkable feature that, for example, He occurs on top of group II, i.e. heading Be, Mg, Ca, Sr,........., because it has a fully filled (outer) s (sub)scale, as holds for the alkaline earth metals.

Intermezzo: The Discoverers of the Periodic System; A First Example of a "Priority Battle"

The discovery of the Periodic Table has many fathers. In the first 60 years of the nineteenth century ideas which can be considered as precursors of the "Periodic Law" emerged. For example, it was recognized that the weights of certain elements of similar properties had specific relationships. Thus, for "trios" of elements of similar properties (like lithium, sodium and potassium; nowadays arranged in the first group of the Periodic Table; cf. Figure 2.9), it was found that the average weight of the lightest element and that of the heaviest element, of such a "trio" of elements, practically equalled the weight of the middle element. This phenomenon, called the "Law of Triads", was discovered by Dobereiner in 1829. Evidently some physical principle must be responsible for such regularity. On the basis of such considerations of pattern finding in the arrangement of the elements, the Periodic System became established in the period 1860–1870.

Names which nowadays are associated with the discovery of the Periodic Table are Meyer (1864, 1868) and Mendeleev (1869, 1870, 1871). In these days communication between scientists was not so easy as nowadays and it was also certainly not a habit for a scientist to inform himself very well about work done at other places. So, when Mendeleev wrote his first major paper on this topic he was perhaps not aware of the work done by Meyer (or, as has been suggested as well, he deliberately did not cite him). And both of them were unaware of what must be called the first example of a Periodic System published earlier by De Chancourtois (1862). Mendeleev and Meyer became involved in a long dispute about "who was first" (within this context, also see the *intermezzo* in Sect. 8.9.2). This may not interest us (also recognizing that De Chancourtois was "first"), but mentioning it serves to illustrate how important human vanity and ambition are as driving forces for scientists; the interested reader is referred to the review provided by Van Spronsen in his book on the history of the Periodic System (1969). There is no doubt that the contributions by Meyer and Mendeleev are seminal and their names are justifiably connected with the Periodic System. Mendeleev should be mentioned in particular because of his many correct predictions of the properties of elements yet to be discovered at the time of publication of the Periodic Table by Mendeleev and for which elements he left open places in his version of the Periodic Table.[4]

This short discussion also illustrates that a great discovery is only very rarely the result of an individual act of a genius occurring like a bolt from the blue in virginal territory. In retrospection one can discern the precursory, gradual developments, which precede the crowning culmination of a period of activity, and which subsequently are forgotten, as their actors are, sometimes unjustifiably.[5]

The "outer" electrons of the atom are relatively weakly bound (to the nucleus, an "attractor") and have a distinct probability to be at locations farthest away from the nucleus. Hence, they play an important role in establishing bonds with other atoms: they can be more or less easily engaged with other "attractors". These "outer", weakly bound electrons are therefore called *"valence electrons"*. Elements with one "outer" electron (group I in the Periodic Table) are said to have valence $= 1$, etc. Hence, the elements in a group (column; also called "family" (of elements)) of the Periodic Table have the same number of valence electrons and consequently similar (chemical) properties.

Note that, in view of the above reasoning, the elements of the transition series should generally have valence $= 2$. However, the valence of the transition elements is less outspoken, which is just a consequence of the closeness of the highest occupied energy levels: e.g. the $3d$ and $4s$ levels for the first transition series. Thus the transition element Cu has an ambivalent nature (see also the discussion regarding the deviating

---

[4] Mendeleev foresaw in 1871 the existence of elements as ekaboron (discovered in 1879 and nowadays called scandium), eka-aluminium (discovered in 1875 and nowadays called gallium) and ekasilicon (discovered in 1886 and nowadays called germanium) and successfully predicted physical and chemical properties of these elements.

[5] The path to the discovery of the elements not yet identified experimentally in, with a view to present-day, previous versions of the Periodic Table, has been an outspoken stage of priority fights. A show case concerns the element technetium (Tc, atomic number 43). In 1925 Noddack, Tacke and Berg claimed to have found Tc in naturally occurring material and called it masurium. Their claim was questioned. In 1937 Perrier and Segrè undoubtedly found Tc in material resulting from nuclear reactions and proposed the now accepted name technetium. However, a discussion running in the literature until recently has been devoted to finding out who the actual discoverers are. And indeed tiny amounts of technetium occur in nature. Various groups of scientists appeared to be capable to establish the validity of the claim made in 1925; i.e. it then cannot be excluded that Noddack, Tacke and Berg were first.... However, the most recent work refutes this assertion (see the account in Materials Research Bulletin, 32 (2007), 857 and Scerri 2013). See also the "epilogue" to this section.

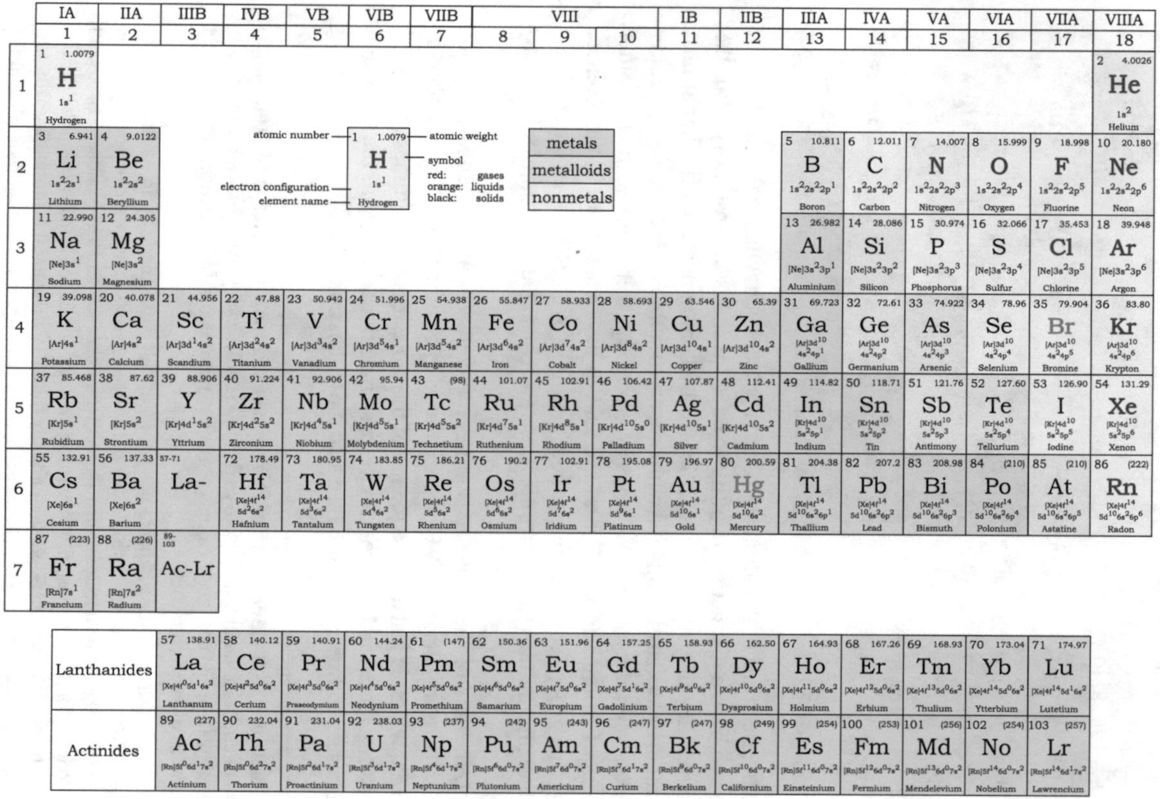

Fig. 2.9 The Periodic Table. Electron configurations as presented in the boxes pertaining to the elements are as given in Table 2.1. All other data presented in the boxes pertaining to the elements have been taken from Landolt-Börnstein, ed. by J. Bortfeld, B. Kramer. Numerical Data and Functional Relationships in Science and Technology, Units and Fundamental Constants in Physics and Chemistry, Subvolume b: Fundamental Constants in Physics and Chemistry (Springer Berlin, 1992) **a** The Periodic Table after Janet (and Dockx) **b** The Periodic Table as recommended by the IUPAC (International Union of Pure and Applied Chemistry).

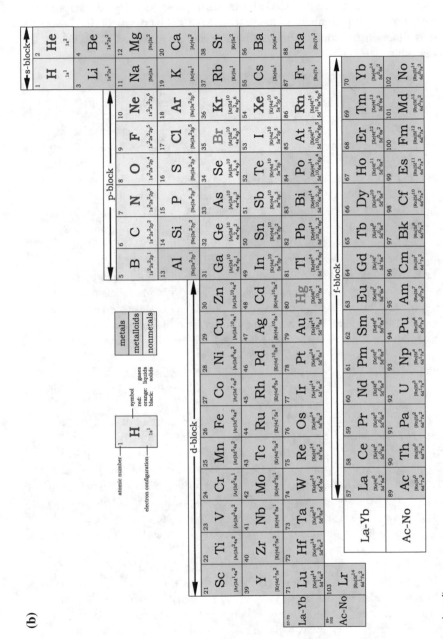

Fig. 2.9 (continued)

electron configuration of Cu above) as exhibited by the occurrence of two oxides: CuO and Cu_2O.

The "Aufbau Prinzip" has provided a beautiful, impressively elegant interpretation of the occurrence of periodicity in the properties of the elements on the basis of a listing as function of atomic number (initially atomic mass) as exhibited by the Periodic Table.

Epilogue: The Extent of the Periodic Table

At this place it appears appropriate to ask how many elements are contained in the Periodic Table. The element with the highest atomic number occurring naturally on earth is uranium (atomic number 92). The elements beyond uranium have to be produced artificially. This can be achieved by nuclear reactions. The occurrence of the heaviest element ever made, the element with atomic number 118, was reported in 2006 (Oganessian et al. 2006). It took until the end of 2015 before this finding was formally accepted, by the International Union of Pure and Applied Chemistry. This element of atomic number 118 was then given a name (in 2016): oganesson, with its symbol Og. Oganesson completes the seventh row/period of the Periodic Table (but the last element to be discovered of the seventh row was the element with atomic number 117 (named tennessine, with its symbol Ts) in 2010, also by Oganessian et al.) Until the writing of this book no further, still heavier elements have been discovered, thereby perhaps inducing the naïve suggestion that a final "rounded-off" state of completion of the Periodic Table has been reached (but see what follows). Note that with element 118, in accordance with the "Aufbau Prinzip", all s, p, d and f orbitals have been filled with electrons. Starting with element 121 a new, further type of orbitals would have to be utilized to describe (the ground state of) the atoms: g orbitals....

Actually, the production of such heavy, beyond uranium, and therefore called "transuranium", elements, has been a matter of strong competition and thus also been subject to controversy: already in 1999 it was claimed that the element now named oganesson had been produced, but later work could not confirm this result and the original data were considered suspect. This remark is made here as another indication that one should be careful in accepting any new, sensational, reported finding; see the footnote in the preface to this book. This holds in particular if some "race" exists, as here in discovering ever-heavier elements.

The lightest elements, hydrogen and helium, emerged soon after the creation of the universe ("Big Bang") upon its cooling: about three quarters of all atoms in the universe (still) are hydrogen atoms; helium constitutes the majority of the remaining quarter.... The "lighter" elements, up to iron (atomic number 26), have been produced by nuclear-fusion processes in stars. These nuclear-fusion processes, leading to element formation, release energy. However, the (equilibrium of the) forces acting within the nucleus (see further below) of atoms implies that the formation of atoms, from their constituents, for atomic numbers beyond 26 costs (does not release) energy. Indeed atoms of atomic number beyond 82 (i.e. beyond lead) are unstable: they decay, i.e. they are radioactive. It is thus thought that the natural elements with atomic number larger than 26 are the result of stellar explosions (supernovae) and/or the collision of neutron stars (having a density of an atomic nucleus: superdense objects). In the universe the natural elements of atomic number larger than 26 comprise only of the order 0.1% of all atomic mass in the universe.

Direct evidence for the existence of the heavy elements produced artificially on earth, as oganesson, is not obtained: the atoms of element 118 live less than a thousandth of a second; their occurrence is deduced from the observation of atomic decay products (see above). The extremely short life of atoms of element 118 is of interest by itself. Quantum mechanics has led to the concept of a shell-like constitution of the nucleus, to be compared with the shell-like constitution for the electron configuration of the atom discussed in this chapter. The positive charge of the protons within the nucleus brings about a strong, in this case repelling, Coulomb interaction, which can overcome the attractive, so-called "strong force" (that, as compared to the Coulomb interaction, acts at extremely short range), leading to nuclear fission. However, the actual outcome of these repulsion and attraction interactions depends on the "arrangement" of the protons and the neutrons within the nucleus. It can be argued that for specific numbers of protons and neutrons in the nucleus *relatively* high stability of the atom/nucleus occurs, which can be compared with the relatively high stability associated with the occurrence of closed shells of electrons (see above). This has led theoreticians to predict "islands of (relative) stability" in the Periodic Table for certain high atomic numbers. However, it is unclear, i.e. it is a matter of debate, which number of protons in the nucleus, beyond 82 (i.e. lead), corresponds with this relatively enhanced stability. It has further been argued that the attractive force acting between the electrons and the nucleus for atoms of atomic number of 173 and larger (the element with atomic number 172 would complete the (until now hypothetical; see above) 8th row of the Periodic Table) has become that large that formation of a neutral atom is impossible…

The idea behind the Periodic System, the occurrence of property periodicity as function of the atomic number (more precisely formulated as similar properties for elements in the same column of the Periodic Table) may be violated by the superheavy elements. As discussed at length in this chapter, the elements in a single column of the Periodic Table are supposed to have similar properties as a result of similar electron configurations in the "outer" shell. Upon increasing the positive charge of the nucleus, i.e. with increasing atomic number, the negatively charged electrons become accelerated. In case of the superheavy elements electron speeds are attained which are a very substantial fraction of the speed of light, implying that relativistic effects (according to Einstein's special theory of relativity) come into play (actually, already also, but less severe, for elements in the sixth row, as Au). As a result a contraction of the (valence) electron orbitals relative to "inner" electron orbitals can occur, as compared to the corresponding orbitals for the lighter elements higher up in the same column. Hence, as a consequence of such relativistic effects, the (chemical) properties of the superheavy elements may significantly deviate from those of the elements above them in the same columns. At present results both confirming and denying such phenomena have been presented (it should be recognized that experiments to this end are extremely difficult in view of the extremely short life time of these superheavy elements). Therefore it remains to be settled if, or to which extent, the Periodic Law is preserved in the realm of the superheavy elements.

As the experiments to be performed for the creation of these heavy elements are based on bombarding targets of one element with ions of another element, with an extremely low yield of the desired fusion product (e.g. bombarding calcium ions (with a nucleus containing 20 protons) onto a target of californium (with a nucleus containing 98 protons; itself not natural and radioactive), which led to three (!) atoms of atomic number 118 (= 20 + 98), i.e. oganesson, upon a bombardment of 10^{19} calcium ions...), only in one or two laboratories on earth one could perform such experiments[6] and thus it may take time to confirm claims of discovery of a new element and thus to extend the Periodic Table further, which, according to current knowledge, in this range of atomic numbers may be considered as of only academic interest....It is then up to mankind to decide if it desires to provide the huge funds for building the machines capable for executing such experiments, to provide answers to such questions.

2.5.1　Atom Size and Ionization Energy

Some important, qualitative statements regarding the dependencies of the size of the atoms and of the ionization energies of the atoms on the atomic number can be made on the basis of the systematization introduced by the Periodic Table (see Figs. 2.10 and 2.11, and (1)–(5) below).

---

[6] The paper referred to above (Oganessian et al., 2006), where it has been claimed that the element with atomic number 118 has been synthesized for the first time, has been authored by 30(!) persons (see under "References" at the end of this chapter). This is not unusual if an enormous experimental/instrumental effort, for example the building of huge (accelerating) machines, and a staggeringly vast experimental expertise is required (as in the field of nuclear/elementary particle chemistry and physics). This is a field of science where, as a consequence, the distance between theoreticians and experimentalists may have grown to an extreme. In materials science this development has not occurred to such extent. Many materials scientists, as the author of this book, feel happy by both performing experiments and developing and testing models (see Chap. 1) which ideally should provide a deep understanding of nature.

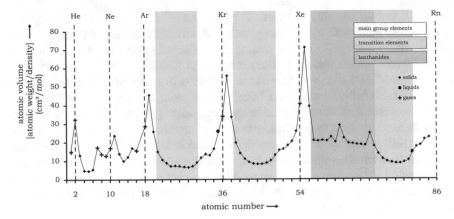

Fig. 2.10   The atomic volume of the elements shown as function of the atomic number. The atomic volume was calculated from the density and molar (atomic) weight, using density values at standard conditions for liquid and solid elements (Data taken from Landolt-Börnstein, ed. by J. Bortfeld, B. Kramer. Numerical Data and Functional Relationships in Science and Technology, Units and Fundamental Constants in Physics and Chemistry, Subvolume b: Fundamental Constants in Physics and Chemistry (Springer Berlin, 1992)) and density values at boiling point and 1 bar for (at standard conditions) gaseous elements. (Data taken from E.W. Lemmon, M.O. McLinden, D.G. Friend, ed. by Eds. P.J. Linstrom, W.G. Mallard NIST Chemistry WebBook, NIST Standard Reference Database 69, June 2005) and atomic weights. (Data taken from Landolt-Börnstein, ed. by J. Bortfeld, B. Kramer. Numerical Data and Functional Relationships in Science and Technology, Units and Fundamental Constants in Physics and Chemistry, Subvolume b: Fundamental Constants in Physics and Chemistry (Springer Berlin, 1992))

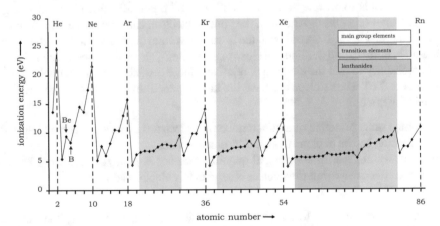

Fig. 2.11   The first ionization energy of the elements shown as function of the atomic number (data taken from W.C. Martin, A. Musgrove, S. Kotochigova, and J.E. Sansonetti, NIST Standard Reference Database 111)

First it should be remarked that the notion "size" of the atom, or size of the "ion" (ion = atom with one or more "outer" electrons taken away by a process of adding energy enough to debind one or more electrons from the atom, which is called "ionization") is unclear. As there is a finite probability to find an electron, thus also an "outer" electron of an atom, at any location in space (apart from at infinity where the probability is zero), the "size" of any atom appears to be of "infinite" nature. However, for example on the basis of considering interatomic distances (that is the distance between the centroid positions of atomic mass) for specific types of bonding between the atoms, one can define "atomic size". A further complication arises because it has been tacitly assumed in the above lines that the atom is a sphere. In a specific kind of chemical bonding the bonding is not of isotropic nature, which holds for so-called "covalent bonding" (see Sect. 3.4). Then "size" refers to a certain direction in space as well.

The dependence of the atomic volume on position in the Periodic Table can now be discussed as follows (see Fig. 2.10):

(1) It is obvious that for the same number of electrons and increasing Z the size of the atom/ion will decrease. This is a straightforward consequence of the Coulomb interaction between the positive nucleus and the negative electrons (see also Eq. (2.2)).

An "outer" electron does not experience the full nuclear charge: the nucleus charge is "screened" by the inner electrons. Because electrons are not confined to specific parts of space outside the nucleus (see discussion in especially Sect. 2.4.1), this screening is not 100% (i.e. for a nucleus of charge $+Ze$ the screening by m "inner" electrons leads to an effective (i.e. "felt" by the "outer" electron) nuclear charge larger than $+(Z - m)e$). And also, the "outer" electron can "penetrate" the "cloud" of "inner" electrons. It appears that the lower the second quantum number l the larger the probability to find the electron close to the nucleus, and at the same time the larger the probablity to find the electron at relatively large distances from the nucleus (see also Sect. 2.6). Thus s electrons have the largest penetrative power and the smallest screening effect:

- for the same principal quantum number n, s electrons experience more Coulomb attraction by the nucleus than p, d, f, electrons do.
- similarly, for the same principal quantum number n, $p, d, f, \ldots$ electrons screen more effectively from the nuclear charge than s electrons do.
  Thus:

(2) Going from the left to the right in a period of the Periodic Table the number of "outer" electrons *in the same shell* increases. Because the screening constant

of these electrons is smaller than 100% and at the same time the nuclear charge increases with one for each electron added upon increasing the atomic number, it follows that, the atomic size decreases going from the left to the right in a period of the Periodic Table. Because s electrons screen less good than p electrons, the size changes between, for example considering the second period, Li, Be and B are larger than those between B, C, N, O and F. Further, considering the transition series, upon increasing the atomic number "inner" electrons of relatively high second quantum number l and thus relatively good"screening" power are added. Hence, the atomic size is practically constant in a transition series.

(3) Going from top to bottom in a column of the Periodic Table the number of "outer" electrons is constant for increasing principal quantum number. The highest probability for finding this "outer" electron occurs at a distance from the nucleus increasing with principal quantum number n (see Sect. 2.6). Thus the atomic size increases going from top to bottom in a column of the Periodic Table. The "inner" electrons "screen" the nuclear charge. But, this "screening" is not 100%. Hence, the effective nuclear charge experienced by the "outer" electrons in a column increases for increasing principal quantum number. This effect causes the increase of atomic size from top to bottom in a column of the Periodic Table to be moderate.

The ionization energies for removal of a first electron (i.e. from the neutral atom) as function of position in the Periodic Table can be discussed in a similar way (see Fig. 2.11):

(4) Going from the left to the right in a period of the Periodic Table the number of "outer" electrons *in the same shell* increases. Because the screening constant of these electrons is smaller than 100% and at the same time the nuclear charge increases with one for each electron added upon increasing the atomic number, it follows that the ionization energy will overall increase going from the left to the right in a period of the Periodic Table. Further, because the d and f electrons added in the transition series have a relatively high "screening" power, it follows that the ionization energies of a transition series are rather equal. Because the p electrons are less "penetrating" than s electrons, their Coulomb interaction with the nucleus is less, their energy level is higher (cf. Fig. 2.8) and consequently they are relatively easily ionizable. This explains the drop in ionization energy experienced going, for example in the second period, from Be to B.

(5) Going from top to bottom in a column of the Periodic Table the number of "outer" electrons is constant for increasing principal quantum number n. The energy of the "outer" electrons increases with n [cf. Eq. (2.3)]. The "inner" electrons "screen" the nuclear charge. But, this "screening" is not 100%. Hence, the effective nuclear charge experienced by the "outer" electrons in a column increases for increasing principal quantum number. The net effect [relative change of effective Z^2 is smaller than the relative change of n^2; cf. Eq. (2.3)] is a decrease of the ionization energy going from top to bottom in a column of the Periodic Table.

2.6 The Shape of the Probability Density Distribution for the Electron

For a one-electron system (hydrogen-like atom) the three-dimensional probability density distribution of the electron in space, around the nucleus, as given by the square of the amplitude, ψ, of the wave associated with the particle as a function of position (x,y,z), can be shown in a picture, thereby presenting an image of the electron distribution. Such a simple pictorial description of the electron distribution is impossible for a two or more electron system, as the probability density for one electron to be in the volume element $\Delta x \Delta y \Delta z$ at a specific location (x, y, z) depends on the coordinates of all other electrons, as remarked in Sect. 2.4.1. Thus Figs. 2.12, 2.13, 2.14, 2.15, 2.16 and 2.17 concern a one-electron system (hydrogen-like atom).

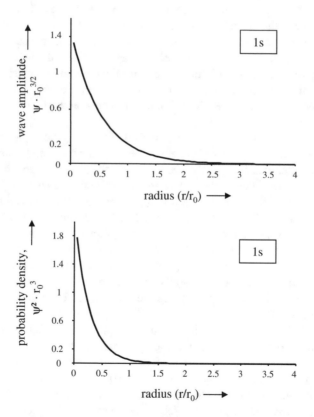

Fig. 2.12 Top part of the figure: The $1s$ wave (probability) amplitude ψ as function of the radial distance to the (centre of the) nucleus for a hydrogen-like atom (i.e. one-electron system). Bottom part of the figure: The $1s$ probability density, ψ^2, as function of the radial distance to the (centre of the) nucleus for a hydrogen-like atom (i.e. one electron).r_0 is the so-called Bohr radius (0.0529 nm) which is the radius of the $1s$ electron orbit in the hydrogen atom according to the Bohr model. The normalization factors $r_0^{3/2}$ and r_0^3 used for the ordinates (make the ordinates dimensionless and) are a direct consequence of the requirement that the probability to find the electron anywhere in space equals 1 (i.e. $\int \psi^2 \, dxdydz = 1$, where the integration ranges for x, y and z cover all space)

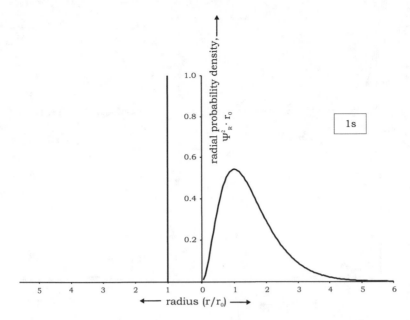

Fig. 2.13 Right part of the figure: the radial probability density distribution; ψ_R^2, according to the wave-mechanical model, for the $1s$ orbital, of a hydrogen-like atom (i.e. one-electron system), shown as function of r/r_0. The normalization factor r_0 used for the ordinate (makes the ordinate dimensionless and) is a direct consequence of the requirement that the probability to find the electron anywhere in space equals 1 (i.e. $\int 4\pi r^2 \psi_R^2 \, dr = 1$, where the integration range for r covers all space). Left part of the figure: the radial probability density "distribution" for the $1s$ electron according to Bohr's model; the electron is at $r = r_0$. r_0 is the so-called Bohr radius (0.0529 nm) which is the radius of the $1s$ electron orbit in the hydrogen atom according to the Bohr model

The result for the $1s$ electron of a hydrogen-like atom is an isotropic distribution of ψ^2 shown as function of the radial distance to the nucleus, r, in Fig. 2.12 (bottom part of the figure). Evidently the probability to find the $1s$ electron in a volume element $\Delta x \Delta y \Delta z$ is the largest close to (at) the nucleus. Often one is more interested in the (average) electron density at a distance r from the nucleus. This involves the calculation of the probability density for the electron to be in the spherical shell of (constant) thickness Δr. One can thus speak of the *radial probability density*, which for distinction with the probability density ψ^2 for the electron in the volume element $\Delta x \Delta y \Delta z$, is indicated by Ψ_R^2:

$$\Psi_R^2 = \psi^2 4\pi r^2$$

Because the function $4\pi r^2$ increases parabolically with r and the function ψ^2 decreases exponentially with r, the net effect is the occurrence of a maximum in Ψ_R^2 at a finite distance from the nucleus; see Fig. 2.13. The location of this maximum agrees with the position of the first orbit in Bohr's model. The distinction between the deterministic model of Bohr and the probability/uncertainty wave-mechanical

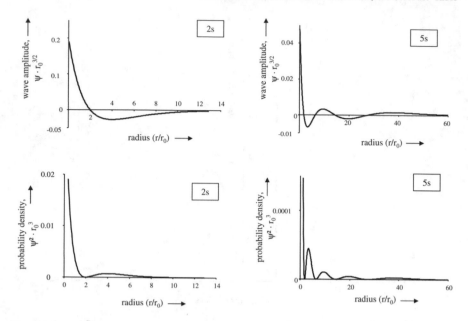

Fig. 2.14 Wave amplitude and probability density for the 2s (left part of the figure) and 5s (right part of the figure) orbitals, of a hydrogen-like atom (i.e. one-electron system), as function of r/r_0. For normalization factors used, see the captions of Figs. 2.12 and 2.13

model should be well appreciated: in the Bohr model the electron can only be at its orbit; in the wave-mechanical model the electron has a finite probability to be in a volume element at any location, apart from at infinite distance from the nucleus where the probability for finding the electron in a volume element must be zero. Most strikingly, the wave-mechanical model predicts that the highest probability for finding the 1s electron is very near to (at) the nucleus (Fig. 2.12 discussed above).

The wave (probability) amplitudes and the radial probability density for the electron in the 2s and 5s orbitals are shown in Fig. 2.14. The largest values for finding the electron at a certain radial distance from the nucleus occur at increasing r for increasing n. Yet, there is an appreciable chance for the s electrons of principal quantum numbers higher than 1 to be at radial distances close to the nucleus (see the subsidiary maxima in Fig. 2.14). This observation remains true also in many-electron systems; it is called "penetration": the s electrons thereby also offer an only modest degree of "screening" from the nucleus charge (see discussion in Sect. 2.5).

For a p electron, two of the three ($l = 1$ with $m_l = -1, 0, +1$) direct solutions of the Schrödinger equation are, in the mathematical sense, complex and cannot be visualized in real space. Provided a coordinate system in real space can or has to be chosen because of the environment of the atom considered (atom in a molecule or in a magnetic field), real space representations, to be obtained by linear combination (i.e. addition and subtraction) of the direct, complex solutions of the Schrödinger equation, make sense. Note that for a differential equation, as the time independent

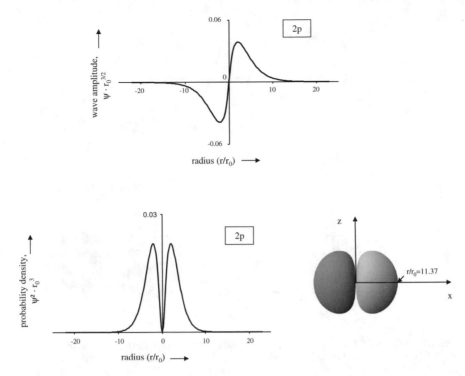

Fig. 2.15 Wave amplitude (upper part of figure) and probability density (left, bottom part of figure) for the $2p_x$ (or $2p_y$ or $2p_z$) orbital, of a hydrogen-like atom (i.e. one-electron system), as function of r/r_0. For normalization factors used, see the caption of Fig. 2.12). The bounding surface for the $2p_x$ orbital, shown in the right, bottom part of the figure (similar for the $2p_y$ and $2p_z$ orbitals), represents the three-dimensional surface where the probability density ψ^2 has decreased to 4×10^{-4} (outside this surface the probability density for the electron is even less, but still finite); the surface has been generated with Orbital Viewer, Version 1.04 (D. Manthey, https://www.orbitals.com/orb)

Schrödinger equation, such linear combinations are automatically also solutions (we will use this property again later, see Sect. 3.4). The three results thus obtained for the p subshell are usually designated by p_x, p_y and p_z. The probability density distributions for these three possible solutions for one $2p$ electron are visualized in Fig. 2.15. The p_x, p_y and p_z functions are concentrated around the x, y and z-axis, respectively. Evidently, spherical symmetry, i.e. isotropy, no longer occurs for the individual solutions. However, if the $2p$ subshell would be half filled (three electrons, one in each of the p functions; all three with parallel spin; cf. the discussion of Hund's rule in Sect. 2.5) or fully filled (six electrons, with two of opposite spin per p function), the resulting, overall electron distribution would be spherical. As compared to the s electron of similar principal quantum number, the p electron has a lesser probability to be found close to the nucleus and also has a lesser probability to be found at large distance from the nucleus: p electrons penetrate less and screen more effectively than s electrons (see Sect. 2.5). Similar results for one $3p$ electron are shown in Fig. 2.16.

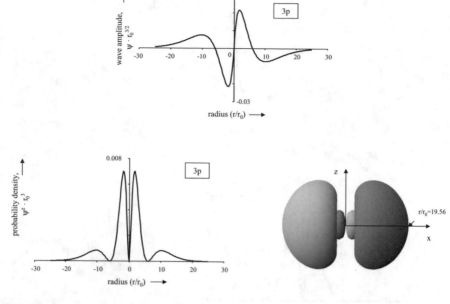

Fig. 2.16 Wave amplitude (upper part of figure) and probability density (left, bottom part of figure) for the $3p_x$ (or $3p_y$ or $3p_z$) orbital, of a hydrogen-like atom (i.e. one-electron system), as function of r/r_0. For normalization factors used, see the caption of Fig. 2.12). The bounding surface for the $3p_x$ orbital, shown in the right, bottom part of the figure (similar for the $3p_y$ and $3p_z$ orbitals), represents the three-dimensional surface where the probability density ψ^2 has decreased to 4×10^{-4} (outside this surface the probability density for the electron is even less, but still finite); the surface has been generated with Orbital Viewer, Version 1.04 (D. Manthey, https://www.orbitals.com/orb)

A similar discussion, as given above for a p electron, is possible for a d electron. Again confining our attention to real space representations only, the five different d functions for the $3d$ subshell, $3d_{xy}$, $3d_{xz}$, $3d_{yz}$ and $3d_{x^2-y^2}$ and $3d_{z^2}$, are visualized in Fig. 2.17 by bounding surfaces (see figure caption). Analagous remarks as for the p electrons regarding the occurrence of spherical symmetry, for a half filled (here 5 electrons) and a fully filled (here 10 electrons) d subshell, and the penetration and screening effects of a d electron can be made.

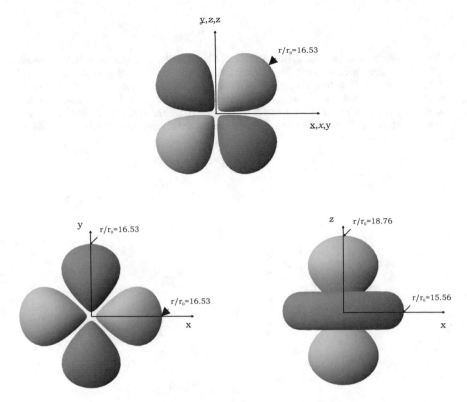

Fig. 2.17 The bounding surfaces for the $3d$ orbitals, of a hydrogen-like atom (i.e. one-electron system), shown represent the three-dimensional surfaces where the probability density ψ^2 has decreased to 4×10^{-4} (outside this surface the probability density for the electron is even less, but still finite); the surfaces have been generated with Orbital Viewer, Version 1.04 (D. Manthey, https:// www.orbitals.com/orb). The upper part of the figure shows the bounding surface for the $3d_{xy}$, $3d_{xz}$ and $3d_{yz}$ orbitals (shape similar; for axis permutation, see the figure). The left, bottom part of the figure shows the bounding surface for the $3d_{x^2-y^2}$ orbital. The right, bottom part of the figure shows the bounding surface for the $3d_{z^2}$ orbital

References

General

W. Finkelnburg, *Einführung in die Atomphysik*, 11. und 12. Auflage (Springer, Berlin, 1967) (in German)

G.F. Lothian, *Electrons in Atoms* (Butterworths, London, 1963).

E.R. Scerri, *The Periodic Table* (Oxford University Press, Oxford, 2007).

E. Scerri, *A Tale of Seven Elements* (Oxford University Press, New York, 2013).

J.W. van Spronsen, *The Periodic System of Chemical Elements* (Elsevier, Amsterdam, 1969).

Specific

Yu.Ts. Oganessian, V.K. Utyonkov, Yu. Lobanov, F. Sh. Abdullin, A.N. Polyakov, R.N. Sagaidak, I.V. Shirokovsky, Yu. S. Tsyganov, A.A. Voinov, G.G. Gulbekian, S.L. Bogomolov, B.N. Gikal, A.N. Mezentsev, S. Iliev, V.G. Subbotin, A.M. Sukhov, K. Subotic, V.I. Zagrebaev, G.K. Vostokin and M.G. Itkis, K.J. Moody, J.B. Patin, D.A. Shaugnessy, M.A. Stoyer, N.J. Stoyer, P.A. Wilk, J.M. Kenneally, J.H. Landrum, J.F. Wild, and R.W. Lougheed, Synthesis of the isotopes of elements 118 and 116 in the ^{249}Cf and ^{245}Cm + ^{48}Ca fusion reactions/ Phys. Rev. **C74**, 044602 (2006)

Chapter 3
Chemical Bonding in Solids; with Excursions to Material Properties

3.1 Attractive and Repulsive Forces; Thermal Expansion and Elastic Constants

Why do atoms stick together? And why do they gather in aggregates exhibiting specific types of three-dimensional (periodic) arrangements? Mankind, on its road to reveal the secrets of nature, time and again returns to these questions in order to develop an ever-growing insight on how matter is formed from its building units.

Bonding requires the existence of an attractive force acting between atoms; one speaks of "chemical affinity". However, atoms can*not* approach each other up till an infinitesimally small distance: "impenetrability of matter". This leads to the recognition that a repulsive force acting between atoms exists too. Whereas the attractive forces range over distances of the order of a number of atomic sizes, the repulsive force is of extremely short-range nature; it acts over a distance of the order of the size of an atom.

Consider two atoms at infinite distance. There is no interaction of these atoms. Hence, the potential energy of interaction is zero. Upon decreasing the distance between the atoms the attractive force is "felt" for the interatomic distance becoming smaller than a number of atom diameters. Consequently, then the potential energy of interaction becomes negative: it costs energy to bring the atoms back at infinite distance, i.e. it costs energy to "debond" the atoms: debonding requires work to be done against the attractive force that drives the atoms together. Alternatively, the action is done by the attractive force by closing the distance between the atoms releases energy. The decrease of potential energy due to the attraction is shown by the dashed curve in Fig. 3.1.

If the distance between the atoms is further decreased, then for distances smaller, they say an atomic diameter the repulsive force is felt: driving the atoms further

© Springer Nature Switzerland AG 2021
E. J. Mittemeijer, *Fundamentals of Materials Science*,
https://doi.org/10.1007/978-3-030-60056-3_3

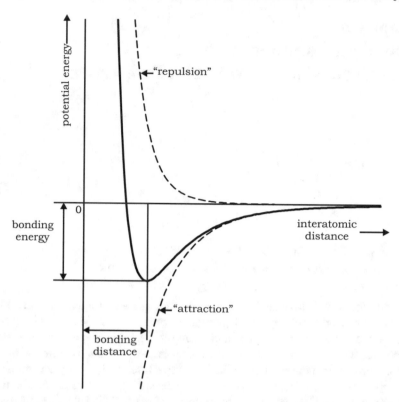

Fig. 3.1 Potential energy of interaction of a pair of atoms as function of the interatomic distance. Result of repulsive and attractive forces (dashed lines) and the total potential energy (solid line). The potential energy curves were calculated according to a Lennard–Jones type equation (see Footnote 5). The optimum, minimal value of potential energy occurs at the bottom of the potential energy well which defines the optimal bonding distance and bonding energy as indicated in the figure

together requires working against the repulsive force, which causes a positive contribution to the potential energy of interaction. See the (other) dashed curve as shown in Fig. 3.1.[1]

The net result of the attractive and repulsive forces is the establishment of an equilibrium situation. There exists an optimal distance between the atoms where the total energy of interaction is minimal. This distance is called the bonding distance

---

[1] The force F acting on a particle in a potential energy field, as due to this field, is determined by the gradient of the potential energy U, i.e. the rate at which the potential energy varies with position: $F = -dU/dr$ (one-dimensional consideration). The minus sign before the gradient immediately makes clear that the force imposed by the field "drives" the particle in a potential energy minimum. Consider the potential energy well in Fig. 3.1 ($r > 0$). If dU/dr is positive, right from the bottom of the well, the force is negative and the particle is driven in the "$-r$" direction, i.e. towards smaller r. If dU/dr is negative, left from the bottom of the well, the force is positive and the particle is driven in the "$+r$" direction, i.e. towards larger r.

and the bonding energy is this minimum value of the (potential) energy of interaction. See the resulting, total (potential) energy of interaction curve (solid line in Fig. 3.1).

The above discussion is restricted to a pair of atoms. For solids, the simultaneous interaction of many atoms needs to be considered. Yet, a consideration on the basis of an energy versus interatomic distance curve, in the above sense, remains possible, at least qualitatively,[2] and we can associate a bonding energy to each atom. Obviously, for solids, the bonding energy is relatively large; for gases, the bonding energy is relatively small and liquids take an intermediate position in this respect. It can be anticipated that the melting point of a solid is a qualitative measure for the bond strength/bonding energy and that an analogous statement holds for liquids with respect to the boiling point, etc.

Because of the short-range nature of the repulsive force and the more long-range nature of the attractive force, the potential energy minimum well has an asymmetrical shape with respect to the position of energy minimum (see Fig. 3.1). This asymmetry has an interesting consequence. The atoms vibrate around their equilibrium positions,[3] with an amplitude that increases with temperature and a frequency that changes only in a minor way with temperature.[4] The vibrational energy contribution has the effect that the system is not at the minimum of the potential energy well, but stays at a somewhat higher level, which is the higher the higher the temperature is; see the horizontal lines indicated in Fig. 3.2. Evidently, if during their vibration the pair of atoms considered becomes separated by a distance smaller that the above-discussed equilibrium distance, the potential energy rises more rapidly than for the situation where, due to the atomic vibration, the interatomic distance becomes larger than the equilibrium distance with the same amount. This is an immediate consequence of the asymmetry of the potential energy well. Hence, for a given level of energy of the system, the allowable decrease of the interatomic distance is smaller than the allowable increase: due to the atomic vibration, the interatomic distance varies between the points where the horizontal line cuts the potential energy minimum well (see the horizontal line segments with arrowheads in Fig. 3.2). Thereby, it follows straightforwardly that *the average interatomic distance increases with temperature*; see the dashed line in Fig. 3.2. Thus, the general, well-known experience that a (solid) body

---

[2] This is a non-trivial statement. The picture used here for the interaction between atoms adopts the action of "central forces": a central force acts along the line joining two atoms and its value depends on the distance between these atoms. The concept of central forces acting between the atoms is certainly an *ina*dequate description for metallic solids, characterized by "non-directional bonding" (see Sect. 3.5), in contrast with the case of covalent bonding which is characterized by directional bonding (see Sect. 3.4). Regarding the limitation/break-down of the central force concept, see also the *"Intermezzo: Short history of the Poisson constant"* in Sect. 12.2. The metallic solid state is insufficiently described as an aggregate of atoms held together by short-range central forces. Yet, a description of even metallic bonding on the basis of pairwise interaction, then consequently of approximate nature, but on the basis of an empirical, i.e. adapted to reality, interaction potential, can provide an at least *qualitatively* realistic picture of bonding in the metallic solid state. The central force approximation is therefore used in the remainder of this subsection as a general approach.

[3] Here, we restrict ourselves to solids, where the thermal disorder is dominated by the thermal vibrations; in a gas, both significant translational motion and significant rotation can occur as well.

[4] Even at 0 K the atoms vibrate (zero point vibrations); an effect predicted by quantum mechanics.

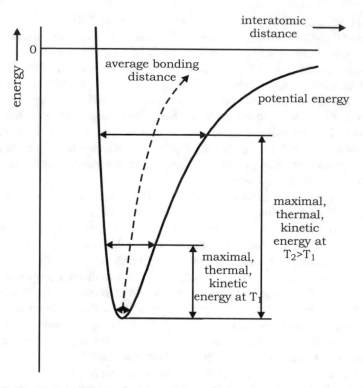

Fig. 3.2 Total potential energy for interaction of a pair of atoms as function of the interatomic distance (cf. Fig. 3.1). The maximal thermal kinetic energy of vibration has been indicated for two temperatures: see the horizontal lines. At a certain temperature, the total energy (potential plus kinetic) is given by the corresponding horizontal line; during a vibration cycle (see the horizontal line segments with the double arrowheads), the kinetic and potential energy vary while the total energy is constant. The thus deduced increase of the *average* bonding distance with temperature has been indicated by the dashed line. (The thermal vibration at 0 K has been suggestively indicated near the bottom of the potential energy well: see the corresponding double arrowheads)

expands upon heating at constant pressure can be understood: the average distance between the atoms becomes larger upon heating. A symmetrical potential energy well would obviously imply that upon heating the interatomic distance and thus the volume of the body remain the same.

The larger the bond strength, the deeper and smaller the potential energy minimum well. It may thus be expected that the linear coefficient of thermal expansion, α (= the relative length increase per K), decreases for increasing melting point (see last sentence of the one but last paragraph), as long as materials of similar atomic arrangement and type of bonding are compared. For example, for Al with melting point 933 K, it holds $\alpha = 23.6 \times 10^{-6}$ K^{-1} and for Ni with melting point 1726 K it holds $\alpha = 13.3 \times 10^{-6}$ K^{-1}, for temperatures at about room temperature. Note that α is not a constant in general, but more or less depends on temperature.

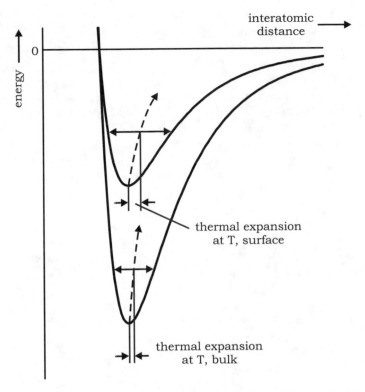

Fig. 3.3 Total potential energy for interaction of a pair of atoms as function of the interatomic distance for both bulk and surface atoms with the potential energy for surface atoms taken as one half of the potential energy of bulk atoms, recognizing the difference in coordination number (see text). The shape of the curves is again according to a Lennard–Jones type equation (see caption of Fig. 3.1). The maximal thermal kinetic energy for vibration at the same temperature T has been indicated by the corresponding horizontal lines for both curves (interaction of bulk atoms and interaction of surface atoms; cf. Fig. 3.2). The resulting thermal expansions at temperature T (as compared to 0 K) for the bulk and surface atoms have been indicated too. The increase of the average bonding distance with temperature has been represented for both the bulk and the surface atoms by the dashed lines (cf. Fig. 3.2)

Intermezzo: The Linear Coefficient of Thermal Expansion of Large and Small Crystals

Atoms at the surface of a crystal are not saturated with respect to their state of bonding: their coordination number (i.e. the number of nearest neighbours; see also Sect. 4.2.4) is less than for bulk atoms. As a consequence, the curve of potential energy per atom versus interatomic distance for a surface atom shows a less deep potential energy minimum well than for a bulk atom (see Fig. 3.3). This can be illustrated for the case of the crystal of a noble, inert gas as Ne, Ar, Kr and Xe. In this case, the potential energy of the crystal can be calculated approximately by considering pair interactions of an atom with only its nearest neighbours (12 in an f.c.c. crystal (cf. Sect. 4.2.1.2), as holds for solid Ne, Ar, Kr and Xe). At the surface, the number of nearest neighbours is smaller than 12. Then, if the pair interaction is given by the same function for surface and bulk atoms, e.g. the so-called Lennard–Jones potential,[5] then a result as shown in Fig. 3.3 is obtained, where the coordination number of a surface atom has been taken as 6. The effect of a similar vibrational kinetic energy for surface and bulk atoms now has consequences of different extent. Evidently, the less deep potential energy minimum for the surface atom, as compared to the bulk atom, leads to a larger atomic position variation due to thermal vibration for the surface atom than for the bulk atom (see horizontal line segments with arrowheads in Fig. 3.3 and see the discussion above with respect to Fig. 3.2). Consequently, in view of the asymmetry of the potential energy minimum wells, the thermal expansion for the surface atoms of the crystal is larger than for the bulk atoms. The smaller a crystal, the larger the ratio of the number of surface atoms and the number of bulk atoms. Hence, the smaller a crystal, the larger its *average* linear coefficient of thermal expansion. Due to severe experimental problems in measuring reliably linear coefficients of thermal expansion of very small crystallites, only recently it has been possible to experimentally prove conclusively this qualitative theoretical prediction (Kuru et al. 2007). The temperature dependence of the interatomic distance (as indicated by the lattice parameter; e.g. see Sect. 4.2.1.2 and Fig. 4.20) of nanocrystalline Ni (in a thin film) was measured. For Ni crystals as small as 35 nm, it was found $\alpha = 13.7 \times 10^{-6}$ K^{-1}, whereas the value for bulk Ni is given by $\alpha = 12.4 \times 10^{-6}$ K^{-1}. By an annealing treatment, the crystal size increased up to 50 nm. Then, the value of the linear coefficient of thermal expansion was determined again and it was found: $\alpha = 12.6 \times 10^{-6}$ K^{-1}, which result is now close to the bulk value. (Conceiving the crystals as spheres, the surface/volume ratio increases with about 43% for the crystal diameter reducing from 50 to 35 nm.)

It can be anticipated that the bond strength is related to the elastic behaviour of a material (more precisely: this holds for so-called linearly elastic behaviour: cf. Sects. 12.2, 12.6 and 12.7). This can be explained more explicitly in the following way. Consider the case that a hydrostatic[6] pressure/force is applied that is either of compressive (leading to compression) or tensile (leading to expansion) nature. As a result of the applied hydrostatic force F, a volumetric strain occurs that leads to a

---

[5] The well-known Lennard–Jones potential energy is of the type: $c_1[(c_2/r)^{12}$(repulsion) $- (c_2/r)^6$(attraction)], with c_1 and c_2 as constants and r as the interatomic distance of the atom pair considered. The Lennard–Jones potential works well for noble gases (in both the solid and liquid state), but is less appropriate for metals, with "squeezable" atoms (see last paragraph of the introductory part of Sect. 4.2).

[6] "Hydrostatic" means that the same force (pressure) acts in all directions, i.e. the force (pressure) is isotropic.

change of the equilibrium interatomic distance: $r_{eq} \rightarrow r_{eq} + dr$, which is associated with a change of the total potential energy: $U_{tot}(r_{eq}) \rightarrow U_{tot}(r_{eq} + dr)$. Using a Taylor's series expansion for $U_{tot}(r_{eq} + dr)$ around r_{eq} and applying the equilibrium condition, given by $dU_{tot}/dr = 0$ at $r = r_{eq}$, it follows:

$$dU_{tot} = U_{tot}(r_{eq} + dr) - U_{tot}(r_{eq}) = \frac{1}{2} \cdot \left(\frac{d^2 U_{tot}}{dr^2} \right)_{r=r_{eq}} \cdot (dr)^2 \tag{3.1}$$

The force that caused the strain associated with the change $r_{eq} \rightarrow r_{eq} + dr$ is calculated from the potential energy according to $F = dU/dr$ (The force considered here is not the force imposed by the potential energy field as considered in Footnote 1. It is an externally imposed force working against the potential energy gradient (the atoms are removed from their equilibrium positions), and therefore the minus sign before dU/dr now has been omitted). Then, it follows for the force F to produce the linear strain $1/r_{eq} \cdot dr$:

$$F = \frac{dU_{tot}}{dr} = \frac{1}{2} \cdot \left(\frac{d^2 U_{tot}}{dr^2} \right)_{r=r_{eq}} \cdot dr \tag{3.2}$$

The linear relation between force/pressure (F) and strain (dr/r_{eq}) is called Hooke's law, with the proportionality constant called elastic constant (see Sect. 12.2). Equation (3.2) provides a validation of Hooke's law and shows that the elastic constants, as the compressibility for the case of hydrostatic loading considered here [see Eq. (11.19)], have a close relationship to the potential energy versus interatomic distance curve describing the chemical bonding (i.e. are determined by $(d^2 U_{tot}/dr^2)_{r=r_{eq}}$).

By straightforward calculus, using equations of the type given by Eqs. (3.3) and (3.4) in Sect. 3.3 for the attractive and repulsive contributions to the total energy and making use again of the equilibrium condition [see above Eq. (3.1)], it can be shown that the elastic constant is mainly determined by the contribution of the repulsive interaction: in general, the short-range repulsive interactions govern the values of the elastic constants.

Hence, macroscopic material properties as the thermal expansion and elastic deformation behaviour can be related in a direct way to the degree and nature of the chemical bonding. A large bond strength associated with a deep and narrow energy minimum well ($d^2 U_{tot}/dr^2$ is large at $r = r_{eq}$) brings about a small coefficient of thermal expansion and large elastic constants.

3.2 Remarks on Model Types of Bonding

The first ideas about the forces between particles (atoms in the present discussion) leading to bonding involved that electrostatic interactions between particles (atoms) govern the chemical bonding (Berzelius in 1812). This picture, in ways much different from the thoughts and concepts of its original proponent, still holds today. In the following, we will be largely, although not exclusively, be concerned with solids. Whatever model for bonding is considered, it always holds that the cohesion of a solid is, after all, explained by the electrostatic interaction between the positively charged atomic cores or nuclei and the negatively charged electrons.

The discussion on the electronic structure of the atom (Chap. 2) has revealed that a completely filled electron shell is very stable relatively; i.e. an electron configuration (of the outer electron shell) as for a (nearest by in the Periodic Table) noble (inert) gas atom is strived for, by giving away one or more valence (i.e. outer and relatively weakly bonded) electrons or accepting one or more of these (cf. the discussion in Sect. 2.5). On this basis, much of the chemical bonding behaviour of the elements can be understood qualitatively. Elements with the tendency to give away valence electrons are called to be *electropositive* and elements with the tendency to take up additional electrons are called to be *electronegative*. Obviously, upon going from left to right in a period of the Periodic Table, the tendency for electronegativity increases and the tendency for electropositivity decreases.

At this place, a remark on the position of hydrogen in the Periodic Table is in order (see also Sect. 2.5). Sometimes, H is not only positioned on top of the column of alkali elements but also placed on top of the column of the "halogens". This reflects that H exhibits properties related to giving away its single and naturally outermost s electron, as holds for the alkali elements, but H also shows the tendency to fill up its $1s$ and naturally outermost shell with one additional electron, in accordance with the behaviour shown by the halogen elements.

In the following, typical bonding types which are met in nature are discussed. It should be realized that these specific types of bonding are (simplified) extremes of reality. Thus, there exists no case of really pure ionic bonding or really pure covalent bonding. It is recognized that often bonding can be well described by one of the bonding types to be discussed, but in many cases distinctly mixed bonding character occurs.

For the formation of solids, a requirement appears to be that the attraction of atoms/ions upon bonding remains "unsaturated": attractive forces operate promoting further addition of atoms/ions, leading to growth of the solid. This contrasts with the formation of molecules as H_2, F_2 or CH_4 (see Sect. 3.4), where after the bonding has been realized "saturation" of bonding has occurred: the molecule formed does not grow. In the latter case, solid formation ("condensation") can only occur if other, often weak(er), attractive forces come into play inducing attraction between the molecules (cf. Sect. 3.6).

3.3 Ionic Bonding; Lattice Energy and The Madelung Factor

The typical system exhibiting this type of bonding is composed of atoms with a few (one or two) valence electrons, as the alkali, alkaline earth and transition metals, and of atoms with a large number (say, seven) of valence electrons, as the halogens. Consider as an example, the compound NaCl (sodium chloride, rocksalt as solid).

The electron configuration of Na is $1s^2 2s^2 2p^6 3s^1$. By giving away its valence electron, Na assumes the electron configuration of Ne. The electron configuration of Cl is $1s^2 2s^2 2p^6 3s^2 3p^5$. By accepting one electron, Cl assumes the electron configuration of Ar. In the combination NaCl, Na can donate its valence electron to Cl, and thereby becomes positively charged, a Na^+ cation occurs, and Cl can accept this electron, and thereby becomes negatively charged, a Cl^- anion occurs. Obviously, this electron transfer is associated with the occurrence of an electrostatic attraction of the Na^+ cation and the Cl^- anion which are oppositely charged. This attractive so-called Coulomb-type interaction induces a potential energy of interaction contribution (taking the ions as spheres) given by[7]:

$$U_{attr} = \frac{1}{(4\pi\varepsilon)} \cdot \frac{(ne)(-me)}{r} \tag{3.3}$$

where n and m are the valences of the cation and the anion, respectively (thus, here $n = m = 1$), e is the charge of the electron, ε represents the permittivity, also called dielectric constant, and r denotes the interatomic (interionic) distance.[8]

As discussed in Sect. 3.1, the interatomic distance cannot be made infinitely small, although this would release the largest amount of potential energy due to the action of the attractive force [cf. Eq. (3.3)]. In this case, the repulsive force can be understood as the consequence of overlapping of the *completely filled* outer electron shells of both ions. This repulsion can also be seen as a consequence of the Pauli exclusion principle: if the two outer electron shells, one of each ion involved, would fuse upon close approach, it would be impossible to assign all now shared electrons to electron energy states of the relatively low energy corresponding to the original shells considered. Excitation of a number of these electrons to a higher energy level would be required. Thereby just another formulation is given for the occurrence of a strong repulsion at such close distance. The discussion implies that if *partly filled* electron shells of approaching atoms would overlap, the outcome can be much different. This is what happens in the case of covalent bonding to be discussed in Sect. 3.4.

The repulsion of the two ions upon close approach leads to a potential energy of interaction contribution which can be written empirically, according to Born and Mayer, as:

---

[7] This equation only holds exactly if the distance between the charged spheres is (much) larger than the sphere radii.

[8] The attractive force is given by (see Footnote 1): $F = -dU_{attr}/dr = -1/(4\pi\varepsilon) \cdot (ne)(-me)/r^2$.

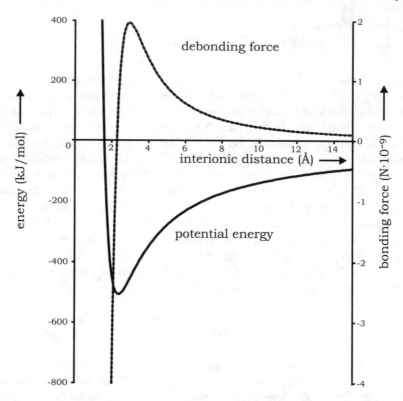

Fig. 3.4 Potential energy (solid line) and bonding force (dashed line) of a Na^+ Cl^- ion pair. The "bonding force" has been calculated as dU/dr, with U as potential energy and r as interionic distance: it is defined as the (external) force needed to debond the ions [cf. discussion below Eq. (3.1)]. The calculation of U has been performed using Eqs. (3.3) and (3.4) given in the text with $a = 1.74428 \times 10^{-16}$ kJ/mol and $b = 3.3 \times 10^{-11}$ m. The constant a has been calculated from the equilibrium distance of the ions taken as 2.36 Å. The constant b has been taken from G. M. Barrow and G. W. Herzog, Physikalische Chemie, 6. bericht. Auflage, Bohrmann, Wien, 1984

$$U_{rep} = a \exp\left(-\frac{r}{b}\right) \tag{3.4}$$

with a and b as constants. The constant b does not depend strongly on the type of ion pair considered and can be taken as 3.3×10^{-11} m; the constant a is ion–pair dependent.[9]

By summing U_{attr} and U_{rep}, to get the total potential energy of the system U_{tot}, taking the derivative with respect to r and equating it to zero ($dU_{tot}/dr = 0$) the bonding distance and the bonding energy are obtained. A result for the Na^+ - Cl^- ion pair is shown in Fig. 3.4. Note that the zero energy level in this figure pertains to the situation where the Na^+ ion and the Cl^- ion are at infinite distance, which should not

---

[9] An alternative, well-known equation for the potential energy contribution due to the repulsion is: $U_{rep} = B/r^n$ with B as an ion pair dependent constant and $7 < n < 12$.

be confused with the case where a Na atom and a Cl atom are at infinite distance, which represents a somewhat lower total potential energy: the ionic description holds for interatomic distances which are sufficiently small (say, smaller than 0.8 nm).

Until now, the discussion was restricted to a single cation–anion pair, with each ion of elementary charge. In an ionic crystal many cations and anions are present and all interactions between all ions have to be considered in order to determine the energy of the aggregate. First, it has to be recognized that matter is electrically neutral: the positively charged cations must be compensated by the negatively charged anions. In particular, this *charge neutrality* must be established at a spatial scale as small as possible. As a consequence, the cations and anions are arranged such in a lattice that they alternate in a regular manner: each cation is surrounded by anions as nearest neighbours and vice versa. It follows that in a lattice of ions, as compared to a single cation–anion pair, not only attractive Coulomb interactions occur, but also repulsive Coulomb interactions: Na^+ - Cl^- Coulomb interactions are favourable (attractive), but Na^+ - Na^+ and Cl^- - Cl^- Coulomb interactions are unfavourable (repulsive). The lattice stability is due to the favourable interactions (cation–anion Coulomb interactions) occurring generally over shorter interatomic (interionic) distances in the crystal lattice than the unfavourable interactions (cation-cation and anion-anion Coulomb interactions); cf. Eq. (3.3). For example, the nearest neighbours of a cation are anions; its next nearest neighbours are cations, and so on.

To calculate the potential energy contribution of the Coulomb interaction for the whole crystal, first a linear crystal is considered:

$$Na^+Cl^-Na^+Cl^-Na^+Cl^-Na^+Cl^-Na^+Cl^-Na^+Cl^-Na^+Cl^- \ldots$$

Select a Na^+ ion. If the distance between a Na^+ ion and the next Na^+ ion is given by the period (of repetition along the linear crystal) a, so that the distance between adjacent Na^+ and Cl^- ions equals $a/2$, the potential energy contribution of all Coulomb interactions of this one Na^+ ion with all surrounding Na^+ and Cl^- ions is given by:

$$-\frac{1}{4\pi\varepsilon}e^2 \left\{ \frac{2}{\frac{a}{2}} - \frac{2}{a} + \frac{2}{\frac{3a}{2}} - \frac{2}{2a} + \cdots \right\}$$

$$= -\frac{1}{4\pi\varepsilon}\frac{e^2}{\frac{a}{2}} \cdot \left\{ 2\left(1 - \frac{1}{2} + \frac{1}{3} - \frac{1}{4} + \cdots\right) \right\}$$

In total, let there be N ($Na^+ + Cl^-$) ions. Then, in order to get the total Coulomb interaction energy for this linear lattice, one may, naively, propose to repeat the above summation for every ion in the structure and thus to multiply the above equation with N. However, thereby it is ignored that every ion pair occurs twice in the series of N summations. Hence, the multiplication factor is $N/2$, leading to the following result for the potential energy of net attractive Coulomb interaction:

$$U_{\text{attr}} = -\frac{1}{4\pi\varepsilon}\frac{e^2}{\frac{a}{2}}\cdot\frac{N}{2}\cdot\left\{2\left(1-\frac{1}{2}+\frac{1}{3}-\frac{1}{4}+\cdots\right)\right\}$$

The term $-1/(4\pi\varepsilon)$ $(e^2/(a/2))$ equals the potential energy contribution by Coulomb interaction of one (isolated) Na^+ - Cl^- pair [cf. Eq. (3.3)]. The factor $N/2$ represents the number of cation–anion pairs in the crystal. The term between braces describes the arrangement of the ions in the lattice and thus is a characteristic of the lattice geometry. It is called the "Madelung factor", M. The value of M indicates the number of times that the potential energy contribution due to the net attractive Coulomb interaction, by the arrangement of the ions in a lattice, is larger than the Coulomb interaction that would occur if the ions would occur as isolated cation–anion pairs. Evidently, for the linear ion arrangement considered here, taking it as infinitely long $(N \to \infty)$, it holds:

$$M_{\text{linear, NaCl}} = 2\left(1-\frac{1}{2}+\frac{1}{3}-\frac{1}{4}+\ldots\right) = 2\ln 2 = 1.386$$

A similar discussion can also be given for Na^+ and Cl^- ions arranged in a plane. For example, consider the two arrangements for the two Na^+ ions and two Cl^- ions are shown in Fig. 3.5. The net Coulomb interaction for the arrangement as shown in Fig. 3.5a is the more desirable one, because the distances for the attractive Na^+– Cl^- Coulomb ion–pair interactions are all smaller than for the repulsive single Na^+–Na^+ and Cl^-–Cl^- ion–pair interactions (cf. the distances for the attractive and repulsive interactions for the arrangement in Fig. 5.3b). The Madelung factor for the arrangement of four ions in Fig. 3.5a equals 1.207.

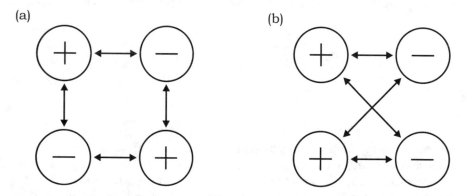

(a)   (b)

Fig. 3.5 Two different possible arrangements of two Na^+ and two Cl^- ions in a plane. The attractive forces have been indicated by arrows. Comparing the distances for the attractive and repulsive pair interactions for the two arrangements, it follows that the configuration in (a) has the lower potential energy and thereby is the more stable one. The relative size of the ions has no resemblance with reality

Fig. 3.6 Crystal structure of NaCl; coordination polyhedra (octahedrons) have been indicated showing that each Na$^+$ ion and each Cl$^-$ ion has six nearest neighbours of Cl$^-$ and Na$^+$ ions, respectively. The Cl$^-$ ions have been represented in black and the Na$^+$ ions have been represented in grey

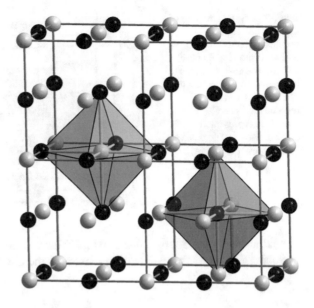

Now, consider the three-dimensional arrangement of Na$^+$ and Cl$^-$ as in the rock-salt structure: see Fig. 3.6.[10] Every Na$^+$ ion has 6 Cl$^-$ ions as nearest neighbours at distances $a/2$ (a is the edge of the unit cell of the f.c.c. Bravais translation lattice pertaining to this material, see Sects. 4.1.1 and 4.1.3), 12 Na$^+$ ions as next nearest neighbours at distances $(a/2)\sqrt{2}$, 8 Cl$^-$ ions as next-next nearest neighbours at distances $(a/2)\sqrt{3}$, 6 Na$^+$ ions as next-next-next nearest neighbours at distances $(a/2)\sqrt{4}$, etc. Then, on the basis of the above reasoning, it appears to follow for the Madelung factor of the rocksalt structure:

$$M_{\text{f.c.c.,NaCl}} = \left(6 - \frac{12}{\sqrt{2}} + \frac{8}{\sqrt{3}} - \frac{6}{\sqrt{4}} + \frac{24}{\sqrt{5}} - \cdots \right) \approx 1.748$$

where the first four terms at the right-hand side directly follow from the consideration above. As a matter of fact, it has been tacitly assumed for the above summation that a (very large) *spherical* NaCl crystal is dealt with, as will be obvious upon reconsidering the text above. This obstructs a truly converging nature of this series. Adopting a (very large) *cubic* crystal (slow,) genuine convergence occurs (Borwein et al. 1985). Obviously, numerical calculation of such summations is required.

Next consider the three-dimensional arrangement of Cs$^+$ and Cl$^-$: see Fig. 3.7.[11] Every Cs$^+$ ion has 8 Cl$^-$ ions as nearest neighbours at distances $(a/2)\sqrt{3}$, 6 Cs$^+$ ions

---

[10] As to be discussed in Chap. 4, it concerns face centred cubic (f.c.c.) arrangements of both the Na$^+$ and the Cl$^-$ ions, with the two substructures shifted over one half of the lattice parameter a (= edge of the f.c.c. unit cell of the f.c.c. Bravais translation lattice; cf. Sects. 4.1.1 and 4.1.3).

[11] As to be discussed in Chap. 4, it concerns primitive cubic arrangements of both the Cs$^+$ and the Cl$^-$ ions, with the two substructures shifted over one half of the diagonal of the primitive cubic unit

Fig. 3.7 Crystal structure of CsCl; coordination polyhedra (cubes) have been indicated showing that each Cs^+ ion and each Cl^- ions has eight nearest neighbours of Cl^- and Cs^+ ions, respectively. The Cl^- ions have been represented in black and the Cs^+ ions have been represented in grey

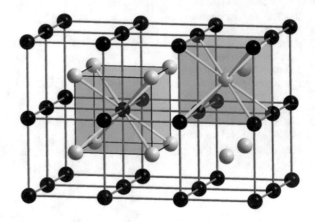

as next nearest neighbours at distances $(a/2)2$, etc. Then, on the basis of the above reasoning, it follows for the Madelung factor of the CsCl structure:

$$M_{\text{prim. cub., CsCl}} = \left(\frac{8}{\sqrt{3}} - \frac{6}{2} + \cdots \right) \approx 1.763$$

The slightly larger value of $M_{\text{prim. cub.,CsCl}}$, as compared to $M_{\text{f.c.c.,NaCl}}$, can be ascribed to the occurrence of 8 instead of 6 (as for the NaCl structure) nearest neighbours of opposite charge in the CsCl structure.

As follows from the above discussion, the Madelung factor is given by the crystal-structure type of the ionic compound considered and thus is independent of the charge of the ions and of the size of the unit cell (i.e. the crystal-structure/translation-lattice dimensions; cf. Sect. 4.1). Some further values of Madelung factors of ionic compounds of specific crystal-structure types are: $M_{\text{ZnS(wurtzite)}} \approx 1.641$, $M_{\text{ZnS(zinc blende)}} \approx 1.638$, $M_{\text{CaF}_2\text{(fluorite)}} \approx 2.519$ and $M_{\text{TiO}_2\text{(rutile)}} \approx 2.408$. If the Coulomb energy is decisive for the crystal-structure energy of the ionic compound considered, the Madelung factor serves as a predictor of the crystal-structure type (but note the ion–pair dependent contribution of the repulsive energy contribution; cf. Eq. (3.4) and see below). The treatment given can have only validity for strictly ionic compounds, as the alkali-metal halides. For example, it should be realized that wurtzite (ZnS, hexagonal) cannot be considered as an ionic compound (rather, wurtzite (ZnS) is covalent; cf. Sect. 3.4), but a compound as BeO has ionic character and has a wurtzite-type crystal structure. If the differences between the values of the Madelung factor for different possible crystal-structure types for a certain ionic compound become small, then, obviously, subtle, tiny energy differences may control the preference for a specific crystal structure. This last situation generally is the rule when different conceivable crystal structures for a solid element, solid solution or

cell of the primitive cubic Bravais translation lattice ($a =$ edge of the unit cell) (cf. Sects. 4.1.1. and 4.1.3).

compound are considered (see the discussion on the (close packed) crystal structures of metals in Sect. 3.5.3).

Interest in calculating Madelung constants ("Coulomb sums") exists until today; their calculation is not a trivial matter, as already indicated by the above discussion of the Madelung constant for rocksalt [see Harrison (2006) and Gaio and Silvestrelli (2009)].

Until now, the role of the repulsive forces in the crystal-structure energy, due to the approaching filled outer electron shells, was ignored. Because of the short-range nature of the repulsive forces (see Sect. 3.1) also for a three-dimensional crystal, only the repulsive interaction with nearest neighbours has to be accounted for. Then, for the total contribution to the crystal-structure energy by the repulsive interactions, it can be proposed to apply the empirical Eq. (3.4), but after replacing the constants a and b by the constants A and B. After adding the crystal-structure-energy contribution due to the total Coulomb interaction and the crystal-structure-energy contribution due to the repulsive interactions, it follows that the change in total crystal-structure energy due to the repulsive interactions is only about 10% of the value of the total Coulomb interaction. The resulting equation for the crystal-structure energy for an ionic crystal composed of singly charged cations and anions thus becomes:

$$U_{\text{crystal}} = -\frac{1}{4\pi\varepsilon} \frac{e^2}{\frac{a}{2}} \cdot \frac{N}{2} \cdot M + A \exp\left(-\frac{\left(\frac{a}{2}\right)}{B}\right) \qquad (3.5)$$

This simple theory has been extremely successful. For example, for the ionic crystal KCl, the theory predicts a crystal-structure energy (bonding energy) of 679 kJ/mole, which agrees very well with the experimental result of 685 kJ/mole. The discussion in Sect. 2.5 implies that the wider the separation of two elements in the Periodic Table, from its lower left corner to its upper right corner, the more ionic the bonding of these two elements is. Hence, recognizing the remark made in Sect. 3.2 about the simplified extreme nature of chemical bonding types, the crystalline solids KCl and CsCl are of more ionic nature than the crystalline solid NaCl.

Only for ionic (alkali-metal halide) crystals such spectacular agreement between theory and experiment can be expected. This is due to the Coulomb description being valid on also the atomic scale. For other types of bonding, such Coulomb interaction is of lesser pronounced importance. Quantum mechanical approaches become necessary, which can be applied to practical cases usually only by imposing reality damaging simplifications.

The potential and the attractive force due to Coulomb interaction are spherically symmetrical [cf. Eq. (3.3)]: the ionic bonding is not orientation dependent. This explains the desire for an ion to be surrounded by as many as possible ions of opposite sign. On the other hand, the size of the ions plays a role; the size of an ion can be interpreted as the distance to the ion at which the repulsive interaction begins to dominate. This implies that, conceiving the ions as hard spheres, the packing of particles of different size has to be considered (cf. begin of Sect. 4.2): a, usually small, cation is in contact with all surrounding (nearest neighbour), usually large, anions.[12] Both considerations lead for singly charged anions and cations to the two most common types of crystal structures for ionic solids:

- the NaCl, face centred cubic type, for which $0.414 < r_{cation}/r_{anion} < 0.732$[13] (indeed for NaCl $r_{cation}/r_{anion} = 0.56$), with the cation at the centre of an octahedron with the six nearest neighbour anions at the corners of the octahedron (see Fig. 3.6), and
- the CsCl, primitive cubic type, for which $0.732 < r_{cation}/r_{anion} < 1.000$ (indeed for CsCl $r_{cation}/r_{anion} = 0.94$), with the cation at the centre of a cube and the eight nearest neighbour anions at the corners of the cube (see Fig. 3.7).

If one of the ions is not singly charged, i.e. one atom accepts the single valence electrons of two atoms of another type or one atom donates two electrons to two atoms of another type, twice as much singly charged ions as doubly charged ions occur. The building structure unit, motif (cf. Sect. 4.1.1), of the crystal structure then is not a cation–anion pair, as for the NaCl and CsCl ionics, but has a basis of three ions, an anion-cation-cation unit or a cation–anion-anion unit. An example of the latter type is

- the CaF_2 (fluorite) type of crystal structure (Fig. 3.8). In this case, $r_{cation}/r_{anion} = 0.75$ and, indeed (see above) the cation (Ca^{2+}) at the centre of a cube is surrounded by eight nearest neighbour anions (F^-) at the corners of the cube. In contrast with the CsCl structure type, the vice versa does not hold in this case: each anion (F^-), here at the centre of a tetrahedron, is surrounded by four nearest neighbour cations (Ca^{2+}) at the corners of the tetrahedron.

The ionic bonding implies that each ion interacts with all other ions in the lattice. Something like a NaCl or a CsCl or a CaF_2 molecule does not exist in the solid state. The whole ionic crystal could be conceived as a single molecule. ...

The lack of directionality of bonding in principle favours plastic behaviour (cf. Sects. 3.4 and 3.5 on covalent bonding and metallic bonding, respectively). Yet, ionic crystals can be quite hard and brittle[14] fracture of ionic crystals can occur: breaking of the crystal implies that strong ionic interactions have to be broken at the interface where fracture occurs. Or said otherwise: local rearrangements of ions, as might

---

[12] Values for the interionic distances can be determined very precisely by X-ray diffraction methods. Numerical values for the ionic radii then follow after, for example, the radius of one well-known ion, e.g. O^{2-}, has been adopted as a standard. This explains why rather varying numerical values for the ionic radii are found in the literature. Such ambiguity does not occur for the atomic radii of metals, which are equal to one half of the interatomic distance, i.e. adopting the close packing of hard, identical spheres as model (cf. Sect. 4.2.1.1), and thus can be determined with high precision by X-ray diffraction methods. I have refrained from providing numerical values for atomic and ionic radii in this book for a number of reasons, one of the more important ones being the problem indicated in this footnote for ionic crystals. An extensive presentation of such atomic radii (for various types of chemical bonding) and of ionic radii has been given by Pauling (1960) in his classical book.

[13] The ranges given here for r_{cation}/r_{anion} follow from straightforward geometrical calculus subject to the above-mentioned constraints.

[14] A material is said to be brittle if it cannot be distinctly deformed without breaking it.

Fig. 3.8 Crystal structure of CaF_2; coordination polyhedra (cubes for Ca^{2+} and tetrahedra for F^-) have been indicated. The cation Ca^{2+} at the centre of a cube is surrounded by eight nearest neighbour anions F^- at the corners of the cube; the anion F^- at the centre of a tetrahedron is surrounded by four nearest neighbour cations Ca^{2+} at the corners of the tetrahedron. The F^- ions have been represented in black and the Ca^{2+} ions have been represented in grey

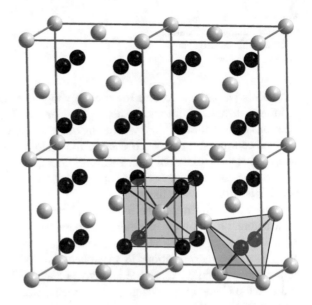

occur by deformation of the crystal, could imply that ions of like charge could occur as nearest neighbours, which of course is energetically very unfavourable. Alternatively, the brittleness has been ascribed to dissolve impurities, as well. Regarding the brittleness, at room temperature, of ionic rocksalt type *poly*crystals, see also Footnote 16 in Sect. 12.12.

3.4 Covalent Bonding

In particular, with a view to the principle of ionic bonding, where bonding occurs between atoms of two elements, the one being electropositive and the other being electronegative, leading to electron "transfer" and bonding due to the thus induced Coulomb interaction of the resulting ions, it seems difficult to apprehend the bonding between like atoms in a condensed state, as between the carbon atoms in diamond.

The overlapping of completely filled electron shells, upon the approach of atoms, leads to strong repulsion (Sect. 3.3). Valence electrons are located in the outermost electron shell which is partly filled. If identical atoms approach they can interact via overlapping valence electron orbitals in partly filled *outermost* electron shells, before strong repulsion due to the completely filled *inner* electron shells can occur. Thereby the occurrence of bonding due to this interaction of the valence electrons is still not made likely.

It is recalled that the (noble gas) electron configuration of a completely filled outer electron shell is strived for by an atom (Sect. 3.2). The point now is, that, by sharing their valence electrons in their partly filled outermost electron shells,

Fig. 3.9 Schematic drawing
of the probability distribution
of the bonding electron pair
in an F_2 molecule

approaching atoms can reach a lower energy together than possible when apart (i.e. at infinite distance from each other). Consider two F atoms. An F atom has the electron configuration $1s^2 2s^2 2p^5$. Adding one electron would realize the (noble gas) electron configuration of Ne. If each of the F atoms of the pair considered would share one of its $2p$ electrons with the other F atom, both F atoms would have realized the electron configuration of Ne, *for a part of the time.* Thereby the occurrence of bonding is still not made likely.

It must be realized that the interaction between atoms is essentially of electrostatic nature, i.e. involving the interaction between electrically charged particles (see Sect. 3.2). Now, if the shared pair of valence (here $2p$) electrons would occupy, most of its time, together, positions in space *between both F nuclei* (see Fig. 3.9), then both electrons would experience attractive Coulomb interactions with *both* nuclei and also the Coulomb repulsion of the F nuclei would be lessened because of the concentration of negative charge in-between. Then, a situation is realized that is of lower energy than corresponding with the energy of the two atoms at infinite separation, i.e. bonding is favoured.[15] It is immediately clear that such advantages would not occur if each of the two valence electrons would spend most of its time outside the internuclear region of space.[16] This simple picture suggests that the covalent bond is realized by a pair of shared valence electrons that spends most of its time (i.e. the probability of finding these electrons; i.e. the squares of the wave (probability) amplitudes (cf. Sects. 2.4.1 and 2.6) are largest) in-between the nuclei of the pair of atoms bonded.

Evidently, the number of valence electrons of an atom is decisive for the number of neighbour atoms that can share such an electron pair with the atom considered. The s and p electrons of the outermost electron shell have the largest probability of being found at large distance from the nucleus (cf. discussion in Sect. 2.5) and thus can be considered as the valence electrons. Therefore, if an atom has N such valence electrons, $8-N$ extra electrons are needed in order to realize a noble gas electron configuration for the s and p electrons in the outermost electron shell. Hence, the atom strives for realization of $8-N$ valence electron pairs to be shared one-to-one with $8-N$ neighbours. So the F atom ($2s$ and $5p$ electrons in the outermost electron shell) needs only one such valence pair and this explains the occurrence of F_2.

---

[15] Here, it is taken for granted that the repulsion energy of the electrons is of lesser importance than the favourable energy contributions indicated; in the language of this section: the electrons may not come too close (for an example demonstrating that electron–electron interactions can govern material behaviour, see the discussion on ferromagnetism in Sect. 3.5.2).

[16] In fact the picture given resembles the one given to explain the stability of an aggregate of cations and anions in a crystal, as compared to the situation where the same total numbers of cations and anions occur as single combinations of cations and anions, in molecules, at infinitely large separation distances (cf. the discussion on the Madelung factor in Sect. 3.3).

An important characteristic of the covalent bond is its directionality: bonding occurs in a specific direction in space (which strongly contrasts with the ionic bond which is not orientation dependent; Sect. 3.3). The background of this phenomenon is the non-isotropic nature of the probability distribution of the p electron (Sect. 2.6). See what follows.

A C atom has the electron configuration $1s^2(\uparrow\downarrow)2s^2(\uparrow\downarrow)2p(\uparrow)2p(\uparrow)$. According to the above "8-N rule" a C atom would strive for four covalent bonds, i.e. four neighbours with which it shares an electron pair. Recognizing that the four valence electrons are composed of two s electrons and two p electrons, it may come as a surprise that the four bonds of C with its four nearest C neighbours in a structure as diamond (see Fig. 3.10) are equivalent. This equivalence is realized by the linear addition and subtraction of the s and p wave functions involved (which combinations, as the individual wave functions, are then also solutions of the Schrödinger equation; cf. Sect. 2.6). These combined wave functions are called hybrids and the process of creating them is called *hybridization*. In the case of C, we start the process with the promotion/excitation of one $2s$ electron to a $2p$ state: $2s^2(\uparrow\downarrow)2p(\uparrow)2p(\uparrow) \rightarrow 2s(\uparrow)2p(\uparrow)2p(\uparrow)2p(\uparrow)$. Evidently this costs energy (the $2p$ state has a higher energy than the $2s$ state; cf. Sect. 2.4). However, this cost in energy is more than compensated by the (subsequent) establishment of the four covalent bonds with the four nearest neighbour atoms. Now the one s and three p wave functions are combined, by linear addition and subtraction in four different ways, to equivalent hybrid wave functions h_i :

(a)   (b)

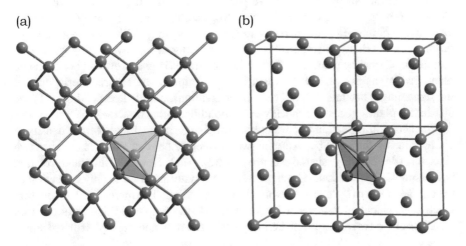

Fig. 3.10 Crystal structure of diamond or silicon or germanium. **a** The covalent bonds, exhibiting tetrahedral coordination, are shown. **b** The crystal as constituted of unit cells is shown. A coordination polyhedron, a tetraeder, has been indicated in both **a** and **b**

$$h_1 \approx \left(s + p_x + p_y + p_z\right)$$
$$h_2 \approx \left(s + p_x - p_y - p_z\right)$$
$$h_3 \approx \left(s - p_x + p_y - p_z\right)$$
$$h_4 \approx \left(s - p_x - p_y + p_z\right)$$

Applying this approach, and using the visualization of the s and p wave functions as discussed in Sect. 2.6, make clear that the four hybrids point to the corners of a tetrahedron, making angles of 109° 28′ between their directions. Because one s and three p functions are combined one speaks of sp^3 hybridization. Next, the overlap of sp^3 orbitals,[17] as given by h_1 through h_4, of approaching C atoms leads to the creation of a valence electron pair establishing a covalent bond between these then bonded C atoms. Hereby the occurrence of four equivalent bonds of one C atom with four neighbouring C atoms as in diamond has been explained. It is also at once clear that the sp^3 hybridization of the neighbour C atoms allows that these neighbour atoms also bond with four neighbours, and so on. Thus, an understanding has been acquired for the crystal structure of diamond, involving a tetrahedral coordination of carbon atoms (see Fig. 3.10).[18] It appears that one diamond crystal can be conceived as one "molecule"; one can speak of a "network solid". It is recalled that a similar remark could be made for an ionic crystal but the distinct differences between the two types of bonding should be appreciated: one ion in the ionic crystal interacts with all other ions in the crystal and the bonding is isotropic, whereas *in the picture given here* (see what follows further below) the bonding interaction of an atom in the covalent crystal is restricted to neighbouring atoms, and thereby the bonding is non-isotropic.

Other variants of hybridization are possible too. After the promotion of one $2s$ electron to a $2p$ state, C can exhibit sp^2 hybridization, i.e. one s and (only) two p orbitals are combined. This pertains to the bonding in graphite; the resulting sp^2 orbitals, by overlap, are responsible for the bonding in the layers of graphite (i.e. parallel to 001 lattice planes; cf. Sect. 4.1.4), characterized by covalent bonds in one plane making angles of 120° between their directions, whereas the overlapping of the remaining p orbitals (left after sp^2 hybridization), oriented perpendicular to the 001 lattice planes, leads to a much weaker bonding in that direction.

The above introduction to covalent bonding suggests that the electron density at the location of the bond between the two covalently bonded atoms is equal to two electrons (the bonding, shared valence electron pair). However, this interpretation is too simple. The amount of charge associated with the bond is only a fraction of the charge of the two electrons involved in the bonding: The more detailed quantum mechanical calculation shows that, as compared to the simple addition of the electron densities of the two electrons in their orbitals of the two, non-bonded, isolated atoms

---

[17] The notion "orbital" is used for electron states with the same values for the n, l and m_l quantum numbers, and thus at most two electrons of different spin quantum number can occupy one orbital, see also below Eq. (2.9).

[18] Yet, it should be recognized that the hybridization has been introduced as a mathematical tool and it should not be interpreted as a process that occurs, in reality, *before* interatomic bond formation. The concept of *directed* orbitals, as explained on this basis, has some arbitrariness.

at, hypothetically, the same distance as bonded, the electron density between the atoms upon bonding is significantly larger: upon bonding a part of the electron density from outside the space between the nuclei has shifted to the internuclear space. However, the amount of charge associated with the bond can be significantly smaller than the full charge of the two electrons involved in the bonding, and yet this fraction is responsible for the large bond strength, as holds for the bond between the carbon atoms in diamond.

In a quantum mechanical description of covalent bonding, it must be realized that if two identical atoms approach each other and two partially filled, originally identical orbitals, each containing a valence electron, begin to overlap, then the valence electrons considered can stay anywhere in the resulting aggregate (molecule) of two atoms and are indistinguishable; the valence electrons can no longer be assigned to the one or the other atom. The Pauli principle prohibits that more than one electron in a system has the same set of (four) quantum numbers. As a result the two original, identical, overlapping orbitals of identical energy are split into two (molecular) orbitals; one of lower energy than the original orbitals and one of higher energy than the original orbitals. The first orbital then is occupied by the two valence electrons considered, has a substantial probability amplitude in the internuclear space and is called the *bonding orbital*; the second orbital remains empty, has a relatively large probability amplitude outside the internuclear space and is called *antibonding orbital*. The energy splitting of the bonding and antibonding orbitals increases with decreasing distance between the two nuclei (see Fig. 3.11). Positioning of the two valence electrons considered in the bonding orbital (subject to Pauli's exclusion principle, so as an electron pair with opposite spins) leads to a substantial electron density in the internuclear space and a reduction of the energy of the system and this released energy is the bonding energy.

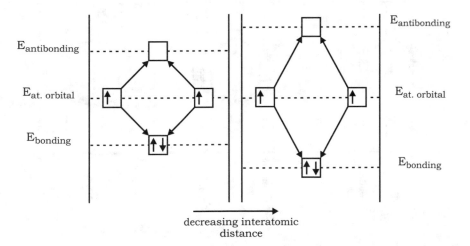

Fig. 3.11 Schematic depiction of the formation of antibonding and bonding orbitals by two approaching carbon atoms, for two different interatomic distances

In accordance with the above discussion and Fig. 3.11, an important rule can now be formulated: *upon bringing individual atoms together into an aggregate, corresponding overlapping atomic orbitals of equal energy split into an equal number of (molecular) orbitals of different energy.*

The electrons responsible for the bonding are the outermost, valence electrons. Upon bonding, in the molecule/aggregate, these electrons are "extended" over the entire molecule/aggregate, but it is recognized that the orbital they occupy still exhibits the nature of the original atomic orbitals. This has led to a description of such *molecular orbitals (MOs)* for the bonding electrons on the basis of a *linear combination of atomic orbitals (LCAO).*

Consider again the diamond crystal composed of sp^3 hybridized C atoms. Bringing the carbon atoms together as in the diamond crystal structure leads to a collection of bonding orbitals closely together in energy and a collection of antibonding orbitals also closely together in energy (for a crystal of N carbon atoms each contributing 4 sp^3 atomic orbitals in the bonding, $2N$ bonding and $2N$ antibonding orbitals occur). The collection of closely spaced (with respect to energy levels) bonding orbitals is called the *bonding band* and the collection of closely spaced (with respect to energy levels) antibonding orbitals is called the *antibonding band* (see Fig. 3.12).

The orbitals in the bonding band are relatively strongly localized, i.e. the probability amplitude of a specific orbital (quantum state) from the bonding band is relatively large between a specific pair of carbon atoms. Evidently, the bonding band is fully occupied: all valence electrons (total $4N$) are positioned on all bonding orbitals (total $2N$) in the bonding band (2 electrons per bonding orbital, with opposite spin as required by Pauli's exclusion principle). This is a general feature for all cases of covalent bonding: the covalent bonding requires the occurrence of electron pairs of

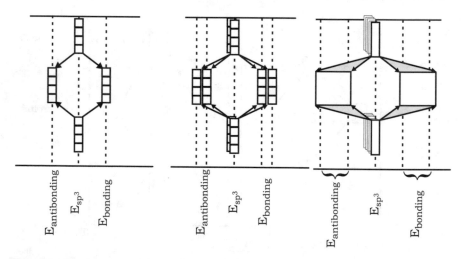

Fig. 3.12 Schematic depiction of the formation of antibonding and bonding orbitals by two, four and an infinite number of carbon atoms. The orbitals merge to a bonding and an antibonding band in the latter case

valence electrons and thus an even number of valence electrons is involved (This situation can be different in the case of the formation of metal crystals where partly filled bonding bands can occur; see next section).

The electrons with energies at the top of the entirely filled bonding band are unable to jump (by thermal agitation) to the empty states of the antibonding band, and thus, in accordance with a more full, similar discussion for the case of metal bonding below, covalently bonded solids are electric insulators: the gap in energy between the bonding and antibonding bands is too large (in case of diamond this gap is about 7 eV).[19]

The occurrence of covalent bonding, characterized by a relatively high electron density between the two atom nuclei in the bond considered (see the discussion immediately above), implies that deformation, involving bond breaking to bring about atom rearrangements locally, is very unfavourable from an energy point of view. Hence, the directionality of the covalent bond induces a usually high hardness[20] of materials where covalent bonding prevails.

3.5 Metal Bonding

Elements which are clearly metals have a small number of valence electrons: one, two, at most three. A noble gas electron configuration is realized by "giving away" the valence electrons. From a chemical point of view, the most outspoken metals are those with one valence electron, i.e. the alkali metals and copper, gold and silver. Aluminium, tin and bismuth, with three, four, and five valence electrons, respectively, exhibit an ambivalent, "amphoteric" character: they can either donate or accept electrons upon interaction with atoms of other elements and thus can show metallic and non-metallic properties.

Metal atoms can also "share" their valence electrons with other metal atoms in an (usually solid) metallic aggregate. Then, if the "8-N rule" would be applied, a metal atom would strive for 7, 6 or 5 neighbours with which it would share an electron pair in covalent bonding. The common crystals structures for metals exhibit 12 nearest neighbours and 8 nearest neighbours (cf. Sect. 4.2). Thereby it becomes apparent that covalent bonding, with electron pairs shared by pairs of atoms, i.e. of the type as described in Sect. 3.4, is not compatible with the metal crystal structure. Obviously, also ionic bonding as discussed in Sect. 3.3 has no relevance to bonding in metals, as for pure metals no interactions between atoms of different electronegativity are involved. Important approaches for the description of bonding in metallic solids have been developed; these will be sketched below.

---

[19] Actually, as follows from the discussion on metal bonding according to the tight binding approach (Sect. 3.5.2), the bonding orbitals and the anti-bonding orbitals originating from the same atomic orbitals comprise *one* band and one could here thus better speak of the bonding and anti-bonding *sub*bands separated by an energy gap. Such an energy gap between the bonding and anti-bonding states in one band is typical for covalent bonding and does not occur in case of metal bonding.

[20] Hardness is a measure for the resistance against plastic deformation; cf. Sect. 12.13.

3.5.1 The Free Electron Models

The metal solid is visualized as an array of positively charged atomic cores of metal atoms, where an atomic core is a metal atom having given away its valence electrons. In the free electron model, the valence electrons move within a "potential well" confined by the surfaces of the crystal, where the (periodic) arrangement of the positively charged metal ions is ignored: the positive charge is assumed to be smoothed out throughout the crystal such that a uniform potential results (see Fig. 3.13).

The potential well is deep: outside the crystal (potential well) the potential is much higher than inside the crystal. It is irrelevant for the model for the potential outside the crystal to be zero and inside the crystal very largely negative, or to be very largely positive outside the crystal and inside the crystal zero. Both descriptions occur in the literature. The point is: it will cost a lot of energy for the electron to be removed from the crystal (it is the potential difference what counts): the electrons stay within the crystal due to their Coulomb interaction with the positive metal ions represented as a constant positive charge within the crystal (see above).

The electrons moving around in the crystal, in between the metal ions, by virtue of their Coulomb interaction with the metal ions, provide a "glue" to keep the repulsive metal ions together in the crystal [cf. the role of the bonding electron pair in covalent bonding (Sect. 3.4)].

Fig. 3.13 Schematic depiction of an infinitely deep one-dimensional potential well

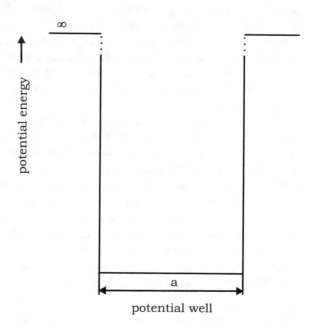

3.5.1.1 The Classical Model; Electron Gas

This first approach is based on ideas by Drude and Lorentz presented at the very beginning of the twentieth century, very soon after the experimental work by (J.J.) Thomson on the nature of the electrons (Thomson was the first to determine values for the charge and the mass of the electron; see also Sect. 2.2). The metal atoms in the crystal loose one or more electrons each. These electrons form a "gas" of negatively charged particles moving within the spatial limits of the metal crystal, and behaving as if they obey the laws of the classical theory of the kinetics of (ideal) gases contained in a vessel. The positively charged atomic cores stay together because of the electrostatic interaction between the atomic cores and the free electrons moving between the atomic cores; i.e. the free electrons act as a "glue".

Electrical conductivity is then explained as follows. In the absence of an external imposed electric field, the free electrons move randomly in all directions; in the presence of an electric field, they are attracted towards the positive side of the field. The collisions with the atoms prevent the electrons from being accelerated continuously, and as a result, a steady state develops with a constant electron current in a constant field, in accordance with Ohm's law.

The theory puts the idea of detaching electrons from their parent atoms to its extreme: the lattice periodicity of the positively charged atomic cores is completely ignored: any special relation/bonding with any specific atom core in the crystal does not occur: the detached electrons are truly free; not even their electrostatic repulsion is taking into account.

In spite of its extreme simplicity, the classical free electron theory has been very successful. Transport of free electrons in the metal crystal is very easy, and thereby high electrical conductivity and thermal conductivity[21] is immediately understood. Even until today, this model is utilized, and, indeed, application of the later, more advanced quantum mechanical approach (see below) has confirmed some basic results of the simple theory. A century after the presentation of the classical theory experimental validation of the relation between the conductivity and the frequency of an alternating electric field, as predicted by the classical theory and confirmed by the quantum mechanical theory, was obtained at last (Dressel and Scheffler 2006). The success of the classical theory implies that the essence for understanding metallic (conductivity) behaviour is the recognition that "free electrons" occur which can move through the entire crystal. This key feature is retained by the later theories.

Yet, serious shortcomings of the theory are apparent as well. It has been mentioned in Sect. 1.2 that the perhaps most typifying property of a metal is that the electrical and thermal conductivity decrease with increasing temperature. This is not what one

---

[21] Heat is not only transferred by the moving free electrons. Also by the vibration of the atoms (atomic cores) thermal kinetic energy can be transported. Concerning metals, the electrons carry only a small part of the heat capacity (see further under (b) in Sect. 3.5.1.2), but they take account of the predominant part of the thermal conductivity. Materials which do not possess free electrons can only utilize the mentioned atomic vibrations to realize heat transfer and this explains why these materials (ionic or covalently bonded materials) are generally bad (electrical and) thermal conductors.

expects if, as in the above discussion, the free electrons are conceived as the carriers of the electrical and thermal conductivity: the thermal conductivity, for example, of an ideal gas (the model for the free electrons) increases with temperature. The discrepancy is represented perhaps most strikingly by the specific heat of a metal. According to classical theory of the kinetics of (ideal) gases the contribution of one mole of "free" particles (electrons) to the specific heat of a system is $3R/2$, with R as the gas constant. Hence, as compared to the heat capacity of an insulator, the heat capacity of a metal should be larger with an amount $n_v 3R/2$ due to the "electron gas", with n_v as the number of free electrons per metal atom. Measurements show that the specific heats of metals are practically equal to those of insulators: the specific heat of a metal is only an amount of the order of a per cent of $n_v 3R/2$ larger than that of an insulator.

Another problem is the low value of the paramagnetic susceptibility of metals. Paramagnetism is related to the number of unpaired electrons with parallel spin (see Sect. 2.5). Truly free electrons would be able to orient their spins in accordance with an applied magnetic field which would lead to large magnetic moments which are not at all observed in reality.

Such problems were the stimulus for the introduction of quantum mechanical considerations in the free electron theory.

3.5.1.2   The Quantum Mechanical Free Electron Theory; "Particles in a Box"

In 1928, Sommerfeld considered free electrons moving in a potential well (Fig. 3.13), subject to the laws of quantum mechanics. Some subtle, but important, differences with the points of departure of the classical theory should be outlined. Firstly, the number of free electrons is taken equal to the number of valence electrons (At the time of development of the classical free electron theory, models for the constitution of the atom as that by J. J. Thomson prevailed [see Sect. 2.2]). Secondly, the array of metal ions (atomic cores) constitute a (very deep) potential well, so that the free electrons are confined between the surfaces of the crystal; yet, the potential is assumed to be constant in the well, and thus, again, the periodicity of the positively charged atomic cores is ignored, i.e. smoothed out, and represented by a constant potential. The positive charge of the metal ions is needed to maintain net zero charge for the metal crystal.

The solutions of the time independent Schrödinger equation depend on the boundary conditions, as was remarked in Sect. 2.4.2. Let us first consider the one-dimensional problem of a (free) electron moving between the walls of an infinitely deep, square potential well as sketched in Fig. 3.14. Boundary conditions, as that there should be a zero chance of finding the electron outside the infinitely deep potential well, that the probability amplitude must be equal to zero at the borders of the infinitely deep potential well and that the chance of finding the electron somewhere

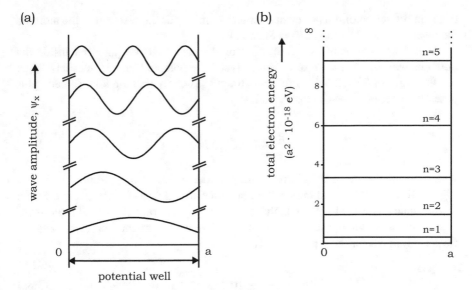

Fig. 3.14 **a** Schematic depiction of standing waves in an infinitely deep one-dimensional potential well of length a, and **b** corresponding energy levels calculated using Eq. (3.7)

inside the potential well equals one, lead to the following results for the probability amplitude within the potential well[22]:

$$\psi_x = \left(\frac{2}{a}\right)^{1/2} \sin\left(\frac{n_x \pi x}{a}\right) \qquad (3.6)$$

with a as the width of the potential well, x as the position coordinate and n_x as a positive integer. Evidently, the only permitted solutions (of the Schrödinger equation) consist of an integral number of half wavelengths fitting to the width of the potential well (see Fig. 3.14a). A useful analogue is a string vibrating between two fixed points: the only possible modes of vibration are those standing waves which have a length of an integral number of half wavelengths with total length equal to the length of the string.

In fact, here the picture of the electron as a wave is adopted and as a consequence the quantum number n_x (in Eq. (3.6)) emerges for the electron in a one-dimensional infinitely deep, square potential well of finite width (cf. the discussion with respect

---

[22] Generally, the boundary conditions for solving Schrödinger's equation imply that (i) the probability amplitude must be finite and continuous as a function of position and that (ii) the derivative of the probability amplitude with respect to position must also be finite and continuous. Strictly speaking, in the above treatment for the *infinitely* deep potential well, a discontinuity occurs for the derivative of the probability amplitude at the borders of the potential well. The general formulation of the boundary conditions is based on *finite* values for the potential. Indeed, introducing a potential well of *finite* depth leads to solutions which comply with the boundary conditions as formulated in this footnote. See the discussion of the "tunnel effect" a little further in this section.

to the introduction of the principal quantum number for the electron in the atom on the basis of Eqs. (2.8) and (2.9) in Sect. 2.4.2).

As a consequence of Eq. (3.6), and for the case that the potential inside the potential wall is set at zero (outside the crystal the potential then is infinitely large; see the begin of Sect. 3.5.1), it follows straightforwardly that the total (kinetic plus potential) energy of the electron satisfies:

$$E = \left(\frac{h^2}{8m_e a^2} \right) n_x^2 \tag{3.7}$$

with m_e as the mass of the electron and a as the width of the potential well (see Fig. 3.14b). It follows that only certain energies are allowed for the (free) electron in the square potential well: quantization occurs. Generalization to three dimensions is possible without being confronted with new fundamental problems and the (expected) result for the total energy is:

$$E = \left(\frac{h^2}{8m_e a^2} \right) (n_x^2 + n_y^2 + n_z^2) \tag{3.8a}$$

$$= \left(\frac{h^2}{8m_e V^{2/3}} \right) (n_x^2 + n_y^2 + n_z^2) \tag{3.8b}$$

where the three different quantum numbers have been designated as n_x, n_y and n_z, a denotes the width of the square potential well in the x, y and z directions and V is the volume of the cube metal considered.

The distance between the energy levels for the electron in the potential well of finite width becomes larger for increasing values of the quantum numbers, which is opposite to the behaviour of the energy levels for the electron in the atom [see Eq. (2.3)]. Evidently, because the crystal dimension [i.e. "a" in Eqs. (3.6–3.8)] is finite and h is very small, the energy levels in the potential well occur very close to each other (they are the closer, the larger the crystal). It is further observed that different combinations of n_x, n_y and n_z give rise to the same energy: *degeneracy of quantum states*, as also discussed for the electron in the atom.

If the potential well is not infinitely deep, the treatment leads to wave functions that constitute a solution of the Schrödinger equation which exhibit finite values for ψ just outside the potential well. Hence, there is a finite chance, i.e. different from zero, to find the electron immediately outside the potential well. This is a result that is impossible according to classical mechanics for electrons of kinetic energies smaller than the potential energy needed to overcome the potential energy barrier. The electrons appear to "tunnel" through the potential energy barrier; therefore, the effect is called "tunnel effect". The size/length parameter, a in Eqs. (3.6–3.8) is relatively large and, in particular for the high energy levels of the free electron in the potential well, which are most relevant for the metallic behaviour (see discussion on metal properties further below), the "tunnelling" of the probability amplitude leads only very close to the surface to minor modifications of the solutions of the

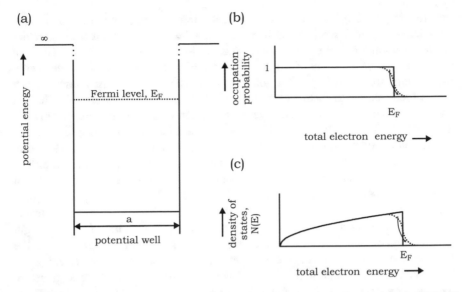

Fig. 3.15 **a** Fermi level in an infinitely deep potential well, **b** Fermi distribution (schematic) for $T = 0$ K (solid) and $T > 0$ K (dashed), **c** density of electron states for $T = 0$ K (solid) and $T > 0$ K (dashed)

Schrödinger equation as obtained for the infinitely deep potential well: only for the analysis of effects occurring near and at the surface of the metal "tunnelling" should be considered.[23]

The filling of the allowed quantum states by the valence electrons should occur in accordance with Pauli's exclusion principle: only two electrons per quantum state with opposite spins, starting with the orbital of lowest energy. According to this "Aufbau Prinzip" all valence electrons in the piece of metal considered are arranged. The orbital filled with the highest energy is still well below the height of the potential well and its energy level is called "Fermi energy" (E_F; see Fig. 3.15a). It is important to realize the difference with the classical model: in the classical model, all free electrons are at the bottom of the potential energy well as then their energy is lowest (there, at 0 K, they are at rest and possess zero kinetic energy). Evidently, according to the quantum mechanical model, at 0 K the valence, free electrons do possess kinetic energy and not all valence, free electrons have the same energy. This recognition has important consequences (see discussion of metal properties further below).

For realistic values of the volume of the piece of metal considered, V, the distances between the energy levels allowed in the piece of metal are that small [cf. Eq. (3.8b)] that one may consider the energy spectrum as if it is continuous. The energy spectrum in terms of the probability that an orbital of energy E is occupied (called *Fermi distribution*) is shown in Fig. 3.15b for $T = 0$ K: all levels are fully occupied, up to

---

[23] Thus, it needs not to surprise that for the initial stages of reactions at surfaces (of metals), as oxidation, this tunnelling effect plays an important role (e.g. see Graat et al. 2002).

the orbital with energy E_F. Thereby all valence, free electrons in the piece of metal considered, say N, have been assigned an orbital/energy level. At a temperature above 0 K, some of the electrons with energy close to E_F can be promoted to an initially unoccupied orbital of slightly higher energy, thereby modifying the energy distribution (shown in Fig. 3.15b for $T > 0$ K). The resulting distribution at $T > 0$ K is such that the probability that an orbital of energy E_F is occupied equals ½ and this can be used as an alternative for definition of the Fermi energy. This last definition can, for example, be used to define the Fermi level of a semiconductor (see the Addendum on "conductors, semiconductors and insulators" later in this section).

The value of E_F at 0 K can be derived as follows. Consider Eq. (3.8b). The energy of a free electron can apparently be represented as a point in the space defined by a three-dimensional Cartesian coordinate system with the three axes representing the values of the three quantum numbers n_x, n_y and n_z (with n_x, n_y and n_z as integers all larger than 0). Quantum states at a distance $r = (n_x^2 + n_y^2 + n_z^2)^{1/2}$ from the origin in this frame of reference all have the same energy. Evidently, the larger r, i.e. the larger E and the larger the degree of degeneracy of the quantum state of energy E. The volume of each point in this (n_x, n_y, n_z) space is unit volume. In total, there are $N/2$ quantum states to be filled up if N is the number of valence, free electrons. These quantum states are given by all points in the octant, pertaining to $n_x > 0$, $n_y > 0$ and $n_z > 0$ of the sphere with radius $r = (n_x^2 + n_y^2 + n_z^2)^{1/2}$. Thus it holds:

$$\frac{N}{2} = \frac{1}{8} \cdot \left(\frac{4}{3}\right) \pi r_{max}^3 \tag{3.9}$$

where r_{max} denotes the radius for the quantum state with the highest energy E_F. Then, using Eq. (3.8b), it is obtained for E_F:

$$E_F = \left(\frac{h^2}{8 m_e}\right) \cdot \left(\frac{3N}{\pi V}\right)^{\frac{2}{3}} \tag{3.10}$$

It is important to realize that E_F is independent of the size of the piece of metal considered: E_F depends on the number of valence, free electrons per unit volume (N/V) and thereby is a characteristic material property.

The degeneracy of a quantum state becomes more pronounced the larger its energy, and thus the density of states as a function of energy increases with energy. The density of states as a function of energy is given by the parameter $N(E)$ such that $N(E)dE$ represents the number of states between the energies E and $E + dE$. $N(E)$ can be calculated as follows. The number of occupied states per unit volume of a metal of volume V up to the quantum states with energy E will be denoted by n. It holds (cf. the above derivation of E_F):

$$n = \frac{1}{8} \cdot \left(\frac{4}{3}\right) \frac{\pi r^3}{V}$$

Then, substituting r according to Eq. (3.8b), it is obtained:

$$n = \left(\frac{4\pi}{3h^3}\right) \cdot (2m_e)^{3/2} \cdot E^{3/2}$$

Straightforward differentiation with respect to E yields:

$$\mathrm{d}n = \frac{3}{2} \cdot \left(\frac{4\pi}{3h^3}\right) \cdot (2\,m_e)^{3/2} \cdot E^{1/2}\,\mathrm{d}E$$

which is nothing else than the number of quantum states per unit volume metal with energies between E and $E + \mathrm{d}E$. Thus, $\mathrm{d}n = N(E)\mathrm{d}E$ and consequently

$$N(E) = \left(\frac{2\pi}{h^3}\right) \cdot (2m_e)^{3/2} \cdot E^{1/2} \tag{3.11}$$

A parabolic relation between E and $N(E)$ occurs (at 0 K) as illustrated in Fig. 3.15c. The effect of temperature on this distribution is sketched as well (see the discussion below on the contribution of the free electrons to the specific heat of the metal).

Finally, the average energy of the free electrons (i.e. kinetic energy; recall that the potential energy of the valence, free electrons has been set equal to zero within the metal which is conceived as an infinitely deep potential well) can now also be calculated simply. The average energy is given by the total energy due to all free electrons divided by the number of free electrons:

$$< E > = \frac{\int_0^{E_F} V \cdot E \cdot 2N(E)\mathrm{d}E}{\int_0^{E_F} 2N(E)\mathrm{d}E}$$

which, after substitution of Eq. (3.11) and using Eq. (3.10), leads to the following result for the average (kinetic) energy per electron at 0 K:

$$< E > = \frac{3}{5}E_F \tag{3.12}$$

In the following, some properties characteristic for the metallic state (cf. Sect. 1.2) are discussed in the light of the quantum mechanical free electron model.

(a) *Electrical conductivity*

The valence electrons with energies at or near to E_F can absorb a small amount of energy and be promoted to orbitals of slightly higher energy, which are initially unoccupied. Such small amounts of energy can be gained by acceleration in an electric field or by absorption of a thermal quantum.

The understanding of electrical conductivity now is as follows. The free electrons behave as a gas: the free electrons move randomly in all directions, but, in contrast with the classical model, their (kinetic) energies correspond with the various occupied quantum states up till the Fermi level. Upon imposition of an electric field gradient, all free electrons will be accelerated: superimposed on the random movements of the free electrons a drift of the electrons, i.e. a current, down the electric field gradient occurs. For a free electron with energy well below E_F and moving in a certain direction, there is always a free electron of the same energy moving in the opposite direction, and hence the free electrons with energy well below E_F do not contribute to a net current. However, upon acceleration in an electric field, electrons with energies close to E_F and moving in the direction of the electric field gradient can occupy orbitals with energies a little above E_F. Then, for the electrons with energies around E_F, moving in the direction of the electric field gradient, there are no free electrons of the same energy in the metal crystal moving in the opposite direction, and compensation as indicated above cannot occur. This leads to the occurrence of a net current. Hence, although all valence, free electrons become accelerated, only those with energies close to E_F (the top of the Fermi distribution) are directly responsible for the occurrence of a net current.

The above analysis about the emergence of a net electric current is incomplete: the acceleration of the electrons would be continuous (leading to superconductivity) and the occurrence of a steady state of constant current has not been explained. One may think that the collisions of the electrons with the positive metal ions in the metal are responsible for the development of a steady state, i.e. the occurrence of resistivity (as was suggested in the introduction of the classical theory; see Sect. 3.5.1.1). However, this statement is too simple. If the arrangement of the metal ions in the crystal is perfect [flawless translational symmetry (cf. Sect. 4.1.1)] and conceiving the electron as a wave, then the individual wavelets scattered by the metal ions upon incidence of the electron wave (the Huygens' principle), interfere coherently and as a result the electron (wave) propagates in an undisturbed manner through the crystal. The deviations of the ideal arrangements of the metal ions lead to destructive interference (incoherent scattering). Such non-idealities are the thermal vibrations of the metal ions and lattice defects as vacancies, dislocations, stacking faults, grain boundaries, etc. (cf. Chap. 5). These thermal vibrations and the crystal imperfections are responsible for the resistivity and lead to a constant current in an electric field according to Ohm's law. Obviously, the thermal vibrations of the metal ions become minimal at 0 K and this explains that the resistivity increases with temperature.

(b)   Specific heat

Small amounts of thermal energy can in principle be transferred to the free electrons by interaction with the thermally vibrating metal ions.[24] However, the only free electrons capable of absorbing such small amounts of energy are those occupying orbitals with energies close to E_F, because they can be promoted to empty orbitals with energies above E_F. Thereby it is immediately clear that the contribution to the specific heat of the metal by the free electrons is much smaller than according to the

---

[24] Actually, from a puristic point of view, this is in conflict with a basic assumption of the free electron model that a constant potential occurs in the metal crystal.

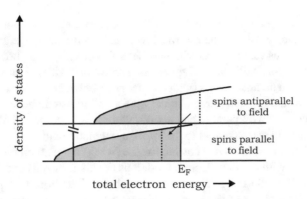

Fig. 3.16 Schematic depiction of the shift of density of states of a material by an external magnetic field, which would lead to two shifted Fermi levels (dashed lines). Then, those electrons with antiparallel spins and energies close to "their Fermi level" can lower their energy by moving to unoccupied states for electrons with parallel spins close to and above the "Fermi level" of the electrons with parallel spins. As a result one, single Fermi level of the metal in the magnetic field occurs (solid line). Consequently, now the number of parallel and antiparallel spins has become unbalanced

classical model (see Sect. 3.5.1.1) and a striking difference between the prediction of the classical theory and experiment is resolved.[25]

(c) *Paramagnetism*

Upon imposing a magnetic field onto a specimen, it can experience a force: the specimen has become magnetized. The ratio of this force and the given magnetic field strength is a material constant: the *magnetic susceptibility*. The magnetic susceptibility is related to the number of unpaired electrons with parallel spin in the specimen. In the presence of a magnetic field, an electron with spin parallel to the field acquires a lower energy, and an electron with its spin antiparallel to the field has an increased energy (the amounts of energy decrease and energy increase for spin parallel and spin antiparallel are the same).

Now consider the free electron model with the density of states as given by Eq. (3.11) (see Fig. 3.15c). Every occupied orbital comprises two electrons with opposite spins. If, upon imposing an external magnetic field, one half of the electrons orients itself with spin parallel to the field, then the other half of the electrons has spins which are antiparallel to the field. Thus, the density of states curve is split in two parts (see Fig. 3.16): one half is shifted to lower energies (corresponding to the electrons with spins parallel to the field) and one half is shifted (with the same amount) to higher energies (corresponding to the electrons with spins antiparallel to the field). This simple picture would imply that now two Fermi levels would occur: one for the electrons with antiparallel spin and one at lower energy for the electrons

---

[25] Also, in other fields of physics the discrepancies between the values for the specific heat as predicted on the basis of classical theories and as obtained by experiments have been resolved only by the application of quantum mechanical approaches.

with parallel spin. Then it is immediately clear that those electrons with antiparallel spins and energies close to "their Fermi level" can lower their energy by moving to unoccupied states for electrons with parallel spins close to and above the "Fermi level" of the electrons with parallel spins. As a result one, single Fermi level of the metal in the magnetic field occurs (with an energy in-between the "Fermi levels" for the electrons with antiparallel and parallel spins, as indicated above), but now the number of electrons with parallel and antiparallel spins is not in balance: there are more electrons with spins parallel to the applied magnetic field.

This imbalance in spins is responsible for the (weak) magnetism for those metals that best comply with the free electron model, as the alkali metals and the metals in group III of the Periodic Table. Because the overshoot of electrons with parallel spins leads to an enhancement of the applied magnetic field, i.e. materials showing this effect are attracted by the magnetic field in the direction of larger field strength, this source of magnetic behaviour is called *para*magnetism, exhibiting a positive susceptibility. The discussed effect on the energy levels of the free electrons is independent of temperature and thus paramagnetism does not depend on temperature.

If repulsion by the magnetic field occurs, i.e. repulsion in the direction away from the larger field strength, one speaks of *diamagnetism*, characterized by a negative susceptibility. Application of the external magnetic field induces the electrons in the (sub)shells of the metal ion to a response such that the magnetic field strength within the ion is reduced. This response involves modification of the movement (velocity) of the core electrons in the subshells.

The resulting, net magnetization of the free electron metals in an applied magnetic field is determined by the outcome of the combined paramagnetism due to the free electrons and diamagnetism of the metal ions. Metals like copper, zinc, silver and gold are diamagnetic (note the filled nature of the subshells of their core electrons (see Table 2.1); thereby the diamagnetic effect is relatively strong, in particular, if only one unpaired valence, free electron occurs that contributes to paramagnetism). Most metals, including the transition metals, with iron, cobalt and nickel above their Curie points, (note the unfilled nature of the $3d/4d$ subshells; see Table 2.1), are paramagnetic. The paramagnetic and diamagnetic effects are small as compared to the ferromagnetic effect which is to be discussed at the end of Sect. 3.5.2.

(d)   *Photoelectric effect*

It is now also clear that the photon capable of "kicking out" an electron from the surface of a metal (see Sect. 2.4) should have an energy $h\upsilon$ at least as large as the difference, Φ, between the potential energy of the electron if placed outside the metal and the Fermi level E_F. Φ is called the *work function* of the metal. Application of incident photons of energies larger than $h\upsilon$ (frequencies larger than υ) allows the electrons at the Fermi energy level E_F to leave the metal with kinetic energy $\frac{1}{2}m_e v^2$ according to:

$$\frac{1}{2} m_e v^2 = h\upsilon - \Phi \tag{3.12}$$

This formulation for the *photoelectric effect*, an energy balance based on the conservation of energy, is the well-known Einstein relation (1905).[26] Note that at a temperature larger than 0 K a few valence, free electrons have energies larger than E_F and can (already) be "kicked out" by photons having a frequency a little lower than Φ/h.

3.5.2 Zone or Band Models

The free electron models are based on the assumption of a constant potential in the metal crystal, i.e. the potential variation due to the positively charged metal atom cores is ignored/smoothed out and the role of the metal ions is restricted to merely guaranteeing that the metal remains electrically neutral. Yet, at one place, in the discussion of the origin of the electrical resistivity, it was necessary to focus on the consequence of the periodic nature of the spatial distribution of the metal ions, and in particular, the deviations from the perfect crystal-structure arrangement. It seems natural that the periodicity of the metal-ion crystal structure is taken into account in more advanced theories of metal bonding. An important result of the thus developed band model is an explanation of the difference in electrical conductivity between conductors and insulators (see also the Addendum to this chapter).

Two approaches will be touched upon below. The so-called *nearly free electron model* is suitable for metals where the valence electrons are nearly free (as holds for the single outershell *s* electrons in the alkali metals), but now propagate in a periodic potential field. The *d* electrons of the transition metals play an important role in the atomic bonding occurring in solids of these elements (note that for increasing atomic number in the Periodic Table the 3*d* subshell becomes filled *after* the 4*s* subshell has been filled; cf. Sect. 2.5 where it was already remarked that the valence of the transition elements is less outspoken). The probability amplitudes of the *d* electrons imply that these (3*d*) electrons are located more closely to the nucleus (than the 4*s* electrons). For these metals, with such more "tightly bound" bonding electrons, the so-called *tight binding model* is appropriate, where orbitals occurring upon bonding are constructed out of the atomic orbitals, as was the case for covalent bonding (cf. Figs. 3.11 and 3.12). Close to the metal ion the orbital formed approaches the atomic orbital; in the region, between the metal ions the orbital resembles the free electron orbital.

---

[26] In 1922, Einstein got the Nobel prize for the year 1921 for his work leading to this equation. Contrary to what is often thought, he explicitly did not get the prize for his work on the theories of special and general relativity, as controversy existed regarding their validity in the absence of experimental confirmation at the time (Pais 1982).

3.5.2.1 The Nearly Free Electron Model; Brillouin Zones

Electrons can be conceived as waves. Waves are diffracted, if certain conditions (pertaining to wavelength and angle of incidence) are satisfied, by scatterers periodically arranged in a lattice. So the lattice (three-dimensional periodic array) of metal ions has a pronounced effect on the propagation of electrons. The effect of diffraction of the conduction electrons by the lattice of metal ions, and thus in the presence of a periodic potential, has been dealt with by Bloch (1928).

It will be assumed that the potential is almost constant: the average value remains zero (as before in the free electron model) and only modest potential maxima (between the metal ions) and modest minima (at the metal ions) occur. For those cases, where the electrons considered cannot diffract, the solution of the Schrödinger equation is practically that of the free electron model implying a uniform electron density. However, if diffraction conditions are satisfied, even a weak potential amplitude of the periodic potential leads to strong scattering and the resulting orbital is vastly different from the one of the free electron model. Therefore, this "nearly free electron" model already is appropriate to highlight the effect of metal ion periodicity. From a technical, mathematical point of view, the small perturbation of the potential due to the small maxima and small minima allows the application of *perturbation theory*, which implies that approximate solutions of the Schrödinger equation can be derived on the basis of the known solutions for the free electron model, which were derived for a constant potential.

Consider electrons moving in a certain direction in the metal crystal. This direction makes an angle θ with a set of crystal planes. Reflection can occur if Bragg's law is satisfied.[27] So, given the angle of incidence, θ, and the spacing of the set of crystal planes, d, reflection can occur if the wave number, defined by $k = 2\pi/\lambda$ where λ is the wavelength of the electron wave, corresponds to:

$$k = \frac{\pm n\pi}{d \sin \theta} \tag{3.14}$$

where the $\pm$ sign indicates that the path difference can be positive and negative and n is an integer. If the Bragg condition is fulfilled, the moving electron is not represented by a travelling wave as holds outside the Bragg condition, but by two standing waves. These waves correspond with probability maxima for the electron at positions in-between the metal ions and at the metal ions, respectively (Fig. 3.17). Thereby, the potential energy of the electron in the field set up by the lattice of metal ions is different for both waves (their kinetic energy is the same). The wave with the highest intensity at the minima of potential energy (at the locations of the metal ions) has the lowest potential energy for the electron. Thus, if the k values, for the moving electrons in the crystal, satisfy Eq. (3.14), two energy levels are possible for the electron. This result has a very important consequence as will be shown next.

---

[27] Bragg's law is usually given by $n\lambda = 2d \sin\theta$ and is discussed and derived in Sect. 4.5.

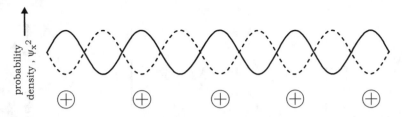

Fig. 3.17 Schematic depiction of the intensity (the probability to find the electron) of the two standing waves which form when the Bragg condition is fulfilled in a one-dimensional chain of positive (metal) ions. The wave represented by the solid line, with intensity maxima at the positions of the positive (metal) ions (indicated with plus sign surrounded by a circle in the figure), has a lower potential energy than the wave represented by the dashed line

In the free electron model, taking the potential in the potential well (i.e. the metal crystal) equal to zero, the total energy of a free electron is composed of kinetic energy only and thus one can write:

$$E = \frac{1}{2} m_e v^2$$

which, using the relation by de Broglie ($\lambda = h/m_e v$; Eq. (2.6)) and $k = 2\pi/\lambda$, can be rewritten as

$$E = \frac{h^2 k^2}{8\pi^2 m_e} \tag{3.15}$$

This equation represents a parabolic relation between E and k (see Fig. 3.18a). Now, according to the nearly free electron model, the results of the free electron model can be adopted if the diffraction condition is not satisfied. However, if the diffraction condition is fulfilled, i.e. k is given by Eq. (3.14), the free electron model fails: two values, instead of one value, occur for the energy of the electron when k satisfies the diffraction condition. This has the consequence as sketched in Fig. 3.18b. Evidently, at those values of k in agreement with the diffraction condition gaps occur in the energies allowed for the electrons[28]: certain ranges of energies are forbidden for the electrons moving in the directions prescribed by Eq. (3.14). On this basis, the development of a *band structure* for the energy spectrum of the electrons in the metal crystal can be understood, where energy ranges allowed for occupation by the electrons are separated by energy ranges which are inaccessible for the electrons (Fig. 3.18c). Generalization of the above argumentation to three dimensions is straightforward. Points in "k space" (with k_x, k_y and k_z components along the three axes of a coordinate system) can be indicated which satisfy Eq. (3.14) and thus define the k_x, k_y and k_z

---

[28] The free electron approximation fails already at k values close to those k values exactly satisfying the diffraction condition. This leads to the dependences of E on k close to the k values given by Eq. (3.14) as sketched in the figure, which deviate from the purely parabolic dependence that holds for truly free electrons [Eq. (3.15)].

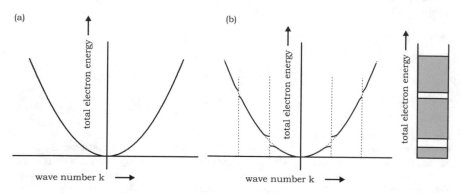

Fig. 3.18 **a** Parabolic relation between total electron energy and wave number, k, for free electrons, as given by Eq. (3.15). In this case, the electron energy is pure kinetic energy, as the potential within the potential well of the crystal has been set equal to zero. **b** Schematic depiction of the effect of a periodic positive (metal-)ion lattice (nearly free electron model) on the total electron energy: splitting up of the total energy at values of the wave number, k, given by Bragg's equation [cf. Eq. (3.14)]; i.e. the position of the borders of the Brillouin zones. Note that at positions away from the Brillouin zone borders the free electron model holds and the total energy is pure kinetic energy, whereas at the Brillouin zone borders, the total energy consists of kinetic plus potential energy (see text). **c** The resulting electron energy bands

values where gaps in the energy band of electron states occur. The regions in k space that comprise the acceptable energy ranges for the electrons are called *Brillouin zones*. One can discern the first, second, third Brillouin zones, etc.

In three dimensions, the boundaries of a Brillouin zone are parallel to the crystal planes that give rise to the reflections considered. Thus, the boundaries of the first Brillouin zone, i.e. the Brillouin zone with the smallest absolute values of k, in a f.c.c. metal are parallel to {111} and {100} planes[29] and in a b.c.c. metal the first Brillouin zone is bounded by {110} planes (see Figs. 3.19a, b). These Brillouin zones are thus polyhedra bounded by the mentioned reflecting planes of the crystal. Boundaries of higher Brillouin zones are determined by other reflections (pertaining to reflections from crystal planes with smaller d values and/or higher values of n, leading to larger absolute values of k [cf. Eq. (3.14)].

Although in a certain direction in k space, i.e. a certain direction of motion for the electron considered, there is always an energy gap in the (E, k) curve (see Fig. 3.18), considering the energy spectrum for *all* electrons in the crystal and recognizing that Brillouin zones in three-dimensional k space are no spheres, there may or may not be a gap in the full range of allowed energies for *all* electrons, irrespective of their directions of movement. A gap for all electrons occurs if the first Brillouin zone is filled completely *before* electrons start to occupy energy levels in the second Brillouin zone. However, if the lowest energy levels in the second Brillouin zone have values below the highest energy levels of the first Brillouin zone (apart from differences in

---

[29] For definition of the Miller indices, denoting a set of lattice planes as {hkl}, characterizing the orientation of the lattice planes considered, and the use of braces, see Sect. 4.1.4.

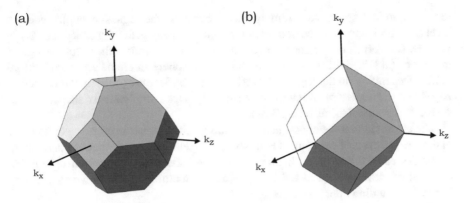

Fig. 3.19 First Brillouin zones of **a** the f.c.c. lattice and **b** the bcc lattice

the directions of the moving, corresponding electrons considered), then the lowest levels in the second Brillouin zone will be occupied already before the first Brillouin zone has been filled completely (this is illustrated in Fig. 3.20). In the last situation, the energy bands (pertaining to the subsequent Brillouin zones) overlap.

In the free electron model, the same absolute value of k corresponds to the same energy [Eq. (3.15)], and hence electrons of the same energy have the same distance

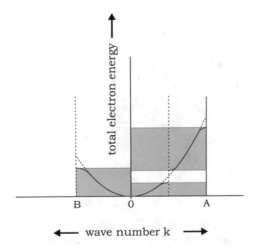

Fig. 3.20 Schematic depiction of energy band overlap caused by the non-spherical shape of Brillouin zones, leading to different ranges for the energy gaps at the borders of the Brillouin zones in different directions in k space. OA and OB correspond with two different directions in k space. The corresponding (E,k) dependencies have been indicated for both directions (here drawn in a colinear fashion: for OA to the right and for OB to the left in the figure). Upon filling the first Brillouin zone with electrons, the highest energy levels (i.e. the border of the first Brillouin zone) have not been reached in the direction OB at the moment the lowest energy levels of the second Brillouin zone in the OA direction have to be occupied (see also Fig. 3.21 and its caption)

to the origin in k space. Accordingly, the electrons at the highest occupied energy level are found at the surface of a sphere in k space that is generally called the *Fermi surface*. The Fermi surface, representing the energy at the limit of the occupied region in k space (at 0 K) will in general no longer be a sphere in the nearly free electron model. Depending on the degree of overlap in energy levels of subsequent Brillouin zones and the degree of filling of the Brillouin zones complicatedly shaped Fermi surfaces can occur (see also Fig. 3.21).

The discussion on the consequences of the whether or not occurrence of overlap of the energy levels of subsequent Brillouin zones (electron energy bands), in relation to the degree of filling of the Brillouin zones with electrons, is postponed until the (alternative) development of the (energy-)band structure has been given according to the tight binding approach (see below).

Finally, at the end of the discussion of (nearly) free electron models, one may wonder if the above has demonstrated that metal bonding must occur, i.e. that the bonding of originally isolated metal atoms in a crystal leads to release of energy. Actually, this is not the case. It has been shown that quantum states of specific energies for (nearly) free electrons in a (nearly) constant potential field exist. Thereby, observed properties of metals could be understood. However, it has not been shown in the above that the resulting system has a lower energy than the collection of isolated metal atoms needed to build up the metal crystal. Such a direct proof for the occurrence of metal bonding is provided by the tight binding approach which is discussed next.

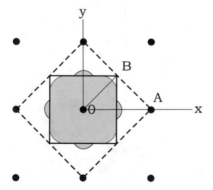

Fig. 3.21 Schematic depiction of complicated Fermi surface shape in a two-dimensional square k lattice as caused by the direction dependence of the discontinuities (gaps) in total electron energy at the borders of the subsequent Brillouin zones (cf. Fig. 3.20). The first Brillouin zone is given by the square indicated by the full lines in the k lattice; the second Brillouin zone is given by the difference of the dashed and full lines in the k lattice. (The points indicated for the k lattice have spacing in the x and y directions equal to $2\pi/a$, with a as the lattice parameter of the two-dimensional square lattice for which the k lattice has been given here; the boundaries of the first Brillouin zone in k space are given by lines that bisect the line segments between the origin of the k lattice and the first points of the k lattice in x and y directions, as measured from the origin; the (E, k) dependencies along OA and OB, as shown in Fig. 3.20, can be interpreted as pertaining to the (E, k) dependencies along OA and OB as indicated in the figure shown here)

3.5.2.2   The Tight Binding Model; The Energy Band Structure

Consider two approaching identical atoms. At a certain interatomic distance, the atomic orbitals of the outermost, valence electrons start to overlap. Then, the valence electrons in these orbitals can no longer be assigned exclusively to the one or the other atom; they can stay anywhere in the aggregate of the two atoms. The original pair of identical atomic orbitals is replaced by two orbitals that extend over the entire aggregate: the electrons of the original pair of identical atomic orbitals move in the field determined by the atomic cores of the aggregate. The two orbitals that replace the pair of identical atomic orbitals are a *bonding orbital* of energy lower than that of the original atomic orbitals and an *antibonding orbital* of energy higher than that of the original atomic orbitals. This was the treatment given for the quantum mechanical description of covalent bonding in Sect. 3.4. As long as the aggregate is restricted to a molecule one speaks of *molecular orbital theory*. A similar approach can now also be followed to indicate what happens with the original atomic orbitals upon the formation of a (metal) crystal.

Thus, the "tight binding method" does not depart from free electrons: the treatment begins with the electrons as in their atomic orbitals associated with isolated atoms. Bringing together, the atoms will lead to (some) sharing of electrons: electrons may spend part of their time with other atoms. One could say these electrons have a certain chance to "hop" from atom to atom using vacant atomic orbitals. This approach is very useful for in particular, but not exclusively, the transition metals. The transition metals have generally partly filled d subshells. The atomic d orbitals only show modest overlap at the interatomic distances of the transition metals occurring in the solid state, which means that the above electron sharing ("hopping") can occur. Yet, these electrons cannot be treated as "free electrons": they still show a relatively strong bonding to the atoms. The atomic nature of the resulting quantum states for these d electrons then naturally suggests an approach as discussed here that can express the relatively "tight binding" with the atoms.

The wave functions for the electrons are constructed out of the atomic orbitals. For a molecule, this development of "molecular orbitals" has often been performed in a variant called *Linear Combination of Atomic Orbitals(LCAO)*, as indicated in Sect. 3.4. Similarly, constructions of "molecular" wave functions for metal crystals out of the atomic orbitals can be performed such that in the regions between the metal ions wave-function characteristics as for free electrons occur, whereas close to the metal ions wave-function characteristics as for the original atomic orbitals prevail. Then, one may also suggest that the crystal could be subdivided in spheres around the metal ions and the space between these spheres and adopt the atomic orbitals within the spheres and the free electron orbitals between the spheres such that continuity in probability amplitude is realized at the sphere surfaces, for the quantum state concerned of the crystal. Indeed, this latter approach has been applied very successfully.

Bringing together N identical atoms will give rise to $N/2$ bonding orbitals and $N/2$ antibonding orbitals out of N identical atomic orbitals. For large values of N, the energy differences between the energy levels of the new (bonding and antibonding) orbitals become (very) small: an energy band of closely spaced energy levels, a "quasi-continuous" band of energy levels, has developed out of the identical atomic orbitals of the same energy. Recall that the number of energy levels in a band is equal to the number of atoms in the aggregate (cf. Sect. 3.4). The width of the band, as the difference between the antibonding state and the bonding state for the two molecular orbitals that occur upon the approach of two identical atoms, depends on the distance between the atomic cores and the type of atomic orbital concerned. The earlier the overlap of identical atomic orbitals of neighbouring atoms occurs the earlier the energy splitting (development of bonding and antibonding "molecular" states) takes place. Further, the separation between the antibonding and bonding states becomes larger upon further approach of the atoms concerned. Hence, upon atomic approach, the outermost valence electrons first experience this phenomenon and the energy bandwidth (energy range from antibonding to bonding states) at the resulting atomic spacing of the crystal will be largest for the band that developed out of the atomic orbitals which "reach most far" from the atom core. This is illustrated in Fig. 3.22. In fact, energy bands also develop for the "core" electrons, but the probabilities to find these core electrons are still largest close to the atoms to which they belong(ed originally).

The resulting energy spectrum for the electrons involved in metal bonding can be illustrated in the following picture (Fig. 3.23). The picture shows a one-dimensional metal crystal. The potential, represented by the full line, is highest outside the crystal. Within the crystal, a periodic variation of the potential occurs with potential maxima in between the atomic cores and potential minima at the atomic cores. The width of the energy bands increases with energy: the energy splitting mentioned above occurs

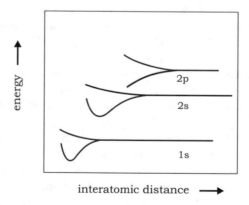

interatomic distance ⟶

Fig. 3.22 Schematic depiction of the formation of energy bands upon decreasing interatomic distance for an infinite number of identical atoms. The earlier the overlap of identical atomic orbitals of neighbouring atoms occurs, the earlier the energy splitting (development of bonding and antibonding "molecular" states) takes place upon decrease of the interatomic distance. Thus, the splitting and band broadening first and hence most pronouncedly occur for the valence electron orbitals upon decreasing interatomic distance; the energy splitting may not occur significantly for the inner, core electrons for interatomic distances as in crystals (see also Fig. 3.23)

Fig. 3.23 Schematic
depiction of valence and
conduction electron bands in
a one-dimensional metal
crystal. The potential,
represented by the full line,
is highest outside the crystal.
Within the crystal a periodic
variation of the potential
occurs with potential
maxima in-between the
atomic cores and potential
minima at the atomic cores

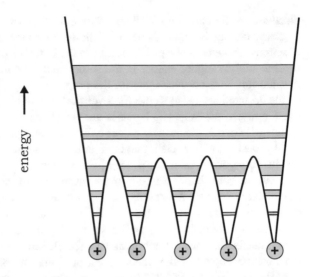

relatively early upon atom approach for the atomic orbitals with distinct probability amplitudes relatively far from the atom core, as holds for the electrons of highest energy, the valence electrons. Bands occur also for the electrons at the atomic core, but these are small and the orbitals have the largest probability amplitudes close to the atom cores. The orbitals in the bands of high energy, where the valence electrons occur, are delocalized, extend over the entire crystal, more or less in the same way as the orbitals in the free electron models. These bands are called the *conduction bands*, which is obvious in view of the explanation for conductivity given above on the basis of the free electron model.

Between the bands energy gaps, *band gaps*, occur, indicating ranges of energy where no acceptable energy levels for the electrons in the system exist. Overlapping of energy bands is possible and, in line with the previous discussion, this will occur as a result of band widening if the distance between the atomic cores is relatively small (cf. the discussion on the occurrence of energy bands separated by energy gaps on the basis of the nearly free electron model).

The electronic configuration of Na is given by $1s^22s^22p^63s$. Each Na atom has one valence electron in the $3s$ orbital. Upon the formation of a Na crystal, a $3s$ band is formed that is half filled by all $3s$ electrons in the crystal. It is immediately understood that Na is a good conductor: for the $3s$ electrons at the Fermi level unoccupied energy levels (quantum states) close to and slightly higher in energy than the Fermi level, and in the same band, are reachable upon acceleration in an electric field (see the discussion on electrical conductivity on the basis of the free electron model). The electron configuration of Mg is $1s^22s^22p^63s^2$. An analogous discussion as for Na cannot explain the metallic behaviour of Mg, as exhibited by its electrical conductivity: the 3s band is full; the Fermi level is at the top of the 3s band. Electrons with energies equal to or a little lower than E_F cannot find an unoccupied level at slightly higher energy in the same band. The reason that Mg is a metal is that the full 3s band overlaps with the empty 3p band; thereby unoccupied levels of the 3p band are close to the E_F level that occurs at the top of the 3s band. The overlapping of the 3s and 3p bands is not very large and, indeed, as compared to the metals with one 3s electron, Mg does exhibit an only modest electrical conductivity. On this basis, it will be clear that the elements of group III in the Periodic Table, as Al, are very good conductors as they possess only one 3p electron.

Now, we come back to a point raised at the end of the discussion of the (nearly) free electron models. An approximate estimate for the binding energy of the metal crystal can be made on the basis of the above band model in particular for those metals which are particularly well described by the concept of free electrons, as the alkali metals with one s electron as valence electron.[30] In that case, it follows that the s band occupied by the valence electrons, and formed upon the approach of the metal atoms, is only half filled (there are as many orbitals (quantum states) in a band as atoms participating in the metal bonding and each orbital can be occupied by two electrons of opposite spin (Pauli's exclusion principle); this all follows from the above).

The bonding states and antibonding states formed upon atom approach have energies below and above, respectively, the energy of the original atomic orbitals (energy splitting; cf. the discussion on covalent bonding in Sect. 3.4). As a crude assumption, it is supposed that the energy of the original atomic orbitals, say E_{atom}, is in the middle of the energy range spanned by the resulting bonding and antibonding states (i.e. the bandwidth). At this stage, it is already clear that a driving force for metal bonding exists: all valence electrons in the band, apart those at the highest filled level, have a lower energy than E_{atom}. Note that it has been tacitly assumed here that the other, core electrons of the metal atoms do not change their energy significantly upon metal bonding.

In the case considered of a single valence electron, the band is half filled (see above) and thus the kinetic energy of the electrons at energy level in the middle of the band equals E_F. Adopting a free electron description for the half filled band, the total energy of an electron in this band is given by its kinetic energy plus the potential energy at the bottom of the band, say V (V was set equal to zero in the free electron model). Hence, it follows from the above for the electron in the middle of the band with total energy $E_F + V$

$$E_F + V = E_{atom} \qquad (3.16)$$

----

[30] The alkali metals are "simple", "open" metals: the ionic cores do not "touch" in the lattice and are practically identical to the cores in the isolated atomic state; the energy levels of the metallic, conducting, nearly "free" electrons in their bands are well above the energy levels of the core electrons.

Now, it has been derived above [Eq. (3.12)] that the average kinetic energy of the free electrons is given by 3/5 E_F. So the average energy of the valence electrons in the band is given by 3/5 E_F + V. Then, the bonding energy for metal bonding is given by [using Eq. (3.16)][31]:

$$E_{bonding} = \left(\frac{3}{5}E_F + V\right) - E_{atom} = -\frac{2}{5}E_F \qquad (3.17)$$

Again, this approximate value for the bonding energy is based on the assumption that only the valence electrons change their energy upon bonding and that the energies of the other, core electrons remains unchanged; i.e. the philosophy behind the free electron model.[32]

The result given by Eq. (3.17) leads to, perhaps surprisingly, very good predictions for the bonding energies of the alkali metals. However, similar predictions for the transition metals are much less good: the absolute values of the bonding energies are underestimated. This is because not only the obvious valence (s) electrons are involved in the bonding, but, also the d electrons in the generally partly filled d subshell play a significant role. The d electrons are more tightly bound to their parent atoms. Yet, unlike for true core electrons, significantly broadened d bands for d electrons occur at interatomic spacing as in the metal crystal. The tight binding approach as sketched above allows a description of this effect recognizing that close to the atoms the orbitals in the d band approximate the original atomic orbitals, whereas at positions in-between the metal ions the orbitals in the d band show free electron-like character.

In the case of the transition metals of the first series (Sc to Zn; cf. the Periodic Table, see Fig. 2.9), (only) two electrons per atom are donated to the rather wide 4s band, whereas a maximum of ten electrons per atom can be donated to the rather small 3d band: the density of states (cf. Fig. 3.24) is much higher in the 3d band than in the 4s band. The wide 4s band and the narrow 3d band overlap. Because of this overlap, the 4s band becomes partly filled *before* the 3d band is filled completely. The electrons of both the 4s band and the 3d band contribute to the metal bonding of the considered transition metals.

---

[31] The reader may be surprised that the bonding energy is negative (recognizing that E_F is positive). This is a matter of definition. It is common in (chemical) thermodynamics to define the difference in energy upon reaction as the difference of the energy of the "product" and the energy of the "parent". This is the rule followed in Eq. (3.17). Hence, the energy *released*, called the "driving force" of a reaction (cf. Sect. 9.1 and its footnotes), is: $-E_{bonding} = 2/5\ E_F$.

[32] Referring to bonding energies it should be clearly indicated how the begin and end states considered have been defined. Here, the parent, begin state is the collection of separate metal atoms at infinite mutual distances and the product, end state is the metal crystal with the atoms at their (equilibrium) lattice-site positions. In the calculation of lattice energies of ionic crystals, in Sect. 3.3, the begin, parent state was the collection of cations and anions at infinite mutual distances and the product, end state was the ionic crystal with the ions at their (equilibrium) lattice-site positions. Alternatively, in the latter case, the bonding energy could have also been calculated starting from neutral atoms. Then, the ionization energy for both atom types has to be added to the bonding energy. In this case, starting from neutral atoms, the bonding energy is also called "cohesive energy".

Fig. 3.24 Schematic depiction of the overlap of a 3d band and a 4s band in a transition metal; density of electron states versus electron energy. The Fermi level has been indicated by the dashed vertical line

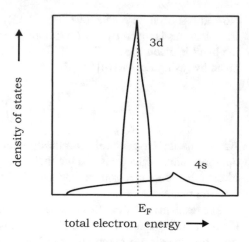

The density of states curve $N(E)$ [see Eq. (3.10)] for the d band is related to the crystal structure of the transition metal. It is interesting to show the shape of the $N(E)$ curve for the d band of the transition metals exhibiting the body centred cubic (b.c.c.; cf. Fig. 3.4) crystal structure (Fig. 3.25). High densities of states occur in the high and low energy regions of the d band; a low density is found in-between. No truly separated subbands (of bonding and antibonding orbitals) occur as in the case of genuinely covalent bonding (see Fig. 3.12). Yet, this picture suggests that a partly covalent nature is associated with the metal bonding of the transition metals. In particular, for those transition metals with less than half filled d shells, this shape

Fig. 3.25 Density of states curve for a bcc transition metal (Cr). The Fermi level has been indicated by a dashed vertical line (data taken from D. G. Laurent, J. Callaway, J. L. Fry and N. E. Brener, Physical Review, B 23 (1981), 4977–4987)

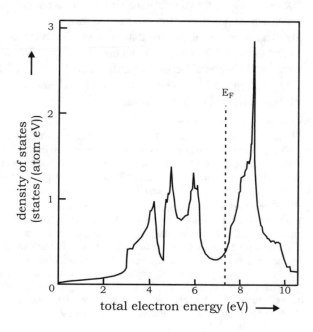

of the $N(E)$ curve indicates that all d electrons can occupy orbitals in the low energy, bottom part of the d band, which appears to favour occurrence of the b.c.c. crystal structure.

The $3d$ band provides no significant contribution to the electrical conductivity. The orbitals in the $3d$ band have large probability amplitudes close to the positions of the metal ions and the chance that an electron in the $3d$ band transfers from one to the other metal ion is relatively small. What is more: electrons from the conducting $4s4p$ band (the $4s$ band overlaps with the $4p$ band) can become "trapped" in the $3d$ band, especially because the density of states in the $3d$ band is relatively high (see above). This explains the relatively low conductivity of transition metals in general. Copper has the electron configuration $1s^2 2s^2 2p^6 3s^2 3p^6 3d^{10} 4s$ (see Sect. 2.5). Hence, in the copper metal, the $3d$ band is fully filled. "Trapping" of electrons in the $3d$ band from the combined, only partly filled, conducting $4s4p$ band cannot occur, and therefore copper (and similarly Ag in the second transition series of the Periodic Table and Au in the third transition series of the Periodic Table; see the electron configurations in Table 2.1) *is* a good conductor.

The generally partly filled nature of the $3d$ band is the cause of some important properties of the transition metals, as for example, the *ferromagnetic behaviour* of iron, cobalt and nickel.

(e)   *Ferromagnetism*

Ferromagnetism implies that, after magnetization has been induced by an imposed magnetic field, distinct magnetization remains after the field has been removed: remanent magnetization. The origin of the ferromagnetic behaviour is sought, as before for paramagnetic behaviour, in the presence of unpaired electrons.[33] Actually, the individual constituents of (only) paramagnetic materials can exhibit permanent magnetization as the result of the presence of unpaired electrons. However, in the absence of an external magnetic field, the directions of these individual magnetizations are randomly oriented in the material and no permanent magnetization can occur. The intriguing question then is which effect brings about the alignment/ordering of the magnetization directions in the absence of an external magnetic field. The following discussion will demonstrate that ferromagnetism is due to subtle electron–electron interactions.[34]

As with paramagnetism, the field-induced (net) parallelization of electron spins upon application of an external magnetic field enhances the imposed magnetic field. The effect is now much larger than for paramagnetism; the susceptibility is both positive and relatively large. The effect is related to the only partially filled core

---

[33] The net magnetic moment is due to the spin of the electron and the orbital magnetic moment. Fully or half filled subshells of atoms have an orbital magnetic moment equal to zero, as simply follows from the addition of the corresponding values of the quantum number m_l (cf. Sect. 2.4.2).

[34] Note that in the considerations until now the electrons were considered to move independently from each other. The only correlation taken into account was due to the compliance of the electrons with Pauli's exclusion principle. Also, an interaction via the effect on the electron's movements by some average effective potential, that can also be determined by the other electrons in the system, was allowed for. This has been called "the independent electron approximation".

subshells: for example, the ferromagnetic materials iron, cobalt and nickel have unfilled $3d$ subshells, containing 6, 7 and 8 electrons, respectively, while the $4s$ subshell is filled completely (see Table 2.1). Note that, in the previous discussion of paramagnetism, the small paramagnetism of the (alkali) metals was due to a fraction of the $4s$ electrons in the $4s$ band corresponding to the small excess of the $4s$ electrons with unbalanced spin, upon imposition of a magnetic field (cf. Fig. 3.16).

The induced/permanent magnetism is due to unpaired electrons and thus pertains to electrons taking part in the metal bonding; the genuinely core electrons are placed in completely filled subshells as electron pairs of opposite spin.

The electrostatic interactions between the charged particles upon (metal) bonding (electrons and ions) depend also on the occurrence of parallel spins or antiparallel spins for the electron configuration. Hund's rule as discussed in Sect. 2.5 implies that electrons in atoms are preferably placed, with parallel spins, in separate orbitals of degenerated atomic orbitals in order to minimize the Coulomb repulsion among them. The condition of parallel spin assures that the electrons are placed in different orbitals (of the same energy) of the atom (cf. Pauli's exclusion principle). One speaks of *spin correlation*. Thus, a correlation energy is associated with the tendency of the electrons to stay away from each other, which is composed of two contributions: the charge correlation, a classical Coulomb interaction energy contribution, and the spin correlation, a purely quantum mechanical energy contribution.

Considering the metal crystal, this can be formulated as that a mechanism, that aligns the spins of similar, unpaired electrons of *neighbouring* atoms (as unpaired $3d$ electrons for the metals considered) in parallel spin configuration upon bonding, causes a lower energy. It is essential that the $3d$ band orbitals have large probability amplitudes close to the positions of the metal ions. Hence, before for these $3d$ electrons strong bonding effects would occur upon further reduction of the interatomic distance, the energy of the system can be lowered for the situation with parallel spins as long as there is an only modest overlap of the electron densities associated with the orbitals of the atoms concerned. Serious overlap of the electron densities of the original atomic orbitals would lead to bonding orbitals where the electrons can only be placed in antiparallel spin fashion according to Pauli's exclusion principle.

The importance of the above-discussed effect of parallel spins for the unpaired $3d$ electrons of neighbouring metal atoms obviously increases if the number of unpaired electrons in the $3d$ subshell increases: the field by the "spinning", unpaired electrons of parallel spin is larger if the number of these electrons per atom is larger, and thereby a larger influence is exerted on the unpaired $3d$ electrons of the neighbouring atoms to have their spins parallel as well; i.e. an arrangement of all these originally unpaired $3d$ electrons in paired, antiparallel fashion in bonding orbitals may be energetically less favourable.

However, it is not easy to understand which mechanism is responsible for the alignment of the magnetizations of the individual atoms upon bonding in a crystal: it has been shown, somewhat unexpected in view of the above discussion, that the magnetic fields of the individual atoms, and thus the magnetic forces they exert on each other, cannot cause the coupling of the magnetizations. And also the energy effects associated with spin correlation cannot explain the extent of the observed coupling. The present situation is unsatisfactory as that there is no complete theory of the origin of ferromagnetism that agrees with physical reality. This is a partly strongly controversial field where the approaches presented often are based on assumptions and approximations that can be in conflict with each other. This just underlines that lot of what is presented as explanatory theory has an ad hoc character.[35]

The occurrence of parallel spin fashion for the originally unpaired $3d$ electrons means that these after bonding still unpaired $3d$ electrons need more orbitals than in the antiparallel fashion, because each unpaired electron needs one orbital. Hence, a further requirement for ferromagnetism is the availability of sufficient empty orbitals in the band considered; i.e. the subshell of the atoms giving rise to the band considered can only be partially occupied, and in this sense, this requirement parallels the one indicated two paragraphs above. Furthermore, because the adoption of parallel spins of electrons in the same (here $3d$) band implies that small changes in kinetic energy associated with this parallel nature of the spins of the electrons exist between the electrons with their spins parallel (consequence of the Pauli exclusion principle), not only sufficient empty orbitals in the band must be available but also they have to be *close together on the energy scale*. Hence, the density of states must be relatively high, which is the case for the $3d$ band as the energy splitting is moderate because of modest overlap of the atomic orbitals in the bonding situation of the metal.

As the temperature increases thermal agitations become increasingly important and above a certain temperature the permanent magnetization is lost: the long-range spin order is destroyed.[36] The temperature above which all permanent magnetization has disappeared is called the *Curie temperature*. Above the Curie temperature, the originally ferromagnetic materials have become paramagnetic. Thus, in view of the spin order, the Curie temperature characterizes an order–disorder transition. The value of the Curie temperature is material specific, does not depend on heating or cooling (rate), i.e. shows no "hysteresis" and therefore is appropriate for temperature calibration as in calorimetry and dilatometry (see Sect. 10.13).

The application of the Curie temperature for calibration in *calorimetry* is based on the discontinuity in the specific heat of the material at the Curie temperature (see Fig. 3.26). The application of the Curie temperature for calibration in *dilatometry* is based on the slight change in specific volume that occurs upon magnetization (cooling) and demagnetization (heating). This effect is called *magnetostriction*. The interaction between the metal ions and the electrons depends on the distance between the ions: the d band becomes wider upon decreasing the distance between the ions, and thereby the energy levels for the d electrons change as well. The magnetostriction effect could thus be understood as a response of the material (slight change of the lattice constants) to modify the band structure such that it becomes most favourable

---

[35] These remarks are made here to highlight with this example that deep, fundamental questions on the properties of solids were *and are* at the heart of materials science. Ferromagnetism is certainly an "old" topic and yet, still very "hot". An illuminating discussion on competing approaches in the field of ferromagnetism can be found in Sects. 3.3 and 3.4 of Aharoni (2000); also the middle part of the preface to the first edition of that book is worth reading against this background.

[36] This can be interpreted as a consequence of the increasing importance of the entropy with increasing temperature in determining the energy of the system (see Sect. 7.3).

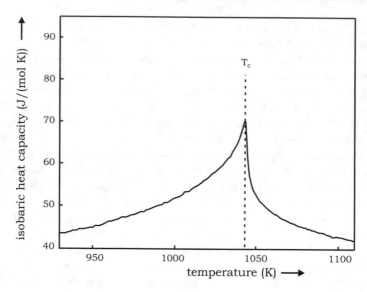

Fig. 3.26 Isobaric specific heat (C_p) of iron as function of temperature. The Curie temperature T_c has been indicated with the vertical dashed line (data from A. T. W. Kempen, F. Sommer, E. J. Mittemeijer, Thermochimica Acta, 383 (2002), 21–30)

for either the occurrence of paramagnetism (above the Curie temperature) or the occurrence of ferromagnetism (below the Curie temperature).

The spontaneous occurrence of ferromagnetism may not be evident from inspection of a macroscopical piece of ferromagnetic material. This is explained by the subdivision of the material in *magnetic domains*. In each domain, the magnetic moments, due to the unpaired electrons of parallel spin, are aligned and each domain has its own direction of magnetization. The oppositely oriented magnetizations of neighbouring domains more or less cancel each other. As a result, the specimen is macroscopically not magnetized. The domain structure should not be confused with the grain structure: each grain can consist of many domains. The origin of the domain structure is the reduction of the *magnetic* energy by the reduced spatial extension of the magnetic field of the domain as compared to the situation where the metal grain would be identical to one domain. Thus, the ferromagnetic grains break up in domains, such that the directions of magnetization across the domain boundaries are opposed (see Fig. 3.27). The lower limit of domain size is related to the cost of energy associated with the generation of the domain boundaries, called *Bloch walls*: the energy balance between reduction of overall magnetic energy and Bloch wall energy[37] is decisive for the domain size. When placed in an external magnetic field

---

[37] Across the Bloch wall the orientation of the spin of the unpaired electrons with parallel spin has to change pronouncedly. This is not achieved abruptly, but occurs gradually in order to maintain a near parallel nature for the spins of the electrons concerned of directly neighbouring atoms. The Bloch wall can thus comprise a few 100 lattice spacings and thereby is much thicker as a "normal" grain boundary which has a thickness of a few lattice spacings (Sect. 5.3).

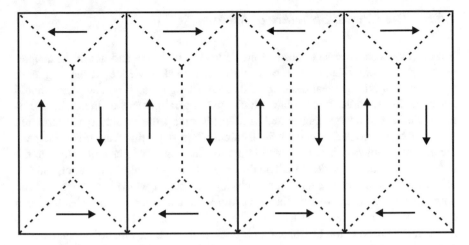

Fig. 3.27 Schematic depiction of domains of different magnetization in a ferromagnetic material. Dashed lines: domain boundaries, solid lines: grain boundaries

the favourably oriented domains increase in volume at the cost of the unfavourably oriented domains and, given a sufficiently strong external field, eventually the state of maximum, saturation magnetization is reached.

(f) *The metallic lustre*

The shiny appearance of a metal, called metallic lustre, has been mentioned as one of the properties typifying a metal (cf. Sect. 1.2). The background of this phenomenon is the relatively strong absorption of visible light by the metal. Visible light (of wavelengths 400–800 nm) can penetrate a metal substrate over a depth up to maximal, say, a wavelength. This absorption occurs, more or less evenly (see below), across the entire visible wavelength spectrum. The absorbed radiation is used to elevate the electrons involved in the metal bonding to unoccupied energy levels. Next, the excited electrons fall back to lower energy levels under reemittance of the energy level differences as visible light. This is observed by the human eye as a strong, bright reflection. The slight differences in the usually greyish colour of the different metals and metallic alloys are a consequence of slight wavelength specificity in the absorption (see above). The yellow-reddish colour of copper is due to a relatively strong absorption of wavelengths shorter than, say, 560 nm, involving excitation of bound $3d$ electrons into the $4s$ band (cf. above remarks on copper).

The reflectivity can be defined as the ratio of reflected intensity to incident intensity. Typical reflectivity values are for gold and silver about 44% and 95%, respectively, which is of the order of the reflectivity of mirrors used in households.

3.5.3   The Crystal Structure of Metals

It may come as a surprise to observe the following text here and not in the chapter on crystallography (Chap. 4). The reason is that our knowledge on the origins of metal bonding, in the ideal case, i.e. if that knowledge would be complete, should allow us, in principle, to calculate which spatial arrangement of the metal atoms gives the largest energy gain upon bonding. Thereby the stable crystal structure would be predicted and explained. Such calculations, departing from first principles, are generally not easy if not impossible; many electron problems still belong to the most difficult ones to be dealt with on the basis of the Schrödinger equation and can only be tackled by approximate approaches, a number of which have been touched upon above. Yet, it can be said that the progress made in the last decades has been enormous, although much of that cannot be dealt with in this book (see also remarks made in the introduction of Chap. 4). Of course, relative statements on the basis of approximate theories have explained successfully certain trends (e.g. in a class of compounds), subject to restrictions corresponding to the method used, have been able to compare the relative stabilities of the one and the other structure and even have provided more or less good predictions of quantitative values of measurable parameters. Indeed, this is a highly interesting and dynamic field of research (e.g. see Martin 2004). Relative stabilities of crystal structures often can only be understood if subtle effects, possibly expressed by tiny energy differences, as, for example, could hold for atomic interactions taking place over long ranges,[38] are taken into account. This recognition is another way of indicating the problems ahead when one desires to predict and explain the crystal structures directly from first principles. The energies corresponding to transitions of one crystal structure to another crystal structure of a solid metal (showing allotropy[39]) are of the order of one per cent of the bonding energy: this is not encouraging for the development of approximate calculation methods which should yield reliable predictions.

Obviously, one approach to predict the occurrence of a certain crystal (metal) structure could involve the calculation of the total energy of the aggregate of atoms considered, for all possible spatial arrangements of the atoms; the arrangement with the lowest energy is the crystal structure we expect to find in equilibrium in nature.

Here, for the crystal structures of metals, one could proceed by considering the energy contributions of (i) the bonding, conduction electrons and (ii) the Coulomb energy of the metal ions in a "sea" of free electrons.

Ad (i). Considering metals it becomes immediately clear that the larger parts of the energy contributions (potential and kinetic energy) of the electrons are structure independent: the kinetic energy of the free electrons [Eq. (3.15)] and their potential energy (set equal to a constant value, zero, in the free electron model) do only depend

---

[38] This remark regarding the significance of long-range interactions, as next-next-nearest neighbour interactions, in fact implies that a central force approximation for atomic interaction can be faithfully used, which is in particular doubtful for metals [see also Footnote 2 in this chapter and Sect. 4.2.4].

[39] Allotropy denotes the occurrence of more than one possible crystal structure for atoms of a single element (see Sect. 4.2.5).

on the volume of the aggregate of metal atoms; indeed, E_F in the free electron model is structure independent as well [Eq. (3.10)]. Hence, only the small correction ("perturbation") to be added to these free electron energy contributions according to the nearly free electron model (Sect. 3.5.2), that expresses the effect of a periodic arrangement of the metal atoms, introduces a structure dependence for the energy contribution of the electrons.

Ad (ii). In order to calculate the electrostatic, Coulomb interaction of the positive point charges (the metal ions) and the "sea" of free electrons, and to be able to demonstrate the sensitivity of the equilibrium crystal structure for tiny energy contributions, first, the concept of the "Wigner-Seitz cell" is introduced.

The crystal, as exhibited by the three-dimensional periodic arrangement of atoms (here, we will restrict ourselves to a single element crystal), can be conceived as a completely space filling arrangement of the following cell:

- take the origin of space at an atom;
- draw lines towards the neighbouring atoms;
- draw the bisections of (i.e. planes perpendicular to and halfway) the line pieces between the atom at the origin and the surrounding atoms;
- the body formed, i.e. the volume enclosed by these bisectional planes, is the cell sought for, called *Wigner–Seitz cell*. Another formulation defining the Wigner–Seitz cell is that it is the locus of all points closer to the origin than to any other lattice point.[40]

The whole crystal can now be constituted by a, completely space filling, stacking of equal polyhedra, each of which is identical to the Wigner–Seitz cell. An example, for the face centred cubic crystal structure, is shown in Fig. 3.28.[41,42]

The structure sensitivity in the electrostatic energy calculation is then due to the "facetting" of the Wigner–Seitz cell, which in a (energy) calculation is uncomfortable to handle. Hence, Wigner and Seitz, recognizing that for the well-known metal structures the overall shape of the Wigner–Seitz cell approaches a sphere (cf. Fig. 3.28), proposed to replace the Wigner–Seitz cell by a sphere of the same volume as the polyhedron. The calculation of the electrostatic, Coulomb interaction of the metal ions and the conduction electrons can then, approximately, be restricted to the calculation of the electrostatic energy of a metal ion in the spherical Wigner–Seitz cell filled with a uniform compensating charge equal to the number of valence electrons per metal ion: there is no interaction of charge neutral, spherical units. Of course, as we wish to find out the relation between structure and energy, we cannot depart from electrically neutral spherical cells; the Coulomb interaction must be calculated for certain

---

[40] The Wigner–Seitz cell also provides an alternative definition of the coordination number of an atom in a crystal structure (cf. Sect. 4.2.4).

[41] Evidently, the first Brillouin zone in three-dimensional k-space (see Sect. 3.5.2 and Fig. 3.21) is nothing else than the Wigner–Seitz cell in k-space.

[42] According to the treatment in Sect. 4.1, the Wigner–Seitz cell is a primitive cell: it contains one motif (here the motif is one atom). However, the Wigner–Seitz cell is not a unit cell as defined in crystallography, because it is not defined by lattice-translation vectors.

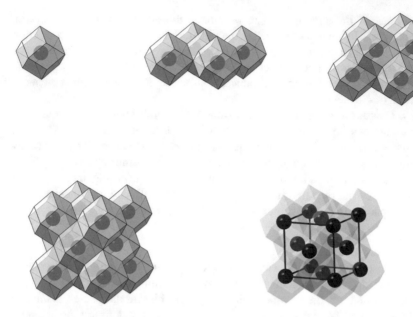

Fig. 3.28 Illustration of how an f.c.c. crystal can be built up from Wigner–Seitz cells. In the last image, the f.c.c. unit cell has been highlighted

three-dimensional, periodic arrangements of metal ions in a "sea" of conduction, free electrons. The Wigner–Seitz cell was introduced here to be able to demonstrate the subtleness of the dependence of the metal-crystal structures observed in nature on tiny energy effects (see what follows).

The calculation of the Coulomb interaction between an array of metal ions in a "sea" of free electrons of uniform density throughout the crystal resembles somewhat the calculation of the Coulomb interaction in an ionic crystal as dealt with in Sect. 3.3. Indeed, this Coulomb interaction is found to be of the type [cf. Eq. (3.3)]:

$$-\frac{(Ze)^2}{d} \cdot \alpha$$

where Z is the number of valence electrons per metal ion (of charge $+Ze$), d is the nearest neighbour distance of the metal ions ($2r_{WS}$, with r_{WS} as the radius of the spherical Wigner–Seitz cell (i.e. the radius of a sphere of one atomic volume), closely resembles d), and α denotes a structure constant that plays the role of a "Madelung" constant. Note the similarity of the above formula with the results presented in Sect. 3.3, where the negative charge was localized (on the anions) and not spread out uniformly as in the case considered here. On the basis of this approach, the following results for α have been obtained (Martin 2004):

$$\alpha_{\text{f.c.c.}} = 1.79175; \alpha_{\text{hcp}} = 1.79168; \alpha_{\text{bcc}} = 1.79186;$$
$$\alpha_{\text{simple cubic}} = 1.76012; \alpha_{\text{diamond}} = 1.67085$$

Evidently, the close packed structures for metal ions in a sea of free electrons exhibit a larger Coulomb interaction than the simple cubic and diamond structures, for the same atomic volume. This can already be considered as an indication that, if metal bonding prevails (which can be formulated as that the concept of metal ions in a sea of (nearly) free electrons holds) that then the close packed structures are favoured. Unfortunately, the differences calculated for the close packed structures and the bcc arrangement appear meaningless. It should be noted that a large number of decimals has been given above for α only to demonstrate that very small differences occur for the f.c.c., hcp and bcc atomic arrangements (which is a striking difference with the values obtained for the Madelung constant for different ionic compounds; cf. Sect. 3.3). Moreover, the result for the spherical Wigner–Seitz cell, which implies ignoring any effect of a spatial (periodic) arrangement of the metal ions (see above), differs only slightly ($\alpha_{\text{WS}} = 1.80$) from the results obtained for the close packed arrangements: for an assessment of the Coulomb interaction the precise metal-crystal translational periodicity plays a minor role. This emphasizes the point made before: subtle, tiny energy differences control the preference for a specific (close packed) crystal structure (for a metal).

As may be anticipated from the above discussion, it must then be the modification of the electron energy by the occurrence of deviations from the free electron model, as exemplified by the occurrence of band gaps, that can be crucial for the preference for the one or the other type of close packed crystal structure.

The above notes can only be considered as an introduction to this topic. The discussion was concentrated on metals that well comply with the nearly free electron model. These are "simple" metals, characterized by core electrons at energies well below those of the "free" electrons which are responsible for the metallic bonding, i.e. the alkali and alkaline earth metals and metals in group III of the Periodic Table. This excludes the transition metals with partly filled d subshells; in that case, the partly localized (i.e. not fully "free") d electrons contribute significantly to the bonding (see the tight binding model). But also copper and the noble metals (silver and gold) with completely filled d shells are excluded, as in the latter case the metal bonding involves something like "hybridization" of the filled d orbitals and higher (s) orbitals, and thus partly localized d electrons contribute to the bonding as well. These (transition) metals deserve special consideration, which is outside the scope of this book (e.g. see Cottrell 1988). As a final remark, regarding the transition metals (as well as the rare earth or lanthanide metals and the actinide metals, with partly filled f subshells; cf. Sect. 2.5), it is mentioned that electron-spin polarization, i.e. lifting the electron-spin degeneracy, can be associated with (tiny) energy differences which can induce preference for the one or the other crystal structure. This implies that magnetism can stabilize a crystal structure, thereby involving that shape-memory effects can be induced by rotation of an externally applied magnetic field (Söderlind and Moore 2008). For the more well-known origin of shape-memory effects, see the *"Intermezzo: Shape-memory alloys"* in Sect. 9.5.2.

In the above discussion on relative stabilities of crystal structures, it has tacitly been assumed that the pressure was constant. As follows from the discussion on the development of bands of orbitals upon closing the distance between the constituent atoms of a (metal) crystal (Sect. 3.5.2.2), both the (energy) width of a band increases and the (energy) gaps between bands decrease if the interatomic separation decreases. Hence, one should expect that the larger the density of the aggregate of atoms, i.e. the more interaction (overlapping) of atomic orbitals occurs, the stronger the metallic character (for example, exhibited by an increasing electrical conductivity) and (close packed) crystal structures typical for metals, may be observed. As remarked in the Preface of this book, any substance may be made *metallic upon densification*. Hydrogen becomes metallic at extremely high pressures, which are not easy to realize at earth. However, it is expected that metallic hydrogen is the dominant component of the cores of massive gas planets as Jupiter and Saturn, where such pressures may prevail. Thus, it may be that, on the scale of the universe, metallic hydrogen is the more common modification of hydrogen. Examples of different type are provided by the difference in metallic character of allotropic forms. Tin as a solid can occur as "grey tin", which shows a diamond-type of crystal structure, with non-metallic properties, but above about 13 °C tin exhibits a body centred tetragonal atom arrangement, "white tin", which has a specific density about 27% larger than "grey tin" and that clearly shows metallic properties (see Sect. 4.2.3.4).

At the end of these remarks about the crystal structure of metals, the explicit difference between the crystal structure of a typical covalently bonded material, e.g. diamond, and the close packed crystal structures of metals is highlighted. The *coordination number*, i.e. the number of nearest neighbours in the crystal, is determined for diamond by the *valency* of carbon: there are four nearest neighbours bonded by directed covalent bonds to the carbon atom considered. The coordination number of metals is to a large extent determined by the *space filling* that assures the largest Coulomb interaction of the positively charged metal ions and the sea of free, valence electrons: the usual coordination numbers are twelve, representative of the face centred cubic (f.c.c.) and hexagonal close packed (h.c.p.) crystal structures (see Figs. 3.29 and 3.30), and eight, representative of the body centred cubic (b.c.c.) crystal structure (see Fig. 3.31).

As a final note, the great deformability of metal crystals can now be understood as follows.

Fig. 3.29 f.c.c. lattice showing the coordination polyhedron (coordination number = 12)

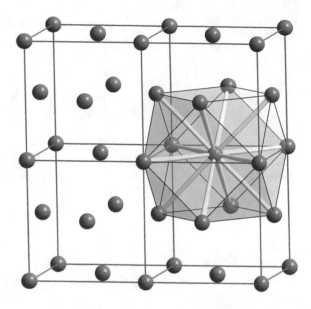

Fig. 3.30 h.c.p. lattice showing the coordination polyhedron (coordination number = 12)

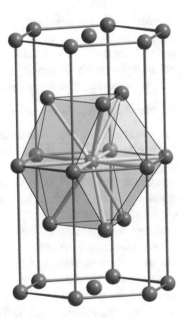

(g)   *Plastic deformation*

The shearing of a metal crystal can be achieved with minor effect on the bonding: the non-directed bonding between the positive metal ions and the "sea" of free, valence electrons is not pronouncedly affected if positive metal ions move with respect to each other (see further Chap. 5, especially Sect. 5.2.5). Thus, metals are malleable. The

Fig. 3.31 b.c.c. lattice
showing the coordination
polyhedron (coordination
number = 8)

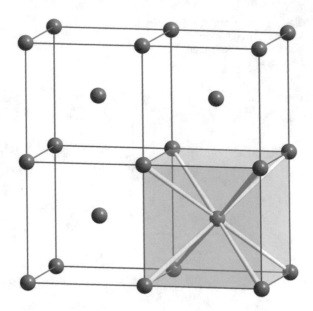

opposite holds for covalently bonded materials as diamond: shearing of diamond implies the disruption of directed covalent bonds between pairs of carbon atoms, which requires a large amount of energy. Thus, covalently bonded materials are not well plastically deformable, i.e. they are very hard (cf. end of Sect. 3.4 and the *"Intermezzo: The hardest materials"* in Sect. 12.13).

Addendum: Conductors, semiconductors and insulators

The discussion on electrical conductivity of metals revealed that the occurrence of partly filled bands (e.g. the alkali metals) or of overlapping of filled and empty bands (e.g. the alkaline earth metals) can be seen as a condition for the material to be a *conductor* (see Fig. 3.32a). It seems appropriate then already here, i.e. even in a chapter on metal bonding, to consider a classification of materials based on their electrical conductivity.

Many materials, other than metals, cannot be considered as conductors. If the valence electrons, the electrons of highest energy, fully occupy a band which in energy lies well below the first higher-in-energy band, the material is no conductor. The full band filled with valence electrons is called *valence band*; the empty band above the valence band is called *conduction band*. Recall that the nearly free electron theory has shown that the band gap depends on direction (in k space; see discussion below Eq. (3.15) and see Figs. 3.20 and 3.21): the band gap is the smallest difference between the lowest energy level in the upper band and the highest energy level in the lower band, irrespective of direction.

The chance that an electron absorbs thermal energy and jumps from the highest energy level of a filled valence band to the lowest energy level of the empty conduction band is given by the Boltzmann factor (see also Sect. 8.6):

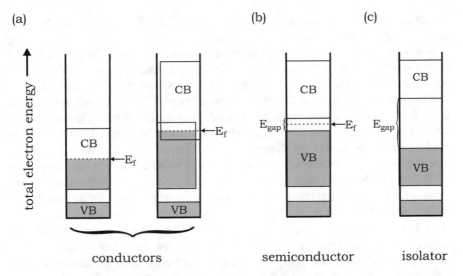

Fig. 3.32 Schematic depiction of band structures of differently well electricity conducting materials. **a** Conductors, left: conduction band half filled (e.g. Na); right: band overlap (e.g. Mg); Fermi level has been indicated by E_F. **b** Semiconductor. The Fermi level according to the alternative definition (see Footnote 43) has been indicated by E_F. **c** Insulator. The energy gap between the conduction band (CB) and the valence band (VB) has been indicated in (b) and (c) by E_{gap}

$$\exp\left(-\frac{E_{gap}}{kT}\right)$$

with E_{gap} as the band gap, k as the Boltzmann constant and T as the absolute temperature. At room temperature, $kT = 25$ meV/atom (corresponding to $RT = 2.44$ kJ/mol, with R as the gas constant).

If E_{gap} is large (say, >5 eV) thermal excitation of an electron of the valence band to the conduction band is negligible, and the material is called an *insulator* (see Fig. 3.32c).

If E_{gap} is moderate (say, around 1 eV), thermal excitation is not negligible and a (modest) conductivity that increases with temperature can be observed. Such a material is called a (n intrinsic) *semiconductor* (see Fig. 3.32b). The conductivity is not only due to the promoted, free electron in the conduction band, but also to the "hole" left in the vacancy band, that, so to speak, has become partially filled due to the excitation of the electron considered.[43] This so-called *intrinsic conductivity* is still relatively small (at room temperature): orders of magnitude smaller than the conductivity exhibited by a true metal, as copper. The striking difference between a metal and a semiconductor should be realized: the electrical conductivity of a metal

---

[43] Recognizing the presence of equal amounts of holes in the valence band and electrons in the conduction band and applying the alternative definition of the Fermi energy for $T > 0$ K as given in Sect. 3.5.1.2, it becomes clear that E_F for an intrinsic semiconductor lies at the middle of its band gap (see Fig. 3.32b).

decreases with temperature [see under (a) in Sect. 3.5.1.2], whereas the electrical conductivity of a semiconductor increases with temperature.

Semiconductors can be forced to show larger electrical conductivities by doping of well-chosen impurities. This impurity-induced conductivity is called *extrinsic conductivity*. The classical example is provided by the semiconductor silicon.

Consider the replacement of one silicon atom by a phosphor atom. Phosphor has five valence electrons, one more than silicon. Four of the valence electrons of the phosphor atom are taken up in the valence band of the silicon crystal. The fifth valence electron will be bonded to the singly positively charged phosphor ion in the silicon parent lattice. In a way, the situation for this fifth valence electron now resembles the bonding of the single electron in a hydrogen atom to the hydrogen atom nucleus. However, the electrostatic interaction is smaller in the present case: the other electrons in the valence band can redistribute in space somewhat, in response to the positive charge on the phosphor core. As a result, the positive charge on the phosphor is shielded to a certain degree, one speaks of "screening", and the fifth valence electron of the phosphor atom does not experience the full charge on the phosphor core. Consequently, the bonding of the fifth valence electron to the phosphor core is rather weak, as also typified by the extent of the orbital of this electron (several tens of Ångstroms). The energy level of the orbital for this valence electron is below but close to the bottom level of the conduction band of the semiconductor (see Fig. 3.33a). The situation described pertains to sufficiently low temperature: the extra electron is trapped at the impurity atom. However, the distance between the energy level of the trapped electron and the bottom of the conduction band is that small [of the order of kT at room temperature, i.e. about 25 meV (see above)] that thermal excitation is easily possible, at room temperature, and thereby the extra electron becomes a free electron in the originally empty conduction band. Given a sufficiently high concentration of such impurities, a significant electrical conductivity is obtained. The impurity discussed donates an extra electron to the silicon, and therefore it is called a *donor*: the energy levels just below the conduction band are called *donor levels* and the resulting semiconductor is called an *n-type semiconductor* ("n" stands for "negative"). Note that the electrical conductivity now is only due to electrons

Fig. 3.33 Schematic depiction of **a** a *n*-type semiconductor with donor levels indicated by dashed lines and **b** a *p*-type semiconductor with acceptor levels indicated by dashed lines

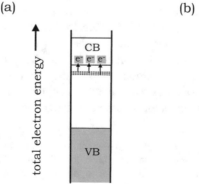
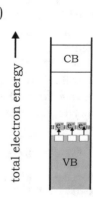

in the conduction band; there are no holes in the valence band as would be due to thermal excitation of valence electrons (see above).

An analogous phenomenon can occur upon introducing an impurity atom that has less valence electrons than silicon, for example, aluminium which has three valence electrons. Obviously, now an electron hole occurs in the valence band that is localized at the impurity atom. If this hole would move away for the impurity atom, this atom would be negatively charged. Electrostatic attraction of this negative charge and the positive hole occurs, which attraction is "screened" as before. As a result, a localized empty orbital occurs at an energy level a little above the upper level of the valence band of the semiconductor (see Fig. 3.33b). At low temperature, the hole remains trapped at the impurity atom. Upon increasing the temperature, an electron from the valence band can jump into this orbital and, given a sufficiently high concentration of impurity atoms, the holes in the valence band allow a significant electrical conductivity. The impurity considered removes electrons from the valence band, and therefore it is called an *acceptor*: the energy levels just above the valence band are called *acceptor levels* and the resulting semiconductor is called a *p-type semiconductor* ("*p*" stands for "positive"). Note that the electrical conductivity now is only due to holes in the valence band; there are no free electrons in the conduction band as would be due to thermal excitation of valence electrons (see above).

It is interesting to remark that even at temperatures that low that neither thermal excitation of electrons from the donor levels to the conduction band (*n*-type semiconductor) nor thermal excitation of electrons from the valence band to the acceptor levels (*p*-type semiconductor) can occur, electrical conductivity is possible if the concentration of donors or acceptors exceeds a certain critical value ("heavily doped" semiconductor): then the relatively extended orbitals corresponding to the donor and acceptor levels (see above) overlap significantly and a band, called "impurity band", occurs. Consequently, electrical conductivity is possible by "hopping" of electrons (or holes) from donor to donor (acceptor to acceptor) atom. Hence, by increasing the impurity concentration, the semiconductor is no longer semiconductive: it has become metallic! This "impurity band" based, induced metallic nature of an originally semiconductor phase, due to increased overlapping of spatially confined (but extended; cf. above) orbitals, parallels the transition of a non-metallic phase to a metallic phase by overlapping of atomic orbitals, realized by densification as discussed in Sect. 3.5.3.

3.6 van der Waals Bonding

Bonding of particles as atoms and molecules is invariably due to electrostatic forces acting between them. One may wonder, then, what causes neutral molecules as CO_2 and F_2, and in particular noble, "inert" gas atoms, as He, Ne and Ar, to condense and eventually to become a solid at a sufficiently low temperature. It is proposed that the weak attractive forces that lead to bonding of the atoms and molecules just mentioned can be conceived as the result of dipole interactions. This type of bonding is called the van der Waals bonding.

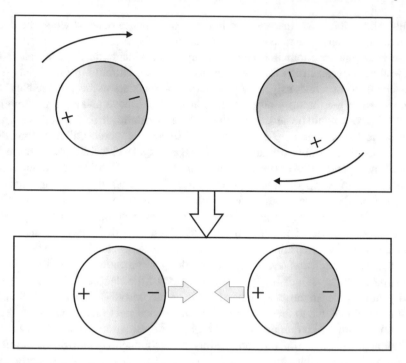

Fig. 3.34 Formation of a van der Waals bond between two atoms/molecules having permanent dipoles

A dipole occurs if a certain separation of the centres of gravity of positive and negative charge exists in a body. Two dipoles can interact and reduce their combined energy if the positive end of one dipole is oriented such that it is close to the negative end of the other dipole (see Fig. 3.34). Such interactions can occur (i) between permanent dipoles of neighbouring particles (polar molecules), (ii) between a permanent dipole of a particle and the induced dipole in its neighbouring particle and (iii) between induced dipoles of neighbouring particles that have or have not permanent dipoles.[44]

Ad (i) Interaction of permanent dipoles. In a molecule like HCl, a concentration of negative charge lies on the chlorine part of the molecule and a concentration of positive charge resides on the hydrogen part (cf. the electronegativities of Cl and H). Similarly, in a molecule of CO, the oxygen part of the molecule is charged negatively relatively and the carbon part is charged positively relatively. In view of the differences in electronegativity between H and Cl and between C and O, one can expect that the resulting dipole for HCl is more pronounced than that for CO. Attractive relative orientations of molecules having permanent dipoles will be more probable than repulsive relative orientations of these molecules, and thus a tendency

---

[44] The interactions of types (i), (ii) and (iii) have been analyzed theoretically first by Keesom (1912), Debije (1920) and London (1930), respectively, and are also often named after them in the literature.

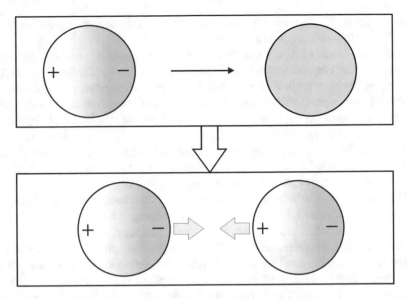

Fig. 3.35 Formation of a van der Waals bond between an atom/molecule having a permanent dipole and an atom/molecule with an induced (by the permanent dipole) dipole

for bonding (attraction) of molecules having permanent dipoles exists. The bonded, condensed state will reflect a favourable (in the above sense) arrangement of the dipoles.

Ad (ii) Interaction of permanent dipoles and induced dipoles. A molecule that has a permanent dipole can induce a dipole in an originally dipole-less molecule/atom having a spherically symmetric electrical charge distribution: the electron distribution of the neighbouring molecule/atom becomes distorted such that an attractive interaction is induced with the (permanent) dipole of the first molecule (Fig. 3.35). Note that the induced dipole can also occur in a neighbouring molecule that already has a permanent dipole; so, for permanent dipoles, the effects discussed under (i) and (ii) have to be combined.

Ad (iii) Interaction of instantaneous and, correspondingly instantaneous, induced dipoles.[45] This effect is thought to be responsible for, also, the bonding exhibited by the inert, noble gases which are composed of atoms with a (time averaged) spherically symmetric electrical charge distribution (closed electron shells); i.e. they do not exhibit permanent dipoles. Consider an isolated hydrogen atom, averaged over time this atom obviously has no dipole. However, at any instant of time, the electron circling the nucleus, is somewhere on its track and an instantaneous dipole occurs. (In fact, according to this picture, we here revert to Bohr's model of the atom; Sect. 2.3). This dipole can, instantaneously, induce a dipole in a neighbouring atom. Thus, fluctuations of the electrical charge distribution in both neighbouring atoms

---

[45] The interaction of type (iii) requires a quantum mechanical treatment.

become correlated. This provides the basis for a weakly attractive, van der Waals force. Thus, the noble, inert gases can solidify.

The three classes of dipole interactions discussed above are summarized under the heading "van der Waals attractions". The energy of the van der Waals interaction depends on the inter-particle (interatomic/inter-molecule) distance, r, according to r^{-6} (for all three types of dipole interactions), and thus the interaction is of short-range nature, as compared to the Coulomb interaction between two ions that depends on the inter-ion distance according to r^{-1} [see Eq. (3.3)] which thereby is of long-range nature. Note that the repulsive interaction (due to overlap of the outermost electron distributions) is of (still) even more pronounced short-range nature (see Sect. 3.3) than the van der Waals attraction. So bonding is possible.

The van der Waals bonding is rather weak: bond energy of magnitude, say, 10 kJ/mol (0.1 eV/atom); to be compared with the ionic, covalent and metallic bond energies which are of magnitude, say, 500 kJ/mol. The van der Waals attraction operates between all atoms and molecules. It takes the stage and provides the predominant source of (*secondary*) bonding between molecules, wherein the atoms are bonded by strong, *primary* (usually of pronouncedly covalent nature) bonds,[46] leading to condensed states (liquid, solid). Thus, the van der Waals bonding is responsible for the bonding leading to most organic crystals. Obviously, because of the weak nature of the van der Waals bond in general, solids of molecules bonded by van der Waals forces are soft, have low melting points and low boiling temperatures.

Between inert gas atoms the van der Waals interaction due to induced dipoles operates. In the gas phase, this interaction causes the non-ideal nature of the gas and is directly related to the occurrence of the correction to be applied to the pressure in the so-called van der Waals equation (equation of state describing the interrelationship of p(pressure), V(volume) and T(temperature) for a real gas). The van der Waals bonding leads at sufficiently low temperature to solid phases for Ne, Ar, Kr and Xe (He is a liquid at $T = 0$ K, at zero pressure). The non-directionality of the van der Waals bonding by induced dipoles (type (iii); see above) suggests that in the solid state the inert, noble gas atoms want to stick together as closely as possible. Indeed, the crystal structures of the solid phases of the inert gases Ne, Ar, Kr and Xe are all cubic close packed (i.e. face centred cubic; see Fig. 3.23 and Sect. 4.2.1). Note that the argument to explain the occurrence of close packed structures for the inert gases differs from that for metals where the Coulomb interaction of positive metal ions with a "sea" of free, negative electrons is considered (Sect. 3.5.3).

A diamond (and also a graphite) crystal can be conceived as one "molecule", implying that, in the above-defined sense, strong, primary bonds occur between all neighbouring carbon atoms; these solids have been called "network solids" (cf. Sect. 3.4). Other types of polymeric carbon were discovered and studied intensively: fullerenes (Kroto 1985) and nanotubes (Iijima 1991). Network solids of fullerenes are not possible, leading to the supposition that, if solids of these materials occur, that then secondary, van der Waals bonding can be predominantly responsible. The perhaps

---

[46] Here, it is recalled that an ionic or metallic solid crystal can be conceived as *one* gigantic "molecule" (see Sects. 3.3 and 3.5).

Fig. 3.36 Geometry of a
molecule of C_{60}
(Buckminister) fullerene

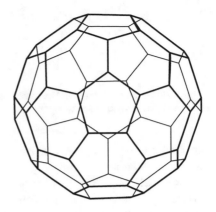

best-known example may be so-called buckministerfullerene which is a substance
composed of C_{60} molecules. Such a molecule has the geometry of a closed, more
or less spherical cage structure of 60 carbon atoms arranged in 20 hexagons and 12
pentagons (Fig. 3.36).[47] Interactions between C_{60} molecules are (always) possible
on the basis of van der Waals forces due to induced dipoles [van der Waals bonding
of type (iii)], just as for the inert, noble gases discussed above. Then, it is no surprise
that the C_{60} crystal exhibits cubic close packing of C_{60} molecules, with a relatively
large lattice parameter of about 1.4 nm.

3.7 Hydrogen Bonding

Another mechanism providing secondary bonding between molecules, composed of
primarily bonded atoms, is the bonding due to "bridging" by hydrogen atoms (actually, by the nuclei, i.e. the protons). Hydrogen is an (the only) element with no core
electrons: its only electron is a valence electron. Having given away its valence electron for bonding, the hydrogen atom nucleus, the proton, is unprotected/unshielded by
core electrons and can directly interact with other electrons (e.g. of another neighbouring molecule): *the proton can be "shared" among neighbouring molecules.*
Hydrogen atoms taken up in a bonding configuration with, for example, oxygen
in a molecule become (partially) positively charged and thereby can interact with
electronegative atoms as nitrogen, oxygen and fluorine present in a neighbouring
molecule. In this sense, the hydrogen bond can be considered as a bonding due to the
interaction of permanent dipoles in neighbouring molecules. (Thereby the hydrogen
bonding could be considered as van der Waals bonding of type (i); cf. Sect. 3.6).

Thus, a simple example of hydrogen bonding is provided by (solid and liquid)
HF (see Fig. 3.37). The molecule HF is significantly polarized. The positively

---

[47] Nanotubes can be conceived as made from graphene planes (i.e. a 001 plane of the graphite
crystal structure; cf. Sect. 4.2.3.3) rolled into a tube. As a result, every carbon atom resides at the
junction of three hexagons of carbon atoms.

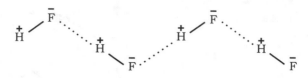

Fig. 3.37 Hydrogen bridges in HF, indicated by dotted lines; covalent bonds designated by solid lines and partial charges shown by "+" and "−"

charged hydrogen nucleus can interact with the negatively charged fluorine atom of a neighbouring HF molecule. Thereby, a hydrogen bridge is formed between the two molecules. Note that the position of the bridging hydrogen needs not to be symmetrical between both molecules: the bridging hydrogen atom is likely more close to the HF molecule it belonged too, already before the association with the neighbouring HF molecule occurred. In this way, solid HF is composed of endless chains of HF molecules. Note that the negatively charged F atoms are less likely to build bridges between the HF molecules: because the F atoms have a substantial volume (due to their core electrons), the F atom cannot approach closely the centre of positive charge; but the very small, practically bare proton (hydrogen atom nucleus) can come close enough to the negatively charged part of the neighbouring molecule. Hydrogen bridges are possible; fluorine bridges do not occur.

The energy of the hydrogen bond usually is a couple of times the energy typical of a van der Waals bond: a few tens kJ/mol.

Many organic compounds as crystalline solids exhibit important contributions to the bonding of the individual molecules by the hydrogen bonds. In fact, the resulting crystal structure, i.e. the spatial arrangement of the atoms, is to a large extent dictated by the occurrence of hydrogen bonds: the molecules arrange themselves in crystals such that the hydrogen atoms of one molecule can bond to electronegative atoms, as oxygen, nitrogen and fluorine, of a neighbouring molecule. This holds, for example, for the "polymeric" structure of crystalline water, i.e. ice (Fig. 3.38). The structure of ice is characterized by tetrahedrons of oxygen. The oxygen is held together by hydrogen bridges between them. The proton at each oxygen–oxygen connection is more closely to one of the two oxygens, implying that the H_2O molecule as entity is identifiable in the crystal structure. Each oxygen atom is surrounded by four hydrogen atoms: two of these are primarily covalently bonded to the oxygen atom (i.e. they together form the H_2O molecule); the other two are "hydrogen bonded". This desire to establish hydrogen bonds between the H_2O molecules in ice leads to the rather open crystal structure of ice, with the consequence that upon solidification of water an increase of volume occurs (see also Sect. 7.5.1).

Biologically active materials are to a significant extent controlled by hydrogen bonding. Proteins are macromolecules produced by polymerization of amino acids and thus are characterized by chains with a –C–C–N repetition unit (Fig. 3.39). The original amino acids used in protein formation have to occur in a specific sequence in the $(-C-C-N)_n$ chain. It is this sequence of amino acids that is crucial for the biological activity of the protein; there are many hundreds of amino acids incorporated in

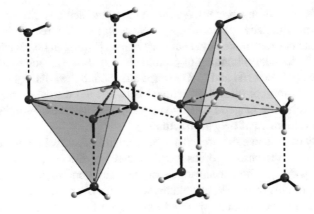

Fig. 3.38 Crystal structure of ice. Covalent bonds have been indicated by solid lines, hydrogen bonds have been designated by dashed lines. The oxygen atoms are represented by relatively large black dots, the hydrogens atoms are shown by relatively small white dots. Coordination polyhedra (tetrahedra) for the oxygen in the water molecules (coordination number = 4) have been indicated. The picture shows only one possible configuration of water molecules. The water molecules can also be oriented differently under certain geometrical constrictions, but, considering the coordination tetrahedron, for each oxygen atom, it holds that always two hydrogen atoms are covalently bonded and always two hydrogen atoms are bonded via an hydrogen bridge

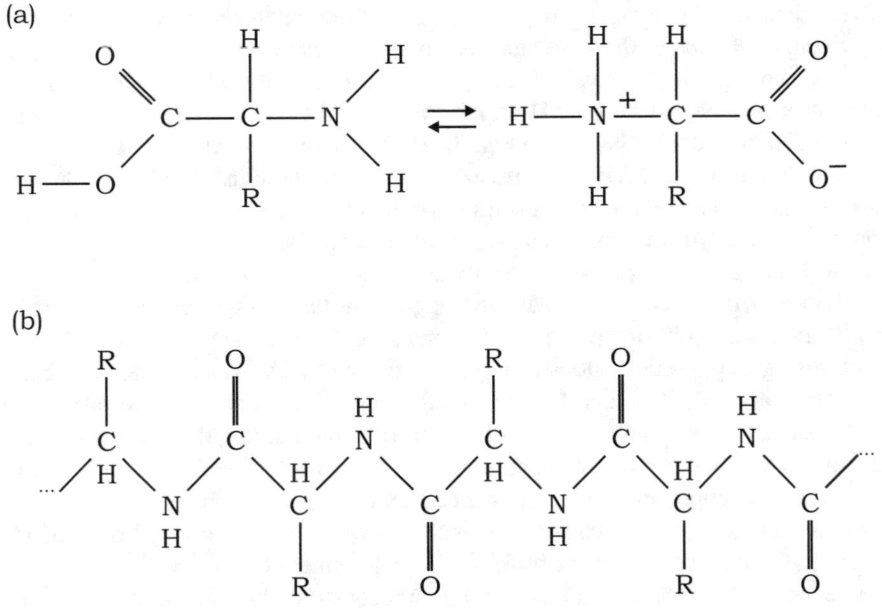

(a)

(b)

Fig. 3.39 a General formula for an amino acid in not-dissociated and dissociated modification and **b** protein chain formed by polycondensation of aminoacids. Amino acids differ in their R groups and polymerization can be achieved by establishing C–N bonds (C and N from different amino acid molecules) under release of H_2O (H from NH_2; OH from COOH)

such a specific manner in a protein. A protein macromolecule is often not composed of only one (polymerized) chain of –C–C–N units: two or three chains can run in parallel and can be cross-linked by secondary bonds. Hydrogen bonds can be responsible for (such secondary) bonding of protein chains. Usually, such hydrogen bonds are established between NH and C=O groups attached to the adjacent chains, where the hydrogen proton constitutes a "bridge" between the nitrogen and the oxygen atoms. The realization of good hydrogen bonding between the protein chains can lead to the development of helical structures.

The variety of protein structure in living organisms is controlled by genetic information carried by the chromosomes in the nucleus of the cell, the smallest living part of a living being. The chromosomes of each cell contain all data necessary to build up the proteins of the living organism. Cell division involves splitting of the chromosomes. The organism is capable to produce a full chromosome on the basis of only a splitted, half of the original chromosome. The question arises how this is achieved.

Chromosomes contain macromolecules: the nucleic (the adjective "nucleic" indicates that the substance is located in the cell nucleus) acids. Deoxyribonucleic acid (DNA) is considered as the carrier of the genetic information ("genetic code"). The DNA molecule is composed of two chains each composed of alternating sugar (S) and phosphate (P) groups: S–P–S–P–S–P–.... Each sugar group is bonded to one of four bases: adenine (A), thymine (T), guanine (G) and cytosine (C) (see Fig. 3.40). The base attached to a sugar group of one of both sugar-phosphate chains is bonded by hydrogen bonds to the base attached to the opposite sugar group of the other chain of both sugar-phosphate chains (see Fig. 3.41). The spatial geometry of the two sugar-phosphate chains of DNA has been found to be that of two intertwined helical chains: a double helix (see Fig. 3.42). The hydrogen bonds indicated above hold the two spirals in position in space. This hydrogen bonding should then be such that adenine from chain one is always opposite thymine from chain two and vice versa, and that cytosine from chain one is always opposite guanine and vice versa, i.e. A–T and G–C base pairs are the only possible base pairs in DNA.

Evidently, this structure immediately suggests the basis of the genetic code and its replication in cell-division processes. Recognizing that there are of the order 10^{10} of such base-group positions along a single S–P chain of a DNA molecule, the precise sequence of the A, T, G and C bases along a single S–P chain can be taken as a code that stores a vast amount of genetical information. Further, the requirement that the opposite bases of the two chains can only occur as A–T and G–C pairs allows a simple replication: after dividing, in a cell-division process, the double helix into its two, then separate S–P chains each with its own bases (i.e. the hydrogen bonds between the opposite bases are broken), the requirement of A–T and G–C pairing immediately makes clear that it is possible to reconstruct the DNA molecule on the basis of only its splitted half part. Thereby out of one DNA molecule two DNA molecules, identical to the original one, have resulted.

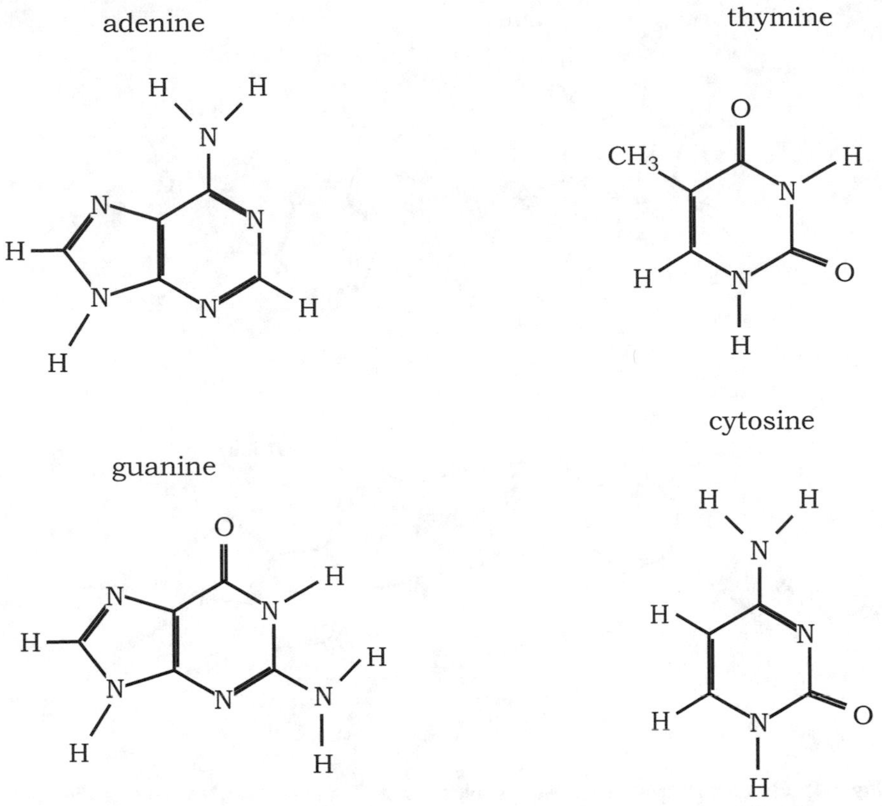

adenine

thymine

guanine

cytosine

Fig. 3.40 Four bases of DNA

Epilogue: "How Science Really Happens"

The discovery of the structure of DNA and, in particular, thereby exposing the replication mechanism of genetic information is without any doubt one of the great discoveries of the twentieth century. Watson and Crick published in 1953 their one-page letter on the structure of DNA with, practically at the end, that one sentence: "It has not escaped our notice that the specific pairing we have postulated immediately suggests a possible copying mechanism for the genetic material". That sentence in fact is the culmination point of their letter which brought them the Noble Prize for Medicine and Physiology in 1962. The reason to dwell upon this here, at the end of this chapter on chemical bonding, that is based on great scientific achievements of others as well, sometimes, but not always mentioned explicitly, is the possibility by this means to make a few remarks about the process of scientific research performed by human beings.

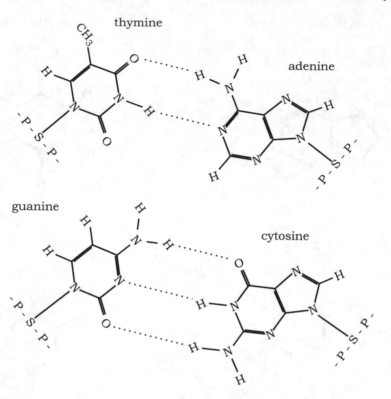

Fig. 3.41 Hydrogen bonding in the two possible pairs (A–T and G–C) of DNA bases (cf. Fig. 3.40), indicated by dotted lines. The –P–S–P– "backbone chain" is composed of alternating sugar and phosphate groups

Fig. 3.42 Schematic depiction of the double helix structure of DNA. The double helix can be described as a spiral staircase with the planar, hydrogen-bonded base pairs as steps

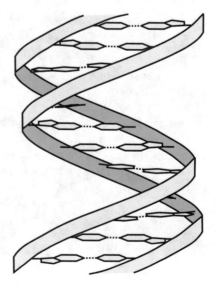

Watson and Crick performed their work on the basis of structure-model building utilizing all relevant experimental results available to them. The important breakthrough occurred when an unpublished X-ray diffraction pattern made by Franklin was shown to Watson by Wilkens, the superior of Franklin. At that moment, Watson and Crick could have decided to propose collaboration to Wilkens and Franklin, work out the possible structure of DNA and publish the result together. They apparently decided otherwise and Wilkens, chasing the structure of DNA by himself, was confronted with the eventual, correct model of DNA after it had been figured out. Then, in the same issue of the journal, where the proposal by Watson and Crick appeared, also, separately, the experimental X-ray diffraction work by the Wilkens group was published. The Nobel Prize committee then decided in 1962 that the prize had to be shared by Watson and Crick and Wilkens. The person left out was Franklin, the coworker of Wilkens, who had done the superb and difficult experimental diffraction work. She perhaps would have been included as Noble laureate, but unfortunately had died at the age of thirty-seven in, already, 1958 and Noble Prizes are not granted posthumously.

The story of the discovery of the double helix structure of DNA has been told by Watson in an exciting book entitled "The double helix", published in 1968. This is a book that in a frank way reveals how science is done in reality by humans. It should be read by anybody becoming involved in scientific research. If fierce competition is felt, if the stakes are high and if the winner takes it all, scientific research is perhaps no longer a noble enterprise but subject to the flaws of human behaviour, as any other activity of mankind. It appears for sure that Watson and Crick did not at all trespass the borders of scientific decency, but reading the book makes clear that a grey zone exists between scientific fairness and scientific abuse. These remarks are the more in order as it has become clear in recent years that, also in the "hard" sciences, the number of cases of flagrant deceit and fraud has increased pronouncedly. One has to be aware of that and realize that there are limits to what is acceptable in scientific research (and pursuing a career). During my career, I have seen unacceptable data manipulation in a laboratory where I have worked, my name has been put on an author list without my allowance and, also, a few papers have been published where my name as author was unjustly left out (to say nothing of those cases where deliberate misleading referencing has been made or referencing to the previous, original work has been omitted at all). Such experiences are not the rule, but not so extremely rare that they should not be mentioned in a world where the number of living scientists is larger than the cumulated number of scientists that lived in the past and where the number of publications can be decisive over a career in science.

Finally, as admitted by Watson himself in the epilogue of his book, Franklin, who may have been not "easy-going", may have suffered from being a female in a male-dominated scientific world. The fate of Rosalind Franklin has developed into a "cause célèbre" in the feminism movement fighting the supposed suppression of female scientists. However, evidence that Franklin has been done basic injustice in the evaluation of her scientific merits lacks. But this is a controversial point.[48]

References

General

A. Aharoni, *Introduction to the Theory of Ferromagnetism*, 2nd edn. (Oxford University Press, Oxford, 2000).

A.H. Cottrell, *Introduction to the Modern Theory of Metals* (The Institute of Metals, London, 1988).

R.M. Martin, *Electronic Structure* (Cambridge University Press, Cambridge, 2004).

L. Pauling, *The Nature of the Chemical Bond*, 3rd edn. (Cornell University Press, Ithaca, New York, 1960).

P. Wilkes, *Solid State Theory in Metallurgy* (Cambridge University Press, Cambridge, 1973).

Specific

M. Gaio, P.L. Silvestrelli, Efficient calculation of Madelung constants for cubic crystals. Phys. Rev. **B79**, 012102 (2009)

D. Borwein, J.M. Borwein, K.F. Taylor, Convergence of lattice sums and Madelung's constant. J. Math. Phys. **26**, 2999–3009 (1985)

P.C.J. Graat, M.A.J. Somers, E.J. Mittemeijer, On the kinetics of the initial oxidation of iron and iron nitride. Zeitschrift für Metallkunde **93**, 532–539 (2002)

---

[48] Although there is no direct link with the scientific content of this book, I may digress, in a footnote to this epilogue on "how science really happens", and thereby emphasizing that science is not an abstract activity but a business run by human beings with ugly flaws, on another, more outspoken, case of (female) discrimination in science. It concerns the career of Lise Meitner. Together with Otto Hahn, she is the discoverer of nuclear fission (1938–1939). Although Meitner received many high honours for her work and was highly regarded by the best scientists of the world in the first half of the twentieth century (notably Einstein and Planck), it was Hahn alone who was awarded in 1944 the Noble Prize for Chemistry for the discovery of nuclear fission. That this Nobel Prize was not shared by Hahn and Meitner is nowadays considered as unfair towards Meitner in view of her contributions. Contemporaries of Meitner, possibly biased by Meitner being a woman, were inclined to consider Meitner as only a coworker of Hahn, working under the intellectual leadership of Hahn, which did not match reality. As late as in 1953, even Heisenberg in a condescending manner still described Meitner as an "assistant" to Hahn. The fate of Meitner was moreover complicated by her being Jewish, which led in fascist Germany to her forced exile by flight from Berlin in 1938. In this sense, there is truth in the statement that Meitner, as a scientist, had suffered from multiple discriminations.

W.A. Harrison, Simple calculation of Madelung constants. Phys. Rev. **B73**, 212103 (2006)

Y. Kuru, M. Wohlschlögel, U. Welzel, E.J. Mittemeijer, Crystallite size dependence of the coefficient of thermal expansion of metals. Appl. Phys. Lett. **90**, 243113 (2007)

A. Pais, *Subtle is the Lord ... The Science and the Life of Albert Einstein* (Clarendon Press, Oxford, 1982)

P. Söderlind, K.T. Moore, When magnetism can stabilize the crystal structure of metals. Scripta Mater. **59**, 1259–1262 (2008)

J.D. Watson, *The Double Helix* (Weidenfeld and Nicolson, London, 1968).

Chapter 4
Crystallography

Asking the laymen what a crystal is, reference most likely will be made to macroscopic solid bodies found in nature (often minerals, possibly presented as gems), more or less or not transparent for visible light, bounded by planar faces (facets) and, thereby, exhibiting regularity. Symmetry may, for example, be apparent as a rotation over a certain angle, e.g. 60°, 90° or 180°, about an axis through the object, leading to the same appearance. The observation of symmetry (not only possible as the result of a rotation as indicated above, but, for example, also as the outcome of a mirroring or an inversion operation) induces a strong emotional stir in human beings: the occurrence of symmetry is experienced as beauty.[1] This sensation may be primarily due to nature and not to nurture.

This felt beauty of matter has led to, partly far-fetched, considerations of the role of symmetry as a bridge between science and art (e.g. see Hargittai 2007 and also Hargittai and Hargittai 1994).

Against the above background it is obvious that already centuries ago mankind tried to find the secret of the regular crystal shape. It was found that the set of occurring angles between the planar faces bounding a crystal is the same for each solid chemical compound or solid element (Stensen's law presented in 1669). Moreover, at about the same time, it was proposed that the regular crystal shape was due to a regular internal arrangement of spherical, cubic or other polyhedric entities (Hooke in 1665 and Haüy in 1784; see Fig. 4.1a, b, for examples). In fact, the last hypothesis, in essence, has been proven right, but this proof had to wait until the discovery of the diffraction of X-rays by crystals in 1912 by Friedrich, Knipping and von Laue (see Sect. 4.5), after the reality of the atoms (and molecules) as smallest building units

---

[1] This statement is too strong. Slight distortions of a symmetrical appearance may be a prerequisite to achieve the strongest appeal: man and women would not feel attracted to perfectly symmetrical women and man, respectively (here is meant that the midplane through the facial front of men and women would be a perfect mirror plane). This effect may also explain the attractivity of the specifically distorted symmetry, but still "regular" appearance, of some two-dimensional patterns designed by Vasarély, as compared to the perfectly translationally symmetrical, two-dimensional patterns created by Escher, examples of both of which decorate walls in my private home.

© Springer Nature Switzerland AG 2021
E. J. Mittemeijer, *Fundamentals of Materials Science*,
https://doi.org/10.1007/978-3-030-60056-3_4

(a) (b)

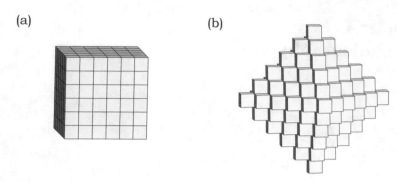

Fig. 4.1 Construction of a macroscopic crystal, by massive arrangement of identical cubes, exhibiting characteristic faces: **a** cube and **b** octahedron

of substances had been convincingly demonstrated in the period 1900–1910 (see the *"Intermezzo: Brownian motion"* in Sect. 8.2).

Intermezzo: Goniometry; the Beginning of Crystallography

Stensen's law may be considered as the beginning of crystallography. The measurement of dihedral angles from the *external* shape (habit) of well developed (facetted), macroscopical crystals, as often exhibited by minerals,[2] characterized especially research in the field of mineralogy for centuries after the formulation of Stensen's law. For the determination of dihedral angles specific instruments were developed: goniometers. This instrumental development, of several hundreds of years, culminated in the devise and construction of the two-circle goniometer after Goldschmidt at the end of the nineteenth century, which instrument allowed an unsurpassed precision. This development has been described by Medenbach et al. (1995), which paper is illustrated with photographs documenting the visual beauty of these instruments (next to the beauty of minerals…) and the impressive technical professionality of the instrument makers (Goldschmidt worked together with the highly skilled technician Peter Stoe (the resulting firm "Stoe", founded in 1887, is still a well-known producer of X-ray diffractometers, also often called X-ray goniometers)). It should be recognized that, for example, the classification of the seven crystal systems (see Sect. 4.1.2) was possible on the basis of such dihedral angle measurements alone. Viktor Mordechai Goldschmidt emerged as one of the most important crystallographers of the time (his nine volume "Atlas der Kristallformen" (1913–1923) is still an important book series of reference). Of course, the discovery of X-ray diffraction in 1912 (see above) ended the importance of performing dihedral angle measurments on macroscopical crystals (minerals). These notes may then also serve as a reminder of the important, scientific contributions of our predecessors, of which we often are no longer aware.

Crystallography then may be described as the science dealing with the *internal* structure, in particular the symmetry, of (ideal, see further below) crystals. The majority of the solid materials are composed of crystals. This explains the importance of crystallography for materials science.

In the past century the determination of crystal structures of specific substances, by diffraction methods (see Sect. 4.5), was usually not a straightforward process. For example, until, say, the third quarter of the twentieth century, it was not uncommon to spend a whole Ph.D. project to the determination of a single crystal structure. The exponential development of crystallographic methods and insights, for which, to emphasize the scientific and technological importance, a number of Nobel prizes have been given through the years, and the enormous increase of computing power, have led to the current situation where the determination of a crystal structure in many cases has become a more or less routine matter that can be handled in a couple of days or much less. It would be misleading to suggest, as a reflection on the last remark, that crystallography as a field of scientific activity thereby has lost its dynamic nature: the discovery of so-called quasicrystals (see Sect. 4.8.2) and their analysis and interpretation is a proof of the opposite.[3]

The description of symmetry, of fundamental importance for the understanding of the regularity of crystal structures, belongs to the realm of mathematics. It turns out that in an n-dimensional space only a finite number of specific combinations of symmetry operations are compatible with periodic, long-range arrangements of building units (e.g. atoms, molecules).[4] Thus, a systematic and complete description

---

[2] Metals can also crystallize as a single crystal of macroscopical dimensions. However, usually, we are confronted with polycrystalline specimens, where the individual crystals exhibit a non-regular morphology, as, for example, forced by the (copious) nucleation of crystals in a melt upon cooling and their unconstrained growth until "hit upon" by neighbouring outgrowing crystals (see the "*Intermezzo: Making Grain Boundaries Visible*" and Fig. 4.2 further on in this section).

[3] Many have suggested, at certain instances of time, the completeness of scientific understanding in a certain field of science, and time and again have been proven to be wrong. A typical contemporary example concerns "thermodynamics": it has often been said that this field has become such mature that nothing of great significance can be added, but current work on the second law of thermodynamics, in systems remote from equilibrium where (local) "order" develops out of "chaos", or the recent development of "interface thermodynamics" represent activities indicating that "thermodynamics" is "hot", also today. Even more dramatic, it is recalled that at the end of the nineteenth century the view was generally held that the entire discipline of "physics" was completed. Then, came relativity and quantum theory; so, one is wise not to condemn a certain field of scientific activity to be "dead" or in a rounded-off state. Unfortunately, science policy makers and scientists over and over again step into this trap. Scientific breakthroughs remain unpredictable and cannot be planned by human beings.

[4] This remark is restricted to crystals as meant usually: crystals exhibiting long-range translational periodicity (see what follows). So-called aperiodic crystals (see Sect. 4.8 and the *epilogue* to this chapter) deserve separate treatment.

of the possible symmetries of crystal structures in three-dimensional space has been achieved (see the "*Intermezzo: A short note on point groups, crystallographic point groups, plane groups and space groups; glide and screw operations*" in Sect. 4.1.2)

Because the spatial distribution of the building units (as atoms/molecules) in crystals is anisotropic (i.e. direction dependent), it may not come as a surprise that many properties of such single crystals are anisotropic as well. Such properties may be the electric or thermal conductivity or the response to mechanical forces/loads (cf. Chap. 12). However, the degree of this anisotropy of a certain property cannot be derived without more ado from the crystal structure. It is even possible that a specific crystal structure for a specific material is accompanied with isotropy of a property that is anisotropic for other materials of the same crystal structure. For example, a crystal of iron (ferrite) has cubic symmetry (body centred cubic; cf. Sect. 4.2.2) and exhibits an anisotropic "mechanical strength" upon uniaxial elastic straining (direction dependent modulus of elasticity; cf. Sect. 12.3), whereas a crystal of tungsten has the same crystal structure but shows practically isotropy for the same property.

Most chemical compounds or alloys become spontaneously crystalline upon solid-ification from the liquid state. The thus formed massive solid will usually not be a *single* crystal, but it will be constituted of many crystals "grown together": it is a *polycrystalline* material. The individual crystals of the polycrystalline material are called the *grains*, having a size which may vary from a few nm to a few mm, or even cm and more. The crystals in the polycrystal can be oriented differently in space, i.e. the arrangements of the building units (atoms/molecules) are the same for each crystal/grain, but these regular arrangements (as exemplified by the "crystal axes") can be oriented differently with respect to the specimen frame of reference. As a result the anisotropy of the polycrystal is not identical with, and in any case less pronounced than, that of the single crystal. As a matter of fact, if the polycrystal consists of very many crystals and the orientation distribution of the crystals in the polycrystal is fully random, the polycrystal as a whole, i.e. on a macroscopic scale, is isotropic for the properties for which the single crystal is anisotropic. In that case, one also says that the polycrystal is *quasi-isotropic*.

Crystalline solids have to be distinguished from *amorphous* solids. Amorphous solids do *not* possess a long-range periodic arrangement of building units (atoms); their atomic structure is characterized by the absence of any long-range order.[5] As may then be expected, amorphous solids show macroscopic isotropy (direction-independent physical properties). Due to the absence of long-range order, the notion "grain" obviously has no relevance for an amorphous solid.

---

[5] However, the atomic arrangement in an amorphous solid, although lacking *long-range order*, can exhibit *short-range order*. For example, in case of amorphous silicon, each individual silicon atom *tends* to be surrounded by four silicon atoms in tetraedron configuration (cf. the discussion on covalent bonding in Sect. 3.4). Further, if an amorphous solid is composed of atoms of more than two elements, then a tendency can occur for the atoms to be surrounded preferentially by unlike atoms (i.e. A atoms in an A/B mixture would tend to have B atoms as nearest neighbours). So, to describe the atomic structure of an amorphous alloy as completely chaotic or structure-less is an overexaggeration.

Fig. 4.2 Etched surface of a cross-section through an aluminium polycrystal. Each differently reflecting region corresponds to a separate grain. In this case, the naked eye allows observation of the grain morphology

---

Intermezzo: Making Grain Boundaries Visible

The *grains* constituting a polycrystalline solid may be visualized with more or less experimental effort. On a clean, flat (polished) surface/cross-section, the assembly of the single crystals constituting a solid are sometimes visible for the naked eye (see Fig. 4.2), but usually a (light-optical) microscope is needed for revealing the individual grains (cf. Sect. 6.6). One usual technique in the analysis of metals (one speaks of *metallography*) is to prepare a very flat surface of the inner part of metal specimen, e.g. a cross-section, which is normally done by cutting the metal specimen, embedding the piece in some material (e.g. an epoxy), followed by grinding and polishing. Next, usually, some etching with some reagent (often an acid) is employed leading to a structuring of the surface revealing the grain morphology in the surface/cross-section. To this end, the etchant can be chosen such that preferentially either grain boundaries or grain faces in the cross-section are etched (cf. Petzow 1999). A result thus achieved is shown in Fig. 4.2.

The internal atomic arrangement of the crystalline solids, i.e. the *crystal structure*, is the focal point of interest in this chapter. The specific type of "regularity" in the crystal structure, which is characteristic for all crystalline materials, is denoted as the *translational symmetry*, which involves a geometric abstraction of the atomic arrangement in *ideal crystals* on the basis of translational symmetry concepts.

The prediction of ideal crystal structures on the basis of an understanding of the principles of chemical bonding is a topic of great interest. The word "prediction" in the preceding sentence can be interpreted as the calculation of the crystal structure by a method which does *not* depart from experimental information, as, for example, unit-cell dimensions (see Sect. 4.1.1 for the notion unit cell). Then, in line with the remarks already made within this context in Sect. 3.5.3, we must conclude that, although enormous progress has been made in recent years, the prediction of ideal crystal structures from first principles remains one of the most difficult problems in solid-state science, which escapes a treatment in this book (for a review, see Woodley and Catlow 2008).

As a final introductory note, it is emphasized that the *ideal crystal*, as considered in this chapter, is infinitely large and does not contain defects/imperfections in the long-range ordered arrangement of the atoms. The very occurrence of a surface of the crystal considered implies that the symmetry operators which can be applied in the bulk do not generally hold for the surface atoms: a symmetry break occurs at the surface, with severe consequences for the properties of the finite crystal (cf. the introduction of Chap. 5). The presence of appreciable densities of defects/mistakes in the long-range ordered atomic arrangement in the bulk of the crystal is partly unavoidable from a thermodynamic point of view: the so-called "equilibrium defects" (cf. Sect. 5.1). In any case, the presence of defects/mistakes can practically always be taken as granted as a consequence of specimen production and handling. These crystal imperfections can determine, often to a very large degree, the properties of practical, crystalline materials and therefore are dealt with extensively in the separate Chap. 5.

4.1 Geometric Description of Crystals

4.1.1 Translation Lattice, Motif and Crystal Structure

The idea that the external shape of crystals derives from the construction of the crystal by a regular, periodic arrangement in a specific way of identical building units was put forward already more than 300 years ago (see the introduction of this chapter). Indeed, simple crystal shapes, as observed in nature, like cubes, tetrahedra or octahedra can be constructed by arrangements of densely packed small cubes as building units (see Fig. 4.1). In our present day language, we call these building units *unit cells* (which are parallelepipeds; a parallelepiped is a prism with parallelograms as faces). These unit cells can be constructed by making use of the *basis translation vectors* of the crystal, which will be introduced now.

In the following, for the reason of simplicity, a number of crystallographic concepts will be introduced with reference to two-dimensional model structures. Crystals of real materials have usually three-dimensional structures; the meaning of the concepts introduced for two-dimensional crystals is straightforwardly extended to three dimensions.

Consider Fig. 4.3a showing a part of an assumedly infinite two-dimensional crystal structure with three types of atoms which may be regarded to be part of a "molecule". The crystal structure in Fig. 4.3a exhibits a regular appearance. How can this regularity be expressed in a scientific fashion? To this end *shifts*, which can be called

(a)

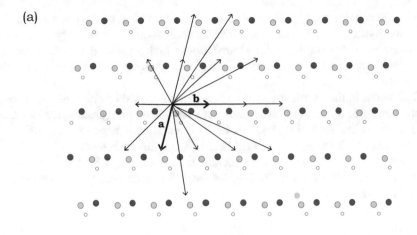

(b)

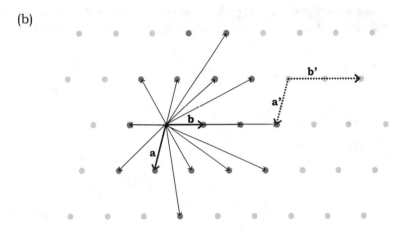

Fig. 4.3 a Two-dimensional model crystal structure with some translation vectors **t** represented by arrows. A shift of the complete crystal structure by any of these vectors leaves the whole crystal structure unchanged (invariant). Due to this definition of the translation lattice, the "starting points" of the arrows representing the translation vectors are arbitrary. **b** Translation vectors **t** from (**a**) with their end points highlighted by dark grey circles. Further (lighter grey) circles have been added to represent the end points of all translation vectors from (**a**), including the null vector. Note that the same arrangement of grey circles will result irrespective of the choice of starting points of the arrows in (**a**). The end points of all translation vectors represent the translation lattice of the crystal structure in (**a**). Two basis vectors **a** and **b** have been chosen from which all translation vectors t can be generated according to $\mathbf{t} = u\mathbf{a} + v\mathbf{b}$ with u and v being integers. The translation vectors $\mathbf{a}'$ and $\mathbf{b}'$ do not form a basis of the complete translation lattice

translation operations, of the crystal structure are identified, which upon their action leave the crystal structure unchanged (invariant operations, i.e. *symmetry operations*). There are an infinite number of such translations which preserve the crystal structure. An atomic structure having such translations vectors, **t**, is said to exhibit *translation periodicity*.

Evidently, in the example considered, a unit composed of one representative of each of the three different types of atoms, of a particular configuration, repeats itself, by specific shifts, i.e. translations, while its orientation is kept. Such a unit of the crystal structure is called a *motif* (see Fig. 4.3a and further below).

For the two-dimensional crystal structure as shown in Fig. 4.3a, two basis vectors **a** and **b** can be indicated (see also Fig. 4.4) such that any arbitrary translation vector **t** can be written as

$$\mathbf{t} = u\mathbf{a} + v\mathbf{b}, \tag{4.1a}$$

where u and v are integers. The end points of all possible vectors **t** form the so-called *translation lattice* of the crystal. Obviously, there are an infinite number of such

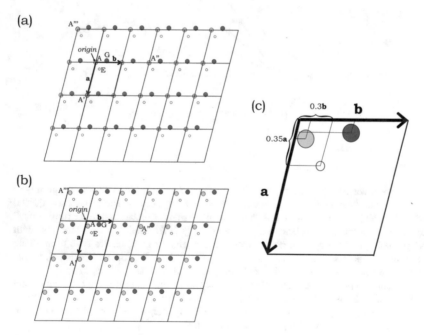

Fig. 4.4 **a** Crystal structure already shown in Fig. 4.3a, depicting a coordinate system spanned by the basis vectors **a** and **b** (a primitive basis), where the origin of the coordinate system was chosen to be located at the centre of gravity of an A atom. **b** The same crystal structure as in (**a**) but with the origin of the coordinate system chosen at another location. Both in (**a**) and (**b**), a unit cell has been indicated by thick lines (fractional coordinates $0 \leq x < 1$ and $0 \leq y < 1$), respectively. In (**c**), the unit cell for (**b**) has been enlarged, demonstrating the way how the fractional coordinates are determined

translation vectors **t** (for an infinite crystal). The basis vectors themselves, **a** and **b**, are translation vectors (with $(u, v) = (1, 0)$ and $(u, v) = (0, 1)$, respectively, to be substituted into Eqs. (4.1a, 4.1b)).

There are different possibilities to choose these basis vectors. The number of possible choices for the basis vectors is infinitely large, even under the constraint for each possible pair of basis vectors that the crystal structure is preserved upon any translation **t** according to Eq. (4.1a), which implies that only specific (pairs of) basis vectors are possible (see discussion below on "basis vectors" **a**′ and **b**′).

The parallelogram spanned by a pair of such basis vectors is the *unit cell* (for the two-dimensional crystal considered here; parallelepiped for three-dimensional crystal; see first paragraph of this section) and has an area given by the vector product **a** × **b**:

$$|\mathbf{a} \times \mathbf{b}| = |\mathbf{a}||\mathbf{b}| \sin \gamma \qquad (4.2)$$

with γ as the angle between **a** and **b**. *As long as the parallelogram spanned by a and b contains only one motif* (in the example considered the motif is a unit composed of one representative of each of the three different types of atoms, arranged in a specific configuration; see Fig. 4.3a), all possible translation vectors can be reproduced by Eq. (4.1a) indeed. For this case, the basis vectors are said to constitute a *primitive basis*. In specific cases, the chosen basis vectors do not provide a primitive basis and then they define a parallelogram (parallelepiped in the three-dimensional case) containing a number of motifs:

The end *points* of all translation vectors represent the *translation lattice* of the crystal structure (see Fig. 4.3b). These end points, mathematical points, are denoted lattice sites or lattice nodes. The vectors **a**′ and **b**′ indicated in Fig. 4.3b form a pair of vectors that does not reproduce all translation vectors **t** (e.g. **t** = 1**a** + 1**b** cannot be constructed from **a**′ and **b**′ according to a recipe similar to Eq. (4.1a)). Yet, **a**′ and **b**′ can be conceived as a pair of basis vectors in the above sense: the translations **t**′ according to the recipe given by **t**′ = u**a**′ + v**b**′ (cf. Eq. (4.1a)) preserve the whole crystal structure, *as long as the atoms contained in the parallelogram spanned by a′ and b′ are reproduced in any similar parallelogram produced by translations t*′. Evidently, the two translation vectors, **a**′ and **b**′, determine a parallelogram containing more than one motif (here the motif is the unit ("molecule") composed of the three different atoms), whereas the parallelogram spanned by the basis vectors **a** and **b** contains only one such motif. Thus, the area |**a**′ × **b**′| is equal to *a number of times* the area |**a** × **b**|. In other words: the vectors **a**′ and **b**′ can be conceived as basis vectors as well, provided the filling of the parallelogram of area |**a**′ × **b**′| with atoms/motifs is prescribed.

Recognizing that Eq. (4.1a) does not generate the entire set of translation vectors **t** if a non-primitive basis has been adopted, a modification of Eq. (4.1a) can be proposed that does describe any arbitrary translation vector **t** of the crystal structure for the chosen non-primitive set of basis vectors:

$$\mathbf{t} = u\mathbf{a}' + v\mathbf{b}' + \mathbf{t}_0 \qquad (4.1b)$$

where $\mathbf{t_0}$ stands for the set of vectors describing/generating the (relative) positions of the motifs in the unit cell. Thus, $\mathbf{t_0}$ comprises a set of N vectors (with N as a natural number) all of which are defined as $u'\mathbf{a'} + v'\mathbf{b'}$ with u' and v' being rational numbers with $0 \leq u', v' < 1$, including always $u' = v' = 0$ (vector $\mathbf{0}$). If $N = 1$, the basis given by $\mathbf{a'}$ and $\mathbf{b'}$ is a primitive one and $\mathbf{t_0}$ includes only $\mathbf{0} = 0\mathbf{a'} + 0\mathbf{b'}$ (i.e. Eq. (4.1b) reduces to Eq. (4.1a)). For the non-primitive basis, $\mathbf{a'}$ and $\mathbf{b'}$ considered in Fig. 4.3b the set $\mathbf{t_0}$ is composed of $\mathbf{0}$ and $\frac{1}{2}\mathbf{b'}$, i.e. $N = 2$. In the following, the basis vectors will be indicated by $\mathbf{a}$, $\mathbf{b}$ and $\mathbf{c}$ (without prime) not only for a primitive basis but also for a non-primitive basis.

The translation lattice, constructed by the operation indicated by Eqs. (4.1a and b), and as spanned by the basis vectors $\mathbf{a}$ and $\mathbf{b}$ (e.g. Fig. 4.3b), provides a geometric abstraction of the crystal structure (e.g. Fig. 4.3a). The positions of the individual atoms are not directly provided by the translation lattice. So *the translation lattice is not identical with the crystal structure.*

Now adopt the basis vectors of the translation lattice as the basis vectors of the coordinate system of the crystal as well. Next, a choice of origin of that coordinate system for the crystal structure has to be made. The choice of origin is in principle arbitrary, but there may be convenient possibilities, e.g. at the position (of the centroid of mass) of a certain atom. First considering a primitive basis only, apparently the crystal structure can then be obtained from the translation lattice upon substitution of each translation-lattice point by a *motif* that by translation according to the translation lattice repeats/reproduces itself at every lattice point. Hence, for example, as shown in Fig. 4.3, the motif is given by the unit composed of the three different atoms (A (grey circle), E (white circle) and G (black circle); cf. Fig. 4.4), of the specific configuration indicated, that repeats itself by specific translations, while its orientation is kept. For example, as shown in Fig. 4.4a, the origin of the primitive translation lattice has been identified with the centroid of mass of an A atom; the configuration of the other atoms of the motif (= the "molecule" composed of the three, A, E and G atoms) with respect to the crystal axes $\mathbf{a}$ and $\mathbf{b}$ has been fixed. Another choice of origin for this two-dimensional crystal structure is shown in Fig. 4.4b, where the origin is chosen "outside" of the motif. This consideration leads to the statement:

$$\text{translation lattice} + \text{motif} = \text{crystal structure}$$

The case considered in Fig. 4.4 pertains to a primitive basis. The above statement does also hold for a non-primitive basis provided the positions of all motifs has been indicated in the translation lattice (see the discussion of the set translations $\mathbf{t_0}$ with respect to Eq. (4.1b)).

With reference to the adopted coordinate system for the crystal structure, the position of each atom can be described by their dimensionless fractional coordinates x and y:

$$\mathbf{r} = x\mathbf{a} + y\mathbf{b} \tag{4.3}$$

Consider the crystal structure shown in Fig. 4.4 described by a primitive basis **a** and **b**. The positions of the atoms designated by A, E, G, A′, A″ and A‴ in Fig. 4.4b, and for the translation lattice indicated, are given by (see also Fig. 4.4c):

$$
\begin{array}{lll}
\text{A}: & x_A = 0.15 & y_A = 0.1 \\
\text{E}: & x_E = 0.35 & y_E = 0.3 \\
\text{G}: & x_G = 0.1 & y_G = 0.5 \\
\text{A}': & x_{A'}, = 1.15 & y_{A'}, = 0.1 \\
\text{A}'': & x_{A''}, = 0.15 & y_{A''}, = 2.1 \\
\text{A}''': & x_{A'''} = -0.85 & y_A''' = -0.9
\end{array}
$$

The first three atoms are those atoms located within the parallelogram defined by the two basis vectors **a** and **b** indicated in Fig. 4.4b; they form the motif in the case considered here. The atoms A′, A″ and A‴ lie outside of that parallelogram. But their fractional coordinates can easily be traced back ("reduced") to those of the corresponding A atom inside the "original" parallelogram, by $x_{A'} = x_A + u$ and $y_{A'} = y_A + v$, with u and v as integers like in Eq. (4.1a). Thus, the translation properties of the crystal structure simply imply that, if at the point r_A an atom A resides, a similar atom A also occurs at the points

$$
\begin{aligned}
r_{A'} &= r_A + t = x_A \mathbf{a} + y_B \mathbf{b} + u \mathbf{a} + v \mathbf{b} \\
&= (x_A + u)\mathbf{a} + (y_B + v)\mathbf{b} = x_{A'}\mathbf{a} + y_{B'}\mathbf{b}
\end{aligned}
\tag{4.4a}
$$

if **a** and **b** form a primitive basis. If **a** and **b** form a non-primitive basis, the discussion leading to Eq. (4.1b) implies that, if at the point r_A an atom A resides, a similar atom A also occurs at the points

$$
\begin{aligned}
r_{A'} &= r_A + t = x_A \mathbf{a} + y_B \mathbf{b} + u \mathbf{a} + v \mathbf{b} + u'\mathbf{a} + v'\mathbf{b} \\
&= \left(x_A + u + u'\right)\mathbf{a} + \left(y_B + v + v'\right)\mathbf{b} = x_{A'}\mathbf{a} + y_{B'}\mathbf{b}
\end{aligned}
\tag{4.4b}
$$

Note that u and v are integers, whereas u' and v' are rational numbers with $0 \le u', v' < 1$.

Similar equations hold for all atoms of the motif as contained in the "original" parallelogram, which is a direct consequence of the translational periodicity of the crystal structure.

It can be concluded that the crystal structure can be built up by an infinite repetition of the "original" parallelogram in identical form, i.e. shifted by all possible **t** according to Eqs. ((4.1a, 4.1b)), to tile, in a massive way (i.e. leaving no "open space"), the whole (here) two-dimensional space. In this sense, the parallelogram spanned by the basis translation vectors is called the *unit cell*.

Thus, in order to construct a complete crystal structure of an ideal crystal, it is sufficient to know the unit cell of the translation lattice, i.e. the basis vectors defining the unit cell and their relative orientation in space, as well the motif and atom content of the unit cell, i.e. the type and fractional position (cf. Eq. (4.3)) of the motifs and the atoms.[6]

Summarizing: For the example considered in Figs. 4.4a and b it holds that the unit cell, i.e. the parallelogram given by the basis vectors **a** and **b**, contains one motif. Such a unit cell is called *primitive cell*; it has translation-lattice points only at the corners. Because every point at a corner of the two-dimensional primitive cell (there are four corners) is shared by four adjacent primitive cells, indeed there is in total one motif per primitive cell (for the three-dimensional case: every corner of the primitive cell, now a parallelepiped with eight corners, is shared by eight adjacent primitive cells and, consequently, again there is in total one motif per primitive cell); cf. the Appendix at the end of this chapter). In especially physics literature often the adjective "*simple*" instead of "*primitive*" is used. In general, the unit cell may contain more than one motif; then its basis vectors are linear combinations of the basis vectors of a primitive cell (here we speak of vectors **a′** and **b′** (as discussed below Eq. (4.2)) as basis vectors; cf. Fig. 4.3b). Evidently, in order to construct the crystal structure from the translation lattice, the positions of all motifs in the unit cell should be indicated. For the example considered in Fig. 4.3b also a unit cell containing two motifs (two "molecules", each composed of the specific combination of the three types of atoms, in the prescribed configuration)[7] has been indicated; the primitive cell is only one of the possible unit cells.

---

[6] Considerable confusion occurs regarding usage of the terms (translation) *lattice* and (crystal) *structure*. As argued in the above text, the translation lattice is a geometrical abstraction to describe the translational periodicity of the crystal considered. Only upon substitution of the translation-lattice points by the motif (sometimes also called "basis"), the crystal structure is obtained. Unfortunately, in many literature sources, the word *lattice* is used, whereas (crystal) *structure* is meant. Against this background, a "highlight" of such conceptual chaos then occurs if one speaks of "lattice defects", whereas "crystal-structure defects" (also often called "imperfections" or "mistakes"; see Chap. 5) are meant. For an extensive discussion of various such abuses of crystallographic notions, with in fact well-defined meanings, see Nespolo (2019). This is the reason that I (try to) use adjectivally and consistently the words "translation" (or "Bravais translation") in connection with the substantive "lattice" (to make clear that the crystal structure is *not* meant), throughout at least this chapter. But I confess that, at other places (where likely no confusion can arise about what is meant; e.g. when the motif is one atom as for many elemental metal crystals), I may have fallen victim to the here discussed sin as well… At many places in the also recent literature, in the fields of materials science, solid-state physics and solid-state chemistry, the word "lattice" in a text can mean either "translation lattice" or "crystal structure". Although the context usually will make clear the intention of the author(s), a pure and correct use of terms well defined in crystallography must be preferred.

[7] Consider the unit cell spanned by the vectors **a′** and **b′** as indicated in Fig. 4.3b. Evidently, the contents of this unit cell are given by (1) the fractional contributions of the motifs at the corners which totals one motif (each motif at a corner of the unit cell is shared by four unit cells; it should be noted that in general the contributions of the motifs at the four corners are unequal, due to the angle γ not being 90°; cf. Fig. A4.1a and Table A4.1 and their full discussion in the Appendix to this chapter, but do recognize that the discussion in this appendix focusses on the number of (each type of) atoms in the unit cell and not on the number of motifs), plus (2) the fractional contributions of the motifs at the middle of the two sides of the unit cell parallel to **b′** (these motifs are shared by two unit cells and the fractional contribution of each of these motifs equals 1/2). Hence, in total, the unit cell contains two motifs.

For the reason of exhibiting the symmetry in the translation lattice more clearly (see further below) often unit cells are defined which are not primitive cells. A well-known example is the so-called face centred cubic unit cell that is employed to describe the translational symmetry of the crystal structure of, for example, a metal as Cu and a salt as NaCl (see also the third paragraph of Sect. 4.2.1.2): see Fig. 4.5 that shows (now for three dimensions) the face centred cubic unit cell and a corresponding primitive cell. Note that in this case, the motif, indicated by a point in the translation lattice, consists of one (copper) atom or one Na^+Cl^- ion pair (see further Sect. 4.1.3).

The unit cell geometry can be defined by its metrics: the lengths of the vectors **a** and **b**, $|a|$ and $|b|$, and the angle γ enclosed by **a** and **b** (cf. Eqs. (4.1a, 4.1b)). At first sight, one may think that description of the metrics of the unit cell, in two dimensions a parallelogram, requires two quantities for each of the vectors **a** and **b** (i.e. their vector components in two-dimensional space), i.e. in total four quantities. However, three quantities suffice as long as the orientation of the crystal in two-dimensional space needs not be specified. Similarly, in the three-dimensional case: one may initially think that description of the metrics of the unit cell, now a parallelepiped, would require nine quantities representing the vector components of the three basis vectors, **a**, **b** and **c**, but, as long as the orientation of the crystal in three-dimensional space needs not be specified, six quantities suffice: the lengths $|a|$, $|b|$ and $|c|$ and three angles, α, β and γ, specifying the relative orientations of **a**, **b** and **c** (see Table 4.1).

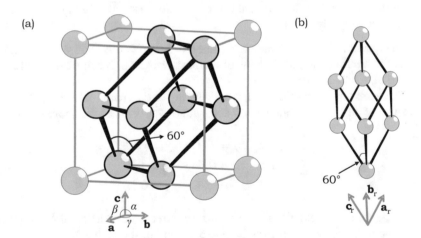

Fig. 4.5 a Translation lattice/crystal structure of copper (Cu) as represented by a face centred cubic unit cell bounded by grey lines (with respect to the usage of the notion "face centred cubic" for, confusingly, both the translation lattice and the crystal structure in case of a metal as Cu, see also the discussion in Sect. 4.2.1.2). Within this cube, a rhombus bounded by eight Cu atoms is shown which can serve as a primitive unit cell, which is shown in a different viewing direction in (**b**). The primitive unit cell constitutes a special rhombohedral one, namely with $\alpha = \beta = \gamma = 60°$. As usual, all atoms with fractional coordinates satisfying $0 \leq x, y, z \leq 1$ are shown. The basis vectors for the face centred cubic unit cell and the rhombohedral primitive unit cell have been indicated without and with subscript "r", respectively

Table 4.1 Description of the crystal structure in two- and three-dimensional space; $a = |\mathbf{a}|$, $b = |\mathbf{b}|$, $c = |\mathbf{c}|$

Concept	Two dimensions	Three dimensions
Translation lattice	$\mathbf{t} = u\mathbf{a} + v\mathbf{b}$	$\mathbf{t} = u\mathbf{a} + v\mathbf{b} + w\mathbf{c}$
Atom position	$\mathbf{r} = x\mathbf{a} + y\mathbf{b}$	$\mathbf{r} = x\mathbf{a} + y\mathbf{b} + z\mathbf{c}$
Fractional coordinates	x, y	x, y, z
Unit cell	Parallelogram spanned by two basis translation vectors, $\mathbf{a}$ and $\mathbf{b}$	Parallelepiped[b] spanned by three basis translation vectors, $\mathbf{a}$, $\mathbf{b}$ and $\mathbf{c}$
Unit-cell dimensions given by unit-cell parameters[a]	a, b, γ	$a, b, c, \alpha, \beta, \gamma$

[a]Also referred to as lattice parameters
[b]Three-dimensional body: a prism bounded by three pairs of parallel parallelograms

The atomic contents of the three-dimensional unit cell can be given by a listing of the fractional coordinates x, y and z of all atoms within the unit cell (see above), i.e. with $0 \leq x, y, z < 1$ (cf. Eq. (4.3)). Within this context, it is remarked that upon considering/drawing unit cells and their fillings it is usual to indicate all atoms in the unit cell for which $0 \leq x, y, z \leq 1$. In the latter case, the fractional contribution of an atom with either $x = 0$ (or 1), or $y = 0$ (or 1) or $z = 0$ (or 1) to the contents of the unit cell is smaller than one. This is discussed in the Appendix at the end of this chapter.

Now, finally, the two-dimensional crystal structure as shown in Figs. 4.3 and 4.4 can be described fully as follows:

1. The (primitive) unit cell parameters are $a = 5$ Å, $b = 6$ Å[8] and $\gamma = 105°$ (here and in the following the lengths of the basis vectors $\mathbf{a}$, $\mathbf{b}$ and $\mathbf{c}$ are denoted by $a \equiv |\mathbf{a}|$, $b \equiv |\mathbf{b}|$ and $c \equiv |\mathbf{c}|$);

2. There are three atoms in the unit cell (listing of the fractional coordinates x, y and z of all atoms within the unit cell with $0 \leq x, y, z < 1$; see above):

$$A: \quad x_A = 0.15, \quad y_A = 0.1$$
$$E: \quad x_E = 0.35, \quad y_E = 0.3$$
$$G: \quad x_G = 0.1, \quad y_G = 0.5.$$

The straightforward extension from two to three dimensions, already partly performed in the above discussion, has been summarized in Table 4.1.

---

[8] Whereas in science distance units as m, cm, mm, μm, nm, pm,... are generally used, on the basis of international agreement, the unit Å (angstrom) = 0.1 nm = 1×10^{-10} m is still normally used in the field of crystallography, in agreement with a recommendation of the International Union of Crystallography (IUCr), recognizing that the size of and distances between atoms in crystal structures are of the order of 1 Å. For example, see the use of distance units in the leading journals of this field, as Acta Crystallographica, Journal of Applied Crystallography and Zeitschrift für Kristallographie.

With respect to the above-described ambiguity in the choice of the unit cell (unit cells containing one or more motifs), the following convention is indicated: The convenient unit cell is (mostly) taken as the smallest possible parallelepiped in three-dimensional space (parallelogram in two-dimensional space) displaying the highest symmetry inherent to the crystal structure and having translational properties yielding the translation lattice. If the angles α, β, γ are not fixed due to symmetry, one should choose them to be closest to 90°.

A few notes with respect to this convention can be made:

- The advantage of choosing the smallest possible unit cell is obvious: a smaller number of different atoms have to be given in terms of their fractional coordinates to describe the unit cell.
- Geometric considerations are easier and more convenient if the angles between the basis vectors are close to 90°; the extension of the unit cell in each crystal-axis direction then is as small as possible.
- Apart from the translation symmetry, other types of symmetry can occur in crystals: e.g. mirror and rotation symmetries (see Sect. 4.1.2). As a consequence, preference may occur for (1) a certain type of origin of the unit cells (such an origin can, but need not be, at the centre of mass of a specific atom) and/or (2) a unit cell larger than the smallest possible one. Such modifications are performed in order to visibly display symmetry properties of the crystal in the unit cell. The non-translational symmetry properties of crystal structures motivate the distinction of two concepts used for the description of crystal structures: *crystal systems* (Sect. 4.1.2) and the *Bravais lattices* (Sect. 4.1.3).

4.1.2 The Crystal System

Mathematically, geometric manipulations which transform a crystal, by motions, into an object (image) indistinguishable from the original are called *symmetry operations*. Symmetry operations are mediated by *symmetry elements*. Translations are shifts of the crystal structure which lead to identical atomic arrangements, and hence translations are symmetry operations with the lattice parameters (unit-cell parameters; cf. Sect. 4.1.1) as the symmetry elements (Fig. 4.6a). Besides translation, other types of symmetry operations can hold for crystal structures: e.g. reflection mediated by a mirror plane, rotation mediated by a rotation axis and inversion mediated by a centre of symmetry (inversion centre: a point (x, y, z) of the crystal transforms into a point $(-x, -y, -z)$ under the constraint that the resulting crystal is indistinguishable from the original). It is beyond the scope of the present text to provide a complete, systematic and rigorous analysis of the symmetry properties of crystals. Here it is attempted to demonstrate how symmetry properties, additional to translational properties, can induce certain additional constraints on the atomic structure of crystals.

For a start, and by omitting translational symmetry, consider symmetry properties of molecules. The geometric structure of a molecule of thionyl chloride is shown in Fig. 4.6b. The atomic arrangement involves the presence of a mirror plane

(a)

(b)

(c)

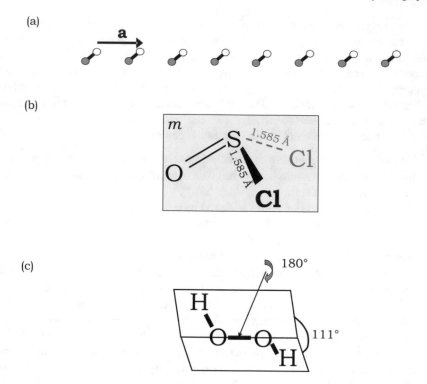

Fig. 4.6 Three types of symmetry operations relevant for atomic structures: **a** translation periodicity as present in crystals with basis vector a (one-dimensional), **b** mirror symmetry in the molecular structure of thionyl chloride $SOCl_2$ (O and S are located on the grey mirror plane indicated by the symbol m, the two Cl atoms are positioned above and beneath this mirror plane), and **c** a two-fold rotation axis in the most stable conformation of the molecular structure of hydrogen peroxide, H_2O_2

(through the O=S bond and bisecting the angle spanned by the two S–Cl bonds). The geometric structure of a hydrogen peroxide molecule is shown in Fig. 4.6c. The atomic arrangement involves that rotation of 180° around the axis indicated leads to an atomic configuration undistinguishable from the original one.

Clearly, symmetry operations provide constraints for the atomic configuration concerned. For example, the mirror plane in thionyl chloride implies that the two bond-lengths S-Cl are the same, i.e. $0.1585\,\text{nm} = 1.585\,\text{Å}$ (see footnote 8). Symmetric structures have relevance for nature: it can be shown, or made likely, that in general atomic configurations exhibiting high symmetry pertain to either a minimum or a maximum value of their energy. In many cases, the high-symmetry states correspond to a minimum value of the energy of the system considered, which explains the preferential occurrence of such atomic configurations in nature, recognizing that systems strive for minimal energy (see also the discussion in Sect. 7.3).

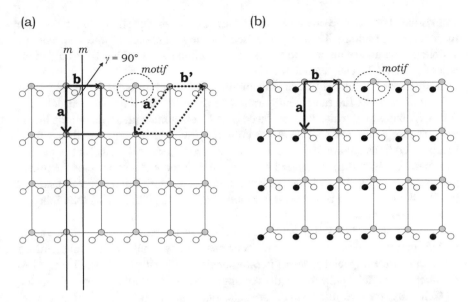

Fig. 4.7 Two-dimensional crystal structures indicating the constraints imposed by symmetry operations on the lattice parameters as given by the translation basis vectors and their relative orientations (i.e. angles between the basis vectors). **a** Crystal structure exhibiting mirror planes (lines) (m). These mirror planes cannot exist if the lattice angle γ between the basis translation vectors **a** and **b** differs from 90°. The alternative unit cell characterized by the basis translation vectors **a'** and **b'** is also primitive, but the lattice angle indicating the angle between these basis translation vectors does not equal 90° and the unit cell by itself, i.e. considered as an isolated object, does not reveal the presence of the mirror planes which do occur in the crystal structure. **b** A related but different crystal structure (different motif, but same translation lattice) that does not show mirror planes as in (**a**)

Symmetry operations corresponding to mirror planes and rotation axes can also occur in the atomic structures of crystals. Consider the two-dimensional crystal structure (not a translation lattice (cf. Sect. 4.1.1)) shown in Fig. 4.7a. Filled (grey) and open (white) circles (dots) may represent two different types of atoms. The intersections/crossing points of the full lines indicate the translation lattice invoked by the basis translation vectors **a** and **b**; the unit cell is primitive, recognizing that the motif consists of one grey and two white atoms in the configuration with the grey atom in the midplane of the two white atoms. Evidently, mirror planes exist in this crystal structure and run perpendicular to the **b** axis (and have been indicated by the symbol m in the figure). Obviously, the occurrence of such mirror planes corresponds with the constraint that the lattice angle γ (i.e. the angle between the **a** and **b** basis vectors; cf. Table 4.1) is exactly 90°. Even an infinitesimal deviation of γ from 90° would remove the mirror symmetry.

An alternative unit cell for the crystal structure as shown in Fig. 4.7a has been indicated with dashed lines. This unit cell is also primitive, but the lattice angle indicating the angle between the corresponding basis translation vectors (indicated with **a'** and **b'** in the figure) does not equal 90° and the unit cell by itself, i.e. considered as an isolated object, does not reveal the presence of the mirror planes which do occur in the crystal structure. Thus, a preference for the first choice of unit cell, making visible the symmetry inherent to the crystal structure (see what has been said at the end of Sect. 4.1.1), can be understood.

A different but related two-dimensional crystal structure is shown in Fig. 4.7b. The motif now has the same configuration as the motif for the crystal structure in Fig. 4.7a, but with a white atom replaced by a black atom. Mirror operations like in the first case would make white atoms coincident with black atoms, and thus these mirror operations are non-existent. It thereby becomes clear that, as for the case as shown in Fig. 4.7b, the translation lattice can exhibit more symmetry (see the mirror planes drawn in Fig. 4.7a, which are mirror planes of the translation lattice considered in Figs. 4.7a and b and of the crystal structure as shown in Fig. 4.7a) than inherent to the crystal structure.

Two-dimensional crystal structures as in Fig. 4.7a, characterized by a lattice angle γ equal to 90° as a consequence of the occurrence of a symmetry operator as a mirror plane, constitute a special group of two-dimensional crystal structures, the so-called *rectangular* crystals, indicated as the rectangular *crystal system*.

Generally, for two-dimensional crystals, one can identify four different types of characteristic constraints imposed by symmetry operations on the unit-cell parameters a, b and γ, leading to four different crystal systems:

1. Quadratic crystals $a = b, \gamma = 90°$;
2. Hexagonal crystals $a = b, \gamma = 120°$;
3. Rectangular crystals $\gamma = 90°$;
4. Oblique (also called monoclinic) crystals   no restrictions

Thus, for rectangular crystals, a in general is different from b, but this needs not necessarily be the case. And for oblique crystals, γ will in general deviate from 90°, but it may, "accidentally", exactly equal 90°. The crystal structure as shown in Fig. 4.7b is oblique, although the lattice angle equals 90°. This crystal structure is not called rectangular: only if the lattice angle is constrained by symmetry operations to be 90°, as holds for the crystal structure as shown in Fig. 4.7a, the crystal structure is genuinely rectangular.

Similarly, for three-dimensional crystals, one can identify different types of characteristic constraints imposed by symmetry operations on the unit-cell parameters a, b, c and α, β, γ, leading to different *crystal systems* (this statement is only largely correct: see the remark on hexagonal/trigonal crystals made below). Hence, for each crystal system, a number of crystal-structure types share the same unit-cell parameter prescriptions (see list below). In total, there are 230 different combinations of specific symmetry elements (operators) possible under the constraint of translational symmetry. These 230 combinations of symmetry elements are called the 230 "space groups" (see, below, the "*Intermezzo: A short note on point groups, crystallographic point groups, plane groups and space groups; glide and screw operations*"); each crystal-structure type complies with one of these space-group symmetries. The seven crystal systems are:

1. Cubic $a = b = c, \alpha = \beta = \gamma = 90°$;
2. Tetragonal $a = b, \alpha = \beta = \gamma = 90°$;
3. Orthorhombic $\alpha = \beta = \gamma = 90°$;
4. Hexagonal $a = b, \gamma = 120°$;
5. Trigonal[9] $a = b = c, \alpha = \beta = \gamma$
                   or $a = b, \gamma = 120°$;
6. Monoclinic $\alpha = \gamma = 90°$;
7. Triclinic   no restrictions

Note that for hexagonal/trigonal crystals, which in principle are all characterized by the constraints $a = b$, $\gamma = 120°$, slightly different categorizations may be found in the literature. Indeed they are also taken together as the hexagonal *crystal family*. The tiny details of definition leading to the different distinctions are beyond the scope of this book.

Again, like for two-dimensional crystals as discussed above, *decisive for the specification of the crystal system is that the constraints on the unit-cell parameters are caused by the symmetry of the atomic arrangement.* For example, for a crystal structure belonging to the tetragonal crystal system a may be equal to c, and yet the crystal structure cannot be assigned to the cubic crystal system. Or, if the crystal system has been specified as tetragonal, one immediately knows that (at least) $a = b$ and $\alpha = \beta = \gamma = 90°$. This leaves unimpeded that (additionally) also $a = c$ can occur in a specific case considered. For an example, see the discussion of the ordered solid solution CuAu as shown in Fig. 4.35 (bottom middle) in Sect. 4.4.1.1.

Intermezzo: A Short Note on Point Groups, Crystallographic Point Groups, Plane Groups and Space Groups; Glide and Screw Operations
Consider a homogeneous (isolated) body, i.e. not necessarily (part of) a crystal, of a certain shape. Determine the symmetry elements compatible with the shape of the body. By operation of the complete collection of symmetry elements pertaining to this body of certain shape, it is found that one point of the body is not transformed by the symmetry operations mediated by the symmetry elements, and all symmetry elements pass through this single point (note that, if the only symmetry elements for a three-dimensional body as considered here, are one or two mirror planes, there is a plane or line, respectively, of such points). Therefore, the complete collection of symmetry elements of this homogeneous body of certain shape is called a *point group*.

For a body as considered above, there are an infinite number of possible point groups. Now, focussing on crystals, the imposition of translational symmetry reduces the number of types of possible rotation axes drastically: the rotation

---

[9] The trigonal crystal system has also been designated as rhombohedral crystal system; but see footnote 10.

must be compatible with the translation lattice, i.e. upon application of the rotation operation to a certain collection of lattice points (as a row of lattice points in one dimension and as a plane of lattice point in two dimensions), the (image) points generated by the rotation must coincide with lattice points. It can simply be shown for two- and three-dimensional lattices that then only one, two, three, four and sixfold rotation axes are possible: fivefold and higher than sixfold rotation axes are impossible (note that the motif applied to the lattice in order to obtain a crystal structure can exhibit such "forbidden", as fivefold, rotational symmetry, but the arrangement of motifs according to the translation lattice cannot exhibit such rotational symmetry elements). The point groups remaining if only the one, two, three, four and sixfold rotation axes are allowed are called the *crystallographic point groups* (also called "crystal classes"). It has been found that in two-dimensional space 10 crystallographic (plane) point groups exist and that in three-dimensional space 32 crystallographic point groups can be discerned.

Next, the symmetry elements in the crystallographic point groups have to be combined with the translations inherent to the specific types of translation lattices (i.e. the translation lattices have been categorized with respect to their symmetry: Bravais translation lattices; see Sect. 4.1.3), in order to determine all possible combinations of symmetry elements in two-dimensional "crystal structures" and three-dimensional crystal structures. There are 5 (Bravais) translation lattices possible in two-dimensional space and 14 (Bravais) translation lattices in three-dimensional space. By straightforward but laborious evaluation, it thus has been found: (1) in two-dimensional space, combining the 10 crystallographic points groups with the 5 (Bravais) translation lattices results in 17 *plane groups*; (2) in three-dimensional space, combining the 32 crystallographic point groups with the 14 (Bravais) translation lattices results in 230 *space groups*. These 17 plane groups and these 230 space groups represent the only possible combinations of symmetry elements in two- and three-dimensional space, respectively, for arrangements of atoms subjected to translational symmetry (i.e. crystals). Yet, an infinite number of two- and three-dimensional crystal structures are possible, because the atomic contents and atomic configuration of the motif have been left unconstrained.

As a final point, it is remarked that combining the translations of the Bravais translation lattices with the crystallographic point groups leads to the recognition that additional symmetry elements can be discerned. In two-dimensional space, the *glide line* can occur. This symmetry element describes a two-step symmetry operation: (1) mirroring with respect to the glide line, plus (2) translation parallel to the glide line over a distance half of the lattice-repeat distance along this line (two successive glide operations applied to the same motif result in a simple lattice translation for this motif parallel to the glide line). In three-dimensional crystals, mirror lines become mirror planes and thus glide lines become *glide planes*. Further, another additional symmetry element is induced for three-dimensional crystals upon combining the translations of the Bravais translation lattices with the crystallographic point groups: the *screw axis*. This symmetry element also describes a two-step symmetry operation: (1) rotation with respect to the screw axis, plus (2) translation parallel to the screw axis over a distance equal to (n/m) times the lattice-repeat distance along the screw axis for an m-fold rotation around the screw axis with n as an integer smaller than m (m successive applications of the screw operation applied to the same motif results in a simple lattice translation for this motif parallel to the screw axis).

Lastly, for the sake of completeness, it is noted that, whereas the glide and screw operators combine a symmetry element of a point group with a translation, and thus pertain to crystals, for a three-dimensional body a combination of two symmetry elements both occurring in a point group is possible: A rotation axis and a centre of symmetry (cf. the begin of Sect. 4.1.2) can be combined: the *inversion axis*. This symmetry operation also describes a two-step symmetry operation: (1) rotation with respect to the rotation axis, plus (2) inversion with respect to the centre of symmetry. This combined symmetry operation is called *rotoinversion* and can occur in bodies not exhibiting translational symmetry as well, in contrast to the glide and screw operations.

4.1.3 The Bravais Categorization of Translation Lattices

Linear combinations of the primitive basis vectors (**a** and **b** for the two-dimensional crystal structure shown in Fig. 4.3a) can also serve as basis vectors (as **a′** and **b′** indicated in Fig. 4.3b and as discussed in Sect. 4.1.1). The lattices defined by the set of lattice points generated by the basis vectors are called translation lattices. For one crystal structure, there are as many different unit cells, spanned by the basis vectors, as there are possible sets of basis vectors: an infinite number of different unit cells. However, it has been made clear in Sect. 4.1.1 that only specific combinations (of which there is an (also) infinite number) of vectors **a** and **b** (**a**, **b** and **c** for a three-dimensional crystal) span a so-called primitive unit cell, i.e. a unit cell that contains only one motif. A categorization of the translation lattices thus obtained, according to their symmetries, leads to the distinction of specific types of translation lattices: the Bravais lattices.

The translation lattice can be more symmetric than the crystal structure: the examples discussed in Sect. 4.1.2 (in particular with respect to Fig. 4.7b) make clear that the translation lattice can posses more symmetry elements than inherent to the crystal structure. Whereas there are 230 possible different combinations of symmetry operations for three-dimensional crystal structures (the so-called space groups, divided over the seven crystal systems listed in Sect. 4.1.2), there are only 14 different three-dimensional Bravais lattices: the Bravais lattices provide a categorization of only the translational symmetry operators of crystal structures.

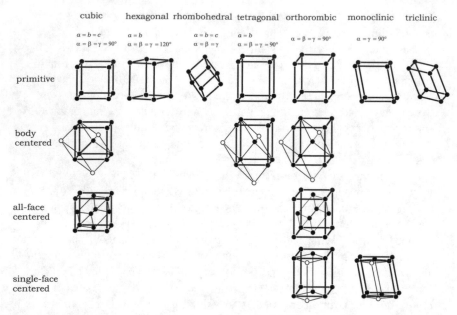

Fig. 4.8 The 14 Bravais translation lattices. Seven types of primitive unit cells can be discerned. Additional constraints on the unit-cell parameters of these primitive unit cells lead to seven additional Bravais translation lattices usually characterized by non-primitive (face centred, body centred or side/end centred) unit cells, as shown; for these latter cases, a possible primitive unit cell has been indicated as well, for which the white lattice points concern lattice points outside the non-primitive unit cell shown. The lattice-parameter characteristics have been indicated on top of the figure

The requirement that the three-dimensional arrangement of unit cells describing the crystal is massive, i.e. fills space completely, restricts the translational symmetry properties to seven categories: cubic, tetragonal, orthorhombic, hexagonal, rhombo-hedral,[10] monoclinic and triclinic. The primitive unit cells used usually to characterize the seven corresponding Bravais lattices are shown in Fig. 4.8.

There are seven more Bravais lattices, which follow from the above-mentioned seven Bravais lattices, with primitive unit cells, each specific for one of the seven categories of translational symmetry, if special additional constraints occur for the unit-cell parameters of the primitive cell considered. The primitive cell for the rhom-bohedral Bravais lattice is characterized by $a = b = c$ and $\alpha = \beta = \gamma$. Now, the additional constraint is imposed that $\alpha = \beta = \gamma = 60°$. The resulting translation lattice, still pertaining to a *primitive rhombohedral* unit cell, can now also be described by a *non-primitive cubic* unit cell with $a = b = c$, $\alpha = \beta = \gamma = 90°$; this is the face

---

[10] Regarding the nomenclature for crystal systems and Bravais translation lattices, the advice of the International Union of Crystallography has been adopted: The adjective *rhombohedral* is used here to designate a specific Bravais *translation lattice*; the adjective *trigonal* is reserved for the crystal-structure types (corresponding to a specific collection of space groups; cf. Sect. 4.1.2) sharing the unit-cell parameter prescriptions as indicated for the trigonal *crystal system* (see also footnote 9). Further, see Hammond (2001) and Schwarzenbach (1996).

centred cubic (f.c.c.) unit cell, containing four motifs (cf. Sect. 4.1.1); see Fig. 4.5. Crystals of the cubic crystal system may pertain to such a cubic translation lattice. Therefore, the cubic translation lattice derived from the primitive rhombohedral unit cell, with the additional constraint that $\alpha = \beta = \gamma = 60°$, is considered as a separate cubic Bravais lattice. On a similar basis, the body centred cubic (b.c.c.) translation lattice is a Bravais lattice of the cubic type; the corresponding primitive unit cell is (again) of the rhombohedral type (now the additional imposed constraint is $\alpha = \beta = \gamma = 109.5°$). In total, by imposing specific constraints on the unit-cell parameters of the primitive cells of the above-indicated group of first, seven Bravais lattices, one thus finds seven additional Bravais lattices, which have the lattice-parameter characteristics $(a, b, c, \alpha, \beta, \gamma)$ for a non-primitive unit cell equal to one of the primitive unit cells of one of the first seven Bravais translation lattices. This second group of Bravais lattices are normally characterized by these non-primitive (face centred, body centred or side/end centred) unit cells (see Fig. 4.8); note that the corresponding primitive unit cells (also indicated in Fig. 4.8) remain of the type of one of the seven in the original, first group of Bravais lattices.

Finally, a special (further; cf. above paragraph) remark has to be made with respect to the rhombohedral Bravais translation lattice, characterized by a *primitive* unit cell. It can also be described, and this is regularly the case in the existing literature, as a hexagonal translation lattice, but thus then is characterized by a *non-primitive* hexagonal unit cell: this non-primitive hexagonal unit cell, apart from (the contributions of; cf. Sect. 4.1.1 and the Appendix to this chapter) the lattice points (motifs) at the unit cell corners, contains in the interior of the cell two additional lattice points (motifs) and is called a "body centred hexagonal unit cell", although these two additional lattice points (motifs) do not occur at the centre of the cell (see Fig. 4.9).

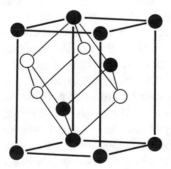

Fig. 4.9 The rhombohedral Bravais lattice, characterized by a primitive unit cell (see Fig. 4.8), can also be described as a hexagonal translation lattice characterized by a non-primitive "body centred hexagonal unit cell" that apart from the (partial) lattice points at the unit-cell corners, contains two additional lattice points (both of which, in contrast with the adjective "centred" used in the name of this unit cell, do not occur at the centre of the unit cell). A rhombohedral primitive unit cell has been indicated as well, for which the white lattice points concern lattice points outside the non-primitive hexagonal unit cell shown

4.1.4 Description of Lattice Planes and Directions; Miller and Miller-Bravais Indices

With respect to the translation lattice, a unified description method for directions in and for orientations of planes in crystals is desired. This allows, for example, to recognize and to identify, with a specific code, the crystal/lattice planes which constitute the planar surfaces of facetted crystals (see the introductory text of this chapter). Or, in this way, the characteristic variation of a certain property as a function of the direction in a crystal can be expressed unambiguously (i.e. independent of the orientation of the crystal in the laboratory frame of reference).

Unsurprisingly, the method adopted in crystallography to describe the orientation of lattice planes and to indicate directions is based on the basis vectors of the Bravais (i.e. translation) lattice, which are parallel to the basis vectors of the crystallographic coordinate system (cf. Sect. 4.1.1). It is this role of the translation lattice for the specification of the orientation of *crystallographic planes and directions* that has led to the usage of the terms *lattice planes* and *lattice directions* in connection with crystal structures.[11]

4.1.4.1 Lattice Planes

The unit cell spanning basis vectors of the three-dimensional translation lattice can be indicated with $\mathbf{a}$, $\mathbf{b}$ and $\mathbf{c}$. Evidently, points on the $\mathbf{a}$, $\mathbf{b}$ and $\mathbf{c}$ axes given by $\mathbf{a}/h$, $\mathbf{b}/k$ and $\mathbf{c}/l$, with h, k, and l as integers, define a plane in the lattice. There is an infinite number of *parallel and equidistant* planes, called a "set of lattice planes" or a "family of lattice planes", which have points of intersection with the $\mathbf{a}$, $\mathbf{b}$ and $\mathbf{c}$ axes given by $n\mathbf{a}/h$, $n\mathbf{b}/k$ and $n\mathbf{c}/l$, with n as integer (see Fig. 4.10a); for $n = 0$, the corresponding member of the family of planes considered runs through the origin of the crystal frame of reference. This family of lattice planes is identified fully with the indices h, k and l. These planes have distances of d_{hkl}, which parameter is also called the interplanar distrance or the lattice-plane spacing. A special situation happens if one or two of the indices h, k, or l is/are equal to 0. For example, consider the case $h = 2$, $k = 3$ and $l = 0$ (Fig. 4.10b). Then, $n\mathbf{c}/l$ can be regarded to take the value $\pm\infty\mathbf{c}$. This implies that the family of planes considered does not intersect the $\mathbf{c}$ axis: these planes run parallel to it (for $n = 0$ the lattice plane contains the $\mathbf{c}$ axis), and at the same time, they intersect the $\mathbf{a}$ and $\mathbf{b}$ axes at points $n\mathbf{a}/2$ and $n\mathbf{b}/3$.

In the above paragraph, h, k and l have been proposed as indices characterizing the family of lattice planes considered. However, an ambiguity occurs: substituting h, k and l by mh, mk and ml with m as an integer, identifies a set of planes that contains also the planes identified above with h, k and l, but that comprises m times more (parallel) planes, which not necessarily are translation-lattice planes. So, in order to restrict the set of parallel planes to only those planes that occur in the translation

---

[11] This convention certainly did and does contribute to the confusing use of the term "lattice" where "(crystal) structure" is meant, as discussed extensively and criticized in footnote 6.

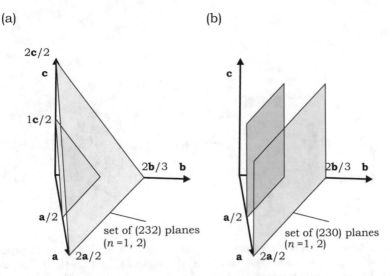

Fig. 4.10 **a** Derivation of the orientation of the set of lattice planes with the Miller indices (232) with points of intersection with the **a**, **b** and **c** axes at n**a**/2, n**b**/3 and n**c**/2, respectively, as shown in the figure for $n = 1, 2$. **b** The same for the set of lattice planes with the Miller indices (230) with points of intersection with the **a**, **b** and **c** axes at n**a**/2, n**b**/3 and "n**c**/0", respectively, as shown in the figure for $n = 1, 2$, where the designation "n**c**/0" effectively means that the planes never intersect the c axis (except for the not shown $n = 0$ plane, within which **c** is located)

lattice, i.e. lattice planes, the so-called *Miller indices* are used. The Miller indices for a lattice plane, i.e. a plane containing translation-lattice points, can be found as follows:

1. Determine the points of intersection of the lattice plane considered with the three axes **a**, **b** and **c**.
2. Take the intercepts as dimensionless numbers, equal to the (rational) number of units a, b and c cut from the **a**, **b** and **c** axes, respectively.
3. Take the reciprocals of these intercepts. For a plane parallel to one of the **a**, **b** and **c** axis, the reciprocal of the intercept is set equal to 0.
4. Multiply or divide these reciprocals by a common factor such that the set of *smallest* integer numbers results; i.e. the set of resulting integers has no common divisor.[12] These resulting integer numbers are called the Miller indices h, k and l, which typify the family of lattice planes considered. The family of lattice planes then is normally specified with the notation (hkl). Thus, one speaks of (100), (110), (111), (211), etc., lattice planes. A bar above a number (e.g. $\bar{2}$) or left of a number (e.g. -2) indicates a negative integer.

On the above basis, crystal faces can now be identified with their Miller indices: some triplet of integer numbers h, k and l presented in the formula (hkl) (Fig. 4.11). Thus,

---

[12] A set of *hkl with a common divisor* has relevance in the discussion of the diffraction by crystals. See the discussion of the so-called Laue indices, with *hkl* replaced by *HKL*, in Sect. 4.5.

(a)   (b)

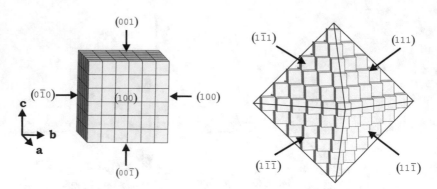

Fig. 4.11 Reproduction of Fig. 4.2 now with indication of the Miller indices for the shown faces of two cubic crystals. **a** A cubic crystal forming a cube bounded by six faces {100}. The unlabelled face on the rear is $(\bar{1}00)$. **b** A cubic crystal forming an octahedron bounded by eight faces {111}. The four unlabelled faces at the rear side are $(\bar{1}\bar{1}\bar{1})$, $(\bar{1}1\bar{1})$, $(\bar{1}11)$, $(\bar{1}\bar{1}1)$

possibly parallel front and back faces of a crystal are denoted by (hkl) and $(\bar{h}\bar{k}\bar{l})$, for the case that the origin of the translation lattice has been chosen somewhere inside the crystal. If only the orientation of the (family of) lattice plane(s) is of interest, it is not necessary to separately consider (hkl) and $(\bar{h}\bar{k}\bar{l})$.

As a general rule it can be said, the higher the density of lattice points in a (each) member of the (hkl) family of planes, the larger the interplanar distance d_{hkl}. Typically, low values of h, k and l imply lattice planes of high lattice-point density and large lattice-plane spacing (see also the lattice-plane spacing formula as given in Table 4.7 in Sect. 4.5).

The Miller indices only indicate the orientation of the concerned family of lattice planes in the translation lattice; the Miller indices thus do not directly provide information regarding their position with respect to the atomic structure of the crystal, because the origin of the crystal coordination system in principle can be chosen freely. Such information is provided for the NaCl-type structure in Fig. 4.12a, where the arrangements of atoms in "atomic" planes parallel to (001) and parallel to (110) are shown.

In many applications, it is relevant to consider "sets" of symmetry-equivalent families of lattice planes (e.g. (hkl), $(h'k'l')$ and $(h''k''l'')$ are equivalent). Such sets have necessarily the same lattice-plane spacing $d_{hkl} = d_{h'k'l'} = d_{h''k''l''}$. Each member of this set comprises a family of lattice planes of specific orientation with respect to the **a**, **b** and **c** axes. An equivalent family of lattice planes then is identified as identical to the first family of lattice planes but oriented differently with respect to the **a**, **b** and **c** axes. In order to indicate in a discussion that a certain statement pertains to the *set of equivalent families of lattice planes*, one uses the notation $\{hkl\}$ instead of (hkl), i.e. $\{hkl\}$ comprises (hkl), $(h'k'l')$ and $(h''k''l'')$.

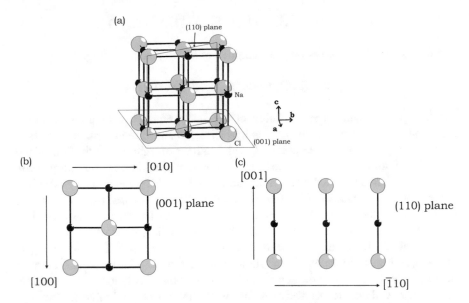

Fig. 4.12 a Unit cell of NaCl showing atoms with $0 \leq x, y, z \leq 1$, indicating atoms which are included in the atomic planes parallel to (001) and (110) as shown in (**b**) and (**c**) (the large grey atoms can be conceived as the Cl⁻ anions and the small black atoms can be taken as the Na⁺ cations. Additionally, in (**b**) and (**c**) directions have been indicated with lie within the shown planes

For most of the relevant cubic crystals, the set of the equivalent families of lattice planes can be directly obtained from the notation/code $\{hkl\}$ by (1) allowing an arbitrary permutation of h, k and l and by (2) additionally allowing each h, k or l to become positive and negative (h and $\bar{h}$). In this way, a maximum of 48 equivalent families of lattice planes belong to the set $\{hkl\}$:

$$(hkl), (klh), (lhk), (khl), (hlk), (lkh),$$
$$(\bar{h}kl), (kl\bar{h}), (l\bar{h}k), (k\bar{h}l), (\bar{h}lk), (lk\bar{h}),$$
$$(h\bar{k}l), (\bar{k}lh), (lh\bar{k}), (\bar{k}hl), (hl\bar{k}), (l\bar{k}h),$$
$$(hk\bar{l}), (k\bar{l}h), (\bar{l}hk), (kh\bar{l}), (h\bar{l}k), (\bar{l}kh),$$
$$(\bar{h}k\bar{l}), (\bar{k}l\bar{h}), (\bar{l}\bar{h}k), (\bar{k}\bar{h}l), (\bar{h}l\bar{k}), (\bar{l}k\bar{h}),$$
$$(\bar{h}k\bar{l}), (\bar{k}l\bar{h}), (\bar{l}h\bar{k}), (\bar{k}h\bar{l}), (\bar{h}l\bar{k}), (\bar{l}k\bar{h}),$$
$$(\bar{h}k\bar{l}), (\bar{k}l\bar{h}), (\bar{l}h\bar{k}), (\bar{k}h\bar{l}), (\bar{h}l\bar{k}), (\bar{l}k\bar{h}),$$
$$(\bar{h}\bar{k}l), (\bar{k}l\bar{h}), (l\bar{h}\bar{k}), (\bar{k}hl), (\bar{h}l\bar{k}), (l\bar{k}\bar{h}).$$

Indeed, the application of the formula for d_{hkl} for cubic crystals (Table 4.7 in Sect. 4.5) yields the same value of d_{hkl} for all of these 48 families of lattice planes.

If certain h, k and l assume the same value, and/or if one or two of these values is 0, some of the 48 families of lattice planes listed above become "degenerated". For example, the set $\{111\}$ ($h = k = l = 1$) comprises only eight families of lattice planes (cf. Fig. 4.11b):

$$(111), (\bar{1}11), (1\bar{1}1), (11\bar{1}), (\bar{1}\bar{1}\bar{1}), (1\bar{1}\bar{1}), (\bar{1}1\bar{1}), (\bar{1}\bar{1}1).$$

and the set $\{100\}$ comprises only six families of lattice planes (cf. Fig. 4.11a):

$$(100), (010), (001), (\bar{1}00), (0\bar{1}0), (00\bar{1}).$$

For hexagonal crystals, the derivation of the set of equivalent families of lattice planes from the notation/code $\{hkl\}$ is less straightforward than for cubic crystals. In order to realize a simple approach, which allows an intuitive derivation of the equivalent families of lattice planes, one introduces an additional, auxiliary index i, to be positioned before the "l" in the notation/code $\{hkl\}$, i.e. a family of lattice planes now is indicated as $(hkil)$ (the so-called *Miller-Bravais indices*). Whereas h, k and l refer to the basis vectors $\mathbf{a} = \mathbf{a}_1$, $\mathbf{b} = \mathbf{a}_2$ and $\mathbf{c}$ in the same way as described above, the additional index i refers to an additional auxiliary basis vector $\mathbf{a}_3 = -(\mathbf{a}_1 + \mathbf{a}_2)$ (which is thus linearly dependent on both $\mathbf{a}_1$ and $\mathbf{a}_2$; note that $|\mathbf{a}_1| = |\mathbf{a}_2| = |\mathbf{a}_3| = a$; cf. Fig. 4.13). It can then be shown that if a family of lattice planes intersect the three axes of the frame of reference of the hexagonal translation lattice at points $n\mathbf{a}_1/h$, $n\mathbf{a}_2/k$ and $n\mathbf{c}/l$, respectively (cf. discussion above), they will intersect an axis along $\mathbf{a}_3$ at $n\mathbf{a}_3/i$ with, always, $i = -(h+k)$ (see the example indicated in Fig. 4.13). Thus, knowing (hkl), one can add i according to $(hkil) = (hk\,(\bar{h}+\bar{k})\,l)$. The advantage of adding the index i is that this index can be taken up into a permutation scheme similar to that described above for (most) cubic crystals (see what follows next).

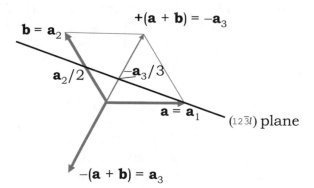

Fig. 4.13 Illustrating the use of Miller-Bravais indices for hexagonal crystals: $(hkil)$. Whereas h, k and l refer to the basis vectors $\mathbf{a} = \mathbf{a}_1$, $\mathbf{b} = \mathbf{a}_2$ and $\mathbf{c}$, the additional index i refers to an additional auxiliary basis vector $\mathbf{a}_3 = -(\mathbf{a}_1 + \mathbf{a}_2) = -(\mathbf{a} + \mathbf{b})$ (which is thus linearly dependent on both $\mathbf{a}_1$ and $\mathbf{a}_2$; note that $|\mathbf{a}_1| = |\mathbf{a}_2| = |\mathbf{a}_3| = a$; cf. the lattice-parameter conditions for the hexagonal crystal system/hexagonal translation lattice given in Sect. 4.1.2 and Fig. 4.8). If a family of lattice planes intersect the three axes of the frame of reference of the hexagonal translation lattice at points $n\mathbf{a}_1/h$, $n\mathbf{a}_2/k$ and $n\mathbf{c}/l$, respectively (cf. Fig. 4.10), they will intersect an axis along $\mathbf{a}_3$ at $n\mathbf{a}_3/i$ with, always, $i = -(h + k)$. (see the intersection points indicated in the figure for the $(12\text{-}3\,l)$ plane with $n = 1$ (cf. Fig. 4.10)). Therefore, on this basis, for each (hkl) one can add i according to $(hkil) = (hk\,(\bar{h}+\bar{k})\,l)$

The set of equivalent families of lattice planes $\{hkil\}$ in (most) hexagonal crystals is obtained by (1) allowing for an arbitrary permutation of h, k and i (*not* involving l) and (2) additionally allowing for h, k and i (simultaneously) being substituted by $\bar{h}$, $\bar{k}$ and $\bar{i}$, and (3) additionally allowing l to become $\bar{l}$. In this way, a maximum of 24 families of lattice planes belong to the set $\{hkil\}$:

$$(hkil), (ihkl), (kihl), (khil), (hikl), (ikhl),$$
$$(\bar{h}\bar{k}\bar{i}l), (\bar{i}\bar{h}kl), (\bar{k}\bar{i}hl), (\bar{k}\bar{h}il), (\bar{h}\bar{i}kl), (\bar{i}\bar{k}hl),$$
$$(hki\bar{l}), (ihk\bar{l}), (kih\bar{l}), (khi\bar{l}), (hik\bar{l}), (ikh\bar{l}),$$
$$(\bar{h}\bar{k}\bar{i}\bar{l}), (\bar{i}\bar{h}k\bar{l}), (\bar{k}\bar{i}h\bar{l}), (\bar{k}\bar{h}i\bar{l}), (\bar{h}\bar{i}k\bar{l}), (\bar{i}\bar{k}h\bar{l}).$$

Indeed, the application of the formula for d_{hkl} for hexagonal crystals (Table 4.7) yields the same value of d_{hkl} for all of these 24 families of lattice planes. Again, as discussed above for cubic crystals, a smaller number of equivalent families of lattice planes arises (i.e. "degeneration" occurs), if certain h, k, i or l are equal or 0.

The advantage of the introduction of the Miller-Bravais indices for hexagonal crystals becomes clear upon deleting the index i at the third position of $\{hkil\}$. Then, the permutability of h, k and i is hidden. This can be illustrated for the set of families of lattice planes indicated by $\{10\bar{1}0\}$. The equivalent families of lattice planes are given in Miller-Bravais indices, following the recipe given above, by:

$$(10\bar{1}0), (\bar{1}100), (0\bar{1}10), (01\bar{1}0), (1\bar{1}00), (\bar{1}010),$$

whereas, the equivalent families of lattice planes are indicated in Miller indices by:

$$(100), (\bar{1}10), (0\bar{1}0), (010), (1\bar{1}0), (\bar{1}00).$$

Thus, although the index i is artificial and is not necessarily needed to unambiguously indicate the orientation of a family of lattice planes, it is a helpful auxiliary index to recognize that a (100) family of lattice planes is equivalent with the $(\bar{1}10)$ family of lattice planes!

4.1.4.2 Lattice Directions

A direction in the translation lattice can be simply given by a linear combination of components along the **a**, **b** and **c** axes of the translation lattice:

$$\mathbf{t} = u\mathbf{a} + v\mathbf{b} + w\mathbf{c} \tag{4.5}$$

where $\mathbf{t}$ is a translation vector, originating from the origin, characterized by the integer numbers u, v, and w. This lattice direction, and lattice directions parallel to it (this is not indicated further), are designated by the notation/code $[uvw]$.

Similar to the agreement of a specific code for equivalent (families of) lattice planes (i.e. (*hkl*) versus {*hkl*}; cf. Sect. 4.1.4.1), a set of equivalent lattice directions can be denoted by a special code, i.e. ⟨*uvw*⟩ denotes all lattice directions equivalent to and including the lattice direction [*uvw*].

With reference to a similar remark made above concerning the meaning of the Miller indices, it should be realized that the indication [*uvw*] for a line/direction in the translation lattice does not directly provide information regarding its position with respect to the atomic structure of the crystal, because the origin of the crystal coordination system in principle can be chosen freely. In Fig. 4.12b, c, "atomic" lines parallel to specific members of the sets of equivalent lattice directions <100> and <110>, in specific "atomic" planes are shown.

For (most) cubic crystals, the set of equivalent directions is obtained in the same way as for the lattice planes, i.e. by arbitrary permutation of *u*, *v* and *w*, and by allowing each of *u*, *v* and *w* becoming (independently) negative. Thus, for the general case (no equal *u*, *v*, and *w*, and neither *u*, *v* and *w* being 0), a set of 48 equivalent directions is obtained.

For hexagonal crystals, a Miller-Bravais-type scheme can be given for the indication of lattice directions that allows an easy derivation of the set of equivalent directions (cf. the introduction of the Miller-Bravais indices for the indication of lattice planes in hexagonal crystals in Sect. 4.1.4.1). Thus, one introduces an auxiliary index T such that the lattice direction is indicated by [$UVTW$] with $T = -(U + V)$ and accordingly

$$\mathbf{t} = U\mathbf{a}_1 + V\mathbf{a}_2 + T\mathbf{a}_3 + W\mathbf{c} \tag{4.6}$$

Although T is only an auxiliary index, like i in the notation for the lattice planes, in this case, it is not that straightforward to obtain this index from the *u* and *v* indices used for indication of the lattice direction in the recipe given by Eq. (4.5), because $u \neq U$ and $v \neq V$. Instead, the U and V, and thus T according to $T = -(U + V)$, have to be derived from the condition of parallelism of the directions **t** as given by Eqs. (4.5) and (4.6). For example, the direction [100] in the notation according to Eq. (4.5) can be indicated as $[2\bar{1}\bar{1}0]$ in the notation according to Eq. (4.6). Similarly, [110] corresponds to $[11\bar{2}0]$ and [210] corresponds to $[10\bar{1}0]$. Then, using the [$UVTW$] indices, equivalent lattice directions can be obtained by application of the same rules used for the lattice planes in Miller-Bravais notation given in Sect. 4.1.4.1.

4.2 Crystal Structures of Elements

Crystal structures are determined by the type and strength of chemical bonding of the constituent atoms. Therefore, some typical crystal structures were already presented in Chap. 3 devoted to "chemical bonding". One typifying parameter in this context is the "coordination number" of an atom in the crystalline state. The coordination number can, in its most simple way, be defined as the number of nearest neighbours of the atom considered; this is the definition used in Chap. 3. As illustrated by Table 4.2, the (range of values for the) coordination number is indicative for the type of chemical bonding. This still leaves a myriad of possible crystal structures also for a single type of chemical bonding.

This section presents a geometric discussion of the (most) simple crystal structures of crystalline solids; i.e. the crystal structures of pure metals are dealt with. A calculation of the crystal structure, departing from first principles, is even today generally still impossible. However, it was argued that those crystal structures for metals are favoured which establish the largest Coulomb interaction of the positively charged metal ions and the "sea" of "free", valence electrons. For the same atomic volume, space fillings which assure this largest Coulomb interaction are the so-called close packed crystal structures (and the body centred cubic crystal structure; cf. discussion in Sect. 3.5.3).

The above consideration immediately suggests to model the atoms as *hard solid spheres ("ping-pong balls")*, packed in order to realize the highest density of intimate contacts: bonding between solid spheres occurs if they touch each other and thus strongest bonding is achieved if as many contacts as possible are realized between the hard solid spheres. This model of hard solid spheres can explain the occurrence of the three most important structure types which can be observed for metallic elements: face centred cubic (also called Cu type), hexagonal close packed (also called Mg type) and body centred cubic (also called W type). Similarly, some of the structures occurring in intermetallic compounds can be understood (see Sect. 4.4). Moreover,

Table 4.2 Types of chemical bonding and the coordination number of the crystalline solid

Type of chemical bonding	Covalent	Ionic	Metallic	van der Waals[b]
Characteristics of interaction[a]	Localized bonds of electron pairs; directional	Attractive and repulsive Coulomb interactions of cations and anions; non-directional	Attractive Coulomb interaction of metal ions and "sea" of (nearly) "free" electrons; non-directional	Electrostatic interaction of instantaneous and correspondingly instantaneously induced dipoles; non-directional
Coordination number[c]	1–6	4–8	8–12	12

[a]The bonding characteristics can only very crudely be indicated here; for a more balanced discussion, see Chap. 3

[b]In Sect. 3.6, three types of dipole interactions, gathered under "van der Waals bonding", have been considered. Here, only the third type of interaction, i.e. between instantaneous and correspondingly instantaneously induced dipoles, is considered. This pertains to the bonding occurring for solid, insert, noble gases

[c]Here, the most simple definition of coordination number is considered: the number of nearest neighbours; see Sect. 4.2.4 for other possibilities

modelling atoms as hard solid spheres makes also sense for ionic compounds dominated by Coulomb interaction of the cations and anions. Apart from these great successes of the model of the atom as a hard solid sphere ("ping-pong ball"; see also Sects. 1.3–1.5), it has its limitations (e.g. see Sect. 4.2.1.3).

Indeed metal atoms cannot be conceived as *hard* spheres. This can be deduced from the change of density occurring upon melting of the solid metal. If the metal atoms upon melting would keep their "size", i.e. they would be "hard", then in the liquid, amorphous state they would assume a so-called dense-random packing with a relative volume density of 0.63, which can be compared with the density of the close packed structures often met for metals (see below) with a relative volume density of 0.74. This would imply an increase of volume upon melting of about 17.5%. Values of only maximally 6% are observed in experiments. Apparently, the metal atoms become "squeezed" in the liquid state as compared with the solid state and thus metal atoms are not *hard* spheres (Egami 2010).

4.2.1 Crystal Structures Derived from Close Packed Arrangements of Hard Spheres

Two of the three most important crystal structures of simple metals, the face centred cubic (Cu type) and the hexagonal close packed (Mg type) crystal structures, are so-called close packed crystal structures. The close packed structures have in common that they constitute the densest arrangements of identical spheres in space (which is not so straightforward to prove; see the story told in Sect. 1.5). In this way, the maximum of contacts (i.e. the highest density of contacts) between the atoms can be achieved (see the introductory text of this Sect. 4.2).

4.2.1.1 The Model of Close Packed Hard Spheres

In the sequel, it is consequently strived for to maximize the number of contacts between identical solid spheres by packing them first in two and then in three dimensions.

If two solid spheres of radius R are brought into contact (as follows from the geometry of the sphere, this is always the closest possible contact), the centres of the spheres assume a distance of $2R$ (Fig. 4.14a). How to bring an additional sphere at shortest possible contact distance (2R) with, simultaneously, the two already contacting spheres? This question is answered by Fig. 4.14b. The new sphere is in contact with both other spheres: there is only one possibility[13] to realize at the same time two new (closest) contacts ("bonds") with the first two spheres. As a result, the centres of the three spheres form an equilateral triangle. This building principle can

---

[13] More correctly: two identical possibilities: the third atom can be attached to either the top (one side) or bottom (opposite side) of the already contacting pair of spheres.

(a)

(b)

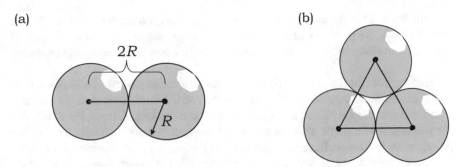

Fig. 4.14 a Two spheres, each of radius R, in a plane perpendicular to the viewing direction. The distance between the sphere centres is $2R$. **b** Addition of a third sphere in contact with both spheres of the pair shown in **a**. The centres of the three spheres form an equilateral triangle

be continued within the plane of this triangle, as shown in Fig. 4.15. Thereby, upon continued addition of spheres, a close packed layer of identical spheres is obtained. In this layer, each atom has six nearest neighbours all at a distance of $2R$. Applying two-dimensional crystallography to this layer, a hexagonal unit cell can be defined with basis vectors $\mathbf{a}_h$ and $\mathbf{b}_h$, and $a_h = b_h = 2R$ and $\gamma = 120°$ (cf. Sect. 4.1.2). Note that upon connecting the centres of the spheres in the close packed layer with

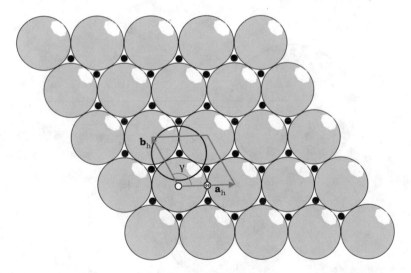

Fig. 4.15 Close packed layer of identical spheres. The vectors $\mathbf{a}_h$ and $\mathbf{b}_h$ are basis vectors ($|\mathbf{a}_h| = |\mathbf{b}_h| = 2R$) of the hexagonal translation lattice and thereby indicate the corresponding unit cell characterizing this *two-dimensional* structure. Different possible positions for placement of additional spheres on top of this original layer have been indicated by the symbols ○, ⊗ and ●. Only the last possibility is realized in *three-dimensional* close packed structures (all ● sites have been indicated in the figure, in contrast with the ○ and ⊗ positions). The open circle represents a sphere added on top of the close packed layer at a ● position

straight lines, a net appears consisting of the equilateral triangles. With a view to what follows, the positions of the centres of the spheres of this first close packed layer, i.e. the nodes of the net of equilateral triangles, are indicated by the character A.

Starting from the first close packed layer, constructed as described above, the next step is extension of the close packing of spheres into the third dimension. To this end, additional spheres are placed in a specific way onto the first, A layer. First consider the placement of a single atom on top of the original A layer. Different imaginable positions for this sphere have been indicated in Fig. 4.15. In the position indicated by O, the additional sphere would be in contact with only the sphere directly underneath, whereas in the position indicated by $\otimes$, two neighbouring spheres would be in contact with the additional sphere. If the position indicated by $\bullet$ is selected, the maximum number of three, simultaneous sphere-sphere contacts is realized. This corresponds to close packed stacking. The resulting local arrangement of the additional sphere, on top of the A layer, touching three spheres of the "A" layer, is shown in Fig. 4.16. The four spheres, in closest contact, form a regular tetrahedron with edges of the length 2R, and with each face of the tetrahedron identical to the equilateral triangle of the original, close packed A layer. The height of the tetrahedron is given by $(2/3)^{1/2}\,2R$.

Having positioned one additional sphere on top of the original, A layer in a $\bullet$ position (Fig. 4.15), one may try to position additional atoms, on top of the A layer, at such positions. It turns out that with respect to a first, single sphere on a $\bullet$ position, the

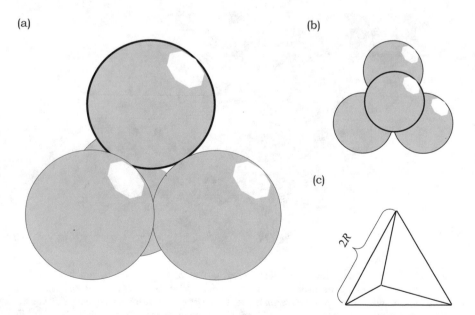

(a)   (b)

(c)

2R

Fig. 4.16 Regular tetrahedron consisting of four hard spheres each of radius R. This tetrahedron results from putting a single sphere (thick margins) on top of a close packed layer (see Fig. 4.15): **a** perspective view, **b** view from the top and **c** edges of tetrahedron formed by connecting the centres of the spheres

three nearest neighbour ● positions cannot be occupied by further spheres, because otherwise the spheres would overlap with the first sphere (see the open circle in Fig. 4.15). In fact, only half of all ● positions can be occupied by spheres. These new spheres then form, again, a close packed layer geometrically identical to the first A layer, but shifted in a direction parallel to the A layer. The spheres of this second layer as shown in Fig. 4.17a are shifted with respect to the spheres of the original layer by a vector $\mathbf{v} = \mathbf{a}_h/3 - \mathbf{b}_h/3$ (having the length of 1/3 of the long diagonal of the unit mesh spanned by $\mathbf{a}_h$ and $\mathbf{b}_h$). The positions of the centres of the spheres of the second layer, with respect to the positions of the centres of the spheres of the first layer (A), have been indicated by the character B.

An alternative lateral position is possible for the second layer, by occupation of the other half of the ● positions provided by the first, A layer; cf. Fig. 4.17a, b.

(a)

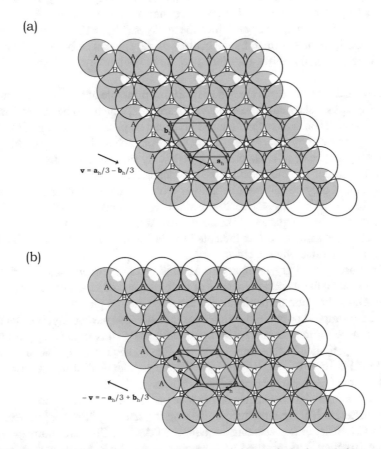

(b)

Fig. 4.17 Two possibilities to place onto a first layer of close packed spheres (sphere centres on position A) a second layer of close packed spheres: **a** second layer in position B, **b** second layer in position C. The vector by which the first layer would have to be shifted to arrive at the lateral position of the second layer (i.e. to make the centres of the spheres of the first and second layer coincident) has been indicated in (**b**) and (**c**)

Table 4.3 Stacking rules for close packed layers of identical spheres

Original layer	Next layer after lateral shift of $\mathbf{v} = \mathbf{a}_h/3 - \mathbf{b}_h/3$	Next layer after lateral shift of $\mathbf{v}' = -\mathbf{v} = -\mathbf{a}_h/3 + \mathbf{b}_h/3$
A	B	C
B	C	A
C	A	B

Starting from a certain layer, layers which can be added (be next in the stacking sequence) are obtained from the preceding layer by one of the two possible lateral shift vectors $\mathbf{v}$ or $\mathbf{v}'$ (cf. Figure 4.17)

Again, a close packed layer of spheres results, which is now shifted by a vector $\mathbf{v}' = -\mathbf{v} = -\mathbf{a}_h/3 + \mathbf{b}_h/3$ with respect to the original, A layer (Fig. 4.17b). The positions of the centres of the spheres in that layer are indicated by the character C. In both cases, i.e. for the B or C layer on top of the A layer (Fig. 4.17a, b), the separation distances of the planes formed by the centres of the spheres of both adjacent close packed layers equal the above-mentioned height of the tetrahedron given by $(2/3)^{1/2} 2R$.

The above discussion leads to a general building principle of close packed arrangements of spheres. Take a close packed layer of spheres, which can be denoted as an A or B or C layer. Then, a next close packed layer can be placed on the top of this layer by shifting it by $\mathbf{v}$ or by $\mathbf{v}' = -\mathbf{v}$ (see Table 4.3). For each of the close packed layers, in a stacking sequence of close packed layers, only three different lateral positions are possible, which are indicated by the characters A, B and C. A given layer, say an A layer, can only be followed by a layer of one of the two other types, so here either a B layer or a C layer. Hence, the stacking sequence of close packed layers can be described by a sequence of the characters A, B and C, with the constraint that two layers of the same type (i.e. indicated by the same character) cannot be direct neighbours in the stacking sequence.

For a close packed crystal structure exhibiting three-dimensional translational periodicity, (also) the stacking sequence of the close packed layers has to be periodic in the stacking direction perpendicular to the close packed planes.

Thus, the simplest possible stacking sequence of close packed layers of identical spheres can be written as AB.AB.AB.etc., with the points indicating the boundaries of the periodically repeating units. Obviously, sequences BA.BA.BA.etc. or AC.AC.AC.etc. and BC.BC.BC.etc. correspond to identical structures (identical arrangements of identical spheres in three-dimensional space).

The next simple stacking sequence of close packed layers pertains to three layers (each of different type) in a periodic unit along the stacking direction: for example, ABC.ABC.ABC.etc. The stacking sequence CBA.CBA.CBA.etc. corresponds to the same structure, but mirrored with respect to ABC.ABC.ABC.etc. by a plane perpendicular to the stacking direction (see discussion of "twinning" in Sect. 5.3).[14]

---

[14] Note that the stacking sequence BA.BA.BA.etc. does *not* produce a mirror structure of the stacking sequence AB.AB.AB.etc., but is fully identical to it.

Table 4.4 Examples of stacking sequences of close packed layers of identical spheres

	Structure type	Jagodzinski symbol
AB.AB.AB	Hexagonal close packed (h.c.p.); Mg type	h
ABC.ABC	Cubic close packed (c.c.p.[a]); Cu type	c
ABAC.ABAC	Double hexagonal close packed (d.h.c.p.); La type	hc
ABACACBCB	Sm type	hhc

[a]Normally described as "face centred cubic" (fcc)

Most of the close packed structures found for pure metals are given by the two simple stacking sequences discussed in the two preceding paragraphs. These two types of crystal structures are dealt with separately in more detail in Sects. 4.2.1.2 and 4.2.1.3. Table 4.4 includes two further stacking sequences occurring in real crystal structures of pure metals, but these usually occur for more "exotic", like rare-earth or lanthanide metals (cf. Sect. 2.5).

The stacking sequence of the close packed layers is one way to characterize the resulting crystal structure. Another approach focusses on the local environment of a sphere in a close packed layer. It appears that for ideal close packed arrangements each sphere is surrounded by 12 nearest neighbours (i.e. the coordination number, defined as the number of nearest neighbours, equals 12): 6 within the closed packed layer containing the sphere considered, and 3 above and 3 below the layer containing the sphere considered. These last 3 plus 3 spheres are incorporated in the close packed layers adjacent to the close packed layer containing the sphere considered. For (ideal) close packed stacking sequences, there are (only) two different possibilities for the positions in space of the nearest neighbours in the adjacent close packed layers, which are exhibited by two geometric structures: the *anticuboctahedron* and the *cuboctahedron* (see Fig. 4.18). If the adjacent layers, of the layer containing the sphere considered, are characterized by the same character (e.g. as it is the case for B in the sequence ....ABA....), the environment of the (each) atom considered in the central (B) layer is that of an anticuboctahedron (set up by the 12 nearest neighbour spheres; Fig. 4.18a). If the adjacent layers are characterized by different characters (e.g. as it is the case for B in the sequence ....ABC....), the environment of the (each) atom considered in the central (B) layer is a cuboctahedron (set up by the 12 nearest neighbours; Fig. 4.18b). Thus, the stacking sequence of close packed layers can be characterized, according to Jagodzinski, by a sequence of the characters h and c indicating the anticuboctahedral and cuboctahedral local environments of the stacked spheres in a close packed layer, respectively. Thus, each layer in the stacking sequence is provided with either the character h or the character c (see Fig. 4.18 and see Table 4.4). Evidently, the simplest close packed stacking sequences, characterized by repeating AB and ABC sequences (or equivalent alternatives; see above) thus contain h layers and c layers only. The symbol h stands for hexagonal, the crystal system of the ABABAB...... stacking sequence, and the symbol c stands for cubic, the crystal system of the ABCABCABC..........stacking sequence (see Sects. 4.2.1.2 and 4.2.1.3).

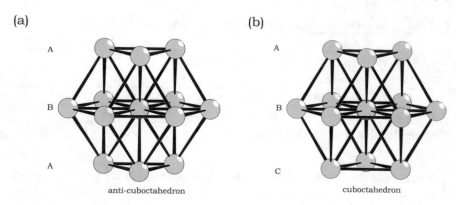

Fig. 4.18 Coordination of one sphere (the sphere in the centre of figures (**a**) and (**b**) indicated with the thicker margin) by its 12 nearest neighbours in close packed structures for the cases of local stacking of the type h (e.g. as in AB.AB.AB.etc. structures) (**a**) and of the type c (e.g. as in ABC.ABC.ABC.etc. structures) (**b**): **a** Anticuboctahedron for a sphere in a B layer in an ABA stack (h) and (**b**) cuboctahedron for a sphere in a B layer in an ABC stack. The shortest sphere-sphere distances (distances equal to $2R$) have been highlighted by thick lines. With reference to this last sentence in particular, it is remarked that the sphere radii in the drawings (**a**) and (**b**) have been made smaller than the true "hard sphere radius" R (as adopted throughout Sect. 4.2) in order to represent more clearly the three-dimensional sphere arrangement

4.2.1.2   The Cubic Close Packed (c.c.p.) or Face Centred Cubic (f.c.c.) Crystal Structure

The crystal structure derived from an ABC.ABC.ABC... stacking sequence of close packed layers of identical spheres is often referred to as the Cu-type crystal structure. One also uses the name cubic close packed (c.c.p.) structure. The most common name is face centred cubic (f.c.c.) structure. The f.c.c. structure is a very common crystal structure type of simple metals, like Cu, Ni, Pd, Pt, Rh, Al, Pb, Ca... and high temperature modifications of Fe (see, in particular Sect. 9.5.2.1) and Co. Also, the noble gases Ar, Kr and Xe solidify, at very low temperatures, according to the f.c.c. structure.

At first sight, it may not be immediately apparent that an ABC.ABC.ABC.... stacking sequence of close packed layers results in a cubic crystal structure. The f.c.c. unit cell has been indicated in Fig. 4.19a for an ABCA.... stack of layers. Note that for the f.c.c. unit cell in this picture a body diagonal of the unit cell (parallel to one of the <111> directions) runs in perpendicular direction, i.e. perpendicular to the close packed planes shown in a perspective way in this figure. The f.c.c. unit cell is shown in untilted, usual way in Fig. 4.19b, c. For crystallographic description of the f.c.c. structure, it suffices to provide a value for the cubic lattice parameter a (e.g. $a = 3.615$ Å for copper at ambient temperature) and to state there is an (e.g. copper) atom at $x = 0$, $y = 0$ and $z = 0$. The further atom positions in the unit cell then are consequently found at $x = 0$, $y = 1/2$ and $z = 1/2$, at $x = 1/2$, $y = 0$ and $z = 1/2$ and at $x = 1/2$, $y = 1/2$ and $z = 0$, bringing about the *face centred nature* of the unit cell.

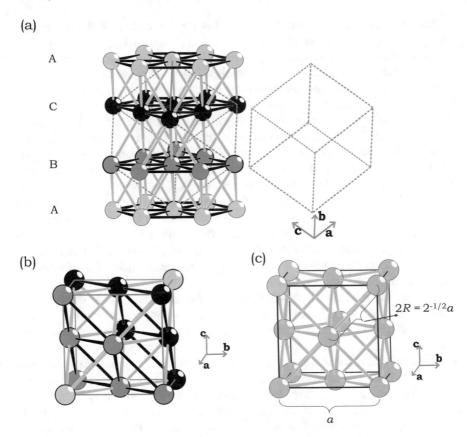

Fig. 4.19 The cubic close packed, face centred cubic, Cu-type crystal structure. **a** Part of an ABC.ABC.ABC… stacking sequence of close packed layers of solid spheres. A face centred cubic unit cell has been indicated by straight dashed lines; see also the repetition of the drawn unit cell at the right side of the picture showing the stacked layers. The atoms with fractional coordinates not belonging to the $0 \leq x, y, z \leq 1$ range, i.e. outside the unit cell indicated (cf. Sect. 4.1.1) have been drawn without margins. The shortest distances between neighbouring atoms (equal to $2R$) in the close packed planes drawn have been indicated in black; the shortest distances between neighbouring atoms in adjacent planes of the stack of close packed planes drawn have been indicated in light grey. **b** "Conventional", untilted drawing of the f.c.c. unit cell showing atoms with $0 \leq x$, $y, z \leq 1$. Colouring of atoms and shortest distances as in (**a**). **c** Same as (**b**) but with uniform colouring of atoms and distances between nearest neighbours. The basis vectors spanning the unit cell have been indicated next to the structures drawn. The spheres have been drawn smaller than corresponding to the sphere radius pertaining to the close packing of spheres model: the spheres of nearest neighbours touch along <110> directions. The relation of the true radius of the solid sphere in the model, R, and the lattice parameter, a, has been indicated in (**c**)

The origin (of the coordinate system and) of the unit cell can be chosen conveniently at the centre of mass of one of the atoms (cf. Sect. 4.1.1). This is what has been done above (see in particular Fig. 4.19b, c). The atoms in the f.c.c. unit cell are located on the origin and on the centre of the faces of the cube making up the unit cell. Thereby the outer appearance of the *crystal-structure* drawing has become identical to that for the face centred cubic *Bravais translation lattice* as shown in Fig. 4.8 (the motif here is one atom; there are four motifs in the unit cell of the f.c.c. Bravais-type translation lattice). Against this background, the Cu-type crystal structure is often referred to without much ado as "the" face centred cubic structure, although this term actually is reserved for the Bravais-type translation lattice of this name. For example, the diamond-type crystal structure or the rocksalt(NaCl)-type crystal structure (see Sects. 4.2.3.2 and 4.4.2.1) are (have a) face centred cubic (Bravais-type translation lattice). In these cases, the distinction between the pictured Bravais-type translation lattice and the pictured crystal structure is evident; not so for metals as Cu. This double use (for metals) of the term face centred cubic for two different concepts (the crystal structure and the translation lattice) has caused a lot of confusion and imprecise statements (see footnote 6).

After having realized the above, it can now be remarked that the drawing of a primitive unit cell of the f.c.c. Bravais-type translation lattice shown in Fig. 4.5 can also be interpreted as a drawing of the primitive unit cell of the f.c.c. Cu crystal structure.

The packing density of the f.c.c. structure, composed of close packed hard spheres of radius R, can be calculated straightforwardly. To this end first the relation of the cubic lattice parameter a with the sphere radius R is determined, recognizing that the nearest neighbouring spheres touch each other along <110> directions (e.g. along the face diagonals of the f.c.c. unit cell). Thus, it follows $2R = 2^{-1/2} a$ or $a = 2 \times 2^{1/2}R$ (for copper: $R = 1.28$ Å). The crystal structure consists of a massive aggregate of identical unit cells. Hence, the packing density is identical to the packing density of a single unit cell. The packing density then is given by the ratio of the volume, V_S, of the spheres (atoms) in the unit cell and the unit cell volume. There are four spheres (atoms) per unit cell (see also Sect. 4.1.1 and the Appendix at the end of this chapter) and therefore V_S equals $4 \times (4/3) \pi R^3$. The volume of the unit cell is a^3. Hence, the packing density, ρ, is given by

$$\rho = 4 \times \frac{4\pi R^3}{3a^3} = \frac{\pi}{3\sqrt{2}} \approx 0.74$$

It follows that, independently of the value of R, 74% of the space in the f.c.c. crystal structure is occupied by the close packed hard spheres. The same result for ρ holds for all stacking variants of the close packed planes (cf. Sect. 4.2.1.1), i.e. also for the hexagonal close packed structure (Mg-type crystal structure; see Sect. 4.2.1.3), provided that the distance between the close packed layers resembles the ideal value of $(2/3)^{1/2} 2R$. Note that the packing density, ρ, as defined here, is a dimensionless quantity, not to be confused with, e.g., the mass density which has the dimension mass/volume.

The f.c.c. crystal structure has a special property with respect to crystal structures derived from other than ABC.ABC.ABC.... stacking sequences of close packed layers of hard spheres. The ABC.ABC.ABC.... stacking sequence does not only occur along the [111] direction of the f.c.c. unit cell depicted in Fig. 4.19a, but it also occurs along the other three of the four body diagonals of the f.c.c. unit cell: the [−1−11], [−11−1] and [1−1−1] directions; for all other types of stacking of close packed layers ("polytypes"), there is only one direction in space perpendicular to which a stack of close packed layers occurs. This equivalence holds as long as the interlayer distances along the different close packed <111> directions are identical, i.e. equal to $(2/3)^{1/2}$ 2R (cf. Sect. 4.2.1.1). If this would not be the case, the principle of close packing of hard spheres is violated and the structure would, furthermore, not be cubic any more. The high symmetry corresponding to the cubic symmetry represents a favourable state of energy (minimum of energy; cf. the third paragraph of Sect. 4.1.2). In fact, from an energetic point of view, this cubic high-symmetry state is therefore favoured by many materials. Yet, crystal structures representing, with respect to the f.c.c. crystal structure, slightly distorted states, exist (see what follows).

A first example is the crystal structure of solid mercury (Hg). This structure can be described using rhombohedral basis vectors: $a = b = c \approx 4.572$ Å and $\alpha = \beta = \gamma = 98.27°$ (this is the deviation: $\alpha = \beta = \gamma = 90°$ for f.c.c.). These lattice-parameter data imply that along the [111] direction the distances between the "close packed" layers are shorter than along the [−1−11], [−11−1] and [1−1−1] directions. The origin for this structural distortion lies in the specific electronic structure of the mercury atoms, which cannot be discussed here.

Another distortion variant of the f.c.c. structure is exhibited by indium (In). In this case, a tetragonal distortion occurs: $c > a$ ($c = 4.94$ Å and $a = 4.59$ Å) and a face centred tetragonal unit cell results. As a consequence of this tetragonal distortion, perpendicular to none of the "close packed" directions, i.e. the [111], [−1−11], [−11−1] and [1−1−1] directions, geometrically perfectly close packed atom layers occur. Again, more detailed consideration of the electronic structure and metal bonding in solid indium is needed to explain these subtleties.

Even in the absence of such deviations as discussed above, the model of close packed hard spheres cannot represent all aspects of the behaviour of a crystalline metal like copper; see the discussion on the use and limitations of deliberately chosen to be simple models, as the "ping-pong ball model" for the crystal structure of metals, in Sect. 1.3. Thus, the lattice parameter of crystalline copper depends on temperature; it increases with temperature (see Fig. 4.20). Evidently, the model of close packed, hard spheres with a constant value for the sphere radius fails to explain this phenomenon. An other model, still very simple, is needed to explain this phenomenon: this model recognizes the short-range nature of the repulsive force and the more long-range nature of the attractive force between atoms, leading to a potential energy minimum well, for the bonding between two atoms as a function of the interatomic distance, which has an asymmetrical shape with respect to the position of energy minimum. But also this model has its limitations, in particular for metals (see further Sect. 3.1 and its footnote 2).

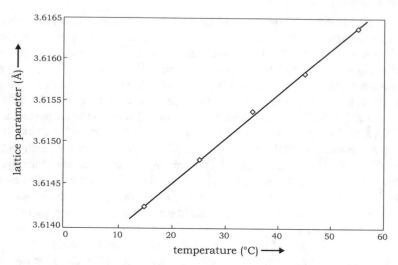

Fig. 4.20 Experimentally determined temperature dependence of the lattice parameter of copper. Data taken from M.E. Straumanis, L.S. Yu, Acta Crystallogr. A 25, 676–682 (1969)

4.2.1.3   The Hexagonal Close Packed (h.c.p.) Crystal Structure

The crystal structure derived from the simplest stacking sequence of close packed layers of identical spheres, AB.AB.AB…, is observed in nature for many metals like Mg, Be, Zn, Ti, Re…, adopting each atom as a hard solid sphere. This crystal structure is often referred to as the Mg-type crystal structure. The more common and more descriptive name for the Mg-type crystal structure is the "hexagonal close packed" crystal structure (abbreviated as h.c.p.). The La type crystal structure (see Table 4.4) also belongs to the hexagonal crystal system.

Like for the cubic close packed structure, the lattice parameters of the hexagonal close packed structure, a and c, can be expressed in terms of the radius of the hard spheres constituting the structure. Two different choices for the unit cell of the Mg-type crystal structure are shown in Fig. 4.21. These two options for the unit cell only differ in the choice of origin. On the basis of Fig. 4.21a it is evident that a corresponds to a_h as defined in Sect. 4.2.1.1, and thus $a = 2R$. The height of the unit cell, as given by c, corresponds to the height of two tetrahedra made up by the considered solid spheres (see Fig. 4.16a). Thus, $c = 2(2/3)^{1/2} 2R$ and the axial ratio $c/a = 2(2/3)^{1/2} \approx 1.633$. Using similar considerations like those for the cubic close packed structure, one obtains a packing density of $\pi/(3 \times 2^{1/2})$, which is exactly the same value as for the cubic close packed structure and for actually all close packed structures (and which corresponds to an occupation of the available space of about 74%).

The unit cell chosen above (two options differing only in the choice of origin) for the h.c.p. *crystal structure* contains two atoms. The *Bravais translation lattice* for the h.c.p. crystal structure is primitive hexagonal (see Fig. 4.8). Hence the motif in this case consists of two atoms and the unit cell contains one motif. Note the difference

(a)   (b)

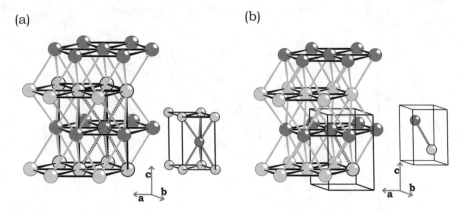

Fig. 4.21 The hexagonal close packed (h.c.p.) structure (Mg-type structure), adopting the atoms as identical hard solid spheres, with two different choices of origin for the unit cell, as represented by grey boxes: **a** with an atom at the origin of the unit cell and in total two fractional atomic positions within the range $0 \leq x, y, z \leq 1$, i.e. at $x = 0, y = 0, z = 0$ and at $x = 1/3, y = 2/3, z = 1/2$, **b** with the origin of the unit cell at a position half-way between two nearest neighbour atoms and in total two fractional atomic positions at $x = 1/3, y = 2/3, z = 1/4$ and at $x = 2/3, y = 1/3, z = 3/4$. Atoms within the $0 \leq x, y, z \leq 1$ range have been indicated with margins, the others without. In order to highlight the close packed layers perpendicular to the stacking direction, the shortest atom-atom distances in these planes have been indicated by fat black lines, whereas the other shortest atom-atom distances have been represented by light grey lines. The two tetrahedra defining the height of the unit cell (c) have been indicated by dashed lines

with the cubic close packed (f.c.c.) crystal structure, where the Bravais translation lattice is identical with the crystal structure with the motif being one atom and four motifs in the unit cell (cf. Sect 4.2.1.2).

In real metals of h.c.p. crystal structure, the axial ratio usually deviates more or less from the ideal value of $2(2/3)^{1/2} \approx 1.633$. This implies that the crystal structure is no longer truly close packed, or, in other words, this is an indication of the limitations of the model of identical hard spheres for understanding the crystal structure (see the end of the introduction of Sect. 4.2).[15] The reasons for such deviations from the ideal axial c/a ratio have to do with details of the metallic bonding in these metals, which are beyond the scope of this book.

In general, both a and c of an h.c.p. metal vary differently with temperature, implying a change of the axial ratio c/a with temperature. Values for a and c are shown in Fig. 4.22 for a series of elements exhibiting the h.c.p. (Mg type) crystal structure. Straight lines of constant c/a and curved lines of constant unit-cell volume have been incorporated in this figure as well. Indeed, although the values occurring in nature for c/a lie around the ideal value of 1.633 (see above), values occur as low

---

[15] As indicated in Sect. 4.2.1.2 for the cubic close packed (Cu type, f.c.c.) structure, a spacing change for the stacks of close packed layers perpendicular to any of the four <111>-directions leads to a distortion rendering the structure non-cubic. In contrast, for the hexagonal close packed (Mg type) structure, a distortion of the ideal (hexagonal) close packed structure by changing c/a does not lead to a change of crystal symmetry.

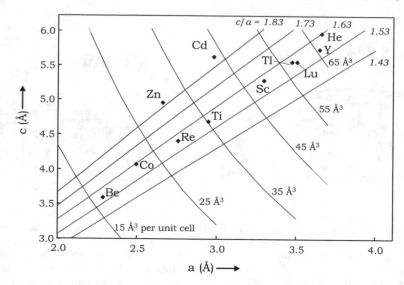

Fig. 4.22 Plot of the lattice parameter c versus the lattice parameter a for a number of elements crystallizing in the hexagonally close packed (Mg type) crystal structure. Points indicated refer to particular elements. Straight, finely dotted lines represent loci of constant axial ratio c/a; curved, full lines represent loci of constant unit-cell volume. Data taken from Inorganic Crystal Structure Base (ICSD), FIZ Karlsruhe

as 1.569 and 1.586, e.g. for Be and Ti, and as high as 1.856 and 1.886, e.g. for Zn and Cd.

Note in particular the relatively large volume of the unit cell pertaining to solid helium (the only noble gas that solidifies, at high pressure (> ≈30 atm) and very low temperature (a few K), according to a h.c.p. stacking; the other noble gases solidify according to a f.c.c. stacking; cf. Fig. 4.22 and Sect. 3.6). With a view to the trend for metals in a period of the Periodic Table, implying that the molar (atomic) volume (unit cell volume, if the same crystal structure is adopted by the elements considered) decreases with atomic number (cf. Sect. 2.5.1), helium is an outsider. This is of course due to the relatively weak van der Waals bonding in solid helium (of the order 10 kJ/mol; Sect. 3.6) as compared to the relatively strong metal bonding (of the order 500 kJ/mol; Sect. 3.5).

4.2.2 The Body Centred Cubic (b.c.c.) Crystal Structure

This third of the three most important structure types does not belong to the close packed stacking variants of hard, solid spheres, although this crystal structure represents an also "quite dense" (see below) packing of identical hard spheres.

The body centred cubic crystal structure (see Fig. 4.23) is also referred to as W-type crystal structure and occurs for the metals W, Nb, V, Fe, Cr, Mo, Na, K…. It can be considered as derived from a body centred cubic Bravais-type translation lattice where the lattice points are identified as atoms, i.e. the motif is one atom; cf. the discussion in Sect. 4.2.1.2 regarding the distinction between the f.c.c. Bravais-type translation lattice and the f.c.c. crystal structure. Thus, the W-type crystal structure is not the only crystal-structure type having a body centred cubic Bravais-type translation lattice (although it is the simplest one; cf. the distinction between the NaCl and Cu crystal structures both having the same f.c.c. Bravais-type translation lattice (Sect. 4.2.1.2)).

In the b.c.c. crystal structure, the shortest distance between two atoms occurs along the body diagonals of the cubic unit cell (i.e. along <111> directions) and has a magnitude of $(3^{1/2})/2 \cdot a$. Hence, $2R = (3^{1/2})/2 \cdot a$ or $a = 4/3^{1/2} \cdot R$, provided the atoms are conceived as hard spheres of radius R. Each atom in the b.c.c. structure has eight nearest neighbours, less than the twelve nearest neighbours for both close packed structures, f.c.c. and (ideal) h.c.p. (see Sect. 4.2.1). However, for the b.c.c. structure, the six next-nearest neighbours occur at distances equal to a ($= 4/3^{1/2} \cdot R \approx 2.3\,R$), which, as compared to the nearest neighbours, thus, are at distances only 15% farther away from the atom considered (see Fig. 4.23), whereas the "gap" between nearest and next-nearest neighbours is considerably larger for the closed packed structures (see also the discussion on the coordination number in Sect. 4.2.4). Against this

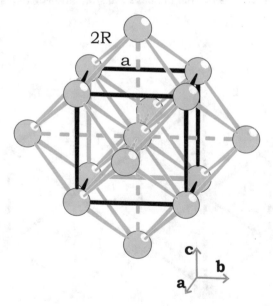

Fig. 4.23 Unit cell of the b.c.c. crystal structure (black box) plus six surrounding atoms; $a = b = c$ and $\alpha = \beta = \gamma = 90°$. Starting from the atom in the centre of the cubic unit cell, eight nearest neighbours occur at distances (indicated by grey, continuous connecting line segments) equal to $(3^{1/2})/2 \cdot a$ (= half of the body diagonal of the unit cell), and six next-nearest neighbours are found at distances (indicated by grey, dashed connecting line segments; i.e. the six atoms drawn outside of the indicated unit cell) equal to a and thus at distances only 15% larger than the nearest neighbour distances

background, the coordination number for the b.c.c. crystal structure is often given as $8 + 6 = 14$.

Adopting the hard sphere model, it can be shown straightforwardly that the b.c.c. structure, although not close packed, represents a quite densely packed structure as well. The b.c.c. unit cell contains two hard spheres (atoms), representing a volume of $2 \times \frac{4}{3} \pi R^3$. Because (see above) $3^{1/2}/2. \, a = 2R$, the unit cell volume in terms of R can be given as $a^3 = \left[\frac{4}{3^{1/2}} R\right]^3 = \frac{64}{3 \times 3^{1/2}} R^3$. Hence, the packing density ρ is given by

$$\rho = \frac{3^{1/2}\pi}{8} \approx 0.68.$$

Indeed the packing density for b.c.c. of 68% is only a little smaller than the packing density for the close packed structures of 74%.

Although the b.c.c. structure is not close packed, relatively densely packed atomic planes can be indicated for the b.c.c. structure, which can be compared with the truly close packed planes in the close packed structures (f.c.c.: {111} planes; h.c.p.: {001} planes). These planes for the b.c.c. structure are the {110} planes (see Fig. 4.24).

It is worthwhile to discuss here a possible, simple orientation relationship of the b.c.c. (W type) structure and the f.c.c. (Cu type) structure. This so-called *Bain orientation relationship* can play a role in phase transformations from an f.c.c. structure

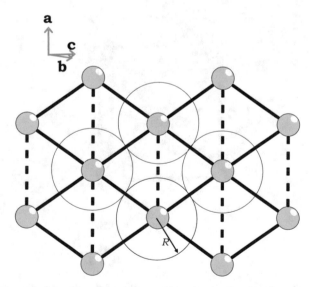

Fig. 4.24 Atom packing in a {110} plane of the b.c.c. crystal structure. Nearest neighbour distances have been indicated with bold connecting line segments; next-nearest neighbour distances have been indicated with dashed connecting line segments (cf. Fig. 4.23). The large circles indicate the size of hard spheres having contact as nearest neighbours along <111> directions, thereby revealing that these most densely packed planes of the b.c.c. structure (the {110} planes) are less densely packed than those in close packed structures: three-sphere contacts do not occur (cf. Figs. 4.14b and 4.15)

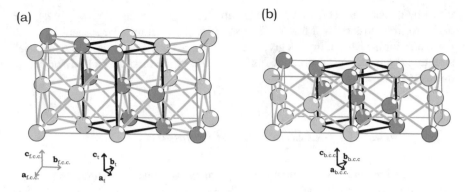

Fig. 4.25 Bain orientation relationship of f.c.c. crystal structure (**a**) and b.c.c. crystal structure (**b**). The dark grey coloured atoms indicate a close packed atomic plane parallel to $\{111\}$ in the f.c.c. crystal structure (**a**) and the corresponding most densely packed atomic plane parallel to $\{110\}$ in the b.c.c. crystal structure ((**b**); see also Fig. 4.24). The body centred tetragonal unit cell indicated with bold solid lines in (**a**) becomes a body centred cubic unit cell indicated with bold solid lines in (**b**) by reducing c_t down to $2^{-1/2}a_{f.c.c.}$, while keeping a_t and b_t fixed (b)

to a b.c.c. structure, and vice versa. For example, this orientation relationship is observed for the precipitation of nitrides as CrN, AlN, VN and TiN (all with f.c.c. translation lattice, as for rocksalt (NaCl)) in ferrite (b.c.c. iron; cf. Figs. 6.19 and 6.22). However, due to misfit-accommodation effects, the orientation relationship observed does not necessarily have to obey this Bain orientation relationship: see the discussion on the austenite (f.c.c.)-martensite (b.c.t.) transformation, for which a Bain *lattice correspondence* holds, but for which the resulting orientation relationship between product and parent phases, as a result of misfit-accommodation processes accompanying the phase transformation, deviates from a Bain orientation relationship (for further, clarifying discussion, see Sect. 9.5.2.2).

Consider two adjacent unit cells of an f.c.c. structure (Fig. 4.25a). Within this pair of adjacent unit cells, an alternative unit cell can be indicated having tetragonal-like unit-cell parameters (subscript "t"), with $a_t = b_t = 2^{-1/2}a_{f.c.c.}$ and $c_t = a_{f.c.c.}$. The new unit cell appears to be body centred, i.e. it contains an atom at the origin of the unit cell and an atom in the middle of the unit cell. Now assume that a_t and b_t remain fixed, and that c_t is reduced continuously. During this "deformation" process, the unit cell remains body centred tetragonal. If c_t has become equal to (i.e. has been reduced to) $2^{-1/2}a_{f.c.c.}$ (Fig. 4.25b), its length has become equal to the length $a_t = b_t$. Thereby the resulting unit cell now has dimensions characteristic for a body centred cubic crystal structure: $a_t = b_t = c_t = a_{b.c.c.} = 2^{-1/2}a_{f.c.c.}$.

However, in the above-described manner of "transformation" (i.e. if the above given relations for the lattice parameters hold), the hard sphere radius R of the atom would be reduced by 15% upon transformation from the f.c.c. structure to the b.c.c. structure. Already only on the basis of this recognition, it can be suggested that in order to compensate this compression (along the c axis), a_t and b_t will increase somewhat, while c_t decreases. Thus, one possibly arrives at a "compromise" structure given by a b.c.c. unit cell with lattice parameter $a_{b.c.c.}$ obeying $2^{-1/2}a_{f.c.c} < a_{b.c.c.} < a_{f.c.c.}$; for this possibility a further discussion of the relation of $a_{f.c.c.}$ and $a_{b.c.c.}$, in particular for the case of iron, is given in Sect. 4.2.5. Or the resulting "compromise" structure is a b.c.t. crystal structure, as in the case of the martensitic transformation discussed in Sect.9.5.2.2.

4.2.3 Further Crystal Structures of Elements

Among the known elements, many metals but in particular, the (few) non-metals exhibit crystal structures, which are different from the f.c.c., h.c.p. (and further close packed structures (cf. Table 4.4 and its discussion in Sect. 4.2.1.1), as well as derivative structures like those of mercury and indium (cf. Sect. 4.2.1.2)) and b.c.c. structures as dealt with in the preceding sections. Often one can understand an occurring deviation from typical metal crystal structures as consequence of a decrease of the degree of metallic bonding in association with a simultaneous increase of the degree of covalent bonding. This leads, in agreement with the overview provided by Table 4.2, to crystal structures exhibiting coordination numbers smaller than 8–12 as holds for typical metals. In particular for elements of group 13 (IIIA) and higher of the Periodic Table (see Fig. 2.9a) complicated and often rather unique crystal structures occur; most of these elements also exhibit allotropy (see Sect. 4.2.5).

In the following subsections, a few examples of such crystal structures of elements are presented, which have relevance for the field of materials science.

4.2.3.1 α-Polonium

The crystal structure of Po is dealt with here because it exhibits the simplest imaginable crystal structure: it is characterized by the primitive cubic unit cell; i.e. with one Po atom in the unit cell positioned according to the fractional coordinates x, y and z obeying $x = y = z = 0$ ($a = 3.359$ Å): α-Po (Fig. 4.26). Apart from α-polonium, some other elements show this crystal structure too but only at high pressures (this holds for, e.g. P, Sb, Bi and Te). Each Po atom in α-Po has six nearest neighbouring atoms (d(Po–Po) $= a = 3.359$ Å) coordinated in the form of an octahedron. The packing density calculated on the basis of a hard sphere model (i.e. the atoms touch each other along <100> directions) is $\pi/6$ ($= 52.4\%$). The lower packing density and the lower coordination number, as compared to the close packed crystal structures and the b.c.c. crystal structure, suggest that the bonding character in the α-Po structure has a covalent contribution.

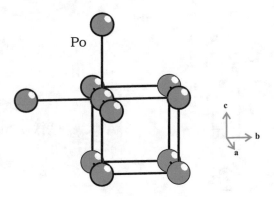

Fig. 4.26 Crystal structure of α-Po. The atoms in the primitive cubic unit cell with $0 \leq x, y, z \leq 1$ have been indicated; the unit cell net contains only one Po atom. Three additional Po atoms have also been depicted in order to show the sixfold coordination of one Po atom by six nearest neighbouring Po atoms (d(Po-Po) $= a = 3.359$ Å; the involved atoms have been highlighted by thick margins in the figure). The shortest Po-Po bonds correspond to the edges of the unit cell

4.2.3.2 Diamond

The diamond-type crystal structure is not only observed for the name-giving form of carbon, but also occurs for the following three elements of group 14 (IVA) of the Periodic Table (see Fig. 2.9a): Si, Ge, Sn (Fig. 4.27a). The *translation lattice* is face centred cubic with, as motif, one atom on (0, 0, 0) and one on (1/4, 1/4, 1/4). As described in Sect. 4.1.1, the fractional coordinates of the further (six) atoms in the unit cell result by adding (1/2, 1/2, 0), (1/2, 0, 1/2) and (0, 1/2, 1/2) to (x, y, z) of both atoms of the motif. So there are altogether eight atoms in the unit cell. The diamond-type crystal structure can also be conceived as generated by two "interpenetrating" f.c.c. *crystal structures* shifted along the body diagonal of the unit cell (parallel to a <111> direction) from (0, 0, 0) to (1/4, 1/4, 1/4).

Each atom is surrounded by four nearest neighbours at distances of $(3^{1/2}/4)a$ (highlighted by thick margins of the involved atoms in Fig. 4.27a). This can be regarded as the distance of two touching hard spheres. On that basis a packing density of $3^{1/2}\pi / 16 (= 34\%)$ can be calculated. This is exactly half of the value for the b.c.c. structure. The geometry of the diamond-type crystal structure is dominated by the optimization of covalent bonding, i.e. in case of carbon (diamond modification), each C atom is sp^3 hybridized (cf. Sect. 3.4) and forms electron-pair bonds with its four nearest neighbours at a distance of d(C–C) $= (3^{1/2}/4)a = 1.55$ Å.

The metallic character of the bonding of the atoms of the group 14 (IVA) elements in their crystal structures increases upon going from lower to higher periods. The most metallic element of group 14 is lead, which simply exhibits an f.c.c. type crystal structure. For Sn, in addition to the diamond-type crystal structure, an allotrope exists (Fig. 4.27b), which is, still, related to the diamond crystal-structure type (see Sect. 4.2.3.4).

(a)

(b)

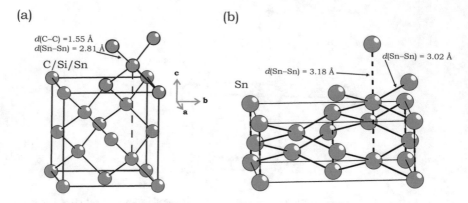

Fig. 4.27 a Diamond-type crystal structure ($a = 3.567$ Å for diamond itself), also observed for e.g. silicon ($a = 5.431$ Å) and α-tin ($a = 6.489$ Å). The atoms in the unit cell with $0 \le x, y, z \le 1$ have been indicated. Three additional atoms have also been depicted revealing the tetrahedral coordination of a given atom by the four nearest neighbouring atoms; the five atoms involved have been indicated by thick margins in the figure. **b** The tetragonal structure of β-tin (face centred tetragonal unit cell with $a = 8.426$ Å and $c = 3.182$ Å) shown selecting analogous atom positions as for (a) (plus an additional atom position) demonstrating the structural analogy. By the drastic change of the c/a ratio upon the transition from α- to β-tin, the coordination tetrahedra of the diamond-type crystal structure are strongly compressed, and, additionally, two atoms along the [001] direction come very close to the central atom in the considered tetrahedron (at a distance of 3.18 Å (the bonds indicated with dashed lines; one of these bonds is also shown in (**a**)) as compared to the interatomic distances pertaining to the bonds within the considered tetrahedron (equal to 3.02 Å, indicated with continuous lines), leading to an increase of the coordination number from 4 (**a**) to 6 (**b**)

4.2.3.3 Graphite

Graphite is another, covalent form of elemental carbon. At ambient temperatures and normal pressure ($= 1$ atm), graphite is more stable than diamond. In graphite, the carbon atoms are sp^2 hybridized (cf. Sect. 3.4), and form a honeycomb-*like* planar network (Fig. 4.28a).

In order to construct a three-dimensional graphite crystal structure, the planar layers described above are stacked parallel to each other in a laterally shifted fashion (Fig. 4.28b). Since the atoms are saturated within the layers with respect to covalent bonding, only relatively weak (van der Waals) bonding forces "glue" the layers together. Consequently, graphite crystals can easily be splitted parallel to and between the planar planes, whereas the forces needed to destroy the planar layers themselves are very much higher.

The most frequent form of graphite exhibits a hexagonal, AB-type of stacking of the planar layers (covalent nets), where similar considerations, concerning stacking sequences, hold for graphite as for closed packed structures (Fig. 4.28b; cf. Sect. 4.2.1.1). Here, the A, B (or C) positions refer to the centres of the hexagons in the planar layers. On this basis a three-dimensional, hexagonal unit cell results,

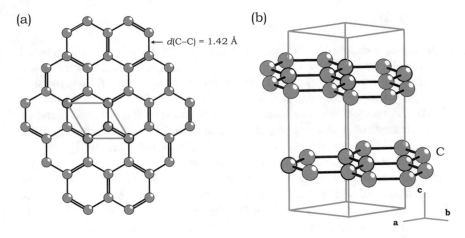

Fig. 4.28 **a** A (001) layer of graphite (covalently bonded carbon atoms; sp^2 hybridized) showing a honeycomb pattern. The unit cell with basis vectors **a** and **b** has been indicated. **b** The crystal structure of hexagonal (AB stacked) graphite ($a = 2.464$ Å, $c = 6.711$ Å). The unit cell has been indicated; the basis vectors **a** and **b** are the same as in (**a**). Atoms within the unit cell with $0 \leq x, y, z \leq 1$ have been highlighted by thicker margins

with, as motif, four carbon atoms at (x, y, z) according to $(0, 0, 1/4)$, $(0, 0, 3/4)$, $(1/3, 2/3, 1/4)$ and $(2/3, 1/3, 3/4)$.

The specific (atomic) volume of graphite (volume per carbon atom, to be calculated as the volume of the unit cell divided by the number of atoms in the unit cell) is significantly larger than that of diamond. This can be explained as a consequence of the weak forces which hold the layers together in the [001] direction (i.e. the direction perpendicular to the planar layers formed by strong covalent bonds in the layer plane). The pronounced difference of the atomic volumes of diamond and graphite $((V_{DIA} - V_{GR})/V_{GR} = -35\%))$ suggests that high pressures may serve to produce diamond from graphite. Diamond has a larger coordination number than graphite, in association with its smaller specific (atomic) volume. Yet, the shortest interatomic distance in diamond is larger than in graphite: d(C–C) in diamond equals 1.55 Å $(= (3^{1/2}/4)a)$; d(C–C) in graphite equals 1.42 Å $(= 3^{1/2}a)$. Regarding this paradox, see Sect. 4.2.5.

In contrast to the colourless, electrically non-conducting and hard diamond, graphite is black, conducting and quite soft. The electrical conductivity is strongly anisotropic: the electrical conductivity parallel to the (001) planes is much higher (of the order 1000 times) than perpendicular to these lattice planes (i.e. along the [001] direction). This is due to the non-localized nature of the $2p$ electrons not taken part in the sp^2 hybridization (cf. Sect. 3.4) of the carbon atoms in a (001) plane.

4.2.3.4 β-Tin

It was already mentioned in Sect. 4.2.3.2 that tin can crystallize in two different modifications (allotropes). Below 13 °C grey tin (α-Sn) is stable, which is a semiconductor with diamond-type structure (Fig. 4.27a). Above this temperature β-Sn is the stable allotrope, called white tin and which is metallic (Fig. 4.27b). The crystal structure of β-Sn can be conceived as obtained by drastic compression of the diamond-like crystal structure of α-Sn along one of the three cubic <100> directions (say, [001]) and by associated extension in directions perpendicular to the (say, [001]) compression direction (cf. Fig. 4.27a, b). As a result, the crystal structure of β-Sn is face centred tetragonal (the crystal structure can also be described by a smaller, body centred tetragonal unit cell; the orientation relationship between the f.c.c. unit cell of α-Sn and the b.c.t. unit cell of β-Sn can then be discussed in a way similar as in Sect. 4.2.2; cf. the face centred and body centred unit cells shown in Fig. 4.25). The compression along a <100> direction strongly deforms the tetrahedron of four nearest Sn neighbours around a given Sn atom in α-Sn (in α-Sn this tetrahedron is ideal) and the distance of the specified Sn atom to these four neighbouring Sn atoms is thereby increased from 2.81 to 3.02 Å (Fig. 4.27a, b). Due to the same compression, two Sn atoms, below and above the specified Sn atom (along [001]), approach this Sn atom at a distance almost equal to the nearest neighbour distance after the compression (3.18 Å vs. 3.02 Å; cf. Fig. 4.27b) so that the coordination number increases from four for α-Sn to six for β-Sn (or, more precisely, 4 + 2; cf. Sect. 4.2.4).

In view of its larger coordination, the density of β-Sn is larger than that of α-Sn. Yet, the shortest interatomic Sn–Sn distance occurs for α-Sn. Regarding this paradox, see Sect. 4.2.5.

As for the allotropes diamond and graphite, the specific (atomic) volume difference of the allotropes β-Sn and α-Sn is large ($(V_\beta - V_\alpha)/V_\alpha = -21\%$). In both cases, the large value of the specific volume difference is due to a change of the type of chemical bonding; in case of α-Sn → β-Sn, the nature of the bonding changes from predominantly covalent to predominantly metallic (see further Sects. 4.2.5 and 3.5).

4.2.4   The Coordination Number

Until now, the coordination number has been conceived as the number of nearest neighbours of an atom in the crystal structure concerned (cf. begin of Sect. 4.2). Thus, values for the coordination number are obtained as 12 for the close packed structures, f.c.c. and h.c.p. (cf. Fig. 4.18), and 8 for b.c.c. (cf. Fig. 4.23); see also Figs. 3.29–3.31. Obviously, as soon as the axial ratio c/a for the h.c.p. structure deviates from the ideal value $(8/3)^{1/2} \approx 1.633$ (cf. Sect. 4.2.1.3), the structure is not truly close packed and, as a consequence, the number of nearest neighbours then is reduced to 6 (these nearest neighbours occur in the same close packed plane as that of the atom considered). For h.c.p. materials with non-ideal c/a axial ratios, one is tempted to ignore the differences between the equal distances with the 6 nearest neighbour distances (in the same close packed plane) and the equal distances with the 3 (close packed plane above) plus 3 (close packed plane below) next-nearest neighbour distances, because these differences are very small in general, and thus, by and large, one accepts the coordination number to be equal to 12 for also non-ideal h.c.p. materials. If that is so, then there is also justification in ignoring the differences between the nearest and next-nearest neighbour distances for the b.c.c. structure, which would lead to a coordination number equal to 14 (instead of 8) for b.c.c. structures.

The above consideration has led to different proposals for the coordination number. For example, the above discussed ambiguity can be removed if a Wigner-Seitz cell is constructed around a considered atom, as described in Sect. 3.5.3.[16] A face of the Wigner-Seitz cell is at equal distances from the considered/central atom and from a nearest neighbouring atom in the direction perpendicular to that face. Thus, one can define the coordination number as the number of such neighbours, i.e. the coordination number is given by the number of facets of the Wigner-Seitz cell. This leads to values for the coordination number equal to 12 for the f.c.c. and h.c.p. structures both with ideal *and non-ideal c/a* axial ratios, and 14 for the b.c.c. structure.

In the past even more evolved concepts for the coordination number have been proposed, which, for example, recognize different types of neighbourship depending on the nearness of other atoms along the straight line connecting the atom concerned and its neighbour considered. However, such further developments are rather useless for crystal structures governed by non-directional types of atomic bonding; see what follows.

With reference to chemical bonding, the use of the notion coordination number tacitly assumes validity of a central force approximation for the interaction between the atoms in the crystal structure. This is a dangerous, in principle wrong, concept for crystals characterized by non-directional bonding (cf. Table 4.2), as holds, for example, for metals (see footnote 2 in Chap. 3). Therefore, in cases where metallic bonding prevails, the prevalence in nature for the f.c.c. or the h.c.p. atomic arrangement, both characterized by the same coordination number according to the above discussed first two concepts, should not be explained on the basis of next-next-nearest neighbour interactions, or so. The central force concept simply fails to explain such fine details for the case of metallic bonding (the energy differences between two crystal structures for the same metal are usually very small: of the order of one per cent of the bonding energy; cf. discussion in Sect. 3.5.3).

---

[16] A Wigner-Seitz cell contains one atom and Wigner-Seitz cells can be arranged such that they fill space completely. Hence, according to conjectures as those by Hooke and Haüy discussed in the begin of this chapter, the Wigner-Seitz cell can be considered as a building unit of a crystal (see Fig. 3.28). Yet, from a crystallographic point of view, a Wigner-Seitz cell is *not* a primitive unit cell because it is not defined by translation vectors of the lattice.

4.2.5 Polymorphism and Allotropy

Solid compounds and elements may occur as crystalline substances of a crystal structure that can be different as function of state variables, as temperature and pressure,
but also as function of the (e.g. thermal) "history" of the specimen concerned. The
occurrence of such different crystalline manifestations of a material is called polymorphism, and the different crystalline phases are called polymorphs. In the special
case that the crystalline material considered is composed of a single element, the terms
allotropy and allotropes can be used, instead of polymorphism and polymorphs. In
Sects. 4.2.3.2–4.2.3.4, the allotropes diamond and graphite, and the allotropes α-tin
and β-tin were discussed.

A well-known example of allotropy of great technological importance occurs in
the case of iron. At normal pressure (= 1 atm) and room temperature, iron has the
b.c.c. crystal structure (also called α-iron or ferrite). At 912 °C (Fig. 4.29), iron
experiences a solid-solid phase transition so that above that temperature iron has the
f.c.c. crystal structure (also called γ-iron or austenite). This γ-iron transforms into
(again) a b.c.c. crystal structure at 1394 °C. This high temperature b.c.c. modification
of iron is called δ-iron, which exists up to the melting temperature of iron at 1538 °C.
The phase transitions from allotrope to allotrope are reversible, i.e. they occur upon
both heating and cooling. Thus, which allotrope occurs (α, γ or δ) depends only on
the temperature (at normal pressure).

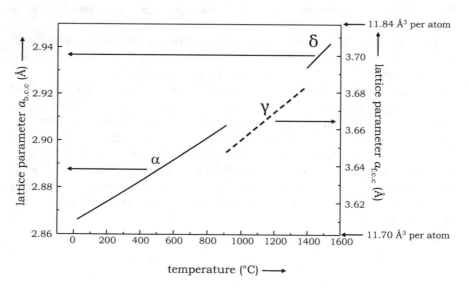

Fig. 4.29 Lattice parameters of the b.c.c. (α and δ) and f.c.c. (γ) allotropes of iron (note the two
different ordinates) as function of temperature at normal pressure (≈ 1 atm). The ranges of both
ordinates were chosen such that they correspond to the same range in specific (atomic) volume,
namely 11.70 to 11.84 Å³. Data taken from A.T. Gorton, G. Bitsianes, T.L. Joseph, Trans. AIME
233, 1519–1525 (1965)

Table 4.5 Changes of the lattice parameter, the shortest interatomic distance and the specific volume (i.e. volume per atom) for pure iron upon the two allotropic transitions b.c.c. $\alpha \to$ f.c.c. γ (at 912 °C) and f.c.c. $\gamma \to$ b.c.c. δ (at 1394 °C) at normal pressure ($\approx$ 1 atm)

	$\alpha \to \gamma$ (912 °C)			$\gamma \to \delta$ (1394 °C)		
	α	γ	Change (%)	γ	δ	Change (%)
Lattice parameter, a (Å)	2.895	3.637	+26	3.680	2.926	−20
Shortest distance, $2R$ (Å)	2.507	2.572	+2.6	2.602	2.534	−2.6
Volume per atom (Å^3)	12.13	12.03	−0.9	12.46	12.53	+0.5

The dependence of the lattice parameter of the three, cubic allotropes of iron is shown as function of temperature in Fig. 4.29. The lattice parameter a increases with temperature for each phase, α, γ and δ, which parallels the behaviour for copper as shown in Fig. 4.20. The specific volumes, expressed as the volume per iron atom (as can be calculated by dividing the unit cell volume by the number of atoms in the unit cell (i.e. four atoms for f.c.c. and two atoms for b.c.c.)), and the lattice parameters of the cubic crystal structures of the allotropes at the temperatures where the transitions $\alpha \leftrightarrow \gamma$ and $\gamma \leftrightarrow \delta$ occur, can now be discussed as follows (see also the discussion at the end of Sect. 4.2.2).

As follows from the data as shown in Fig. 4.29 and gathered in Table 4.5, the lattice parameters a for the α (b.c.c.) and γ (f.c.c.) phases, and the γ (f.c.c.) and δ (b.c.c.) phases, at the respective transition temperatures, differ considerably: differences of 20% or more. This numerical finding is by itself meaningless. A discussion of geometrical parameters of the crystal structures of allotropes, that provides physical insight, should recognize the numbers of atoms in the respective unit cells. The f.c.c. unit cell contains four atoms and the b.c.c. unit cell contains two atoms. Then it is no surprise that the f.c.c. unit cell is "larger" than the b.c.c. unit cell. Indeed, the consideration in Sect. 4.2.2 already indicated that $a_{b.c.c.} < a_{f.c.c.}$. The discussion in Sect. 4.2.2 also suggested that, for the allotropes involved, the hard sphere radius of the atoms should not change "too much" upon structure change. Indeed, the shortest interatomic distances in the respective crystal structures change only modestly upon structure change: only about 2.6%. An even smaller difference occurs for the specific volume, expressed as the volume per iron atom (see above): the average volume per iron atom equals $a_{f.c.c.}^3/4$ for the f.c.c. allotrope and $a_{b.c.c.}^3/2$ for the b.c.c. allotropes (these volume per atom data should not to be confused with the hard sphere volume data used in the calculation of the packing densities in Sects. 4.2.1.2 and 4.2.2). It follows that the differences in the specific volume for the allotropes involved are less than 1%. It can be concluded that in order to predict a value of $a_{b.c.c.}$ from a given $a_{f.c.c.}$, and vice versa, a viable approach is adoption of equality of the specific volumes (volumes per (iron) atom) of the allotropes involved.

In fact, for assessing magnitudes of lattice parameters of different polymorphs, the adoption of equal specific (atomic or molar) volumes always provides reasonable estimates of (relative values of) lattice parameters of different polymorphs, *as long as the type of chemical bonding does not change considerably upon the transformation* (in this context, consider again the allotropic transformation examples graphite $\leftrightarrow$ diamond in Sects. 4.2.3.2 and 4.2.3.3 and α-tin $\leftrightarrow$ β-tin in Sect. 4.2.3.4; in these cases the type of bonding changes upon allotropic transformation and the specific volume difference is large).

The very small changes in specific volume for the allotropic $\alpha \leftrightarrow \gamma$ and $\gamma \leftrightarrow \delta$ transitions for pure iron can be attributed to the higher packing density of the close packed, f.c.c. γ phase in comparison with the not so close packed, b.c.c. α and δ phases. The specific volume of f.c.c. γ-iron is smaller than that of b.c.c. α-iron and that of b.c.c. δ-iron (Table 4.5). However, surprisingly and counter-intuitively, for many metals which can crystallize in f.c.c. (/h.c.p.) and b.c.c. modifications, the not close packed b.c.c. modification exhibits the smaller specific (atomic) volume (see also footnote 6 in Chap. 8).

Perhaps at first sight (also) surprisingly, the smaller specific volume of the f.c.c. γ phase, as compared to both b.c.c. phases, is associated with larger shortest interatomic distances. This is a paradox, which can hold for polymorphic/allotropic phase transitions where the coordination number (defined as the number of nearest neighbours; i.e. 8 for α-iron and δ-iron and 12 for γ-iron) changes. An *increase* in coordination number can be associated with a *decrease* of atomic volume, but then is accompanied with an increase of the nearest neighbour distance. Apparently, if a larger number of bonds is established by an atom (i.e. increase of the coordination number), the strength of each single bond may decrease, which can lead to an increase of the bond length. Examples of the phenomenon discussed here are also provided by the allotropes graphite and diamond (Sects. 4.2.3.2 and 4.2.3.3) and the allotropes α-tin and β-tin (Sect. 4.2.3.4).

4.3   The Notions Alloy, Solid Solution, Ordered Solid Solution and Compound

A liquid or solid substance can consist of a mixture of atoms of different elements, which atoms can be in intimate contact in more or less specific arrangements. If the nature of this mixture is metallic, the substance is called an *alloy*. Such a liquid or solid substance/alloy can be homogenous or heterogeneous. If the substance is a crystalline solid, it may be composed of grains of the same crystal structure possibly but not necessarily of the same composition, or it can be composed of grains exhibiting different crystal structures possibly but not necessarily of different compositions.

Different types of atoms can occur together in a crystal structure. If atoms of more than one element can be present randomly distributed on the atomic sites of the same crystal structure, for a certain range in composition, the corresponding substance is called a *solid solution*. If for this crystal structure such random distribution does not occur and, instead, the different types of atoms occupy preferably element specific sites, for a certain range in composition, a so-called *ordered solid solution* results.

The degree of order is one (i.e. the ordering is perfect) if *all* atoms reside *only* at their preferred sites.

The notion *compound* indicates a substance of specific, so-called stoichiometric composition. If the compound is a crystalline solid, it may be composed of grains of the same crystal structure of the same composition, or it can be composed of grains exhibiting different crystal structures but of the same composition.[17] For a crystal structure of a compound, it holds that the different types of atoms all (ideally) reside on only element specific sites of the crystal structure.

True "line compounds", i.e. compounds of a prescribed, exact composition (in the binary case represented by a line in the phase diagram, e.g. see Figs. 7.17 and 7.19) do not occur in nature: it is invariably found that some compositional variation, albeit for a very or extremely small compositional range, always occurs. Thereby the distinction between a compound and a solid solution disappears.

Ionic crystalline compounds like NaCl (f.c.c. translation lattice; cf. Fig. 3.6 in Sect. 3.3) and CsCl (primitive cubic translation lattice; cf. Fig. 3.7 in Sect. 3.3) have already been introduced and discussed. In the following, the crystal structures of crystalline solid solutions and compounds will be discussed, which are built from two metals, or from a metal and a metalloid like N, C or O, and which in particular have crystal structures derived from the three main crystal-structure types presented above, i.e. face centred cubic, hexagonally close packed and body centred cubic.

4.4   Crystalline Solid Solutions and Compounds

4.4.1   Substitutional Solid Solutions

At room temperature and at 1 atm, pure solid iron has a body centred cubic crystal structure and pure solid aluminium exhibits a face centred cubic crystal structure. Now, at sufficiently high temperature and at 1 atm, add a small amount of aluminium to a melt of pure iron such that a liquid of a homogeneous mixture of iron and aluminium atoms results. Upon cooling down this liquid alloy, solidification occurs. The solid obtained (at room temperature and at 1 atm) does not consist of a small volume fraction of f.c.c. aluminium and a large volume fraction of b.c.c. iron. Instead, it is composed of one single b.c.c. solid solution. This can be demonstrated by performing an X-ray diffraction experiment (see Sect. 4.5): the diffractogram does not show the reflections of solid aluminium (face centred cubic) besides those of pure iron (body centred cubic), but only the reflections characteristic of a body centred cubic substance are observed. Analysis of the reflection positions reveals that the lattice parameter a is somewhat larger than the value expected for pure b.c.c. iron. This already suggests that the small amount of aluminium atoms has been incorporated in

---

[17] The last described situation happens at so-called phase boundaries and triple points of a single component system, with the combination of atoms in the chemical formula for the compound considered taken as the "component" (cf. Sect. 7.5.1).

the crystal structure pertaining to pure iron (at the same temperature and pressure) at sites otherwise occupied by iron atoms: the size difference between the aluminium and iron atoms causes the lattice parameter of the b.c.c. crystal to be different from that of pure b.c.c. iron (see below). This phenomenon is called a solid solubility of, in this case, aluminium in iron; iron is referred to as the solvent and aluminium as the solute.

Recognizing that in the crystalline solid solution the aluminium atoms reside at the same sites in the crystal structure as available for the iron atoms, it can be said, for, for example, the solid solution containing 5at.% Al (and consequently 95 at.% Fe) that there is a probability of 5% to find an aluminium atom at a specific site in the crystal structure and a probability of 95% to find an iron atom at the same site. In other words, the aluminium (and iron) atoms are "statistically distributed": a disordered distribution of the aluminium (and iron) atoms on the sites of the b.c.c. crystal structure occurs. Hence, the aluminium atoms can *substitute* the iron atoms in the crystal structure; one speaks of a *substitutional* solid solution.

The hard sphere radius of an aluminium atom is larger than that of an iron atom. Then, the size mismatch of the aluminium and iron atoms could be thought to have an effect on the positions of the atoms surrounding an aluminium atom, dissolved substitutionally on a site of the parent iron crystal structure, as indicated in Fig. 4.30b: the surrounding atoms are slightly pushed away from the aluminium atom considered. For a random distribution of the dissolved aluminium atoms on the parent iron crystal structure, an increase of the (average) lattice parameter would thus occur with increasing aluminium content. This agrees with the experimental finding (see above and Fig. 4.31). If the substitutional solute atom is smaller than the iron atom, a decrease of the (average) lattice parameter would occur (as holds for the solute Si, Fig. 4.30c and see also Fig. 4.31).

For certain combinations of elements, complete substitutional solid solubility on the same crystal structure may occur. Empirical, so-called Hume-Rothery rules actually formulated for alloys, can be given for the occurrence of complete solid solubility:

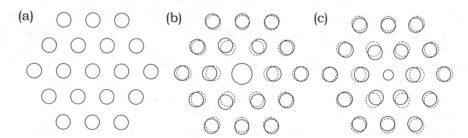

Fig. 4.30 Substitutional solid solutions. **a** Single element crystal, i.e. with identical atoms. **b** Substitution by a larger atom. **c** Substitution by a smaller atom. In the imaginary cases sketched, the circles drawn with full lines indicate the positions of solvent atoms surrounding the dissolved solute atom; the circles drawn with dashed lines represent their ideal positions before substitutional replacement of the solvent atom in the centre of the figure

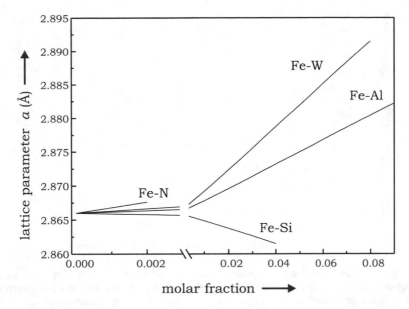

Fig. 4.31 Dependence of the lattice parameter of ferritic (i.e. b.c.c.) iron-based binary solid solutions as function of molar fraction solute for substitutionally dissolved tungsten, aluminium and silicon (cf. Fig. 4.30), and for interstitially dissolved nitrogen (cf. Fig. 4.37. Data taken from B. Predel, Landolt-Börnstein—Group IV Physical Chemistry, Numerical Data and Functional Relationships in Science and Technology (Springer, Berlin, 1994)and H.A. Wriedt, L. Zwell, Transa. AIME 224, 1242 (1962)

1. The crystal structures of the pure solid elements involved must be identical.
2. The size of the atoms of the elements involved should not differ more than, say, 15%.
3. The elements should not differ too much chemically (i.e. their electronegativities (cf. Sect. 3.2) should be not too different; else compound (e.g. intermetallic) formation with a crystal structure deviating from those of the crystalline pure elements involved may occur).

Whereas the first condition is a "conditio sine qua non" for the realization of complete solid solubility, satisfying the size-factor condition (2) may, in case of violation of condition (1), already lead to extended solid solubility observed for the so-called terminal solid solutions (i.e. the composition ranges of these solid solutions extend from the pure element (solvent) to some significant fraction of solute; see the discussion of "terminal solid solutions" in Sect. 7.5.2). Note that for this last possibility the size-factor condition (2) is only a necessary, not a sufficient condition.

In the case of complete solid solubility, the lattice parameter a can be determined for the whole range of composition. An example is shown for the f.c.c. Cu–Pd solid solution in Fig. 4.32. Evidently, the dependence of the lattice parameter on solute content is about linear. Such linearity has been observed for a number of systems where complete (mutual) solubility occurs, as well as for many terminal solid

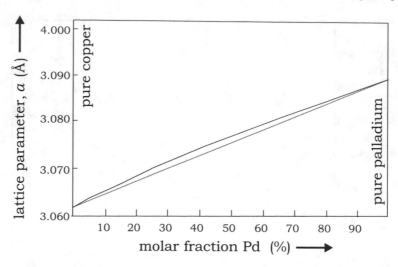

Fig. 4.32 Dependence of the lattice parameter of the solid solution Cu–Pd (face centred cubic crystal structure) on solute content. Note that the true relationship (bold line) is not exactly linear (dashed line) as prescribed by Végard's law. Data taken from B. Predel, Landolt-Börnstein—Group IV Physical Chemistry, Numerical Data and Functional Relationships in Science and Technology (Springer, Berlin, 1994)

solutions; it is called: *Végard's law*, which can be expressed for the solid solution A(B) as:

$$a_{A(B)} = a_A + \text{const.}\, x_B \tag{4.7}$$

with $a_{A(B)}$ as the lattice parameter of the solid solution, a_A as the lattice parameter of pure A (solvent) and x_B as the mole fraction B (solute). Végard's relation can be interpreted as the expression of the assumption that the lattice parameter of the solid solution is linearly dependent of the number of solute atoms in the unit cell. Indeed, for substitutional solid solutions, this leads to a linear dependence of the lattice parameter on the mole fraction solute (this becomes different for interstitial solid solutions, see Sect. 4.4.2). In view of the suggestion due to Fig. 4.30 and its discussion, it might be proposed that the constant in Eq. (4.7) (the slope of the straight line in plots as provided by Fig. 4.32) can be derived on the basis of the elastic deformation (cf. Chap. 12) induced in the parent crystal structure by the difference in size (mismatch) of the dissolved solute atoms and the solvent atoms (see the different positions of the circles with full lines and those with dashed lines in Fig. 4.30). However, this elastic, mechanistic approach to explain Végard's law quantitatively has been found to be in vain: elasticity theory cannot be applied to misfitting inclusions of atomic size[18]: electronic interactions play an important role (as well).

---

[18] However, elasticity theory has been applied successfully to describe quantitatively the change of the (average) lattice parameter of, and its variation in, a matrix containing misfitting precipitates

4.4.1.1 Ordering in Substitutional Solid Solutions; Occurrence of Superstructures

In the above discussion of substitutional solid solutions the distribution of atoms of the different elements simultaneously present in the same crystal structure was taken to be random, i.e. a so-called disordered substitutional solid solution was considered.

In a disordered solid solution, $A_x B_{1-x}$ (with $0 < x < 1$) each atom A and each atom B can have various different local surroundings, which combinations of atom considered with its surroundings (thus) can be of different energy (e.g. because of the differences in local distortions due to the different atomic radii of A and B; cf. Fig. 4.30b, c; however, as indicated above, on the atomic scale electronic interactions (usually) dominate elastic misfit effects). It may therefore be expected that certain, energetically favourable local arrangements are formed preferentially. In other words, a preferred occupation can occur of specific (types of) atom sites of the crystal structure by the atoms of each element concerned: an ordered solid solution is formed (see Fig. 4.33a, b). The structure resulting after such ordering is also called a *superstructure*. Some symmetry operations possible for the disordered structure are no longer possible after ordering (see further below).

The ordering process, involving that distinction (now) has to be made between the positions of the atoms of different elements in the crystal structure, implies that the size of a *primitive* unit cell of the Bravais translation lattice, pertaining to the disordered state, has to be increased in order to arrive at a primitive unit cell of the Bravais translation lattice of the ordered state: the unit cell of the superstructure thus is a *supercell* of the original Bravais translation lattice. This is the origin of the name *superstructure* (see also the description of a superstructure as a "commensurate

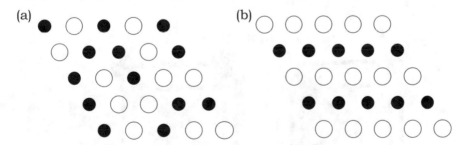

Fig. 4.33 Schematic presentation of **a** a disordered, substitutional solid solution and **b** an ordered, substitutional solid solution

(i.e. inclusions much larger than a single atom; see E.J. Mittemeijer, P. van Mourik, ThH de Keijser, Philos. Mag. A 43, 1157–1164 (1981), E.J. Mittemeijer, A. van Gent, Scripta Metallur. 18, 825–828 (1984) and J.G.M. van Berkum, R. Delhez, ThH de Keijser, E.J. Mittemeijer, Phys. Status Solidi (a) 134, 335–350 (1992)). An overview of changes of the overall lattice parameter in specimens by elastic accommodation of misfitting, coherent and incoherent (see at the end of Sect. 5.3) precipitates is provided by T. Akhlaghi, S.R. Steiner, E.J. Meka, Mittemeijer, J. Appl. Crystallogr. 49, 69–77 (2016). See also the *"Intermezzo: Coherent and incoherent interfaces versus coherent and incoherent diffraction"* at the end of Sect. 5.3

(compositional and/or positional) modulation" in Sect. 4.8). At the same time, it becomes clear that the Bravais translation lattice of the superstructure then is a *sublattice* of the original Bravais translation lattice.

A similar statement needs not to hold for *non-primitive* unit cells: in the following ordering is described on the basis of a non-primitive unit cell for the disordered state that allows specification of the type of ordering induced without changing the unit-cell parameters (i.e. the lattice parameters; cf. Table 4.1).

Various types of ordering occur in substitutional solid solutions. A (most) simple one pertains to β-brass (a Cu–Zn alloy of about 50 at.% Zn, hence, characterized by the chemical formula CuZn): At high temperature, β-brass is a disordered substitutional solid solution having a body centred cubic crystal structure, different from the crystal structures of pure copper (face centred cubic) and pure zinc (hexagonally close packed). Below 460 °C ordering occurs (the ordered state is also called β'-brass): then the atom positions at the origin (corner position(s)) of the unit cell and in the middle of the unit cell become occupied in an ordered fashion by Cu and Zn, respectively, or vice versa; see Fig. 4.34b (note that each corner position of the unit cell contributes 1/8 atom, thereby preserving the chemical formula CuZn; cf. Sect. 4.1.1 and the Appendix at the end of this chapter). In the ordered state, the unit cell is not body centred any more: it has become a primitive unit cell; indeed, the translation by (1/2, 1/2, 1/2) does not leave the structure unchanged as holds for a b.c.c. unit cell.

Three other well-known ordered arrangements (superstructures) occur departing from a face centred cubic crystal structure, as can be observed for Au-Cu alloy at the compositions Cu_3Au, CuAu and $CuAu_3$. The corresponding ordered atomic arrangements can be presented by starting with a unit cell as for the face centred cubic crystal structure and superimposing a specific distribution of the copper and gold atoms over the atomic sites of this crystal structure (see Fig. 4.35).

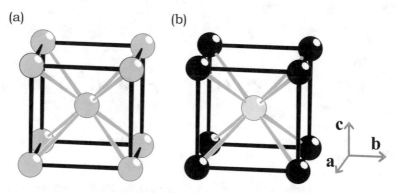

Fig. 4.34 a Unit cell of the disordered solid solution of β-brass (β-CuZn): b.c.c. crystal structure (b.c.c. Bravais translation lattice). The probability for each atom position to be occupied by either a copper atom or a zinc atom is 50%. **b** Unit cell of ordered state called β'-brass (β'-CuZn): CsCl crystal-structure type: primitive cubic Bravais translation lattice. The atom positions at the origin (corner position(s)) of the unit cell and in the middle of the unit cell are occupied in an ordered fashion by Cu and Zn, respectively, or vice versa

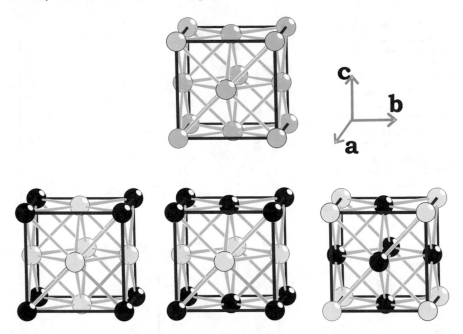

Fig. 4.35 Unit cells of the disordered, face centred cubic solid solution $Cu_{1-x}Au_x$ (top), and the ordered, primitive cubic solid solution Cu_3Au (bottom left), the ordered, primitive tetragonal solid solution $AuCu$ (bottom middle) and the ordered, primitive cubic solid solution $CuAu_3$ (bottom right)

The crystal structures of the ordered solid solutions $AuCu_3$ and $CuAu_3$ are very similar: their Bravais translation lattices are primitive cubic and they can be converted into each other by simply replacing copper atoms by gold atoms, and vice versa (cf. unit cells at bottom left and bottom right in Fig. 4.35). In these cases, i.e. the ordered solid solutions $AuCu_3$ and $CuAu_3$, the unit cell is not face centred cubic any more, as for the disordered case, it has become primitive cubic by the ordering: indeed, the translation of a gold atom at $x = 0$, $y = 0$, $z = 0$ by (1/2, 1/2, 0) (or by (1/2, 0, 1/2), or by (0, 1/2, 1/2)) does not leave the crystal structure unchanged: by these translations one arrives at a copper atom.

The Bravais translation lattice for the ordered solid solution $CuAu$ is primitive tetragonal. As drawn in Fig. 4.35 (bottom middle), the unit cell has cubic geometry: i.e. $a = b = c$ and $\alpha = \beta = \gamma = 90°$. This thereby is an example of the case discussed at the end of Sect. 4.1.2: for the tetragonal crystal system, with $a = b$ and $\alpha = \beta = \gamma = 90°$, it may happen that $a (= b) = c$ and yet the crystal structure cannot be assigned to the cubic crystal system. Only if the additional constraint $a (= b) = c$ is imposed *by the symmetry of the atomic arrangement*, cubic crystal symmetry occurs. Against this background, for the example discussed here, one often speaks, somewhat confusingly, of *pseudo-cubic* crystal symmetry. In reality, for the ordered solid solution, $CuAu$ $a (= b)$ is not exactly equal to c.

Fig. 4.36 X-ray diffraction patterns (calculated) of (top) disordered Cu_3Au (face centred cubic) and (bottom) ordered Cu_3Au (primitive cubic); Cu $K\alpha_1$ radiation. Note the emergence of additional reflections, the so-called superstructure reflections, which are extinguished in the disordered state (cf. footnote 22 in Sect. 4.5)

(a)

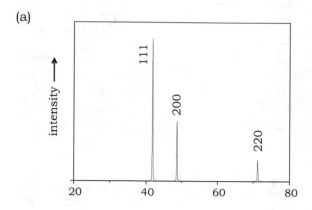

(b)

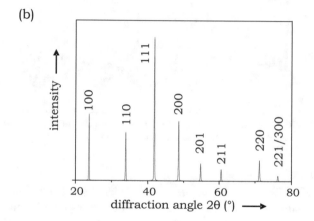

In (X-ray) diffraction patterns (cf. Sect. 4.5), the occurrence of ordering of the atoms of various elements in a crystal structure leads to the emergence of additional reflections, so-called *superstructure reflections*, as compared to the disordered state. For example, (100) and (110) (superstructure) reflections are observed for the ordered, primitive cubic solid solution $AuCu_3$, which reflections are absent for the corresponding, disordered, face centred cubic solid solution (cf. footnote 22 in Sect. 4.5); see Fig. 4.36.

4.4.2   Interstitial Solid Solutions

Iron can also form solid solutions/alloys by dissolving carbon or nitrogen. Since nitrogen and carbon atoms are distinctly smaller than iron atoms, the above discussion on substitutional solid solutions may induce the expectation that a decrease of the lattice parameter a of ferrite (b.c.c. iron) occurs upon dissolving carbon/nitrogen. However, the reverse is true: an increase of a is observed upon dissolving carbon/nitrogen, as shown for nitrogen dissolved in b.c.c. iron (ferrite) in Fig. 4.31. The explanation for this phenomenon is that nitrogen and carbon do not replace iron atoms, i.e. they do not substitute for iron atoms, but instead occupy *interstices* of the W type (b.c.c.) arrangement of iron atoms: such inclusion of carbon/nitrogen atoms expands the crystal structure of ferrite (cf. Fig. 4.37 for a schematic illustration of interstitial insertion of atoms into a host structure).

Also, for interstitially dissolved solutes, Végard relations for the dependence of the lattice parameter on solute content are often good approximations of reality (cf. Fig. 4.32 for nitrogen in ferrite). However, a subtlety should be recognized here. Assuming that the lattice parameter is linearly dependent on the number of solute atoms in the unit cell, leads for substitutional solid solutions to a linear dependence of the lattice parameter on the mole fraction solute (cf. Eq. (4.7)), but for interstitial solid solutions to a linear dependence of the lattice parameter on the number of solute atoms *per* solvent atom and therefore Végard's law should in the latter case be written as:

$$a_{A(B)} = a_A + \text{const.}\, x_B^r \qquad (4.8)$$

where x_B^r can be expressed as, for example, and as done in the literature, the number of interstitial, solute (B) atoms per 100 solvent (A) atoms. In many cases, the distinction of Eqs. (4.7) and (4.8) may be irrelevant in view of the first-order nature of the assumption that the lattice parameter is linearly dependent on the number of solute atoms in the unit cell. However, lattice parameters can be measured by diffraction methods (cf. Sect. 4.5) with very high precision and thus, for interstitial solid solutions, distinction of application of Eq. (4.7) and of Eq (4.8) can be of significance.

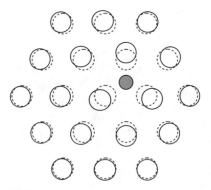

Fig. 4.37 Interstitial solid solution (cf. analogous figures for substitutional solid solutions: Fig. 4.30). The relatively small atom (grey in the figure) is positioned at an interstitial site of the single element parent crystal structure (cf. Fig. 4.30a). In the imaginary case sketched the dark circles drawn with full lines indicate the positions of solvent atoms surrounding the interstitially dissolved solute atom; the circles drawn with dashed lines represent their ideal positions before insertion of the solvent atom into the concerned interstitial site of the parent crystal structure

Interstitial solid solutions are commonly formed by transition metals upon dissolving metalloids as N, C, O and H. Not all binary solid phases constituted of a transition metal with one or more of the mentioned metalloids can be classified as interstitial solid solutions. For example, zirconium may dissolve oxygen up to a composition of about $ZrO_{0.4}$; this material is metallic and can be conceived as an interstitial solid solution. However, the oxidic compound ZrO_2 exists as well, is largely ionic, colourless and an electric insulator, and cannot at all be classified as an *interstitial* solid solution/compound.

Often the insertion of interstitials at interstices of the host (metal) crystal structure changes the crystal-structure type the host (metal) assumes. In the following interstitial phases based on closed packed and W type (b.c.c.) crystal structures are dealt with separately.

4.4.2.1 Interstitial Solid Solutions Based on Close Packed Crystal Structures

Two main types of interstitial sites can be discerned in close packed crystal structures: octahedral and tetrahedral interstitial sites. The location of these interstices can by illustrated considering two adjacent layers of close packed spheres (see Fig. 4.38). An octahedral interstitial site is surrounded by six spheres: three spheres of each of two adjacent close packed layers. A tetrahedral interstitial site is surrounded by four spheres: one sphere of one close packed layer and three spheres of the adjacent close packed layer. The overall three-dimensional arrangement of the octahedral and tetrahedral interstices, of course, depends on the stacking sequence of the close packed layers (cf. Sect. 4.2.1.1). However, for all closed packed structures, there are one octahedral interstitial site and two tetrahedral interstitial sites per closed packed sphere (atom).

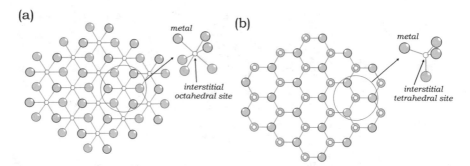

Fig. 4.38 Two close packed layers of spheres (grey) viewed along the layer-normal direction. The spheres have been depicted with reduced size with respect to the hard sphere radius to enhance clarity. The "upper" layer of spheres has been drawn with thicker margins. The small white spheres indicate the interstitial sites. **a** Octahedral interstitial sites. **b** Tetrahedral interstitial sites

The arrangement of the octahedral and tetrahedral interstices in the cubic closed packed f.c.c. structure is shown in Fig. 4.39 (see also Table 4.6).

If all octahedral sites are occupied by "solute" atoms a compound of NaCl rocksalt-type crystal structure results (per host, "solvent" atom/ion there is one octahedral interstice); in rocksalt the Cl anions constitute a cubic close packed arrangement with the Na cations occupying all octahedral interstices (or vice versa, because the arrangement of octahedral interstitial sites itself forms a Cu type f.c.c. crystal structure).

In an interstitial solid solution, the occupation of the octahedral interstices is only partial: TiN principally obeys the rocksalt crystal-structure prescription, but for a composition corresponding to about $TiN_{0.5}$ a partial occupation (of about 50%) of the octahedral interstices formed by the Ti f.c.c. host crystal structure is observed. In other cases, like for C or N austenite (f.c.c. Fe host crystal structure) only occupancies up to about 9% occur (cf. Fig. 9.22), corresponding to the compositions $FeC_{0.09}$ and $FeN_{0.09}$ (see also Sect. 9.5.2.1). To achieve larger amounts of dissolved interstitials in austenite, see the last part of the "*Intermezzo: Thermochemical surface engineering; nitriding and carburizing of iron and steels*" (at the end of this Sect. 4.4.2).

Occupation of the tetrahedral interstitial sites is much less common than occupation of the octahedral interstitial sites. If all tetrahedral interstitial sites are occupied, the CaF_2 type crystal structure results (with the Ca cations constituting the f.c.c. host crystal structure with the F anions occupying all, four tetrahedral interstices of the unit cell; per host, "solvent" atom/ion there are two tetrahedral interstices).

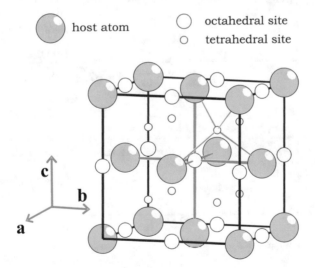

Fig. 4.39 Octahedral and tetrahedral interstitial sites (indicated by large and small white spheres, respectively) in an f.c.c. crystal structure (with atoms indicated by large grey spheres). For one octahedral interstice and one tetrahedral interstice, the distances to the nearest surrounding six (octahedral interstice) or four (tetrahedral interstice) surrounding host atoms have been indicated (see also Table 4.6)

Table 4.6 Geometrical features of the f.c.c., h.c.p. and b.c.c. crystal structures, and their interstitial sites, conceiving the host (metal) atoms as hard solid spheres

	Cu type (f.c.c.)	Mg type (h.c.p.)	W type (b.c.c.)
Lattice, lattice parameters and fractional coordinates	Face centred cubic a M: 0 0 0 (M: 1/2 1/2 0)[a] (M: 1/2 0 1/2) (M: 0 1/2 1/2)	Primitive hexagonal a, c M: 0 0 0 M: 1/3 2/3 1/2[b]	Body centred cubic a M: 0 0 0 (M: 1/2 1/2 1/2)[a]
Lattice parameters in terms of the radius of ideal, hard solid spheres	$a = R \times 2\sqrt{2}$	$a = 2R$ $c = R \times 4\sqrt{2/3}$	$a = R \times 4\big/\sqrt{3}$
Packing density of hard solid spheres	$\sqrt{2}\pi\big/6 \approx 0.74$	$\sqrt{2}\pi\big/6^{c} \approx 0.74$	$\sqrt{3}\pi\big/8 \approx 0.68$
Number of octahedral interstices per close packed atom	1	1	3
Coordinates of octahedral interstices	1/2 0 0 (0 1/2 0) (0 0 1/2) (1/2 1/2 1/2)	2/3 1/3 1/4 2/3 1/3 3/4	1/2 0 0 (x) 0 1/2 0 (y) 0 0 1/2 (z) (1/2 1/2 0 (x))[a] (1/2 0 1/2 (y)) (0 1/2 1/2 (z))
Shortest distance between two octahedral interstices	2R	$R \times 2\sqrt{2/3} = c/2$ $\approx 1.63R$	$R \times 2\big/\sqrt{3} \approx 1.15R$
Distance of the centre of an octahedral interstice to next host atoms	$R \times \sqrt{2}$ $\approx 1.41R$ (6×)	$R \times \sqrt{2} \approx 1.41R$ (6×)	$R \times 2\big/\sqrt{3} \approx 1.15R$ (2×) $R \times 4\big/\sqrt{6} \approx 1.63R$ (4×)
Number of tetrahedral interstices per closed packed atom	2	2	6

(continued)

Table 4.6 (continued)

	Cu type (f.c.c.)	Mg type (h.c.p.)	W type (b.c.c.)
Coordinates of tetrahedral interstices	1/4 1/4 1/4 3/4 3/4 3/4 (..)[a]	0 0 3/8 0 0 5/8 1/3 2/3 1/8 1/3 2/3 7/8	1/4 0 1/2 3/4 0 1/2 1/2 1/4 0 1/2 3/4 0 0 1/2 1/4 0 1/2 3/4 (...)[a]
Shortest distance between two tetrahedral sites	$R \times \sqrt{2}$	$c/4 = R \times \sqrt{2/3}$ $\approx 0.41R$	$R \times \sqrt{2/3} \approx 0.41R$
Distance of the centre of a tetrahedral interstice to next host atoms	$\sqrt{6}/2\,R \approx 1.22R$ (4×)	$\sqrt{6}/2\,R \approx 1.22R$ (4×)	$\sqrt{5/3}\,R \approx 1.29R$ (4×)

[a]The sites in brackets are automatically generated by the information about the non-primitive Bravais lattice type
[b]Note that here for convenience one M atom was set on the origin of the unit cell. The more common choice is to set the metal atoms on 1/3 2/3 1/4 and 2/3 1/3 1/4, and the octahedral sites on 0 0 0 and 0 0 1/2
[c]For ideal axial ratio (cf. Sect. 4.2.1.3)

Interstitial compounds with complete occupation of the tetrahedral interstitial sites are, for example, UN_2 and ScH_2 (These phases can also exhibit partial occupation of the tetrahedral interstices corresponding to interstitial solid solutions UN_{2-x} and ScH_{2-x}).

Whether metalloid atoms occupy tetrahedral or octahedral interstices in a metal host crystal structure depends to a large extent on the size of the interstitial atoms relative to the size of the metal atoms (again it is referred here to the ambiguity met in defining sizes for atoms and ions; see footnote 12 in Chap. 3). The following considerations and values apply to all close packed structures. The size of the interstitial sites (voids) can be quantified in terms of the maximum radius r of an interstitial atom which can be placed into such a void, without distorting it, i.e. while the host (metal) atom spheres (of radius R) still touch each other. It follows that the ratio of the radius of an interstitial atom on an *octahedral* interstitial site and the radius of the host (metal) atom amounts to (see Fig. 4.40):

$$\frac{r}{R} = \sqrt{2} - 1 \approx 0.41$$

and similarly it is obtained for the ratio of the radius of an interstitial atom on a *tetrahedral* interstitial site and the radius of the host (metal) atom

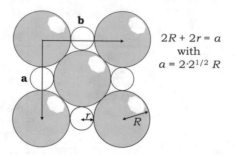

Fig. 4.40 One atomic layer parallel to (001) (unit-cell face) of the f.c.c. crystal structure (close packed; touching solid spheres (grey circles) of radius R) with the octahedral sites occupied by spheres (white circles) of maximum radius equal to $r = \left(\sqrt{2} - 1\right) R \approx 0.41 R$ (cf. Fig. 4.39)

$$\frac{r}{R} = \frac{\sqrt{6}}{2} - 1 \approx 0.22.$$

The above results suggest that "smaller" interstitial atoms could tend to occupy tetrahedral interstices, whereas "larger" interstitial atoms could tend to occupy octahedral interstices. Nitrogen, carbon and oxygen usually occupy octahedral interstitial sites in metallic interstitial solid solutions, whereas in particular hydrogen occupies (also) tetrahedral interstitial sites. Only for uranium (large R!), an interstitial solid solution UN_{2-x} is known for which tetrahedral interstitial sites are occupied by nitrogen (see above). Note that the corresponding composition already indicates that it cannot be realized by only occupation by nitrogen of octahedral interstitial sites (only one octahedral interstitial site per host atom; see above). However, a NaCl rocksalt-type UN exists as well (i.e. occupation of all octahedral interstitial sites of the host f.c.c. crystal structure constituted by the U atoms). These observations indicate that uranium has an atomic radius at about the border where apart from octahedral (for smaller R) also tetrahedral interstitial sites (requiring larger R) can be occupied by nitrogen.

If the interstitial sites are only partially occupied (i.e. the lattice formed by only the interstitial sites is occupied by interstitial atoms and *vacancies*) ordering of the interstitial atoms, on the interstitial site lattice, can occur, similar to the ordering observed for substitutional solid solutions. An example is shown in Fig. 4.41: the compound Fe_4N can be conceived as derived from an f.c.c. crystal structure of iron with occupation of the octahedral interstices in an ordered manner by nitrogen such that only 25% of the octahedral interstices is occupied. In Fig. 4.41, the nitrogen atom has been put at the octahedral interstice in the centre of the unit cell (there are four octahedral interstices per unit cell). Evidently, the ordering of the nitrogen atoms causes the f.c.c. translation lattice of the iron host lattice to change into a primitive cubic translation lattice.

The three-dimensional arrangement of the octahedral interstitial sites in the h.c.p. crystal structure is shown in Fig. 4.42. If all octahedral interstitial sites are occupied, the NiAs structure type results: the arsenic atoms form an h.c.p. crystal structure and the nickel atoms occupy all octahedral interstitial sites (Fig. 4.42). Like NaCl,

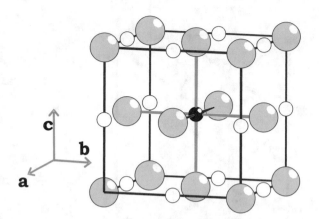

Fig. 4.41 Ordered occupation by nitrogen atoms of 1/4 of the octahedral interstitial sites in an f.c.c. arrangement of iron atoms pertaining to the compound γ'-Fe$_4$N. The nitrogen atom (black circle) has been put at the octahedral interstice in the centre of the unit cell. The iron atoms have been indicated by grey circles; the unoccupied octahedral interstices have been depicted by white circles. Note that due to the ordering of the nitrogen atoms the crystal structure no longer has a f.c.c. translation lattice, but has a primitive cubic translation lattice, like ordered Cu$_3$Au (see Fig. 4.35)

as derived from an f.c.c. host crystal structure, NiAs, as derived from an h.c.p. host crystal structure, cannot be regarded as an interstitial solid solution; it is a compound.

The three-dimensional arrangement of the octahedral interstitial sites in the h.c.p. crystal structure exhibit a feature different from the f.c.c. crystal structure. Many transition-metal nitrides, carbides and some oxides form crystal structures related to the NiAs crystal structure, but in these cases, only up to 50% of the octahedral interstices are occupied. This can be understood as follows. Whereas, in terms of the host atom size, in the f.c.c. structure, the shortest distance between two octahedral interstitial sites corresponds to $2R$ (equal to the distance between two host atoms), the shortest distance between two octahedral interstitial sites in the h.c.p. structure, occurring along the [001] direction, is much smaller: $c/2 = 2\sqrt{2/3}R \approx 1.63R$ (cf. Table 4.6). This latter distance between two neighbouring octahedral sites can in many cases be too small for simultaneous occupation of these neighbouring octahedral interstices. If the octahedral interstices, in the "chains" of octahedral sites running parallel to the [001] direction, are alternately occupied and unoccupied (cf. Figure 4.42), the resulting, frequently observed, occupancy of the octahedral interstitial sites is 50% (corresponding to a composition indicated by MX$_{0.5}$ or M$_2$X, with M being a transition metal and X a metalloid).

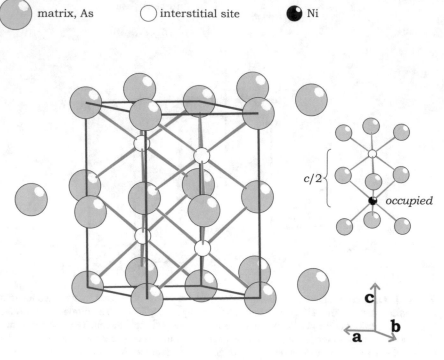

Fig. 4.42 Octahedral interstitial sites in an h.c.p. crystal structure. The unit cell drawn contains two (host) atoms and two octahedral interstices. Full occupation of the octahedral interstitial sites occurs in case of the NiAs crystal structure: the As atoms form an h.c.p. crystal structure and the Ni atoms occupy all octahedral sites. The relatively short distance between neighbouring octahedral interstices along the [001] direction (given by $c/2 \approx 1.63R$; see text and Table 4.6) brings about that for many transition-metal carbides and nitrides the octahedral interstices, in the "chains" of octahedral interstices running parallel to the [001] direction, are alternately occupied and unoccupied, as indicated at the right side in the figure

The tetrahedral interstitial sites in the h.c.p. crystal structure can be occupied in particular by hydrogen atoms (interstitial hydrides), but similar restrictive occupation rules as discussed above for the octahedral interstitial sites hold for the tetrahedral interstitial sites as well.

4.4.2.2   Interstitial Solid Solutions Based on the b.c.c. Structure

The geometry of the interstitial sites in the b.c.c. structure is more complicated than for the close packed structures. Firstly, the interstitial sites are not *perfectly regular* octahedra or tetrahedra: the octahedral and tetrahedral interstices are *distorted* (see Fig. 4.43), in contrast with the perfectly regular octahedral and tetrahedral interstices occurring in the close packed structures. Secondly, there are three octahedral and six

(a)   (b)

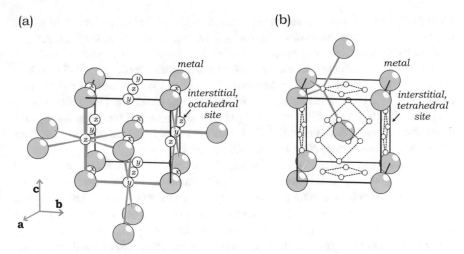

Fig. 4.43 a Octahedral interstitial sites (types x, y, and z) and **b** tetrahedral interstitial sites in a b.c.c. structure. In particular note in (**a**), the very short distances between two (opposite) host atoms of the six host atoms surrounding an octahedral interstice, which have been highlighted in the figure by thicker "bonds" for some selected octahedral interstices; these shortest distances are equal to $a/2 = (2/\sqrt{3})R \approx 1.15R$ (cf. Table 4.6)

tetrahedral interstitial sites per host atom, in contrast with one octahedral and two tetrahedral interstitial sites per host atom in the close packed structures (see also Table 4.6). This leads to smaller shortest distances between neighbouring octahedral and tetrahedral interstitial sites than for the close packed structures (apart from the shortest distances between neighbouring tetrahedral interstices for the h.c.p. and b.c.c. structures, which are equal). This already suggests, in the sense of the discussion in Sect. 4.4.2.1, that simultaneous occupation of such neighbouring interstitial sites is less likely. Indeed, no b.c.c.-derived crystal structures are known, in which all the octahedral or all the tetrahedral sites are occupied simultaneously. On the contrary, usually only a very small fraction of interstitial sites is occupied in solid solutions based on the b.c.c. structure.

Upon inspection of the *shortest* distances from the centres of the interstitial sites to the centres of the surrounding host atoms in the b.c.c. structure, it becomes evident that, in this sense, the tetrahedral sites are larger than the octahedral interstitial sites, which contrasts with the close packed structures: for the b.c.c. structure, the shortest distances to the host atoms are $1.29R$ for the tetrahedral interstices but only $1.15R$ for the octahedral interstices (allowing maximum interstitial atom radii r of $0.29R$ for tetrahedral interstitial sites and of $0.15R$ for octahedral interstitial sites, see Table 4.6). Yet, metalloids like C and N usually occupy the octahedral sites of the b.c.c. metal (M) host structure. This can be understood by the fact that insertion of a metalloid X on an octahedral site is possible by significant displacement of only two (of original distance X-M of $1.15R$) of the six surrounding metal atoms constituting the octahedron; the other four surrounding metal atoms (of original distance X-M of 1.63R) come a little closer to the interstitial atom (see Fig. 4.44); for further discussion, see Sect. 9.5.2.1.

For a b.c.c. crystal, three kinds of octahedral interstitial sites can be distinguished: the x, y and z sites differ by the orientation of the axis formed by the pair of closest host (metal) atoms constituting the octahedron (i.e. the M atoms with the original distance X-M equal to 1.15R); this axis can be oriented along either the [100] direction (*x site*) or the [010] direction (*y site*) or the [001] direction (*z site*) (see Fig. 4.43a; such distinction of octahedral interstitial sites does not occur for close packed structures where all octahedral interstices are perfectly regular).

This distinction of three types of octahedral interstitial sites for a b.c.c. crystal plays a special role for octahedral interstitial site occupation in the presence of many interstitials and leads to tetragonality of the crystal structure as follows. The distortion field around an interstitial in the considered b.c.c. crystal is of tetragonal nature. In the presence of many interstitials, the elastic strain energy can be minimized by alignment of the tetragonal distortion fields of neighbouring interstitial atoms. As a result, upon realization of a high concentration of interstitials, the initially b.c.c. lattice cannot maintain its cubic nature and becomes on average tetragonal: a body centred tetragonal lattice containing a relatively large amount of interstitial atoms on preferably one of the three types of octahedral interstices. The preference of the interstitial atoms for only one of the three types of octahedral interstitial sites is an ordering phenomenon, albeit it is recognized that the interstitials are distributed randomly on the preferred type of interstitial sites. This type of ordering is referred to as *Zener ordering*. In case of an iron host lattice and carbon and/or nitrogen as interstitials, the resulting b.c.t. crystal structure is called *martensite*. For further discussion, see Sect. 9.5.2.1.

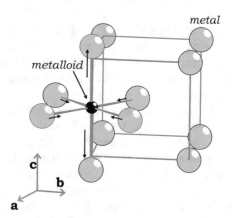

Fig. 4.44 Characteristic displacements of the host (metal) atoms around a metalloid (like carbon or nitrogen) occupying a *z*-type octahedral interstitial site in a b.c.c. metal structure (cf. Fig. 4.43a). Note that these displacements make the octahedron less irregular (see also Sect. 9.5.2.1)

Intermezzo: Thermochemical Surface Engineering;
  Nitriding and Carburizing of Iron and Steels

Interstitial solid solutions on the basis of iron (as solvent; host lattice) and carbon and nitrogen (as solutes; at interstices of the host lattice), and compounds derived thereof, play a great role in technology.

The importance of martensite as a bulk phase for the hardenability of (carbon) steels is discussed in Sect. 9.5.2 (see, especially, the *"Intermezzo: The hardness of iron-based interstitial martensitic specimens"* and the *"Intermezzo: Tempering of iron-based interstitial martensitic specimens"*).

Often the surface of a workpiece is most severely loaded (mechanically and/or chemically) during application: wear, corrosion and fatigue properties are strongly dependent on the surface quality. This recognition gives rise to the thought that it makes sense to enhance the (mechanical and/or chemical) strength of, selectively, the surface region of the workpiece. Accordingly one speaks of surface engineering and surface engineered materials. Thermochemical methods widely used to this end are nitriding and carburizing involving that from a surrounding (e.g. gaseous, salt or plasma (a plasma is an ionized gas)) medium, nitrogen or carbon can diffuse into (and possibly react with components in) the surface region of the iron-based workpiece.

Nitriding is applied to ferritic iron-based materials (ferrite = α-Fe = b.c.c. Fe; cf. Sect. 4.2.5), usually at temperatures below 600 °C and at a pressure of 1 atm. Upon nitriding a *compound layer* of iron nitrides can form at the surface$_x$. These nitrides are based on close packed iron lattices: γ'-Fe$_4$N$_{1-x}$, based on a f.c.c. arrangement of iron atoms, with an ordered distribution of nitrogen atoms on octahedral interstices (the ideal crystal structure of this compound for the composition Fe$_4$N has been discussed above (see Fig. 4.41)), and ε-Fe$_2$N$_{1-z}$, based on a h.c.p. arrangement of iron atoms, with nitrogen atoms distributed over the octahedral interstices in a more or less ordered way. This rather hard compound layer can improve the resistance against wear and corrosion appreciably.[19] Under the compound layer, a so-called *diffusion zone* develops during nitriding, where nitrogen has diffused into the ferritic matrix. Upon nitriding of ferritic steels containing alloying elements with (chemical) affinity for nitrogen, as Cr and Al, the nitrogen in the diffusion zone will not stay in solid solution, but it will precipitate: in the presence of Cr as the nitride CrN (rocksalt crystal-structure type), in the presence of Al as the nitride AlN (rocksalt crystal-structure type (cubic) or wurtzite crystal-structure type (hexagonal); see the *"Intermezzo: Nucleation of AlN in Fe-Al alloy"* in Sect. 9.2), and in the presence of both Cr and Al as the mixed nitride Cr$_x$Al$_{1-x}$N (rocksalt crystal-structure type). The rocksalt crystal-structure type nitrides, at least initially, are largely coherent (cf. Sect. 5.3) with the matrix and give rise to very pronounced hardening. Because of the tendency to volume expansion by the precipitation of the nitrides in the diffusion zone, a distinct compressive, internal, residual (cf. Sect. 12.18) stress parallel to the surface develops in the diffusion zone. Both the combination of high hardness and the compressive nature of the internal, residual stress parallel to the surface cause a large increase of the fatigue strength, which effect is discussed in more detail at the very end of this book (Chap. 12; *"Epilogue: The essence of materials science; optimizing the fatigue strength of ferritic steels by nitriding"*). The possibility for great improvement of material properties as diverse as tribological, corrosion and fatigue properties, by application of tuned nitriding treatments, has made nitriding the most versatile surface engineering method of our times.

Carburizing is applied to ferrous alloys to enhance the carbon content in the surface layer, usually at temperatures around 900 °C and at a pressure of 1 atm where the matrix is austenitic (austenite $= \gamma$-Fe $=$ f.c.c. Fe; cf. Sect. 4.2.5). Upon quenching to low temperature (e.g. room temperature or below that), a high-carbon martensite (see text immediately above this *intermezzo*) develops in the surface layer which has a high hardness (the hardness of Fe–C martensite increases with carbon content (see the *"Intermezzo: The hardness of iron-based interstitial martensitic specimens"* and in particular Fig. 9.33 in Sect. 9.5.2)). As the above discussion suggests, and in contrast with nitriding, carburizing involves the development of a diffusion zone only; a compound layer does not occur (i.e. the composition of the carbon-delivering medium is such that no iron-carbon compound, e.g. cementite (Fe_3C), develops at the surface). The favourable properties of the carburized zone adjacent to the surface are not only due to its high hardness. As holds for the diffusion zone produced by nitriding, a distinct, compressive, internal, residual stress parallel to the surface occurs in the carburized zone, due to the tendency to volume expansion of the surface layer upon martensite formation from austenite upon quenching (cf. Sect. 12.18). Both the high hardness and the compressive nature of the internal, residual stress parallel to the surface in the carburized zone can cause a pronounced increase of the fatigue strength.

In recent years, considerable attention has been paid to metastable solid solutions of C and/or N in austenite containing substitutionally dissolved elements having (chemical) affinity for C and/or N, as Cr (i.e. these elements tend to form carbides and/or nitrides). The amount of interstitials incorporated in these materials can be as large as 25 at.% (the maximal carbon and nitrogen solubilities in pure f.c.c. iron are about 9 at.%; cf. Sect. 4.4.2.1); this strongly enhanced interstitial solubility has led to the name "expanded austenite" for the phase produced, recognizing the pronounced increase of lattice parameter by dissolved interstitial components (cf. Eq. (4.8)). Such interstitial solid solutions can be produced upon carburizing/nitriding of Fe-based austenitic alloys containing Cr, in gaseous atmospheres, or plasmas thereof, comprising carbon and nitrogen containing components. The treatment temperature should be sufficiently low (say, about 450 °C; see what follows), and therefore, to stabilize the austenite at such low temperatures (and at a pressure of 1 atm), additional, austenite stabilizing, alloying elements, as Ni, should be present, as holds for austenitic steels. Apparently, at the treatment temperature, the diffusitivity of the substitutional solute, as Cr, is that small, also as compared to the diffusitivity of the interstitial solutes C and/or N, that carbide/nitride formation does not occur (if the treatment time does not surpass some limiting value): the C and/or N atoms and the Cr atoms remain in solid solution, but they may show some form of association, e.g. as exhibited by a short-range ordering (i.e. the Cr atoms are surrounded/associated with more C and/or N atoms than expected on the basis of a random distribution of the interstitials (cf. footnote 5 in this chapter)). The resulting interstitial solid solution exhibits (very) hgh hardness and possesses a high corrosion resistance. The large uptake of interstitials in the diffusion zone adjacent to the surface induces a strong tendency to volume expansion of the surface layer and a high compressive, internal, residual stress parallel to the surface can develop: compressive stresses as large as a few GPa can occur. Such stresses make likely that plastic accommodation processes in the diffusion zone can occur. Indeed the surface adjacent "expanded austenite" zone exhibits a high density of stacking faults, dislocations and (micro)twins (for description of these microstructural phenomena, see Chap. 5). This surface treatment has been successfully applied to austenitic steels. The discovery of these metastable solid solutions and in particular their (commercial) application is due to B. H. Kolster in the first half of the eighties of the passed century. This surface engineering method has accordingly been named *Kolsterizing* and more recently is denoted as *S-phase surface engineering* (the unfortunate name "S-phase" is a remnant of the time where no satisfactory understanding of the nature of this phase existed in the scientific literature. As a fine point, it can be remarked that Kolster, already in the first half of the eighties of the passed century, had a clear and correct, albeit crude, interpretation of the "expanded austenite" developing upon "Kolsterizing", as demonstrated, at the time, in discussion with the author of this book). Thermochemical surface engineering is discussed at more length in a recent book edited by Mittemeijer and Somers (2015).

One reason to include this *intermezzo* at this place in this book is to make clear, already at this stage, how important in materials science and engineering non-equilibrium states of matter, and their control, are (see, especially, footnote 19 and the above paragraph; further see, in particular, Chap. 9).

---

[19] Here, the following remark is in order. At the applied nitriding temperature (<600 °C) and pressure (usually 1 atm), these iron-nitride phases are not equilibrium phases: they are prone to decomposition in iron and nitrogen gas (see the *"Intermezzo: The Fe–C and Fe–N phase diagrams"* at the end of Sect. 9.5.2.1). As a consequence these iron-nitride compound layers can contain an amount of porosity due to the precipitation of nitrogen gas during nitriding, particular in the "oldest" part of the compound layer (i.e. the surface adjacent part). Such porosity generally has a negative effect on the mechanical properties (perhaps with exception for the case of friction under lubrication). This leads to dedicated nitriding treatments to minimize or even to avoid this porosity, or one removes

4.4.3 Crystal Structures of Further Materials

The focus in the preceding Sect. 4.4.2 has been on the crystal structures of rather simple, but very important, solid solutions and on the crystal structures of a few related compounds. Of course, a myriad of crystal structures occurs and has been found for different types of chemical compounds and solid solutions.

Rocksalt, NaCl, of which the crystal structure has been referred to and discussed at several places before, belongs to the large group of ionic compounds exhibiting a great variety of crystal structures. The crystal structures of ionic compounds can to a large extent be understood by an optimization of the Coulomb interactions of the constituting ions (atomic ions, like Na^+ or O^{2-}, or molecular ions, like NH_4^+ or SO_4^{2-}); this has been discussed to some extent in Sect. 3.3, where, apart from the rocksalt crystal structure also a few other crystal structures of simple ionic compounds have been given.

Ceramic compounds as Al_2O_3 and SiO_2, and materials derived thereof, can have chemical bonding intermediate between ionic and covalent. Their crystal structures are characterized by the attempt to optimize fulfilment of both local, directed (covalent) bonding of the atoms and undirected (orientation independent) Coulomb interactions (cf. Sects. 3.3. and 3.4).

Then, there is the large group of intermetallic compounds characterized by a largely metallic bonding and exhibiting a great variety of crystal structures, which often can be understood on the basis of geometric principles (cf. the treatment in Sect. 4.2.1).

The crystal structures of materials belonging to the above classes may be relatively simple, but may reach as well a high degree of complexity, as expressed by the occurrence of unit cells containing up to more than 1000 atoms. Anyhow, the formal concepts described in Sect. 4.1 can be applied to all crystal structures, irrespective of their complexity. More information on the crystal-structure building principles and, more generally, crystal chemistry (i.e. the relation between crystal structure and chemical bonding) for different types of material classes is offered in, e.g. Chaps. 7–9 in Giacovazzo et al. (2002) and in Müller (2007).

4.5 Determination of the Crystal Structure; (X-Ray) Diffraction Analysis

The three-dimensional periodicity of the arrangement of the atoms in a crystal implies that a crystal can act as a three-dimensional diffraction grating. In order to generate a diffraction pattern from a grating, comprising a number of diffraction maxima, the wavelength of the radiation incident to the grating should be of the same order as the periodicity inherent to the grating (as expressed by the distance of the scatterers of

---

mechanically the porosity affected surface adjacent part of the compound layer after the nitriding treatment.

the grating (i.e. the slit distance of one-dimensional gratings in classical diffraction experiments using visible light)). Hence, in order to possibly observe a diffraction pattern from a crystalline solid, the wavelength of the incident radiation should be of the order of 1 Å = 0.1 nm, as the distance between the atoms in a solid is of this order of magnitude. As indicated in the introduction to this chapter, it required the acceptance of a periodic arrangement of "building units", governing the regularity of crystals, and the recognition that, accepting atoms/molecules as realities and as the basic entities for the constitution of solid bodies, the regularity of crystals is due to the periodic arrangement of atoms/molecules, in order that the thought developed that a radiation of the required wavelength could be provided by X-rays. X-rays were discovered in 1895 by Röntgen (see Sect. 1.5) and the Brownian movement had proven, in the years 1900–1910, that atoms are the smallest building units of materials (see the introduction to this chapter). Against this background, the idea for and the execution of the first diffraction experiment with a crystal was developed and performed, respectively, in 1912 by Friedrich, Knipping and von Laue.

Here, the following side remark is in order. The occurrence of a diffraction pattern exhibiting sharp interference maxima (peaks), upon subjecting a crystal to a beam of incident electro-magnetic radiation of appropriate wavelength (as, for example, provided by X-rays), was long considered as the *ultimate* proof for the *translational* symmetry of crystals. This is in line with the spirit of the reasoning in the third paragraph of the introductory part of this chapter. However, the use of the adjective "ultimate" in the above has led to an overstatement. Research in the last part of the twentieth century has made clear that so-called *aperiodic crystals*, which do *not* possess translational symmetry, upon diffraction give rise to diffraction patterns exhibiting sharp interference maxima (peaks) as well (see Sect. 4.8; the only part of this chapter where crystals devoid of translational symmetry, in real, "physical" space, are considered).

The condition that an interference maximum occurs for a crystal upon diffraction of incident monochromatic light (electro-magnetic radiation) was most simply formulated by Bragg (1912),[20] and his line of reasoning is followed in principle below.

Consider Fig. 4.45, a set of (*hkl*) lattice planes in a crystal is exposed to an incident beam of parallel X-rays "hitting" the set of (*hkl*) lattice planes of the crystal

---

[20] W. L. Bragg ("Sir Lawrence Bragg") was the son of W. H. Bragg. Both, father and son, have contributed, separately and in cooperation, enormously to the field of the diffraction of X-rays by crystals. However, it was the son who first derived what is now known as "Bragg's law". That it had to be him, and not his father, may be due to the initial inclination of the father to focus on a particle-like, rather than a wave-like character of the X-rays. The joint results by father and son Bragg constitute an impressive example of the fruitful effect of family ties for the progress of science. Another such example is provided by the Burgers brothers, W. G. and J. M. (see the *"Intermezzo: a historical note about the Burgers vector"* in Chap. 5). The fascinating, early history of X-ray crystallography has been recorded in two books: (1) J.M. Bijvoet, W.G. Burgers, G. Hägg (eds.) Early Papers on Diffraction of X-rays by Crystals (Published for the International Union of Crystallography by A. Oosthoek's Uitgeversmaatschappij N.V., Utrecht, The Netherlands, Volume I, 1969 and Volume II, 1972) and (2) P.P. Ewald, Fifty Years of X-ray Diffraction (Published for the International Union of Crystallography by A. Oosthoek's Uitgeversmaatschappij N.V., Utrecht, The Netherlands, 1962)

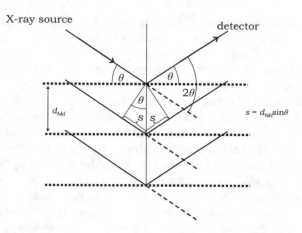

Fig. 4.45 Derivation of Bragg's law. If the path difference between the rays scattered by consecutive (*hkl*) planes (the dashed lines in the figure), i.e. 2 s = $2d_{hkl}\sin\theta$, with d_{hkl} as the (*hkl*) lattice spacing, equals $n\lambda$, with λ as the wavelength (of the X-rays) and n as a positive integer, then constructive interference occurs and a diffraction maximum can be observed in the diffraction pattern

at an angle θ. The atoms (actually the electrons) in each of the individual lattice planes act as scatterers for the incident X-rays in all directions. Position a detector, at "infinite" distance from the crystal, such that the X-rays (possibly) scattered in the "reflected" direction, i.e. also at an angle θ with the set of (*hkl*) lattice planes, are recorded. As follows from the geometric construction in the figure, for the X-rays scattered by the top lattice plane at the (central) position indicated in the figure and by the lattice planes beneath and at positions directly below this (central) position, constructive interference of the scattered rays occurs if the path difference between the rays scattered by consecutive (*hkl*) planes, i.e. $2d_{hkl}\sin\theta$ with d_{hkl} as the (*hkl*) lattice spacing, equals $n\lambda$ with λ as the wavelength of the X-rays and n as a positive integer. This reasoning holds for all positions on the top lattice plane and the corresponding positions on the lattice planes underneath. Hence, a diffraction maximum (interference maximum), reflection, is observed for the geometry indicated if

$$n\lambda = 2d_{hkl}\sin\theta \qquad (4.9)$$

This is the famous Bragg law. Hence, for fixed λ and fixed d_{hkl}, a diffraction maximum, reflection, is recorded if θ obeys Eq. (4.9).

In the last sentences of the preceding paragraph, the notions "diffraction maximum" and "reflection" have been used for the same. Apparently, the derivation of Bragg's law and the pictorial assistance (Fig. 4.45) give rise to the use of the word "reflection", although the diffraction phenomenon is not a reflection phenomenon: reflection would occur for any value of θ, whereas diffraction is only possible for specific values of θ prescribed by Eq. (4.9). In all directions corresponding to values of θ incompatible with Eq. (4.9), no (diffracted) intensity is recorded by the detector.

Yet, as in the literature, in the following next to "diffraction maximum", the term "reflection", and also the term "peak", will be used as equivalents.

Intermezzo: The von Laue theory

Von Laue (1912) had derived a complete equation for the intensity distribution of the diffraction maximum. This equation is of the type as expressed by Eq. (6.10), derived in Sect. 6.2 and which holds for diffraction by a one-dimensional grating: recognizing the three-dimensional nature of the diffraction grating provided by a crystal, the Laue result then occurs as the product of three such "$\sin^2/\sin^2$" functions as in Eq. (6.10), one for N_1, one for N_2 and one for N_3, where N_1, N_2 and N_3 stand for the number of "scatterers"/"slits" in each of the three dimensions of the crystal, respectively (cf. N in Eq. (6.10)). It can then be shown that the (2θ) position of the maximum of this three-dimensional intensity distribution (approaching a Dirac (δ) function for N_1, N_2 and N_3 approaching infinity (infinitely large crystal)) is compatible with Bragg's law (Eq. (4.9)). Thus, this probably not too transparent discussion, to explain the occurrence of diffraction maxima, serves to understand the much greater popularity of the Bragg equation, as compared to the Laue result. On the other hand, the treatment by von Laue, also called the kinematical diffraction theory, in a general way allows calculation of the amount of intensity contained in a diffraction maximum and also of the shape of the intensity distribution of a diffraction maximum, which last aspect is of great importance to analyse the crystal imperfection from the shape/broadening of diffraction lines/maxima, which is discussed in Sect. 6.9.

The angle between the incident X-rays and the diffracted X-rays equals 2θ (cf. Fig. 4.45). This angle, 2θ, is called the *diffraction angle*. To understand the significance of this definition, consider the following. The usual interpretation of Fig. 4.45, tacitly but not always recognized, involves that the set of (hkl) lattice planes is parallel to the surface of the specimen/crystal with the top plane of the (hkl) set of lattice planes at the surface. However, the experiment needs not be arranged in this way: the specimen may be tilted such that the set of (hkl) lattice planes possibly giving rise to diffraction (i.e. the occurrence of an interference/diffraction maximum) makes an angle with respect to the surface (see Fig. 4.46a, b; such tilting of the specimen is crucial for determining the macrostress in the specimen by diffraction, as discussed in Sect. 6.10.2). For the tilted specimen, as long as the X-ray source and the detector remain fixed (in the laboratory frame of reference), obviously the angle between the incident/diffracted X-rays and the surface of the specimen does not equal θ, whereas this holds for the untilted specimen. However, in both cases, the angle between the incident and diffracted X-rays is the same: 2θ. Therefore, a more correct form of Bragg's law would be: $n\lambda = 2d_{hkl}\sin[(2\theta)/2]$. This explains that the diffracted intensity is usually plotted as function of 2θ, rather than θ.

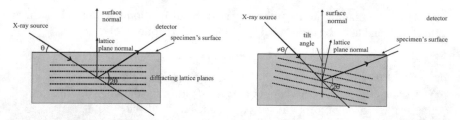

Fig. 4.46 Diffraction by a set of (*hkl*) lattice planes can occur if the diffraction angle 2θ complies with Bragg's law, for example, for fixed value of the wavelength λ. (a) The geometry of the experiment can be such that the diffracting (*hkl*) planes are parallel to the surface of the specimen. Then, the incident and diffracted X-ray beams are oriented symmetrically with respect to the surface of the specimen and the angle of incidence with respect to the surface of the specimen equals θ. **b** If the specimen is tilted, by rotation around an axis in the surface of the specimen, while keeping the X-ray source and the detector fixed in the laboratory frame of reference, then a set of (*hkl*) lattice planes can diffract that is not parallel to the surface of the specimen: the angle between the incident/diffracted X-rays and the surface of the specimen no longer equals θ. Such tilting of the specimen is crucial for the analysis of stress in the specimen, as discussed in Sect. 6.10.2

To predict the position (2θ value) of the possible reflections for crystal structures according to the seven crystal systems (cf. Sect. 4.1.2), the lattice spacings d_{hkl} have to be calculated from the corresponding lattice parameters (also indicated in Sect. 4.1.2). The necessary formulas to perform these calculations have been given in Table 4.7.

For a crystal, with the set of (*hkl*) lattice planes parallel to the surface, and for the diffraction geometry, where the incident, monochromatic X-rays make an angle of θ with the surface ("untilted" specimen; cf. above discussion), it follows from Eq. (4.9) that only for specific values of θ diffraction (positive interference) occurs for this crystal (for fixed d_{hkl}: for each value of n, one value of θ; θ < 90°). Now consider a polycrystalline specimen, of a single element or compound, containing many crystallites in various (possibly randomly distributed) orientations. For the different values of d_{hkl} (pertaining to the different sets of lattice planes parallel to the surface in the different crystals), different values of θ can be indicated which provide compliance with Eq. (4.9). In the presence of very many crystals in the volume irradiated by the incoming X-rays, it appears that, for any θ compatible with a d_{hkl}, there will always be a number of crystallites in the specimen which fulfil the requirement according to Eq. (4.9). Hence, for such, a polycrystalline specimen a plot of diffracted intensity versus 2θ yields a "fingerprint" of the material considered: at specific values of 2θ diffracted intensity can be observed (Fig. 4.47).

By convention, the various reflections observed in the diffractogram (intensity versus 2θ) are denoted by *HKL*, without brackets or braces, where $H = nh$, $K = nk$ and $L = nl$. (*H, K* and *L* are called "Laue indices", to be distinguished from the "Miller indices", h. k and l (which are numerically equal to the Laue indices for n = 1); cf. Sect. 4.1.4). Thus, the 100 reflection is the "first-order" (n = 1) reflection originating from the (100) family of lattice planes of a single crystal or from the

Table 4.7 Relation between the lattice spacing d_{hkl} and the lattice parameters (unit-cell parameters; cf. Table 4.1) for the seven crystal systems

Cubic:

$$d_{hkl}^{-2} = \frac{h^2+k^2+l^2}{a^2}$$

Tetragonal:

$$d_{hkl}^{-2} = \frac{h^2+k^2}{a^2} + \frac{l^2}{c^2}$$

Orthorhombic:

$$d_{hkl}^{-2} = \frac{h^2}{a^2} + \frac{k^2}{b^2} + \frac{l^2}{c^2}$$

Hexagonal:

$$d_{hkl}^{-2} = \frac{4(h^2+hk+k^2)}{3a^2} + \frac{l^2}{c^2}$$

Rhombohedral/trigonal (see footnotes 9 and 10):

$$d_{hkl}^{-2} = \frac{(h^2+k^2+l^2)\sin^2\alpha - 2(hk+kl+hl)\cos\alpha(1-\cos\alpha)}{a^2(1-3\cos^2\alpha+2\cos^3\alpha)}$$

Monoclinic:

$$d_{hkl}^{-2} = \frac{h^2}{a^2\sin^2\beta} + \frac{k^2}{b^2} + \frac{l^2}{c^2\sin^2\beta} - 2\frac{hl}{ac\sin^2\beta}\cos\beta$$

Triclinic:

$$d_{hkl}^{-2} = \frac{\left[\begin{array}{l}\frac{h^2}{a^2}\sin^2\alpha + \frac{k^2}{b^2}\sin^2\beta + \frac{l^2}{c^2}\sin^2\gamma + 2\frac{hk}{ab}(\cos\alpha\cos\beta - \cos\gamma) \\ +2\frac{kl}{bc}(\cos\beta\cos\gamma - \cos\alpha) + 2\frac{hl}{ac}(\cos\alpha\cos\gamma - \cos\beta)\end{array}\right]}{1 - \cos^2\alpha - \cos^2\beta - \cos^2\gamma + 2\cos\alpha\cos\beta\cos\gamma}$$

{100} set of equivalent families of lattice planes[21] in a polycrystalline specimen (cf. Sect. 4.1.4.1 for definition of "family" and "set"). The 200 reflection then is the second-order ($n = 2$) reflection originating from the same (100) family of lattice planes of a single crystal or from the {100} set of equivalent families of lattice planes in a polycrystalline specimen (cf. Fig. 4.47).

Note that the formulas in Table 4.7 can be used as well after replacing all h, k and l by H, K and L; the $d_{HKL}(= d_{hkl}/n)$ values, thus obtained utilizing these formulas, can be substituted into Bragg's law in the form $\lambda = 2d_{HKL}\sin\theta$ in order to determine, for an experiment applying monochromatic X-rays of wavelength λ, the peak positions (2θ values) of the HKL reflections.

On this basis, phase (compound/element) identification is possible, by using the measured 2θ values of diffraction peaks and comparing these with peak-position values for the enormous body of compounds and elements contained in a database, as the one maintained and continuously augmented by the ICDD (= International Centre of Diffraction Data).

---

[21] For example, for a cubic crystal, the (100), (010), (001), (−100), (0−10) and (00−1) sets of lattice planes are equivalent, the precise (*hkl*) notation being dependent on how the crystal axes have been defined for the individual crystal. With the notation {*hkl*}, i.e. the use of the braces, it is indicated here that any set of equivalent lattice planes of {*hkl*} type can contribute to the reflection concerned.

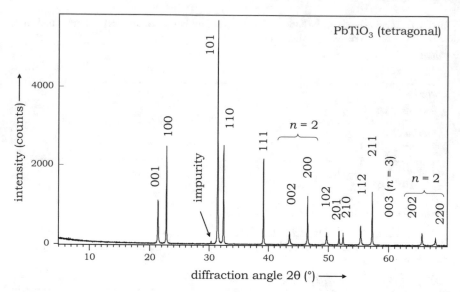

Fig. 4.47 Example of an experimental X-ray diffraction pattern. Recorded for a polycrystalline powder specimen of PbTiO₃ (tetragonal), using Cu Kα₁ radiation. At specific values of the diffraction angle, 2θ, diffracted intensity can be observed. Consequently, the diffraction pattern can be conceived as a "fingerprint" of the diffracting substance. Such phase identification concerns one of the main applications of X-ray diffraction in materials science (measurement by Dr. A. Leineweber, Max Planck Institute for Metals Research)

In fact, the "fingerprint" nature of the (X-ray) diffraction pattern is based not on the (2θ) peak positions alone. The intensity values of the *HKL* reflections are important as well as indicative parameters (and these have also been incorporated in the ICDD data files). To make this clear, it is recalled that the crystal can be considered as a three-dimensional diffraction grating. In the Laue (kinematical) theory of diffraction, as discussed in the *intermezzo* above, a periodic, three-dimensional arrangement of "scatterers" is considered. According to the treatment in this chapter, the crystal is constituted of a three-dimensional, massive, periodic arrangement of unit cells. The unit cell then can be conceived as the "scatterer". It will be obvious that the scattering power of a unit cell depends on its filling: the number and type of atoms and their fractional coordinates (cf. Eq. (4.3)). This scattering power differs from *HKL* to *HKL*.[22] and in different ways for different crystalline substances. Hence, the

---

[22] The scattering power of a unit cell can even be zero for a certain *HKL*; i.e. this *HKL* reflection does not occur in the diffraction pattern. One then speaks of "systematic extinction". In a sense, this effect is artificial: it depends on the unit cell chosen and occurs only for non-primitive unit cells in which more than one motif (cf. Sect. 4.1.1) is present such that the waves scattered by the various motifs in the unit cell interfere destructively for specific *HKL*'s (again: note that the *HKL* notation of a reflection depends on the choice of unit cell) and as a consequence the unit cell has zero scattering power for these reflections. Example: consider the f.c.c. unit cell relevant for many metals (Cu, Al, etc.). This is a non-primitive unit cell containing 4 motifs (= 4 identical metal atoms; cf. Sect. 4.2.1.2). For this choice of unit cell, it can be shown that the 100 reflection

intensity contained in a diffraction maximum depends on the crystal structure, i.e. the filling of the unit cell.

Hence, information about the crystal structure is contained in both the positions and the intensities of the diffraction peaks.

The first crystal structures determined by X-ray diffraction were, of course, relatively simple. Thus, the crystal structures of simple metals, rocksalt, zincblende and related materials could be deduced on the basis of primarily the positions of the diffraction peaks. However, for crystal structures of greater and enormous complexity, it is imperative to utilize the intensity information as well. A straightforward procedure cannot be indicated and special methods for more or less special cases were developed. This was the time, referred to in the beginning of this chapter, where one Ph.D. student could devote a whole dissertation to the determination of a single crystal structure. Due to the colossal, exponential growth of computer power and also further theoretical developments, one can nowadays determine an also complicated crystal structure in a couple of days. Moreover, whereas in former days, crystal-structure determination required the availability of a single crystal (large enough) allowing the recording of its diffraction pattern, nowadays also methods have been developed that allow the refinement and even the direct determination of the crystal structure from diffraction patterns recorded from finely polycrystalline specimens/powders (e.g. see review by Cerny and Favre-Nicolin 2007).

Until now, the focus was on the diffraction of X-rays, as the electro-magnetic radiation of appropriate wavelength to give rise to diffraction effects by crystals. However, electrons of suitable energy (wavelength) can also be diffracted by crystals (cf. Sect. 2.4) and, indeed, electron-diffraction methods have also been and are successfully used to determine crystal structures, or to at least provide crystal-structure information. A similar remark can be made about neutrons. The penetrative power of electrons and neutrons is very different: whereas neutrons can be used to generate diffraction patterns from material relatively deep (cm) under the surface of a specimen, electrons can be used for very thin specimens (up to 100 nm) or for structure investigations of the surface. X-rays take an intermediate position: penetration of materials up to a few μm.

4.6 The Stereographic Projection

The description of directions in and orientations of planes in crystals has been discussed in Sect. 4.1.4. Both the algebraic character of the treatment given and perspective images, i.e. projected images, of crystals do not provide an easy visualization of the angular relations between directions, between planes and between

---

is extinguished. This is no longer true, if, for example, departing from the f.c.c. unit cell, with a random distribution over the atomic sites of the crystal structure of Cu and Au atoms of a solid solution of the composition $AuCu_3$, the distribution of the Cu and Au atoms becomes ordered (cf. bottom left of Fig. 4.35). Then, the 100 reflection is no longer extinguished (see Fig. 4.36a, b).

directions and planes in crystals/lattices. To this end, the stereographic projection is often used.

The stereographic projection is a graphical method to represent the orientation of a crystal. Consider Fig. 4.48, a crystal is positioned with its centre at the centre of a sphere which is very large as compared to the crystal. Planes of the crystal can be extended hypothetically and intersect the sphere. Since the crystal is infinitely small, the plane of intersection runs through the centre of the sphere and the sphere is intersected by a circle; because the crystal is infinitely small, all planes of a family of lattice planes (cf. Sect. 4.1.4.1) reduce to a single plane of intersection in the construction considered. The diameter of these circles is equal to the diameter of the sphere, and therefore these circles are called *great circles*. Any plane of the crystal can thus be represented by a great circle. An alternative way of representing the orientation of a plane of the crystal considered is by erecting a normal to the plane at the centre of the sphere and determining the point of intersection with the sphere. This point of intersection is called the *pole* of the plane considered. A pole thus represents (the orientation of) a family of lattice planes (see above).

The angles between the planes of the crystal are equal to the angles between the corresponding poles (and equal to the angles of intersection of the corresponding great circles). Now to arrive at a two-dimensional representation of these angular relationships, the so-called stereographic projection (further denoted as SGP) is realized as follows. The horizontal plane through the centre of the sphere in the above discussion is taken as the plane of projection. The projection is performed by drawing a connection line, for all poles on the upper hemisphere of the sphere, from the pole to the "bottom point", B, of the sphere: the points of intersection of these lines with the horizontal plane of projection provide the projections of the poles considered.

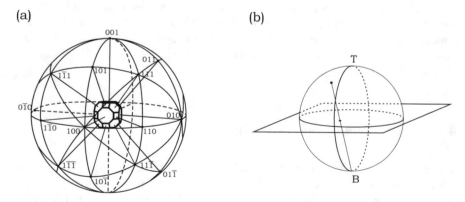

Fig. 4.48 Stereographic projection (SGP) of a crystal. **a** The crystal is put at the centre of a sphere which is very large as compared to the crystal. Hypothetical extension of crystal planes, passing through the centre of the sphere in view of the smallness of the crystal, intersect the sphere according to *great circles*. Normal to crystal planes erected at the centre of the sphere intersect the sphere at points called *poles*. **b** Lines connecting the poles on the upper hemisphere with the "bottom point" B of the sphere intersect the horizontal plane, i.e. the plane of projection, at points which thereby are the projections of the poles concerned

Note that it suffices to consider (project) poles on the upper hemisphere only: if the pole of (*hkl*) occurs on the upper hemisphere, the pole of (−*h*−*k*−*l*) occurs on the bottom hemisphere, but both poles represent the same crystallographic direction; so all orientation information (angular relationships) is (are) represented by the poles of the upper hemisphere already. The plane of projection is enclosed by a circle which is the great circle of the plane with pole at the "top point", T, of the sphere. This external boundary of the SGP is called basic circle or equator.

The main virtue of the SGP is that the angular relationships between the poles, and thus the corresponding great circles/crystal planes, are preserved in the SGP. For measurement of these angles in the SGP, see further below.

An example of the SGP for a cubic crystal with the plane of projection equal to the (001) plane is shown in Fig. 4.49. The points shown in the SGP are the

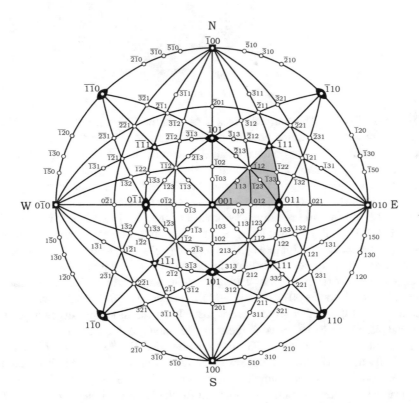

Fig. 4.49 Standard (001) SGP of a cubic crystal, i.e. the plane of projection (cf. Fig. 4.48) is equal to the (001) plane. The points shown in the SGP are the projections of poles. The (curved) lines shown in the SGP all are the projections of great circles. Note that the fourfold rotational symmetry of the [001] axis perpendicular to the plane of projection is exhibited by the SGP. For cubic crystals, it holds that the standard (001) SGP contains all non-equivalent directions in the crystal in the so-called *standard stereographic "triangle"*, which is enclosed by the projected great circle connections of the poles 001, 011 and −111 in the standard (001) SGP (see the grey area in the figure)

Fig. 4.50 Wulff
stereographic net. Longitude
circles (arcs) and latitude
circles (arcs) are shown in
gradations of, in this case,
2°. The longitude circles are
great circles

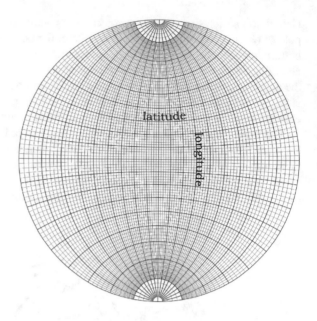

projections of poles. The (curved) lines shown in the SGP all are the projections
of great circles: great circles become circular arcs intersecting the basic circle at
two diametrically opposite points (obviously, great circles oriented perpendicularly
to the plane of projection are projected as straight line segments; e.g. see the NS
and EW lines in the SGP). A property of the SGP is that the symmetry properties
pertaining to the crystallographic axis perpendicular to the SGP are exhibited by the
SGP: hence, the fourfold rotational symmetry of the [001] axis is exhibited by the
standard[23] (001) SGP is shown in Fig. 4.49. The high symmetry of the cubic crystal
system, in particular as exhibited by the standard (001) SGP, leads to the finding, for
cubic crystals, that all non-equivalent directions in the crystal are contained in the
so-called *standard stereographic "triangle"*, delineated by the projected great circle
connections of the poles 001, 011 and −111 in the standard (001) SGP (see the grey
area in Fig. 4.49). In other words: every possible crystal direction (every normal to
a crystal plane) can be indicated in the standard stereographic triangle.

To facilitate working with a SGP often, a so-called *Wulff stereographic net* can
be fruitfully used. The Wulff net shows longitude circles (arcs) and latitude circles
(arcs) in gradations of, for example, 2° (see Fig. 4.50). The longitude circles (arcs)
are great circles (arcs), which does not hold for the latitude circles (arcs) (with the
exception of the EW connecting line; cf. Fig. 4.49).

One of the important applications of the Wulff stereographic net is the measurement of the angles between poles (= the angle between the corresponding crystal
planes) from the poles as projected in the plane of projection. In a practical way,

---

[23] The adjective "standard" refers to a SGP of a low index plane of the crystal. Such standard SGPs
are available for the various crystal systems (e.g. see Johari and Thomas 1969), but, of course, can
also be generated (simply) by available (commercial and free) software.

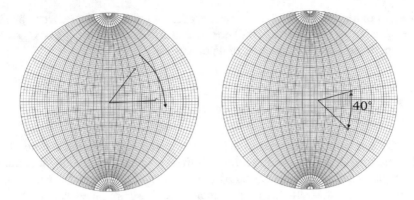

Fig. 4.51 Measurement of the angles between the poles of two crystallographic planes (i.e. the angle between the two crystallographic planes) using a SGP, exhibiting the projections of the poles, and the Wulff net. The SGP shown in the left part of the figure is rotated on top of a Wulff net such that the two poles concerned fall on a great circle (longitude circle; cf. Fig. 4.50). The result after the rotation is shown in the right part of the figure. The difference in latitude then is the angle between the two poles (i.e. the angle between the two crystallographic planes)

one then may proceed as follows. The SGP is plotted on a transparent paper (tracing paper). This transparent paper with the indicated SGP is laid on top of the Wulff net (with coinciding centres), such that it can be rotated, without restrictions, relative to the Wulff net. Now, the transparent paper with the SGP is rotated such that the two poles considered fall on the same great circle (longitude circle). The angle between the poles then is given by the difference in latitude, which is simply read off from the Wulff stereographic net (see Fig. 4.51).

For the materials scientist, the SGP plays an important role in transmission electron microscopical analysis (see Sect. 6.8) of phase transformations (orientation relationships between product and parent phases; cf. footnote 20 in Chap. 9), twinning (Sect. 5.3) and plastic deformation (slip system; Sect. 5.2.5); for all these phenomena identification of crystallographic directions and planes is imperative and to this end the SGP, as a tool to interpret diffraction patterns and images recorded by transmission electron microscopy, is indispensable (see Johari and Thomas 1969).

4.7 The Texture of a Polycrystal

A crystalline, massive material very often does not consist of a single crystal but of many crystals. Then, it is said to be polycrystalline and one speaks of a polycrystal. The crystals in the polycrystal can be, and usually are, oriented differently with respect to the specimen frame of reference. If the distribution of the orientations of the crystals in the specimen is not random, one speaks of *preferred orientation* or that the specimen exhibits a *texture* (see also the discussion in the introductory part of this chapter). Already only because of the (intrinsic) anisotropy of many properties of a single crystal, it thus appears of great practical importance to be able to characterize the (volume weighted) orientation distribution of the crystals (grains), in the specimen frame of reference, of a polycrystal.

4.7.1 The Pole Figure

The SGP can be fruitfully used to exhibit the preferred orientation/texture of the grains in a (massive) polycrystal. To this end, the dependence of the amount (volume) of crystals of corresponding crystallographic orientation (see at the end of next paragraph) on direction in the specimen is measured and presented as follows.

The integrated intensity of a selected reflection HKL (= area enclosed by the HKL diffraction profile, i.e. area under the peak; cf. Sect. 4.5; for the distinction between Miller indices, hkl, and Laue indices, HKL, see Sect. 4.5 as well) is measured for a range of ψ and φ values, where ψ denotes the specimen-tilt angle, and φ represents the angle of rotation around the specimen normal (cf. Fig. 4.52). The integrated intensity of a reflection is proportional with the amount of diffracting material contributing to the measured reflection (Sect. 4.5). Hence, the integrated intensity measured in a direction characterized by a certain (ψ, φ) combination is proportional with the amount (volume) of crystals oriented with a $\{hkl\}$ plane perpendicular to the direction indicated by the specific (ψ, φ) combination. Note that the orientations of the crystals contributing to the measured integrated intensity are only such restricted

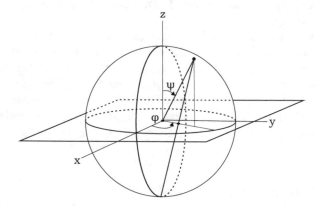

Fig. 4.52 Definition of the angle ψ, the specimen-tilt angle, and the angle φ, the angle of rotation around the normal of the surface of the specimen. A (ψ, φ) combination denotes a direction in the specimen ($\psi = 0$ denotes the surface-normal direction). Every (ψ, φ) combination indicates a certain direction in the specimen which intersects the sphere. The points of intersection on the upper hemisphere can be projected according to the stereographic projection (Fig. 4.48) onto the plane of projection which is the horizontal plane of the sphere (surface of the specimen). The intensity of a HKL reflection recorded for the (ψ, φ) direction can be indicated at every (ψ, φ) point in the SGP. The result is called a *pole figure*; see Figs. 4.53, 4.54 and 4.55

that their diffracting $\{hkl\}$ planes are oriented perpendicular to the (ψ, φ) direction: the rotation around the (ψ, φ) direction (i.e. the direction perpendicular to the diffracting lattice planes) is unrestricted. In this sense, the crystals contributing to the measured integrated intensity in the (ψ, φ) direction are *not* of identical crystallographic orientation and this explains (1) why we spoke of "crystals of corresponding crystallographic orientation" at the end of the previous paragraph and (2) how the adjective "corresponding" has to be understood.

For $\psi = 0°$, and variable φ, crystals with $\{hkl\}$ planes parallel to the surface diffract; for other values of ψ and specific values of φ $\{hkl\}$ planes with a specific orientation with respect to the surface of the specimen diffract (Fig. 4.52).

Consider the sphere as shown in Fig. 4.52. The polycrystalline specimen is positioned at the centre of the sphere. The horizontal plane is identified with the surface of the specimen. Every (ψ, φ) combination indicates a certain direction in the specimen which intersects the sphere. The points of intersection on the upper hemisphere can be projected according to the stereographic projection on the plane of projection which is the horizontal plane of the sphere (surface of the specimen).

Every (ψ, φ) combination corresponds with a specific value of measured (integrated) intensity of the *HKL* reflection; i.e. corresponds with a certain amount (volume) of crystals with $\{hkl\}$ lattice planes oriented perpendicularly to the direction specified with the considered (ψ, φ) combination. This value of integrated intensity is indicated at every (ψ, φ) point in the SGP for which the integrated HKL intensity has been measured. The obtained representation of spatial distribution of integrated HKL intensity, and thus preferred orientation of crystals in the specimen, is called the *hkl pole figure*, where hkl denotes the Miller indices of the diffraction lattice planes.[24] The integrated intensities can be indicated in the two-dimensional SGP by iso-intensity (contour) lines (see Figs. 4.53 and 4.54) or in a three-dimensional representation by plotting the integrated intensities in the direction perpendicular to the plane of projection at their corresponding (ψ, φ) locations in the SGP (see Fig. 4.55). For crystallographic interpretation and manipulation, the two-dimensional presentation of the pole figure in a SGP as in Figs. 4.53 and 4.54, rather than the perspective three-dimensional representation in Fig. 4.55, is much more appropriate (cf. the first paragraph of Sect. 4.6).

For a polycrystalline specimen with a random distribution for the orientation of the constituting crystals (grains), the intensity distribution in a pole figure seems to be uneven/inhomogeneous: close to the equator of the pole figure the intensity appears, seemingly, relatively low. This is just a consequence of the type of projection of the intensity distribution, recorded as function of ψ and φ, in a SGP. The intensity

---

[24] A pole figure should not be referred to using the Laue indices, *HKL*, of the reflection used for measurement of the pole figure (See Sect. 4.5 for the distinction between Miller and Laue indices). For example, for a f.c.c. crystal structure, the notion "200 pole" ($HKL = 200$) is meaningless, in contrast with "$\{100\}$ pole" ($hkl = 100$). In case of the example considered, a $\{100\}$ ($hkl = 100$) pole figure can be measured employing a 200 ($HKL = 200$) reflection (as the 100 ($HKL = 100$) reflection is extinguished; see footnote 22 in Sect. 4.5). Violations of this ruling occur frequently in the literature, i.e. within the framework of the example discussed, often "$\{200\}$ pole figures" are published....

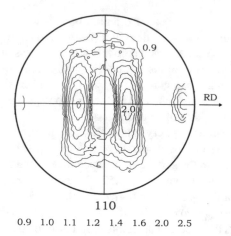

110

0.9 1.0 1.1 1.2 1.4 1.6 2.0 2.5

Fig. 4.53 {110} pole figure, measured using the 110 reflection of the ferrite (b.c.c. Fe) matrix, of the surface of a hardened and tempered (cf. the *intermezzi* at the end of Sect. 9.5) and subsequently cold rolled (80% thickness reduction) SAE 52100 steel specimen. The rolling direction has been indicated with RD. The texture is relatively weak and not sharp: the iso-intensity lines indicate relatively moderate intensity maxima and a large spread in orientation. The values given for the levels of the iso-intensity lines (contours) shown have been indicated as (**a**) *x* random, where "random" pertains to the intensity level in the absence of preferred orientation and with (**a**) as the number indicated beneath the pole figure. Taken from A.P. Voskamp, E.J. Mittemeijer, Metallur. Mater. Trans. A 27A, 3445–3465 (1996)

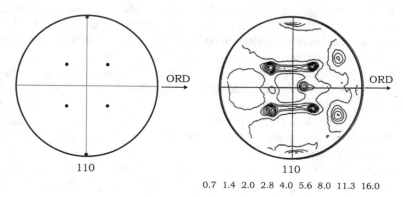

110   110

0.7 1.4 2.0 2.8 4.0 5.6 8.0 11.3 16.0

Fig. 4.54 {110} pole figure, measured using the 110 reflection of the ferrite (b.c.c. Fe) matrix, for the plane parallel to the surface at depth 0.2 mm below the lowest point in the deep groove of an endurance tested ball bearing inner ring prepared from hardened and tempered SAE 52100 steel (the inner ring had experienced 2.10^8 rotations under a maximal stress of 4.9 GPa at 6000 rpm at an operation temperature of 55 °C). The overrolling direction has been indicated with ORD. The {100}<110> texture component has been indicated in the {110} standard SGP at the left-hand side. Comparison with the measured pole figure at the right hand side reveals that part of the crystals in the specimen more or less are oriented according to the {100}<110> texture component. However, in the present case, there is also an other texture component in the specimen which has been identified as a {221}<411> component (for indication at the bottom of the figure of the levels of the iso-intensity lines, see the caption of Fig. 4.53). Taken from A.P. Voskamp, E.J. Mittemeijer, Metallur. Mater. Trans. A 27A, 3445–3465 (1996)

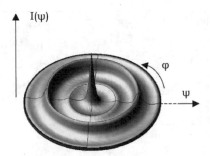

Fig. 4.55 {110} pole figure of the surface of a b.c.c. Nb film of thickness 500 nm prepared by magnetron sputter deposition on a 500 μm thick single crystalline Si wafer, with a (510) surface, covered with a 50 nm thick amorphous SiO_2 layer. The preferred orientation is described as a {110} fibre texture, with the fibre axis parallel to the normal of the film surface. The intensity maximum at the centre of the {110} pole figure corresponds to crystals with {110} planes parallel to the surface. These same crystals also give rise to intensity maxima in the pole figure at those tilt angles ψ where the other members of the set of equivalent families of {110} planes diffract; further see Fig. 4.56. Taken from B. Okolo, P. Lamparter, U. Welzel, E.J. Mittemeijer, J. Appl. Phys. 95, 466–476 (2004)

distribution on the (upper hemi)sphere as shown in Fig. 4.52 is homogeneous for a random distribution, i.e. the intensity contained in an area of specific size on the sphere is the same for every area of the same size on the sphere. The SGP is not "area true": it is simply seen that a specific area on the sphere close to the top of the (upper hemi)sphere is projected as a relatively small area near the centre of the pole figure, whereas an identical area on the sphere remote from the top of the (upper hemi)sphere is projected as a relatively large area close to the equator. This disparity between visual impression and reality for the orientation distribution is avoided if so-called equal-area projection is applied: then, the difference with a possibly random distribution of the orientations in the specimen is directly revealed visually. It has been proposed that "equal-area projection" should be preferred for the representation of preferred orientation (Kocks et al. 1998). However, the SGP has distinct advantages for crystallographic analysis: angular relationships are clearly revealed (see Sect. 4.6).

For the case as considered in Fig. 4.55, the preferred orientation does not depend on φ but only on ψ. Then one speaks of the occurrence of a *fibre texture*, implying that the texture is characterized by a (strong) tendency to orient a specific {*hkl*} plane perpendicular to the fibre axis independent of the angle of rotation around this axis. Often the fibre axis is oriented perpendicular to the surface, as frequently observed for thin polycrystalline films: for the Nb film pertaining to Fig. 4.55 {110} planes prefer to be parallel to the surface. The intensity maximum at the centre of the {110} pole figure corresponds to crystals with {110} planes parallel to the surface. Of course, these same crystals also give rise to intensity maxima in the pole figure at those tilt angles ψ where the other members of the set of equivalent families of {110} planes diffract (see Sect. 4.1.4.1). These are not the only intensity maxima which can be discerned in this pole figure (see below).

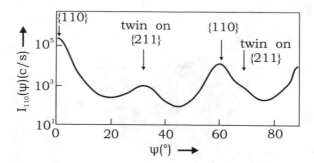

Fig. 4.56 ψ scan, also called pole figure section, is shown for the same b.c.c. Nb film with {110} fibre texture considered in Fig. 4.55. Apart from the intensity maxima pertaining to those crystals with {110} planes parallel to the surface, additional intensity maxima can be discerned, which are due to twinning along {211} planes of the crystals with {110} planes parallel to the surface. Taken from B. Okolo, P. Lamparter, U. Welzel, E.J. Mittemeijer, J. Appl. Phys. 95, 466–476 (2004)

Evidently, in case of a fibre texture with the fibre axis parallel to the normal of the specimen, as in Fig. 4.55, it suffices for the description of the texture to consider the intensity dependence on tilt angle ψ only. An example of such a ψ *scan* or *pole-figure section* is shown in Fig. 4.56 for the same Nb film with {110} fibre texture considered in Fig. 4.55. The ψ scan reveals the intensity maxima pertaining to those crystals with {110} planes parallel to the surface. Additionally, intensity maxima due to twinning, along {211} planes of the crystals with {110} planes parallel to the surface, can be discerned (note that crystalline Nb has a b.c.c. structure; for twinning, see Sect. 5.3).

The {110} pole figure depicted in Fig. 4.53 shows the preferred orientation of the crystals in the tempered martensitic (b.c.c. ferritic) matrix (see the "*Intermezzo: Tempering of iron-based interstitial martensitic specimens*" in Sect. 9.5.2.4) of cold rolled steel. The rolling direction has been indicated in the pole figure. The figure shows the typical rolling texture occurring for b.c.c. Fe. The texture is rather weak and not sharp: the iso-intensity lines indicate relatively moderate intensity maxima and a large spread in orientation. A different situation occurs for the texture developing in deep groove ball bearings upon overrolling. In this case, the texture is relatively strong and sharp (cf. Figs. 4.54 and 4.53).

The so-called texture components are indicated as {hkl}<uvw>, where {hkl} denotes the crystallographic plane that is preferably parallel with the specimen surface (rolling contact surface, for the case under consideration) and <uvw> denotes the crystallographic direction that is preferably parallel with a specific specimen direction (the overrolling direction in the inner bearing ring groove, for the case under consideration). The pole figure as shown in Fig. 4.54 reveals the presence of a {100}<110> texture component. For a single crystal oriented according to the ideal {100}<110> texture component the positions of the {110} poles in the SGP are shown in the left part of Fig. 4.54. Comparison with the measured pole figure in the right part of Fig. 4.54 indeed shows that part of the crystals in the specimen more or less are oriented according to this texture component. However, there are also intensity maxima in the pole figure incompatible with the {100}<110> texture component: in the present case, there is also an other texture component in the specimen which has been identified as a {221}<411> component. Depending on the loading conditions predominance for one or an other texture component can occur.

4.7.2 The Orientation Distribution Function

The intensity at a (ψ, φ) location in a *hkl* pole figure represents all crystals, in the volume of material that can diffract, which have the *hkl* pole in the direction (in the specimen frame of reference) specified by ψ and φ, implying that the rotation of the crystal around the *hkl* direction is unconstrained: all crystals in the specimen with their *hkl* lattice planes perpendicular to the direction considered are able to diffract. Indeed, the direction considered in a pole figure is defined by two angles, ψ and φ (see above and Fig. 4.52), whereas the definition of the orientation of a crystal in the (Cartesian, orthogonal) frame of reference of the specimen (or of the laboratory) requires three angles: The orientation of a crystal (with an orthogonal crystal frame of reference, which can always be given) with respect to the orthogonal frame of reference of the specimen is given by the Euler angles Ψ, θ and Φ, as shown in Fig. 4.57. The *orientation distribution* (in the specimen frame of reference) for all crystals in the specimen thus is given by the amount (volume) of crystals oriented according to a (Ψ, θ, Φ) combination as function of Ψ, θ and Φ (the space defined by the coordinates Ψ, θ and Φ is called *Euler space*). The problem then is: how to

Fig. 4.57 Orientation of a crystal in a specimen with respect to the frame of reference of the specimen can be specified by the three Euler angles Ψ, θ and Φ. Both the crystal coordinate system (x, y, z) and the coordinate system of the specimen (X, Y, Z) are orthogonal

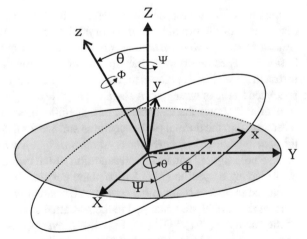

specimen frame of reference (X,Y,Z)
crystal frame of reference (x,y,z)
Euler angles (Ψ,θ,Φ)

deduce the orientation distribution (OD) function (ODF), in Euler space coordinates, from experimentally determined pole figures?

The above paragraph indicates that a pole figure is obtained by integration along a path in the OD involving a full rotation (i.e. over 2π radians) about the *hkl* pole direction. This calculation is possible in a straightforward, direct way. However, to determine the OD from pole figures, the inverse problem must be solved, which evidently is a less directly solvable problem. It becomes immediately clear that one pole figure will generally not suffice to extract, in a unique, i.e. unambiguous way, the OD. Elaborate procedures, using several pole figures recorded for "geometrically independent" hkl poles, have been developed in the last 70 years, which usually impose a number of physical and also intuitive conditions to reach a unique solution for the OD (Kocks et al. 1998). The number of independent pole figures needed depends on (1) the crystal symmetry of the material investigated and (2) the number of coefficients used for the (possibly) applied series development of the ODF. Thus, for material of cubic crystal symmetry only two geometrically independent hkl pole figures can be sufficient for successful determination of the ODF.

Evidently, knowledge of the OD is a prerequisite to be able to calculate, i.e. to predict, the behaviour of a polycrystalline aggregate. For example, accounting for the so-called grain interaction in calculations of the mechanical response of massive polycrystalline specimens (as described in the *"Intermezzo: Grain interaction"* in Sect. 6.10.2) calls for the ODF as input.

4.7.3   The Inverse Pole Figure

In the above discussion the focal point of interest was the characterization of the distribution of the orientation of the crystals/grains of a phase in the specimen *with respect to the specimen frame of reference*. Thus, a pole figure exhibits the distribution of a specific crystallographic direction, i.e. the direction perpendicular to the diffracting {*hkl*} lattice planes for the crystals/grains of a phase in the specimen, in the specimen frame of reference. A so-called *inverse pole figure* presents a pendant representation: it shows the distribution of a specific direction in the specimen frame of reference, e.g. the normal of the specimen surface for the crystals/grains of a phase in the specimen, *with respect to the crystal frame of reference*. Thus, a "surface-normal inverse pole figure" shows a representation of the (volume) fraction of a crystalline phase having a specific {*hkl*} lattice plane perpendicular to the surface normal. As discussed in Sect. 4.6, for cubic crystals, all non-equivalent (crystallographic) directions in the crystal are contained in the so-called *standard stereographic "triangle"* in the standard (001) SGP (see Fig. 4.49): every possible crystal direction (every normal to a crystal plane) can be indicated in the standard stereographic "triangle". For the case discussed, in the standard stereographic "triangle", the surface-normal inverse pole figure shows (see Fig. 4.58), either by means of "iso-frequency" lines or by means of colour/grey contrast, the relative fraction, in the volume analysed,

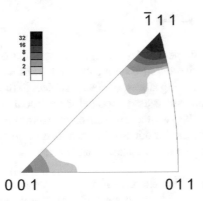

Fig. 4.58 Surface-normal inverse pole figure for a cold rolled (from more than 20 mm thickness down to smaller than 1 mm thickness) and recrystallized (at 800 °C for 2 h) sheet of a ferritic (b.c.c.) Fe-0.2at.%Cr-0.1at.%Ti alloy. The crystallographic orientations of surface adjacent volume elements in the specimen analysed (mean grain size of about 90 μm) were determined by electron backscatter diffraction (EBSD) with a lateral resolution of about 1 μm. The frequency of the volume elements in the analysed surface region with a specific crystallographic direction parallel to the normal of the surface of the specimen has been indicated by grey scale contrasting. The values given for the levels of the iso-frequency contours shown have been indicated in the *top left* of the figure as (**a**) *x* random, where "random" pertains to the frequency level in the absence of preferred orientation and with (**a**) as the number indicated. Evidently, in the case shown a majority of the crystals have a {111} lattice plane about parallel to the surface; there is also a minor preference for the crystals to have a {001} plane parallel to the surface of the specimen (measurement by Dr. E. Bischoff, Max Planck Institute for Metals Research)

of crystals of the phase considered with variable {*hkl*} perpendicular to the specimen surface normal. Electron backscatter diffraction (EBSD) experiments, from the surface of a specimen[25], can allow the full determination of the crystallographic orientation of surface adjacent volume elements of each crystal/grain in the surface adjacent region of a polycrystalline specimen with respect to the specimen frame of reference. Thereby a direct representation of such obtained results in a surface-normal inverse pole figure suggests itself. An example of a surface-normal inverse pole figure determined by EBSD is shown in Fig. 4.58.

---

[25] Electron backscatter diffraction (EBSD) is usually carried out in a scanning electron microscope (SEM; see Sect. 6.9). The electrons impinging on the surface (tilted with respect to the incident electron beam) and penetrating surface adjacent material of a crystalline solid, after backscattering may be diffracted, according to Bragg's law (Eq. (4.9)), by lattice planes inclined with respect to the surface of the specimen. Such diffracted electrons may escape from the surface of the material over a depth ranging till, say, 10–40 nm. These thus back-scattered and diffracted electrons can produce a diffraction pattern on a detector (screen). This diffraction pattern reveals so-called Kikuchi bands where each band corresponds to one set of diffracting lattice planes. The (mostly computerized) interpretation of this diffraction pattern (pattern of Kikuchi bands) leads to determination of the crystal orientation in the specimen frame of reference.

4.8   Aperiodic Crystals

An ideal crystal presents perfect long-range translational order of the constituting atoms. A completely disordered (chemically and translationally), chaotic spatial distribution of the constituting atoms, of course subject to the constraint of excluded, occupied volume, could serve as a model for the ideal amorphous solid. These descriptions represent the two extreme cases of atomic arrangements in a solid. As already remarked in the introductory part of this chapter, both extremes do not comply with reality: crystals can contain defects and amorphous solids can exhibit short-range order.

In the final section of this chapter, and with reference to the above paragraph, the focus is on, in a way, intermediate types of atomic arrangements which are not compatible with perfect long-range *translational* order and yet do give rise to diffraction patterns exhibiting more or less sharp diffraction peaks: *aperiodic crystals*. These crystals exhibit, ideally perfect, long-range, geometrical order, but do not possess translational periodicity. Translational periodicity is only one way to establish long-range geometrical order....

4.8.1   Incommensurately Modulated Atomic Structures

Starting with a perfect crystal, characterized by a perfect, translationally periodic structure, a modulation, also of translational periodicity, can be superimposed on this parent periodic structure. The modulation can pertain to the composition (type of atom) or the position of the atoms. The modulation period can be *commensurate*, i.e. equal to a rational number of periods of the underlying parent, translation lattice (implying that x times the period of the parent, translation lattice equals y times the translational period of the modulation), or the modulation period can be *incommensurate*, i.e. equal to an irrational number of periods of the underlying parent, translation lattice (implying that x times the period of the parent, translation lattice can never equal y times the translational period of the modulation). As an illustration, see Figs. 4.59 and 4.60.

A one-dimensional crystal composed of a random mixture of two types of atoms, A and B is shown at the top of Fig. 4.59: the chance to meet an atom A at a specific atomic site (here the sites/nodes of the translation lattice are taken as the sites available for the atoms) is given by the atom fraction of A in the crystal and one speaks of a disordered solid solution (see Sect. 4.4.1.1); the crystal exhibits perfect long-range translational order. A compositional modulation occurs as soon as specific sites are occupied preferably by atoms of type A. The preference may be that strong that only atoms of type A can occur at the sites specified, or the preference is less outspoken and now and then also atoms or type B occur at the sites preferred by atoms of type A. This dependence of atom-type preference on position in the crystal can be described by a modulation function, the amplitude of which is a measure for the preference of atom type A for the site at the position concerned. As an example, a sinusoidal compositional modulation function is shown in the figure that has a period equal to an integer number of the translation period of the parent translation lattice and that, in the specific case shown, peaks at each third lattice/atomic site (positive, maximal amplitude). In the extreme case, all A atoms are only at their most preferred sites and a completely ordered solid solution has been obtained ("degree of order" = 1); then, of course, the in reality occurring compositional variation, in contrast with the above modulation function describing the dependence of atom-type *preference* on position in the crystal, is not sinusoidal. Whether the degree or order equals one or not, the ordering has led to the development of a "superstructure" (see Sect. 4.4.1.1). The crystal structure now has a (primitive) unit cell larger than in the unmodulated case (see the discussion in Sect. 4.4.1.1). In this case of a modulation function that is superimposed on the parent, translation lattice such that the period of the modulation function equals a *rational (e.g. integer) number* of periods of the parent, translation lattice, one speaks of a "*commensurately* modulated crystal structure".

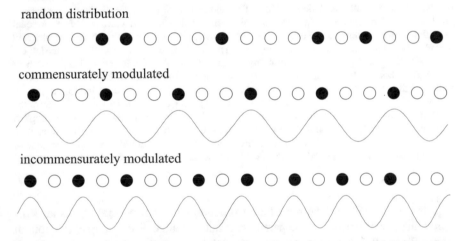

random distribution

commensurately modulated

incommensurately modulated

Fig. 4.59 Compositional modulation. The top of the figure shows a one-dimensional crystal, with translation period a, of a binary system with the constituting A (black) and B (white) atoms distributed in a random manner. The middle part of the figure represents the case that lattice-site preference for occupation by the A (black) atoms is described by a modulation function of translation period $3a$ (see the sinusoidal function drawn below the one-dimensional crystal). In the case shown, the A atoms have chosen the sites where the modulation function exhibits a maximum. The period of the modulation function equals a rational (here: integer) number of periods of the parent, translation lattice. A commensurately modulated crystal structure has formed: the translational periodicity is maintained, albeit with a different period (a "superstructure" has formed). If, as shown in the bottom part of the figure, the period of the compositional modulation equals an irrational number of times the original translation period (in the figure the period of the compositional modulation is given by $(3/\sqrt{2})a$), then translational periodicity is lost. An incommensurately modulated crystal structure has formed, which is characterized by two translation periods: the one of the parent, translation lattice and the one of the compositional modulation

constant interatomic distance

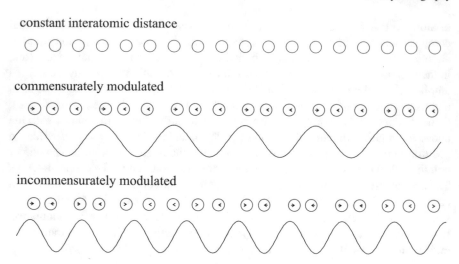

commensurately modulated

incommensurately modulated

Fig. 4.60 Displacive modulation. The top part of the figure shows a one-dimensional crystal of one type of atoms with translation period a. The middle part of the figure represents the case of positional displacement for each atom according to a modulation function with a period $3a$, i.e. equal to a rational (here: integer) number of periods of the parent, translation lattice. A commensurately modulated crystal structure has formed. The bottom part of the figure shows a case where the lattice-site modulation function is equal to an irrational number of times the original translation period (here $(3/\sqrt{2})a$). As holds for the compositional modulation shown in the bottom part of Fig. 4.59, thereby translational periodicity is lost: an incommensurately modulated crystal structure has formed, which is characterized by two translation periods: the one of the parent, translation lattice and the one of the displacive modulation. The arrows shown in the "atoms" represent the displacement vector applied to the centre of the atom, such that the arrows originate from the original, not displaced position of the atom and end with the arrow head at the actual displaced position; sign and magnitude of the displacements are prescribed by the sinusoidal modulation drawn below the one-dimensional crystal

It is conceivable that the modulation function has a period that equals an *irrational* number of periods of the parent, translation lattice. For example, the bottom part of Fig. 4.59 shows a compositional modulation with a period that equals $(3/\sqrt{2})$ times the period of the parent, translation lattice. Now it is impossible to define a unit cell (coincidence of x times the period of the parent, translation lattice and y times the period of the modulation function never occurs): the structure has lost its translational periodicity! One then speaks of an "*incommensurately* modulated crystal structure". The incommensurately modulated crystal structure is fully characterized by two periodicities. Thereby the "chaos" introduced by the modulation in the parent crystal structure is still of highly regular, periodic nature, and therefore the diffraction pattern still reveals a collection of sharp diffraction maxima.

The modulation function needs not pertain to a modulation of the composition: it may indicate a position displacement. In the last case, the atoms at their sites have been shifted as compared to the sites prescribed by the parent, perfect, translationally periodic structure. A one-dimensional crystal composed of one type of atoms is shown at the top of Fig. 4.60. A displacive modulation function can be defined such that its amplitude is a measure for the displacement to be applied to the atom on the concerned site of the parent structure. In the figure, two such modulation functions are shown, as well as the resulting atomic arrangements: A modulation function with a period equal to a rational (e.g. integer) number of periods of the translation period of the parent, translation lattice leads to a *commensurately* modulated crystal structure: a "superstructure". A modulation function with a period equal to an irrational number of periods of the translation period of the parent crystal lattice leads to loss of periodicity and an *incommensurately* modulated atomic structure results.

Another type of incommensurately modulated atomic structure results by the interpenetration of two crystal structures of different atoms; each crystal structure possesses its specific, own periodicities. Such an "intergrowth compound"/"composite structure" can exhibit incommensurability if at least along one direction in the resulting atomic structure incommensurability of corresponding periodicities of both crystal structures occurs (Fig. 4.61).

In nature, the situation can be more complex than sketched above. For example, the modulations can be of three-dimensional nature. But the essence of the discussion remains unaffected.

To find out if an atomic structure is commensurately or incommensurately modulated may not be trivial. The period of modulation for a commensurately modulated atomic structure and the period of the parent, translation lattice may come into register only over an enormous distance. Then, from a practical point of view,

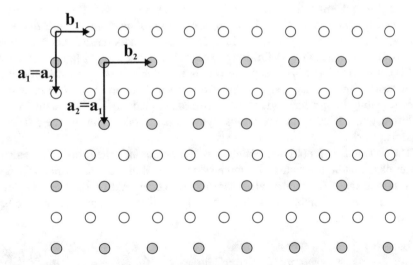

Fig. 4.61 Two-dimensional illustration of an "intergrowth compound"/"composite structure". Two interpenetrating crystal substructures, here with their atomic sites occupied by white and grey atoms, respectively. Such a composite structure exhibits incommensurability if along at least one direction in the resulting atomic arrangement incommensurability of periodicities of both constituting crystal substructures occurs. In the example shown: $\mathbf{a}_1 = \mathbf{a}_2$ and $\mathbf{b}_1 = 1/2 \sqrt{2} \mathbf{b}_2$

distinction between such commensurate modulations and genuinely incommensurate modulations cannot be made.

An incommensurately modulated atomic structure in three-dimensional, "physical" space, itself lacking translational periodicity, can be described as an imaginary, translationally periodic structure in higher dimensional space. Thus, for one-, two- and three-dimensionally incommensurate modulations, the corresponding translationally periodic structures (density functions) occur in four-, five- and six-dimensional space, respectively. This concept of higher dimensional *superspace* was introduced by de Wolff in 1974 and led to the development of superspace *groups* (cf. Sect. 4.1.2 for the notion "space groups") to describe the symmetry of aperiodic, incommensurately modulated atomic structures in three-dimensional space (Janssen et al. 1999). The real, incommensurately modulated atomic structure is obtained as the "cut" ("intersection") of the imaginary, translationally periodic "crystal" in higher dimensional space with (three-dimensional) "physical" space.

Incommensurately modulated atomic structures may be much more common than possibly perceived. Papers reporting such results can be found in the literature from 1950 onwards with a frequency increasing with time.

4.8.2   Quasicrystals

It can be shown that three-dimensional translational symmetry excludes the occurrence of fivefold and more than sixfold rotational symmetry (see the *"Intermezzo: A short note on point groups, crystallographic point groups, plane groups and space groups; glide and screw operations"* in Sect. 4.1.2). For a long time, this was considered as, almost, a (negative) definition of a crystal. Starting with the famous discovery by Shechtman et al. (1984), this picture changed dramatically. Many metallic alloys upon quenching (= very fast cooling) from the melt develop solid atomic structures revealing long-range order (sharp diffraction peaks occur) but with fivefold, eightfold, tenfold and 12-fold rotational symmetries, as deduced from the diffraction patterns.

The occurrence of rotational symmetries of the types indicated above immediately makes clear that no translational symmetry can prevail in these materials, which yet possess long-range (orientational) order (note the occurrence of a diffraction pattern exhibiting (a finite number of) sharp diffraction peaks). These materials have been called *quasicrystals*.

Intermezzo: A Revolution in Crystallography; "Young" versus "Old"

The observation of "forbidden" (icosahedral) symmetry in *long-range geometrically ordered* atomic structures was experienced by the crystallographic community as nothing less than a shock. It may be no surprise that Shechtman initially experienced great problems in convincing, amongst others, his peers and colleagues (thereby including coauthors of his eventual 1984 paper), in particular because it was well known that multiply twinned (cf. Sect. 5.3) structures can exhibit icosahedral symmetry. After publication of the paper, the finding was controversially discussed: one of the opponents being nobody else than Noble-prize winner Linus Pauling. The resistance offered by Pauling, thereby, in effect, representing crystallographic establishment, is (very) remarkable, as Pauling, through his career, has acted rather as a revolutionist (in this context, it is also noted that he was a fierce competitor of Watson and Crick in the race for the clarification of the structure of DNA; see the *"Epilogue: how science really happens"* of Chap. 3 and in particular the book by Watson referred to there). However, in retrospect, the original results as published in the 1984 paper, and later work, convincingly demonstrated the genuinity of the long-range orientational ordering allowing (icosahedral) symmetry incompatible with translational periodicity. In 2011, Shechtman was awarded the Nobel Prize for Chemistry for his discovery made about 30 years earlier.

This story thus reflects a "happy ending" for a new and revolutionary finding and idea proposed by an, at the time, relatively unknown scientist. It is typical for the progress of science that an established scientific community only reluctantly, and delayedly, adopts the consequences of a "break-through". Indeed, the road of science is paved also by scientists who, disappointedly, leave a scientific field, and even abandon science as a profession, because the "break-through" nature of their work has not been recognized soon enough (if at all).

Quasicrystals often reveal icosahedral symmetry (are orientationally long-range ordered; an icosahedron, composed of 20 triangular faces, having 12 vertices where five triangular faces meet, is shown in Fig. 4.62). Liquids frequently exhibit, locally, atomic arrangements of, more or less, icosahedral structure (cf. the *"Intermezzo: Entropy of fusion and the structure of liquids"* in Sect. 7.5.1). Thus, upon rapidly cooling, it may be conceivable that "freezing in" of such local structures may occur. For an appropriate range of cooling rate (of the liquid), the icosahedral entities are preserved in the solid state, and (had the chance to become) aligned such that they have the same orientation, but an ordering according to a Bravais *translation* lattice has not been realized. This picture suggests that the structure of a quasicrystal is based on a collection of icosahedral "units" of more or less identical orientation, not exhibiting any translational ordering, interspersed with "disordered" material. This model is described as the "icosahedral glass" model.

A more explicit interpretation of the structure of quasicrystals derives from, partly ancient, ways of tiling a plane. A tile is a planar object of a specific shape. Subject to the condition that the tiling of the plane considered is "massive" (one then also speaks of "tessellation"), i.e. without gaps between the tiles and without overlapping of tiles, the plane can be tiled in (yet) an infinite number of ways.

Fig. 4.62 Icosahedron, composed of 20 triangular faces, having 12 vertices where five triangular faces meet. Note that an axis, revealing by rotation about it the occurrence fivefold rotational symmetry, passes through a vertex

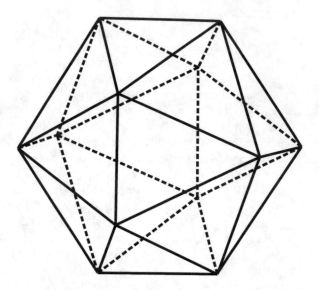

A *periodic* tiling, exhibiting long-range translational symmetry, occurs if a (two-dimensional) unit cell can be designated that by translation fills the plane completely; the unit cell itself can be composed of, for example, a set of identical tiles related by rotation/reflection operations. One type of massive tilings (i.e. tessalations; see above), of *non-periodic* nature but of outspoken long-range *orientational* order, are the so-called Penrose tilings, named after the mathematician Penrose who described such tilings in 1974. A Penrose tiling can be constructed from two tiles of different, specific shapes. For example, by taking the two rhombic unit cells indicated in Fig. 4.63a, i.e. two lozenges with equal edges: a "fat" lozenge with angles of 72° and 108° and a "skinny" lozenge with angles of 36° and 144°, the plane can be filled massively, applying certain "matching rules" (i.e. the tiles are joined in a particular fashion), such that a case of high orientational long-range order occurs (Fig. 4.63b). Evidently, fivefold rotational symmetry occurs in this tiling. The tiling can be considered as representing a specific non-periodic (two-dimensional) atomic structure after having put atoms at, for example, the nodes in the tiling.

As follows from inspection of Fig. 4.63b, at specific locations in the two-dimensional tiling, fivefold rotation axes, with the rotation axes perpendicular to the plane of the tiling, can be indicated. It is important to remark that upon operation of such a rotation over (a multitude of) 72°, a coincidence with the original pattern/tiling is only established for a region immediately surrounding the rotation axis concerned. This is a striking difference with crystals exhibiting translational symmetry: in that case, such rotations (if allowed; see the beginning of Sect. 4.8.2) lead to coincidence for the entire crystal. Further consideration of Fig. 4.63b shows that the building units to construct this tiling, the rhombic "unit cells" shown in Fig. 4.63a, in the tiling take only orientations out of a limited set of possibilities: see the arrangement of identical lozenges around a number of the rotation axes. Thereby

(a)

"fat" lozenge   "skinny" lozenge

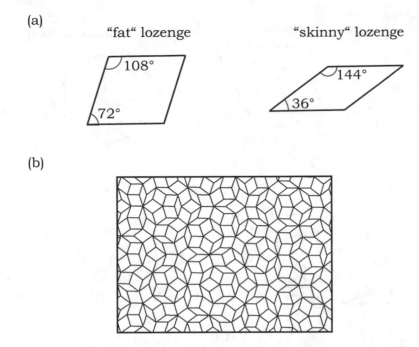

(b)

Fig. 4.63 **a** Example of two basic tiles necessary and sufficient to construct a massive tiling ("tessellation") showing long-range orientational ordering. In this case, two rhombic unit cells are shown with equal edges: a "fat" lozenge and a "skinny" lozenge. **b** By application of certain "matching rules" to join the basic tiles shown in (**a**) in a particular fashion, the plane can be filled massively such that a tiling of long-range orientational order occurs: a Penrose tiling

the occurrence of (long-range) orientational order, and the absence of translational order, has been illustrated.

In three dimensions, a corresponding Penrose "tiling" of space is possible. The two "tiles" then are two rhombohedra: a "prolate" rhombohedron and an "oblate" rhombohedron (see Fig. 4.64a). A specific non-periodic (three-dimensional) atomic structure results after having put atoms at specific positions in the two "tiles", for example, at the nodes/vertices in the tiling (see Fig. 4.64b).

The true atomic structure of three-dimensional quasicrystals may be given by some intermediate of both extreme models discussed above, i.e. in-between the icosahedral "glass" and the three-dimensional Penrose "tiling".

"Icosahedral" quasicrystals, as discussed above, exhibit no translational symmetry at all. Quasicrystals which do preserve translational symmetry in one or two dimensions have been observed.

The construction of quasicrystals, out of the individual atoms, is subjected to some rules (Janot 1994). One condition reads: A quasicrystal should be "quasi-periodic", i.e. it should be possible to express the atomic density function as a finite sum of periodic functions; the periods of a few of these periodic functions should be incommensurate (a similar "quasi-periodicity" holds, of course, for the incommensurately

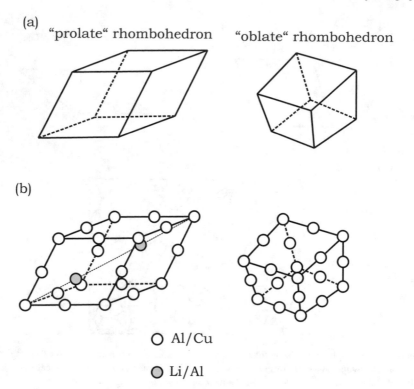

Fig. 4.64 a Two basic "tiles" for filling massively three-dimensional space with a Penrose "tiling" (cf. Fig. 4.62) are two rhombohedra: a "prololate" rhombohedron and an "oblate" rhombohedron. **b** A specific non-periodic three-dimensional atomic structure results if in the corresponding three-dimensional Penrose "tiling" atoms are put at specific sites of the two basic "tiles", as indicated in the figure for "icosahedral" Al_6CuLi_3 (see van Smaalen 1995)

modulated atomic structures discussed in Sect. 4.8.1). Indeed, it can be shown that "Penrose tilings", as considered above, comply with this rule. As a consequence, the real, three-dimensional atomic structure of a quasicrystal can be conceived as the "cut" ("intersection") of an imaginary, translationally periodic structure in high-dimensional space (cf. van Smaalen 1995) with (three-dimensional) "physical" space. Such superspace description was already introduced for the incommensurately modulated atomic structures in Sect. 4.8.1. The description of a quasicrystal of icosahedral orientational symmetry as a translational periodic structure requires an (at least) six-dimensional "superspace".

Finally, one may wonder which types of interatomic forces govern the occurrence of incommensurately modulated crystals and, in particular, quasicrystals. In the last case, two geometrically different types of "clusters" of atoms (the notion "unit cell" is not allowed) appear to be preferred, which, moreover, are subjected to highly specific "matching rules" upon constituting the quasicrystal (if the Penrose "tiling" approach is adopted). In view of the degree of complexity involved in the atomic structural arrangement in, especially, quasicrystals, it may be suggested that stabilization by a contribution of configurational entropy (entropy of mixing) plays a lead role, then of course especially at higher temperatures (cf. Sects. 7.3 and 5.1.1 for the notion "(configurational) entropy" and understanding of this remark), which rises doubts about the thermodynamic stability of quasicrystals at (very) low temperatures (see, for quasicrystals, an overview by Steurer (2018)). It may be that local bonding requirements, which are difficult to be accommodated in a translationally periodic structure, induce such aperiodic crystals. Recognizing the subtlety of minor energy effects in controlling the prevalence for a certain, normal (i.e. translationally symmetric) and (even) relatively simple crystal structure, as discussed in Sect. 3.5.3 and at the end of the introductory part of this chapter, it does not come as a surprise to remark that the current state of knowledge does not comprise understanding profound enough to answer such questions detailedly.

Epilogue: The Notion Crystal Revisited

In the beginning of this chapter, a crystal has been defined as the regular, periodic, three-dimensional space filling arrangement of unit cells. However, in the preceding, Sect. 4.8 *incommensurately modulated atomic structures* and *quasicrystals* were introduced, which structure types have been gathered under the heading *aperiodic crystals*. A definition of the type just reiterated apparently does not comply with these last mentioned structural arrangements of atoms: these atomic arrangements do not display three-dimensional translational periodicity and yet diffraction patterns occur with well-defined intensity maxima, as for the usual crystals exhibiting translational periodicity (cf. Sect. 4.5). If then the focus is on the whether or not occurrence of well-defined (researchers of aperiodic crystals here often speak of "essentially discrete") diffraction maxima, an ideal crystal may be defined through its diffraction pattern, thereby comprising the aperiodic crystals. The "Online Dictionary of Crystallography", maintained by the Commission for Crystallographic Nomenclature of the International Union of Crystallography, (thus) provides, at the time of writing this text, a definition of a crystal as a substance that "has essentially a sharp diffraction pattern (see: http://reference.iucr.org/dictionary/Crystal) However, also this definition has flaws: it is, after all, based on an experimental image and these cannot be made of perfect, involving infinitely large as well, crystals. It has thus been proposed to define an ideal crystal as a solid body having long-range *positional* order (Ben-Abraham 2007). But this, in turn, can be criticized: what actually is long-range positional order, after all?[26] The problem is even more complicated if one strives for also including real, i.e. imperfect crystals in the definition to be given for a crystal.

---

[26] It can be shown that the (X-ray) diffraction pattern of a crystal can be conceived as the Fourier transform of its (electron) density. (The mathematical operation "Fourier transform" cannot be

What we touch upon here concerns an on-going debate in the field of crystal-lography. Moreover, a trend exists to extend the field of crystallography beyond its classical borders (the inclusion of aperiodic crystals is only a first example): any material exhibiting a certain, small or large degree of ordering of atomic entities can be subject of investigation (to describe the atomic arrangement) by crystallographers.

Far from being a conclusive discussion, the above presents a message. Sharp, watertight definitions about natural phenomena and objects are very difficult to formulate in the natural sciences, as materials science, in contrast with defini-tions of phenomena and objects in the isolated (here is meant "detached from the real world") field of mathematics. Another such, problematic definition is encountered in Chap. 7, where the concept "phase" is introduced.

Appendix: How to Deal with Atoms at Unit-Cell Boundaries

The composition and the density of a unit cell are equal to the composition and density of the whole crystal. Thus, the information about geometry ($a \equiv |a|$, $b \equiv |b|$ and $c \equiv |c|$, α, β, γ) and contents (fractional coordinates $0 \leq x, y, z < 1$; compare also Table 4.1) of the unit cell suffices to calculate the composition and the density of massive, solid crystalline material, provided the grain-boundary density is insignificant (Thus, this statement need not hold for nanocrystalline materials; i.e. for crystal/grain sizes less than 100 nm).

A crystal structure can be represented graphically by a unit cell of that crystal structure, displaying atoms with fractional coordinates in the range $0 \leq x, y, z \leq 1$ (note that for drawing the unit cell the second "$\leq$" is not a "$<$" like above). Upon counting the atoms on the basis of a unit-cell drawing, a problem arises for atoms located at fractional coordinates with either $x = 0$ (or 1) or $y = 0$ (or 1) or $z = 0$ (or 1), i.e. for atoms somewhere at the bounding faces or edges or at corners of the unit cell. Two examples of such cases are shown in Fig. 4.65a, b. Table 4.8 lists the fractional coordinates of the atoms observed in these unit-cell drawings.

The drawn unit cell of the CsCl structure in Fig. 4.65b shows 8 Cs atoms but only 1 Cl atom (see Table 4.8). The drawn unit cell may thus suggest erroneously that the composition of this ionic compound is given by the formula Cs_8Cl, whereas the

---

introduced in this book, but that does not obstruct understanding the essence of this footnote) Thus, it has been proposed to define "long-range positional order" as equivalent with the occurrence of sharp peaks in the Fourier spectrum of the object. In this way, one would have given an operational definition of "long-range positional order", and by avoiding a reference to an experimental diffrac-tion pattern but instead relying on a mathematical operation to be applied on the object considered, some obscurity in the definition of a crystal in the sense discussed here could be avoided (Lifschitz 2007).

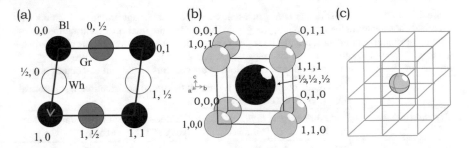

Fig. 4.65 a Unit cell of the two-dimensional crystal structure of the hypothetical compound "BlGrWh". **b** Unit cell of the three-dimensional crystal structure of the physical compound CsCl. The fractional coordinates x and y in (**a**) and x, y and z in (**b**) have also been indicated. **c** A Cs atom, of the crystalline compound CsCl, at a corner of the unit cell considered in (**b**), contributes to the eight unit cells sharing this corner

true composition is represented by the formula CsCl. The discrepancy is resolved if one recognizes that the Cs atoms, having their centres of mass located at the corners of the unit cell, also contribute to/are part of the adjacent, neighbouring unit cells. Hence, adopting the Cs atoms as solid spheres, only a fraction 1/8 of each sphere is a part of the drawn unit cell, as every corner of the unit cell is equally (cf. footnote a to Table 4.8!) shared by 8 unit cells (see Table 4.8 and Fig. 4.65c). The fractional contribution of an atom that would lie on an edge of the unit cell is 1/4, because every edge of the unit cell is shared by 4 unit cells. The fractional contribution of an atom that would lie on a face of the unit cell is 1/2, since every face of the unit cell is shared by 2 unit cells.

Thus, it now simply follows that the unit cell as drawn in Fig. 4.65b contains 8 × 1/8 Cs = 1 Cs atom and 1 Cl atom, in agreement with the true composition of the compound.

In general, if unit cells are considered, as in drawings, with all atoms with fractional coordinates in the range $0 \leq x, y, z \leq 1$, then, for counting the atoms in the unit cell, and if any of their fractional coordinates equals 0 or 1, one has to sum them according to their partial contributions to the unit cell considered. This complication is obviously avoided if one only considers/draws all atoms with fractional coordinates in the range $0 \leq x, y, z < 1$ (note the use of "<" instead of "≤"); then all atoms considered/drawn contribute fully to the unit cell considered/drawn. But such graphical presentations are not made usually.

Table 4.8 Two ways of listing the atoms in a unit cell. The first variant provides all atoms with fractional coordinates in the range $0 \leq x, y, z \leq 1$. In this variant, the fractional contribution of each atom to the contents of the unit cell can be smaller than one. The second variant provides only the atoms with fractional coordinates in the range $0 \leq x, y, z < 1$ (note use of "<" instead of "≤"). In this variant, the fractional contribution of each atom to the contents of the unit cell equals one

| Crystal structure | Fractional coordinates x, y, (and z); fractional contribution, w, of atom at $(x, y(, z))$ to the contents of the unit cell | |
	Two-dimensional crystal structure of "BlGrWh" (Fig. 4.65a)	Three-dimensional crystal structure of CsCl (Fig. 4.65b)
All atoms in unit cell considered/drawn with $0 \leq x, y, z \leq 1$ (usual consideration)	Bl: 0, 0 $w^a = \gamma/360°$ Bl: 1, 0 $w^a = 1/2 - \gamma/360°$ Bl: 0, 1 $w^a = 1/2 - \gamma/360°$ Bl: 1, 1 $w^a = \gamma/360°$ Gr: ½, 0 $w = 1/2$ Gr: ½, 1 $w = 1/2$ Wh: 0, ½ $w = 1/2$ Wh: 1, ½ $w = 1/2$ unit cell content: 1 Bl + 1 Gr + 1 Wh	Cs: 0, 0, 0 $w = 1/8$ Cs: 1, 0, 0 $w = 1/8$ Cs: 0, 1, 0 $w = 1/8$ Cs: 0, 0, 1 $w = 1/8$ Cs: 1, 1, 0 $w = 1/8$ Cs: 1, 0, 1 $w = 1/8$ Cs: 0, 1, 1 $w = 1/8$ Cs: 1, 1, 1 $w = 1/8$ Cl: 1/2, 1/2, 1/2 $w = 1$ unit cell content: 1 Cs + 1 Cl
All atoms in unit cell considered/drawn with $0 \leq x, y, z < 1$	Bl: 0, 0 $w = 1$ Gr: 1/2, 0 $w = 1$ Wh: 0, 1/2 $w = 1$ unit cell content: 1 Bl + 1 Gr + 1 Wh	Cs: 0, 0, 0 $w = 1$ Cl: 1/2, 1/2, 1/2 $w = 1$ unit cell content: 1 Cs + 1 Cl

[a]The fractional contribution, w, of the Bl atoms depends on the value of the angle γ. If a rectangular (or quadratic; cf. Sect. 4.1.2) two-dimensional unit cell would occur, i.e. $\gamma = 90°$, the weights for all four Bl atoms would be 1/4. An analogous complication occurs for atoms at the corners of three-dimensional unit cells which are not "rectangular" (i.e. not cubic, tetragonal or orthorhombic; cf. Sect. 4.1.2); but it always holds that the sum of the fractional atom contributions of all corners of the unit cell equals one. For three-dimensional unit cells, similar complications also occur for atoms at edges, but do *not* occur for atoms at faces (i.e. the w values for atoms at faces are always 1/2 for three-dimensional unit cells). For two-dimensional unit cells, the w values for atoms at edges are always 1/2

References

General

C. Giacovazzo (ed.), *Fundamentals of Crystallography*, 2nd edn. (Oxford University Press, Oxford, England, 2002)

C. Hammond, *The Basics of Crystallography and Diffraction*, 2nd edn. (Oxford University Press, Oxford, England, 2001)

U. Müller, *Inorganic Structural Chemistry*, 2nd edn. (Wiley, West Sussex, England, 2007)

D. Schwarzenbach, *Crystallography* (Wiley, West Sussex, England, 1996)

R. Tilley, *Crystals and Crystal Structures* (Wiley, West Sussex, England, 2006)

Specific

S.I. Ben-Abraham, What is a crystal? Z. Kristallogr. **222**, 310 (2007)

R. Cerny, V. Favre-Nicolin, Direct space methods of structure determination from powder diffraction: principles, guidelines and perspectives. Z. Kristallogr. **222**, 105–113 (2007)

T. Egami, Understanding the properties and structure of metallic glasses at the atomic level. JOM **62**, 70–75 (2010)

I. Hargittai, M. Hargittai, *Symmetry: A Unifying Concept* (Shelter Publications Inc., Bolinas, California, USA, 1994)

M. Hargittai, Symmetry, crystallography, and art. Appl. Phys. A **89**, 889–898 (2007)

C. Janot, *Quasicrystals; A Primer*, 2nd edn. (Clarendon Press, Oxford, England, 1994)

T. Janssen, A. Janner, A. Looijenga-Vos, P.M. de Wolff, Incommensurate and commensurate modulated structures, in *International Tables for Crystallography*, vol. C, 2nd edn. ed. by A.J.C. Wilson, E. Prince (Kluwer Academic Publishers, Dordrecht, The Netherlands, 1999), pp. 899–947

O. Johari, G. Thomas, *The Stereographic Projection and Its Applications* (Interscience Publishers, New York, 1969)

U.F. Kocks, C.N. Tomé, H.R. Wenk (eds.), *Texture and Anisotropy* (Cambridge University Press, Cambridge, United Kingdom, 1998)

R. Lifschitz, What is a crystal? Z. Kristallogr. **222**, 313–317 (2007)

O. Medenbach, P.W. Mirwald, P. Kubath, Rho und Phi, Omega und Delta—Die Winkelmessung in der Mineralogie. Mineralien-Welt **6**, 16–25 (1995)

E.J. Mittemeijer, M.A.J. Somers (eds.), *Thermochemical Surface Engineering of Steels* (Elsevier, Woodhead Publishing, Cambridge, United Kingdom, 2015)

M. Nespolo, Lattice versus structure, dimensionality versus periodicity: a crystallographic Babel? J. Appl. Crystallogr. **52**, 451–456 (2019)

G. Petzow, *Metallographic Etching* (ASM, Materials Park, Ohio, USA, 1999)

D. Shechtman, I. Blech, D. Gratias, J.W. Cahn, Metallic phase with long-range orientational order and no translational symmetry. Phys. Rev. Lett. **53**, 1951–1953 (1984)

S. van Smaalen, Incommensurate Crystal Structures. Crystallogr. Rev. **4**, 79–202 (1995)

W. Steuer, Quasicrystals: what do we know? What do we want to know? What can we know? Acta Crystallograph. A **A74**, 1–11 (2018)

S.M. Woodley, R. Catlow, Crystal structure prediction from first principles. Nat. Mater. **7**, 937–946 (2008)

Chapter 5
The Crystal Imperfection; Structure Defects

Idealized presentations of atomic arrangements exhibiting long-range translation symmetry, i.e. idealized crystal structures, have been presented and discussed in the previous chapter. Very many properties of crystalline materials cannot be understood merely on the basis of such perfect atomic arrangements. As a matter of fact, defects in the atomic arrangement, as compared to the idealized ordering, strongly determine material properties as mechanical strength, diffusion, electrical conductivity and so on.

Not all deviations of the perfect arrangement can be named defects. Thermal vibrations of the atoms, with their ideal lattice sites as centroids of these vibrations, occur, with amplitudes increasing with temperature and frequencies of the order 10^{13}/s. The frozen-in configuration of atoms is not considered as a defect structure: lattice defects are defined with respect to the *time averaged* atomic configuration. Further, atomic arrangements modified by elastic strains are also not considered as defect: if a crystal structure can be rendered perfect by applying a purely elastic deformation, the crystal considered is said to be perfect, still.

At this place it is appropriate to remark that the finiteness of a crystal should be considered as a defect: a perfect crystal is infinitely large. The presence of a surface implies that the crystal contains atoms (in the surfaces) with incomplete bonding (as compared to atoms in the bulk): at the surface a symmetry break occurs.[1] These surface atoms thus have a higher energy than the other (bulk) atoms. In fact, this is a way to make likely that the generation of a surface is associated with the introduction of an extra, so-called surface, energy. Also X-ray (electron, or....) diffraction experiments indicate that the finite size of a crystal is a defect, as the diffraction lines/peaks become broadened due to the finite size (smallness) of crystals (see Sect. 6.10.1).

---

[1] This phenomenon of incomplete bonding (unsatisfied bonds) at the surface of a crystal is also responsible for the occurrence of a coefficient of linear thermal expansion that is larger for small crystals (with a relatively large ratio of surface to bulk atoms) than for the corresponding bulk material (cf. Sect. 3.1).

© Springer Nature Switzerland AG 2021
E. J. Mittemeijer, *Fundamentals of Materials Science*,
https://doi.org/10.1007/978-3-030-60056-3_5

The word "defect" has a negative aura. This is not generally justified within the context of materials science, in particular as dealt with in this chapter. Defects can be essential in realizing profitable material properties and their handling is a cardinal, if not the dominant part of the manipulation of the microstructure by the materials scientist and engineer. This will be made clear in this and, especially, the forthcoming chapters of this book.

In the following, a generally applied classification of structure defects is adopted, where the dimensionality of the defect concerned is considered to be a distinguishing feature.

5.1 Point Defects (Zero-Dimensional)

5.1.1 Singular Point Defects: Thermal and Constitutional Defects; Vacancies; Interstitial, Substitutional and Antistructure Atoms

Consider a translation lattice, where the lattice points (lattice sites/lattice nodes; see Sect. 4.1.1) all are occupied by atoms of a single element. If at an atom site of the resulting crystal structure an atom of this element lacks, the defect resulting, i.e. an unoccupied crystal-structure site, is called a *vacancy* (see Fig. 5.1a). Such a vacancy can be generated by thermal excitation (see Eq. (5.1) below) and then is called a *thermal vacancy*.

Thermally induced vacancies should be considered as stable (=equilibrium) defects. It costs energy to introduce a vacancy into the crystal (introducing a vacancy can be considered as "alloying", i.e. an atom of a foreign element is dissolved into the infinitely large, perfect crystal of element A); this is called the formation enthalpy of a vacancy, ΔH_{vac} (>0). Yet, the very many ways to realize this vacancy (the vacancy can be positioned at any site of the lattice), is a stabilizing factor for the crystal considered and thus associated with a decrease of energy (equal to the product of absolute temperature and the entropy of mixing (i.e. (the change in) configurational entropy; for background, see Sect. 7.3). As a result, applying the condition of minimal Gibbs energy for the system, the following equation describes the relation between the equilibrium fraction of vacancies, c_{vac}, and the temperature:

$$c_{vac} = \exp\left(\frac{-\Delta G_{vac}}{RT}\right) \tag{5.1}$$

where ΔG_{vac} denotes the Gibbs energy (free enthalpy; see Sect. 7.3) of formation of a mole of vacancies, apart from the contribution of the entropy of mixing (= change in configurational entropy; cf. Sect. 7.3): $\Delta G_{vac} = \Delta H_{vac} - T \Delta S_{vib}$, where ΔS_{vib} represents (largely) the change in vibrational entropy of the crystal (in particular due

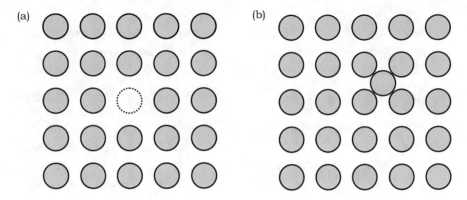

Fig. 5.1 Vacancy (**a**) and interstitial atom (**b**) in a simple cubic structure (shown here for the two-dimensional simple square crystal-structure analogon)

to change in the vibrational frequencies in the neighbourhood of the vacancies) upon introduction of the vacancies (for ΔH_{vac} see above; further see Sect. 8.6.1).

This is an equation of a type one will encounter more often: this so-called Boltzmann-type equation indicates the probability that a (thermally activated) process, that costs an amount of energy (here ΔG_{vac}), occurs. Recognizing that ΔG_{vac} can be of the order of 100 kJ/mol, it follows that close to the melting point of, for example, a metal, the equilibrium fractional vacancy concentration, c_{vac}, is of the order 0.1%.[2] At lower temperature, the equilibrium amount of vacancies is, of course, (very) much smaller (cf. Eq. 5.1). As will be shown in Chap. 8, vacancies, even at such low concentrations, often play a dominant role in diffusion in crystalline materials.

Instead of a missing atom, an extra atom can be positioned at an interstitial site of the above considered crystal structure. Such an atom is called an *interstitial* atom (see Fig. 5.1b).

Further, atoms of a foreign element B can be incorporated in the parent crystal structure of element A considered. Exchange of an atom of the parent element on a site of the crystal structure considered with an atom of the foreign element, leads to so-called *substitutional* dissolution of an atom of element B in the matrix of element A (Fig. 5.2a). Similarly, an atom of element B can also occupy an interstitial site (Fig. 5.2b). Obviously, an atom that can be dissolved interstitially is most likely

---

[2] The values reported in the literature for the equilibrium vacancy concentration of metals (near the melting point) are rather diverse. This has been discussed controversially. In a personal book, devoted to only this problem (!), points in favour of and points detracting opposing points of view have been elaborated in substantial detail, with as a conclusion the emergence of a clear "winner" (Y. Kraftmakher, *Lecture Notes on Equilibrium Point Defects and Thermophysical Properties of Metals*, World Scientific, Singapore, 2000). This is mentioned here to illustrate (again; cf. footnote 35 in Sect. 3.5.2.2) that fundamental questions connected with basic properties of materials, as "simple" as the equilibrium vacancy concentration, cannot be answered satisfactorily and definitively until today. The book referred to here thereby also provides another example of the progress of science, not as a smoothly proceeding development, but rather as characterized by battles of conflicting conceptions of nature, fought by their proponents.

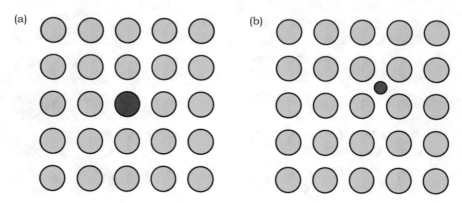

Fig. 5.2 Substitutional (**a**) and interstitial (**b**) solute atom in a simple cubic lattice (shown here for the two-dimensional simple square crystal-structure analogon)

smaller than the atoms of the parent crystal structure which reside on the sites of the parent crystal (see above), as holds for C or N in ferrite (α-Fe) or austenite (γ-Fe); see Sects. 4.4.2 and 9.5.2.1.

Consider the so-called B2-ordered intermetallic compound AB (e.g. NiAl and CoAl). This is a CsCl-type compound where the A atoms reside on the lattice nodes of a primitive cubic translation lattice, which holds as well for the B atoms, and where the two primitive cubic translation lattices and thus the two primitive cubic *substructures* are displaced with respect to each other according to $\frac{1}{2}a<111>$ of the "CsCl" unit cell (see Fig. 3.7, Sect. 4.4.1.1 and in particular Fig. 4.34). A perfectly ordered structure, with all sites of the combined translation lattices occupied, can be realized only at the stoichiometric composition and at absolute zero temperature. At 0 K *a deviation in composition* from the stoichiometric composition, while maintaining the, in this case B2-type, crystal structure, requires the introduction of point defects: the so-called *constitutional defects* (see next paragraph). The number of constitutional point defects is no direct function of temperature.

For an ordered compound the ratios of the numbers of atomic sites on the various substructures are fixed. Thus, for the B2 crystal structure, the numbers of atomic sites on both substructures are equal. Two types of point defects are typical for ordered compounds: antistructure atoms and vacancies. If, for the B2 structure, A atoms are ideally placed on their A substructure and B atoms on their B substructure, then *antistructure atoms* are A atoms on the B substructure and B atoms on the A substructure. If the A and B atoms have about the same size, antistructure atoms can occur on both substructures and one speaks of antistructure defect compounds. If the B atoms are considerably larger than the A atoms, the B atoms may not occur as antistructure atoms. Then, to realize a deviation from the ideal composition, an excess of A atoms involves the presence of vacancies on the B substructure and/or the presence of A antistructure atoms on the B substructure (cf. Fig. 5.3a, b) (vice versa, in case of A atoms considerably larger than the B atoms, an excess of B atoms

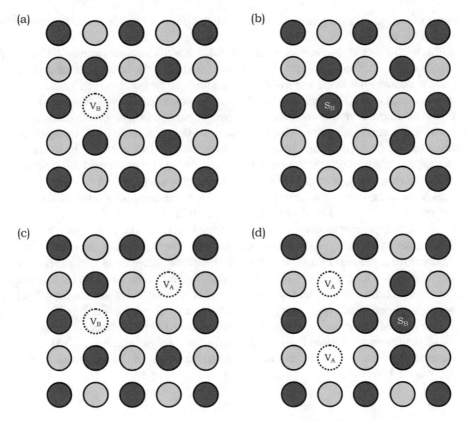

Fig. 5.3 Point defects in an ordered B2-type, binary, CsCl-type compound, where the A atoms (black in the figure) reside on a primitive cubic substructure, which holds as well for the B atoms (grey in the figure), and where the two primitive cubic substructures are displaced with respect to each other according to $\frac{1}{2}a\langle111\rangle$ of the "CsCl" unit cell (shown here for the two dimensional square crystal-structure analogon). **a, b**: *constitutional defects*. In both cases, deviation from the 1:1 stoichiometry towards increasing fraction of A atoms occurs. **a** A vacancy (V_B) on the B substructure. **b** A substitutional A-atom (S_B) on the B substructure; this A atom is called an antistructure atom. **c, d**: *thermal defects*, which occur under the constraint of preservation of the composition, i.e. here the 1:1 stochiometry has to be maintained. **c** Same number of vacancies on the A and B substructures ($V_A - V_B$ vacancy pairs), **d** A triple defect composed of two vacancies on the A substructure (V_A) and one substitutional A atom on the B substructure, i.e. one antistructure atom (S_B). The triple defect is found if the formation of vacancies on one substructure (here: the B substructure) is not possible (The triple defect shown can be conceived as originating from a vacancy pair as shown in **c** by transfer of an A atom to the empty atomic site on the B substructure)

involves the presence of vacancies on the A substructure and/or the presence of B antistructure atoms on the A substructure).

Point defects in ordered compounds can also be thermally activated (cf. Eq. 5.1) while *maintaining the composition*: the so-called *thermal defects* (see Sect. 5.1.2). The same type of point defect can be of constitutional or thermal nature. Usually, in ordered compounds the number of constitutional point defects is much larger than that of the thermal point defects.

In fact, the pictorial presentations in Figs. 5.1, 5.2 and 5.3 are too simplistic. Generally, the introduction of point defects is associated with the occurrence of crystal-structure distortions in the immediate surroundings of the point defect. These *static* displacements should be distinguished from the so-called *dynamic* displacements due to the thermal vibrations of the atoms (see begin of this chapter). In accordance with the above treatment, two cases of occurrence of such static displacements can be indicated: (i) An interstitially dissolved atom usually does not fit in the interstice of the perfect parent crystal structure: the atoms surrounding an interstitial atom are displaced (e.g. "pushed aside") from their ideal lattice-site positions (see Fig. 4.37). This leads to the well-known tetragonal distortion of the octahedral interstitial site in the ferrite (α-Fe) crystal structure upon incorporation of a carbon atom or a nitrogen atom, which atoms are too large for the octahedral interstice offered; see Sect. 9.5.2.1. (ii) Relaxation of the crystal structure surrounding a vacancy occurs: the surrounding atoms move somewhat into the cavity left by the atom "taken away", and as a result the volume of a vacancy is significantly smaller than the atomic volume (as a rule of thumb: the volume of a vacancy is about ½ that of the original atom; see Sect. 8.6.1 where an experimental route to arrive at such a result has been indicated).

These local distortions can be quantitatively assessed by their effect on the average lattice parameter as deduced from (X-ray) diffraction analysis (see Sect. 8.6.1) and/or the analysis of the so-called diffuse scattering around (X-ray) diffraction maxima (e.g. see Warren 1969).

5.1.2   Combined Point Defects: Vacancy Pairs and Triple Defects; Schottky and Frenkel Defects

Again consider the B2-ordered intermetallic compound AB and recall that the ratios of the numbers of atomic sites on the various substructures are fixed (see Sect. 5.1.1). Now, if the composition has to be maintained, as is the case for thermal defects, and if only vacancies occur as point defects, then only *vacancy pairs*, composed of one vacancy on the A substructure and one vacancy on the B substructure, can occur (Fig. 5.3c). Or, also as a consequence of maintaining the composition, the occurrence of an A antistructure atom, on an additional site of the B substructure, has to be balanced by the introduction of two A vacancies: one at the site of the A substructure left by the A atom, now on the B substructure, and one generated as an additional, balancing site of the A substructure: such a combination of point defects that maintains the composition (one antistructure atom in one substructure and two vacancies on the other substructure) is called a *triple defect*, which is a thermal defect (Fig. 5.3d).

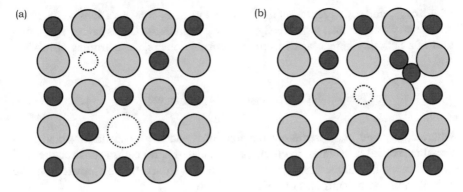

Fig. 5.4 Point defects in an ionic crystal of NaCl-type (or CsCl-type). Cations: small black circles. Anions: large grey circles. Charge neutrality has to be preserved. **a** Schottky defect: one vacancy on the anion substructure and one vacancy on the cation substructure. **b** Frenkel defect: one vacancy on the cation substructure and one interstitial cation

A complication with the introduction of point defects occurs for ionic crystals. Consider the f.c.c. structure of NaCl (cf. Figs. 3.6 and 4.12). The f.c.c. crystal structure is composed of two f.c.c. substructures: one occupied by cations (Na^+) and one occupied by anions (Cl^-), where the two f.c.c. substructures are displaced with respect to each other according to $\frac{1}{2}a<100>$. The creation of (only) a vacancy in the bulk on the anion substructure would violate the charge neutrality of the crystal; a build up of electrostatic energy would occur. Therefore, nature prefers to maintain charge neutrality (no build-up of space charge).

This charge neutrality can be achieved with the simultaneous formation of a vacancy on also the anion substructure. The resulting pair of vacancies (one on the cation substructure and one on the anion substructure) is called a *Schottky defect* (proposed in 1930; Fig. 5.4a). Compare this discussion with the one given above for the occurrence of constitutional vacancies, where the focus was on maintaining the compositional homogeneity.

Another way to preserve charge neutrality would involve that the formation of a vacancy on, say, the cation substructure is associated with the simultaneous formation of an interstitial cation. The resulting pair of a vacancy and an interstitial ion is called a *Frenkel defect* (proposed in 1926; Fig. 5.4b).

Schottky and Frenkel defects are thermally activated, equilibrium defects. Applying thermodynamical equilibrium considerations, it follows for pure, ionic materials (no impurities are present), similarly as for thermally induced vacancies in the crystal of a single element (see above):

$$\text{for Schottky defects:} \quad c_{\text{vac/an}} = c_{\text{vac/cat}} = \exp\left(\frac{-\Delta G_S}{2RT}\right), \quad (5.2)$$

where $c_{vac/an}$ and $c_{vac/cat}$ represent the fractional vacancy concentrations on the anion and cation substructures, respectively, and ΔG_S denotes the Gibbs energy for formation of a Schottky defect, apart from the contributions of the entropies of mixing, and

$$\text{for Frenkel defects}: \quad c_{vac/cat} = c_{int/cat} = \exp\left(\frac{-\Delta G_F}{2RT}\right), \qquad (5.3)$$

where $c_{vac/cat}$ and $c_{int/cat}$ represent the fractional vacancy concentration and fractional interstitial concentration on the (as considered here) cation substructure, and ΔG_F denotes the Gibbs energy for formation of a Frenkel defect, apart from the contributions of the entropies of mixing.

5.2   Line Defects (One-Dimensional)

Long before the first dislocations could be visualized directly, as by application of transmission electron microscopical analysis,[3] the concept of the linear, one-dimensional defect called dislocation was introduced:

- The (shearing) force required to deform a crystalline solid can be a factor of about 10^4(!) smaller than the theoretical (shearing) force necessary for deformation of a perfect crystal. This dramatic discrepancy was resolved by the, at the time, hypothetical, assumed presence of structure defects called dislocations (Orowan, Taylor and Polanyi, independently(!) in 1934, applied the concept of edge dislocations building on ideas introduced earlier by, also, Dehlinger and Kochendörfer (1927) who were led by their analysis of X-ray diffraction-line broadening of deformed crystalline material and had introduced the notion "Verhakungen" for structure distortions related to those called dislocations later).[4]
- Experimental observations of the growth rate of crystals with smooth faces from a supersaturated solution indicate that the degree of supersaturation compatible with nucleation of new solid on these smooth faces, in order to establish (further) growth of the crystals, is at least an order of magnitude larger than the experimentally needed supersaturation. Defects as (screw) dislocations can be associated with the

---

[3] The first observations of (edge) dislocations, made by using a transmission electron microscope (see Sect. 6.8), were published in 1956.

[4] Analogous to the discussion with respect to the discoverers of the Periodic System (see the corresponding *intermezzo* in Sect. 2.5), the dislocation concept did not came as a thunder bolt without warning: precursors can be found in the literature. The first forerunner of the dislocation concept was proposed shortly after Friedrich, Knipping and von Laue had shown in 1912 by X-ray diffraction that crystals consist of a periodic arrangement of the constituting atoms (see the introductory part of Chap. 4). Prandtl, as early as in 1913, recognized that discrepancies between mechanical properties observed in reality and those expected for hypothetical perfect crystals necessitate the presence of crystal-structure imperfections in real crystals. See, in particular, the first part of the personal retrospective by Seeger published in International Journal of Materials Research, 100 (2009), 24–36.

occurrence of steps (ledges) in otherwise smooth surfaces (see Fig. 5.6 discussed below). Nucleation of new solid at such steps in surfaces is energetically less costly than nucleation on truly flat faces and thus the low values of supersaturation for growth could be explained (Frank in 1949).

In the following, the main geometrical properties of the two basic types of dislocations are discussed.

5.2.1 The Edge Dislocation

One way to "produce", in a purely imaginary way, an edge dislocation in a perfect crystal runs as follows. Consider the primitive (also called "simple") cubic crystal block (with one atom of a single element on each site of the primitive cubic translation lattice) shown in the left-hand part of Fig. 5.5a. Make a cut along ABCD. Shift (by shearing) the upper, "loosened" part of the block with respect to the bottom part

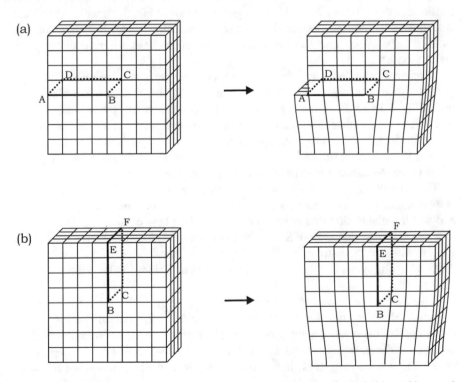

Fig. 5.5 Two hypothetical views on formation of an edge dislocation (in a primitive cubic crystal structure). **a** A planar cut is made along ABCD in a crystal block and the upper, "loosened" part of the crystal block is sheared over one lattice spacing to the right. **b** An extra plane is inserted along the planar cut EBCF. In both cases, an edge dislocation with its dislocation line along BC results

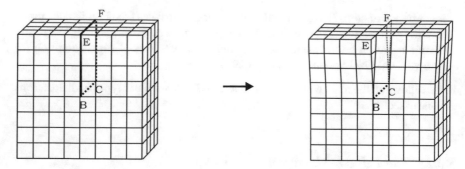

Fig. 5.6 Formation of a screw dislocation (in a primitive cubic crystal structure). A planar cut is made along EBCF of a crystal block, and the upper right half of the crystal block is sheared over one lattice spacing along this plane with respect to the upper left half of the crystal block

of the block, along the plane ABCD and perpendicular to the line BC, over one atomic distance. This leads to the occurrence of the step in the left face of the crystal block left of the line AD in the plane defined by ABCD. As a result, an atomic configuration results as sketched in the right-hand part of Fig. 5.5a, which shows the deformation of (crystal-) lattice planes[5] parallel and close to the line BC, which is called the *dislocation line*. Evidently, the largest structure distortions occur near the dislocation line.

Another way to "produce", in a purely imaginary way, an edge dislocation in a perfect crystal is as follows. Consider the (same) crystal block shown in the left-hand part of Fig. 5.5b. Make a cut along EBCF. Introduce an extra *half-plane* along the cut. The half-plane terminates within the crystal at an *edge*, which is a characteristic of an edge dislocation. The result obtained and shown in the right-hand part of Fig. 5.5b is equal to the picture shown in the right-hand part of Fig. 5.5a for the deformation of the (crystal-) lattice planes parallel to the dislocation line BC.

The symbol "⊥" is used for the edge dislocation shown in Fig. 5.5, where the symbol indicates that the extra half-plane has been inserted from the top (Fig. 5.5b) and that the dislocation line is perpendicular to the plane of drawing. One speaks of a *positive* edge dislocation. Similarly, the extra half-plane can be inserted along a cut made from the bottom (cf. Fig. 5.5b) and one speaks then of a *negative* edge dislocation: the symbol "⊤" indicates that the extra half plane has been inserted from the bottom.

From the "action" discussed above to generate an edge dislocation, by "pushing-in" a half-plane, it immediately follows, by intuition (see, in particular Fig. 5.5b), that the part of the crystal having accepted the half-plane, and as a consequence of the cohesion of the entire crystal, must experience a state of stress that has a compressive nature: the surroundings of the half-plane in this part of the crystal (the upper part of

---

[5] As indicated at the start of Sect. 4.1.4, the description of directions and of orientations of planes in crystals is realized with respect to the translation lattice. Therefore, we do (yet; cf. footnote 6 in Chap. 4) use the terms "lattice planes" and "lattice directions" for describing the orientations of crystallographic planes and directions in crystals.

the crystal block shown in Fig. 5.5) counteract the insertion of the extra half-plane; in other words: the local hydrostatic component of the stress field is compressive. Similarly, the other part of the crystal (the bottom part of the crystal block shown in Fig. 5.5) experiences the presence of an extra half-plane in the first part as well: it is strained to adapt to the larger dimension of the first part, while the crystal remains cohesive, i.e. it is strained in a tensile way; in other words: the local hydrostatic component of the stress field is tensile. All stress components are proportional with the reciprocal of their distance to the dislocation line.

5.2.2 The Screw Dislocation

The crystal block with the cut made along EBCF (cf. Fig. 5.5b) is shown again in the left-hand part of Fig. 5.6. Now displace the upper part of the crystal block right from the plane EBCF relative to the upper part of the crystal block left from the plane EBCF, over one lattice spacing parallel to the line BC. This leads to the atomic configuration and the deformation of the (crystal-) lattice planes (see footnote 5) around the dislocation line BC as sketched in the right-hand part of Fig. 5.6: the screw dislocation. To explain the notion "*screw*", the following thought experiment is carried out. Make yourself as small as an atom, position yourself at E in the crystal block shown in the right-hand part of Fig. 5.6 and perform a "walk" with BE as radius vector. After having made one full circle by anti-clockwise rotation around BC, while maintaining contact with the (crystal-) lattice plane you are walking on, you will find yourself not in E again. Instead you have moved one lattice spacing parallel to BC in the direction of BC. Upon continuing this operation, it follows that a spiral surface (helicoid) is followed along the spiral axis BC.

Displacing the upper part of the crystal block right from the plane EBCF relative to the upper part of the crystal block left from the plane EBCF, over one lattice spacing parallel to the line BC, *but now in a direction opposite to the one pertaining to the case shown in* Fig. 5.6, also leads to a screw dislocation with BC as dislocation line. Evidently, making a similar "walk" as discussed in the previous paragraph, now the direction of net movement parallel to the dislocation line has reversed. Thus, one can speak of left-hand and right-hand screw dislocations (cf. positive and negative edge dislocations discussed in Sect. 5.2.1). For a right-hand screw dislocation, a *clockwise* rotation by a "walk" as described above leads to advancement of one lattice-plane distance parallel to and downwards along the dislocation line. An advancement of one lattice-plane distance parallel to and upwards along the dislocation line holds for a left-hand screw dislocation upon such clockwise rotation. Thus, a left-hand screw dislocation is shown in Fig. 5.6.

The hydrostatic component of the stress field of a screw dislocation is nil; only two shear components occur: a shear component in planes perpendicular to the dislocation line and in circumferential direction, and a shear component in planes through the dislocation line and parallel to the dislocation line.

Although the geometric discussion and in particular the three-dimensional representations of the associated atomic arrangements given above for edge and screw dislocations suggest otherwise, in fact the screw dislocation is the more simple one of the two linear defects, as is exemplified by considering the dislocation-strain field, which is more complex for the edge dislocation.

5.2.3   Dislocation Line and Burgers Vector; Dislocation Density

The dislocation line indicates the core of the distortion in the crystal structure due to this defect. Thus, for an edge dislocation the dislocation line merely indicates the end of the extra half-plane (cf. Sect. 5.2.1; line BC in Fig. 5.5). One defines the vector $\mathbf{l}$, which is a vector of unit length that indicates the orientation of the dislocation line.

The magnitude and orientation dependence of the distortion associated with a dislocation is characterized by the so-called Burgers vector. The Burgers vector can be determined by the use of a so-called Burgers circuit. A Burgers circuit involves a closed loop, "atom (lattice site)-to-atom (lattice site)" path. For such a closed-loop path made in a perfect crystal, i.e. the Burgers circuit does not enclose a dislocation, in order that one returns at one's starting point in the crystal, the total number of lattice-spacing steps (atom-to-atom distances) made to the right must be equal to the total number of lattice-spacing steps to the left, and, similarly, the total number of steps made upwards must be equal to the total number of steps made downwards (see Fig. 5.7a). Now consider the situation that the Burgers circuit encloses an edge dislocation (In Fig. 5.7b, c, the Burgers circuit is made in a plane perpendicular to the dislocation line of an edge dislocation and a screw dislocation, respectively). The path followed in this real (i.e. containing the dislocation) crystal must be made through "good" crystal, i.e. remote from the dislocation line. It then immediately follows that if the same step (atom-to-atom) sequence is followed as for the Burgers circuit in the case of the perfect crystal (Fig. 5.7a), then the path followed now leaves a gap, i.e. the loop does not close (Fig. 5.7b, c). The extra vector that is needed to close the loop, and thereby to establish the Burgers circuit in the real crystal, characterizes the magnitude of the distortion brought about by the dislocation and its orientation dependence. This vector is called the Burgers vector, usually indicated by the symbol $\mathbf{b}$.

The definition of the direction of the Burgers vector has not yet been accomplished in this way. One possible convention for fixing the direction of the Burgers vector is as follows. First the positive direction of the dislocation line has to be established. This is a completely arbitrary choice (for the examples shown in Fig. 5.7b, c this direction is taken perpendicular to the plane of the paper towards larger "depths" below the plane of the paper). Then the Burgers circuit has to be made around the dislocation, in clockwise sense with respect to the positive dislocation-line direction. The (size and) direction of the Burgers vector is then determined by defining the Burgers vector

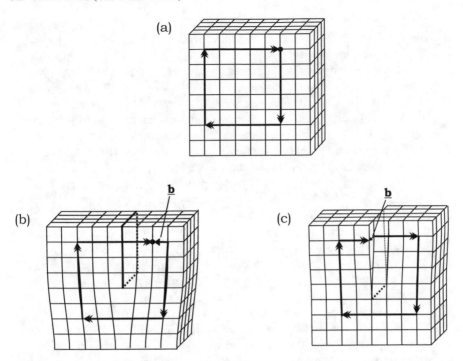

Fig. 5.7 Burgers circuit in **a** a perfect, primitive cubic crystal, **b** a primitive cubic crystal containing an edge dislocation and **c** a primitive cubic crystal containing a screw dislocation. In case of **b, c** the Burgers vector, **b**, has been taken as the vector pointing from starting point to end point of the path followed in the real crystal (see text)

as the vector pointing from starting point to end point of the path traced in the real crystal in clockwise sense. This is the approach followed in Fig. 5.7b, c to determine the (size and) direction of the Burgers vector.

Evidently, the direction of the Burgers vector is reversed (its "sign" is changed) by making an anti-clockwise Burgers circuit or by taking the Burgers vector as the vector pointing from end point to starting point of the path followed in the real crystal. It is also possible, as an equally valid procedure, to first make a, now closed-loop, Burgers circuit in the real crystal and then make the same atom-to-atom path in the perfect crystal; the thus occurring closing failure in the perfect crystal then defines the Burgers vector, etc. There is no generally adopted convention. Upon reading the literature one should be aware of this; of course, consistency requires that a single definition is adopted throughout one work.

Intermezzo: A Historical Note About the Burgers Vector

The Burgers vector (and the concept of the screw dislocation, and the concept of (low-angle) grain boundaries as arrays of dislocations (cf. Sect. 5.3) is due to the Dutch scientist J.M. Burgers, an authority on fluid dynamics, who should not be confused with his younger brother W.G. Burgers, a well-known physical metallurgist. J.M. Burgers was "dragged" to dislocation theory in its infancy by his brother, made his seminal contribution (1939,[6] 1940) and left the field thereafter, leaving to others, as his brother, to apply his results. It strikes that in some textbooks on materials science and, even sometimes, on dislocation theory, the original work by J.M. Burgers is often not cited or cited incorrectly, in contrast with other important, original, initial work as due to e.g. Taylor, Orowan and Polanyi. Moreover, many materials scientists erroneously attribute the invention of the Burgers vector to W.G. Burgers (who acquired a world reputation especially because of his work on recrystallization; cf. Sect. 11.2.1). This personal note (the author occupied "Burgers Chair" at the Delft University of Technology during 12 years) seems especially useful, not only to restore historical correctness where needed, but in particular as an illustration of how occasional family ties or friendships can lead to very distinct, original contributions in a specific field of science by the fertilization by a relative *outsider*, who may leave the field immediately after having made his/her "discovery". That in this case it had to be J.M. Burgers, and not his brother W.G. Burgers, who made the definitive breakthrough, may have a lot to do with J.M. Burgers being the by far most mathematically gifted and experienced of the two brothers. The "story" behind the "discovery" has been told by W.G. Burgers (to the author and) in his very last publication: *"How my brother and I became interested in dislocations"* (Proceedings of the Royal Society (London), A **371** (1980), 125–130), a good read for any beginning materials scientist.

Evidently, it follows from the exercise performed in Fig. 5.7b, c, that

$$\text{for an edge dislocation: } \mathbf{b} \perp \mathbf{l}, \tag{5.4}$$

$$\text{for a screw dislocation: } \mathbf{b} \,//\, \mathbf{l}. \tag{5.5}$$

The Burgers vector of a dislocation is fixed. However, the dislocation-line vector is not necessarily a constant quantity. Dislocations of mixed character can and very often do occur. At those places along such dislocation lines where the conditions (5.4) or (5.5) are satisfied, one speaks of edge and screw character at these locations of the dislocation of mixed character considered (see Fig. 5.8).

---

[6] J.M. Burgers published this original work in 1939 first in the "Proceedings of the Royal Society of Sciences (Amsterdam)" (usually referred to as Proc. K. Akad. Wet. Amst.): two contributions (in English) in volume 42, starting at pages 293 and 378, respectively. The paper published in 1940, taken up in the list of references at the end of this chapter, can be considered as (and was meant by Burgers to be) a summary and an extension of these preceding papers. This lucid paper has been written very well, is particularly instructive and is a pleasure to read, also by students, even after, now more than, 80 years.

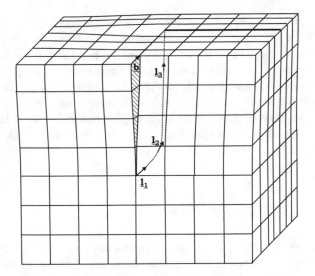

Fig. 5.8 Dislocation of mixed character. The line vector has been indicated with l_x; the Burgers vector with b. To avoid any confusion in the figure regarding the vector nature of the line vector and the Burgers vector, they have been indicated by a line under the characters "l" and "b" in the figure, which is a generally accepted, alternative way of vector indication (otherwise realized in the text by bold characters for "l" and "b"). At the location where the line vector is given by l_1, $l \parallel b$ and the dislocation has pure screw character. At the location where the line vector is given by l_3, $l \perp b$ and the dislocation has pure edge character. At the location where the line vector is given by l_2 the dislocation has mixed character. See also text

Apparently, for the cases considered above, where it could be said that the focus actually was on the simple cubic crystal structure (Figs. 5.5, 5.6, 5.7 and 5.8), the Burgers vector can only be between two atom sites of the crystal, i.e. between two nodes of the translation lattice (see begin of Sect. 5.2.1), and hence the Burgers vector is a lattice vector. Dislocations for which this holds are called *perfect dislocations*. (for imperfect, partial dislocations, see Sect. 5.2.8).

Dislocations cannot simply start or end at some arbitrary position within a crystal. They can end and start at surfaces and grain boundaries/interfaces. The way out of this is manifested by dislocation loops (see Fig. 5.12, discussed below) or by branching/dissociation of dislocations at some location within the crystal.

The density of the dislocations in a crystal is defined as the length of dislocation line divided by the volume of the crystal considered. Thus, the dimension of dislocation density is [length]/[volume] $=$ [area]$^{-1}$. This can also be interpreted as the number of intersection points of dislocation lines through a cross section of unit area. A usual estimate for the average distance between dislocations in a random distribution of dislocations with density ρ_d then is $1/\sqrt{\rho_d}$. (The area "confined to" one dislocation in the above cross section of unit area thus is taken as $(1/\sqrt{\rho_d})\,(1/\sqrt{\rho_d})$ $= 1/\rho_d$).

The dislocation density can be very variable. The dislocation density of an annealed metal is $(10^{10}-10^{12})$ m^{-2}. Deformed, e.g. cold-rolled, metals exhibit dislocation densities as large as 5.10^{15} m^{-2}. By special precautions single crystals (of Si and Ge) can be grown with dislocation densities of the order 10^6 m^{-2}, which should be considered as a very small value. Moreover, growth of tiny, needle-shaped crystals, called "whiskers", of virtually zero dislocation density can occur. However, normally the presence of dislocations in crystalline materials cannot be avoided as they result from the production route followed (solidification and plastic deformation due to mechanical action or thermal loading).

5.2.4 Strain Energy of a Dislocation

Obviously, the presence of dislocations in a crystal involves the introduction of strain energy. Close to the dislocation line, i.e. close to the so-called *core* of the dislocation, (linear, i.e. Hooke's law) elasticity theory (cf. Sect. 12.2) no longer holds. Therefore, the total strain energy of a dislocation is written as:

$$E_{strain} = E_{core} + E_{elastic} \tag{5.6}$$

The contribution $E_{elastic}$ can be given for a single dislocation, per unit length of dislocation line, along the axis of a cylindrical crystal as

$$E_{elastic} = \text{const.} \, G\, b^2 \ln\left(\frac{R}{r_0}\right) \tag{5.7}$$

where the constant is (about 20–50%) larger for an edge dislocation as compared to a screw dislocation. G denotes the shear modulus (elastically isotropic material is considered; cf Sect. 12.3 and Eq. 12.6), b is the size of $\mathbf{b}$, R is the radius of the (cylindrical) crystal and r_0 represents the radius of the core. This equation for $E_{elastic}$ immediately makes clear that there is no characteristic "dislocation line energy": $E_{elastic}$ becomes infinite for R approaching infinity (and r_0 becoming zero); the dislocation line energy depends on the size of the crystal containing the dislocation. The need for a lower boundary for the integration of strain energy through space, in order to arrive at a result as given by Eq. (5.7), represents that the distortion of the material within a cylinder around the dislocation line of radius r_0, the "core radius" which is of the order of the size of the Burgers vector, cannot be described by linear elasticity theory (cf. Sect. 12.2). Estimates for E_{core} suggest that its value is about 10–30% of $E_{elastic}$. It appears possible to include the contribution of E_{core} by (artificial) adjustment or r_0 in Eq. (5.7), thereby rendering $E_{elastic}$ into E_{strain}. The thus adjusted values of r_0 are in the range $1/4b$–$2b$, also depending of the type of material (e.g. ionic or metallic) considered (cf. Hirth and Lothe 1982).

Evidently, for the single dislocation considered above, $E_{elastic}$ depends on the size of the crystal (cf. R). If many dislocations are present, cancellation of the long-range parts of the elastic strain fields of the individual dislocations can occur. Further, the logarithmic dependence of $E_{elastic}$ on R (cf. Eq. 5.7) makes $E_{elastic}$ insensitive for the precise choice of R. Then R can appropriately be taken as half of the average spacing between the dislocations, which can be assessed by $1/\sqrt{\rho_d}$, with ρ_d as the dislocation density (cf. Sect. 5.2.3).

For practical purposes one may write for the elastic strain energy introduced by a dislocation per unit length of dislocation line:

$$E_{elastic} = \text{const. } G \, b^2 \tag{5.8}$$

with the constant having values between 0.5 and 1.0.

One might think that a dislocation might be generated by thermal excitation. Such dislocations then would be "equilibrium defects" as the "thermal vacancies" (cf. Sect. 5.1). The chance of formation of such "thermal dislocations" may be estimated by considering a "Boltzmann-type" equation as Eqs. (5.1)–(5.3). A crude estimate obtained on the basis of Eq. (5.8) for the energy *per atom* along a dislocation line is 5 eV ($= 8 \times 10^{-19}$ J). The thermal (kinetic) energy of an atom at room temperature (kT) is about 0.025 eV ($= 0.04 \times 10^{-19}$ J). Application of these numbers in a "Boltzmann-type" equation leads to a probability of $\exp(-200) \approx 10^{-87}$ that dislocation line can be produced by thermal activation. Hence, the generation of thermal dislocations is very unrealistic.

The strain energy of a dislocation obviously is proportional with its length. The energy increase per unit length increase of a dislocation can be conceived as the *line tension* of the dislocation (in a way similar to the concept surface tension; cf. Sect. 11.3.1). This line tension explains the tendency of any curved dislocation to straighten and thus reduce its length. Under the action of a shear stress, a dislocation may become curved (for example, if the dislocation is pinned at two locations; cf. Sect. 5.2.6). Then it can be derived that the shear stress τ_0 necessary for maintenance of a certain radius of curvature of the (pinned) dislocation, obeys

$$\tau_0 = \text{const. } \frac{Gb}{r} \tag{5.9}$$

with r as the radius of curvature of the dislocation concerned (and the same constant as in Eq. (5.8)).

5.2.5 Glide of Dislocations; Slip Systems

Consider Fig. 5.9. By imposing the shear stress τ, the upper part of the primitive cubic crystal block (as considered until now in this Sect. 5.2) may be shifted with respect to the bottom part of the crystal block over, say, one atomic distance (for discussion of shear stress and shear strain, see Sect. 12.2). The plane along which this shearing has occurred is called the slip plane. If the crystal would have been

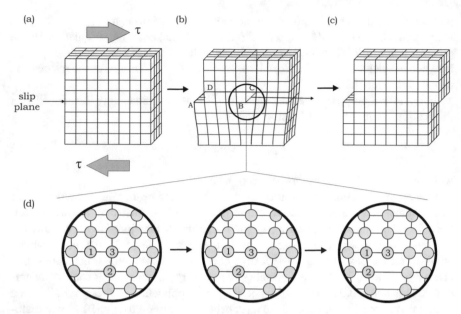

Fig. 5.9 Shear deformation accomplished by motion of an *edge* dislocation. **a** Unsheared crystal (shear stress τ has been indicated by grey arrows). **c** Sheared crystal. **b** Transitional state with edge dislocation moving along the slip plane; the direction of motion of the dislocation line, perpendicular to the dislocation line and *parallel* to the Burgers vector (τ is parallel to **b**; if τ is not parallel to **b**, the component of τ in the direction of **b** controls the occurring slip) has been indicated by the arrow. In this case (edge dislocation), the direction of slip is *perpendicular* to the dislocation line and, as always, parallel to the Burgers vector. **d** Enlarged view (encircled regions; the plane of drawing in **d** is the same plane of drawing as for **a–c**, showing the change of the atomic arrangement at the dislocation line during the motion of the dislocation

perfect, the occurrence of such shearing requires that the atoms in the upper part of the crystal and adjacent to the slip plane have to move *at the same time in a coordinated way*. This is an extremely difficult to realize process that would require a very high value of shear stress τ. In reality shear stress values a factor of, say, 10^4 smaller are needed. This led to the concept of (initially only edge) dislocations to explain the occurrence of relatively easy slip (see the introduction of Sect. 5.2).

Dislocations can glide over a slip plane. The slip plane possible for a dislocation is defined by the Burgers vector **b** and the dislocation line vector **l**. Thus for an edge dislocation, the slip plane is perpendicular to the half-plane and runs through the edge of the half-plane. In the above sense, for a screw dislocation the slip plane is undefined, because **b** and **l** are parallel to each other (cf. Sect. 5.2.3). Yet, in reality, screw dislocations, as edge dislocations, normally glide over the most densely packed planes.

Return to Fig. 5.5a. After having made the cut along ABCD and having performed the displacement of the upper part of the crystal block with respect to the bottom part of the crystal block, along the plane ABCD and perpendicular to the line BC, over one atomic distance, an edge dislocation has been introduced with dislocation line along BC. One can now say: the dislocation along BC designates the boundary of the (left, upper) part of the crystal block that has slipped over one atomic distance, with respect to the unslipped (left, bottom) part of the crystal block. Now, exert (continue to exert) the shear stress τ (parallel to **b**). The crystal responds by moving the dislocation at an already relatively small value of τ. Minor arrangements of the atomic arrangements around the dislocation line suffice to achieve this, which can be described in a crude way as follows: By establishing the bond between the atoms 1 and 2 and disrupting the bond between the atoms 3 and 2, under the action of shear stress τ, the dislocation (the half-plane) has moved effectively one atomic position to the right; see Fig. 5.9d. This process continues and at the end, after the dislocation considered has traversed the entire crystal block along the slip plane concerned, the whole upper part of the crystal block has slipped over a distance as large as the Burgers vector, and in a direction parallel to the Burgers vector, as exhibited by the resulting step at the surface (Fig. 5.9c). The point is: the same plastic deformation, as resulting, in the absence of a dislocation, from a coordinated, simultaneous translation of all atoms in the upper part of the crystal block, with respect to the bottom part of the crystal block (see above), has now been realized by the *sequential rearrangement* of a few atoms close to the moving dislocation line. One may intuitively and correctly presume that the latter process requires much smaller values of shear stress τ.

The same plastic deformation of the crystal block considered can also be realized by slip of a single screw, instead of edge (as above), dislocation. In this case, the dislocation line is oriented parallel to the shear stress τ (parallel to **b**; see Fig. 5.10). Under the action of the shear stress, the dislocation line now moves in a direction perpendicular to τ. In the end, for a shear stress acting in a specific direction, exactly the same plastic deformation has been realized (compare Figs. 5.9c and 5.10c for the slip realized *in the direction of the acting shear stress* τ).

Summarizing it can be said that for an edge dislocation **l** moves in the direction parallel to **b** and that for a screw dislocation **l** moves in the direction perpendicular to **b** (with, in the above discussion, τ parallel to **b**; if τ is not parallel to **b**, the component of τ in the direction of **b** controls the occurring slip). The slip occurs always in the direction of **b**.

Intermezzo: The Peierls Stress

The shear stress required to move a dislocation line from atomic position to atomic position, in the sense of the above discussion, can be derived from the (slight) dependence of the dislocation-line energy on the position of the dislocation line in the crystal structure, which energy will be a periodic function of the length of the Burgers vector/position in the crystal structure. The shear stress applied must make it possible to overcome the potential energy barrier, for movement of the dislocation line, which occurs for the position of the dislocation line in-between two atomic positions (see Fig. 5.11). Simple treatments have been based on elasticity theory (cf. Chap. 12) and have led to

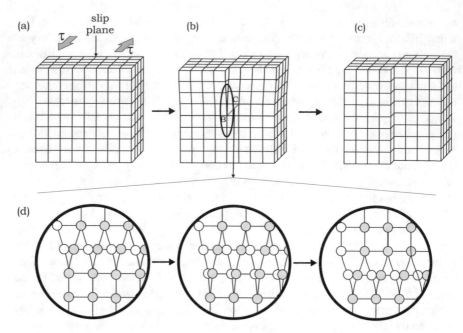

Fig. 5.10 Shear deformation accomplished by motion of a *screw* dislocation. **a** Unsheared crystal (shear stress τ has been indicated by grey arrows). **c** Sheared crystal. **b** Transitional state with screw dislocation moving along the slip plane; the direction of motion of the dislocation line, perpendicular to the dislocation line and *perpendicular* to the Burgers vector (τ is parallel to **b**; if τ is not parallel to **b**, the component of τ in the direction of **b** controls the occurring slip), has been indicated by the arrow. In this case (screw dislocation), the direction of slip is *parallel* to the dislocation line and, as always, parallel to the Burgers vector. **d** Enlarged view (encircled regions; the plane of drawing in **d** is perpendicular to the plane of drawing in **a**–**c**; grey atoms: atoms above plane of drawing, and white atoms: atoms below plane of drawing), showing the change of the atomic arrangement at the dislocation line during the motion of the dislocation

an expression for the shear stress required to move the dislocation; this stress is called the Peierls stress (e.g. see Hull and Bacon 2001). Although this concept is widely used, the thus determined Peierls stress has limited validity: the stress field at and in the immediate surroundings of a dislocation line (this part of the material is called the "dislocation core"; see Sects. 5.2.3 and 5.2.4) cannot be described by elasticity theory: quantum–mechanical interactions come into play. Indeed, the Peierls stress as calculated for simple metals is much too large. It has been argued, supported by the experimental data, that the Peierls stress for metals can be negligibly small (Gilman 2007). This can be expressed in other words: the energy of the dislocation core in metals does not depend strongly on its precise atomic structure/arrangement and thus does not change distinctly upon movement of the dislocation line from atomic position to atomic position. This statement is of course closely related to the recognition that the energy

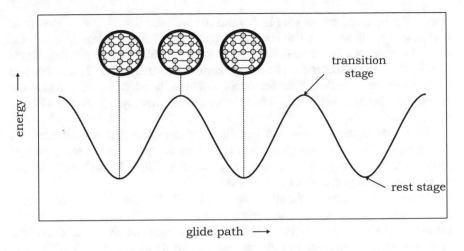

Fig. 5.11 Energy profile for glide of an edge dislocation (cf. Fig. 5.9)

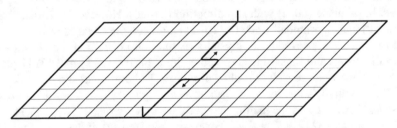

Fig. 5.12 A kink displaces the dislocation line locally by a unit distance in the slip plane. Here a double kink is shown for an edge dislocation. Lateral movement of the kink (i.e. parallel to the original dislocation line) can already be realized by a relatively small stress (as compared to the movement of the edge dislocation line as a whole perpendicular to the dislocation line, i.e. in the direction of **b**; see Fig. 5.9)

of the metal crystal does only slightly depend on its precise structure (e.g. see the very small differences between the values for the "Madelung" constant α for the f.c.c., h.c.p. and b.c.c. atomic arrangements, and the corresponding discussion, given in Sect. 3.5.3).

Here it should be remarked that a dislocation line need not be a straight line. As a result of thermal activation, so-called kinks can occur: A *kink* displaces the dislocation line locally by a unit distance in the slip plane (see Fig. 5.12; steps in dislocation lines not within the slip plane, so-called jogs, are discussed in Sect. 5.2.6). Lateral movement of the kink, i.e. parallel to the original dislocation line, can already be realized by a relatively small stress, as compared to the movement of the dislocation line as a whole, i.e. perpendicular to the dislocation line by the action of a (the Peierls) stress (as considered above; see Fig. 5.9 and the *intermezzo* immediately above). In particular if the dislocation line tends to be parallel to a close packed direction, the potential energy barrier for motion of a kink parallel to this close packed direction is relatively small. Such lateral movement of double kinks (see Fig. 5.12) can therefore be an important ingredient of the mechanism of glide, because the movement of the dislocation line as a whole, i.e. as a perfectly straight line, would require a higher value of stress.

Glide of dislocations occurs such that the distortion associated with it is minimal. This implies that dislocation glide normally occurs along the most densely packed planes (which have the largest interplanar spacings; see Sect. 4.1.4.1) and along the most closely packed directions in these planes.

Hence, for f.c.c. crystals the slip planes are the {111} planes. In each of the four distinguishable (i.e. unique), close packed {111} planes of the f.c.c. crystal the three distinguishable (i.e. unique) <110> directions are the most closely packed directions. A combination of slip plane and slip direction is called a *slip system*. Thus for f.c.c. crystals 12 (= 4 × 3; see above) slip systems occur.

Similarly, considering h.c.p. crystals the preferred slip plane is the most densely packed, basal plane (0001) with slip direction [11–20]. However "prismatic" glide on (10–10) and "pyramidal" glide on (10–11) has been observed as well.

In the not close packed b.c.c. crystal structure, slip occurs in the close packed <111> directions and thus glide is possible along {110}, {211} and {321} planes, which planes contain this close packed direction. The dominant slip plane depends on the temperature. Thus, for b.c.c. metals, as a rough rule, slip along {211} occurs at $T < T_m/4$ (with T_m as the melting temperature in K), along {110} at $T_m/4 < T < T_m/2$, and along {321} at $T > T_m/2$. For b.c.c. iron (ferrite), it has been found that at room temperature slip occurs on all three glide planes indicated along a common <111> direction; this phenomenon is called "pencil glide".

As follows from Table 5.1, the f.c.c. and b.c.c. crystal structures possess 12 or more slip systems, whereas the h.c.p. crystal structure has much less slip systems at its disposal. Consequently, the possibility to respond to severe loading by plastic deformation (yielding) is most pronounced for f.c.c. and b.c.c. materials; h.c.p. material is relatively brittle.

5.2.6 Dislocation Production: Frank–Read Source, Cross-Slip and Vacancy Condensation

A straight dislocation (segment) is pinned at the positions A and B and lies on a slip plane. The "pinning points" A and B can be, for example, solute atoms, or particles of a different phase, or dislocations lying on planes which are not the slip plane of the dislocation (segment) considered here. Upon exerting a shear stress τ parallel to the Burgers vector of the gliding dislocation line (segment), the dislocation intends to glide along the slip plane. Because of the pinning at A and B, the dislocation line

Table 5.1 Number of slip planes, slip directions and number of independent slip systems for f.c.c., b.c.c. and h.c.p. structures

Lattice type	Slip planes	Number of slip planes (multiplicity)	Slip directions	Number of slip directions per slip plane	Number of *independent* slip systems[a]
f.c.c	{111}	4	<110>	3	5
b.c.c	{110}	6	<111>	2	5
	{211}	12	<111>	1	5
	{321}	24	<111>	1	5
h.c.p	{0001}	1	<11–20>	3	2
	{10–10}	3	<11–20>	1	2
	{10–11}	6	<11–20>	1	4

[a]See Sect. 12.12

bows out: see the sketch in Fig. 5.13, where the plane of the figure is taken as the slip plane and the shear stress acts parallel to AB and thereby parallel to the plane of the

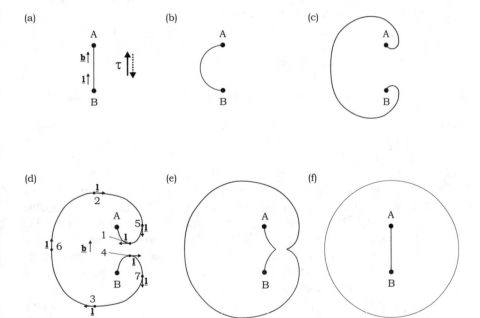

Fig. 5.13 Formation of a dislocation loop at a Frank–Read source. The dislocation line is pinned at the locations A and B. A shear stress has been applied as indicated by the arrows (τ). In **d**, the Burgers vector has been indicated by **b** and the line vectors **l** have been indicated at the locations 1–7 with **l**. At points 5, 6 and 7, the dislocation has pure screw character. At points 1, 2, 3 and 4, the dislocation has pure edge character. Note that the dislocation at points 1 and 4 is of opposite sign, which leads to an attractive force and recombination of the dislocation line at these points

figure (i.e. the initial dislocation (segment) has pure screw character in the example considered in the figure; note that the dislocation line of a pure screw dislocation upon slip moves perpendicular to **b**; the slip occurs always in the direction of **b**; cf. Sect. 5.2.5). This bowing out of the dislocation implies that the dislocation-line length increases and thus the line tension requires a minimal value for τ in order that the curvature corresponding to the bowing out can be achieved/maintained: see Eq. (5.9). The largest minimal shear stress occurs for the smallest radius of curvature, which occurs if the dislocation has become a half-circle with AB as diameter. This largest minimal shear stress thus equals [taking const. in Eq. (5.9) as ½; see Eq. (5.8)]:

$$\tau_0 = \frac{Gb}{d} \tag{5.10}$$

with $d = $ AB.

For a shear stress τ larger than τ_0 the dislocation proceeds to bow out beyond the half-circle shown in Fig. 5.13b. Successive possible shapes for the lengthening dislocation line are schematically shown in Fig. 5.13b–f. In the present discussion the original state of the dislocation characterized by the line segment AB has screw character, i.e. **b** is parallel to **l** (= parallel to AB). The Burgers vector of a dislocation is a fixed quantity, whereas the dislocation-line vector is not (Sect. 5.2.3). In an advanced stage of glide (Fig. 5.13d–f), only at certain locations along the dislocation line pure screw and pure edge character occurs. For example, at the locations denoted 1 and 2 and 3 and 4 pure edge character happens, as, there, **b** $\perp$ **l** (see Fig. 5.13d). It should be realized that the edge dislocations at 1 and 2 and those at 3 and 4 have opposite signs: for example, making clockwise Burgers circuits around the parts of the dislocation line at 3 and 4 leads to Burgers vectors of opposite sign. As a result, the pair of dislocations at 1 and 4 can combine and thereby annihilate. Thereby a closed dislocation ring is established and a line segment as the starting one, AB, remains (Fig. 5.13e, f). Still under the action of the applied shear stress, the process discussed can continue and thereby a series of ever widening dislocation rings is produced. This process could proceed in principle until an obstacle, for example a grain boundary, occurs on the slip plane against which the dislocations generated will tend to "pile up". The later arriving, newer dislocations (loops) interact with the earlier, older ones and as a result a so-called back stress occurs that works against the applied (shear) stress and the dislocation (Frank–Read) source becomes eventually inactive (See Sect. 12.14.2 for the possible role of dislocation pile-ups in explaining the effect of grain-boundary density on strengthening).

According to the above picture, dislocation multiplication (here a developing series of dislocation loops) takes place in a single slip plane (the plane of drawing in Fig. 5.13). Now recall that for a screw dislocation, the slip plane is undefined as **b** // **l** (Sect. 5.2.3). Yet, the screw dislocation tends to glide in specific crystallographic planes only (as {111} planes in f.c.c. materials; cf. Sect. 5.2.5). Because **b** and **l** do not define a slip plane, it is possible for the screw dislocation to change slip plane (say, in f.c.c. material from (111) to (11–1), or in b.c.c. material from (1–10) to (−321), etc.; see Table 5.1 and at the end of Sect. 5.2.5) under the influence of a (very localized) change of the state of stress. This phenomenon is called *cross-slip*. Considering the dislocation ring produced and as shown in Fig. 5.13, it follows that at the locations 5, 6 and 7 pure screw character prevails (for **b** parallel to the original line segment AB). At these locations, the expanding dislocation ring can leave the original glide plane (i.e. the plane of the drawing) if promoted by the local state of stress. Thus, very complicated three-dimensional dislocation configurations can develop as the result of multiple cross-slip. Because the applied, overall state of stress implies a shear stress acting parallel to the plane of the drawing, i.e. the primary glide plane, a band of "dislocated" material parallel to the primary glide plane develops in the material: the shear stress components will on average be the largest parallel to the primary glide plane and the dislocation ring will have the largest tendency to expand parallel to the primary glide plane. On this basis, a dislocation line that is continuous over many parallel glide planes, as connected by relatively short dislocation-line segments on other glide planes inclined with respect to the primary glide plane, can develop. The above discussion describes the development of so-called *glide bands* in materials where the plastic deformation of the specimen/workpiece considered can be concentrated (see also Sect. 12.9.2).

The above-mentioned "relatively short connecting dislocation line segments" are called jogs: A so-called *jog* displaces the dislocation line locally from one slip plane to the next one. As holds for kinks (cf. Sect. 5.2.5): the occurrence of jogs is a thermally activated process.[7]

Condensation of *excess* vacancies can be an origin of dislocation loops as well. If a crystalline material is quenched from elevated temperature with sufficiently high quench rate, then many or all of the thermal vacancies present in the material at high temperature may be retained at low temperature after the quench. This state of the material is not an equilibrium one. If the final temperature is high enough or if, subsequently, the quenched material is annealed at a moderate temperature, so that the vacancies are sufficiently mobile, clustering of these vacancies on specific (close packed) planes may occur. The part of the plane thus occupied with a disc of vacancies is mechanically unstable: collapse can occur and the filled (close packed) planes originally neighbouring the disc of vacancies become neighbours,[8] which leads to the formation of the dislocation loop (see Fig. 5.14). In this case, the Burgers vector of the dislocation ring formed is oriented perpendicular to the ring and hence the dislocation ring is of pure edge character. Glide is unlikely, as the slip "plane" (given by **l** and **b**) is not a closed packed plane. (Another consequence of quenched-in

---

[7] The movement of kinks (cf. Sect. 5.2.5) and jogs in dislocation lines obeys the same rules as described for edge and screw dislocations in Sects. 5.2.5 and 5.2.7. Example: a jog in a screw dislocation is an edge dislocation-line segment. Thus this jog can glide *along* the dislocation line (cf. Sect. 5.2.5). The same jog can move *with* the screw dislocation only by climb (cf. Sect. 5.2.7). As compared with glide, climb is a relatively slow process and thereby the movement rate of a largely gliding screw dislocation is slowed down.

[8] Thereby, if this collapse of vacancies has occurred for vacancies originally clustered on a closed packed, (111) plane of an f.c.c. crystal, a stacking fault (see Sects. 5.2.8 and 5.3) has been realized at this location.

(a)

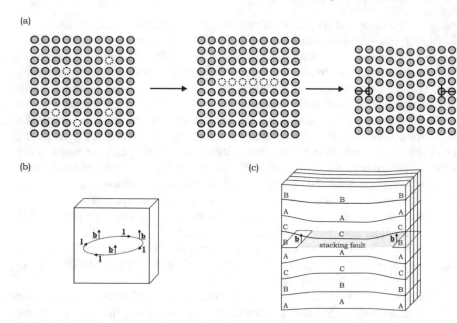

(b)   (c)

Fig. 5.14 Formation of a dislocation loop by condensation of vacancies on a crystal plane. **a** Primitive cubic crystal (with one atom of a single element on each site of the primitive cubic translation lattice). The dislocation has pure edge character all along the dislocation loop, as can be seen from Burgers and line vectors **b** and **l** in **b**. **c** Face-centred cubic crystal (with one atom of a single element on each site of the face-centred cubic translation lattice). In this case, the dislocation is a (Frank) partial dislocation and a stacking fault (cf. Sect. 5.2.8) results. A, B, C indicate the different stacking positions in the close packed structure (cf. Sect. 4.2.1)

vacancies: they can have a large influence on (substitutional) diffusion processes in crystalline solids, as discussed in Sect. 8.6.2).

Note that in case of the dislocation-ring formation process discussed in relation with the Frank–Read source (cf. Fig. 5.13), the Burgers vector was in the plane of the dislocation ring. In general, dislocation rings or loops may have Burgers vectors inclined with respect to the plane of the dislocation ring/loop; in case of the example discussed in the preceding paragraph, the Burgers vector is oriented perpendicular to the plane of the dislocation ring/loop. In the latter case glide, as discussed in this section, as occurring in the plane of the ring/loop, leading to expansion or shrinkage of the ring/loop in the plane of the ring/loop, is not possible. Expansion or shrinkage in the plane of the ring/loop then is only possible by a process called climb, which is discussed next.

5.2.7 *Climb of Dislocations*

Climb of an edge dislocation is defined as the move of the dislocation out of its plane of glide, as defined by the Burgers vector and the dislocation-line vector. If, on the basis of a vacancy–atom exchange mechanism (see Sect. 8.5.2), a vacancy diffuses to the dislocation line in the primitive cubic crystal shown in Fig. 5.15a, an atom at the half-plane can be replaced by the vacancy. This has as immediate consequence that the half-plane at this location has moved upward; i.e. a pair of jogs cf. Sect. 5.2.6) has been formed (Fig. 5.15b). Continuation of this process causes the edge dislocation to move upwards (Fig. 5.15c). Similarly, the edge of the dislocation, i.e. the end of the half-plane, can emit a vacancy. Continuation of this process causes the half-plane to move downwards. These processes are called positive climb and negative climb, respectively. As implied by the above explicitly mentioned formation of jogs: climb normally does not occur along the entire length of the dislocation line.

Pure screw dislocations cannot climb: there is no extra half-plane to be extended or to be consumed. However, at the location of a jog on the dislocation line of the screw dislocation, climb may be initiated, in agreement with the above discussion (cf. footnote 7).

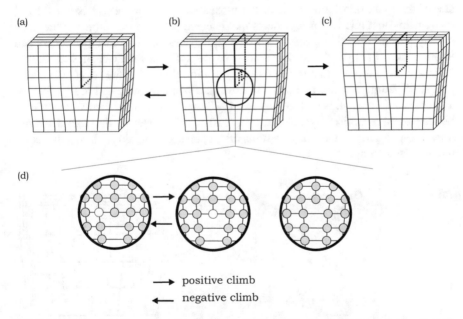

positive climb

negative climb

Fig. 5.15 Climb of an edge dislocation (again a primitive cubic crystal block is considered). Both bottom **d** and top **a–c** can be read from left to right (positive climb) and from right to left (negative climb). **a–c** Motion of the dislocation line by formation of jogs. **d** Enlarged view of encircled area, showing (going from left to right in the figure) the diffusion of a vacancy to the dislocation core (the emission of a vacancy from the dislocation core, for going from right to left in the figure)

The above discussion suggests that the climb rate can be controlled by the transport of vacancies. Then it follows that climb in general is a thermally activated process, as the number of thermal vacancies depends on temperature according to Eq. (5.1). Note that the presence of quenched-in vacancies (see Sect. 5.2.6) would reduce the activation energy of climb as the formation energy of a vacancy drops out (the activation energy for migration of a vacancy remains; see discussion in Sect. 8.6.2). Hence, at relatively low temperatures the only process of plastic deformation/shear by dislocation movement is due to glide.

As a final note, it is remarked that in principle climb can also be established by the transport of interstitial atoms to and from the edge of the edge dislocation. This is a less likely mechanism, as long as a single-element system is considered.

5.2.8 Partial and Sessile Dislocations

Until now the focus was on *perfect dislocations* where the Burgers vector is a lattice vector (cf. Sect. 5.2.3). Such dislocations can occur in primitive (simple) cubic crystal structures. Considering the f.c.c. crystal (with one atom of a single element on each site of the f.c.c. translation lattice), the most likely Burgers vector is $\frac{1}{2}a<110>$, recognizing that this vector represents the distance between two atoms in the most closely packed direction (there are 3 equivalent ones) in the most closely packed lattice plane ({111}) (cf. the discussion on slip systems for f.c.c. in Sect. 5.2.5). It immediately follows that for an edge dislocation then there are two (110) lattice planes per Burgers vector and thus the extra half plane as indicated in Fig. 5.5, for a simple cubic crystal, is replaced by two (110) half-planes (see the two (110) type lattice planes, comprised by the vector $\frac{1}{2}a<110>$, as indicated by the dashed and full lines in Fig. 5.16a, b). One may presume that, upon shearing, this pair of half-planes will not easily move dependently, as a pair.

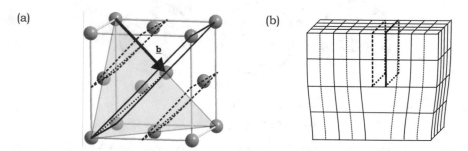

(a)   (b)

Fig. 5.16 a Full edge dislocation in an f.c.c. crystal. The corresponding full Burgers vector is indicated by an arrow, the (111) slip plane in grey. The solid and dashed black lines indicate (1–10) lattice planes. **b** As the full Burgers vector comprises two (1–10) lattice-plane spacings, a full edge dislocation in the f.c.c. crystal can be taken as two inserted (1–10) half-planes

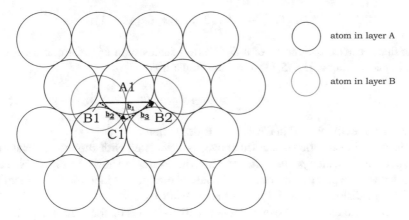

Fig. 5.17 Atomic path of shear along a (111) plane in an f.c.c. crystal. The atoms drawn with solid lines are below the slip plane, the atoms drawn with dashed lines are above the slip plane. The upper layer, represented, e.g. by the atom at B1, is first translated according to $\mathbf{b}_2$ to C1 (first Shockley partial), then translated according to $\mathbf{b}_3$ to B2 (second Shockley partial). The vector sum $\mathbf{b}_2 + \mathbf{b}_3$ provides the displacement according to the perfect dislocation $\mathbf{b}_1$. A, B, C indicate the different lateral stacking positions of close packed planes (cf. Sects. 4.2.1.1 and 4.2.1.2)

Slip occurs along close packed planes with (net; see what follows) shifts along close packed directions in these planes. A top view on a close packed plane in the f.c.c. crystal ($\{111\}$) is shown in Fig. 5.17. As discussed in Sects. 4.2.1.1 and 4.2.1.2, if the layer shown is denoted by A, the next two close packed layers (say, on top) are positioned such that their atoms reside at sites corresponding to the pits B and C of layer A. Thus, the sequence ABCABC… for the close packed planes of the f.c.c. crystal results. Now suppose, upon applying an appropriate shear stress (parallel to the (111) plane considered), that layer B has to be shifted with respect to layer A in order that a displacement is caused in the slip direction. A shift from pit B1 to pit B2 along a straight line, and over a distance corresponding to $\frac{1}{2}a<110>$, requires that the B atom concerned has to move over the "top" of the A1 atom. This picture serves to make likely that the shift of the B atom rather occurs in two steps: (i) from pit B1 to pit C1 and then (ii) from pit C1 to pit B2, which path clearly is associated with less distortion of the crystal structure than would occur along the straight line from pit B1 to pit B2. As a result, the net shift from positions B1 to B2 occurs in a zig-zag mode. In terms of the passage of dislocations to realize shear, this zig-zag mode implies that the unit displacement according to the Burgers vector $\frac{1}{2}a<110>$ is replaced by two partial displacements of type $\frac{1}{6}a<211>$. In other words, apparently two dislocations have passed, one after the other: the perfect, unit dislocation with Burgers vector $\mathbf{b}_1$ has split, one says "dissociated", into two partial dislocations with Burgers vectors $\mathbf{b}_2$ and $\mathbf{b}_3$. It holds:

$$\mathbf{b}_1 \rightarrow \mathbf{b}_2 + \mathbf{b}_3 \quad \text{with} \quad \mathbf{b}_1 = \mathbf{b}_2 + \mathbf{b}_3 \tag{5.11}$$

which in this case can be made explicit according to (see Fig. 5.17):

$$1/2a[110] \rightarrow 1/6a[211] + 1/6a[12-1]$$

Note that, because the energy of a dislocation scales with b^2 (cf. Eq. 5.8), for the reaction according to Eq. (5.11) to occur it must hold

$$b_1^2 > b_2^2 + b_3^2 \tag{5.12}$$

which is evidently obeyed in the case considered here.

The partial dislocation discussed above, as associated with slip in f.c.c. material, is called the *Shockley partial*. Considering the example shown in Fig. 5.16, each half-plane of the pair of "inserted" (110) half-planes now is identified as a Shockley partial edge dislocation, with Burgers vector b_2 or b_3.

For the same reason that two perfect edge dislocations on the same slip plane and with the extra half-plane at the same side of the slip plane repel each other (cf. the discussion in Sect. 5.2.6 on the Frank–Read source where the attraction of a pair of edge dislocations on the same slip plane with their half-planes at opposite sides of the slip plane is discussed), also the pair of (110) half-planes considered here push away each other. In general, the Shockley partial dislocations that form a pair repel each other. As a consequence a *stacking fault* (see also Sect. 5.3) occurs between the two Shockley partials. It already follows directly from the discussion given with respect to Fig. 5.17 that the shift according to b_2, i.e. from pit B1 to pit C1, results in a stacking fault: the order ABCABCABC… is replaced by ABCA**C**ABCA… (the plane where the stacking fault occurs has been indicated by the bold character). Of course, the introduction of a stacking fault causes an increase of the energy of the system proportional with the area of the stacking fault. Hence, an equilibrium situation develops where the energy gain reached by a further increase of the distance between the two Shockley partials is less than the energy increase due to the corresponding extension of the stacking fault (see Fig. 5.18). It is obtained for the optimal, equilibrium distance D_S of the two Shockley partials:

$$D_S = G \frac{b_2 \cdot b_3}{2\pi\gamma} \tag{5.13}$$

where γ represents the stacking fault energy per unit area. Upon glide the two Shockley partials with the stacking fault in-between move as an entity, so to speak as a ribbon of constant width.

The above discussed formation of two Shockley partial dislocations separated by a stacking fault is not restricted to edge dislocations: the similar phenomenon occurs for screw and mixed dislocations as well. This has an important consequence. A perfect screw dislocation has no well-defined glide plane, as b and l are parallel (cf. Sect. 5.2.5). Once the screw dislocation has become dissociated, the very existence of a planar, stacking fault in-between the two Shockley partials now prescribes that glide of the screw dislocation has to occur along the plane of the stacking fault (a {111} lattice plane). A further consequence is that a dissociated screw dislocation cannot cross slip (cf. Sect. 5.2.6). In order that a dissociated screw dislocation

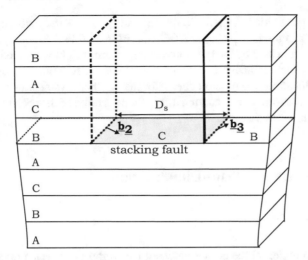

Fig. 5.18 Stacking fault (grey area) between both inserted (1–10) half-planes in an f.c.c. crystal upon dissociation of a full, i.e. perfect dislocation. The two (Shockley) partial dislocations have been indicated by their Burgers vectors $\mathbf{b}_2$ and $\mathbf{b}_3$. The separation distance between the two partial dislocations has been denoted by D_s. A, B and C indicate the different lateral stacking positions of close packed planes (cf. Sects. 4.2.1.1 and 4.2.1.2)

can cross-slip, the two partials have to recombine at one or more locations along the dislocation (where local barriers to dislocation glide may occur); one speaks of the formation of constrictions (at these locations the ribbon width reduces to a line width) and the perfect dislocations formed at these locations can then cause local occurrences of cross-slip. Evidently, on the basis of the above discussion, the formation of a constriction costs energy and thereby becomes thermally activated: the higher the temperature the more likely the occurrence of cross-slip. The formation of a constriction will be easier, if the equilibrium distance between the two partials is relatively small. Thus (cf. Eq. 5.13) for materials with high stacking fault energy, than aluminium (166 mJ/m^2), cross-slip will be more than for materials with low stacking fault energy, as silver (16 mJ/m^2).

If the Burgers vector of a dislocation does not lie in a close packed plane, glide of the dislocation is impossible. Such a dislocation is called a *sessile* dislocation. Consider the collapse of the disc of vacancies (monolayer of vacancies of finite lateral dimensions) on a close packed plane of an f.c.c. crystal, as illustrated in Fig. 5.14 (Sect. 5.2.6). The resulting dislocation loop has edge character: the Burgers vector of magnitude 1/3a<111> (corresponding to the thickness of one {111} plane) is oriented perpendicular to the dislocation line (plane of the dislocation loop). This sessile dislocation is called a *Frank partial dislocation*.[9] As holds for the Shockley partial dislocations, the Frank partial dislocation, always of edge character, and to be conceived in general as the consequence of the insertion or removal of one half-plane of a close packed plane, lies at the border of a stacking fault: left of the Frank

---

[9] The Shockley partial dislocation obviously is a *glissile* (i.e. able to glide) dislocation.

partial at the right side in Fig. 5.14c the stacking order of the close packed planes (perpendicular to the half-plane) is ABCACABCA… (the location of the stacking fault has been indicated by the bold character), whereas right from the Frank partial the ideal order (perpendicular to the half-plane) ABCABCABC…, occurs.

Partial dislocations are possible for materials of other crystal structures (as h.c.p. and b.c.c., etc.) as well. For discussion thereof, one is referred to specialized literature given at the end of this chapter.

5.3   Planar Defects (Two-Dimensional)

5.3.1   Interfaces

The transition region at the border between two tightly connected crystals, as in a massive polycrystalline specimen, has a thickness (lateral size) of a couple of atomic sizes. A general name for the transition region is *interface*. If the boundary exists between two grains of the same crystal structure and composition but of different crystal orientation (with respect to the specimen frame of reference), one speaks of a *grain boundary*. It the boundary occurs between two grains of different crystal structure (or between two grains of the same crystal structure but of different composition), one speaks also of an *(inter)phase boundary*.

Naively one may expect that the arrangement of the atoms in an interface is of chaotic, amorphous nature in general, recognizing that the atoms in the interface region would tend to comply at the same time with two prescriptions for their positions (as given by the crystal structures of the two adjacent crystals), or, in other words, that the state of chemical bonding for the atoms at the interface is less ideal than in the bulk of each of the crystals. This approach to the interface structure, i.e. the lack of ordered arrangement of atoms at the interface, has been called the "amorphous cement theory". However, many interfaces do reveal a more or less regular arrangement for the atoms in the interface. A few of these cases are dealt with below.

Confining ourselves to the interface of two crystals of the same crystal structure (and of the same composition) but of different crystal orientation, the grain boundary is defined by the orientation relation of the two crystals and the orientation of the grain-boundary plane. The orientation relationship between two crystals of the same crystal structure can be described by a (smallest) rotation of say, θ ("misorientation angle"), around a specific axis. The orientation of the axis (in space) is determined by two (Eulerian) angles. The normal of the boundary plane is also determined by two (Eulerian) angles. Hence, a set of five independent variables, also called degrees of freedom (cf. Sect. 7.4), is required to define the grain boundary in a *macroscopic* sense, i.e. without bothering about the precise positions of the atoms at the grain boundary.

In reality, the atoms at the grain boundary may take positions which are determined (also) by relaxation processes at the boundary, as a consequence of which these atoms occupy positions in space incompatible with the prescriptions of both crystal lattices. Thus, the above definition of the grain boundary does not provide an atomic picture of the grain-boundary structure. It could be said that the five independent variables indicated above define the boundary conditions for such relaxation processes. It is found that a set of four, further independent variables is needed to describe the outcome of the relaxation processes at the grain boundary. These four, so-called *microscopic*, parameters involve a possible relative translation of the crystals with respect to each other (the so-called *rigid body translation*) and the position of the grain-boundary plane in the direction of the grain-boundary plane normal. Additionally, local shuffles of atoms close to the boundary plane can occur. Against this background it can be understood that a grain boundary has a certain thickness (say, about two atoms thickness): outside the grain-boundary transition region the perfect crystal structures of both crystals constituting the grain boundary occur.

For special low-angle grain boundaries, the atomic structure at the grain boundary can be directly guessed. Examples of these are the following.

Tilt boundaries are the class of boundaries for which the orientation relation between the two crystals is defined by a rotation around an axis lying in the boundary plane (see Fig. 5.19a). The special case of a low-angle *symmetrical tilt boundary* is shown in Fig. 5.20 ("symmetrical" means that the boundary is positioned symmetrically with respect to the orientation of both crystals). Evidently, the atomic arrangement at the low-angle symmetrical tilt boundary can be conceived as a (vertical) wall of edge dislocations at a constant distance D according to

$$D = \frac{b}{2\sin(\theta/2)} \approx \frac{b}{\theta} \tag{5.14}$$

For an *infinite* wall of edge dislocations the long-range nature of the dislocation strain fields of the individual edge dislocations has been annihilated (Read and Shockley 1952). This is no longer true, to that extent, for a *finite* wall of edge dislocations, as

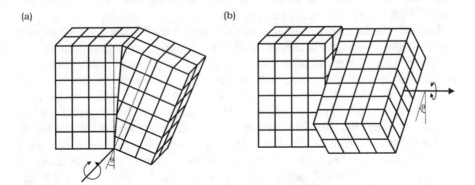

(a)   (b)

Fig. 5.19 Schematic representations of **a** a tilt grain boundary and **b** a twist grain boundary. The misorientation angle has been indicated by θ; the rotation axis has been represented by the solid black line

Fig. 5.20 Representation of a symmetrical small angle tilt boundary by a regular, vertical arrangement of a single set of edge dislocations along the boundary, a dislocation wall (here for a simple cubic lattice). The parameters D (distance between the edge dislocations in the dislocation wall/the boundary), b (length of the Burgers vector) and θ (angle of misorientation) pertain to Eq. (5.14)

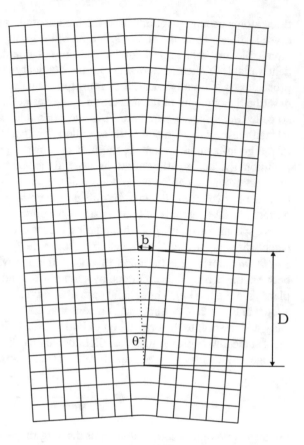

occurs in practice (Beers and Mittemeijer 1978). This configuration of edge dislocations is rather stable; the only possible mobility for edge dislocations thus arranged is by climb. The symmetrical tilt boundary has a relatively low energy. See also the remarks about polygonization in Sect. 11.1.1.

The concept of the low-angle symmetrical tilt boundary described above has been important in the history of materials science, as it provided an indirect proof for the existence of dislocations. It follows from Eq. (5.14), that, for reasonable values of b and sufficiently small values of θ, a value of D results (for numerical values of b and θ equal to 0.3 nm and 0.05°, D equals about 0.34 μm), which is larger than the minimal distance that can be resolved by light optical microscopy (which is about 0.2 μm; cf. Sect. 6.5.2). Since the distortion of the crystal structure is largest at the location of the dislocation line, etching of a cross section of the specimen (perpendicular to the grain-boundary plane) by a chemical etchant will be most severe at the locations in the cross section where dislocation lines end in the cross section. This leads to the emergence of "pits" in the cross section. According to the above discussion, these pits will occur at the low-angle symmetrical tilt boundary at distances which can be resolved by light optical microscopy. Indeed, in a famous experiment, Vogel et al.

(1953), on this basis were able to demonstrate that the concept of the low-angle tilt boundary as a wall of edge dislocations, as shown in Fig. 5.20, made sense.

Twist boundaries are the class of boundaries for which the orientation relation between the two crystals is defined by a rotation around an axis perpendicular to the boundary plane (see Fig. 5.19b). The special case of a low-angle *symmetrical twist boundary* is shown in Fig. 5.21. The atomic arrangement at the boundary can be conceived as a square net of two sets of screw dislocations. With reference to the above discussion on the wall of edge dislocations, it is remarked that a single set of screw dislocations does exhibit a pronounced long-range stress field, but that the presence of the second set annihilates this long-range nature (for infinitely extended sets of screw dislocations). The distance between the dislocations in each of both sets of screw dislocations obeys Eq. (5.14) as well.

At this place, it is appropriate to remark that Burgers (1940; see Sect. 5.2.3) was the first to propose the general conception that low-angle "boundary surfaces" can be made up by "sets of parallel dislocation lines" and the famous pictorial presentation of a symmetrical low-angle tilt boundary (Fig. 5.20) originates from his work [see Fig. 7 in Burgers (1940)].

Considering Eq. (5.14), it follows that the energy of the low-angle boundary increases with increasing θ: the distance between the individual dislocations in the boundary decreases. However, this reasoning does not recognize the interaction of the dislocation-stress fields. Read and Shockley (1952) have derived the following equation for the energy per unit area of low-angle tilt and twist boundaries, γ:

$$\gamma = \gamma_0 \, \theta (A - \ln \theta) \tag{5.15}$$

where γ_0 and A are constants. Whereas the first term in this equation, $\gamma_0 \theta A$, simply expresses that the total energy in the boundary increases if the dislocation density in the boundary increases (θ is proportional with the dislocation density in the boundary; cf. Eq. 5.14), the second term, $-\gamma_0 \theta \ln \theta$, accounts for the effect of the interaction of the stress fields of the dislocations: the long-range nature of the dislocation-strain fields of the individual dislocations becomes annihilated upon arrangement of the dislocations as in these small angle boundaries (this is the driving force for an effect called polygonization discussed in Sect. 11.1.1). Thus, although the total grain-boundary energy increases with increasing θ, the *energy per dislocation* in the boundary decreases with increasing θ. Notwithstanding that Eq. (5.15) is often quoted, there has been remarkably little experimental verification; but there is no reason to doubt the general philosophy behind the model.

For high-angle grain boundaries, say $\theta > 10°$, the cores of the dislocations tend to overlap, one might also speak of the occurrence of a certain "delocalization" of the dislocation cores, and in general the concept of simple dislocation configurations to model the atomic structure of the grain boundary, and thus an equation like Eq. (5.15), looses its significance.

For high-angle grain boundaries (between crystals of the same crystal structure), the concept of the *coincidence site lattice* (CSL) has been proposed (Friedel, 1926), which comprises the lattice points common to the two translation lattices (imagined

(a)

(b)

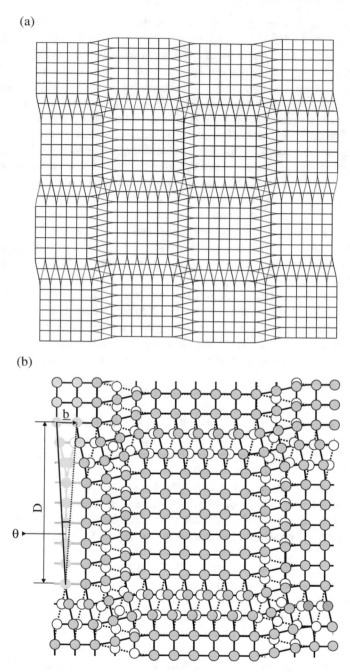

Fig. 5.21 **a** Representation of a symmetrical small angle twist boundary by the regular arrangement of two sets of screw dislocations into a square net in the boundary plane (here for a simple cubic lattice). **b** Enlarged view showing the atomic positions. Grey atoms: atoms above the boundary plane; white atoms: atoms below the boundary plane. At the left, the parameters D (distance between the screw dislocations of a set), b (length of the Burgers vector) and θ (angle of misorientation) pertain to Eq. (5.14)

to extend infinitely) of the two grains which are separated by the grain boundary considered (in this discussion it is assumed here that all lattice points are occupied with one atom of a single element on each site of both translation lattices). The orientation relationship of the two crystals separated by the grain boundary can then be indicated by the notation "Σn", where n, an integer number, denotes the ratio of the unit cell of minimum volume of the CSL and the volume of a primitive unit cell of the single crystal. Hence, a low value for "n" involves that the density of coincidence lattice sites is high. It may be proposed that a grain boundary has a low energy if the grain boundary corresponds with a high density of coincidence lattice sites. Thus a (coherent) twin boundary as discussed below for f.c.c. crystals is a $\Sigma 3$ {111} boundary (see Fig. 5.23).

An outcome of the CSL concept is that a periodicity occurs for coincident lattice sites at the (high-angle) grain boundary. For small deviations from an orientation relationship determined by a high densities of coincidence lattice sites, i.e. corresponding to a low value of "n" in "Σn", a network of so-called *secondary dislocations* can be introduced in the grain boundary, which secondary dislocation networks can be well described by the O-lattice theory due to Bollmann (1982); this description holds for deviations from these ideal coincidence orientations up to, say, 15°.

A description of the structure of grain boundaries on the basis of (only) a consideration of the corresponding CSL can only be of limited significance: the CSL model considers the geometrical matching of the two translation lattices of the bicrystal concerned and thus possesses most relevance for crystalline materials with one-atom motifs (cf. Sect. 4.1.1), as holds for many metals (see the restrictive remark in the one but last paragraph above); in case of multicomponent systems with motifs composed of more than one atom the CSL approach cannot be expected to provide a full description of the grain-boundary structure, if at all. Moreover, the orientation of the grain-boundary plane (two degrees of freedom; see above) is ignored...

Further models for the structure of grain boundaries are those based on the arrangement of polyhedral units. According to these models, the structure of a grain boundary is described by a two-dimensional periodic arrangement of one or more basic types of atomic unit arrangements. For further discussion, see Sutton and Balluffi (1995).

An energetic, i.e. thermodynamic approach is possible as well. Such treatments depart from the surface energies of the faces of the two crystals to be "glued" together and their interaction energy (= energy released upon forming the grain boundary). If one of both faces has a relatively low surface energy, as holds for a low index (densely packed) plane, then the grain-boundary energy may be relatively minimal too. This can be related to the possibility offered by a densely packed plane at the boundary to allow better fit (via minor atom-position adjustments) to its counterpart and thereby establish a larger decrease of free volume than possible with a high index (less densely packed) plane (Randle 2010). Interestingly, it can be noted that the energetic approach, as referred to here in general, has rather recently led to the theoretical and experimental finding that for specific categories of materials a very thin region at and comprising the interface of two phases A and B can be an AB alloy/compound, of *amorphous* (i.e. not crystalline) nature, as the *equilibrium* structure for the interphase (see *"Intermezzo: Interface Stabilized Microstructures"* in Sect. 11.3.1). This may be interpreted as (late) support for the "amorphous cement theory", mentioned above, that initiated serious research on the structure of grain boundaries and interphase boundaries, now about a century ago (see Brandon 2010).

The general picture of the energy of high-angle grain boundaries involves that this energy can be taken as about constant, with the exception of special orientations where a high density of coincidence-lattice sites may occur and the grain boundary may be described as a periodic arrangement of regions of good and bad fit between the adjacent grains; these special orientations have a low energy. As a result, a picture for the dependence of grain-boundary energy on rotation angle θ emerges as shown in Fig. 5.22, with energy minima at orientations corresponding to specific values of Σ. Note that the case shown deals with a grain boundary for variable misorientation in *copper*, that has an f.c.c. translation lattice with, indeed (see above argumentation regarding the limitations of the CSL concept), a motif of one atom.

If the grain-boundary energy is isotropic (as would hold for large ranges in θ for high-angle boundaries; see immediately above), a massive arrangement of columnar, parallel grains (i.e. actually a "two-dimensional" grain-boundary network) would strive for a honeycomb configuration for its grain boundaries, as an angle of 120° between the grain boundaries provides a (metastable) equilibrium configuration (in genuine equilibrium no grain boundaries would be present; i.e. a single crystal represents the equilibrium situation). Note that this "ideal" of 120° for the angle between grain boundaries cannot be realized for a truly three-dimensional, massive arrangement of grains with flat faces: moderate curvatures of (part of) the grain boundaries at

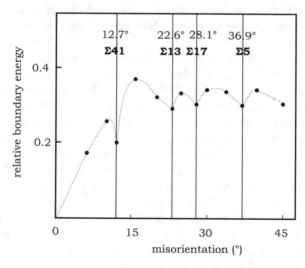

Fig. 5.22 Experimentally determined relation between grain-boundary energy and misorientation angle for [001] twist boundaries in Cu. For several orientations characterized by low values for Σn, cusps in the grain-boundary energy-misorientation curve can be detected. The cusps have been indicated by the black lines and labelled with their misorientation angles and the corresponding Σn-values [taken from T. Mori, H. Miura, T. Tokita, J. Haji, M. Kato, Philos. Mag. Lett. **58**, 11–15 (1988)]

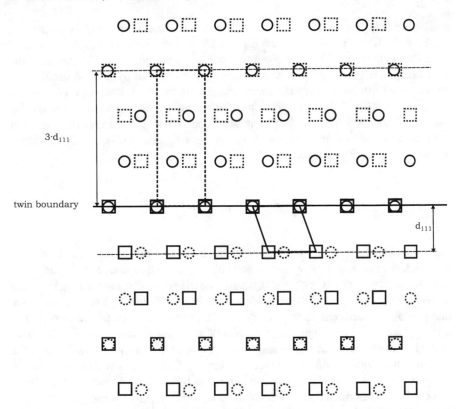

Fig. 5.23 Twinning on an {111} plane in an f.c.c. crystal (with one atom of a single element on each site of the f.c.c. translation lattice). The (coherent) twin boundary has been indicated by the black line. Real (occupied) atomic sites have been drawn with solid lines; virtual (unoccupied) atomic sites have been drawn with dashed lines. The circles indicate atomic sites of the original, parent crystal; the squares indicate atomic sites of the twinned crystal. Above the twin boundary, only sites of the original crystal are occupied; below the twin boundary only sites of the twinned crystal are occupied. The figure also illustrates the concept of the coincidence site lattice (CSL) (here it holds that all translation lattice points are occupied with one atom of a single element on each site of both translation lattices). Coincidence sites are given by the superposition of sites of the original translation lattice and the twinned translation lattice (here: coincidence of circle and square; the unit cell of this CSL lattice, indicated by the dashed rectangle in the figure, is three times as large as the unit cell of the f.c.c. translation lattice, indicated by the rhomboid in the figure. Note that "unit cell" here refers to a unit cell in the two-dimensional projection of the crystal structures (translation lattices) on the plane of drawing. Evidently the orientation relation of "parent" and "twin" can be indicated with $\Sigma 3$ and the coherent twin boundary is a $\Sigma 3$ {111} boundary [see discussion of the CSL concept below Eq. (5.15)]. The interplanar spacing (lattice spacing) of the f.c.c. {111} planes has been denoted by d_{111}

grain-boundary junctions have to occur (for a less crude discussion and explanation, see Sect. 11.3.1; in particular below Eq. (11.13) and see Sect. 11.3.2).

It can be concluded that simple interpretations may not be valid generally: the macroscopic, overall geometry of a boundary, as expressed by a value for Σn (see above), need not be a straightforward indicator of the grain-boundary energy: the microscopic, relaxed, atomistic structure of the boundary, the amount *and* distribution of free volume over the boundary, may be of decisive importance for the energy of the high-angle grain-boundary and its properties (e.g. see Bos et al. 2007). This is an area of high research activity.

5.3.2   Twin Boundaries and Stacking Faults

A special example of a high-angle grain boundary of low energy is the (coherent; see Sect. 5.3.4) twin boundary, where all sites of the translation lattice (indicating here as well the atom sites of the crystal structure; see the above text introducing the CSL) at the boundary are sites of the coincidence site lattice. Twinning has occurred when one part of two adjacent parts of a crystal is a mirror image of the other part. The mirror plane is the twin boundary; the two parts of the crystal constitute the twin; in the literature often, but erroneously, with "twin" is meant the twinned part of the crystal only. Twins are often observed in annealed f.c.c. metals with a low stacking fault energy; such twins then are called *annealing twins*. In f.c.c. materials, twinning can be simply described on the basis of the stacking order of the close packed {111} planes. For perfect f.c.c. material, the order is (cf. Sect. 4.2.1.2):

$$ABCABCABCABC\ldots$$

Let us suppose that the second plane B in this series acts as twinning plane. Then the stacking order becomes:

$$ABCABACBACBA\ldots$$

A coherent twin boundary and an illustration of the concept of the Coincidence Site Lattice (CSL; see above) is presented in Fig. 5.23.

Twinning is not restricted to f.c.c. materials. Twinning, for example, occurs also in b.c.c. and h.c.p. metals, in particular as a result of plastic deformation; such twins are then called *deformation twins*. Twinning thus is the response of the crystal to an externally applied stress and realizes a *homogeneous* shearing of the crystal. Thereby twinning is an other way of shearing than realized by slip/glide of dislocations, which latter process establishes *inhomogeneous* shearing (cf. Sect. 5.2.5).

The occurrence of an error in the stacking order of (close packed) planes causes a planar fault called *stacking fault*. For f.c.c. materials with the ideal stacking order of the closed packed {111} planes according to:

ABCABCABCABC...

the introduction of a stacking fault at the location of the second plane B leads to:

ABCACABCABCA...

It is obvious that the stacking order *after* the occurrence of the fault is identical with the original one. In the sense of this discussion, as a result of twinning, the first plane that is "faulted", as compared to the original stacking order, is the first plane *beyond* the twinning plane, and all other planes in the twinned region are "faulted" as well, with reference to the original stacking order. And also the following statement can be made: the stacking fault can be regarded as one single layer of twinned material. Whereas the stacking fault then is a very localized error, the twinning thus affects macroscopically large volumes.

One can distinguish intrinsic and extrinsic stacking faults. Removal of (part of) a closed packed plane leads to an *intrinsic* stacking fault (cf. Fig. 5.14); insertion of (part of) a closed packed plane leads to an *extrinsic* stacking fault.

Two ways have been indicated in Sect. 5.2.8 how a stacking fault can be terminated within a crystal: (i) two Shockley partial dislocations occur at the extremities of the stacking fault (Fig. 5.18), or (ii) the boundary of the stacking fault is formed by a (sessile) Frank partial dislocation (cf. Fig. 5.14).

5.3.3 Antiphase Boundaries

Without destroying the symmetry of the arrangement of the atom sites of the crystal structure, as happens with twinning and the introduction of stacking faults, the distribution of the various kinds of atoms over the available atom sites for a single crystal can lead to the emergence of a special kind of planar faults called *antiphase boundaries*.

At relatively high temperatures, the distribution of the atoms of the various components can be random. Upon cooling ordering may occur. For example, consider the alloy Cu_3Au. At elevated temperatures, the Cu and Au atoms occupy the atom sites of an f.c.c. crystal in a random fashion, i.e. each atom site has a chance of 25% to be occupied by an Au atom, and consequently a chance of 75% to be occupied by a Cu atom. During cooling, and starting at a certain temperature, the Au atoms wish to take, exclusively, atom sites of only one of the four simple cubic substructures which compose the f.c.c. crystal (cf. Sect. 4.2.1.2). As a result, an ordered structure develops, a *superstructure* called here the $L1_2$ structure, that can be described by taking an f.c.c. unit cell and putting the Au atoms at the corner sites of the cube, and thus the Cu atoms are left with the sites at the centre of the faces of the cube. The structure (*superstructure*) thereby becomes primitive cubic, i.e. it is no longer f.c.c. As a consequence in the (X-ray) diffraction pattern extra reflections appear, so-called superstructure reflections (cf. Sect. 4.4.1.1 and Figs. 4.35 and 4.36).

As there are four simple cubic substructures in the f.c.c. structure, the Au atoms can take any of these four simple cubic substructures. If this ordering occurs upon cooling, it is conceivable that ordered *domains* nucleate at about the same time at different locations in the same crystal. It is well possible that the simple cubic substructure chosen by the Au atoms at nucleus 1 in the crystal considered differs from the simple cubic substructure chosen by the Au atoms at nucleus 2. Then, by growth of the ordered nuclei in the crystal they will eventually "hit" upon each other. At the regions of contact evidently the orderings of the two domains are not compatible, since different simple cubic substructures have been chosen by the Au atoms in the two domains. The "boundaries" which occur between such domains in a single crystal are called antiphase boundaries (APBs).

Considering glide along an {111} plane of the original f.c.c. crystal, glide with $\frac{1}{2}a<110>$, i.e. after passage of a perfect dislocation which realizes a shift/shear from pit B1 to pit B2 for layer B (cf. Fig. 5.17 and its discussion in Sect. 5.2.8), restores the ideal packing if the Cu_3Au crystal is disordered. In the case that the $L1_2$ type ordering has been established, this no longer holds: at the glide plane, for the region where the perfect dislocation has passed, an APB has been formed. The ideal atomic arrangement for the ordered structure reappears at the location of the APB if a second perfect dislocation, that produces an additional slip according to $\frac{1}{2}a<110>$, has passed. The pair of perfect dislocations is called a *superstructure dislocation* (in the literature often designated as *superlattice dislocation*; but see footnote 6 in Chap. 4): After passage of a superstructure dislocation, the original atomic arrangement has been retrieved in the ordered structure, in the same way as after passage of a perfect dislocation the original atomic arrangement has been retrieved in the disordered structure. Now, in analogy with the stacking fault (SF) in f.c.c. crystals as discussed above, it can be concluded that an APB can be terminated within a crystal by two perfect dislocations which are separated by an APB. Because each of the two perfect dislocations is dissociated in two Shockley partial (S. partial) dislocations separated by a SF (cf. Sect. 5.2.8), the following defect configuration corresponding to the superstructure dislocation is obtained in the superstructure:

$S. partial|$ **SF + APB** $|S. partial|$ **APB** $|S. partial|$ **SF + APB** $|S. partial$

The separations between the four partial dislocations in the structure sketched above obviously depend on the values which hold for the SF and APB energies.

5.3.4 Coherent and Incoherent Interfaces

With respect to the above, a discussion of the notions *coherent*, *semi-coherent* and *incoherent* for the character of interfaces appears appropriate as a concluding remark on planar faults. If we make ourselves as small as an atom and take a walk on a (crystal-) lattice plane, we may cross a grain boundary or interphase boundary. Generally, the plane on which we walk does not remain continuous across the boundary, i.e. we have to make a step to accommodate a difference in (crystal-) lattice plane "height" at the boundary. The special case of continuity occurs if the atomic arrangements at the interface, as prescribed by the crystal structure of both grains adjacent to the grain boundary, are the same. Then the interface is called *coherent*. This leaves unimpeded that the plane we walk on may be somewhat distorted/curved close to the interface due to coherency, misfit strains which occur if the equilibrium values for the spacings of the corresponding (crystal-) lattice planes at both sides of the interface are somewhat different. Evidently, in this last case the difference between the equilibrium values of the (crystal-) lattice spacings of the corresponding (crystal-) lattice planes at both sides of the interface considered has been accommodated fully elastically by the system (composed here of two crystals). Such coherent interfaces, where the plane formed by the atoms at the interface is part of the crystal structure of both adjacent grains, can occur for, possibly, any orientation of the interface if, for example, the two crystals correspond to phases of the same crystal structure with identical orientation but with slightly different lattice parameters. However, coherent interfaces can also occur between two phases if only for a specific {hkl} the plane of the atoms at the interface is common to both phases.

An obvious example of a fully coherent interface is the twin boundary discussed above: the atomic distances parallel to the interface at both sides of the interface fit perfectly (Fig. 5.23). A coherent interface is often met in the initial stage of precipitation of a second phase in a matrix of the first phase. Consider the vanadium-nitride (VN; rocksalt-type crystal structure) particle precipitated in the ferrite (α-Fe; b.c.c.) matrix shown in the high-resolution transmission electron micrograph given in Fig. 5.24. The orientation relationship between α-Fe and VN is such that {001} planes of the α-Fe matrix are parallel to {001} planes of the VN platelet, which planes are also parallel to the platelet/matrix interface, i.e. the {001} α-Fe plane also acts as so-called *habit plane* for the platelet precipitate. It appears that in the micrograph the set of (110) lattice planes in the α-Fe matrix continues as a set of (111) planes in the VN platelet, as indicated by the black–white line contrast in the micrograph which traverses the matrix and the particle, a thin platelet, in a continuous way. This illustrates the coherent nature of the interface between the matrix and the faces of the platelet. The misfit between VN and α-Fe leads to a distortion of the lattice planes, exhibited by curvature close to the interfaces between the faces of the platelet and the matrix. At the extremities of the platelet misfit dislocations have developed; the misfit in directions perpendicular to the platelet faces, as experienced at the platelet circumference, is (very) much larger than parallel to the platelet faces.

The misfit between the precipitate phase and the matrix can be accommodated fully elastically only for small misfit and small extents of the interface (small particles). Here it should be recognized that, whereas on a macroscopical scale strains can only be accommodated elastically up to strain values of a few tenths of a percent (cf. Sect. 12.9), on a microscopical scale strains can be accommodated elastically up to strain values of, say, 10–20% (e.g. see Mittemeijer et al. 1981).

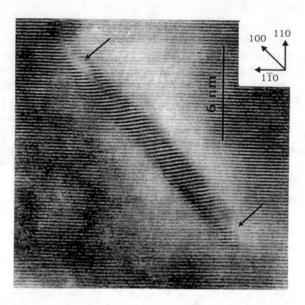

Fig. 5.24 Vanadium-nitride precipitate (rocksalt-type crystal structure) in an α-Fe (b.c.c.) matrix (high-resolution TEM; cf. Sect. 6.8.6). At the top right corner, crystallographic directions referring to the b.c.c. crystal orientation of the α-Fe matrix are shown. The set of (110) planes in the α-Fe matrix continues as a set of (111) planes in the VN platelet, as indicated by the black–white line contrast in the micrograph which traverses the matrix and the particle, a thin platelet, in a continuous way: the interface between the matrix and the faces of the nitride platelet is coherent. The curvature of the lines (often called lattice fringes) is due to elastic accommodation of the misfit between matrix and platelet. Misfit dislocations occurring at the platelet's extremities have been indicated by arrows; these can be conceived as a consequence of the misfit in directions perpendicular to the platelet faces, as experienced at the platelet circumference, being (very) much larger than parallel to the platelet faces [Fe-2.2 at.%V alloy nitrided for 25 h at 913 K (= 640 °C); T.C. Bor, A.T.W. Kempen, F.D. Tichelaar, E.J. Mittemeijer, E.A. van der Giessen, Philos. Mag. A, **82**, 971–1001 (2002)]

   A well-known case of coherent-incoherent transition is provided by the growth of a thin, crystalline layer A on a crystalline substrate B. As long as layer A is very thin, *epitaxial* (i.e. with fixed orientation relation between A and B) growth of a coherent layer A on top of B can occur. Because the substrate B is usually very thick as compared to the layer A, the substrate can be conceived as being rigid and the possible misfit between B and A is fully accommodated by layer A. As long as layer A is very thin, layer A may accommodate the misfit fully elastically: the B/A interface is fully coherent. This type of growth of layer A on substrate B is called *isomorphous*. Upon increasing thickness of layer A, the elastic strain energy incorporated in layer A increases. Beyond a limiting, critical thickness the overall strain energy can be reduced if so-called misfit dislocations are introduced at the B/A interface: the dislocation-strain energy plus the remaining elastic strain in layer A then are smaller than the elastic strain energy in the case where the whole layer A would be strained elastically to accommodate the misfit. The theory predicting and quantifying the emergence of such misfit dislocations is due to Frank and van der Merwe (1949). This process leads to the introduction of series of so-called *misfit dislocations* at the interface with (ideally) constant spacing. Because of the two-dimensional nature of the problem, misfit dislocation networks (comprising two or more sets of misfit dislocations running in different directions) can occur. The B/A interface, as characterized by a network of regularly spaced misfit dislocations, thus consist of regions of good, coherent fit between the dislocations and of the dislocation network with the misfit concentrated at the dislocation lines. Such an interface is called *semi-coherent*. See the schematic picture shown in Fig. 5.25, where an array of edge dislocations accounts for most of the misfit between layer A and substrate B. This type of interface can be considered as an analogue of the low-angle grain boundaries shown in Fig. 5.20 and discussed above. Note that the misfit dislocations in Fig. 5.25 are edge dislocations with the extra half-planes oriented perpendicular to the boundary/interface plane, which contrasts with the low-angle symmetrical tilt boundary where the (misfit) dislocations are also of edge character but with the extra half-planes oriented parallel to/within the boundary plane (see Fig. 5.20).

In case of large misfit, regular networks of dislocations do not occur generally at the B/A interfaces. Such interfaces can be considered as analogues of high-angle grain boundaries and are called *incoherent* interfaces.

One should not make the mistake of identifying incoherency with necessarily non-elastic, plastic accommodation of misfit (as by misfit dislocations). Incoherent interfaces can occur with full elastic accommodation of the misfit. For example, the initial stage of precipitation of tiny Si particles, of diamond-type crystal structure (cf. Sect. 4.2.3.2), in the Al matrix of f.c.c. type crystal structure (a case of volume misfit of 23%), likely leads to incoherent Si particles but, as shown experimentally, with fully elastic accommodation of the misfit (Mittemeijer et al. 1981).

Twin boundaries can also be of incoherent nature. Then the boundary and the mirror/twinning plane are not identical, as holds for the coherent twin boundary (see

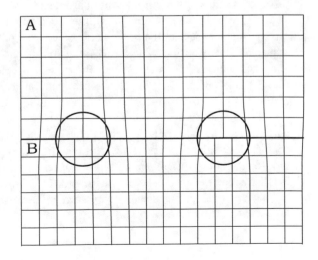

Fig. 5.25 Semi-coherent interface. Schematic illustration of misfit dislocations (circles) at the interface between layer A and substrate B, both of primitive cubic crystal structure (with one atom of a single element on each site of the translation lattice). The lattice spacing of A is slightly larger than the corresponding lattice spacing of B

above). With reference to the twinning process discussed above for the f.c.c. crystal structure, in fact the twinning operation can be conceived as the result of the repeated introduction of stacking faults for all closed packed planes constituting the twinned part of the twin. A stacking fault is caused by the passage of a Shockley partial along a closed packed plane (see Sect. 5.2.8). Then the incoherent twin boundary can be conceived as composed of all Shockley partial dislocations that have passed along the closed packed planes of the twinned part in order to realize the twinning for the twinned part of the crystal considered.

Intermezzo: Coherent and Incoherent Interfaces versus Coherent and Incoherent Diffraction

The adjectives coherent and incoherent have also been used frequently in association with the application of (X-ray) diffraction analysis. It should be recognized that *coherency/incoherency of diffraction* by second phase particles (as precipitates), in a matrix, with the diffraction by the matrix, may but need not coincide with occurrence of *coherency/incoherency of the second phase/matrix interface*. This can be made clear as follows.

Constructive interference, i.e. coherent diffraction, occurs if the waves scattered by separate parts of the diffracting material have a (more or less) fixed phase difference (reduced modulo 2π). Destructive interference, i.e. incoherent diffraction, occurs if the phase difference between waves scattered from one part of the diffracting material (say matrix phase) and from another part of the diffracting material (say second, precipitate phase) takes any value between o and 2π *with equal probability*.

Thus, because of the variability of the phase jump at an *incoherent* boundary between a second phase particle and the matrix, due to its irregular structure, it appears likely that in this case the second phase particles and the matrix diffract independently, i.e. incoherently, and in the diffraction pattern separate diffraction peaks of the second phase particles and the matrix occur.

For *coherent* (and semi-coherent) interfaces, more complicated diffraction effects can be expected. If a misfit between the second phase particles and the matrix exists, distortions, due to elastic accommodation, occur in the matrix and the second phase particles, especially close to the particle/matrix interfaces (cf. Fig. 5.24). The phase difference between waves scattered by the matrix and waves scattered by the second phase particle depends on both the position (difference) vector from one scatterer (in the matrix) to the other (in the particle) and the value of the diffraction angle (i.e. the length of the diffraction vector). Then, given the imperfect (strained) but (semi-)coherent crystal structure for the entity matrix/second phase particles, it depends on the length of the diffraction vector if coherent or incoherent diffraction occurs. This effect has been recognized and discussed by van Berkum et al. (1996).

Fig. 5.26 Evolution of the ferrite 211 diffraction-line profile (Co K_α radiation) of thin-foil Fe–Cr alloy specimens (**a**) and Fe-V alloy powder specimens (**b–d**) upon nitriding and denitriding. The shown diffraction-line profiles result after subtracting from the measured profile the $K_{\alpha2}$ component as well as removing the instrumental profile. **a** Fe-2.0 at.%Cr thin foil nitrided at 723 K (450 °C) with a nitriding potential of 0.1 atm$^{-1/2}$ in steps of 72 h to a total of 792 h (only some of the intermediate nitriding steps are shown here). **b–d** Fe-V powders of indicated compositions (in at.%), nitrided for 4 h at 773 K (500 °C) with a nitriding potential of 0.1 atm$^{-1/2}$. Upon (prolonged) nitriding the diffraction-line profiles shift to lower diffraction angles. The results after a denitriding treatment at 673 K (400 °C) for 16 h in flowing H_2 atmosphere are shown as well: the diffraction-line profiles reveal a partial shift back upon denitriding (Akhlaghi et al. 2015).

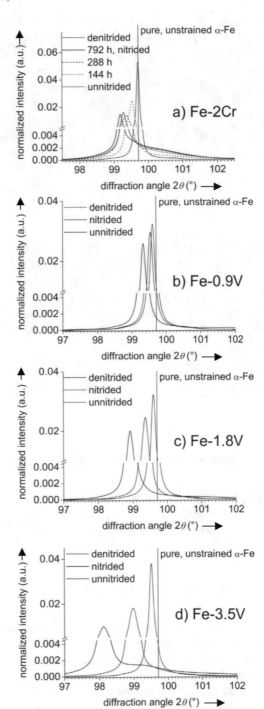

For an example, exhibiting complex diffraction effects in case of precipitates diffracting coherently with the matrix, we consider fully (i.e. homogeneously) nitrided Fe–Cr (thin foils) and Fe-V (powders) specimens. The nitriding process was performed at relatively low temperatures (450 and 400 °C, respectively) so that the platelike CrN and VN precipitates, which develop upon nitriding, are fully coherent with the α-Fe (ferrite) matrix. The X-ray diffraction patterns observed (Fig. 5.26) only exhibit (broadened and shifted towards lower diffraction angles) ferrite reflections, i.e. *separate CrN and VN reflections do not occur: the CrN and VN precipitate particles diffract coherently with the ferrite matrix.* The shift observed upon nitriding of the ferrite reflection to lower diffraction angle implies that the overall lattice parameter has increased, which could be fully quantitatively ascribed to the elastic accommodation of the precipitate/matrix misfit (see footnote 18 in Sect. 4.4.1). The ferrite reflections are severely broadened due to the misfit strains in the specimen (Fig. 5.26; for an introduction to diffraction-line broadening due to (micro)strains, see Sect. 6.10.1). The observed, characteristically shaped broadening is the consequence of the nature of the misfit-strain field surrounding the nitride platelets in association with the nitride platelets diffracting coherently with the matrix: the strong, *tetragonally shaped* strain fields surrounding the misfitting, tiny, coherent precipitates lead to an intensity tail at the right-hand side (high diffraction-angle side) of the ferrite-peak maximum (Akhlaghi et al. 2015).

5.4   Volume Defects (Three-Dimensional)

The volume defects considered pertain to three-dimensional objects contained within a matrix. Three-dimensional structures composed of zero-, one- or two-dimensional defects are not considered here.

5.4.1   Second Phase Particles

Second phase particles, precipitated within, as a consequence of a thermal treatment, or taken up, as a consequence of a material processing route, into a matrix of the first, dominant phase, disrupt, more or less (as possibly associated with the occurrence of incoherent or coherent interfaces; see Sect. 5.3), the long-range translation symmetry of the matrix. They may induce considerable misfit-stress fields and thus can influence material properties pronouncedly. Such stress fields surrounding the second phase particles can be due to misfit between the volume occupied by the second phase particle when unconstrained and the space ("hole") put at its disposal by the matrix. Such misfit can arise due to specific volume differences induced by precipitation, or by different thermal expansion or shrinkage upon heating or cooling the specimen.

A possibly favourable effect of second phase particles is a contribution to the enhancement of mechanical strength. Considering yielding of a material as related to glide of dislocations (Sect. 5.2.5), any mechanism obstructing dislocation glide improves the mechanical strength. In the discussion of the Frank-Read source for dislocation (-line) production (Sect. 5.2.6) it was made clear that second phase particles can serve as obstacles for dislocation migration: the stress fields surrounding the second phase particles can be of "antagonistic" nature and "block" propagation of the stress field of a migrating dislocation: the second phase particle acts as "pinning point". It was already indicated that in order that a dislocation can pass two pinning points (A and B in Fig. 5.13; see Sect. 5.2.6) a critical shear stress is needed that depends on the distance between the obstacles (which can be second phase particles):

$$\tau_0 = \frac{Gb}{d} \tag{5.10}$$

where d represents the distance between A and B and thus reflects the dependence of the critical shear stress τ_0 on the second phase particle density and distribution. This mechanism for hardening is designated as the *Orowan process* (with τ_0 as the *Orowan (shear) stress*; see also Sect. 12.14.4). As a result of the Orowan process, upon passage of the pinning points by a series of gliding dislocations, a system of concentric loops is formed around the second phase particles (see Fig. 5.27). Consequently, the effective average distance between the second phase particles has decreased to d', which implies a necessary increase of the maximal value of shear stress required for continuation of dislocation glide (cf. Eq. 5.10).

5.4.2 Pores

Pores are important defects in particular for sintered ceramic materials. Sintering implies the increase of the density of powder material by the shortening of the distances between the centroid points of (mass) gravity between adjacent powder particles. Originally, the pores in the not yet sintered powder occur between the particles. In a later stage, the pores will occur at the grain boundaries between the, originally separated, individual powder particles. In a final stage of sintering, the pores may have been left behind by (in the wake of) the moving grain boundaries, i.e. they have become detached from the powder-particle boundaries, and the pores then are incorporated within the matrix of the grains (see Fig. 5.28). These pores, in this final stage of sintering, are rather stable, since their elimination now requires volume diffusion of atoms (for pores situated at grain boundaries, as in an earlier stage of sintering, the relatively fast grain-boundary diffusion is rate determining for pore elimination). Pores need not occur as hollow spheres. In particular for sufficient

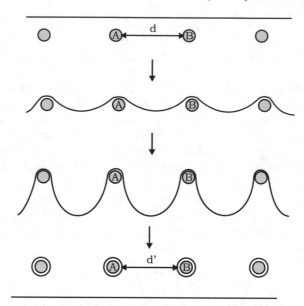

Fig. 5.27 Schematic depiction of the Orowan process of work hardening (from top to bottom). Upon application of a shear stress, a dislocation line (black line) starts to move (recall that for an edge dislocation the dislocation line moves in a direction parallel to **b** and that for a screw dislocation the dislocation line moves in a direction perpendicular to **b**; in both cases, the component of τ parallel to **b** controls the occurring slip; cf. Sect. 5.2.5). The dislocation gets pinned at second phase particles (grey). The average distance between these particles is d. The dislocation bows out upon increasing the shear stress. The maximal value of shear stress corresponds with half-circles of the curved dislocation line between the pinning points (cf. Eq. 5.10). After passage of the dislocation, concentric dislocation rings are left around the second phase particles, which decreases the effective average distance between the second phase particles to d', and thereby implies an increase of the maximal value of shear stress required for continuation of dislocation glide (cf. Eq. 5.10)

Fig. 5.28 Pores during the sintering process of powders. **a, b** Pores (black) are the relicts of the spaces between the original powder particles (white) and thus normally located at grain boundaries. **c** In late stages of sintering, grain growth may occur, during which the grain boundaries move away from the pores, leaving them in the interior of the grain

time at sufficiently elevated temperature, facetting of the pore inner surface can occur as consequence of anisotropy of the surface energy. One then speaks of the development of "negative crystals", possibly having in mind the development of facets on the surface of a machined spherical single crystal upon annealing, also as a consequence of an anisotropic [i.e. depending on (hkl)] surface energy.

References

General

F. Agullo-Lopez, C.R.A. Catlow, P.D. Townsend, *Point Defects in Materials* (Academic Press, London, 1988).
J.P. Hirth, D. Lothe, *Theory of Dislocations*, 2nd edn. (Wiley, New York, 1982).
D. Hull, D.J. Bacon, *Introduction to Dislocations*, 4th edn. (Butterworth-Heinemann, Oxford, 2001).
A.P. Sutton, R.W. Balluffi, *Interfaces in Crystalline Materials* (Clarendon Press, Oxford, 1995).

Specific

M. Akhlaghi, T. Steiner, S.R. Meka, A. Leineweber, E.J. Mittemeijer, Lattice-parameter change induced by accommodation of precipitate/matrix misfit; mitfitting nitrides in ferrite. Acta Mater. **98**, 254–262 (2015)
A.M. Beers, E.J. Mittemeijer, Dislocation wall formation during interdiffusion in thin bimetallic films. Thin Solid Films **48**, 367–376 (1978)
J.G.M. van Berkum, R. Delhez, Th.H. de Keijser, E.J. Mittemeijer, Diffraction-line broadening due to strain fields in materials. Acta Crystallographica, A **52**, 730–747 (1996)
W. Bollmann, *Crystal Lattices, Interfaces, Matrices: An Extension of Crystallography* (W. Bollmann, Geneva, 1982).
C. Bos, F. Sommer, E.J. Mittemeijer, Atomistic study on the activation energies for interface mobility and boundary diffusion in an interface-controlled phase transformation. Phil. Mag. **87**, 2245–2262 (2007)
D. Brandon, Defining Grain Boundaries: An Historical Perspective. Mater. Sci. Technol. **26**, 762–773 (2010)
J.M. Burgers, Geometrical considerations concerning the structural irregularities to be assumed in a crystal. Proc. Phys. Soc. (Lond.) **52**, 23–33 (1940)
F.C. Frank, J.H. van der Merwe, One-dimensional dislocations. II. Misfitting monolayers and oriented overgrowth. Proc. R. Soc. **A198**, 216–225 (1949)
E.J. Mittemeijer, P. van Mourik, Th.H. de Keijser, Unusual lattice parameters in two-phase systems after annealing. Philos. Mag. A **43**, 1157–1164 (1981)
J.J. Gilman, The "Peierls Stress" for pure metals (evidence that it is negligible). Philos. Mag. **87**, 5601–5606 (2007)
V. Randle, Role of grain boundary plane in grain boundary engineering. Mater. Sci. Technol. **26**, 774–780 (2010)
W.T. Read, W. Shockley, *Dislocation models of grain boundaries*, in ed. by W. Shockley, J.H. Hollomon, R. Maurer, F. Seitz *Imperfections in Nearly Perfect Crystals* (Wiley, New York, 1952), pp. 352–371

F.L. Vogel, W.G. Pfann, H.E. Corey, E.E. Thomas, Observations of dislocations in lineage boundaries in Germanium. Phys. Rev. **90**, 489–490 (1953)
B.E. Warren, *X-Ray Diffraction* (Addison-Wesley, Reading, Massachusetts, 1969).

Chapter 6
Analysis of the Microstructure; Analysis of Structural Imperfection: Light and Electron Microscopical and (X-Ray) Diffraction Methods

Materials are substances that have now, or are expected to find in a not too distant future, practical use (see Chap. 1). The microstructure of a material (beautifully described by the untranslatable German word "Gefüge") is a notion that comprises all aspects of the atomic arrangement in a material that should be known in order to understand its properties. Mostly, we are concerned with crystalline materials. The conception microstructure then narrows to the description of the so-called crystal imperfection (cf. Chap. 5).

The notion microstructure encompasses a long list of specificities: the compositional inhomogeneity, the amount and distribution of the phases in the material, the grain size and shape and distribution functions of the grain-size parameters, the grain(crystal)-orientation distribution function (texture), the grain boundaries/interfaces and the surface of the material, the concentrations and distributions of crystal defects as vacancies, dislocations, stacking and twin faults, and lattice distortions.

The microstructure to a very large extent determines the properties of a material. As already indicated in the Preface of this book, materials science boils down to: *the development of models which provide the relation between the microstructure and the properties.* To this end, characterization methods of the microstructure of a material are a prerequisite.

Every materials scientist involved in research has to spend a considerable amount of time on methodological development. This in particular holds if top-level expertise should be acquired on the analysis of the microstructure of materials. In particular, application and thorough knowledge of basic aspects of, on the one hand, image-forming, microscopical techniques, which provide local information, and, on the other hand, (X-ray) diffraction analysis methods, which give statistically averaged information on the (defect) structure.

Practically, all microstructural analyses of any specimen of a material start, or should start, with a light optical microscopical examination. The light microscope can be fruitfully applied in some stage of investigation of practically all applied and fundamental research projects in materials science. There nowadays exists a

© Springer Nature Switzerland AG 2021
E. J. Mittemeijer, *Fundamentals of Materials Science*,
https://doi.org/10.1007/978-3-030-60056-3_6

tendency to overlook the possibilities of the light microscope in favour of the scanning electron microscope in particular. Without ignoring the advantage of enhanced resolving power offered by, for example, the scanning electron microscope, there is no competitor for the versatility and the relative ease of application of the light microscope. In fact, it can be advised at least to perform light-microscopical analysis before more evolved, but unavoidably more constrained, (electron) microscopical techniques are applied.

On the one hand, it is often taken unjustly for granted that interpretation of light optical and in particular (transmission) electron micrographs is an activity for specialists only. On the other hand, unknowingly one may blunder grossly: for example, ask if one is aware of the difference between a dark field image in light optical microscopy and in transmission electron microscopy. Any materials scientist should be familiar with basic knowledge on image formation and the principle of the functioning of a microscope. This knowledge must and can be transferred in the beginning of any study on materials science, i.e. before or at the time that one for the first time looks at a microstructure as imaged by a light optical or an electron microscope. Therefore, some space in this book, on fundamentals of materials science, is devoted to this topic, without that a course on microscopy is given.

Against the above background in the following essential aspects of light optical microscopical techniques are discussed first. Along the way, main elements of image-formation theory are introduced, thereby facilitating a subsequent discussion on "supermicroscopy", transmission (and scanning) electron microscopy and X-ray diffraction methods.

6.1 The Lens

The lens consists of a transparent (for visible light) body bounded by two (curved) surfaces. In practice, lenses are employed with flat or spherical refracting surfaces such that rotational symmetry occurs with respect to the optical axis (=line connecting the centres of curvature); only this kind of (centred) lenses can be simply made in series with sufficient accuracy.

The action of the lens should ideally be to distort the light (wavefronts)[1] propagating from a point of the object, in order that they converge into a single point of the image, such that there exists a one to one correspondence between all points in the object plane perpendicular to the lens axis and all points in the image plane also perpendicular to the lens axis.

---

[1] Light is considered here as an electromagnetic wave propagation, which can be characterized by its amplitude and phase. With reference to the discussion on the dualistic nature of light in Sect. 2.4, and noting that there is no such medium, as "ether", through which the "wave" would propagate, the only observable quantity of the light is the (time averaged) intensity which is proportional to the squared amplitude.

6.1.1 The Paraxial Approximation

In the limiting case of (i) infinitely small inclination of the light ray considered to the lens axis and (ii) thin lenses, the so-called *paraxial imaging equation* holds:

$$\frac{n_1}{v} + \frac{n_2}{b} = \frac{n - n_1}{r_1} - \frac{n - n_2}{r_2} = F \tag{6.1}$$

where n_1 and n_2 are the indices of refraction for object and image space, respectively; n represents the index of refraction of the lens; v and b denote the distances of object and image, respectively, to the point of intersection of the (infinitely thin, centred) lens with the lens axis (the lens is represented by its so-called principal plane H in Fig. 6.1) and r_1 and r_2 are the radii of curvature for the lens surfaces adjacent to object and image space, respectively. If $n_1 = n_2 = 1$ (object space and image space are vacuum), the well-known formula is obtained:

$$\frac{1}{v} + \frac{1}{b} = \frac{1}{f} = (n - 1)\left(\frac{1}{r_1} - \frac{1}{r_2}\right) \tag{6.2}$$

By agreement v and b are taken as positive quantities if object and image are real, whereas they are taken as negative quantities if object and image are virtual. Further, r_1 and r_2 are considered as positive quantities if the lens surface considered is convex for the incident light, whereas they are considered as negative quantities if the lens surface considered is concave for the incident light.

From Eq. (6.1), it immediately follows for the refraction of light by a lens:

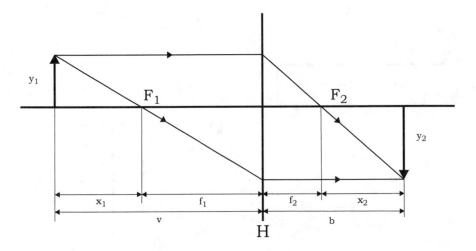

Fig. 6.1 Schematic illustration of the paraxial imaging equation. The infinitely thin lens is represented by the principal plane H

(i)   rays parallel to the lens axis ($v = \infty$) after refraction pass through a single point called focus ($b = f_2 = n_2/F$);

(ii)   rays passing through the (other) focal point ($v = f_l = n_1/F$) after refraction propagate parallel to the lens axis ($b = \infty$).

If the focal distances in object and image space are denoted by f_1 and f_2 and the distances of object y_1 and image y_2 to the corresponding focal points are given by x_1 and x_2 (cf. Fig. 6.1), then it follows for M_l = lateral (transverse) magnification = ratio of (linear) sizes of image and object:

$$M_l = \frac{y_2}{y_1} = \frac{x_2}{f_2} = \frac{f_1}{x_1} = \frac{f_1}{f_2}\frac{b}{v} \qquad (6.3)$$

In case $n_1 = n_2$, $f_1 = f_2$. Then the ray through the centre of the lens passes unrefracted and $M_l = b/v$.

Further with $f_1 > 0$ it follows (see Fig. 6.2a–e):

– $f_1 < x_1 < \infty$: real image smaller than object;
– $x_1 = f_1$: real image of same size as object;

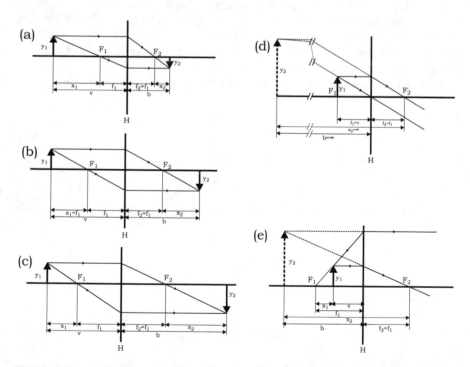

Fig. 6.2 Real and virtual image formation for different distances, x_1, of object y_1 to focal point f_1 in case of a symmetric lens ($n_1 = n_2, f_1 = f_2$). **a** $f_1 < x_1 < \infty$: real image smaller than object, (**b**). $x_1 = f_1$: real image of same size as object, **c** $0 < x_1 < f_1$: real image larger than object, **d** $x_1 = 0$: virtual image at infinite distance, **e** $-f_1 < x_1 < 0$: virtual image larger than object

- $0 < x_1 < f_1$: real image larger than object;
- $x_1 = 0$: virtual image at infinite distance;
- $-f_1 < x_1 < 0$: virtual image larger than object.

From Eqs. (6.1) and (6.3), it follows with $n_1/n_2 = f_1/f_2$:

$$\frac{db}{dv} = -\frac{f_2}{f_1} M_l^2 = -\frac{n_2}{n_1} M_l^2 \tag{6.4}$$

It is concluded that the longitudinal (axial) magnification is proportional with the square of the lateral magnification. Hence, upon image formation, a large spatial distortion occurs in general. In view of the mostly very limited depth of focus this does not normally constitute a serious problem.

6.1.2 The Compound Lens

Image formation employing a thick or compound lens can be described in an analogous, paraxial manner. The principal plane H representing the thin lens (Fig. 6.1) is now replaced by two principal planes, H_1 and H_2 (Fig. 6.3). The focal distance is measured starting from the point of intersection of the principal plane concerned with the lens axis. The rules (i) and (ii) mentioned in Sect. 6.1.1 remain valid.

Further two points, K_1 and K_2, the so-called nodal points, can be indicated at the lens axis such that (cf. Fig. 6.3):

$$F_1 K_1 = f_2; \quad F_2 K_2 = f_1 \tag{6.5}$$

The ray of incident light, travelling through K_1, after refraction passes through K_2 parallel to its original direction (angular magnification $= 1$). Because, $f_1/f_2 = n_1/n_2$

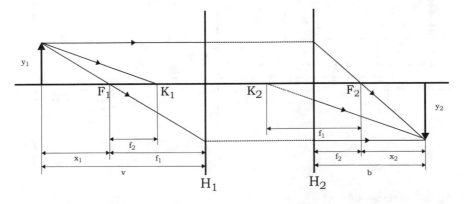

Fig. 6.3 Schematic illustration of the paraxial imaging equation for a compound lens

(cf. Sect. 6.1) it is concluded in case $n_1 = n_2$ that both nodal points coincide with the points of intersection of the principal planes with the lens axis.

6.2   Image Formation

In the sequel, the refractive indices of object and image space are set equal to one: Image formation is thought to occur in vacuum, or, which is practically the same, in air. This does not impose an essential restriction.

Consider an object illuminated by a monochromatic pencil of parallel rays. Diffraction of the light will occur at angles determined by the wavelength of the light, and the spacing and orientation of the microstructural features of the object. The diffracted rays of corresponding structural features (think of grating) are parallel to each other and converge in a point in an image plane at infinite distance. Thus, a diffraction pattern of the object is produced in this image plane. In this case of source of light and image screen at infinite distances (*Fraunhofer diffraction*), the diffraction pattern can be produced at a finite distance in the (back) focal plane of a lens or lens system. Superposition of light originating from the diffraction maxima in the diffraction pattern (focal plane) leads to an image of the object in the image plane (Fig. 6.4). Image formation was considered in the above manner by Abbe.

A grating with spacing p is coherently illuminated by a monochromatic pencil of rays parallel to the lens axis (Fig. 6.4). Suppose that in the (back) focal plane, a maximum occurs for rays diffracted at an angle φ with the axis (The location of this maximum is determined as the point of intersection, S, of the ray making an angle φ with the lens axis and passing through the nodal point K_2, with the focal plane; cf. Sect. 6.1.2). From Eqs. (6.5) and (6.1), it follows:

$$F_2 K_2 = f = \frac{F_2 S}{\tan \varphi} \tag{6.6}$$

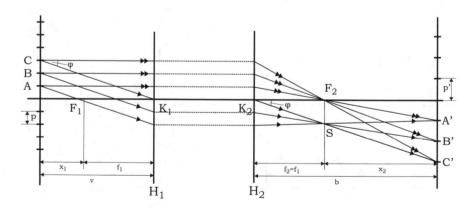

Fig. 6.4 Image formation after Abbe

A diffraction maximum occurs if the path difference between neighbouring rays equals $m\lambda_o$ with $m = 0$ (principal diffraction maximum), $\pm 1, \pm 2 \ldots$ and where λ_o represents the wavelength (in vacuum). Then[2]

$$\sin \varphi = \frac{m\lambda_o}{p} \tag{6.7}$$

For sufficiently small φ, $\sin \varphi \approx \tan \varphi$ and from Eqs. (6.6) and (6.7) it is obtained

$$F_2 S = \frac{f m \lambda_o}{p} \tag{6.8}$$

The distance between the adjacent diffraction maxima equals

$$p' = \frac{f \lambda_o}{p} \tag{6.9}$$

As a consequence of the *finite* number of slits of the grating giving rise to the diffraction maxima in the (back) focal plane (which, in absence of the lens system, corresponds to observation of the diffraction maxima at infinite; see discussion above), these maxima have a certain lateral extension. The smaller the number of slits of the grating, the broader the intensity maxima in the focal plane. As a result of a rigorous calculation, the intensity distribution in the focal plane, I_p, can be determined as a function of φ (i.e. as a function of the location in the focal plane) and as a function of the number of slits N contributing to the diffraction. It is obtained:

$$I_p = \text{const.} \frac{\sin^2 \left\{ \frac{\pi}{\lambda_o} N p \sin \varphi \right\}}{\sin^2 \left\{ \frac{\pi}{\lambda_o} p \sin \varphi \right\}} \tag{6.10}$$

Examples of the intensity distribution in the focal plane calculated according to Eq. (6.10) are presented in Fig. 6.5 for slits of infinitesimal width and a total number of slits, N, of 1, 2, 4 and 8. The following remarks can be made:

- Primary maxima, proportional to N^2, occur at values $p \sin \varphi = m \lambda_o$ with $m = 0$, $\pm 1, \pm 2, \ldots$ (Eq. 6.7).
- Secondary maxima are found between the primary maxima.
- The ratio of the heights of the primary and secondary maxima increases as the number of slits, N, increases. In the limit $N \to \infty$ a set of "point" maxima appears.
- The distance between the primary maxima is *inversely* proportional to the spacing p of the grating (cf. Eq. 6.9).

---

[2] Note that in case in object space $n_1 \neq 1$: $\sin \phi = m\lambda_o/(n_1 p)$.

Fig. 6.5 Formation of intensity maxima in the focal plane for diffraction of light waves by a grating with N slits

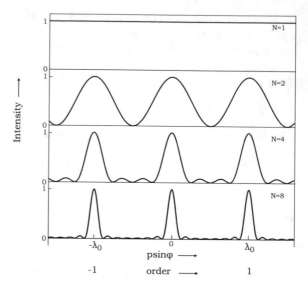

- The principal effect of a finite width of the slits of the grating considered would be that, roughly speaking, the height of the primary maxima rapidly decreases for increasing values of m.

The crucial part of the image-formation process follows now: With respect to image formation, the diffraction maxima in the focal plane can be considered as coherent sources of (secondarily diffracted) light. For large values of N, the diffraction pattern in the focal plane can be conceived as a grating composed of $2M + 1$ slits (numbered as: $-M, -(M-1), \ldots, 0, \ldots (M-1), M$; cf. Fig. 6.5) of infinitesimal width with a spacing p' according to Eq. (6.9). Hence, interference phenomena observed at an image screen at an arbitrary distance, l, from the focal plane, very much larger than the spacing p', can be described completely analogous to the above discussion for the primarily diffracted light interfering at the focal plane. In analogy with Eq. (6.10), the obvious result for the intensity distribution in the image plane, $I_{p'}$, as a function of φ' (i.e. as a function of the location in the image plane) and as a function of the number of diffraction maxima (of infinitesimal extension, i.e. N is very large), $2M + 1$, contributing to the (secondary) diffraction, is given by:

$$I_{p'} = \text{const.} N^2 \frac{\sin^2\left\{\frac{\pi}{\lambda_o}(2M+1)p'\sin\varphi'\right\}}{\sin^2\left\{\frac{\pi}{\lambda_o}p'\sin\varphi'\right\}} \tag{6.11}$$

Analogous to the remarks made above with respect to the intensity distribution in the focal plane, it is concluded for the intensity distribution in the image plane:

- A set of primary maxima is observed with a spacing p'' which conforms [following a derivation which parallels Eqs. (6.6)–(6.9)]

$$p'' = \frac{l\lambda_o}{p'} = \frac{lp}{f} \tag{6.12}$$

and thus p'' is *directly* proportional to the spacing p of the grating (see also Eq. (6.9) and the discussion above regarding the intensity distribution in the focal plane).

- The intensity distribution can only be considered as a faithful "image" of the grating if $2M + 1$ ($=$ number of diffraction maxima contributing to "image" formation) is large: see Fig. 6.5.

The distance l is not subject to any restriction, apart from $l \gg p'$. Therefore, in this special case, an "image" of the grating is not only obtained in the image plane as prescribed by the paraxial image construction (cf. Sect. 6.1). Now, if also plane waves making various angles with the lens axis are incident to the object, then a series of corresponding diffraction patterns is generated in the focal plane. For each diffraction pattern, the distance between adjacent maxima equals p'; but the diffraction patterns are shifted with respect to each other.

Interference of light originating from these diffraction patterns now only gives rise to a set of interference fringes constituting an image of the object in the image plane if $l = x_2$ (cf. Figs. 6.1, 6.3 and 6.4), because then the optical path length from object point to image point is the same no matter via which diffraction maximum the diffracted ray travels.[3] Thus, the image of the grating is observed in the image plane as prescribed by Eq. (6.1), and it holds (cf. Eq. 6.3)

$$p'' = x_2 \frac{p}{f} = M_l p \tag{6.13}$$

The information content of diffraction pattern and image is in principle the same, but the data are differently distributed: in the diffraction pattern for each maximum an averaging occurs over the entire object, whereas in the (perfect) image a point-to-point correspondence exists with the object (see Fig. 6.4). This statement describes the distinguishing characteristics of microscopic and diffraction methods for microstructure analysis.

Hence, diffraction methods generally yield information averaged over the illuminated part of the specimen (e.g. the crystal structure as determined by X-ray diffraction), whereas image-forming methods provide local information (e.g. the location of a precipitated particle as determined by light optical microscopy and the location of a dislocation as determined by transmission electron microscopy).

---

[3] Fermat's principle says that the light follows always the course of minimal optical path length. The optical path length is defined by the product of path length and index of refraction. Constant optical path length then implies that the number of wavelengths corresponding to the length of the path followed is the same, thereby accounting for possible variations of the value of the index of refraction along the path followed.

A lens for image formation, in the sense discussed here, does not always exist. In the sense of the present discussion, this is, for example, the case for X-rays; then only the diffraction pattern can be studied.

6.3 The (Reflected) Light Optical Microscope

6.3.1 The Magnifier ("Loupe")

As indicated in Sect. 6.1, a double-convex lens (with $f \equiv f_1 = f_2$) invokes a virtual image y_2 of object y_1 for object distances $-f < x_1 < 0$ (Fig. 6.6a), which image can be transformed by a converging lens (e.g. the human eye) into a real image (Fig. 6.6b).

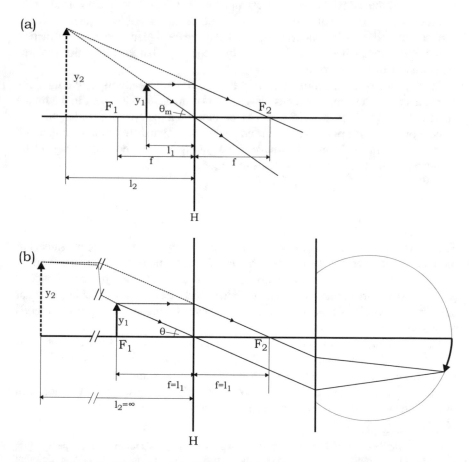

Fig. 6.6 a Schematic illustration of the formation of a virtual image y_2 in case $-f < x_1 < 0$ (case(e) in Fig. 6.2). **b** Image formation on the eye's retina using a magnifier, with a fully relaxed eye lens and with the object in focal point F_1 (image y_2 at infinite distance)

Using a fully relaxed eye lens (then the distance eye lens to retina = focal distance eye lens) objects at infinite distance are sharply imaged on the retina. In order to apply a fully relaxed eye lens on observing through a magnifier, the rays "originating" from a point of the virtual image should be parallel to each other (then they converge into a single point in the focal plane = retina). Hence, then (i) the object should be placed in the focal plane (F_1) and (ii) the virtual image occurs at infinite distance. The objects are observed larger by reducing the object distance in combination with accommodating the eye lens (= reduction focal distance eye lens). By definition, a reference distance $l_r = 250$ mm is taken as the smallest distance from the eye allowing "easy" observation. The magnifying power of the magnifier is expressed by the angular magnification, M_a, defined as the ratio of the tangent of the angle subtended at the eye by the virtual image of the object and the tangent of the angle subtended at the naked eye by the object when placed at the reference distance, l_r.

For the situation of Fig. 6.6, it follows for the angle θ_m subtended at the eye by the virtual image:

$$\tan \theta_m = \frac{y_1}{l_1} = \frac{y_2}{l_2} \tag{6.14}$$

Using $v = l_1$ and $b = -l_2$ (cf. definitions in Sect. 6.1.), it follows from Eqs. (6.2) and (6.14):

$$\tan \theta_m = y_1 \left\{ \frac{1}{f} + \frac{1}{l_2} \right\} \tag{6.15}$$

For the angle θ subtended at the naked eye, it follows in consideration of the definition made above

$$\tan \theta = \frac{y_1}{l_r} \tag{6.16}$$

Thus, it is obtained for the angular magnification

$$M_a = \frac{\tan \theta_m}{\tan \theta} = l_r \left\{ \frac{1}{f} + \frac{1}{l_2} \right\} \tag{6.17}$$

Two extreme situations can be considered:

(i) The instrument is adjusted such that the rays originating from the object seemingly come from infinite distance (fully relaxed eye lens). Then $l_2 = \infty$ and

$$M_a = \frac{l_r}{f} \tag{6.18}$$

(ii)   The instrument is adjusted such that the rays originating from the object seem-
        ingly come from the smallest distance for "easy" observation. Then $l_2 = l_r$
        and

$$M_a = 1 + \frac{l_r}{f} \qquad (6.19)$$

In this case, the angular magnification is equal to the ratio of the (linear) sizes of
image and object (y_2/y_1).

Equation (6.18) is normally used in practice as a definition for the magnifying
power, M_p, of all optical instruments (e.g. a magnifier) producing virtual images (at
distances between 250 mm and infinite) of nearby objects:

$$M_p = \frac{l_r}{f} \qquad (6.20)$$

Then, in contrast to the lateral magnification, M_l, the magnifying power is a characteristic quantity of the optical system. To distinguish M_p and M_l for a "magnification"
of "a" one writes: $M_p = ax$ and $M_l = a{:}1$ (e.g. $M_p = 30x$ and $M_l = 30{:}1$).

6.3.2   The Compound Microscope

The microscope optically consists of two lens systems, the objective and the eyepiece,
which are separated by a distance larger than the sum of their focal distances. For a
schematic presentation, both lens systems are replaced by a single thin positive lens.
The paraxial image construction (cf. Sec. 6.1.) has been performed in Fig. 6.7.

In practice, the microscope is adjusted such that the object under examination, y_1,
is placed just outside (below) the outer(lower)-focal plane of the objective. Then, the
objective provides a real, upside-down, image, y_2, of the object in the inner(lower)-
focal plane of the eyepiece or just passed (above) that. This real image is called
primary or *intermediate image*. The eyepiece, functioning as a magnifier, transforms
the intermediate image into a virtual image, y_3, without further image inversion (cf.
Fig. 6.7), which image is located somewhere between infinite and l_r.

For the magnifying power of the microscope, M_p^{micro}, it follows (cf. Fig. 6.7)

$$M_p^{micro} = M_l^{ob} M_p^{ep} \qquad (6.21)$$

where the superscripts "*ep*" and "*ob*" indicate eyepiece and objective, respectively.

In accordance with Eq. (6.3)

$$M_l^{ob} = \frac{t}{f_{ob}} \qquad (6.22)$$

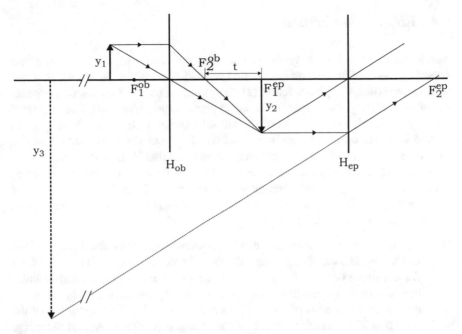

Fig. 6.7 Image formation in a compound microscope

where t represents the *tube length* (=distance between the inner focal points of objective and eyepiece) and where it has been assumed that the intermediate image exactly coincides with the inner(lower)-focal plane of the eyepiece (Fig. 6.7). M_p^{ep} can be taken in accordance with Eq. (6.20). Then a practical definition for the magnifying power of the microscope can be given as

$$M_p^{micro} = \frac{t l_r}{f_{ob} f_{ep}} = \frac{l_r}{f_{micro}} \qquad (6.23)$$

From Eq. (6.23), it follows that the effective focal distance of the compound microscope, $f_{micro} = f_{ob} f_{ep}/t$, can be made very small (and thus the magnification large) by applying a large tube length (e.g. $t = 250$ mm). Because lenses of very small focal distance are difficult to produce and to adjust, Eq. (6.23) immediately makes obvious the use of the compound microscope.

On changing of objective, the tube length of the microscope remains fixed and accordingly the objectives are fully characterized by M_l^{ob} (Eq. (6.22)), just as the eyepieces are characterized by M_p^{ep}. These values are normally indicated at the respective lens settings using the symbolism designated at the end of Sect. 6.3.1.

It is remarked that the larger M_p^{micro}, the smaller the field of view, since the diameter of the field lens of the eyepiece, which determines the exit pupil, is constant (e.g. 22 mm or 33 mm).

6.4 Köhler Illumination

Non-luminous objects for study under a microscope have to be illuminated. The oldest known method of illumination is so-called *critical illumination*. In this method, the light source is imaged by a condenser (a lens system to produce a satisfactory concentration of light) on the specimen in the object plane of the microscope. Two problems accompanying this technique are (i) the inhomogeneity of the source leading to an uneven illumination of the specimen, and (ii) the extension of the source which should be sufficiently large to illuminate a substantial part of the specimen.

A method of illumination where every point of the specimen surface receives light of the entire light source eliminates the above mentioned objections and was developed by Köhler. The so-called Köhler illumination (1893) led to a revolution in microscopy. It has the following basis (see Fig. 6.8):

(i)   The source is imaged by means of a *collector*-lens on the focal plane of the condenser. Hence, after passage through the condenser, all rays originating from a single point of the source are parallel to each other. Hereafter interaction with the specimen (surface) occurs. After passage of the objective, the non-diffracted pencils of parallel rays are converged in the focal plane of the objective where a second image of the source is produced. At this place, the light diffracted by the specimen yields a diffraction pattern of the specimen; see Sect. 6.2.

(ii) The condenser simultaneously images an iris diaphragm, the *field stop* positioned nearby the collector, on the specimen surface. In this way, the illuminated part of the specimen can be restricted to the field of view, thus avoiding interplay with "false" light diffracted by structural features of the specimen outside the field of view and by microscope parts passed through by the light. Finally, an iris diaphragm, the *aperture stop*, is placed in the focal plane of the

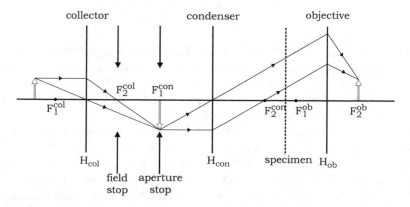

Fig. 6.8 Schematic illustration of Köhler illumination (transmitted light microscopy; transparent specimens). The light source is symbolized by a white arrow in contrast to the object in previous figures (black arrow)

condenser, which allows control of the angular aperture of the light striking the specimen. The resolving power and the depth of field are dependent on the angular aperture.

Köhler illumination was originally developed for transmitted light microscopy (transparent specimens); see Fig. 6.8. For reflected light microscopy (opaque specimens, e.g. metallic specimens), an adaptation is necessary (see Fig. 6.9):

The light is introduced into the tube by means of, for example, a prism or a semi-transparent mirror making an angle of 45° with the optical axis of the microscope. The objective now has a double function: it also acts as the condenser. However, the condenser diaphragm cannot be placed in the focal plane of the condenser (=inner-focal plane of the objective), since the image formation of the specimen would be affected. Therefore, a set of auxiliary lenses is applied.

The source is imaged by the collector at the place of the aperture stop which is positioned in the focal plane, F_1^{L1}, of the (positive) lens 1. So lens 1 images source and aperture stop at infinite. The (positive) lens 2 has a position such that the rear focal plane, F_2^{L2}, coincides with the inner (upper) focal plane, F_2^{ob}, of the objective (=condenser). Hence, source and aperture stop are imaged by lens 2 in the focal plane of the condenser, as required by Köhler illumination.

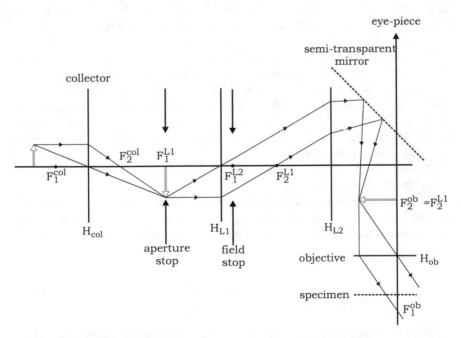

Fig. 6.9 Schematic illustration of Köhler illumination (reflected light microscopy; opaque specimens). The light source is symbolized by a white arrow in contrast to the object in previous figures (black arrow)

Further, the field stop has been placed in the focal plane F_1^{L2} of lens 2. Lens 2 accordingly generates an image of the field stop at infinite. Subsequently, the objective images the field stop in the outer (lower) focal plane, F_1^{ob}, just above the specimen (see discussion in Sect. 6.3.2).

Note that in the optical path, the order of field and aperture stop in reflected light microscopy is reversed as compared to transmitted light microscopy.

6.5   Resolving Power

6.5.1   *Minimal Image Construction*

The resolving power of a microscope is determined by the objective; the eyepiece merely enlarges the primary image (empty magnification; assuming that the eyepiece presents all information present in the primary image). Details not resolved in the primary image are not observed.

The first diffraction maximum of a grating with spacing p, which is taken as a model for an actual specimen, conforms to Eq. (6.7) with $m = \pm 1$ (cf. Sect. 6.2 and Figs. 6.4 and 6.5). (Intermediate) Image formation occurs by superposition (interference) of light originating from diffraction maxima in the diffraction pattern. For a minimal form of image construction, at least two diffraction maxima should contribute, e.g. at least one of the first-order diffraction maxima ($m = \pm 1$) together with the zeroth-order diffraction maximum should be encompassed by the aperture of the objective (cf. Fig. 6.5) In such a case, no faithful image is obtained in general: Consider the image resulting for a grating of slits of infinitesimal slits if only two maxima in the focal plane contribute to image formation. This image is indicated by the case $N = 2$ in Fig. 6.5, recognizing that this figure characterizes the intensity distribution not only in the focal plane but also in the image plane (cf. Eqs. (6.10) and (6.11) and their discussion). Obviously, the broad diffraction maxima can only be considered as a strongly distorted representation of point maxima which is the "ideal" image of such a grating.

Hence, considering an incident pencil of rays parallel to the lens axis, if the most branched-off diffracted ray, which is just transmitted by the objective, makes an angle u with the lens axis, then it follows for minimal image construction (cf. Eq. (6.7) and its footnote)

$$A_{ob} = n_1 \sin u = \frac{\lambda_o}{p_{min}} \tag{6.24}$$

Note that for this case (incident pencil of rays parallel to the lens axis; but see below) in fact three maxima contribute: $m = -1$, $m = 0$ and $m = +1$ (cf. Fig. 6.5). By definition A_{ob} is called the *numerical aperture* of the objective. The spacing, p_{min}, which is just resolved by the objective is given by Eq. (6.24). The resolving power, RP, can be defined as

$$RP = \frac{1}{p_{min}} = \frac{A_{ob}}{\lambda_o} \qquad (6.25)$$

In general, a real specimen is no grating, but p_{min} will be of the same order of magnitude as the value indicated by Eq. (6.24).

From Eqs. (6.24) and (6.25) it follows that the resolving power by application of immersion liquids with $n_1 > 1$, which fill object space between object and lens, is larger than in air (n_1 very close to 1).

The resolving power can also be enhanced by giving the incident pencil of rays an angle of inclination, u, with respect to the lens axis, such that just the zeroth diffraction maximum and one of the first-order maxima is transmitted by the objective. Then [4]

$$RP = 2\frac{A_{ob}}{\lambda_o} \qquad (6.26)$$

For this case of *oblique illumination*, the resolving power according to Eq. (6.26) is experienced only if the direction of the inclined incident pencil of rays is perpendicular to the lines of the grating. This phenomenon is called *azimuth effect*. It can be avoided by using a central-closed aperture stop only permitting illumination by an annulus close to the edge of the condenser: *annular-oblique* or *hollow-cone illumination*.

In deriving Eqs. (6.24)–(6.26) an incident plane wave coming from one direction (point source; see Sect. 6.4) was considered. In practice, a finite angle of aperture is normally applied, as determined by the aperture iris diaphragm (see Sect. 6.4 and Figs. 6.8 and 6.9). Hence

$$RP \leq \frac{(A_{ob} + A_d)}{\lambda_o} \leq \frac{2A_{ob}}{\lambda_o} \qquad (6.27)$$

where A_d represents the aperture of the diaphragm applied ($A_d \leq A_{ob}$).

6.5.2 Maximal Magnification

The eye can just distinguish details which subtend an angle of $1'$ (i.e. one minute ($'$) $= 1/60$ of a degree ($°$)) at the eye. The minimal spacing, p_{min} (cf. Eq. 6.25), which has to be observed by the eye, has to be (angularly) magnified up to at least this measure. For "easy" observation, it can be even stated that the angle subtended should be $4'$.

If, after magnification, the smallest observable spacing subtends an angle α at the eye, then it follows that the maximal magnifying power, M_p^{max}, should be (see Sect. 6.3.1)

---

[4] This is the same resolving power as achieved from self-luminous objects.

$$M_p^{\max} = \frac{\tan \alpha}{p_{\min}/l_r} \cong \frac{\alpha l_r}{p_{\min}} = \alpha l_r \mathrm{RP} \tag{6.28}$$

For $\alpha = 4'$ and $\lambda_o = 550$ nm and with $A_{ob}/\lambda_o \leq \mathrm{RP} \leq 2A_{ob}/\lambda_o$ (cf. Eqs. (6.25)–(6.27) it is obtained

$$500 A_{ob} \langle M_p^{\max} \langle 1000 A_{ob} \tag{6.29}$$

A magnification larger than about $1000A_{ob}$ has no significance in practice for the resolving power. Such a case is denoted as "empty" magnification, i.e. the smallest spacing resolvable by the objective, $p_{\min}$, subtends an angle larger than $4'$ at the eye.

Practical maximal values for the numerical aperture of the objective are $A_{ob} = 0.92$ (in air with $n_1 = 1$ and $u = 67°$) and $A_{ob} = 1.40$ (in oil with $n_1 = 1.52$ and $u = 67°$). Thus, it is obtained for the maximal magnifying power with $\alpha = 4'$, $\lambda_o = 550$ nm and $l_r = 250$ mm:

$$M_p^{\max} = 2\alpha l_r \frac{A_{ob}}{\lambda_o} = 975x \text{ (in air) and } 1480x \text{ (in oil)}$$

and the corresponding smallest resolvable spacing equals

$$p_{\min} = \frac{\lambda_o}{2A_{ob}} = 300 \text{ nm (in air) and } 195 \text{ nm (in oil)}$$

It should be recognized that the morphology of the structural details just resolvable is not at all imaged faithfully (see the discussion in the second paragraph of Sect. 6.5.1).

6.6 Bright and Dark Field and Other Imaging Techniques by Light Optical Microscopy

6.6.1 Bright Field Microscopy

Bright field microscopy can be described as image formation on basis of the "natural" diffraction pattern: Principal ($m = 0$) and secondary maxima ($m = \pm 1, \pm 2, \pm 3, \ldots$; cf. Sect. 6.2) are unaffected. Using polychromatic light, the following mechanisms contribute to contrast in the bright field image:

(i)   Diffraction occurs at discontinuities in the specimen surface (e.g. grain boundary; scratch). Consequently, a local decrease of reflected intensity is observed.
(ii) Selective absorption of one or more wavelengths by a particular phase in the microstructure leads to a coloured appearance (e.g. Ti(C, N) appears pink).

(iii)   Absorption throughout the entire spectrum offered by a particular phase in the microstructure causes a grey-tinted appearance (e.g. Si in AlSi alloys).

An etched cross section of a two-phase, Cu_3P (hexagonal)-Cu (cubic), specimen (of hyper(=above)eutectic (cf. Sect. 7.5.2) P content is shown in Figs. 6.10a–f according to various light microscopical imaging techniques. Upon solidification of the alloy, first Cu_3P phase has solidified (primary phase). Then, upon continued cooling, eutectic solidification of the remaining liquid at the eutectic temperature has occurred eventually (cf. Sect. 7.5.2, discussion around Eq. (7.19)), implying at that stage the coupled precipitation of Cu phase and Cu_3P phase under the development of a lamellar two-phase microstructure (cf. Sect. 7.6 and Fig. 7.31, in particular for solidification for an alloy of alloying element B content larger than the eutectic composition, x_{eut}, but smaller than the B content of the β phase in Figs. 7.13 and 7.31). The two microstructures (primary phase and the two-phase eutectic) are clearly discernable in the etched cross section.

Mechanisms as described above contribute to the contrast in the bright field micrograph made with "white" (i.e. polychromatic) light shown in Fig. 6.10a. The grain boundaries in the eutectic microstructure are revealed by the height difference induced by the polishing (Cu is relatively soft and therefore it can be assumed that the Cu_3P lamellae protrude from the surface; the cross section was not etched), and scratches can be observed as dark lines. The Cu phase in the eutectic microstructure appears with a yellow-reddish shine, and the Cu_3P phase appears green.

6.6.2 Dark Field Microscopy

The purpose of dark field microscopy is to analyse the light diffracted by the specimen in the absence of the not-diffracted, reflected light. The not-diffracted light is gathered in the principal maximum in the focal plane of the objective ($m = 0$; cf. Sects. 6.2 and 6.3). The quintessence of the dark-field technique is to avoid that the principal maximum contributes to image formation. This could, for example, be realized by introducing an absorbing (opaque) plate at the central order maximum in the focal plane: *central dark field*. In practice normally *oblique dark field* is applied. Then the incident light makes an angle with the lens axis such that the not diffracted, reflected light does not pass through the objective. In the reflected light microscope, all-sided oblique dark field illumination can be realized by employing an annular (lens or mirror) condenser all round the objective by which the specimen is illuminated (Fig. 6.11). As is evident from Fig. 6.11 and as compared to bright field microscopy, dark field microscopy requires relatively large object distances.

The above description indicates that, applying dark field microscopy, diffracting objects in the specimen surface are observed as "self-luminous" against a dark background; as compared to bright field microscopy (=image formation employing the diffraction pattern with also the principal maximum), the contrast for the observation of the diffracting objects is considerably improved. For example, scratches are

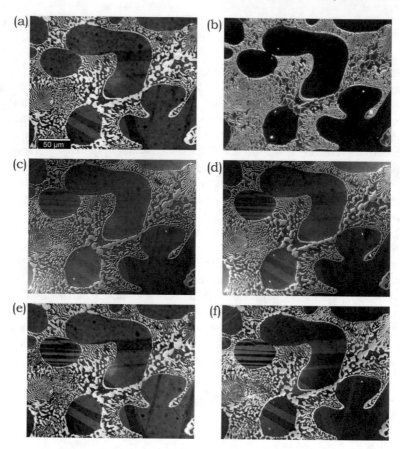

Fig. 6.10 Examples of different light microscopical image-formation techniques applied to an etched cross section of a two-phase Cu_3P-Cu specimen. The larger crystals are (tetragonal) Cu_3P primarily solidified phase; the lamellar arrangement of (cubic) Cu and (tetragonal) Cu_3P phases is the result of the upon continued cooling, eventual eutectic solidification at the eutectic temperature (micrographs made by Dr. E. Bischoff, Max Planck Institute for Metals Research). **a** Bright field, using "white" (i.e. polychromatic) light. **b** Dark field, applying obliquely incident light (see text). **c**, **d** Differential interference contrast with, from **c** to **d**, sign reversal for the phase difference of waves 1 and 2 (see text). The differential interference contrast is generated in the light microscope applied by using a polarized beam of light which is led through a double-refracting prism, thereby producing two planar, parallel wavefronts of different index of refraction and different polarization (always mutually perpendicularly, linearly or elliptically polarized; see Footnote 6 in this chapter). The different indices of refraction lead to an optical path (i.e. phase) difference for the two wavefronts upon passage through the prism. The thus produced two wavefronts hit the surface of the specimen with a small lateral displacement (see text). If the two wavefronts are reflected from (laterally nearby) parts of the specimen different in height or of different optical activity, a further phase difference is added to both (now reflected) wavefronts. The reflected wavefronts reenter the prism (implying that a further phase difference is added) and thereafter are recombined in one plane by an analyser. The intensity (colour) variation in the image made with the reflected light passed through the analyser depends on the phase differences induced locally. **e** and **f** Polarized light contrast with, for **e** and **f**, different rotations of the analyser with respect to the polarizer (see text)

Fig. 6.11 Schematic illustration of the principle of dark field microscopy

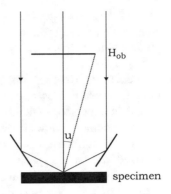

sharply outlined in dark field. Even the presence of objects smaller than the minimal resolvable spacing (cf. Sect. 6.5.1) can be pursued by virtue of their diffracted light. Obviously, in the latter case nothing can be said about the morphology of these objects.

An example of a dark field micrograph is shown in Fig. 6.10b for the cross section of the Cu_3P-Cu specimen described above. Evidently, the scratches now appear as white lines against a dark background (cf. the bright field micrograph in Fig. 6.10a where the scratches are imaged as black lines). The grain boundaries between the Cu phase and the Cu_3P phase in the etched eutectic microstructure, corresponding with height changes/discontinuities in the surface of the cross section and thereby giving rise to relatively pronounced diffraction, are now very well delineated as bright lines against a dark background.

Other light optical microscopical techniques are sometimes very useful; see the following Sects. 6.6.3–6.6.5.

6.6.3 Phase-Contrast Microscopy

Phase-contrast microscopy involves the transformation of a non-observable phase (see first footnote in Sect. 6.1) difference, between light (transmitted or reflected) from parts of a transparent (transmitted light microscopy) or opaque (reflected light microscopy) specimen, into an observable intensity difference. This is realized by influencing the diffracted light and the (transmitted/)reflected light separately, recognizing that in the (back) focal plane the reflected light is gathered in the principal, central order maximum ($m = 0$) and the diffracted light is concentrated in the side maxima ($m = \pm1, \pm2, \pm3, \ldots$). In phase-contrast microscopy a phase plate is inserted at the place of the central order maximum ($m = 0$) in the focal plane, which plate (i) changes the phase of the reflected light by one quarter of a period ($=\pi/2$) and (ii) reduces the amplitude of the reflected light.

In practice the phase-contrast technique is put into effect by using an annular, central-closed aperture stop (cf. Figs. 6.8 and 6.9 and Sect. 6.4); only an annular zone of the objective is utilized. Therefore, the central order maximum is also of annular shape and consequently an annular phase plate is required. This all-round oblique illumination not only enhances the contrast but also improves resolution (see Sect. 6.5.1).

From the above discussion it follows that, from the point of view of image formation, dark field microscopy is an extreme form of phase-contrast microscopy, as in dark field microscopy the intensity of the reflected light is reduced to zero.

This technique allows to reveal qualitatively, by contrast difference, height differences in the surface (thickness for transmitted light microscopy) of, e.g., a polyphase polished and etched specimen or crystal-orientation differences, as, e.g., a result of twinning, for non-cubic materials. The smallest phase differences which can be detected by phase-contrast microscopy are about $6°$, which would imply for light optical microscopy that height differences of about 10 nm ($6°/360° \times 550$ nm) can be visualized by a contrast difference.

Phase-contrast microscopy plays an important role as imaging technique (transmitted light microscopy) in biology and biotechnology for the investigation of (living) cells which act predominantly as phase changing objects upon passage of visible light (e.g. see Horn and Zantl 2006).

6.6.4 Interference Microscopy

Interference microscopy[5] involves the branching of an initial light ray into two (or more) coherent ones, which are separately influenced before they are recombined in an image plane. Just as with phase-contrast microscopy, interference microscopy is intended to visualize phase differences as evoked by differences of refractive indices and/or of height (thickness for transmitted light microscopy). In contrast with phase-contrast microscopy, interference microscopy allows a quantitative analysis of phase differences (not in all variants; see below). The following notes pertain to the case of a two-wave interference microscope.

Suppose that wave 1 is influenced by the (reflecting and diffracting) specimen and wave 2 is left unaffected, by appropriate guidance in the microscope. The two rays can then be (re)combined in the image plane such that they (the two wavefronts) make a small angle with respect to each other. As a result, a set of fringes occurs in the image plane. Each fringe represents the locus of points of constant optical path difference (see Footnote 3 in Sect. 6.2) between the two wavefronts in the image plane. The deviation of a straight line for these fringes, at the location where a phase changing object occurs in the surface of the specimen, can be utilized to calculated quantitatively the phase change which, for example, may be due to a height difference.

---

[5] In principle the entire field of light microscopy can obviously be described as interference microscopy, because image formation implies the interaction (interference) of transmitted/reflected and diffracted beams of light (cf. Sect. 6.2)

The two wavefronts can also be made parallel. A phase changing object then reveals itself by a contrast different from its surroundings in the image.

One could also let interact both parallel wavefronts with the specimen surface and in the image plane realize a lateral shift of both parallel wavefronts. If, in this "*shearing method*", the lateral image shift is as small as the resolving power of the microscope (see Sect. 6.5) one speaks of "differential image shift". Then the two "images" are observed as coinciding. Small phase differences can yet occur in the field of view, most distinctly in the regions where boundaries are imaged which occur between two objects giving rise to different phases upon reflection (e.g. due to a height difference of the two phases in the surface of the specimen). Using monochromatic light such a boundary is observed, say *brighter* than the (average) brightness of the surroundings, whereas then consequently, the opposite boundary of the phase changing object, as emerging in the surface of the specimen, is observed *darker* than the surroundings (or vice versa), because the phase differences at both boundaries considered have opposite sign. For very small phase changing objects, it is this nearby occurring contrast reversal which leads to enhanced visibility; a very small total phase shift can still be discerned.

In case polychromatic light is applied, the boundaries of the phase changing object are observed with a colour/tint different from the surroundings.

It follows from the above discussion that parts of the specimen with identical relief (flat surfaces of phase changing object B and of matrix A) are observed with identical brightness (monochromatic light) or colour/tint (polychromatic light), because the phase difference between wavefronts 1 and 2 is the same (but note the additional effects caused by the use of polarized light in case of optically anisotropic (i.e. not cubic) materials; see Sect. 6.6.5).

The above discussed *differential interference-contrast (DIC) method* is especially suited to reveal the surface relief of specimens. It should however be borne in mind that the spatial impression of the image does not resemble a stereoscopic observation: "Ridges" may look like "canals" and vice versa by a reversal of sign for the imposed phase difference of wavefronts 1 and 2.

Differential interference-contrast (DIC) micrographs for the etched cross section of the $Cu–Cu_3P$ specimen considered in this section are shown in Figs. 6.10c, d. Indeed, the contrast reversal mentioned above can be observed at many places: if a grain boundary of a Cu lamella with the adjacent Cu_3P phase appears bright, then the appearance for the grain boundary at the opposite side (in the cross section) of the Cu lamella is dark. The phase difference of wavefronts 1 and 2 as pertaining to Fig. 6.10c has been subjected to sign reversal and the image has been recorded again with the result shown in Fig. 6.10d. As expected, for those opposite grain boundaries of a Cu lamella where the above discussed bright–dark sequence is observed in Fig. 6.10c, a dark–bright sequence is observed in Fig. 6.10d.

6.6.5   Polarized Light Microscopy

Polarized light microscopy employs linearly (plane) polarized incident light. Linear polarization is achieved by utilizing a *polarizer*. By agreement the polarizer is oriented in the microscope such that the plane of vibration in the image is parallel to the tilt axis of the illuminating mirror/prism (cf. Sect. 6.4 and Fig. 6.9).

In case of perpendicular incidence of linearly polarized light (and if no depolarization within the instrument occurs), the following situations can be distinguished:

(i)   The specimen is optically isotropic (this holds for all crystallographically cubic materials). Then, the reflected light is linearly polarized with the same plane of vibration as the incident light. Note that for *absorbing* isotropic material (e.g. cubic metals) the reflected light is only linearly polarized for the case of perpendicular incidence.

(ii)   The specimen is optically anisotropic.[6] Then, the reflected light can be:

- linearly polarized with a rotated plane of vibration. This situation is met with transparent media. It also occurs with absorbing uniaxial or, in special cases, absorbing biaxial media. Note that for absorbing anisotropic material this statement only holds for the case of perpendicular incidence;
- elliptically polarized. This situation is met with absorbing biaxial media in general.

For the analysis of the reflected light a polarizer called *analyser* is employed which is usually oriented such that the light passed through is linearly polarized in a plane of vibration perpendicular to that of the polarizer. Then, i.e. employing "crossed" polarizer and analyser, in case (i) no light will be transmitted by the instrument, in contrast with case (ii). This forms the basis of a powerful method to distinguish between optically isotropic and optically anisotropic material.

Polarized light micrographs for the etched cross section of the Cu–Cu$_3$P specimen are shown in Fig. 6.10e, f. The Cu$_3$P phase is optically anisotropic; the Cu phase is optically isotropic. The Cu$_3$P phase, in particular the primary grains, as these are relatively large as compared to the Cu$_3$P lamellae, shows the presence of regions within the grains of contrast differing with their immediate surroundings. These are regions in twin orientation with respect to the grain matrix. Because the effect on the polarization of the incident light depends on the orientation of the Cu$_3$P crystal with respect to the incident light, regions in twin orientation appear with different contrast. By rotation of the analyser, the contrast differences between twin and matrix vary and can become of opposite nature (cf. Fig. 6.10e, f).

---

[6] Propagation of a planar wavefront in an anisotropic, possibly absorbing, crystalline medium leads to splitting into two parallel planar wavefronts of different index of refraction and different polarization (always mutually perpendicularly, linearly or elliptically polarized). This is the phenomenon of "double refraction", also called "birefringe". There are, however, in general two directions for the incident planar wavefront, with respect to the crystal axes, for which such splitting does not occur. These directions are called "optical axes". Both optical axes may coincide. Thus, one distinguishes biaxial and uniaxial materials.

As a side remark, it should be noted that, for optically anisotropic material, also in differential interference-contrast micrographs the effect of crystal orientation on the contrast generated is visible if polarized light is used, which is the case for the DIC images shown in Fig. 6.10b, c (see instrumental details described in the caption of these figures). Indeed, bands of different contrast can be seen in (some of) the primary Cu_3P grains (Fig. 6.10b, c), which bands are of the same morphology and orientation as observed in the polarized light micrographs (Fig. 6.10e, f).

Structural details making an angle with the specimen surface give rise to "anisotropic" phenomena in the sense discussed above, i.e. considering a specimen of optically isotropic material, they appear "light" against a dark background in case of crossed polarizer and analyser. These phenomena are the result of a local occurrence of non-perpendicular incidence, which for absorbing, isotropic material (e.g. many metals) gives rise to elliptic polarization. Typical examples of such structural details are (polishing) scratches and protruding phase boundaries after etching (e.g. in case of a finely lamellar microstructure) in the specimen surface.

As a final note to Sects. 6.3, 6.4 and 6.6, which deal rather exclusively with light microscopy (the other sections up till here have a more general bearing for diffraction and image formation, as will become clear from what follows in this chapter), one should recognize that even today light microscopy is an important (first) tool of microstructural analysis: see what has been said in the introduction of this chapter. Moreover, the development of the light microscope continues even today. Thus the introduction of the digitized recording of images has led to new designs of light optical lens systems [see, for example, Drent (2005)]. As a further illustration of this point, in the following Sect. 6.7 attention is paid to (comparatively) recent developments to circumvent the limit to the resolving power as expressed by Eqs. (6.25) and (6.26).

6.7 Resolution Beyond the Diffraction Limit: "Supermicroscopy"

The constraint imposed on the resolution inherent to microscopy is in fact a "diffraction limit" (see Sect. 6.5.1). Abbe presented this result in 1873. Although he was not the first to derive an equation for the maximally possible resolving power, nowadays this criterion [especially Eq. (6.26)] is also denoted as the "Abbe limit". As a consequence, the laterally smallest resolvable spacing (in other words: the smallest distance between two microstructural features just allowing their observation) according to the diffraction limit, $p_{min} = \lambda_0/(2A_{ob})$, is about 200 nm (see Sect. 6.5.2). For a long time this boundary condition was considered as unbreakable and, as a matter of fact, it is. However, in recent years "tricks" have been developed to bypass (it is incorrect to say "to overcome") the diffraction limit. One class of methods, allowing attaining resolution larger than the diffraction limit, involves numerical *calculation* using multiple, recorded diffraction-limited images (cf. the development of (admittedly entirely different) computational models for the image-formation process in high-resolution transmission electron microscopy as discussed in Sect. 6.8.6; see especially there the text under the subheading *Image simulation*). Here the focus is on *experimental* "image" (perhaps better denoted as "response map"; see footnote 16 in Sect. 6.8.5) recording methods, employing special microscopic devices, providing lateral/depth resolution (considerably) better than indicated by the diffraction limit. The basis for a very promising such method, "*stimulated emission depletion (STED) microscopy*", originates from (i.e. is a variant of) older "*confocal microscopy*". Therefore in the following firstly the principle of confocal microscopy is discussed and secondly attention is devoted to STED.

6.7.1   Confocal Microscopy

The microscopic techniques discussed so far can all be called wide-field methods, i.e. a beam of parallel rays of incident light illuminates evenly the entire specimen (apart from possible confinement by a field stop; see the Köhler illumination discussed in Sect. 6.4). This contrasts with a [here discussed reflected light (the nowadays mostly used variant of the)] confocal microscope, which employs point illumination (see Fig. 6.12):

Light from a source is spatially confined, e.g. by a very small circular opening, called pinhole, in a stop. The light passing through this pinhole (here stop with pinhole outside the microscope tube) is focussed (here after reflectance by a semi-transparent mirror; cf. Fig. 6.9) by the objective lens on a plane located in the specimen, i.e. an image of the pinhole (smaller than the pinhole) is produced there. In fact, as a consequence of diffraction, the image of the pinhole is not a point: the intensity in the image "point" is spread out (according to the so-called point spread function). The image of the original point thus is a diffraction pattern characterized by a bright central region (the so-called Airy disc) and a set of light and dark concentric rings around the bright central region, forming, together with the Airy disc, the Airy pattern.

Confocal microscopes are most often used with more or less transparent objects (biological objects, as (possibly living) cells) and then, also to reveal three-dimensional structures/morphologies, the image plane, i.e. the plane where the sharp image of the pinhole is generated, can be located in a controlled manner at variable depth underneath the surface of the specimen, which explains why in the above consideration the plane with the sharp image of the pinhole is explicitly located *in* the specimen, not necessarily at the surface, as would be the case with more or less opaque specimens, as metals.

The light originating from the illuminated "point" now passes through the objective (here double function of the objective; cf. Köhler illumination for reflected light microscopy) and is focussed as a "point", actually a diffraction pattern (see above),

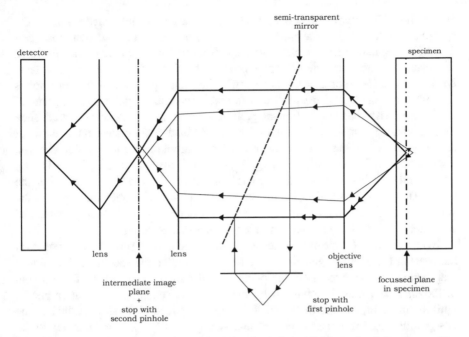

Fig. 6.12 Schematic illustration of the principle of a (reflected light) confocal microscope. The lenses are represented by their principal planes (see Sect. 6.1.1). Light from the aperture (here a very small circular opening: a *pinhole*, positioned on the optical axis of the auxiliary lens) in the first stop is focused by the objective on a plane in the specimen, where an image of the pinhole in the first stop thereby is generated. The image of the pinhole in the stop is not a point but spread out due to diffraction. The light reflected from this illuminated "point" is focused by the objective on the intermediate image plane where a second stop with a pinhole (positioned on the optical axis of the microscope) resides, positioned such that only a central part of the image "point" in the intermediate image plane passes through the pinhole in this stop. Light originating from planes below (as sketched in the picture) or above the plane in the specimen where the sharp image of the pinhole in the first stop occurs, arrives in the intermediate image plane as a disc of intensity and therefore is largely blocked by the second stop

on an intermediate image plane. The illuminated "point" in the image plane in the specimen and its image "point" in the intermediate image plane are thus in a conjugated, *confocal* (=with a common focus) relationship, which explains the origin of the name of this microscopical technique.

In the intermediate image plane, a second stop with pinhole resides. And now the essence of this confocal microscopy is met: this pinhole is such located that it encompasses only *part of* the image/diffraction pattern of the illuminated "point" in the specimen: typically the diameter is such limited that only the Airy disc of the intermediate image fits in the pinhole, i.e. the circumference of the pinhole runs through the first minimum of the diffraction (Airy) pattern. The advantage of this technical feature can be explained as follows. Light from the specimen not only originates from a mathematical point in the image plane in the specimen, but in fact a small volume in the specimen can be indicated giving rise to diffracted light (or emitted light if fluorescent material is contained in the small illuminated volume in the specimen; often fluorescence microscopy is performed with a confocal microscope). The light originating from a plane different from the image plane in the specimen obviously does not converge to a single point in the plane of the intermediate image, but in a plane above or below this intermediate image plane. This in turn implies that this (cone of) light causes a disc of intensity (and not a "point") in the plane of the intermediate image plane (see the path of the rays of light sketched with dashed lines in Fig. 6.12 and cf. the origin of discs in the back focal plane (in that case not an image plane!) of the objective in case of convergent beam transmission electron microscopy; cf. Figure 6.19). The largest part of this disc of intensity is blocked by the second stop in the intermediate image plane: the detector positioned after the second stop with pinhole does largely not record this light. Thereby the depth (in the specimen) resolution of the signal recorded by the detector is pronouncedly enhanced as compared with conventional light microscopes discussed before.

Evidently, at this stage of the discussion, the detector only provides a response on the basis of the intensity generated by a single "point" in the specimen. A response map pertaining to the extent of the plane with the sharp image of the first pinhole in the specimen must be obtained by rastering, i.e. scanning "pixel after pixel", with the incident, focussed beam of light (e.g. produced by a laser, leading to the designation "confocal laser scanning microscopy (CLSM)"), in the plane with the sharp pinhole image. The thereafter constructed response map of the intensity per measured "point"/pixel is usually called an "image" of the specimen at the depth analysed (cf. Footnote 16).

Largely excluding, by the action of the second stop with pinhole, the light originating from locations in the specimen below and above the image plane in the specimen, which contains the sharp image of the first pinhole, especially decreases strongly the depth of focus (see above) and increases the contrast in the response map, as compared to what holds for image formation with a conventional light microscope. The improvement of the lateral resolution, by confocal microscopy, however, is only modest.

The perhaps most important applications of confocal microscopy occur in biological and life sciences: by varying the depth of the plane with the sharp image of the first pinhole, "optical sectioning", of even more or less undamaged (parts of) living beings can be performed, leading to reconstructions of three-dimensional structures/morphologies in such specimens.

Confocal microscopy was proposed and initially demonstrated in the nineteen fifties. Its widespread use became possible only after the later development of laser systems and computational advances in the nineteen eighties, which allowed practical application.

6.7.2 Stimulated Emission Depletion Microscopy (STED)

It should be noted that at this stage of the development, still no distinctly larger lateral resolution, than possible with conventional light microscopy, was obtained (see above). This situation changed in the nineties of the past century. The basic idea runs as follows.

The lateral resolution in confocal microscopy is governed by the lateral size of the illuminated "point" (image of the first pinhole) in the plane of the sharp image of the first pinhole. If this lateral size can be made smaller, then, obviously, the lateral resolution becomes improved. Thus, methods to reduce the size of the "point" image of the first pinhole are sought for: "point spread function *engineering*" (see above). Perhaps the most well known of these techniques is "stimulated emission depleted microscopy (STED)". STED essentially is based on the fluorescence response of a specimen. Then, to understand the functioning of STED, first some effort is needed to describe the fluorescence phenomenon (see Fig. 6.13).

Fluorescence is initiated with the excitation of an electron from a ground state to an excited state upon absorption of a photon of energy equal to the difference in energy of the excited and ground states. Thereafter (e.g. vibrational), relaxation

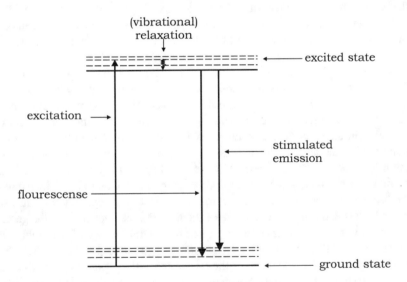

Fig. 6.13 Fluorescence and stimulated emission. *Fluorescence*: Upon excitation of an electron from its ground state to an excited state, (vibrational) relaxation (without emittance of radiation) occurs: the electron in the excited state transfers to a (sub)state of lower energy. Thereafter, the electron returns to the ground state emitting radiation of somewhat larger wavelength than that of the incident radiation inducing the excitation. *Stimulated emission:* Subjection the system simultaneously to a second radiation of wavelength different but about the same as the wavelength of the incident radiation causing the excitation, induces the electron at its relaxed, excited state to return to the ground state at a higher (vibrational) energy level under emission of radiation of same wavelength as that of the second, imposed radiation

of the excited stage occurs, i.e. the excited atom/molecule loses energy without that radiation is emitted (heat dissipation). Then, only a few nanoseconds after the excitation, the excited electron falls back to its ground state under emittance of a photon which then, consequently, is of lower energy [i.e. of longer wavelength ("red shift")], than that of the incident photon inducing the excitation. The fluorescence effect can be suppressed by, simultaneously to the first (laser) beam of light, that causes the excitation of the electrons as discussed above, subjecting the specimen additionally to a second, highly intense (laser) beam of light of energy (wavelength) about the same as that of the first beam. Then the atoms/molecules subjected to both radiations are induced upon excitation to return, after (vibrational) relaxation as discussed above, to the ground state, *but at a higher energy (vibrational) level* under emittance of a photon of the same energy (wavelength) as that of the second (laser) beam of light. Hence, as compared to the photon inducing the excitation, and the fluorescence photon ("red shifted"), a further "red shift" occurs for the "stimulated emission" photon (cf. Fig. 6.13). Hence this "stimulated emission" and the genuine fluorescence radiation can be distinguished on the basis of their energy (wavelength) difference.

Fluorophores (also called fluorochromes) are fluorescent chemical compounds often used to stain/mark cells, tissues or, in a more general way, regions in materials, on the basis of forming a covalent chemical bond with (macro)molecules. In STED, such fluorophores are used to mark certain regions in the specimen to be analysed.

In order to reduce the lateral size of the illuminated "point" in the specimen, i.e. the image of the first pinhole, the activity of fluorophores for "normal" fluorescence in an outer part of the illuminated "point" can now be suppressed by imposing there a "stimulated emission" process as described above, so that a smaller inner part of the illuminated "point" remains, where fluorescence can and should still occur. To this end, by involvement of a second (laser) beam, a second "point spread function" is generated in the plane in the specimen where the "point spread function" of the first pinhole is generated by the first (laser) beam of light. This second point spread function must be of toroidal shape ("doughnut" shape); it is overlaid on the image of the first pinhole, with coinciding midpoints (Fig. 6.14). Complicated interference methods have been devised in order that the second "point spread function" has zero intensity in the central part of the image of the first pinhole. Hence, only the genuine fluorescent light from this small inner part of the "point" image of the pinhole can generate the "point" image in the intermediate image plane (see Sect. 6.7.1). This leads to a significant enhancement of the lateral resolution upon rastering the plane with the sharp image of the first pinhole in the specimen (see above); the Airy disc of the intermediate image is much reduced in lateral size as compared with conventional confocal microscopy (cf. Sect. 6.7.1). The lateral resolution now is in the range 30–90 nm, i.e. a factor 2 till 7 better than possible with the common light microscope (cf. Sect. 6.5.2). In some special cases a lateral resolution of about 3 nm has been reported.

The application of this type of fluorescence microscopy especially concerns the exposition of structures/macromolecules in biological material, as in particular within *living* cells. Of course, as compared with the practically atomic resolution offered

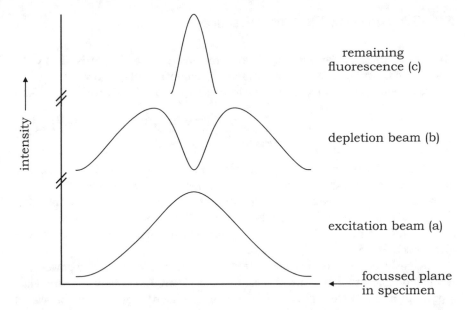

Fig. 6.14 Essence of STED: suppression of fluorescence from the outer part of the image of the first pinhole, i.e. the first point spread function, in the specimen (cf. Fig. 6.13). This is realized by *superimposing* on this image, generated by a first laser beam, a second point spread function, generated by a second laser beam of toroidal (doughnut) shape. The picture shows schematic intensity distributions, in the plane of the specimen where the sharp image of the pinhole occurs, along a line containing the centre of the intensity distributions: **a** for the fluorescence radiation occurring if only the first laser beam would be active, **b** for the depletion, fluorescence suppressing radiation, occurring if only the second laser beam would be active, and **c** for the remaining, laterally confined fluorescence radiation if both laser beams are active

by high-resolution transmission electron microscopy (HR(T)EM; see Sect. 6.8.6), the gain in resolution, as compared to light microscopy and conventional confocal microscopy, is not impressive. However, TEM and its variants as HR(T)EM require ultra-high vacuum and electron transparent (very thin, of the order of, say, 100 nm and less) foils as specimens to be investigated and thus are inappropriate to operate with living material (apart from the radiation damage possibly induced by the incident electrons) and to reveal morphological changes as function of depth in a single cell. Against this background one can understand why in 2014 the Nobel Prize in chemistry was awarded for the above sketched developments of "supermicroscopy" as a tool allowing practical application, albeit, at the time of writing this text, especially to realize its more or less unconstrained application in the life sciences, some distinct problems still have to be dealt with (Vicidomini et al. 2018).

6.8 Transmission Electron Microscopy

The development of the transmission electron microscope (TEM) is simply the outcome of the striving for higher resolution, i.e. higher resolving power.

The resolving power of a microscope in vacuum can be given by (cf. Sect. 6.5.1):

$$RP = \frac{1}{p_{min}} = \frac{A_{ob}}{\lambda_0} = \frac{\sin(u)}{\lambda_o} \qquad (6.25)$$

where p_{min} is the spacing that is just resolved by the objective lens, u is the angle with the lens axis made by the most branched-off diffracted ray, which is just transmitted by the objective, and λ_o is the wavelength of the radiation used in vacuum. It should be recognized that the definition of the resolving power of an objective lens is somewhat arbitrary (cf. the treatment in Sect. 6.5.1), but the usual definitions are of the type given by Eq. (6.25).

Obviously, decrease of wavelength leads to increase of the resolving power. The recognition that a stream of material particles, as electrons, not only has particulate aspect but also a wave aspect (cf. Eq. 2.6) immediately suggests the application of an electron beam as light source for a microscope, because the wavelength of accelerated electrons can be made very small: for example, the wavelength for an accelerating voltage in the range 100–400 kV[7] is in the range 0.0037–0.0016 nm. Recognizing that the limiting wavelength of visible light is about 400 nm, it would follow that the resolving power of a TEM could be a factor 100000–250000 larger than that of a light microscope. If this would be true p_{min} values of a few percent of an Ångstrom (the typical size of an atom is a couple of Ångstrom; one Ångstrom = 0.1 nm[8]) would be attainable.

The (ideal) action of a lens has been indicated in Sect. 6.1. A beam of visible light can be refracted and focussed by glass. A beam of electrons can be refracted and focussed by a magnetic field. Electron microscopes are supplied with magnetic electron lenses [the magnetic field is generated by an electrical current through a (copper) wire surrounding a core of soft magnetic material (soft iron)]. An important distinction between the glass lens for visible light and the magnetic lens for electrons is that the (position of the) focal planes of the magnetic lenses can be varied by varying the current through the wiring of the lens and that the focal planes of glass lenses are fixed: the action of magnetic lenses can be considered as similar to that of convex glass lenses but with controllable, variable focus. As a consequence, in the light microscope the glass lens systems are moved, for example to achieve "focussing", whereas in the TEM the lenses are fixed, but their foci can be changed (i.e. their "strength" can be changed).

---

[7] The currently commercially available transmission electron microscopes have accelerating voltages in this range.

[8] In crystallography, the "Ångstrom" is still often used as a distance/length unit and has even been formally sanctioned as such by the International Union of Crystallography (IUCr); cf. Footnote 8 in Sect. 4.1.1.

Now, unfortunately, the aperture of a magnetic lens is in no way comparable with that of a glass lens: the aperture angle u can be estimated at about $1°$. This very limited (allowable) aperture angle is such small because of severe (in particular spherical[9]) aberrations of the magnetic lens. Combining this constraint with the increase in resolving power due to the decrease in wavelength indicated above, it can be concluded that the TEM, as compared to the light microscope offers a resolution that is a factor 1000–3000 better than the light microscope, i.e. the minimal spacing discernable is about 0.1–0.3 nm. This implies that (rows of[10]) atoms can be resolved.

The penetrative power of the accelerated electrons is limited. This leads to a major limitation in the application of TEMs: the specimens to be investigated must be electron transparent and therefore must be thin: foils of thicknesses not larger than 500 nm (depending on the acceleration voltage and the (average) atomic number of the material to be investigated) and preferably less then, say, 100 nm have to be made.

Using Bragg's law (see Sect. 4.5), it immediately follows from the small wavelength associated with the accelerated electrons that very small diffraction angles occur.[11] This implies that the diffracting crystallographic planes in the specimen/foil are oriented practically parallel to the incident electron beam and thereby are practically perpendicular to the surface of the specimen/foil considered.

6.8.1 Basic Constitution and Action of the TEM; Imaging and Diffraction Modes

The TEM essentially is composed of a number of consecutive lens systems. In a way its construction resembles that of the compound light optical microscope when used with a photographic plate to record the real image (see Sects. 6.3.2 and 6.4). Drawing a ray diagram is performed as for the light microscope, but now the optical axis is not

---

[9] Spherical aberration occurs when, on imaging an object point on the lens axis, the rays refracted by different lens zones do not converge in a single image point on the lens axis: the edge zone of the lens refracts too strongly: an (already) curved wavefront becomes (even) more curved. Abbe was the first to demonstrate in 1872 that in light optical microscopy spherical aberration in principle can be eliminated by use of a compensating, composite lens system: a doublet consisting of a convergent lens and a divergent lens. For transmission electron microscopy, with magnetic lenses, it took until 1998 before such first, spherical-aberration-corrected lens systems were used for the objective lens system. However, this important step forward still does not lead to direct (atomic structure) image formation of the structure analysed, in a way as holds for image formation in the bright field mode in a light optical microscope: phase information in the diffracted electron waves has to be converted into amplitude information (see first footnote in Sect. 6.1 and the discussion on phase-contrast microscopy in Sect. 6.6.3). A corresponding discussion is beyond the scope of this book (see an overview by Urban 2007).

[10] The "viewing direction" in the electron microscope is perpendicularly through the specimen/foil. Atoms on top of each other, arranged in a row in the "viewing direction", are projected on top of each other in the image.

[11] For example, for a wavelength of 0.003 nm and a lattice spacing of 0.3 nm, the diffraction angle, 2θ, is less than $0.6°$.

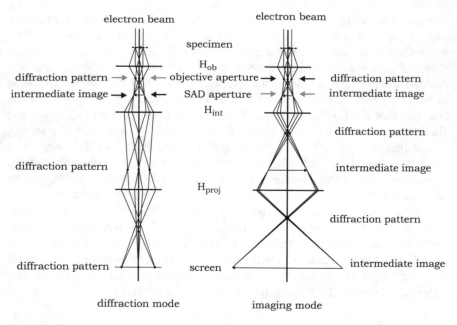

Fig. 6.15 Optical paths in a transmission electron microscope operating either in diffraction mode or in imaging mode

drawn horizontally but vertically, thereby representing reality for a TEM: with the electron source on top of the column/microscope and the final real image (of either the specimen or the diffraction pattern) plane at the bottom of the column/microscope. Four lens systems and two apertures can be discerned in the column, and these are discussed below (see also Fig. 6.15).

The illumination system (not shown in Fig. 6.15). The conventional way of operating a TEM involves that the specimen is illuminated by a (practically) parallel beam of electrons. This can be achieved by the action of a system of condenser lenses: either an underfocussed image of the (light) electron source is produced on the specimen, involving the application of an almost parallel beam of electrons, or the principle of Köhler illumination[12] can be employed. Normally, one restricts the illuminated part of the specimen/foil to that part that is really investigated ("viewed"), implying that a truly parallel beam does not occur; some "convergence" is introduced.

The objective lens system. Employing the non-diffracted and diffracted rays emanating from the specimen positioned in its object plane, the objective lens produces a diffraction pattern in its back focal plane and an image of the specimen/foil in the image plane (cf. Sect. 6.2). A unique feature of the TEM is that it allows the investigation of both the diffraction pattern and the image of the same part

---

[12] An image of the source (also called "cross-over") is produced at the front focal plane of a final condenser lens which then produces a truly parallel electron beam hitting the specimen (cf. Sect. 6.4)

of the specimen. This can be achieved by variation of the "strength" of the next lens system.

The intermediate lens system. The intermediate lens produces an image in its (fixed) image plane of either the diffraction pattern of the specimen in the back focal plane of the objective lens or the image of the specimen/foil in the image plane of the objective lens. The transition from imaging the diffraction pattern to imaging the specimen/foil is achieved by decreasing the focal distance of the intermediate lens (see Fig. 6.15).

The projector lens system. The projector lens serves to generate a final image from the image produced by the intermediate lens (which is either an image of the diffraction pattern or an image of the specimen/foil) on the viewing screen/plate/detector.

The objective aperture. This aperture selects the part of the diffraction pattern that one wishes to operate in the image formation (cf. Sect. 6.4 and the action of the "aperture stop" in the light microscope). In contrast with the light microscope (cf. Figs. 6.8 and 6.9), the real objective aperture, not a virtual one, is introduced in the back focal plane of the objective lens.

The selected area aperture. This aperture selects a part of the specimen of which one wishes to analyse the diffraction pattern and therefore it is called "selected area diffraction aperture" (abbreviated by "SAD" aperture). Its action can be compared with the field stop in a light microscope discussed in Sect. 6.4. The selected area aperture is inserted in the image plane of the objective lens, implying the application of a virtual aperture in the specimen/foil plane (this is a pendant of the "trick" performed with the field stop in the light microscope; cf. Sect. 6.4 and Figs. 6.8 and 6.9). The diffraction pattern originating from the part of the specimen selected by the selected area aperture is accordingly called "selected area diffraction pattern" (usually abbreviated by "SADP"). The smallest "size/length" of the virtual selected area aperture in the plane of the specimen is usually a few tenths of a μm. To analyse diffraction patterns of smaller specimen/foil areas one could apply convergent beam electron diffraction, discussed in Sect. 6.8.5.

In the case of applying the TEM in diffraction mode, the intermediate lens has a "strength" adjusted such that the back focal plane of the objective lens acts as object plane for the intermediate lens (cf. Fig. 6.15). The presence of an objective aperture at the location of the back focal plane of the objective lens would obstruct the image formation of the diffraction pattern by the intermediate lens. Hence:

- in case of the diffraction mode, the objective aperture has to be removed (and the selected area aperture has to be introduced), and also
- in case of the imaging mode, the selected area aperture is removed (and the objective aperture has to be introduced).

6.8.2 The Diffraction Pattern; the Zone Law

As made likely at the end of the introduction to this Sect. 6.8, the diffracting crystal-lographic planes in the specimen/foil are oriented practically parallel to the incident electron beam and thereby are practically perpendicular to the surface of the specimen/foil considered. A *zone axis* is defined as the direction common to a number of (*hkl*) families of crystallographic planes (cf. the introduction of Miller indices and the corresponding symbolism in Sect. 4.1.4). Evidently, the incident electron beam direction is the direction of the zone axis of the diffracting crystallographic planes! If the direction of the incident electron beam is designated as [*uvw*] *in the crystal coordinate system* concerned, it holds for all (*hkl*) families of crystallographic planes belonging to the zone axis [*uvw*] and giving rise to *HKL* diffraction spots in the diffraction pattern[13]:

$$hu + kv + lw = 0 \text{ and } Hu + Kv + Lw = 0 \tag{6.30}$$

which so-called *zone relation* is valid for all crystal classes. The *HKL* diffraction spots (see Footnote 13) in the electron-diffraction pattern must satisfy this relation.[14]

The (geometrical) relation between the diffracting crystal and its diffraction pattern as recorded by a TEM is illustrated in Fig. 6.16. A b.c.c. crystal is considered which has been oriented with one of its <100> axes, here chosen to be indicated as the [001] axis (the zone axis for the case considered), parallel to the electron-beam direction. It is easy to verify (see the total diffraction pattern shown in the centre of the top part of Fig. 6.16) that Eq. (6.30) is obeyed.

6.8.3 Diffraction Contrast Images; Bright Field and Dark Field "Imaging"

Superposition of "light" originating from the diffraction maxima in the diffraction pattern in the back focal plane of the objective lens leads to an image of the object in the image plane of the objective lens. It has been demonstrated in Sect. 6.2 that a faithful image of the specimen is only obtained if the number of diffraction maxima contributing is very large ("infinitely large"). Evidently, already the light microscope does not provide an ideal image as the numerical aperture of the light microscope is limited, i.e. only a part of the diffraction pattern is involved in the image-formation process. The most minimal (distorted) image (of a grating) is formed if only two diffraction maxima contribute (see Fig. 6.5 and its discussion in Sect. 6.5.1).

---

[13] According to convention (see Sect. 4.5), the reflection observed from the (*hkl*) family of crystal-lographic planes is designated with Laue indices as *HKL* (with $H = nh$, $K = nk$ and $L = nl$; $n =$ order of reflection), without brackets or braces, in the (X-ray, or electron, or …) diffraction pattern.

[14] Actually, this statement only holds for the so-called Laue zone of order zero (further, see Williams and Carter 1996).

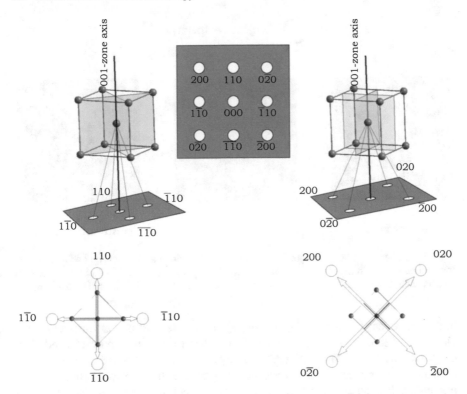

Fig. 6.16 Formation of a diffraction pattern originating from a b.c.c crystal in a TEM in diffraction mode. The crystal is oriented with [001] parallel to the electron-beam direction. Left part of figure: diffraction by {110} planes in perspective view and top view. Right part of figure: diffraction by {100} planes in perspective view and top view. The small grey spots in both partial diffraction patterns (top views: bottom parts of the left and right parts of the figure) indicate the orientation of the crystal relative to the position of the diffraction spots (large white circles) concerned (cf. the upper parts of the left and right parts of the figure). The addition of both contributions (partial diffractograms) leads to the total diffraction pattern shown in the centre of the top part of the figure

In the above sense, i.e. considering the faithfulness of the image produced, the electron microscope usually is a much more limited instrument than the light microscope: the very small, allowable aperture of the objective lens (see the begin of Sect. 6.8), which decides the upper limit of achievable image quality, obstructs the cooperation of many diffraction maxima in the image-formation process. Keeping this in mind, the usual bright field and dark field "imaging" modes will be discussed.

Bright field. The SADP is composed of a, usually very intense, central spot due to the non-diffracted electrons and a number of spots each corresponding to electrons diffracted in a specific direction (see Fig. 6.17). The central spot can be selected in the diffraction pattern by introducing the objective aperture and positioning it such that only the non-diffracted electrons contained in the central spot can propagate (left part of Fig. 6.17). Then, by removing the SAD aperture and strengthening the

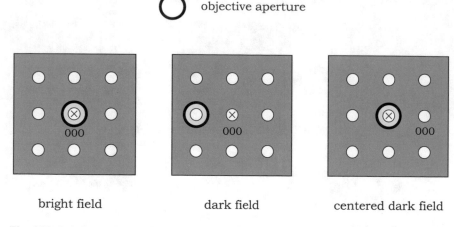

Fig. 6.17 Relative positions of objective aperture and diffraction pattern for bright field, dark field and centred dark field modes in operating a TEM. The optical axis of the TEM in all cases runs perpendicular to the plane of drawing and through the spot in the centre indicated with a cross

intermediate lens (cf. Fig. 6.15), the image plane of the objective is imaged onto the viewing screen/detector. This "image" then is due to only the non-diffracted electrons. Obviously, this "image", generated by only one diffraction maximum, is not even a minimal image, and therefore, the word "image" has been put between quotation marks. Instead one speaks of *"bright field diffraction contrast image"*: the parts of the specimen/foil where little or no diffraction of the incident electrons occurs, *irrespective of the direction of diffraction*, appear relatively *light* in the "image"; the parts of the specimen/foil where pronounced diffraction of the incident electrons occurs, *irrespective of the direction of diffraction*, appear relatively *dark* in the "image".

Dark field. The objective aperture can also be used to select a spot of diffracted electrons implying that only these diffracted electrons can propagate (middle part of Fig. 6.17). Analogous to the bright field case discussed above, a *"dark field diffraction contrast image"* thus is generated: the parts of the specimen/foil where little or no diffraction of the incident electrons occurs, *in the specific direction selected by the objective aperture*, appear relatively *dark* in the "image"; the parts where pronounced diffraction of the incident electrons occurs, *in the specific direction selected by the objective aperture*, appear relatively *light* in the "image".

Hence, the bright and dark field diffraction contrast images are not fully complementary: what is relatively dark in the dark field diffraction contrast image, appears relatively light in the bright field diffraction contrast image. However, parts of the specimen/foil, which appear relatively dark in the bright field diffraction contrast image, appear relatively light in the dark field diffraction contrast image only if for the dark field diffraction contrast image a diffraction spot has been selected [by the appropriate positioning of the objective aperture in the diffraction pattern, as it appears in the back focal plane of the objective lens (see above discussion)] that pertains to strong diffraction in the direction corresponding to that diffraction spot by *those* parts of the specimen/foil (see also Fig. 6.18, discussed in Sect. 6.8.4).

Notwithstanding the above stipulation of the distinction between *image* and *diffraction contrast*, it is customary, e.g. in textbooks on transmission electron microscopy, to speak about bright and dark field *images*, and this will be the case in the sequel as well. As a final remark, it then is crucial to recognize the difference with the same notions as used for light optical microscopy, where a bright field image implies the image generated by the maximum due to the non-diffracted rays *and* all maxima of diffracted rays in as far as allowed by the numerical aperture and a dark field image implies the image generated by *all* maxima of diffracted rays in as far as allowed by the numerical aperture (cf. Sects. 6.6.1 and 6.6.2).

Now reconsider the manipulation performed to establish a dark field image in the TEM according to the above discussed procedure. Positioning of the objective aperture around some spot due to diffracted electrons implies that the electrons contributing to the dark field image follow a route in the TEM relatively remote from the optical axis (cf. Figure 6.15, imaging mode). As suggested by the discussion in the first paragraphs of Sect. 6.8, "off-axis" electrons suffer from the magnetic lens aberrations. To remedy the associated disadvantageous effects, it is usual to provide a tilt to the incident beam of electrons such that the incident beam hits the specimen/foil surface under an angle with the optical axis that equals the diffraction angle pertaining to the diffraction spot considered. As a result, the concerned diffracted electrons travel along the optical axis, i.e. the diffraction spot is located at the centre of the diffraction pattern and the objective aperture can be positioned at this centre (as for the bright field imaging mode) to select this diffraction spot (right part of Fig. 6.17). This procedure is called *centred dark field* ("*CDF*") imaging and is the preferred way to make a dark field image in the TEM.

6.8.4 Examples of Bright and Dark Field TEM Images

A situation which is often met in practice concerns the analysis of a polycrystalline, polyphase material. The power of TEM analysis by separate imaging of phases and individual grains of a phase is illustrated in the schematic Fig. 6.18.

The cross section (thin foil; cf. the one but last paragraph of the introduction of Sect. 6.8) shown in the upper left corner of the figure comprises a b.c.c. matrix grain containing three second phase, h.c.p. particles/grains, which are considered here to be the result of precipitation out of the supersaturated matrix (cf. Chaps. 7 and 9). The diffraction patterns originating from the four grains (matrix plus three precipitate grains) have been indicated within the grains in the cross section. Evidently, the matrix and two of the three precipitate grains are in diffracting condition (i.e. have an appropriate crystal orientation with respect to the incident electron beam). Note that the two diffracting precipitate grains have a different orientation with respect to the matrix, as follows from the orientation of the diffraction patterns (here identical

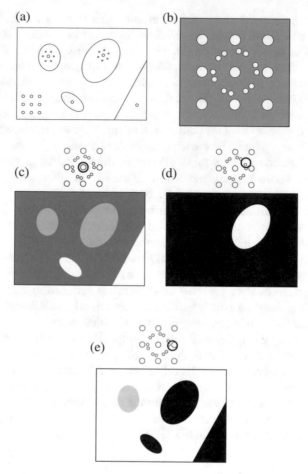

Fig. 6.18 Illustration of bright and dark field image formation in a TEM. **a** Schematic depiction of an examined microstructure: the b.c.c. matrix; two h.c.p. precipitate grains in diffracting orientation; a third precipitate which is not in diffracting orientation. The lower right corner represents the edge of the electron-transparent cross section/foil, i.e. the hole in the foil as could be due to the foil-preparation procedure. The respective contributions, partial diffraction patterns, to the total diffraction pattern are indicated within each microstructure constituent (top left). **b** The total diffraction pattern of the region described under **a** (top right). **c** Bright field image formed using the primary, not diffracted, central, 000 spot, as indicated schematically above the image (bottom left). **d** Dark field image formed using one of the diffracted beams/diffracted spots originating from one of the diffracting precipitate grains, as indicated schematically above the image (bottom middle). **e** Dark field image formed using simultaneously a diffraction spot of the matrix grain and one diffraction spot of one of the precipitate grains, as indicated schematically above the image (bottom right)

apart from rotation around the normal of the cross section/foil) of these two precipitate grains. One precipitate grain does not diffract: its diffraction pattern only shows the central, not diffracted, 000 spot (cf. Figs. 6.16 and 6.17). Obviously, only the central, 000 spot occurs for the hole in the foil analysed, as observed for the bottom right corner of the cross section/foil. The total diffraction pattern resulting from the various constituents of the microstructure is shown in the upper right corner of Fig. 6.18, i.e. Fig. 6.18b.

If the objective aperture (indicated in the figure by an open circle with bold circumference (as also in Fig. 6.16) is positioned around the central, 000 spot, a bright field image is obtained as shown in the bottom left corner of Fig. 6.18, i.e. Fig. 6.18c: The not diffracting third precipitate grain appears bright, as compared to the matrix grain and the two diffracting precipitating grains, which may be of different darkness in the image, recognizing that the matrix grain and the two diffracting precipitate grains give rise to different amounts of diffraction out of the incident electron beam. Obviously, the hole in the foil appears bright in the bright field image.

If the objective aperture is positioned around one of the diffraction spots of one of the diffracting precipitate grains, then a dark field image is obtained that shows the diffracting precipitate concerned as appearing bright against a dark background that forms the remainder of the image (see the middle of the bottom part of Fig. 6.18, i.e. Fig. 6.18d).

If the objective aperture is positioned such that it encompasses both a diffraction spot of the matrix and a diffraction spot of one of the diffracting precipitate grains, then a dark field image is recorded that shows the not diffracting precipitate grain and the diffracting precipitate grain, of which no spot is included by the objective aperture, and the hole as appearing dark against the bright matrix including the bright appearing diffracting precipitate grain of which the spot is included by the objective aperture (see the right bottom part of Fig. 6.18, i.e. Fig. 6.18e).

Experimental examples of corresponding bright and dark field images are provided by Fig. 6.19. Evidently, the diffracting microstructural constituents, of which a diffraction spot is utilized for obtaining a dark field image in the micrographs shown, appear bright against a dark background in the dark field images, in agreement with the treatment of Fig. 6.18: see cases (a) and (b) in Fig. 6.19 (bottom part of the figure). As compared to bright field images, dark field images often provide more clearly microstructural information; relative variations in diffracted intensities are shown with larger contrast. Thus, the "broken up"/"fragmented" nature of (Cr,Al)N and VN precipitate platelets is clearly revealed in the dark field images of cases (a) and (b) shown in Fig. 6.19.

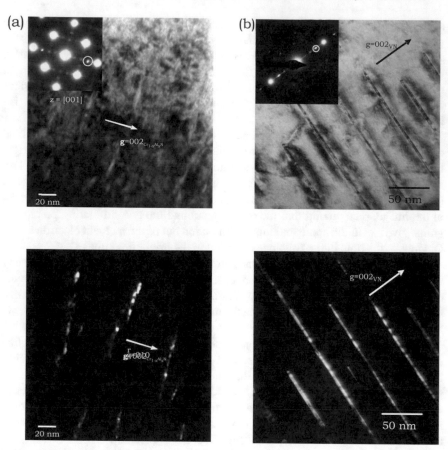

Fig. 6.19 Experimental examples of bright field (shown in *top* part of the figure) and dark field (shown in the *bottom* part of the figure) image formation in a TEM. The diffraction spot used for dark field imaging has been indicated for the cases **a** and **b** below by a white circle in the corresponding diffraction patterns shown in the insets in the bright field images. **a** (Al,Cr)-nitride precipitates in a nitrided Fe-1.5 wt% Al-1.5 wt% Cr alloy, nitrided for 15 h at 570 °C. The dark field image at the bottom is obtained employing a $Cr_{1-x}Al_xN$ nitride 002 diffraction spot (taken from A. Clauss, E. Bischoff, R. Schacherl and E.J. Mittemeijer, Metallurgical and Material Transactions A, 40A (2008), 1923–1934). **b** VN precipitates in a an Fe-2.23 at.% V alloy nitrided for 10 h at 580 °C and subsequently annealed at 750 °C for 10 h. The dark field image at the bottom is obtained employing a VN nitride 002 diffraction spot (taken from N.E. Vives Díaz, S.S. Hosmani, R.E. Schacherl and E.J. Mittemeijer, Acta Materialia, 56 (2008), 4137–4149)

6.8.5 *Convergent Beam Electron Diffraction (CBED); Microdiffraction; Scanning Transmission Electron Microscopy (STEM)*

In the case of selected area diffraction, the lateral size of the area on the specimen/foil surface is at least a few tenths of a μm (Sect. 6.8.1). Obviously, structural defects in the specimen, as dislocations, and second phase particles, as precipitates, can have dimensions much smaller than this size. Hence, the desire to obtain diffraction patterns of much smaller parts of the specimen than possible by SAD. The best known, mostly used "microdiffraction" technique is "convergent beam electron diffraction (CBED)".

Application of a convergent electron beam hitting the specimen/foil surface allows minimization of the volume of the specimen/foil, that gives rise to the diffraction pattern that one wishes to analyse. In fact an image of the light source (a so-called cross-over) is made on the plane of the specimen/foil, which resembles the classical case of "critical illumination" for light microscopy (see Sect. 6.4). The convergent electron "probe" that thus hits the specimen/foil has a lateral size which can be as small as 10 nm; in combination with a field emission gun as electron source regions of lateral size even smaller than one nm can be investigated (for sufficiently thin foils).

In the case of SAD, a (practically) parallel beam of electrons hits the specimen/foil surface. The diffracted rays of similar structural features in the specimen-surface plane are parallel and as a result they all converge in a single point (diffraction maximum) of the diffraction pattern generated in the back focal plane of the objective lens: the diffraction maxima are sharp (Fig. 6.20a and see also Fig. 6.4). If a convergent beam hits the specimen surface, this is no longer true. Consider Fig. 6.20b. The incident convergent beam has a convergence angle of, say, φ. If the two limiting, converging rays drawn in Fig. 6.20b, which differ in orientation by rotation over the angle φ, hit the specimen/foil surface and are diffracted with the same diffraction angle, then obviously the corresponding diffracted, diverging rays also are rotated over the same angle φ. Following the paraxial approximation (Sect. 6.1.1), a diffracted ray intersects the back focal plane of the objective lens at a location that is determined by the point of intersection of the line drawn through the lens centre parallel to the diffracted ray considered and the back focal plane. Performing this construction for both rays considered, as a result in the back focal plane of the objective lens the similarly diffracted rays do not converge in a single point, as for the parallel incident beam, but (and for all rays in the convergent beam) give rise to a disc of intensity (Fig. 6.20b). Hence, the diffraction maximum considered is represented by a disc of intensity the lateral size of which is determined by the convergence angle of the incident convergent beam.

Employing a very thin specimen/foil, the discs in the CBED pattern are rather uniform of intensity. These patterns can be fruitfully applied for crystallographic analysis on the basis of diffraction patterns in the same way as holds for SADPs, with the difference that the diffraction pattern now originates from a very small

(a)

(b)

(c)

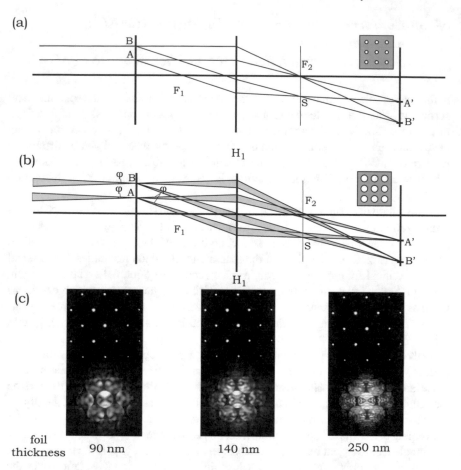

foil
thickness   90 nm   140 nm   250 nm

Fig. 6.20 Schematic illustration of **b** convergent beam electron diffraction in comparison with **a** conventional electron diffraction. In case of CBED, the diffraction spots are enlarged to discs (cf. Fig. 6.4). **c** Experimental examples. The SAD patterns [(practically) parallel beam of electrons hitting the foil surface] and the corresponding CBED patterns (convergent beam of electrons hitting the foil surface) are shown for a silicon foil with <110> (as zone axis; cf. Sect. 6.8.2) parallel to the foil normal. The SAD and CBED patterns have been arranged in order (left to right) of increasing foil thickness: 90, 140 and 250 nm, as determined by computer simulation (SAD and CBED patterns recorded by and CBED patterns simulated by Dr. W. Sigle, Max Planck Institute for Metals Research)

area/volume and thus very local crystallographic information can be obtained. For thicker specimens/foils, contrast phenomena appear in the disc, which information can be used for example to determine the full three dimensional crystal symmetry (including the space group; cf. in Sect. 4.1.2 the "*Intermezzo: A short note on point groups, crystallographic point groups, plane groups and space groups; glide and screw operations*"). Examples of a CBED pattern (and the corresponding normal SADP) are provided by Fig. 6.20c for a silicon foil at three different foil thicknesses.

Superposition of the light originating from the diffraction maxima in the diffraction pattern leads to an image of the object in the image plane provided coherency of the incident light prevails. This is the case for an incident (on the specimen/foil) parallel beam of light (see Fig. 6.4 and Sect. 6.2). Coherency is lost if the incident beam is convergent. So in the way discussed above for bright and dark field image (contrast) formation from crystalline materials, one cannot expect in general that a useful image (contrast) is produced if a convergent electron beam is employed. This problem is circumvented in a special mode of transmission electron microscopy where a convergent electron beam *scanning the specimen/foil* is used (scanning transmission electron microscopy, STEM) and "image" formation is realized without using an imaging lens (see, in particular, also the operating principle of scanning electron microscopy (SEM) discussed in Sect. 6.9; see also Fig. 6.24).

In STEM mode, the foil is scanned by the focussed electron beam and simultaneously, in the diffraction pattern, the intensity in a certain angular range of the scattered electrons is recorded.[15] The thus measured "bright field" intensity (measured in the diffraction pattern at the location of the transmitted, non-scattered electron beam) or "dark field" intensity (measured for a certain angular range in the diffraction pattern) is used to vary the intensity of a *separate* electron beam that scans a television/cathode-ray-tube screen in the same way as and synchronously with the first, specimen-foil scanning electron beam. Thereby a contrast appears on the display that has a one-to-one relation with the specimen foil. As a result, a "mapping" of the foil is obtained.[16] It is important to be aware of the meaning of the concepts "image", "diffraction contrast image" and "map", as discussed here, because normally the single word "image" is used for all these notions.

To select either the non-diffracted or a specific type of diffracted electrons, one might propose to introduce a detector at the appropriate position in the back focal plane of the objective lens. However, in imaging TEM mode, the objective aperture is located there (see Sect. 6.8.1 and Fig. 6.15). Therefore, to use a TEM in STEM mode, the detector is positioned in the image plane of the projector lens, where an image is produced from the diffraction pattern (implying that the TEM, in order to

---

[15] Note that electrons scattered in same directions, but originating from different locations in the specimen foil, converge in the back focal plane of the objective lens in a single point: parallel diffracted rays intersect the back focal plane of the objective lens at a location that is determined by the point of intersection of the line drawn through the lens centre parallel to the diffracted rays considered and the back focal plane. Hence, although the probe in STEM mode moves (it scans the specimen foil surface), the generated CBED pattern in STEM mode is stationary. Normally, for CBED application, e.g. to analyse the diffraction pattern of a very small specimen volume, the electron beam is convergent but it is not moved, i.e. it is *not* scanning the foil surface!

[16] Scanning (modes of) microscopes in general do not produce images in the sense as discussed in Sect. 6.2; such (modes of) microscopes produce "response maps" of the specimen with respect to one specific type of response to the action of a scanning probe (here a focussed electron beam, or, as holds for STED (Sect. 6.7), a laser beam), as, for example, the amount of (specifically) diffracted or of not diffracted electrons (as in STEM), the amount of secondary or of back-scattered electrons (as in SEM; cf. Sect. 6.9), or the amount of generated X-rays originating from a specific element, as in composition analysis (EPMA and EDS; cf. Sects. 6.8.7.1 and 6.9.3) or the amount of fluorescence (STED; cf. Sect. 6.7).

use as a STEM, has to be in diffraction mode; cf. Sect. 6.8.1). The bright field detector is usually positioned centrally in the diffraction pattern to simply transmit the non-diffracted, "direct" electrons. For the diffracted/scattered electrons, often an annular detector is used that only transmits the intensity from electrons diffracted/scattered in a certain angular range. STEM images are in particular useful if incoherent elastic scattering is important, i.e. in cases where the contrast of the specimen foil is dominated by atomic mass (differences) and thickness (variations) as can also hold in particular for non-crystalline, amorphous materials. Diffraction contrast due to the coherent elastic scattering of electrons by crystalline materials is by far best analysed in TEM images.

A convergent electron beam is a very useful means for local composition analysis. The focussed and thus very localized (see above) electron beam that hits and passes through the foil generates X-rays on its way through the specimen. The energy of these X-rays is element specific, and thus, energy dispersive spectroscopical analysis (EDS) of these X-rays leads to very local composition analysis of the foil (see also Sect. 6.9.3).

6.8.6 High-Resolution Transmission Electron Microscopy (HRTEM)

Application of an objective aperture large enough to enclose more than one diffraction spot gives rise to (minimal) image formation in the sense of Abbe's theory (Sect. 6.2). To achieve a faithful image, the involvement of a large number of diffraction maxima (corresponding to the application of an objective aperture of relatively large diameter) is required in the image-formation process. As discussed in the introduction of this Sect. 6.8, the magnetic electron lenses of a TEM have such flaws that a faithful image cannot be produced if a relatively large objective aperture is used. However, instrumental/lens improvements and the availability nowadays of algorithms to calculate the instrumental effects on the not so faithful image produced make it possible, on the basis of a model of the (atomic) structure of the specimen, to simulate the (flawed) image, and, if a good agreement with the experimentally observed image occurs, one may conclude that the proposed model of the specimen provides a realistic description. This operation mode of TEM is called high-resolution transmission electron microscopy (HRTEM, also abbreviated as HREM): although the image produced is affected by instrument/lens aberrations, these are accounted for by the computer calculation of the affected image and in the end the high resolution corresponding to the diameter of the objective aperture utilized has in fact been achieved.[17]

---

[17] In this sense, the future of high resolution in microscopy *in general* needs not in the first place lie in hardware instrumental advancements leading to nicer, i.e. sharper and contrast richer, images but rather in the development of computational models for the image-formation process in non-ideal microscopes, leading to algorithms for the processing of the enormous quantities of data contained in recorded non-ideal images.

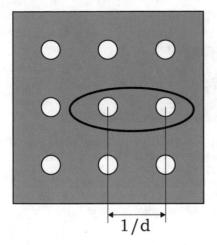 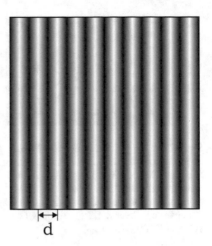

Fig. 6.21 Illustration of the formation of lattice fringes (right part of the figure) by interference of two diffracted beams as indicated in the diffraction pattern (left part of the figure)

Lattice-fringe imaging. If two beams, say the non-diffracted beam and one diffracted beam, are enclosed by the objective aperture, a minimal form of image formation is possible in principle (see Sect. 6.2 and, in particular, Sect. 6.5.1). The result is a set of parallel fringes: parallel lines of maximal and minimal intensity can be observed: see Fig. 6.21 and see Fig. 6.5 for $N = 2$. The spacing between these lines of corresponding (as maximum or minimum) intensity is equal to the spacing of the crystallographic planes giving rise to the diffracted beam.[18] The observed fringes (of maximum or minimum intensity) do *not* represent the (atomic) positions of the crystallographic planes concerned in the (image of the) specimen; only the period-icity corresponding to these crystallographic/lattice planes[19] is visualized. Note that because the diffracting crystallographic planes are oriented practically perpendicular to the surface of the specimen/foil (see the last paragraph of the introduction to this Sect. 6.8), the fringes in the image produced can only be due to crystallographic planes oriented practically perpendicularly to the foil surface (i.e. these planes are oriented "edge on"). Now, if a number of spots of diffracted beams, in addition to the spot of the non-diffracted beam, are enclosed by the objective aperture, fringes in various directions can occur in the image produced. These crossing fringes may lead to patterns of intensity maxima and minima in the image which do not directly hint at the fringes which are the origin of the image produced. Again: the crossing fringes cannot be identified as representing the positions of the atomic planes in

---

[18] Actually, the fringe spacing is given by the reciprocal of the distance between the two diffraction spots in the diffraction pattern as enclosed by the objective aperture. Only if one of the two diffraction spots is the non-diffracted one and the other diffraction spot is a first order reflection, the fringe spacing is equal to the spacing of the crystallographic planes giving rise to the diffraction spot.

[19] Regarding the usage of "lattice planes" versus "crystallographic planes", see (again) the begin of Sect. 4.1.4.

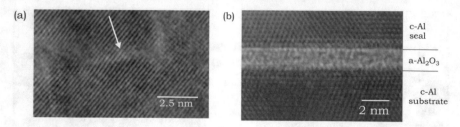

Fig. 6.22 a VN precipitate in an α-Fe (b.c.c., ferrite) matrix. The 110-lattice fringes become curved due to lattice distortion caused by coherency stresses (taken from N.E. Vives Díaz, S.S. Hosmani, R.E. Schacherl and E.J. Mittemeijer, Acta Materialia, 56 (2008), 4137-4149). **b** Amorphous Al_2O_3 layer grown onto Al substrate with a {111} surface, formed during initial stage of oxidation in oxygen. The sample was afterwards sealed by an Al layer on top for protection of the oxide layer (taken from F. Reichel, L.P.H. Jeurgens, G. Richter and E.J. Mittemeijer, Journal of Applied Physics, 103 (2008), 093515)

the image of the specimen, and thus, the image produced is *not* a direct image of the (structure of the) specimen: merely, the relative orientation (here relative rotation) of the sets of crystallographic planes concerned and the corresponding lattice spacings, i.e. the structural periodicity, can be determined from the image produced. These lattice-fringe images are powerful means to study the local crystal structure, e.g. the local deformation around a (coherent) precipitate (see Fig. 6.22a; see also Fig. 5.24)[20] or the structure (presence of dislocations) at an interface or the amorphous or crystalline nature of a thin layer or precipitate (see Fig. 6.22b). Any further interpretation of these images, e.g. in terms of atomic positions, requires extensive computer calculation (see what follows).

Image simulation. To achieve the highest (atomic) resolution in structure analysis using the TEM, the experimentally observed image has to be simulated on the basis of (i) knowledge of the effect of instrumental/lens parameters on the process of image formation in the TEM and (ii) an atomistic model of the specimen. To this end, a number of elaborate methods contained in (also commercially) available computer programmes exist. Then, to conclude that the model of the specimen is satisfactory, it is usually required that the match of the simulation with the experimental reality persists over a range of defocus values, i.e. a set of experimental images of the same part of the foil is recorded for a number of defocus values, which is called a "through-focus series". An example is shown in Fig. 6.23. It should not be forgotten that such agreement between calculation and experiment does not yet provide a guarantee of uniqueness regarding the structural details found. A detailed account on (TEM) image (processing and) simulation is beyond the scope of this book. Textbook treatments have been provided by Williams and Carter (1996) and De Graef (2003).

---

[20] Conventional TEM images also reveal the presence of such distortions, albeit at a spatially less resolved scale. See Fig. 6.19 case (b), the bright field image shown in the top part of the figure: the nitride (VN) platelets in the ferrite (b.c.c. iron) matrix are surrounded by dark contrast along their faces. This contrast is caused by precipitate/matrix misfit strains inducing local bending of lattice planes in the matrix leading to the contrast observed.

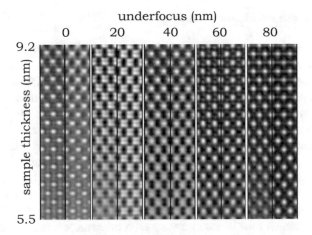

Fig. 6.23 Through-focus series. The specimen is a silicon foil with <110> as zone axis parallel to the foil normal (cf. Sect. 6.8.2). Experimental images and simulated images are shown for a series of defocus (underfocus) values, indicated at the top of the figure. For each defocus (underfocus) value, a simulated image (at the left) and the experimental image (at the right) are shown. At the left side of the figure, the variation of foil thickness along the ordinate has been indicated (experimental images recorded by K. von Hochmeister; image simulations by K. Du; see K. Du, K. von Hochmeister and F. Philipp, Ultramicroscopy, 107 (2007), 281–292; Dr. F. Philipp, Max Planck Institute for Metals Research)

6.8.7 Analytical Electron Microscopy (AEM); Chemical Composition Maps

The interaction of the incident electrons with the specimen/foil gives not only rise to elastically scattered electrons. The electron beam/solid interactions also lead to the generation of X-rays, Auger electrons and inelastically scattered electrons, exhibiting an energy loss with respect to the incident electrons.[21] The energies of these X-rays, Auger electrons and inelastically scattered electrons are element specific. Using a scanning electron probe, their analysis in principle allows the determination of chemical composition maps. Transmission electron microscopes equipped with apparatus allowing such measurements are called "analytical electron microscopes".

---

[21] Elastic scattering involves the interaction of the incident electrons with the nuclei of atoms in the specimen, which is associated with no loss of energy but a change of direction (momentum). Diffraction is an elastic scattering process. Inelastic scattering involves the interaction of the incident electrons with the electrons of atoms in the specimen, which is associated with both energy loss and change of direction (momentum). An incident electron can eject an electron out of a near core orbital of an atom, leaving a "hole" in the near core orbital. Next, an electron in a higher orbital of the atom concerned can jump into the near core orbital with the "hole". This process is associated with the emittance of the energy difference between the higher orbital and the lower orbital in the form of either X-rays of characteristic energy (in this way the X-rays from X-ray tubes used in X-ray diffraction analysis are produced; cf. Sect. 4.5 and Sect. 6.10) or, less frequently, in the form of a so-called Auger electron of characteristic energy which is ejected from a relatively high atomic orbital.

6.8.7.1   Electron Probe Micro-analysis (EPMA/EDS)

The interaction of the incident electron beam, "probe", with the material irradiated (also) gives rise to the emittance of X-rays of energies (wavelengths) which are element specific. Measuring the energy(wavelength) of specific X-rays across a scanned surface then leads to maps revealing the distribution of the elements in the surface. This is a powerful analysis, called electron probe micro-analysis (EPMA),[22] for investigating the local compositional variations in a specimen, e.g. due to the presence of precipitate particles or concentration profiles across an interface induced by a diffusion process. Performing EPMA in a STEM (of course, with a *not* scanning electron beam; the convergent nature of the incident electron beam is the desired feature) with an electron probe size smaller than 1 nm (which requires the use of a field emission gun (FEG) as electron source) and the associated use of thin electron-transparent specimen foils, leads to a very high lateral resolution for the chemical analysis: the X-rays generated originate from specimen-foil areas as small as the electron probe diameter (this is different for EPMA performed on bulk specimens, as carried out in a SEM, see Sect. 6.9.3). However, the presence of a minimal amount of the element to be quantified in the specimen volume analysed is a prerequisite: there is a detectability limit. In fact, due to the smallness of the volume analysed (probe size smaller than 1 nm is possible), the presence of a few atoms of the element considered in the volume analysed can be established. The analysis of the X-ray radiation generated is usually performed by energy-dispersive spectroscopy (EDS) applying a solid-state detector. (For use of wavelength dispersive spectroscopy (WDS) in EPMA, see Sect. 6.9.3).

6.8.7.2   Electron Energy-Loss Spectroscopy (EELS)

The energy losses experienced by those incident electrons which experience inelastic collisions with electrons of atoms of the specimen are indicative of the nature and bonding state of these atoms of the specimen. Measuring the spectrum of electron energies thus can provide a lot of local chemical information. This technique is called electron energy-loss spectrometry (EELS). EELS is especially useful for the analysis of light elements, where EPMA in TEM [in the EDS variant (see Sect. 6.8.7.1)] becomes problematic. As for EPMA in the TEM (see Sect. 6.8.7.1), for highest spatial resolution of EELS (a resolution of 0.1 nm is possible), the TEM is preferably operated in STEM mode (the TEM then is in diffraction mode (see Sect. 6.8.5); again (see above) with a not scanning electron beam; the convergent nature of the incident electron beam is the desired feature). The EELS detector is often positioned in the back focal plane of the projector lens (i.e. underneath the CBED pattern viewed on

---

[22] In fact, the methods based on using a scanning electron probe and measuring the energies of, e.g., the emitted Auger electrons or the inelastically scattered electrons (as in EELS) could have also been called EPMA, but because of the historical development this is not usual and the designation EPMA is reserved for the analysis of the X-ray radiation induced by a scanning electron probe.

the image screen of the TEM). A typical EELS spectrum reveals (i) the zero-loss peak (predominantly) due to elastic (forward) scattering of electrons, (ii) the low energy-loss region till about 50 eV energy loss, representing the incident electron interactions with the weakly bonded outer electrons of the atoms in the specimen and (iii) the high energy-loss region, where the incident electron interactions with relatively strongly bonded inner electrons of the atoms in the specimen can also lead to ionization of these atoms exhibited by characteristic element-specific ionization edges in the EELS spectrum.

Comparing EELS with EPMA in AEM (the TEM in STEM mode; see above), it can be said that EELS offers a somewhat higher spatial resolution (beam spreading affects EPMA (even for the electron-transparent foil; see, in particular, Sect. 6.9.3 for EPMA of bulk specimens) but is not of similar consequence for EELS as only electrons in a narrow angular range are collected by the spectrometer) and also the detectability limit is better (in favourable situations the presence of even only a single atom of a specific element in the volume analysed can be demonstrated). The major limitation of EELS in AEM is the requirement of very thin foils (less than a number of tens of nm thick), to avoid multiple scattering effects which obscure the energy spectrum and can render the ionization edges invisible.

With special detectors it is also possible to filter electrons in a specific energy range out of the entire spectrum of electron energies. Thereby *energy-filtered* images or diffraction patterns can be displayed. Thus, composition maps (elemental distribution maps) can be made[23] and in diffraction patterns the contribution of the inelastically scattered electrons (diffuse background scattering) can be removed.

6.9 Scanning Electron Microscopy (SEM)

The scanning electron microscope (SEM) provides a picture of the surface (region) of the specimen. In this way, and because of its high resolution (details 0.5 nm apart laterally can be resolved), a SEM is complementary to the light optical microscope [that can resolve details in (the surface of) the specimen 200 nm apart laterally (Sect. 6.5.2)]. A SEM is generally no competitor for a TEM in view of the richness of structural details revealed by TEM, e.g. the analysis of a dislocation network in an interface is beyond reach for a SEM. The popularity of SEM is undoubtedly to a large extent due to the relative ease of specimen preparation as compared to TEM and, also, light optical microscopy.

As compared to a light microscope, a SEM does not produce an image of the (surface of the) specimen in the sense of Sect. 6.2, rather it is an instrument that "maps" the (surface of the) specimen (see Footnote 16). Its function principle can be described as follows (see Fig. 6.24).

---

[23] It is even possible to make maps of the state of bonding and type of chemical environment of atoms by selective energy filtering, i.e. (again) selecting those electron energy-loss ranges which are particularly sensitive to the effects of interest.

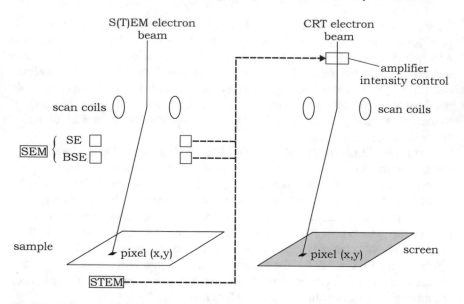

Fig. 6.24 Schematic illustration of the operation principle of a SEM (or a STEM). While the S(T)EM electron beam scans the sample, a cathode-ray-tube (CRT) electron beam is scanned in an analogous manner over a screen, while its intensity is determined by the signal of the STEM detector in case of STEM and by the BSE (backscattered electrons) or SE (secondary electrons) detector in case of SEM

A focussed electron beam scans the surface of the specimen. The interaction of the incident electron beam with the (surface region of the) specimen generates so-called secondary electrons originating from the specimen, and part of the incident electrons are "back-scattered". The intensity of the secondary or back-scattered electrons is measured while the incident electron beam, the "scanning probe", scans the specimen surface. This measured intensity is used to vary the intensity of a *separate* electron beam that scans a television/cathode-ray-tube (CRT) screen in the same way as and synchronously with the first, specimen-surface scanning electron beam. Thereby a contrast appears on the display that has a one-to-one relation with the surface of the specimen. This contrast picture is called the image of the specimen (surface) in the sequel (see Fig. 6.24 and compare with the action of a STEM described in Sect. 6.8.5).

If the size of the area scanned by the second, image generating electron beam on the display is $b \times b$ and the size of the area scanned by the first, (electron) radiation generating electron beam on the specimen surface is $a \times a$, it follows for $M_l =$ lateral (transverse) magnification = ratio of (linear) sizes of image and object ([cf. Eq. (6.3)]:

$$M_l = \frac{b}{a} \tag{6.31}$$

The scanned areas are actually divided in a number of lines, with each line composed of a number of scan points. Such a scan point is called a "pixel". Obviously, no detail smaller than the pixel size can be resolved, even if the lateral incident beam size is smaller than the pixel size. (The above consideration pertains to STEM as well, of course; cf. Sect. 6.8.5) Hence, if b has a length of, say, 10 cm and corresponds to 10^3 pixels, it follows from Eq. (6.31) that M_1 takes a value of 10^5 in order that one pixel in the image corresponds to 1 nm on the specimen surface. At present, a minimum probe size of 0.5–1 nm is attainable with a SEM using a field emission gun as electron source. The smallest detail resolvable can be larger than that because of electron-interaction effects in the surface region of the specimen causing that for one pixel radiation is received from a lateral area larger than the pixel/beam size.

The incident electron beam originates from a source and is then accelerated to attain an energy typically in the range (even well below[24]) 1–30 keV. This accelerated electron beam is focussed by a condenser/objective lens system on the specimen surface (cf. Sect. 6.8.1 and 6.8.5). Scan coils realize the rastering/scanning of a preset area on the specimen surface. With modern SEMs, equipped with a field emission gun and lens-aberration correctors, in particular the spherical aberration corrector (see footnote 9 in this chapter and related text), lens aberrations and other instrumental parameters are of lesser importance for the limiting resolution than the above-mentioned electron-interaction effects in the specimen.

6.9.1 Secondary Electron Images

The incident electron beam has a penetration depth into the specimen of the order of 1 μm, depending on its energy. The incident beam generates the emittance of secondary electrons. These secondary electrons have energies of only up to about 50 eV and therefore can "escape" from the specimen only from depths beneath the surface not larger than, say, some nm. If an incident electron during penetration is back-scattered, it can after back scattering also induce emittance of a secondary electron. Also these secondary electrons can escape from the specimen only if generated at depths less than some nm. This last type of secondary electrons can originate from a region of lateral size larger than corresponding to the incident beam (because they are induced by electrons back-scattered in variable directions from the incident beam) and thereby they give rise to a background intensity in the image. The first type of secondary electrons are generated from a region of lateral size equal to the lateral size of the incident beam and thus are responsible for the high-resolution information in the image.

Secondary electron images give a very "plastic" impression (*topographic contrast*) of the surface morphology of a specimen. Light and dark (shadowing)

---

[24] So-called low-voltage SEM (LVSEM) allows high-resolution imaging of delicate biological structures sensitive to electron radiation induced damage.

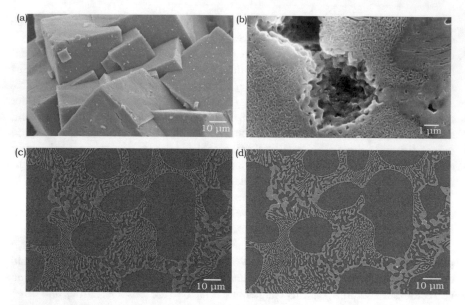

Fig. 6.25 Examples of image formation using a SEM. **a** Secondary electron (SE) image: MOF (metal-organic framework) crystals. **b** Secondary electron (SE) image: Pore in a ZrO_2 ceramic. **c** Secondary electron (SE) image: a two phase, Cu_3P-Cu specimen (cf. Fig. 6.10). **d** Backscattered electron (BSE) image: a two phase, Cu_3P-Cu specimen (cf. Fig. 6.10) (micrographs made by Dr. E. Bischoff, Max Planck Institute for Metals Research)

contrast effects occur that emphasize a three-dimensional impression: see in particular Fig. 6.25a, b). The occurrence of such effects can be understood as follows. The amount of secondary electrons generated depends on the angle of the surface irradiated with respect to the incident electron beam: a surface area at a large angle with respect to the incident electron beam appears relatively bright (relatively large yield of secondary electrons); a surface area normal to the incident electron beam appears relatively dark (relatively small yield of secondary electrons). Protrusions and edges in the surface morphology appear with bright contours with respect to their surroundings because secondary electrons can escape through more than one, protrusion/edge defining, surface. These few remarks serve to explain the contrast phenomena observed in most secondary electron images as a result of the surface topography. This relative ease of interpretation of secondary electron images, next to the relative ease of specimen preparation (see above), has contributed pronouncedly to the enormous popularity of SEM as well.

6.9.2 Back-Scattered Electron Images

The energy of back-scattered electrons from the incident beam can of course be much higher than the energy of the secondary electrons: their energy spans the large range from at most the energy of the electrons in the incident beam down to, say, 50 eV. Scattered incident electrons are detected for scattering angles close to 180° and are then called back-scattered electrons.

Because of their high energy, back-scattered electrons can escape from relatively large depths beneath the surface: depths of the order of 1 μm. As a consequence, an image produced by back-scattered electrons is not in the first place an image of the surface (as holds for the secondary electron image). Further, the lateral resolution of the image due to back-scattered electrons is less than that of a secondary electron image.

The interest in back-scattered electron images is largely due to the dependence of the amount of back-scattered electrons on the atomic number of the material irradiated: the larger the atomic number (and the larger the atomic density), the larger the amount of back-scattered electrons. Thus, the back-scattered electron image becomes composition sensitive (*material contrast*). An example is shown in Fig. 6.25d (compare with the secondary electron image shown in Fig. 6.25c). Further, the back-scattered electron intensity is sensitive to the magnetic domain structure of magnetic materials and the orientation of the irradiated crystal with respect to the incident electron beam, both effects giving rise to special contrast effects.

6.9.3 Chemical Composition Maps; Electron Probe Micro-analysis (EPMA)

The classical application of a scanning electron probe to analyse the element specific, characteristic X-ray radiation generated is realized in a SEM. As compared to EPMA performed in a STEM (Sect. 6.8.7.1), it should now be realized that, due to multiple scattering of the incident electrons within the specimen investigated, for a bulk specimen the X-rays generated originate from a volume beneath the surface which is of a lateral size (much) larger than the lateral size of the incident electron beam which in the case of a SEM is of the order of, say, 10 nm to 1 μm. Therefore, the lateral resolution of such composition maps recorded from bulk specimens cannot be better than of the order 0.1 μm (also here (cf. the AEM discussed in Sect. 6.8.7.1) the best resolution is obtained applying a field emission gun (FEG) as electron beam source). The energy of the emitted, characteristic X-rays can be analysed by an energy-dispersive spectroscopic (EDS) system, as is usual in AEM (Sect. 6.8.7.1), but now also a wavelength dispersive spectroscopic (WDS) system can be applied. WDS is based on the Bragg reflection of the characteristic X-ray radiation, originating from the specimen, by an analysing crystal. To cover various wavelength ranges various analysing crystals have to be applied which involves a large instrumental occupation

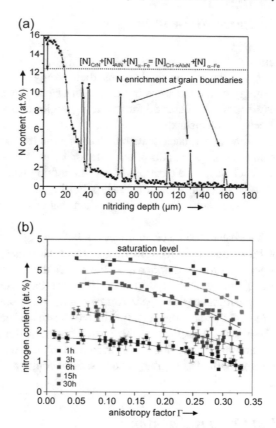

Fig. 6.26 Examples of electron probe microanalysis with wavelength dispersive spectroscopic analysis (EPMA with WDS) yielding *quantitative* composition data: **a** Nitrogen concentration-depth profile recorded from the cross section of a recrystallized Fe-4.3 at.%Cr-8.1 at.%Al specimen nitrided in a NH_3–H_2 gas mixture at 450 °C for 144 h. The dashed horizontal line represents the nitrogen level expected for (i) the development of the cubic CrN and AlN or cubic NaCl-type mixed $Cr_{1-x}Al_xN$ nitrides plus (ii) the N dissolved in the ferrite matrix at the applied nitriding conditions. The nitrogen taken up above the dashed line (indicated by the line piece with two arrowheads in **a**) is called excess nitrogen. In the deeper region of the nitrided zone of the recrystallized specimen, relatively large N contents occur at the grain boundaries due to the presence of nitride precipitates there (taken from M. Akhlaghi, S.R. Meka, E.A. Jägle, S.J.B. Kurz, E. Bischoff and E.J. Mittemeijer, Metallurgical and Materials Transactions A, 53 (2016), 4578–4593). **b** Nitrogen content measured on the surface of various, crystallographically differently oriented grains of an Fe-4.5 at.% Cr specimen as a function of the orientation factor of the corresponding grains, as measured from a specimen nitrided for different times in a NH_3–H_2 gas mixture at 450 °C. To exemplify the error of the measured averaged N content for each grain, error bars for the EPMA measurements at two nitriding times (1 and 3 h) have been shown. The polynomial fits have been given only to guide the eye. The orientation factor, Γ, indicated along the abscissa, is given by $(h^2k^2 + h^2l^2 + k^2l^2)/(h^2 + k^2 + l^2)$; hence Γ is minimal for the (100) plane ($\Gamma = 0$) and Γ is maximal for the (111) plane ($\Gamma = 1/3$). Evidently, the surface nitrogen content decreases with increasing orientation factor; an effect not observed and explained before (taken from M. Akhlaghi, M. Jung, S.R. Meka, M. Fonović, A. Leineweber and E.J. Mittemeijer, Philosophical Magazine, 95 (2015), 4143-4160)

of space and makes this variant of EPMA less suited for AEM. The advantages of WDS, as compared to EDS, are a much higher energy (wavelength) resolution, a higher count rate, a better (i.e. lower) detection limit and, in particular, a more efficient analysis of light elements (as C and N). This is a very powerful technique for quantitative analysis on an absolute basis of composition profiles (also of, especially, light elements) with high compositional accuracy. For examples, see Figs. 6.26a, b and Sato et al. (2007).

6.10 X-ray Diffraction Analysis
of the Imperfect Microstructure

X-ray diffraction analysis was introduced in Chap. 4 as the classical and nowadays still most important method to determine the idealized crystal structure of a material, i.e. the filling of the unit cell is determined from the position and integrated intensities of the reflections (also called "peaks" or "line profiles") in a diffraction pattern as recorded from the material to be analysed (The integrated intensity of a reflection is given by the area under the diffraction-line profile). The position of a HKL reflection (cf. Sect. 4.5 for the nomenclature used here) is given by Bragg's law [cf. Sect. 4.5 and Eq. (4.9)]:

$$n\lambda = 2d_{hkl} \sin \theta \qquad (6.32)$$

The information contained in the shape of the diffraction peak is thereby ignored. Indeed, for specimens composed of large crystals of perfect atomic arrangement, the reflections are of infinitesimal structural width (but of finite area (integrated intensity), i.e. the *structural* line profiles are mathematically speaking Dirac (δ) functions) and the observed line broadening then is only due to instrumental effects [as finite slit widths and the spectral line broadening due to the applied wavelength distribution (instead of truly monochromatic radiation)]. However, in reality the imperfectness of the materials investigated, containing grain boundaries, defects as dislocations and stacking faults and exhibiting stresses, in other words "the microstructure" (see the beginning of this Chap. 6), induces the occurrence of so-called *structural* line broadening and possibly line-profile position shifts, as compared to the idealized case indicated above.

Diffraction analysis is perhaps the most powerful technique for investigating the microstructure of materials by exploiting, especially, its sensitivity for (variations in) the atomic arrangement and also the element specificity of the scattering power of an atom [for books providing an overview of this research field, see Mittemeijer and Scardi (2004) and Mittemeijer and Welzel (2013)].

Each line profile in the diffraction pattern represents an average over the diffracting material; in the case of conventional X-ray diffractometry the diffracting volume is usually of the order 1 mm^3. This indicates the strength and at the same time the limitation of diffraction analysis: average values for structure/microstructure parameters are obtained (e.g. the dislocation density, the internal stress) which have a close bearing on the properties on mesoscopical and macroscopical scale, but the atomic arrangement around an individual, isolated defect cannot be revealed in this way. Here one is referred to the discussion on image formation in Sect. 6.2 (see text below Eq. (6.13) in particular). X-ray diffraction as considered here represents a case of Fraunhofer diffraction (see the begin of Sect. 6.2): the diffraction pattern is studied at "infinite" distance from the diffracting object. The same diffraction pattern would occur in the (back) focal plane of a lens capable of refracting X-rays sufficiently (Fig. 6.4).

In the diffraction experiment, the diffracted X-rays originate from lattice planes oriented symmetrical with respect to the incident and diffracted X-rays (Fig. 6.27; constructive interference according to Bragg's law (cf. Eq. 6.32). Hence, from the peak position, a value is obtained for the lattice spacing in the direction perpendicular to the diffracting planes; similarly, the information contained in the diffraction-line profile shape (the structural line broadening) represents the (micro)structure in that direction.

Fig. 6.27 Course of incident and diffracted beams during an X-ray diffraction measurement **a** for $\psi = 0$, i.e. the surface of the specimen is oriented symmetrical with respect to the incident and diffracted beams and **b** for $\psi \neq 0$, i.e. the surface is not oriented symmetrical with respect to the incident and diffracted beams. Only in case **a** the angle between surface and incident/diffracted beam is θ. In **b**, this is not the case. However, in all cases the diffraction angle equals 2θ

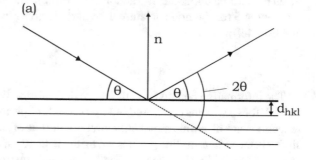

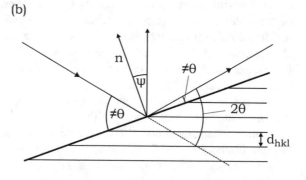

6.10.1 Determination of Crystallite Size and Microstrain

From a fundamental point of view, smallness of size of an otherwise perfect crystal should be considered as a "defect": the long-range atomic arrangement is disrupted at the interface with another crystal (grain) or at the surface (see also the introductions of Chaps. 4 and 5). This view is corroborated by the occurrence of "defect" line broadening due to the smallness of crystallite[25] size. In this section, we will present the most simple approach to determine "crystallite size" and "microstrain" parameters from occurring (X-ray) diffraction-line broadening to illustrate the unique possibilities of this technique.

The following result is obtained for the intensity distribution (shape of the line profile) due to the finite crystallite size[26]:

$$I(h_3) = \text{const.} \, \frac{\sin^2(\pi N_3 h_3)}{\sin^2(\pi h_3)} \qquad (6.33)$$

where N_3 is the number of lattice planes in the direction perpendicular to the diffracting lattice planes and $h_3 = 2d_{hkl}^{\text{ref}}(\sin\theta)/\lambda$, with d_{hkl}^{ref} as the (constant) chosen reference value of the lattice spacing in the direction perpendicular to the diffracting lattice planes (the subscript "3" pertains to the direction perpendicular to the diffracting lattice planes; the subscripts "1" and "2" would indicate principal directions parallel to the diffracting lattice planes, but are irrelevant here).

Using Eq. (6.33), the intensity distribution can be calculated as a function of the so-called diffraction angle, 2θ, which is the angle enclosed by the directions of the diffracted and incident X-rays (Fig. 6.27). Note that only for the case that the surface of the specimen is oriented symmetrical with respect to the incident and diffracted X-rays (Fig. 6.27a), the angle between the incident X-rays and the surface equals θ, in all other orientations for the surface this is not true, whereas the diffraction angle remains 2θ (Fig. 6.27b; Fig. 6.27 is another, more explicit, version of Fig. 4.46 in Chap. 4, and thus provides an introduction to Fig. 6.32 discussed in Sect. 6.10.2). Therefore, in a diffractogram the diffracted intensity is plotted as function of 2θ (or as function of h_3) and not as function of θ.

---

[25] Often the word "crystallite" instead of "crystal" is used to allow the finite size of the coherently diffracting crystalline domain, giving rise to the observed "size broadening", to be smaller than the size of a grain in a polycrystalline specimen. This can be relevant if, for example, specific defect (e.g. dislocation) arrangements occur in an otherwise perfect crystal which induces incoherency of diffraction at the location of such a defect arrangement. Then the "size" leading to the size broadening in the measured diffraction-line profile is smaller than the grain size.

[26] Equation (6.33) to a large extent parallels Eq. (6.10). However, there occurs a significant difference. The path difference of two rays diffracted by two neighbouring slits of the grating in Fig. 6.4 equals $p\sin\varphi = d_{hkl}^{\text{ref}}\sin(2\theta)$, recognizing the similar roles of p and d_{hkl}^{ref} and of φ and 2θ. However, the path difference between two (X-) rays diffracted from two neighbouring lattice planes equals $2d_{hkl}^{\text{ref}}\sin\theta = 2p\sin(\varphi/2)$. This difference is a consequence of the grating in Fig. 6.4 being oriented not symmetrical with respect to the incoming and diffracted rays, whereas the lattice planes in the X-ray diffraction experiment are oriented symmetrical with respect to the incident and diffracted X-rays.

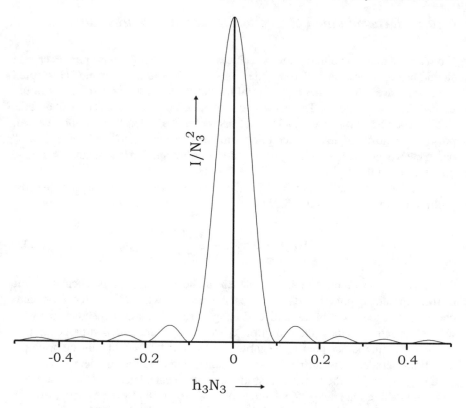

Fig. 6.28 (X-ray) diffracted intensity distribution according to Eq. (6.33) for $N_3 = 10$. Note that this intensity distribution repeats itself for the various orders of reflection (cf. Fig. 6.5)

A plot of I versus h_3 is shown in Fig. 6.28. The peak maximum is const. $(N_3)^2$ and the area under the peak equals const. N_3. Against this background, normalized coordinates have been employed in Fig. 6.28: the intensity values have been divided by $(N_3)^2$ (the ordinate); the h_3 values have been multiplied with N_3 (the abscissa). To characterize the width of the peak, the so-called *integral breadth*, β, is introduced which is given by the ratio of peak area (const. N_3) and peak maximum (const. $(N_3)^2$), and thus, the contribution to the integral breadth due to finite crystallite size, β_{size}, is proportional to $1/N_3$ and is given by:

$$\text{on } 2\theta \text{ scale}: \quad \beta_{\text{size}} = \frac{\lambda}{\left\{\left(N_3 d_{hkl}^{\text{ref}}\right)\cos\theta\right\}} = \frac{\lambda}{\{D_{hkl}\cos\theta\}} \tag{6.34a}$$

$$\text{on } h_3 \text{ scale}: \beta_{\text{size}} = \frac{1}{N_3} \tag{6.34b}$$

with $D_{hkl} = N_3 d_{hkl}^{\text{ref}}$ as the crystallite size in the direction perpendicular to the diffracting lattice planes. Hence, the integral breadth of the only size-broadened peak provides a direct measure of the crystallite size in the direction perpendicular to the diffracting lattice planes.

The intensity distribution according to Eq. (6.33) was first given by von Laue in 1912 (in the paper by Friedrich, Knipping and von Laue reporting the discovery of the diffraction of X-rays by crystals); it is often referred to as "Laue function". Already shortly thereafter Scherrer (1918) realized that the breadth of a reflection can be fruitfully used as a measure of the average finite size of the diffracting crystals: Eq. (6.34a) is usually called "Scherrer equation".

Consider the case that lattice-spacing variations occur within the diffracting crystallites, as caused by the presence of crystal imperfections as, for example, dislocations, faulting and misfit-stress fields around, e.g., precipitates. As a crude approach to the effect on diffraction-line broadening by the presence of such lattice-spacing variations Bragg's law, in the form $\sin \theta = n\lambda/2d_{hkl}$ (cf. Eq. 6.32), can be differentiated:

$$\frac{\partial \sin \theta}{\partial d_{hkl}} = (\cos \theta)\frac{\partial \theta}{\partial d_{hkl}} = -\frac{n\lambda}{2d_{hkl}^2} = -\frac{\sin \theta}{d_{hkl}}$$

and thus, in difference form:

$$\text{on } 2\theta \text{ scale: } \Delta(2\theta) = -2\left(\frac{\Delta d_{hkl}}{d_{hkl}}\right)\tan \theta \qquad (6.35a)$$

$$\text{on } h_3 \text{ scale: } \Delta(h_3) = -2\left(\frac{d_{hkl}^{ref}}{\lambda}\right)\left(\frac{\Delta d_{hkl}}{d_{hkl}}\right)\sin \theta = -h_3\left(\frac{\Delta d_{hkl}}{d_{hkl}}\right)\sin \theta \quad (6.35b)$$

Equation (6.35a) expresses that a homogeneous change of lattice spacing Δd_{hkl} leads to a shift of the peak position $\Delta(2\theta)$ on the 2θ scale, or $\Delta(h_3)$ on the h_3 scale.

Now consider the case that the specimen is constituted of crystallites, each of constant lattice-parameter value, but that the lattice-parameter values differ from crystallite to crystallite. Then, the diffraction-line broadening due to lattice-spacing variations *between* the diffracting crystallites can be expressed in terms of a contribution to the integral breadth due to microstrain, $\beta_{\text{microstrain}}$, by an equation on the basis of Eq. (6.35a):

$$\text{on } 2\theta \text{ scale: } \beta_{\text{microstrain}}(2\theta) = \langle \Delta(2\theta) \rangle = 4e_{hkl}\tan \theta \qquad (6.36a)$$

$$\text{on } h_3 \text{ scale: } \beta_{\text{microstrain}}(h_3) = \langle \Delta(h_3) \rangle = 4\left(\frac{d_{hkl}^{ref}}{\lambda}\right)e_{hkl}\sin \theta = 2h_3 e_{hkl}\sin \theta$$

$$(6.36b)$$

where

$$e_{hkl} = \left\langle \left|\frac{2\Delta d_{hkl}}{d_{hkl}}\right| \right\rangle \qquad (6.37)$$

is a measure for the lattice-spacing variation between the diffracting crystallites, considering that the lattice spacing varies over a range from $d_{hkl} - \Delta d_{hkl}$ to $d_{hkl} + \Delta d_{hkl}$.

The derivation leading to Eq. (6.35a, b) is in general inappropriate to describe the effect on the shape of a diffraction-line profile of a lattice-spacing variation *within* a "coherently diffracting" crystallite. Yet, as a gross approximation it has been proposed that Eq. (6.36a, b) can still be used. Only for the case that the microstrain distribution in the specimen is Gaussian,[27] it can be shown that a simple equation describes the relation between the microstrain parameter describing the line broadening, e_{hkl} [cf. Eqs. (6.36a, b) and (6.37)], and the root mean square of the (local) strain, ε_{hkl} (see Delhez et al. 1983)[28]:

$$e_{hkl} = \frac{1}{2}(2\pi)^{1/2}\langle \varepsilon_{hkl}^2 \rangle^{1/2} \tag{6.38}$$

To establish a relation between the "size" and "(micro)strain" parameters, as derived from diffraction-line broadening analysis, and microstructure parameters more common to characterize "the solid state", as the dislocation density of a cold-worked metal, can be difficult. An example of an often used relation between the dislocation density, ρ_d, the root mean square of the local strain ε [to be derived from e (e.g. by applying Eq. (6.38))] and the crystallite size D (cf. Eq. 6.34a)[29] reads (Williamson and Smallman 1956):

$$\rho^{1/2} = \frac{2(3)^{1/2}\langle \varepsilon^2 \rangle^{1/2}}{Db} \tag{6.39}$$

with b as the length of the Burgers vector (cf. Sect. 5.2.3). Considering Eqs. (6.34a, b) and (6.36a, b) it is seen that β_{size} and $\beta_{microstrain}$ depend differently on 2θ or h_3 (e.g. on h_3 scale β_{size} is independent of h_3, whereas $\beta_{microstrain}$ increases linearly with h_3). These different dependences of β_{size} and $\beta_{microstrain}$ on 2θ and h_3 can be used to determine "size" (i.e. D_{hkl}) and "strain" (i.e. e_{hkl}) separately from the structural line broadening measured for two orders of reflection, e.g. hkl and $2h2k2l$, recorded from the same crystals in the specimen.

Assuming that the total structural line broadening, β_{total}, can be written as a sum of the "size" and "strain": broadenings[30]:

---

[27] More precisely: the microstrain distribution must be Gaussian for all correlation distances. The correlation distance is the distance between two points in the specimen, in a direction perpendicular to the diffracting lattice planes, for which the strain is considered.

[28] The *local* strain is the strain for which the correlation distance is nil (see Footnote 27).

[29] Here it is assumed that no hkl dependence of e, ε and D occurs, so the subscripts "hkl" have been omitted.

[30] The total structural line broadening cannot be equated with the measured diffraction-line broadening because instrumental broadening occurs as well and is included in the measured line profile. Various more or less exact approaches exist to correct for the instrumental broadening. Within the

$$\beta_{\text{total}} = \beta_{\text{size}} + \beta_{\text{microstrain}} \tag{6.40}$$

it follows that, for β_{total} as measured on a 2θ scale, plotting of $\beta_{\text{total}}(\cos\theta)/\lambda$ versus $\sin\theta$ results in a straight line ["Williamson–Hall plot" (1953)]:

$$\frac{\beta_{\text{total}}(\cos\theta)}{\lambda} = \frac{1}{D_{hkl}} + \left(\frac{4e_{hkl}}{\lambda}\right)\sin\theta \tag{6.41}$$

From the slope, a value for e_{hkl} is obtained, and from the part cut from the ordinate, a value for D_{hkl} results. By repeating the analysis for various hkl, the shape of the diffracting crystallites and the anisotropy of the microstrain can be determined. In case isotropy prevails, the results of all hkl reflections can be combined in a single "Williamson–Hall plot". Although this last variant is often applied, such an approach is not generally justified.

Examples of such analysis of diffraction-line broadening are shown in Fig. 6.29.

The so-called Williamson–Hall plot (Eq. (6.41)) is often used in line-profile analysis to get a semi-quantitative description of the microstructure. The linear addition of β_{size} and $\beta_{\text{microstrain}}$ (Eq. (6.40)) is an approximation; it only holds exactly if the "size-broadened" and "strain-broadened" component line profiles are Lorentzian functions (i.e. of the type: $1/(1 + x^2)$), which is not valid in general. More advanced line-profile analyses are available, where the full shape (instead of only the integral breadth β) of the measured profiles is taken into account. In fact the dependences on 2θ or h_3 and the interpretation of the parameters "size" and "strain" can be rather complicated (Berkum et al. 1994). But the principle remains the same: the different 2θ or h_3 dependences of the size and strain broadenings are used to determine the size and strain parameters.

The above discussion, including the examples, has been devised to introduce the power of diffraction-line broadening analysis for characterizing the microstructure; thus there is no other technique that can provide quantitative information on simultaneously the crystallite size and microstrain distributions in a specimen. The advanced methods for these analyses, albeit part of commercial software packages sold by diffractometer producers, are more complex than the most simple one illustrated here [see a classical textbook (Warren 1969) and an overview paper (Mittemeijer and Welzel 2008)].

---

context of the discussion in this section, the following procedure is indicated. The instrumental line broadening is measured using a standard specimen that does not show structural line broadening. Then the integral breadth of the total structural line broadening may approximatively be obtained according to: $\beta_{\text{total}} = \beta_{\text{measured}} - \beta_{\text{instrumental}}$.

(a)
(b)

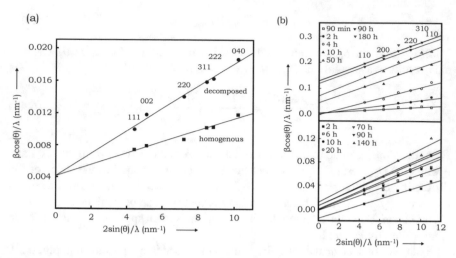

Fig. 6.29 Examples of Williamson–Hall plots. **a** Decomposition of a Pd(B) solid solution. Upon annealing and decomposition of the Pd(B) solid solution into a B-rich and a B-poor solid solution, pronounced microstrains develop additionally to the microstrains present due to dislocations in the initial state. Consequently, the slope of the fitted straight line increases upon annealing/decomposition, while the ordinate intercept (grain size) remains constant (taken from M. Beck and E.J. Mittemeijer, Zeitschrift für Metallkunde, 92 (2001), 1271–1276). **b** Ball milling of molybdenum in a attritor mill (upper diagram) and in a planetary mill (lower diagram). The grain size decreases with milling time (increasing ordinates intercepts), whereas the microstrains increase (increasing slopes). The effects are more pronounced in case of the attritor mill (taken from I. Lucks, H.P. Lamparter and E.J. Mittemeijer, Journal of Applied Crystallography, 37 (2004), 300–311)

6.10.2 Determination of (Residual) Macrostress

In thin films and in the surface regions of bulk materials residual,[31] internal macrostresses are usually present. In general, a state of stress can be either internally imposed (as indicated by the previous sentence) or externally imposed (by an acting external load). The analysis of the state of stress to which a material/workpiece is subjected, is of great technological importance, because stresses can be beneficial or detrimental with respect to, in particular, mechanical properties. For example, it is imaginable that a residual tensile stress parallel to the surface of the specimen would promote crack development at the surface, for example during fatigue where the specimen is subjected to very many cycles of compressive and tensile loading stresses. The presence of a residual compressive stress in such a situation would counteract the development of the crack (see Fig. 6.30).

The (macro)stress or (macro)strain as discussed here, which is taken as a constant over distances covering many grains in the specimen, should be distinguished from

---

[31] Residual stresses are present as internal stresses in a material body without that an external load acts on the body. Residual stresses can result after some treatment the body has been subjected to (see the introduction of Chap. 12 and, in particular, Sect. 12.18).

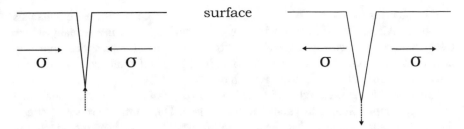

surface

Fig. 6.30 Effect of (residual, internal) stress parallel to the surface of a piece of material (workpiece) on the propagation/growth of cracks perpendicular to the surface. Whereas a compressive stress promotes closure of the crack (retards further growth), as indicated in the left part of the figure, a tensile stress promotes crack growth, as indicated in the right part of the figure

the "microstrain" dealt with in Sect. 6.10.1, where the "microstrain" parameter, as deduced from the diffraction-line broadening, is a measure for the *local* (i.e. within the grains) *variation of the strain* due to strain fields associated with dislocations and/or other mistakes, misfitting precipitates, etc. (see Fig. 6.31).

In view of their importance, materials engineers have always been keen in developing and applying methods for measuring such internal, residual macrostresses. The stress is a not measurable quantity, but it exhibits itself by the stress-invoked strain. Then, considering Eq. (6.35a, b), it immediately becomes clear that the X-ray diffraction technique allows a straightforward method for determining the (macro)strain in a specimen from the homogeneous change of lattice spacing Δd_{hkl} which can be determined by measuring the corresponding shift of the peak position $\Delta(2\theta)$ on the 2θ scale (or $\Delta(h_3)$ on the h_3 scale).

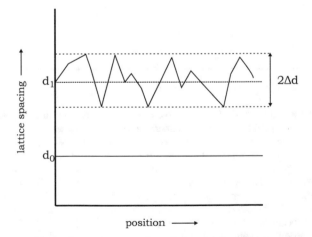

Fig. 6.31 Distinction of macro- and microstrain. Schematic variation of lattice spacing, d, as function of a position (distance) parameter within the stressed specimen. While macrostrain changes the overall lattice spacing from d_0 to d_1 over ranges covering many grains/the whole specimen, microstrains cause local (on an atomic scale) variations in lattice spacing

On this basis the method of stress determination by (X-ray) diffraction analysis was born. Already in 1927, a textbook by Glocker appeared in Germany where (also) the essential elements of the X-ray diffraction method for stress determination have been presented (The paper by Friedrich, Knipping and von Laue, reporting the discovery of diffraction by X-rays, is from the year 1912).

The basic idea is sketched in Fig. 6.32. A polycrystalline specimen is subjected to a, say, compressive stress parallel to the surface. Due to the presence of stress, the lattice spacing of the *hkl* lattice planes in a crystallite depends on the orientation of the crystallite with respect to the surface of the specimen. By X-ray diffraction experiments, and by varying the orientation of the specimen with respect to the (fixed,

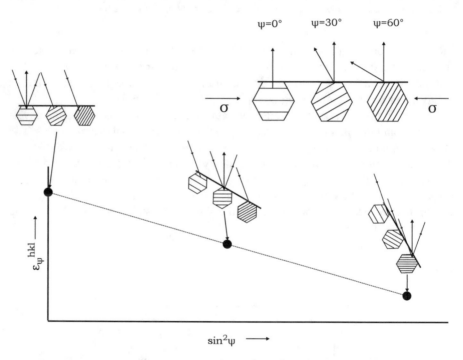

Fig. 6.32 Schematic illustration of the principle of the $\sin^2 \psi$ method. In the upper right corner, three grains, in a polycrystalline, massive specimen, with different orientations ψ with respect to the surface of the specimen/workpiece have been indicated and the different effects of a compressive stress, parallel to the surface, on the lattice spacing d_{hkl} of a *hkl* set of lattice planes (of different orientation with respect to the surface of the specimen in each grain) has been shown. In case of a compressive stress parallel to the surface, as shown, d_{hkl} will be larger than the unstressed value for the grain with the *hkl* planes parallel to the surface; d_{hkl} will be smaller than the unstressed value for a grain with the *hkl* planes strongly inclined to the surface. By measuring the *hkl* reflection upon varying the specimen-tilt angle ψ, different grains, with sets of *hkl* planes differently oriented with respect to the surface, come into diffraction and the value of d_{hkl} derived from the *hkl* peak position will depend on ψ (see the graph at the bottom of the figure). Elasticity theory shows that, for elastically isotropic specimens, the strain ε_ψ^{hkl}, as derived from d_{hkl} and the reference value for d_{hkl}, depends linearly on $\sin^2 \psi$. (For a more complete discussion, see text)

in the laboratory frame of reference) directions of the incident and diffracted X-rays (cf. Fig. 6.27), the direction dependence of the (macro)strain can be determined, where the (macro)strain in a certain direction,[32] ε, is derived from the lattice spacing of the hkl reflection, d, as measured in that direction, from the peak position according to:

$$\varepsilon = \frac{d - d_0}{d_0} \qquad (6.42)$$

The analysis of the direction dependence of ε provides the means to determine the value of the stress parallel to the surface of the specimen (see what follows).

To relate a measured strain to the stress, it is now necessary to first present the basic relations between the stress and strain components (see Chap. 12). Suppose that in the surface region of the specimen, two equal principal stress components parallel to the surface occur, $\sigma_{//}$. Perpendicular to the surface, the stress must be equal to zero (requirement of mechanical equilibrium). Then for intrinsically elastically isotropic material, it follows (cf. Eq. 12.20 in Sect. 12.4):

$$\sigma_{//} = \left(\frac{E}{(1-\upsilon)}\right)\varepsilon_{//} \qquad (6.43)$$

In this case, the following relation holds:

$$\varepsilon_\psi^{hkl} = \left(\frac{-2\upsilon}{E}\right)\sigma_{//} + \left\{\frac{1+\upsilon}{E}\right\}\sigma_{//}\sin^2\psi \qquad (6.44)$$

where ψ denotes the angle of specimen tilt (See Footnote 32). For the case considered, there is no dependence for the measured lattice spacing, and thus lattice strain ε^{hkl}, on the rotation of the specimen as characterized by the angle φ (See Footnote 32).

Hence, a plot of ε_ψ^{hkl} versus $\sin^2\psi$ results in a straight line (see Fig. 6.32), the slope of which provides a value of $\sigma_{//}$, provided the value of the elastic constant $(1+\upsilon)/E$ is known. The type of equation represented by Eq. (6.44) has led to the denomination "$\sin^2\psi$" method for stress determination.

The situation expressed by Eq. (6.44), i.e. the case of a biaxial, rotationally symmetric state of stress, is, by far, the one investigated most often. Evidently, whereas the value of the lattice strain ε_ψ^{hkl} is seriously affected by an error in the value used for d_0 (Eq. 6.42: small difference of d_ψ and d_0), this holds to a very much lesser extent for the slope of the straight line in Fig. 6.32.

Not only materials engineers have a strong and obvious interest in stress analysis (see also Chap. 12). Adopting a purely scientific, fundamental point of view, it is

---

[32] The orientation of the measured strain (lattice spacing), in a certain coordinate system called here the "specimen frame of reference", is fully described by two angles: φ representing the rotation angle of the specimen about the specimen surface normal and ψ representing the tilt (inclination) angle of the specimen surface normal.

obvious that diffraction-stress analysis can teach us a lot about pressing problems in materials science, as illustrated by the examples discussed in the three "intermezzi" below.

Intermezzo: Grain Interaction

If a massive, polycrystalline specimen, constituted of crystals (grains) which are intrinsically elastically anisotropic,[33] is subjected to a macrostress, each grain (crystal) is not free to deform as if it were alone/"standing free": the grain is constrained by its surroundings. As a consequence of the resulting "grain interaction" a distribution of stresses and strains occurs over the crystallographically differently oriented crystallites composing the specimen.

About a century ago already two approaches have been proposed to describe extremes of grain interaction: it is assumed that all crystals of the specimen exhibit either identical strains [the Voigt model (1910)] or identical stresses [the Reuss model (1929)] in the specimen frame of reference. In view of the intrinsic elastic anisotropy of the individual grains, both models imply that incompatible elastic behaviour would occur at both sides of the individual grain boundaries (discontinuity in stress Voigt model) or strain (Reuss model), and thus reality may be anticipated to be somewhere in between both extremes types of grain interaction. More or less cumbersome models have been developed to assess such "intermediate" types of grain interaction.

Now, the diffraction analysis of stress allows a sensitive testing of grain interaction: each diffraction line contains information on the elastic strain of crystallites only for such crystallites in the specimen which have their $\{hkl\}$ planes oriented perpendicular to the measurement direction,[34] i.e. only the elastic strain of this subgroup of crystallites composing the polycrystalline specimen is measured. Therefore, the mechanical strain taken in the same direction in the specimen, representing an average over *all* crystallites in the sample, is not equal to the strain measured by (X-ray) diffraction, which represents an average of only a subgroup of the crystallites composing the specimen. For this reason, the diffraction-strain measurement is much more sensitive to grain interaction than the mechanical (macroscopic) strain measurement.

---

[33] Intrinsic elastic anisotropy means that if a constant uniaxial (state of) stress is applied to a single crystal of the material considered, then the resulting strain in the direction of the stress depends on the orientation of the crystal with respect to the direction of the stress. Elastic isotropy involves that the strain in the direction of the stress is the same independent of the orientation of the crystal, for the same value of applied stress.

[34] The diffraction experiment as described above relies on the determination of the lattice spacing of the diffracting lattice planes and thus the "measurement direction" is perpendicular to the diffracting lattice planes (cf. the above discussion of the $\sin^2 \psi$ method and Fig. 6.32).

In accordance with the above discussion, the effect of grain interaction can express itself in the diffraction-strain measurement through hkl-dependence of, e.g., the slope and the part cut from the ordinate by the straight line in the $\sin^2\psi$ plot (Eq. 6.44) and also curvature in the $\sin^2\psi$ plot may occur. The analysis or need of taking into account of grain interaction is a topic of debate until today [Noyan and Cohen (1987), and Hauk (1997)]. In particular, recent diffraction-strain measurements performed with unprecedented accuracy have provided direct experimental evidence for mechanisms of grain interaction in real specimens more complicated than those described by the simple models discussed above (Welzel et al. 2005); see also the next *intermezzo*.

Intermezzo: Surface Anisotropy and Thin Films

The classical approaches to grain interaction involve that the grain interaction is conceived to be isotropic (i.e. independent of the direction in the specimen frame of reference), which, for example, holds for the classical Voigt and Reuss models (see *intermezzo* above). Even if the individual crystallites of a polycrystal are intrinsically elastically anisotropic, the body as a whole can still exhibit macroscopic elastically isotropic behaviour, provided a random distribution of the orientation of the crystallites prevails, which case is usually called quasi-isotropic behaviour (cf. the introduction of Chapter 4). However, it can intuitively be understood that the grain interaction becomes anisotropic (i.e. dependent on the direction in the specimen frame of reference) if crystallographic texture (see Sect. 4.7) occurs.

Now consider the nearby presence of a surface. For the crystallites adjacent to the surface of a polycrystal, not all directions are equivalent and the (mechanical and diffraction) elastic properties of these crystallites can be at most transversely (i.e. parallel to the plane of the surface) isotropic, reflecting a rotational symmetry of the specimen with respect to the surface normal. This "surface anisotropy" is a source of direction-dependent grain interaction and thus macroscopic (mechanical) elastically anisotropic behaviour: the grain interaction perpendicular to the surface can be different from the grain interaction parallel to the surface, recognizing that straining perpendicular to the surface can be relatively unconstrained at the surface as compared to the bulk.

Especially for the analysis of the state of stress in thin films, the occurrence of direction-dependent grain interaction may be expected to have severe effects. Although the possibility of surface anisotropy has been discussed already by Stickforth in 1966, the effect remained unproven until recently. High-precision diffraction-stress analysis of thin films has now demonstrated unambiguously that the effect occurs: see Fig. 6.33. A $\sin^2\psi$ plot (see Fig. 6.32 and its discussion) is shown in Fig. 6.33 for an (ultra-)thin (50 nm thickness) Pd film. The lattice strain is measured for various angles of specimen tilt ψ using the 200 reflection. It can be shown that even if crystallographic texture occurs, that then the $\sin^2 \psi$ plot for the lattice-strain data obtained for the 200 reflection should lie on a straight line according to the classical isotropic grain-interaction models. The pronounced curvature seen in the plot of Fig. 6.33 is an unambiguous demonstration of the occurrence of direction-dependent grain interaction in thin films (and surface regions of bulk materials). The full, curved line drawn in the figure is the result of a model of anisotropic grain interaction applied to the data shown. It could even be shown experimentally that the extent of the anisotropy of the grain interaction depends on the distance from the free surface: the grain interaction becomes of more isotropic nature at an increasingly larger distance from the surface (Kumar et al. 2006). Thus, diffraction-stress analysis provides deep insight into the mechanical behaviour of solid bodies.

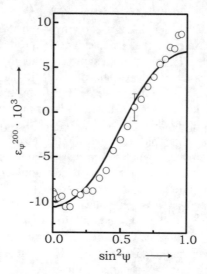

Fig. 6.33 Effect of surface anisotropy. Residual stress in a 50 nm thick Pd film. The strain ε_ψ^{200}, as calculated from the lattice-spacing values derived from the 200 reflection recorded for varying tilt angle ψ, is shown as a function of $\sin^2 \psi$ (cf. Fig. 6.32 and its discussion). In the absence of surface anisotropy, the ε_ψ^{200} values should fall on a straight line in this plot. The occurrence of pronounced curvature demonstrates the existence of surface anisotropy as a genuine effect. The full line drawn presents the predicted result as the consequence of the occurrence of surface anisotropy (taken from U. Welzel, A Kumar and E.J. Mittemeijer, Applied Physics Letters, 95 (2009), 111907)

Intermezzo: Colossal Stress and Texture Gradients
in (Even Ultra-)Thin Films

As if the above difficulty does not suffice to complicate (X-ray) diffraction-stress analysis (in thin films), the finding that, on top, pronounced microstructural depth gradients can occur in especially thin films, poses a further, serious problem for the experimentalist. Whereas in case of relatively thick films, it may often justifiably be assumed that over the depth of penetration of the X-rays a stress-depth gradient is marginal, this may not be an acceptable approach for thin films (see also Kumar et al. 2006).

In this sense, the situation may be dramatic if ultra-thin films are studied. It was shown for a tungsten film of only 50 nm thickness, produced by magnetron sputtering in ultra-high vacuum, that an extremely steep stress-depth profile prevails in the specimen: the (compressive) stress changes from -2800 MPa $(=-2.8$ GPa$)^{35}$ near the surface to about -1000 MPa $(=-1$ GPa$)$ at a depth of only 25 nm (see Fig. 6.34). Also the degree of preferred orientation changes very strongly with increasing depth beneath the surface: the $\{110\}$ fibre texture is sharp and strong at the surface and is absent near the interface with the substrate, i.e. over a depth of 50 nm the tendency for preferred orientation is completely lost. A straightforward diffraction analysis of the state of stress in the film, with such colossal stress- and texture-depth gradients, is no longer possible in the way as described in Sect. 6.10.2: The measured data of lattice strain have to be compared with the outcome of simulations.

In the very many papers published in the literature presenting diffraction analyses of stress in thin films, on a more or less routine basis, as one of the tools utilized to characterize the microstructure, the possible effect of complications as discussed in this and the preceding *Intermezzo* is usually not considered until now. Such superficiality is regrettable, as consequently, at least in some cases, meaningless results are presented. Notwithstanding this observation, in the spirit of the concluding remark of the preceding *Intermezzo* it is concluded that very detailed insight on the microstructure of nanomaterials, as (ultra-)thin films, can be extracted from diffraction-stress analyses.

---

[35] The compressive nature of a stress component is, by convention, expressed by deliberately positioning a minus sign before a corresponding stress value (cf. Sect. 12.2).

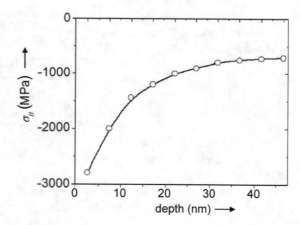

Fig. 6.34 Stress, parallel to the surface, $\sigma_{//}$, in an ultra-thin (50 nm thickness) tungsten film, in a state of biaxial, planar, isotropic stress, as a function of depth beneath the surface of the film. A very steep stress-depth profile happens in the first 25 nm beneath the surface (a monotonous change of stress of about 2 GPa (=2000 MPa) over 25 nm) (taken from Y. Kuru, U. Welzel and E.J. Mittemeijer, Applied Physics Letters, 105 (2014), 221902)

References

General

M. de Graef, *Introduction to Convential Transmission Electron Microscopy* (Cambridge University Press, Cambridge, 2003)

D.B. Williams, C.B. Carter, *Transmission Electron Microscopy*, vols. I–IV (Plenum Press, New York, 1996)

S. Amelinckx, D. Van Dyck, J. Van Landuyt, G. Van Tendeloo, *Handbook of Microscopy*, vol. 1–3 (VCH Verlagsgesellschaft, Weinheim, 1997)

E.J. Mittemeijer, P. Scardi (eds.), *Diffraction Analysis of the Microstructure of Materials*. Springer Series in Materials Science, vol. 68 (Springer, Berlin, 2004)

E.J. Mittemeijer, U. Welzel (eds.), *Modern Diffraction Methods* (Wiley-VCH, Weinheim, 2013)

B.E. Warren, *X-ray Diffraction* (Addison-Wesley, Reading, Massachusetts, 1969)

Specific

J.G.M. van Berkum, R. Delhez, ThH de Keijser, E.J. Mittemeijer, Diffraction-line broadening due to strain fields in materials. Acta Crystallographica, A **52**, 730–747 (1996)

R. Delhez, ThH de Keijser, E.J. Mittemeijer, Determination of crystallite size and lattice distortions through X-ray diffraction line profile analysis. Recipes, methods and comments, Fresenius Zeitschrift für Analytische Chemie **312**, 1–16 (1982)

P. Drent, Properties and selection of objective lenses for light microscopical applications. Microsc. Anal. **19**, 5–7 (2005)

V. Hauk, *Structural and Residual Stress Analysis by Nondestructive Methods; Evaluation-Application-Assessment* (Elsevier, Amsterdam, 1997)

E. Horn, R. Zantl, Phase-contrast light microscopy of living cells cultured in small volumes. Microsc. Anal. **20**, 15–17 (2006)

A. Kumar, U. Welzel, E.J. Mittemeijer, Depth dependence of elastic grain interaction and mechanical stress: analysis by X-ray diffraction measurements at fixed penetration/information depths. J. Appl. Phys. **100**, (2006)

E.J. Mittemeijer, U. Welzel, The "state of the art" of the diffraction analysis of crystallite size and lattice strain. Z. Kristallogr. **223**, 552–560 (2008)

I.C. Noyan, J.B. Cohen, *Residual Stress; Measurement by Diffraction and Interpretation* (Springer, New York, 1978)

A. Sato, N. Mori, M. Takakura, S. Notoya, Examination of Analytical Conditions for Trace Elements Based on the Detection Limit of EPMA (WDS). JEOL News **42E**, 46–52 (2007)

K.W. Urban, The New Paradigm of Transmission Electron Microscopy. MRS Bull. **32**, 946–952 (2007)

G. Vicidomini, P. Bianchini, A. Diaspro, STED super-resolved microscopy. Nat. Methods **15**, 173–182 (2018)

U. Welzel, J. Ligot, P. Lamparter, A.C. Vermeulen, E.J. Mittemeijer, Stress Analysis of Polycrystalline Thin Films and Surface Regions by X-ray Diffraction (Review). J. Appl. Crystallogr. **38**, 1–29 (2005)

G.K. Williamson, R.E. Smallman, III. Dislocation densities in some annealed and cold-worked metals from measurements on X-ray Debye-Scherrer spectrum. Phil. Mag. **1**, 34–46 (1956)

Chapter 7
Phase Equilibria

The appearance of a system can be homogeneous or heterogeneous. Even in equilibrium situations, involving that no further (net) changes in the system occur and are possible, provided the boundary conditions remain constant, heterogeneity can prevail: for example, in an Al-Si alloy at room temperature (and at 1 atm pressure), in equilibrium an Al-rich part of the system (f.c.c. crystals with very little Si dissolved) and a Si-rich part of the system (crystals of diamond-type structure with very little Al dissolved) can be distinguished. These, generally dispersed, parts of the system, which are in equilibrium with each other, will be called phases. Obviously there is a great scientific, fundamental interest and, even greater, practical/technological interest, to know and understand these "heterogeneous, phase equilibria". This has led to:

(i)   the publication of huge data files providing compilations of phase diagrams describing, on a largely phenomenological (i.e. experimental) basis, these phase equilibria (e.g. for binary systems, Massalski et al. (1996) and, for ternary systems, Petzow and Effenberg (from 1988 onwards);
(ii)   great theoretical development in a field called "materials thermodynamics" in order to arrive at fundamental understanding of these phase equilibria;
(iii)   models and algorithms to predict phase diagrams for cases where they have or cannot be measured [e.g. the CALPHAD (= calculation of phase diagrams) approach and corresponding software package; see Saunders and Miodownik (1998) and Lukas et al. (2007)].

Among scientists the interest for producing highly accurate data for material-property databases, as those pertaining to phase diagrams, has waned considerably in recent years. This in part has certainly to do with the lack of scientific status for that type of work. This development is regrettable and worrying. Much scientific research and, especially, very many practical, engineering applications of materials depend strongly on the availability of reliable phase-diagram data. We know, on the basis of own bitter experience, as materials scientists, how unpleasant it is in the course of a research project to be confronted with observed effects, i.e. phenomena, which are

© Springer Nature Switzerland AG 2021   357
E. J. Mittemeijer, *Fundamentals of Materials Science*,
https://doi.org/10.1007/978-3-030-60056-3_7

simply a consequence of the nature of the phase diagram, that was, as published and consulted, evidently incomplete or incorrect. Then, not only the work performed is ill-fated, one subsequently has to carry out oneself this phase-diagram-determination, experimental work, which involves an enormous amount of extra research at an unplanned moment of time.

7.1   The Notion Phase

The conventional definition of a phase is: *a phase is a macroscopically, homogeneous body exhibiting uniform physical and chemical properties*. Often systems consist of more than one phase and then are said to be heterogeneous, whereas the constituting phases are homogeneous themselves. An example of a heterogeneous system is a polycrystalline specimen where the grains (crystals) pertain to one of two crystal systems and where these two types of grains (crystals) have possibly different but in themselves homogeneous compositions. The system is said to be constituted of two phases. This holds, for example, for the two-component specimen composed of the Al-rich f.c.c. phase and the Si-rich diamond-like phase in the Al-Si specimen at room temperature (and at 1 atm pressure) discussed above, or for a one-component Fe specimen at 1184 K (and at 1 atm pressure) where the ferrite (α-Fe, b.c.c.) phase and the austenite (γ-Fe, f.c.c.) phase coexist.

Note that the parts of one phase considered here can be different in shape and dispersed through the whole system; it remains one, single phase.

For the definition of a phase, one should not look at the atomic scale: a system composed of two components, like a solid solution of elements A and B, would be heterogeneous if considered on the atomic scale.

The word phase is also used in non-equilibrium situations, where the "phase" considered may, for example, exhibit a concentration profile and may show non-uniform physical and chemical properties. It appears that the notion phase thus is restricted actually to states of equilibrium, and designation of system parts in such non-equilibrium situations by the notion "phase" is strictly not correct. However, in practice, and also in this book, one is not puristic down to this level.

The three forms of aggregation, solid, liquid and gas, should not be identified as three phases: although all gases are completely miscible, and thus indeed there is only one gas phase possible if gas components are in contact, liquids composed of two components can decompose, leading to the presence of a number of liquid phases in a system, and already even for a one-component solid system the occurrence of various (crystal) modifications can lead to as many phases (but, in case of simultaneous presence in a state of equilibrium, subject to the phase rule; see Sects. 7.4 and 7.5.1): e.g. the α-Fe (ferrite, b.c.c.), the γ-Fe (austenite, f.c.c.) and the δ-Fe (b.c.c.) modifications/phases of solid iron.

7.2 The Notion Component

At first sight the definition of the number of components appears simple: to constitute the phases of the A-B alloy the two atomic elements A and B can be designated as the components of the alloy which suffice to form all possible phases. However, this definition is not economical: water as vapour (one phase), liquid (one phase) and solid (various phases, since a number of modifications of ice exist) is composed of the elements H and O. However, the different phases of water can all be constituted of one component: H_2O. Similarly, at the melting point minimum of a binary A-B alloy (cf. Sect. 7.5.2), both the liquid phase and the solid phase have the same composition and hence the A-B alloy in this situation is a one-component system. The definition of the number of components for the cases considered here thus is: the *minimum* number of different chemical species, i.e. the number of *mathematically independent* components to build up all phases in the system. Fixed ratios for certain atomic elements for all phases in a system (as holds for the two examples given above) reduce the number of components as compared to the total number of atomic elements (see further Sect. 7.4).

7.3 The Notions Equilibrium and Stationary State; Internal Energy, Entropy, (Helmholtz) Free Energy and Gibbs Energy

The isolated part of the world that comprises the interacting phases is called the *system*. The state of the system is described by *state variables*, as energy, pressure, mass, temperature, etc. One distinguishes between *extensive state variables*, which depend on the size of the system, as energy, mass and volume, and *intensive state variables*, which are independent of the size of the system, as pressure, temperature and composition variables. In a discussion of phase equilibria, it is usual taken for granted that the amount of the phases present does not play a role (which does not strictly hold if, for example, the relative amount of interface (grain-boundary) and surface area depends on system size). Then, to specify a state of a system it suffices to focus on intensive state variables only.

If, for the observer, and under certain controlled and constant conditions, no net changes in the amounts and distribution of the phases present and their structure (including defects as discussed in Chap. 5) and chemical composition occur, the system is often said to be in equilibrium. However, such an observation of permanence is not sufficient to define equilibrium. In the thermodynamic sense equilibrium occurs if, under the specified constant conditions, no further minimization of the energy content of the system is possible.

At this place some effort is needed to explain what types of energy one may consider in order to define a state of equilibrium.

The *internal energy* of a system, indicated by the symbol U, can be conceived as the sum of the potential and kinetic energies of the particles the system is composed of. A potential energy depends on the position of the particle (atom, molecule) in the field of force which acts on it (e.g. the electric field associated with the Coulomb interaction in an ionic crystal; cf. Chap. 3). A kinetic energy results from the motion of the particle (in a crystal lattice the atoms vibrate around their equilibrium positions; cf. begin of Chap. 5). Equilibrium cannot be defined by the internal energy alone. For example, the internal energy of a system, initially a solid, increases upon melting or vaporization; so minimization (see previous paragraph) of internal energy alone cannot explain a state of equilibrium.

Considering the various ways A and B atoms can be distributed over the sites of a fixed crystal structure, it becomes immediately apparent that the occurrence of a disordered distribution is much more likely than the occurrence of an ordered distribution (cf. Chap. 4; there are very many more disordered distributions possible than ordered distributions). Hence, given a certain mobility of the atoms on the sites of the crystal structure, even if a given ordered distribution would occur, it would be transferred, by random movements of the atoms over the sites of the crystal structure, into a disordered state and it would be very unlikely that the ordered state would reoccur. Hence, on the basis of this statistical argument only, a disordered state would represent the equilibrium situation. Thus, according to this reasoning, melting and vaporization of a solid are favoured because the disorder increases, very drastically as compared to ordering on a crystal structure, as a result of these processes. The degree of disorder of a certain state could be expressed by the number of corresponding distributions. This degree of disorder, that in equilibrium situations is as large as possible, is expressed by a quantity called *entropy*, indicated by the symbol S.

Note that, for a given crystalline phase, the entropy of mixing of the various types of atoms and/or vacancies on the crystal structure (this is the example discussed above; the corresponding entropy is called *configurational entropy*) does not comprise the total entropy: also the atomic vibrations, at a certain temperature, contribute to the entropy (the corresponding entropy is called *vibrational entropy*); also see discussion of Eq. (5.1) in Sect. 5.1.1.

A certain distribution (of atoms A and B on the crystal structure in the example considered here) corresponds with a certain value of the internal energy. Distributions with a low value of the internal energy are preferred by nature. This holds for many ordered distributions. Hence, this explains the following antagonistic effect: ordering is preferred because of a low value of internal energy (this occurs if unlike (here A and B) atoms attract each other), but disorder is preferred because of its high probability (i.e. there are very many more disordered distributions than ordered distributions). The probability of a state, p_{state}, can thus be taken as proportional to the product:

$$p_{\text{state}} \sim w \cdot \exp(-U/kT) \tag{7.1}$$

where w denotes the number of distributions of internal energy U and $\exp(-U/kT)$ represents the probability that a state of internal energy U occurs at temperature

T (cf. Eq. 5.1) in Sect. 5.1.1; evidently, a state of relatively low internal energy is more likely to occur at relatively low temperature). The constant $k = R/N_{Av}$, where N_{Av} represents Avogadro's number, is called Boltzmann's constant. *Equilibrium* is defined by the *highest* possible value of p. Thermodynamics has expressed this as corresponding to a *minimum* value of the *free energy*, also called Helmholtz energy, F, defined as:

$$F = U - TS = -RT \ \ln(p_{state}) \tag{7.2}$$

with, as follows from Eq. (7.2), $S = k \ln(w)$, which is called the entropy. Hence, a (thermally and materially isolated) system becomes more stable for a smaller value of internal energy, U, and/or a higher value of the entropy, S. Equation (7.2) makes clear the role of the temperature: at low temperatures, the internal energy U may dominate F and thus ordered distributions may occur (see above discussion), whereas at elevated temperatures the term TS may become dominant and thus disordered distributions may be preferred. Thus, as one obvious consequence of this statement: for a certain material, the solid, crystalline state is generally stable at a relatively low temperature, whereas the liquid state is generally stable at a relatively high temperature.

In fact, the minimum of F describes equilibria if the temperature and the volume of the system do not change. If a change of volume, dV, occurs upon a transformation associated with a decrease of F at constant temperature, then the system can perform an amount of work against the prevailing pressure, p, which is given by pdV and at most is equal to the occurring decrease of F. Then, at equilibrium at constant temperature and pressure, the so-called *Gibbs energy* (often designated as "*free enthalpy*" in Germanic languages) has a minimum value, with:

$$G = F + pV = U + pV - TS = H - TS \tag{7.3}$$

where $H = U + pV$, which is called *enthalpy*. A change of G of a system, $\Delta G \equiv G_{end} - G_{begin}$, can at constant temperature and pressure be written as:

$$\Delta G = \Delta H - T\Delta S \tag{7.4}$$

where, if, with respect to ΔU, $p\Delta V$ is small because ΔV is small (e.g. for very many solid–solid and solid–liquid transformations), ΔH is predominantly given by ΔU.

Normally, phase equilibria are considered at constant temperature and pressure and thus we are concerned with the Gibbs energy as the energy parameter that by its minima prescribes the possible equilibrium situations. Moreover, then the deviation of G from its minimum can be considered as a "driving force" acting on the system to transform it into equilibrium and thereby the deviation of G from its minimum can be incorporated in kinetic theories for describing the rate of phase transformations (see Sect. 9.1 and Chap. 10).

Now, consider an energy landscape for the system considered as sketched in Fig. 7.1, which shows the energy as function of some state variable. The state of lowest energy occurs at position I. If at position I the system experiences a small

Fig. 7.1 Energy landscape for a system: the energy as function of some state variable. Stable (I), unstable (II) and metastable (III) states have been indicated

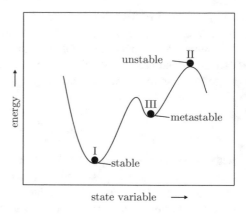

variation in energy, due to a variation in one or more variables defining its state, invariably a minor increase in energy of the system occurs and as a result the system is driven back to its equilibrium state, recognizing that nature strives for a state of minimal energy. This is immediately understood if the system shown in Fig. 7.1 is conceived as a ball in a landscape of hills and valleys: at I the ball is at the bottom of the deepest valley; a small variation in its position (by a push) moves the ball upwards on a slope; after the push the ball will roll down (again) to its bottom position of lowest energy. Therefore, at position I the system is said to be in equilibrium. A position on a slope of the energy landscape will lead to unforced movement of the ball to a position of lower energy. Now consider the system at the top (crest of the hill) position II. If no action is exerted on the ball, it will stay at rest. Any push will cause the system to move to either position I or position III. Evidently position II does not correspond to a state of (stable) equilibrium. Although at position III equilibrium occurs in the above sense (for moderate pushes at III the ball will roll back to III), in an absolute sense the equilibrium at III (corresponding to a side minimum for the energy of the system) is only of *metastable* nature: if given the chance (a strong push at III, so that the ball gains enough kinetic energy to overcome the potential energy barrier), the system will transfer to position I, as there its energy is at (absolute) minimum.

The above discussion may suggest the following definition of equilibrium:

$$\frac{d(\text{energy})}{d(\text{state variable})} = 0 \qquad (7.5)$$

However, this requirement is fulfilled not only in positions I and III, but also in position II. Hence, although often applied, this characteristic of equilibrium does not provide a sufficient definition of (stable or metastable) equilibrium. Gibbs, who laid the foundations for the thermodynamics of phase equilibria (around 1875), expressed the variational principle of the equilibrium condition by stating that

$$\text{Var (energy)} = 0 \qquad (7.6)$$

and applied the mathematics of variational calculus (beyond the scope of this book) to derive the equilibrium conditions. The above discussion on the ball at the bottom of the valley, that got a small push and was "naturally" driven back to its position of minimal energy, can be considered as a precursor of the formalism expressed by Eq. (7.6).

In general the type of equilibrium dealt with in this book is of *dynamic* nature, which is usually observed as the outcome of competing thermally activated processes: for example, per unit of time the amount of atoms of a certain type passing through an interface in one direction is equal to the amount of atoms of the same type passing through the same interface in the reverse direction. As a net result no change is apparent.

Now consider again the phase equilibrium for an Al-Si alloy at room temperature and at 1 atm pressure: in equilibrium an Al-rich phase (f.c.c. crystals with very little Si dissolved) and a Si-rich phase (crystals of diamond-type structure with very little Al dissolved) can be distinguished (cf. Sect. 7.1). The Gibbs energy for a phase i of the two-component system A–B, G^i_{A-B}, can always (i.e. also in a not equilibrium situation) be formally written as:

$$G^i_{A-B} = c^i_A G^i_A + c^i_B G^i_B \qquad (7.7)$$

where G^i_A and G^i_B are called the *partial Gibbs energies* of A and B in phase i, c^i_A and c^i_B are the fractions A and B in phase i ($c^i_A + c^i_B = 1$) and where Al and Si in the example considered take the roles of A and B. It is noted that G^i_A and G^i_B generally depend on the composition of phase i. The requirement of minimal Gibbs energy for the system in dynamic equilibrium immediately makes clear that the (partial) Gibbs energies for Al must be equal in both phases, else a net transport of Al would occur to that phase where G^i_{Al} would be smaller. A similar statement can be made for component Si. Hence, and this conclusion is needed in Sect. 7.5, *for a phase equilibrium it holds that the partial Gibbs energies of each component are equal for all phases.*

A term often used for partial Gibbs energy is *chemical potential* which is denoted by the symbol μ. Then equilibrium in a system composed of two phases α and β constituted from two components A and B requires that two conditions are satisfied:

$$\mu^\alpha_A = \mu^\beta_A \quad \text{and} \quad \mu^\alpha_B = \mu^\beta_B$$

In the experimental investigation of phase equilibria one of the greatest problems is the occurrence of stationary states: no net changes in the system occur, although the underlying dynamics of the system can be pronounced, and yet, an (dynamic) equilibrium situation does not occur. An example is sketched in Fig. 7.2.

At the left and right sides of the solid foil a gas is present that contains a component that can in principle dissolve in the foil. Suppose that the concentrations in the gas of this dissolvable component are kept at constant values c_1 and c_2 at the left and right sides of the foil, respectively (this holds, for example, for endlessly large reservoirs of gas at both sides of the foil). As a result a stream of this dissolved gas

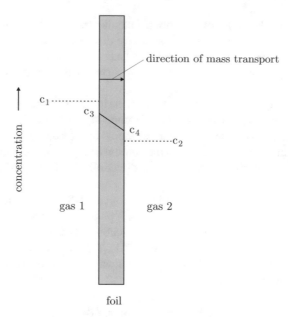

Fig. 7.2 A stationary state: mass transport through a thin foil for a component that can be dissolved in the foil and that is present with different constant concentrations in gases 1 and 2 at the left and right sides of the foil, respectively. The rates of absorption and desorption are of similar order of magnitude as the rate of diffusion through the foil, leading to a stationary state. Local equilibrium at both foil surfaces could have been approached very closely if the rates of dissolution and desorption are very much larger than the diffusion rate in the foil

component occurs through the foil, from left to right if c_1 is larger than c_2. At the left interface between gas and solid a competition occurs between the rate of dissolution and the rate of inward diffusion that results in a stationary value, say c_3, for the concentration of dissolved gas component in the foil at the left interface. Similarly, at the right interface between solid and gas a competition occurs between the rate of outward diffusion and the rate of desorption that results in a stationary value, say c_4, for the concentration of dissolved gas component in the foil at the right interface. Considering this situation, it is clear that the whole system is not in equilibrium, although for the observer no (net) changes occur as a function of time: equilibrium would require that the concentration of the dissolved gas component in the foil is constant throughout. Even more important: at the interfaces between foil and gas even no local (near) equilibrium occurs: *local equilibrium* would have required that c_3 and c_4 equal the values of dissolved gas component in the foil corresponding to chemical equilibrium with the gases of composition c_1 and c_2, respectively. Local equilibrium could have been approached very closely if the rate of dissolution and the rate of desorption are very much larger than the diffusion rate in the solid. The real occurrence of mass flux through the foil immediately makes clear that true local equilibrium can never occur exactly, because no net transport of material is possible in a genuine equilibrium situation.

The occurrence of stationary states at interfaces between solid phases (as in solid (A)-solid (B) diffusion couples, where product phases develop between the original A and B parent phases; cf. Chap. 8) and at surfaces (i.e. at interfaces between gas and solid phases) has often been interpreted erroneously as the happening of local (near) equilibria at these interfaces/surfaces. Thus, many phase boundaries, as determined on the basis of measurements of these interface adjacent compositions occurring in such experiments and as accordingly published in (supposedly equilibrium) phase diagrams, are simply wrong. The Fe-N phase diagram as published in the compilation provided by Massalski et al. (1996) provides an example of such affected phase-boundary data (see Mittemeijer and Somers 1997), and this is no exception.

7.4   Degrees of Freedom; The Phase Rule

To define the state of a system (a number of) the values of its state variables must be known. Obviously, (at least one of) the extensive state variables must be known in order that the size of the system is specified. However, phase equilibria can usually be defined using intensive state variables only (see first paragraph of Sect. 7.3).

Apart from the composition variables, usually pressure and temperature are chosen as the state variables of interest. However, this is no restriction: it is imaginable that electric, gravitational field strengths, etc., have to be considered as well and then the following derivation has to be modified accordingly. Here we confine ourselves to pressure, temperature and composition variables. Not all of these variables have to be specified in order to describe a phase equilibrium. The number of intensive state variables which can be varied independently for the phase equilibrium considered is called the number of degrees of freedom.

Suppose r phases interact and constitute an equilibrium. The number of components needed to build up these phases is n (see further below). Then, at first sight, the total number of variables would be:

$$rn + 2$$

where the "2" stems from the variables pressure and temperature.

For each phase, composed of n components, only $(n - 1)$ composition variables are independent, because the restraining condition per phase is $\sum c_i = 1$, with c_i as the fraction of component i in the phase considered. There are obviously as many of such restraining conditions as there are phases, so this already reduces the number of variables, to be specified in order to fix the phase equilibrium, to

$$rn + 2 - r$$

Next, the condition that the system considered is in equilibrium has to be imposed. It has been made clear in Sect. 7.3 that in equilibrium the partial Gibbs energies of each component must be equal for all phases in the system (discussion below Eq. 7.6).

Thus for component j:

$$G_j^1 = G_j^2 = \cdots = G_j^r$$

For component j equilibrium involves imposition of $(r - 1)$ conditions (equations), as follows from inspection of the series of equalities given above. There are n components and thus $n(r - 1)$ equilibrium conditions.

Finally, it follows for the number of independent intensive state variables which are sufficient to determine a phase equilibrium, i.e. the number of degrees of freedom, f:

$$f = rn + 2 - r - n(r - 1) = n - r + 2 \tag{7.8}$$

which is Gibbs' famous phase rule to be applied in the next section.

At this place it is appropriate to recall that the number of components to be substituted in the phase rule is not necessarily equal to the total number of constituents (or species) in the system (see the discussion in Sect. 7.2): $n =$ the *minimum* number of different components [constituents (or species)], i.e. the number of *mathematically independent* components to build up all phases in the system. The number of *independent components* is given, in a more general way than as indicated in Sect. 7.2, by the number of dependent components [constituents (or species)], minus the number of relations (or constraints) linking the components [constituents (or species)] mathematically (such as atom balances), minus the number of *independent* chemical reactions involving the (active) components. An independent reaction (in equilibrium) is one that contains one or more active components [constituents (or species)] not yet taking part in other reactions already specified and which components [constituents (or species)] have not yet been constrained/defined by earlier specifications of degrees of freedom (specified intensive state variables, given constraints, etc.). All independent reactions together must comprise all active components.[1]

Interestingly, also the number of phases of a system to be substituted in the phase rule is not in all circumstances obvious. $r =$ the number of *independent* phases. Interdependency of phases occurs if their Gibbs energies are equal; then r is smaller than the total number of phases actually present in the system. However, phase-energy degeneration appears to be of largely academic interest only (Chen et al. 2008).

---

[1] The text presented in this paragraph is perhaps not easy to penetrate at first glance. However, much confusion occurs in the determination of n for cases (much) more complicated than as considered until now (i.e. as in Sect. 7.2). The above given, general formulation to determine the minimum number of necessary components, n, has been taken, practically one-to-one, from a literature source where a focus is on gas and gas–solid reactions (J.T. Slycke, E.J. Mittemeijer and M.A.J. Somers in E.J. Mittemeijer and M.A.J. Somers (Eds.), Thermochemical Surface Engineering of Steels, Elsevier (Woodhead), 2015, p. 18). In that work the general approach to determine n has to be followed and has been illustrated by a number of examples of variable, distinct complexity. The treatment in the present book, for the determination of n, deals with cases not more complicated than as discussed in Sect. 7.2.

7.5 Phase Diagrams

Phase diagrams present fields of stability for phases as a function of intensive state variables (as pressure, temperature, composition). At the boundaries between these phase-stability fields phase transformations occur (e.g. a solid–liquid transformation which changes a solid into a liquid, or a solid–solid transformation of one crystal structure to an other one: *polymorphism*, which in case this happens for a single element is called *allotropy*; cf. Sect. 4.2.5).

7.5.1 One Component (unary) Systems

Applying the phase rule, it follows $f = 3 - r$. Hence in a single phase region two degrees of freedom occur: pressure and temperature can be chosen independently. At a transition from one phase to another phase only one degree of freedom remains: either the temperature or the pressure can be varied independently. An equilibrium of three phases corresponds with fixed values for both pressure and temperature: the so-called *tripel point*. These results can be illustrated as follows.

The phase diagram of a substance composed of atoms of a single element at constant pressure (say, 1 atm) reduces to a single line: the temperature axis (Fig. 7.3). An allotropic transformation in the solid, melting of the solid and boiling of the liquid occur at fixed temperatures.

Without specifying one of the intensive state variables, the fields of phase stability can be fully presented in a two-dimensional diagram: a p–T diagram. A famous example is the p–T diagram of water (Fig. 7.4). The lines a, b and c in this diagram represent the sublimation-pressure curve (the equilibrium between ice and water vapour), the melting-pressure curve (the equilibrium between ice and (liquid) water) and the boiling-pressure curve (the equilibrium between (liquid) water and water

Fig. 7.3 Phase diagram for a unary system with an allotropic transition in the solid state

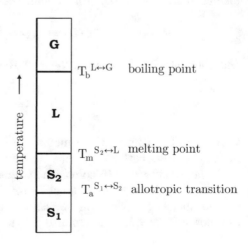

$T_b^{L \leftrightarrow G}$ boiling point

$T_m^{S_2 \leftrightarrow L}$ melting point

$T_a^{S_1 \leftrightarrow S_2}$ allotropic transition

Fig. 7.4 Pressure
(P)–temperature (T) diagram
for water

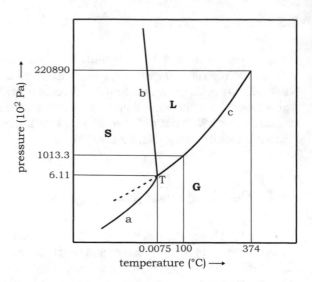

vapour), respectively. The tripel point has been indicated by "*T*". Note that the tripel point occurs at a temperature (0.0075 °C) slightly higher than the melting temperature of ice at 1 atm, but at the rather different pressure of 611 Pa. This diagram also provides the opportunity to hint at the occurrence of metastable eqlibria (cf. Sect. 7.4). Upon cooling water below the freezing point it may not immediately solidify, for example because no sites for easy nucleation of ice are available (cf. Chap. 9). Then *supercooled* water occurs. The dashed line in Fig. 7.4 describes metastable equilibrium between supercooled (liquid) water and water vapour. The metastability of this equilibrium should be understood in the sense of Fig. 7.1: given the chance (e.g. nucleation on an ice crystal put into the supercooled water) it will solidify immediately, because this releases energy.

At a two-phase boundary two phases are in equilibrium. Consider, for example, melting in the single component system. Equilibrium of solid and liquid implies:

$$\Delta G_f = G_{melt} - G_{solid} = 0 \tag{7.9}$$

where the subscript "*f*" denotes "fusion". Hence (cf. Eq. 7.4):

$$\Delta S_f = \frac{\Delta H_f}{T_m} \tag{7.10}$$

with T_m as the temperature of melting. Both ΔH_f and T_m can differ widely for different systems, but both are related to the bonding energy of the solid such that, as a rule, large values of ΔH_f correspond with large values of T_m and small values of ΔH_f correspond with small values of T_m (cf. Sect. 3.1). Hence, in view of Eq. (7.10), the corresponding variations in ΔS_f can be much smaller. This can be discussed also as follows. Since ΔS_f is determined by the change in entropy upon melting and, as already remarked in Sect. 7.4, the disorder upon melting of a solid increases very drastically, as compared to the ordering on a crystal lattice of a solid, differences in entropy between solids are less relevant for ΔS_f. Indeed, considering a series of *homologous* materials,[2] as close packed metals, ΔS_f is practically constant: ΔS_f equals about 8.5 J/(mol K) for close packed (f.c.c. and h.c.p.) and, but slightly different, for b.c.c metals. It has been possible to give a quantitative understanding for this numerical, practically constant value for the entropy of fusion of the elements: see the following *intermezzo*.

Intermezzo: Entropy of Fusion and the Structure of Liquids

A liquid strives for a structure that maximizes the *local* density, whereas in closed packed structures of solid substances the *overall, global* density is maximized. The configuration of highest density of hard spheres, as models for the atoms of the element considered, is a tetrahedron with the spheres at the vertices (cf. Sect. 4.2.1.1). This leads to the proposal to describe the structure of liquids as "polytetrahedral". Such tetrahedra can be packed, locally, around a common edge or common vertex. If this is done local structures occur which exhibit fivefold rotational symmetry (which type of rotational symmetry is impossible in case of *long-range* translational symmetry, as in crystals; cf. Chap. 4, in particular Sect. 4.8.2). For example, 20 tetrahedra can be packed around a common vertex leading to an icosahedral local structure. However, taking the individual tetrahedra as rigid, the packing cannot be accomplished

Fig. 7.5 Icosahedron formed by 20 regular tetrahedra packed around a common vertex. The packing cannot be perfect: gaps at some edges of the icosahedron must occur: one such resulting configuration for the geometrically unavoidable gaps is shown in the figure

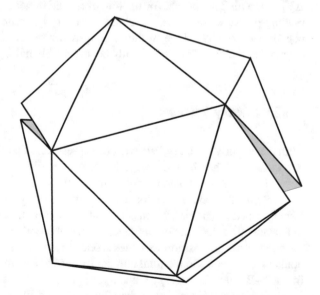

---

[2] Homologous materials are materials of similar (chemical) structures and related physical properties, as in the present case ΔH_f through T_m.

such that a perfect icosahedron results: gaps, at edges of the icosahedron, occur in the structure (Fig. 7.5). For the location of such gaps in the icosahedron a number of equivalent possibilities exist. The "configurational" entropy associated with this "degeneracy" [i.e. the number of ways to distribute the gaps over the icosahedron; cf. discussion around Eqs. (7.1) and (7.2)] can be conceived as predominating the configurational entropy *increase* upon melting (Spaepen 2005; the configurational entropy of the crystalline solid is neglected with respect to the configurational entropy of the liquid). The straightforward calculation shows that about 5/6 of the entropy of fusion, ΔS_f, can be conceived as due to this configurational entropy. The remaining about 1/6 of ΔS_f is ascribed to the increase in vibrational entropy upon melting, recognizing that the liquid has a lower density than the crystalline solid.

The slopes of the phase-boundary lines in the p–T phase diagram for the one component system are given by the *Clausius-Clapeyron equation*, which for the case of melting reads:

$$\frac{dp}{dT} = \frac{\Delta S_f}{\Delta V_f} = \frac{\Delta H_f}{T_m \Delta V_f} \tag{7.11}$$

where use has been made of Eq. (7.10). Evidently it holds (see above discussion) $\Delta S_f = S_{melt} - S_{solid} > 0$ (or $\Delta H_f > 0$ and $T_m > 0$). On this basis a peculiar feature of the p–T phase diagram of water can be explained: because upon melting of ice $\Delta V_f < 0$ (which is unusual for melting of a solid in general), the slope dp/dT of the melting-pressure curve, i.e. the phase boundary between ice and (liquid) water, is negative. A negative slope of the melting-pressure curve is a rare observation: for example in the case of metals it only occurs for Sb and Bi.

7.5.2   Binary Systems

With the adjective "binary" we wish to express here that the system is composed of two elements. Thereby the system is not necessarily a two-component system (see Sects. 7.2 and 7.4 and below), although mostly it is.

Applying the phase rule for a two-component system it follows: $f = 4 - r$. Usually phase diagrams are considered at $p = 1$ atm, thereby reducing the number of independently selectable intensive state variables with one. For a binary A–B phase the number of composition variables is one, e.g. the mole fraction of B, x_B, because it holds $x_A + x_B = 1$. So the total number of intensive state variables to be considered for an A–B phase is three: p, T and x_B.

The above consideration makes clear that the number of degrees of freedom within the field of stability of a single binary phase equals $f = 4 - 1\ (=r) - 1$

(because $p =$ fixed) $= 2$: temperature and composition are independently variable. If two binary phases are in equilibrium it follows that only one independent state variable remains (at $p =$ fixed): either the temperature or the composition of one of the phases (the composition of the other phase is thereby fixed). This is called a *univariant* equilibrium. Finally, an equilibrium of three binary phases has nil degrees of freedom (at $p =$ fixed): the temperature and the compositions of the three phases are fully determined. This is called a *nonvariant* equilibrium.

Phase diagrams for binary systems are usually presented as $T - x_B$ diagrams at $p = 1$ atm. For the remainder of this section, we will be concerned with this case. The temperature is usually expressed in either K or °C. The composition variable can be expressed as atom fraction or mass fraction. If x_B is given as atom fraction, say x_B^a, the corresponding mass fraction, say x_B^m, follows from

$$x_B^m = \frac{x_B^a \cdot A_B}{x_A^a \cdot A_A + x_B^a \cdot A_B} \tag{7.12}$$

with A_A and A_B as the atom masses of A and B, respectively. Similarly, the atom fraction x_B^a can be derived from the mass fractions according to

$$x_B^a = \frac{x_B^m / A_B}{x_A^m / A_A + x_B^m / A_B} \tag{7.13}$$

Both composition scales are often indicated in $T - x_B$ diagrams as representations of the abscissa at the top and bottom of the diagram, with the more fundamental, linearly presented atom-fraction (-percentage) scale usually at the bottom.

7.5.2.1 Isomorphous System; The Lever Rule

A system with complete solubility in the liquid and solid phase (i.e. the solid and liquid phases can be stable for any value of $x_B = 1 - x_A$) is the Nb-Ta system. The phase diagram is shown in Fig. 7.6b[3]; a schematic presentation of such a phase diagram is given in Fig. 7.6a. Such a system is called an *isomorphous* system. The

---

[3] At a number of places, in especially this chapter, for the purpose of illustration, phase diagrams of some binary systems are presented, which have been redrawn from the compilation provided by Massalski et al. (1996). The numerical composition and temperature data, as indicated for specific points/lines in these diagrams, have been adopted as given in this compilation. These numerical data, as presented in some cases, can suggest an accuracy, which from an experimental point of view, is surprisingly high. For example, see the Al-Si phase diagram shown in Fig. 7.12. The melting point of Al (at 1 atm) has been indicated as 660.452 °C. Such an indication implies that the true melting point of Al would likely be in the range 660.4515–660.4525 °C. An experimentalist knows that knowledge of the *relative* temperature in an experiment with a precision of 0.01 °C (0.01 K) is already a very good achievement. Apart from the melting points of the pure elements, the numerical values given in these phase diagrams can be based on the outcome of a computational model description/evaluation of the thermodynamics of the system, indeed derived from experimental data, but the significance of the computed/evaluated values can never be better than the inaccuracy corresponding with the

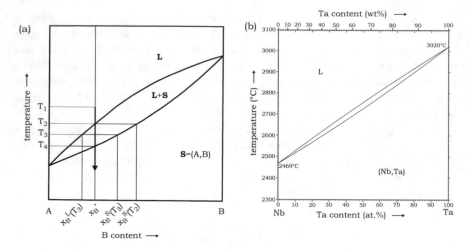

Fig. 7.6 a Schematic binary phase diagram of an isomorphous system (see text) and **b** the Nb–Ta phase diagram (redrawn from Massalski et al. 1996)

regions where the single-phase liquid and the single-phase solid are stable have been indicated with the symbols "L" and "S", respectively. The more or less lenticularly shaped region between L and S represents a two-phase region where a liquid phase and a solid phase of different compositions are in equilibrium and has been indicated with "L + S". The line separating the L and L + S regions is called *liquidus*; the line separating the L + S and S regions is called *solidus*. The way to read this diagram can best be illustrated by carrying out the following thought experiment.

Suppose at the start of the experiment a liquid of composition x'_B exists at a temperature T_1 (see Fig. 7.6a). Upon cooling nothing more than that the liquid gets a continuously lower temperature occurs (see arrow in Fig. 7.6a). At the moment the liquid has reached the temperature T_2 (the liquidus temperature for the alloy composition x'_B) the system cannot longer exist as a single liquid phase: solidification starts with the development of an infinitesimally small amount of a solid of composition $x^S_B(T_2)$. Hence, at this temperature the liquid of composition $x'_B = x^L_B(T_2)$ is in equilibrium with the solid of composition $x^S_B(T_2)$. The system has become univariant (see above discussion). The horizontal line in a two-phase region connecting the two phases in equilibrium is designated as *tie line* or, in Germanic languages, *Konode*. Upon further cooling it is obvious that more solid phase will develop and that the amount of liquid phase will decrease, which occurs under adaptation of the compositions of the liquid phase and the solid phase in order to satisfy the equilibrium requirements as indicated by the courses of the liquidus and solidus. Thus at T_3 a liquid of composition $x^L_B(T_3)$ is in equilibrium with a solid of composition $x^S_B(T_3)$. When the temperature is infinitesimally smaller than T_4 (the solidus temperature for the alloy composition

experimental errors inherent in the data used, although a computer can produce a practically endless list of decimals.

x'_B) the system cannot longer exist as a two-phase (liquid + solid) system: the system consists of a single solid phase of composition x'_B, implying that solidification has been completed. Still further cooling (follow the arrow in Fig. 7.6a) only causes a continuous decrease of the temperature of the solid.

It goes without saying that one is interested to know the amounts of the two phases in equilibrium at a certain temperature in a two-phase region. The calculation of these amounts is simply performed on the basis of two mass, or number of atoms, conservation expressions (see also Fig. 7.7). In the two-phase region the AB alloy of composition c_0 (we omit here the subscript B to simplify the notation) cannot exist as a single phase: it decomposes in a phase α (e.g. a liquid) of composition c_α and a phase β (e.g. a solid) of composition c_β. The total mass, or number of atoms, in the system remains constant and thus:

$$N_\alpha + N_\beta = N_0 \tag{7.14}$$

where N_α, N_β and N_0 denote mass, or number of atoms of the phases α and β and the alloy, respectively. It also holds that the mass or the number of atoms B remains constant, implying:

$$c_\alpha N_\alpha + c_\beta N_\beta = c_0 N_0 \tag{7.15}$$

From Eqs. (7.14) and (7.15), it is derived straightforwardly:

Fig. 7.7 Partial, binary, A–B phase diagram, showing a part of a two-phase, $\alpha + \beta$ region. Illustration of the lever rule; see Eqs. (7.14)–(7.17)

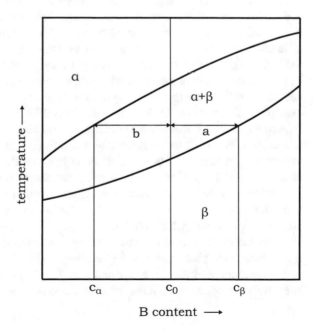

$$\frac{N_\alpha}{N_0} = \frac{c_\beta - c_0}{c_\beta - c_\alpha} = \frac{b}{a+b} \qquad (7.16)$$

and

$$\frac{N_\beta}{N_0} = \frac{c_0 - c_\alpha}{c_\beta - c_\alpha} = \frac{a}{a+b} \qquad (7.17)$$

So the amounts of phases α and β are proportional to the parts, line-lengths, a and b, of the tie line (see Fig. 7.6a). In other words: the amount of one of the phases (α or β) is proportional to the absolute value of the difference in concentration between the other phase (β or α) and the gross composition of the alloy. The tie line takes the role of a lever with the gross alloy composition as the fulcrum. For this reason, the result expressed by Eqs. (7.16) and (7.17) is called *the lever rule*.

The reader can now verify that indeed the amount of solid phase increases and the amount of liquid phase decreases upon cooling in the two-phase, $L + S$ region in Fig. 7.6a.

Substitutional dissolution of a solute atom of element B in a solid crystalline phase of element A can be associated with the generation of (local) distortions and thus introduces strain energy in the phase that opposes the chemical energy promoting the mixing. Systems where the A-B atom-size difference is only a few percent (say, less than 5%), as holds for Ag-Au and Ta-Ti, can accommodate this size difference relatively effortless. In these cases, phase diagrams of the type shown in Fig. 7.6 occur. For larger atom-size differences (say more than 10%), as holds for Au-Ni and Cr-Mo, the introduction of strain energy becomes relatively more important, in particular for mixing in the solid phase. In the liquid phase such an atom-size difference is easier to accommodate, since no lattice sites fixed in space occur. Thus, the liquid phase becomes relatively energetically preferred. This can be expressed by the phase diagram of the system: the liquid phase region becomes extended at the cost of the solid phase region: the system may reveal a melting point minimum. An example is shown by the Cs-Rb phase diagram (Fig. 7.8a).

At the melting point minimum (and also at a melting point maximum) the liquidus and solidus lines must meet (and at the position of the melting point minimum (or maximum) the tangent of both the liquidus line and the solidus line is nil). If this coincidence of the minima of liquidus and solidus lines was not the case, it would be possible to identify temperatures at which, *within* the two-phase $L + S$ region, two solid phases of different composition would be in equilibrium in the additional presence of a liquid, which obviously is physical nonsense (see Fig. 7.8b). The number of degrees of freedom at the melting point minimum is nil, which follows by application of the phase rule: $n = 1$ (the compositions of the liquid and solid phases in equilibrium are identical), $r = 2$ and thus $f = 1$, and since p is fixed (1 atm) all intensive state variables are fixed (cf. discussion in Sect. 7.2).

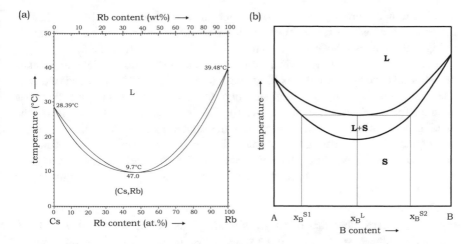

Fig. 7.8 a Cs-Rb phase diagram exhibiting a melting point minimum (redrawn from Massalski et al. 1996). A melting point minimum may occur as the result of a size difference of atoms A and B which is easier to accommodate in the liquid than in the solid. **b** Illustration of the requirement that at the melting point minimum the liquidus and solidus lines must meet. If this were not the case it would be possible to identify a temperature at which, *within* the two-phase L + S region, two solid phases would be in equilibrium in the additional presence of a liquid, in violation of Gibbs' phase rule (see text)

7.5.2.2 From Mixing to Demixing: Emergence of the Miscibility Gap and the Eutectic System

Defining the change in Gibbs energy, at constant temperature and pressure, upon mixing the amount of atoms A and the amount of atoms B on the sites of the crystal structure concerned as ΔG_{mix}, one can write [cf. Eq. (7.4)]:

$$\Delta G_{\mathrm{mix}} = \Delta H_{\mathrm{mix}} - T \Delta S_{\mathrm{mix}} \tag{7.18}$$

Of course, mixing is preferred if $\Delta G_{\mathrm{mix}} < 0$. If the atoms A and B do not interact favourably, for example because of a relatively large introduction of strain energy upon mixing, ΔH_{mix} can be larger than zero and thus oppose mixing. The contribution of the entropy of mixing (related to the degree of disorder, cf. the discussion in Sect. 7.3) then can still favour mixing (ΔS_{mix} is always larger than zero). This contribution, $T \Delta S_{\mathrm{mix}}$, obviously increases with temperature. ΔS_{mix} is relatively, say per % dissolved element, large for a small amount of dissolved solute. Hence, it is conceivable that at sufficiently low-temperature decomposition takes place: a two-phase region (also called "miscibility gap") occurs in the phase diagram where two crystalline solid phases are in equilibrium at a fixed temperature: an A-rich phase and a B-rich phase; for both phases it holds that the amounts of solute decrease with decreasing temperature. At sufficiently high temperature the mixing promoting effect of $T \Delta S_{\mathrm{mix}}$ (>0) may overcome the mixing counteracting effect of ΔH_{mix} (>0):

a critical temperature, T_c, occurs above which an A–B solid solution is possible over the entire composition range. The resulting phase diagram is shown schematically in Fig. 7.9a; see also the phase diagram of the Ni-Rh system (Fig. 7.9b).

The two effects discussed, which involve responses of nature to demixing trends in A-B alloys, occurrence of a melting point minimum (Sect. 7.5.2.1) and of a two-phase, solid–solid region in the phase diagram, can occur together: see the phase diagram for the Au-Ni system (Fig. 7.10). The larger ΔH_{mix} (>0), the lower the melting point minimum and in particular the higher T_c may be. This suggests a development in the appearance of the phase diagram for increasing ΔH_{mix} as sketched in Fig. 7.11a–c. The resulting phase diagram (Fig. 7.11c) is called a *eutectic* diagram and is observed, for example, for the Al-Si system (Fig. 7.12; cf. the begin of this Chap. 7 and Sect. 7.1).

The schematic presentation of the eutectic diagram in Fig. 7.11c (see also Fig. 7.13) makes clear that three two-phase regions occur: solid $\alpha + L$, solid $\beta + L$ and solid α + solid β, and three single-phase regions: solid α, solid β and L. The boundary between the single solid phase (α) region and the two-phase, solid (α)–solid (β), region is called a *solvus*. The composition ranges of the solid solution phases α and β are limited by the extreme compositions, pure A and pure B, respectively, and therefore the α and β phases are also called "*terminal solid solutions*". As a general rule, it can be stated that in these binary phase diagrams *between two single-phase regions always a two-phase region occurs* (see also Sect. 7.5.3 and the "*Epilogue: The Topology of Phase Diagrams; some Rules*").

A remarkable feature takes place at the so-called eutectic temperature, T_{eut}, where three phases are in equilibrium: the solid phases α and β and the liquid phase L. This is the only place in the diagram where such a three-phase equilibrium occurs; this

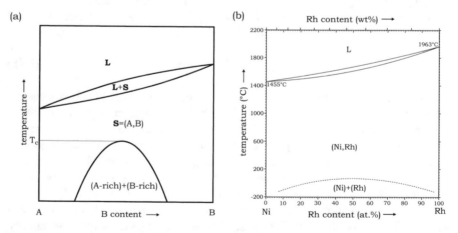

Fig. 7.9 a Occurrence of a miscibility gap in the solid state for a binary system, as a deviation from an isomorphous system (see Fig. 7.6), for example caused by the need to avoid the introduction of a relatively large strain energy (due to a size difference of atoms A and B) upon mixing. **b** The phase diagram of Ni-Rh (redrawn from Massalski et al. 1996)

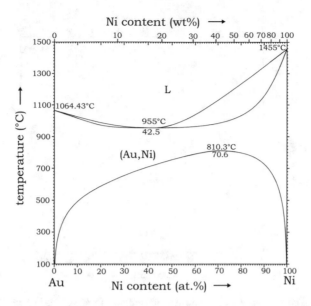

Fig. 7.10 Au-Ni phase diagram showing a melting point minimum and a miscibility gap (redrawn from Massalski et al. 1996)

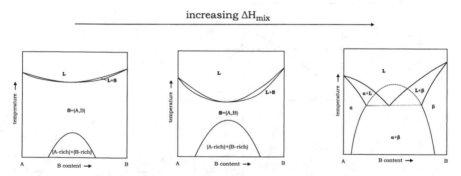

Fig. 7.11 Evolution of a eutectic system from a system with melting point minimum and a miscibility gap. With increasing mixing enthalpy, the melting point minimum and the critical temperature of the miscibility gap approach each other

equilibrium is nonvariant (see the begin of Sect. 7.5.2): the temperature and the compositions of the three participating phases are fixed [in case of fixed pressure (1 atm)]. Evidently, upon cooling a liquid phase L of the eutectic composition, solidification of the solid phases α and β starts upon reaching the temperature T_{eut} (see Fig. 7.13) and the system stays at this temperature until all L has transformed:

$$\text{eutectic reaction: } L \leftrightarrow \alpha + \beta \tag{7.19}$$

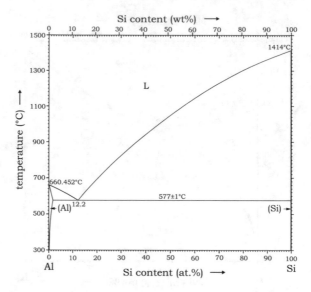

Fig. 7.12 Al–Si phase diagram; a eutectic system (redrawn from Massalski et al. 1996)

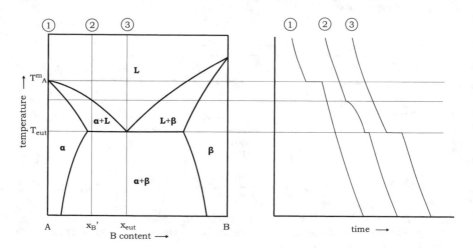

Fig. 7.13 Schematic presentation of a eutectic binary system (left part of the figure) with cooling curves (right part of the figure), starting from the liquid, for alloys of compositions 1, 2 ($= x'_B$) and 3 ($= x_{eut}$)

It follows from the extensive discussion in Sect. 7.5.2.1 with respect to the thought (cooling) experiment performed for the isomorphous system sketched in Fig. 7.6, that upon cooling a binary liquid, normally solidification occurs for a range of temperature, which contrasts with the solidification of a pure element that takes place at a fixed temperature (Sect. 7.5.1). It is shown above that a liquid of the eutectic composition also solidifies completely at one temperature (while extracting heat from the system). Adopting the eutectic phase diagram given in Fig. 7.13, consider the case of continuous cooling of a liquid of initial composition x'_B smaller than the eutectic composition. Then, solidification starts with the development of the (primary) solid phase α at a temperature higher than T_{eut} and the equilibrium between the solid phase (α) and the liquid phase changes upon continued cooling, similarly as discussed with respect to Fig. 7.6. Hence, the liquid phase becomes enriched continuously in B and the developing solid phase α becomes continuously enriched in B as well, while the temperature decreases. At the moment the eutectic temperature is reached, the liquid phase has the eutectic composition and the formation of solid phase β starts, while the formation of solid phase α is continued: reaction (7.19) runs until completed while the system stays at T_{eut}. Thereafter continued cooling maintains the two-phase system $\alpha + \beta$, while continuously adjusting their compositions according to the solvus lines.

The above discussion provides a qualitative understanding for the cooling curves (temperature versus time) as could be observed for the eutectic A–B system by continuously extracting heat from the system (see Fig. 7.13). The pure elements solidify at fixed temperature, i.e. during solidification the cooling rate is nil. The liquid alloy of composition deviating from the eutectic composition experiences a temperature range over which solidification occurs; the release of heat during the solidification will cause a decrease of the cooling rate as compared to the fully liquid alloy. When the eutectic temperature has been reached the residual liquid of eutectic composition will solidify into the two solid phases while the temperature remains constant (zero cooling rate).

The occurrence for a eutectic alloy of a fixed melting/solidification temperature, instead of a melting/solidification temperature *range* and, furthermore the relatively low value of the eutectic temperature, as compared to the melting/solidification temperatures of the pure elements, makes an eutectic system attractive for technological application as soldering alloy. The classical alloy for soldering purposes was the Pb–Sn alloy of composition about 40wt%Pb–60wt%Sn, with a low eutectic temperature of about 185 °C (see Fig. 7.14). Because Pb is poisonous and may not be applied to this end anymore, other much more complicated (higher than binary) alloys are investigated nowadays for soldering applications (e.g. see Sommadossi et al. 2002).

A special type of demixing and associated miscibility gap in the solid state occurs for systems showing development of ordering and thus "superstructures". As discussed in Sect. 7.3 and below Eq. (7.18), the internal energy (enthalpy) may control the phase occurrence at relatively low temperature, whereas at elevated temperature the entropy may be dominant. If for a binary system the A–B atom interaction is favoured over A–A and B–B (pairwise) atom interactions, a tendency occurs for unlike atoms to be nearest neighbours on the sites of the crystal structure considered. Thus, at sufficiently low temperatures, ordered distributions ("superstructures") can develop (cf. Sect. 4.4.1.1). The region in the phase diagram separating the stability region of the superstructure phase from those of the terminal solid solution phases cannot be given by a single line, as holds for the border between the miscibility gap

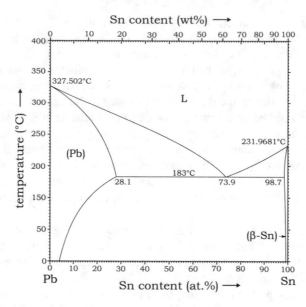

Fig. 7.14 Pb–Sn phase diagram; eutectic system with technical relevance of eutecticum (redrawn from Massalski et al. 1996)

discussed above and the terminal solid solutions. In accordance with the rule indicated above, between two single phase regions (as a terminal solid solution phase and the superstructure phase) a two-phase region must occur, where the "superstructure" phase is in equilibrium with the terminal solid solution phase. This suggests a form of the phase diagram as illustrated in Fig. 7.15a: in the region left from the

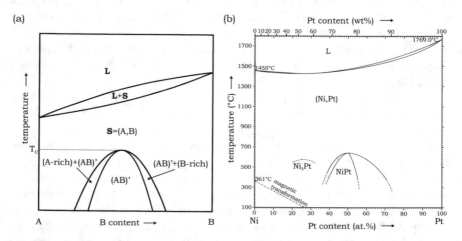

Fig. 7.15 **a** Schematic phase diagram for a binary, A–B system, exhibiting a superstructure phase, (AB)′ in the solid state. **b** The Ni–Pt phase diagram (redrawn from Massalski et al. 1996)

single superstructure phase, $(AB)'$, the superstructure phase is in equilibrium with the A-rich phase and in the region right from the single superstructure phase, $(AB)'$, the superstructure phase is in equilibrium with the B-rich phase. An example of such a phase diagram is shown in Fig. 7.15b for the Ni-Pt system.

7.5.2.3 Monotectic System

The size difference between the atoms of the two elements, A and B, may be that large that even in the liquid the tendency to decomposition becomes strong (cf. Sect. 7.5.2.2). Then a two-phase, liquid–liquid, phase region can occur. Obviously and in line with the previous discussion, if this happens strong demixing behaviour in the solid region of the phase diagram is for sure. An illustration of such a diagram is provided by Fig. 7.16a, b.

Consider cooling of a liquid of composition x_B' starting at a temperature in the single phase field L (Fig. 7.16a). Upon reaching the border between the single-phase field L and the two-phase field $L_1 + L_2$, decomposition of the liquid phase L into two liquid phases L_1 and L_2 occurs. Further cooling will lead to changes of the compositions of the phases L_1 and L_2 according to the phase boundary. Upon reaching the temperature indicated with T_{mon} the phase L_1 must transform into solid phase α and liquid phase L_2, while remaining at the same temperature until all L_1 has been transformed. This is called a *monotectic* reaction, which is nonvariant:

$$monotectic\ reaction: L_1 \leftrightarrow \alpha + L_2 \qquad (7.20)$$

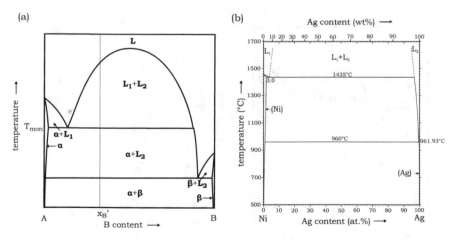

Fig. 7.16 **a** Schematic phase diagram for a binary, monotectic system. **b** The Ni–Ag phase diagram as example (redrawn from Massalski et al. 1996)

7.5.2.4 Congruently and Incongruently Melting "Line" Compounds;
The Peritectic System

Certain phase regions can be such narrow, i.e. limited in composition range, that they usually are represented by a vertical line in the phase diagram. This can happen for intermetallic compounds (cf. Chap. 4). If these "line" compounds melt, then they may do so at a certain, specific temperature under formation of a liquid of the same composition. This can occur if the melting temperature of the compound is higher than the melting temperatures of the elements constituting the compound. Such compounds are called *congruently melting compounds*. An example is shown in Fig. 7.17a, b.

If the stability of the compound considered is limited, as for the case where the melting point of one of the elements is pronouncedly higher than that of the compound, congruent melting of the compound cannot occur. Instead the compound melts under formation of a solid and a liquid of compositions different from that of the "line" compound: *incongruently melting compound*. An example is shown in Fig. 7.19a, b.

One can conceive the phase diagram shown in Fig. 7.19a as the result of a continuous increase of the melting point of component B in the phase diagram shown in Fig. 7.17a (see Fig. 7.18): the two-phase region $L + \beta$ becomes more extended and the two-phase region $C + L$ shrinks until, eventually, it has disappeared; thereby the phase diagram shown in Fig. 7.17a has transformed into the phase diagram shown in Fig. 7.19a.

Consider cooling of a liquid of composition equal to that of the incongruently melting compound, x_{comp}, starting at a temperature in the single phase field L

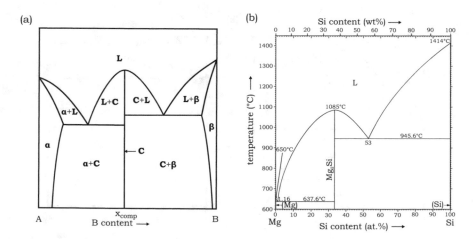

Fig. 7.17 a Schematic phase diagram for a binary system with a congruently melting compound indicated with C. **b** The Mg–Si phase diagram as example (redrawn from Massalski et al. 1996)

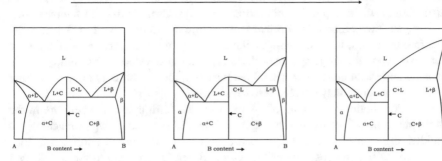

Fig. 7.18 Evolution of a peritectic system from a binary system with a congruently melting compound (cf. Fig. 7.17a). With increasing melting point of component B the two-phase region L + β becomes more extended and the two-phase region C + L shrinks until, eventually, it has disappeared

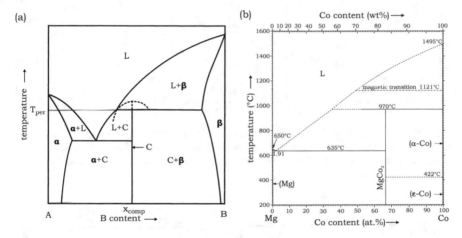

Fig. 7.19 a Schematic phase diagram of a binary, peritectic system with an incongruently melting compound C. b The Mg-Co phase diagram as example (redrawn from Massalski et al. 1996)

(Fig. 7.19a). Upon reaching the liquidus, the formation of the solid phase β is initiated. Further cooling will lead to changes of the compositions of the phases β and L according to the solidus and liquidus. Upon reaching the temperature indicated with T_{per} the phase (compound) C must form from solid phase β and liquid phase L, while remaining at the same temperature until all β and (in this case also) all L have been transformed. This is called a *peritectic* reaction, which is nonvariant:

$$\text{peritectic reaction: } \beta + L \leftrightarrow C \qquad (7.21a)$$

Another type of peritectic phase diagram can be conceived as another manifestation of the increase of the atomic size difference between A and B (cf. Sects. 7.5.2.2

and 7.5.2.3). The eutectic phase diagram resulted if, upon increasing atomic size misfit, both a melting point minimum and a two-phase, solid–solid region ("miscibility gap") with increasing critical temperature developed (see Fig. 7.11). If the system considered, upon increase of the atomic size difference, can maintain the lenticular shape for the L + S two-phase region, but develops a miscibility gap in the solid state with increasing critical temperature, then a phase diagram exhibiting a peritectic reaction may develop (see Fig. 7.20).

Consider cooling of a liquid of composition x'_B starting at a temperature in the single phase field L of the phase diagram shown in Fig. 7.21a. Upon reaching the liquidus, the formation of the solid phase β is initiated. Further cooling will lead to changes of the compositions of the phases β and L according to the solidus and

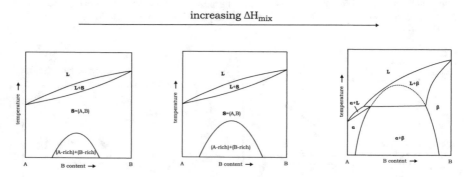

Fig. 7.20 Evolution of a peritectic system from a system with miscibility gap, but without melting point minimum (note this difference with the situation sketched in Fig. 7.11 pertaining the development of a eutectic system). With increasing mixing enthalpy, the miscibility gap and the solidus approach each other

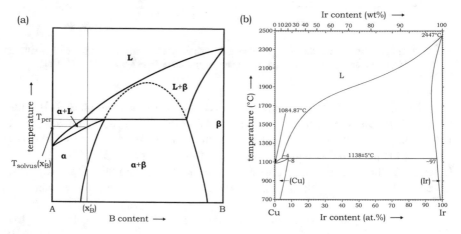

Fig. 7.21 **a** Schematic phase diagram of a binary, peritectic system. **b** The Cu–Ir phase diagram as example (redrawn from Massalski et al. 1996)

liquidus. Upon reaching the temperature indicated with T_{per} the phase α must form from solid phase β and liquid phase L, while remaining at the same temperature until all β has been transformed. Upon continued cooling more phase α is formed, until at temperature $T_{solvus}(x'_B)$ all L has been transformed in α. The peritectic reaction for this example reads:

$$\beta + L \leftrightarrow \alpha \tag{7.21b}$$

7.5.2.5 Eutectoid, Monotectoid and Peritectoid Systems

Reactions of types analogous to those described by Eqs. (7.19–7.21) can occur also with only solid reactants and products. Thus, one can discern:

$$\text{eutectoid reaction:} \quad \gamma \leftrightarrow \alpha + \beta \tag{7.22}$$

$$\text{monotectoid reaction:} \quad \beta' \leftrightarrow \alpha + \beta \tag{7.23}$$

where β and β' denote solid phases of the same crystal structure but different composition, and

$$\text{peritectoid reaction:} \quad \alpha + \beta \leftrightarrow \gamma \tag{7.24}$$

Note that the peritectoid/peritectic and eutectoid/eutectic reactions are pendants: the concerned parts of the phase diagrams are mirror images (inverted temperature axis).

Examples of these nonvariant, solid–solid phase transformations can be found in the phase diagrams shown in Figs. 7.22[4], 7.23 and 7.24.

This Sect. 7.5.2 ends as it begins: with an application of the phase rule. Consider the terminal solid solutions in the phase diagram for the eutectic system. If the solubility of B in the α phase is very small and the solubility of A in the β phase is very small, one is tempted in a phase diagram drawing to let coincide the solvus lines with the left

---

[4] The Fe–C phase diagram is of great practical importance (steels!). It returns at a number of places in this book in different fashions: Fig. 7.22b, Fig. 9.9, Fig. 9.22a (based on the same literature source as for Fig. 7.22b, but differently presented; see what follows), and finally Fig. 9.24 (the latest version). Thereby the ongoing efforts to determine and understand the underlying thermodynamics of the Fe–C phase diagram are illustrated. Also see the "*Intermezzo: The Fe–C and Fe–N phase diagrams*" at the end of Sect. 9.5.2.1.

Finally it is recalled that in phase–diagram drawings the composition parameter along the abscissa *at the bottom* of the figure usually is "the more fundamental, linearly presented atom-fraction (-percentage)", whereas, simultaneously, the weight(mass) fraction is usually shown, then not linearly, along the abscissa *at the top* of the figure [see below Eq. (7.13) in Sect. 7.5.2]. However, in case of Figs. 9.9 and 9.22a reverse policy has been followed: the, then linearly presented, weight(mass)-fraction scale has been placed *at the bottom* of the Fe–C phase diagram. This reflects the preferred use by engineers in industrial practice of weight(mass) fractions (e.g. in the steel making process).

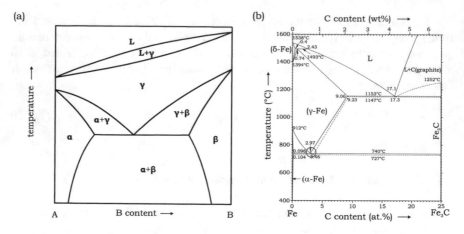

Fig. 7.22 a Schematic phase diagram of a binary system with a eutectoid reaction. **b** The Fe–C phase diagram as example (redrawn from Massalski et al. 1996). The stable diagram (with the phase graphite) is given by the solid line. The metastable diagram (with cementite, Fe_3C) is given by the dashed line); the eutectoid reaction, $(\gamma\text{-Fe}) \leftrightarrow (\alpha\text{-Fe}) + Fe_3C$, has been marked by a circle. See also Fig. 9.9 and the *"Intermezzo: The Fe–C System; Steels and Cast Irons"* in Sect. 9.4.2

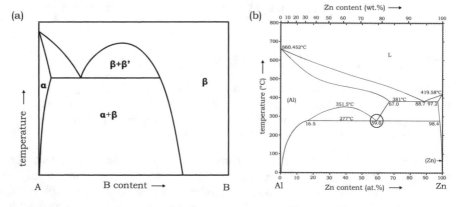

Fig. 7.23 a Schematic phase diagram of a binary system with a monotectoid reaction. **b** The Al-Zn phase diagram as example (redrawn from Massalski et al. 1996); the monotectoid reaction, $(Al)'$ $\leftrightarrow (Zn) + (Al)$, has been marked by a circle

and right ordinates: see Fig. 7.25. Although this type of phase-diagram drawing has been often performed for the case considered and for similar cases where marginal solubilities for solutes occur, this is a misleading representation of the phase diagram as it is in conflict with the phase rule: For the solid solution phase (α or β) the number of degrees of freedom is 2 (at fixed pressure). Hence, temperature and composition are independent, intensive state variables. The sketched diagram erroneously suggests that only one degree of freedom remains (the temperature). The phase rule thus indicates an important thermodynamic consequence: in equilibrium there is always

(a)

(b)

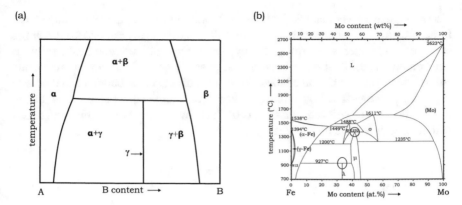

Fig. 7.24 **a** Schematic phase diagram of a binary system with a peritectoid reaction. **b** The Fe-Mo phase diagram as example (redrawn from Massalski et al. 1996); two peritectoid reactions, $R + \sigma \leftrightarrow \mu$ and $(\alpha\text{-Fe}) + \mu \leftrightarrow \lambda$, have been marked by circles

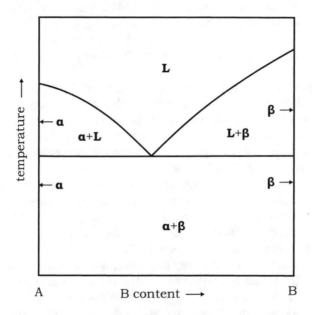

Fig. 7.25 Schematic depiction of a binary, eutectic phase diagram with very small solubility ranges for the terminal solid solutions: the solvus lines practically coincidence with the ordinates. However, in principle the solvus lines cannot coincide with the ordinates of the phase diagram. (Very) Close to the ordinates there have to be single phase fields corresponding with two degrees of freedom (B content and temperature, at fixed pressure; cf. Gibbs' phase rule)

a certain (possibly very small, yet finite) solubility of any component in any phase. In accordance with the discussion in Sect. 7.3 and beneath Eq. (7.18), it follows that even if the internal energy (enthalpy) change upon dissolution of the solute is unfavourable (>0), then there is always a, possibly very small, amount of solute atoms which become dissolved, because as a result the contribution of the entropy of mixing becomes that large relatively (larger for smaller amounts of solute) that $-T\Delta S_{mix}$ (<0) overcompensates ΔH_{mix} (>0). It is obvious that the solubility increases with temperature (role of T in $T\Delta S_{mix}$).

7.5.3 Ternary Systems

Applying the phase rule for a three-component system it follows: $f = 5 - r$. Again considering phase diagrams at fixed pressure ($p = 1$ atm), the number of independently selectable intensive state variables is reduced with one. For a ternary A–B–C phase the number of composition variables is two, e.g. the mole fraction of B, x_B, and the mole fraction of C, x_C, because it holds $x_A + x_B + x_C = 1$. So the total number of intensive state variables to be considered for an A–B–C phase is four: p, T, and x_B and x_C.

The above consideration makes clear that the number of degrees of freedom within the field of stability of a single ternary phase equals $f = 3$ (recognizing $p =$ fixed): temperature and two composition parameters are independently variables. If two ternary phases are in equilibrium it follows that only two independent state variables remain (at $p =$ fixed): either the temperature and one composition variable of one of the phases, or two composition variables of one of the phases (the composition of the other phase is thereby fixed). Analogous remarks can be made for an equilibrium of three ternary phases. Finally, an equilibrium of four ternary phases has nil degrees of freedom (at $p =$ fixed): the temperature and the compositions of the four phases are fully determined: a *nonvariant* equilibrium, also called "ternary eutectic point", occurring at the "ternary eutectic temperature".

Evidently, a visual display of ternary phase diagrams requires a three-dimensional representation (T, x_B and x_C are the variables to be considered at $p =$ fixed (1 atm); see above). This is achieved by representing the compositions of the phases occurring at constant temperature in a plane and drawing the temperature axis perpendicular to these *isothermal sections* of the phase diagram.

To indicate the composition of a phase of a ternary system in an isothermal section one usually applies the so-called *composition triangle*, also called Gibbs' triangle. This is an equilateral triangle where the corners represent pure (100%) A, B and C. The composition of a phase P can always be described by a point within this triangle. Draw lines parallel to the sides of the triangle through P (see Fig. 7.26a). Line a has a constant distance to side BC and represents all phase compositions with a same amount of A. This amount of A can be read from the side AC of the triangle where the fraction of A has been indicated. Similarly, lines b and c through P parallel to sides AC and AB are lines of constant amounts B and C, respectively, and can be used to determine the amounts of B and C in phase P.

The geometrical background for the above construction is the recognition that the sum of the lengths of the lines of projection of P onto the sides of the equilateral triangle is constant, i.e. independent of the position of point P within the triangle (which constant is equal to the height of the equilateral triangle). The lines drawn

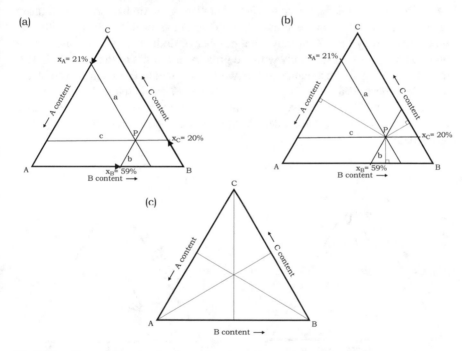

Fig. 7.26 **a** Composition triangle of a ternary phase diagram (constant temperature; constant pressure). The arrows indicate the contents of A, B and C for a phase of composition P. **b** Same composition triangle, showing (additionally) the lines of projection of P onto the sides of the triangle. The sum of the lengths of these lines of projection is constant, independently of the location of P. The lines a, b and c are lines at constant distance from the sides of the triangle and thus lines representing constant A content (line a, here 21%), B content (line b, here 59%) and C content (line c, here 20%). **c** Composition triangle with three lines of constant ratio (here 1:1) of two of the three components

through P parallel to the sides of the triangle then are lines at distances to the sides equal to the lengths of the corresponding lines of projection of P (Fig. 7.26b).

Other special lines in the composition triangle are straight lines through the corners. Suppose such a line emanates from corner A. Then it holds that at this line phase compositions occur with a constant ratio of the amounts of B and C. Similar statements hold for straight lines through B and C (Fig. 7.26c).

In a two-phase region, it is necessary to know the compositions of the two phases which are in equilibrium. Such a pair of two composition points, located at the boundaries of the two-phase field concerned in the composition triangle (i.e. at constant temperature [and at constant pressure)], can be connected by a straight line, a tie line as for the binary systems. In contrast with the tie lines in the two-phase regions of a binary phase diagram (at constant temperature and at constant pressure), the position of the tie lines in the two-phase regions of a ternary phase diagram cannot be drawn without more ado; i.e. it cannot be known beforehand which compositions the two phases in equilibrium have (but the tie lines in the two-phase region cannot intersect): to describe the possible equilibrium states completely the tie lines in the two phase regions should be given in the ternary phase diagram as well. A schematic illustration is provided by Fig. 7.27a showing a number of tie lines in a L + γ two-phase region. The tie lines must comply with the gross composition of the ternary alloy. If the tie line is known, the amounts of the two phases in equilibrium are given by the lever rule as derived for the binary system (cf. Sect. 7.5.2.1). Note that, at constant temperature (and at constant pressure), only one tie line is possible in a two-phase field of a binary system, independent of the gross composition of the binary alloy considered.

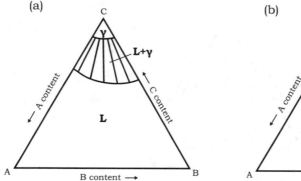
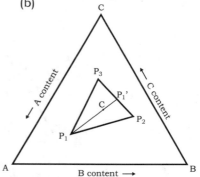

Fig. 7.27   a Two-phase field (L + γ) with tie lines within a composition triangle of a ternary alloy system (constant temperature; constant pressure). **b** Within the composition triangle shown a three-phase field occurs. This three-phase field is not shown. The points P_1, P_2 and P_3 indicate the compositions of the three phases in equilibrium. Connecting the points P_1, P_2 and P_3 by straight lines a triangle is constructed within the composition triangle. The relative amounts of phases P_1, P_2 and P_3 are given by $C P_1'/P_1 P_1'$, $C P_2'/P_2 P_2'$ and $C P_3'/P_3 P_3'$, respectively (see text)

A similar construction of a lever rule for a three-phase region is a bit more complicated. In an isothermal section (i.e. after selection of the temperature [and the pressure)] no degree of freedom is left for the three-phase equilibrium (i.e. the three-phase
equilibrium at a selected temperature is nonvariant): the compositions of the three
phases have become fixed as well. Hence the three-phase equilibrium in a composition triangle (for the temperature and pressure selected) can be represented by three
points (here P_1, P_2 and P_3) indicating the fixed compositions for the three participating phases (Fig. 7.27b). By connecting these three points, a triangle is constructed
within the composition triangle. The gross composition of the ternary alloy is indicated by the point C, the centre of gravity of the triangle defined by the compositions
P_1, P_2 and P_3. The amounts of the three phases can now be obtained by application
of a variant of the lever rule, as follows:

$$\frac{N_{P_1}}{N_0} = \frac{CP_1'}{P_1P_1'}; \quad \frac{N_{P_2}}{N_0} = \frac{CP_2'}{P_2P_2'}; \quad \frac{N_{P_3}}{N_0} = \frac{CP_3'}{P_3P_3'} \tag{7.25}$$

where N_{P_1}, N_{P_2} and N_{P_3} denote the numbers of moles in phase P_1, P_2 and P_3,
respectively, and N_0 represents the number of moles in the whole specimen. P_1'
is determined by extending line piece P_1C until its intersection with side P_2P_3.
Similarly for P_2' and P_3'.

As a conclusion to this section some experimentally determined isothermal
sections for the ternary system Ag–Cu–Ni are presented in Fig. 7.28 for the
temperature range 1440–700 °C and at $p = 1$ atm, as an exercise to read and interpret.

7.6 Microstructure Development
with Reference to the Phase Diagram

The phase diagram can be applied not only to find out which phases are in equilibrium
at given pressure and temperature. Responses of the system to some action exerted
from outside leading to phase changes may be predicted (qualitatively). An example
is provided by the discussion on the nature of the cooling curves in Sect. 7.5.2.2
(Fig. 7.13). Also, the microstructural development (even its non-equilibrium nature)
can be conjectured.

First, turn to the binary, isomorphous system illustrated in Fig. 7.29 (cf. Fig. 7.6).
By cooling the liquid of composition x_B' (due to the extraction of heat), solidification
starts upon entering the two phases, L + S region. If equilibrium is maintained
during the entire cooling process, then the composition of the developing solid should
move along the solidus line as discussed in Sect. 7.5.2.1. However, in practice the
adaptation of the composition of the solid is hindered by a relatively slow diffusion in
the solid state (solute atoms, B atoms in the example considered, have to move from
the inside to the surface of the developing solid particle and transfer to the liquid
phase there). As a result new solid, of the composition given by the solidus at the

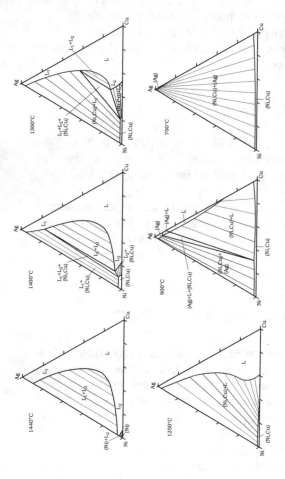

Fig. 7.28 Isothermal sections of the Ag–Cu–Ni system for the temperature range 1440–700 °C and at p = 1 atm. Note the three phase regions: $L_1 + L_2 +$ (Ni,Cu) at 1400 °C, $L_1 + L_2 +$ (Ni, Cu) at 1300 °C and (Ag) + L + (Ni,Cu) at 900 °C . (taken from G. Petzow and G. Effenberg (eds.), Ternary Alloys: A Comprehensive Compendium of Evaluated Constitutional Data and Phase Diagrams, Vol.1 Ag-Al-Au to Ag-Cu-P, Wiley-VCH Verlag, Weinheim, 1988).

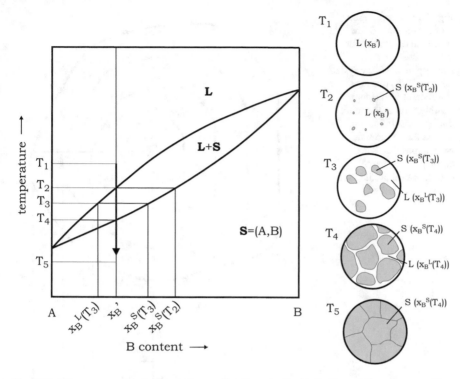

Fig. 7.29 Microstructure evolution during cooling from the liquid of a binary alloy of composition x'_B in an isomorphous system

prevailing temperature, precipitates at the surface of the already existing particle as grown at a higher temperature and with a composition richer in solute. Eventually, after completion of the solidification, massive material results with in each grain a composition gradient in solute (Fig. 7.30). This is the reason that after casting normally a homogenization treatment is performed with the solidified, as cast material by subsequent annealing at elevated temperature, but at a temperature where the material remains solid, in order that the diffusion in the solid state is that fast that compositional homogenization can be achieved.

This type of consideration can be applied to a eutectic system as well. Consider Fig. 7.31 where cooling of a liquid alloy is performed for four different alloy compositions: x'_B, x''_B and x'''_B and x_B^{eut}. For the alloy of composition x'_B the discussion given above can be copied.

For alloy composition x''_B, it follows that, after completed solidification of solid α phase, entering the two-phase, $\alpha + \beta$ field upon continued cooling implies that the solid α phase becomes supersaturated with respect to element B. If the driving force is large enough to overcome a nucleation barrier (cf. Sect. 9.2), precipitation of the solid β phase must occur. Two cases can be considered: (i) the driving force is very large (cooling at high rate so that the formation of β phase occurs at a relatively low

Fig. 7.30 Composition
profile in a grain of the
specimen resulting after
solidification in an
isomorphous, binary system
(cf. Fig. 7.29) due to
insufficient diffusional
equilibration in the solid
during the solidification. The
dashed lines represent
iso-composition lines
corresponding to the
concentration profile in the
grain shown below

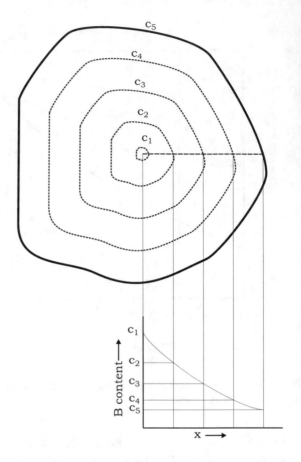

temperature in the $\alpha + \beta$ phase field) and thus homogeneous nucleation within the
solid grains of the initial α phase is possible, or (ii) the driving force is relatively
low (cooling at moderate rate so that the formation of β phase occurs at relatively
high temperature in the $\alpha + \beta$ phase field) and thus heterogeneous nucleation may be
predominant as at grain boundaries (and dislocations; see the "*Intermezzo; nucleation
of AlN in Fe-Al alloy*" in Sect. 9.2).

For the alloy of composition x_B^{eut} solidification starts and is completed at T_{eut}. Both
phases, α and β, have to solidify simultaneously. A lamellar structure, composed of
alternating α and β lamellae, develops (this morphology occurs often but not always).
The liquid immediately in front of a growing α lamella will be relatively rich in B.
This excess in B has to be removed by lateral diffusion of B to the neighbouring β
lamellae, where the adjacent liquid is relatively poor in B. Similarly, the excess in A
in front of a β lamella diffuses laterally to the neighbouring α lamellae. Hence, the
development of a lamellar structure composed of alternating α and β lamellae makes
short diffusion paths possible to realize the desired redistribution of A and B atoms
during the, necessarily isothermal, eutectic solidification. Indeed, the higher the rate

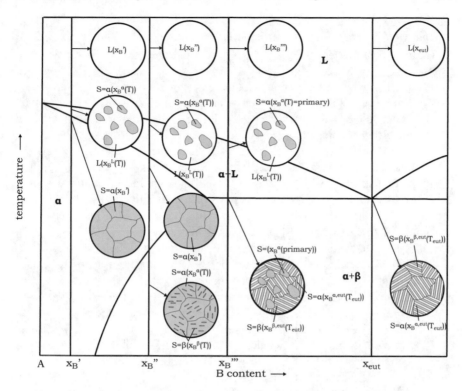

Fig. 7.31 Microstructure evolution during cooling of initially liquid, binary alloys of different compositions as indicated in the phase diagram for a eutectic system (cf. Figs. 7.12 and 7.13)

of heat extraction, the finer the lamellar structure is, to shorten the diffusion paths of A and B in the liquid at the solidification front (see also Sect. 9.4.2 on eutectoid transformations).

Finally, for the alloy of composition x_B''' solidification starts with the formation of (primary) α crystals in the melt, as long as the system has a temperature within the two phase, $\alpha + L$ field. Upon continued cooling the composition of both the solid α phase and the liquid phase L become enriched in B. When the temperature has reached the value T_{eut}, solid α phase of composition $x_B^{\alpha(eut)}$ and liquid phase L of composition x_B^{eut} occur (assuming that equilibrium is realized at all temperatures during cooling; but see the discussion above). Upon further extraction of heat, first all L has to solidify, while the temperature stays at T_{eut}. In accordance with the above discussion on eutectic solidification, the final microstructure will be composed of a matrix of α and β lamellae colonies with dispersed (primary) α crystals (which had solidified before the development of the α/β lamellar structure). Note that the occurrence of nonequilibrium, segregation phenomena, as discussed for the precipitation of α phase in an isomorphous system (Fig. 7.30), will lead to dispersed α crystals and α lamellae different in (overall) composition.

Epilogue: The Topology of Phase Diagrams; Some Rules

The method followed in Sect. 7.5, to present a manifold of phase-diagram types, concerned the geometrical description of the shape and location of phase fields in a 1, 2, 3,... dimensional space put up by 1, 2, 3,... intensive state variables. Such an approach thus can be denoted as a topological treatment. Qualitative arguments were used to arrive at phase diagrams of stepwise becoming higher complexity (especially in Sects. 7.5.2.2–7.5.2.4). The technique followed involved the design of a schematic sketch of the type of phase diagram dealt with, followed and compared with an example found in nature of that type of phase diagram, which was taken from the literature. Already simple descriptions of the Gibbs energies of phases suffice, by variation of only one (!) parameter to characterize the Gibbs energies of different phases, to present a range of phase-diagram types as presented here (e.g. see Lele, 2011). Such treatments also provide the demonstration of some simple rules and boundary conditions which should be obeyed in the construction (e.g. on the basis of experimental data) of phase diagrams:

(1)   *Upon crossing the border line (phase boundary) between two adjacent phase fields in a phase diagram the number of phases can only change by one, with, moreover, the other phase(s) remaining the same.* This rule holds independent of the number of components. The reader can simply check the validity of this rule considering the given examples of phase diagrams of binary and ternary systems. Of course, the phase rule has to be obeyed as well. Thus, the number of phases cannot be increased beyond the maximum possible for the given number of selected intensive state variables: in a two component system, at fixed pressure, the

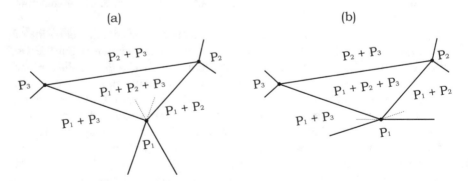

Fig. 7.32 A three-phase (P_1, P_2 and P_3) field in a composition triangle (at fixed pressure and fixed temperature) of a ternary system, demonstrating the only two possible cases of extrapolation of the two one-phase/two-phase boundaries (i.e. here the $P_1/(P_1 + P_2)$ and $P_1/(P_1 + P_3)$ phase boundaries) confining a one-phase (here P_1) field: **a** both into the three-phase field and **b** both outside the three-phase field (i.e. one into each of the two two-phase fields adjacent to the one-phase field). The extrapolations have been indicated by dashed lines

maximum number of phases in equilibrium is three; in a three-component system it is four at fixed pressure and it is three with fixed pressure and fixed temperature as in the composition triangle.

In fact this discussion provides the generalization of the rule given in Sect. 7.5.2.2 for binary systems that "between two single phase regions always a two-phase region occurs". The applicability of this rule can be tested for a real, ternary system represented by Fig. 7.28.

(2) *Pure elements are not stable phases in multi-element systems.* There is always some, although possibly very small, mutual solubility for the elements in the system, as a consequence of the increase of the configurational entropy upon solution formation (see Sect. 7.3). The number of degrees of freedom in binary and ternary systems for (terminal) solid solutions is 2 and 3, respectively (at fixed pressure), implying that the temperature and one (binary systems) or two (ternary systems) composition parameters are independent intensive state variables. Phase diagram sketches showing no compositional extent of the phase field for terminal solid solutions are misleading, as they suggest that the temperature is the only remaining degree of freedom for such phases. Such an in principle incorrect, but more often than rarely, presented drawing of a binary phase diagram was already discussed in a note at the end of Sect. 7.5.2 (Fig. 7.25). Preferably, the occurrence of such minute, terminal solid solution ranges is indicated yet in the phase diagram by use of arrows pointing at the ordinates (see " $\leftarrow$ (Mg)" and "(Si) $\rightarrow$ " in Fig. 7.17b). A similar remark can be made for so-called line compounds (see Sect. 7.5.2.4 and Figs. 7.17, 7.18 and 7.19). The composition range, that must occur for any and thus also this type of compound, because of the thermodynamic background indicated above, is too small to be resolved for the composition scale employed in the drawn phase diagram and thereby a conceptually wrong interpretation is possible (see above).

(3) Extrapolation of phase-boundary lines in phase diagrams is often applied to consider/find out possible metastable equilibria: For example, in case of solvus lines, see the dashed line in Fig. 7.11 at the extreme right, or, for supercooled water in metastable equilibrium with water vapour, see the dashed line in Fig. 7.4.

Consider the phase diagram of a binary, eutectic system, as shown in Fig. 7.11 at the extreme right and in Fig. 7.13. Thermodynamics dictates that *the tangents of the $\alpha/(\alpha + \beta)$ phase boundary and the $\beta/(\alpha + \beta)$ phase boundary, at the eutectic temperature, must lie within the two-phase fields $\alpha + L$ and $\beta + L$, respectively.* In other words, the metastable continuations of these phase boundaries must lie within the two-phase fields mentioned. It can be verified that the experimental phase diagrams shown in Figs. 7.12 and 7.14 satisfy this constraint.

In case of possible extrapolations of phase-field boundaries, confining a one-phase field touching a corner of a three-phase field in a composition triangle of a ternary system, the following constraint for such extrapolations can be useful (thus, for example a possible incompatibility of the lines drawn in a composition triangle, e.g. on the basis of experimental data, can be exposed):

Consider a one-phase (P_1) region in an isothermal section (composition triangle) that neighbours on one side with a two-phase $(P_1 + P_2)$ region and on an other side with an other two-phase $(P_1 + P_3)$ region, as sketched in both Figs. 7.32a, b. Apparently the two phase boundaries of the P_1 phase field can intersect at one of the corner positions of the triangle set up by the compositions of the three phases P_1, P_2 and P_3 of the (here, because at fixed pressure and fixed temperature, nonvariant) three-phase equilibrium of P_1, P_2 and P_3. Note that the sketches in Fig. 7.32 satisfy the rule that upon crossing a phase-boundary line the number of phases can only change by one, while the other phases remain the same (see under (1) above).

Now it must hold that *extrapolations of the $P_1/(P_1+P_2)$ and $P_1/(P_1+P_3)$ phase boundaries must follow either (i) tracks within the triangle set up by the compositions of the three phases P_1, P_2 and P_3 (i.e. the three-phase field) or (ii) tracks outside the three-phase field, i.e. one extrapolation runs in the (P_1+P_2) two-phase field and one extrapolation runs in the (P_1+P_3) two-phase field (see the dashed lines in Fig. 7.32a, b).*

It can now be verified that these constraints hold for the $L_1 + L_2 + $ (Ni, Cu) three-phase equilibrium, shown in Fig. 7.28 at 1300 °C, regarding the extrapolations of the $L_2/(L_2 + $ (Ni,Cu)) and $L_2/(L_1 + L_2)$ phase boundaries (extrapolations *within* the three-phase field), and for the (Ag) $+ L +$ (Ni,Cu) three-phase equilibrium, shown in Fig. 7.28 at 900 °C, regarding the extrapolations of the (Ni,Cu)/((Ni,Cu) $+$ (Ag)) and (Ni,Cu)/((Ni,Cu) $+ L$) phase boundaries (extrapolations *outside* the three-phase field into the bordering two-phase fields).

Generalizations and extensions of the above rules can be given, to a large extent based on theoretical, thermodynamic analyses of more than a century ago. The interested reader is referred to the general literature indicated at the end of this chapter and references cited therein.

References

General

Y.A. Chang, Phase diagram calculations in teaching, research and industry. Metall. Mater. Trans. A **37A**, 273–305 (2006)

S. Lele, Phase diagrams—Past Present and Future. IIM Metal News **14**, 22–41 (2011)

H.L. Lukas, G.S. Fries, B. Sundman, *Computational Thermodynamics* (Cambridge University Press, Cambridge, 2007).

T.B. Massalski (ed.), *Binary Alloy Phase Diagrams*, 2nd edn. (ASM, Metals Park Ohio, 1996)

B. Predel, M. Hoch, M. Pool, *Phase Diagrams and Heterogeneous Equilibria* (Springer, Berlin, 2004).

N. Saunders, P.A. Miodownik, *CALPHAD Calculation of Phase Diagrams* (Pergamon Press, Oxford, 1998).

G. Petzow, G. Effenberg (eds.), *Ternary Alloys: A Comprehensive Compendium of Evaluated Constitutional Data and Phase Diagrams* (Wiley-VCH Verlag, Weinheim) (a series of volumes published starting 1988)

Specific

S.-L. Chen, R. Schmid-Fetzer, K.-C. Chou, Y. Austin Chang, W.A. Oates, *A note on the application of the phase rule*. Int. J. Mater. Res. **99**, 1210–1212 (2008)

E.J. Mittemeijer, M.A.J. Somers, Thermodynamics, kinetics, and process control of nitriding. Surf. Eng. **13**, 483–497 (1997)

S. Sommadossi, W. Gust, E.J. Mittemeijer, Characterization of the reaction process in diffusion-soldered Cu/In-48 at.% Sn/Cu joints. Mater. Chem. Phys. **77**, 924–929 (2002)

F. Spaepen, A survey of energies in materials science. Phil. Mag. **85**, 2979–2987 (2005)

Chapter 8
Diffusion

Transport of material by migration of atoms or molecular entities, i.e. diffusion, is one of the most fundamental, elementary processes in materials and thus of great importance to the materials scientist and engineer. Firstly, a desired redistribution of the atoms of the elements in a solid/workpiece can be evoked by subjecting the material to a thermal treatment giving possibly rise to the development of new phases and microstructure (see Chaps. 9 and 10), leading to optimum, desired properties. The rate (i.e. the kinetics) of such processes is, often next to nucleation processes, in many cases determined by the necessary diffusion processes. Obviously, reactions between a solid and a liquid and/or a gas involve diffusion processes as well. Secondly, restricting ourselves to diffusion in solids, understanding the mechanism of diffusion processes in solids can lead to deep insight into the nature and density of defects exhibited by the atomic arrangement, as in crystals (e.g. vacancies and dislocations; cf. Chap. 5).

One should distinguish between the net, *macroscopic* flow of material and the movements, more or less haphazard jumps, of the individual atoms which provide the *atomistic* mechanism of nature at the background of the diffusion phenomenon. This sentence introduces the two ways/levels to describe diffusion: the continuum approach (Sect. 8.1) and the atomistic approach (Sect. 8.2).

8.1 The Continuum Approach to Diffusion; Fick's First and Second Laws

A system strives for a state of minimal energy in order to be in equilibrium. If the spatial distribution of the components of a system does not correspond to equilibrium, a tendency exists to realize by material transport such a state of equilibrium. It seems natural to assume that the energy deviation from equilibrium is adopted as the "driving force" for material flow. Then, to first-order approximation, it can be proposed that the local flux is proportional with the local gradient in energy (i.e. the derivative with

© Springer Nature Switzerland AG 2021
E. J. Mittemeijer, *Fundamentals of Materials Science*,
https://doi.org/10.1007/978-3-030-60056-3_8

respect to position; cf. first term of a Taylor series expansion). This approach already guarantees satisfying the boundary condition that no (net) material flow occurs if the energy gradient has become nil. This philosophy leads for one-dimensional diffusion to:

$$J = -\text{constant.}\frac{\text{d}(\text{energy})}{\text{d}x} \tag{8.1}$$

where the flux J represents the amount of transported material per unit of time and per unit of area of the cross section perpendicular to the diffusion direction (e.g. unit of J: kg/(m^{2}s) or number of moles (atoms)/(m^{2}s)). The position coordinate x denotes the direction along which diffusion takes place. The minus sign at the right-hand side of Eq. (8.1) expresses that net material flow takes place in the opposite direction of the energy gradient, recognizing that the diffusion brings about energy release. In the sense of the discussion given in Chap. 7, it can be suggested that at constant temperature and pressure, for the diffusion of a single component, the "energy" parameter in Eq. (8.1) can be interpreted as the partial Gibbs energy, i.e. the chemical potential, of the component considered (cf. Eq. (7.7) in Sect. 7.3); then, d(energy)/dx in Eq. (8.1) is written as dμ/dx.

If diffusional mixing of (ideal) gases is considered, it becomes immediately clear that the diffusion is driven by the increase in entropy (which decreases the (Gibbs) energy; see the discussion in Sect. 7.3, where the notion of (configurational) entropy was introduced). In more condensed systems, as liquids and solids, and at decreasing temperature, where the interaction between the diffusing entities of different kinds becomes increasingly important, additional, enthalpy effects obstruct descriptions of diffusion that at the same time are both simple and fully rigorous.

Equation (8.1) is perhaps the simplest proposal to describe diffusional flow one could conceive. For example, we could have included higher order terms, in accordance with the Taylor series expansion, and/or similar terms describing the dependences of the flux on (for diffusion of even a single component) all (i.e. of all components in the system) chemical potential gradients, and also on (possible) electric potential gradients, mechanical stress gradients, etc.

Historically a different route was followed. It was observed that upon annealing an otherwise homogeneous system exhibiting compositional heterogeneity often the compositional variations decreased and eventually vanished. From now on, we will largely focus on binary (two component) systems in which one-dimensional diffusion takes place. It was accordingly proposed that the diffusional flux would be proportional to the concentration gradient of one (of both) diffusing components and thus:

$$J = -D\frac{\partial c}{\partial x} \tag{8.2}$$

In this expression for the flux J, the proportionality constant has the name *diffusion coefficient*. Recognizing that J is expressed as quantity per unit area of cross section and per unit of time and c is expressed as quantity per unit of volume, it follows directly from Eq. (8.2), that the dimension of D is (length)^2/time, and thus a usual unit for D is m^2/s.

The last expression is the one commonly used in descriptions of diffusion. It was first proposed by Fick (1855).[1] Because, within the context discussed here (but see Sect. 8.2), it cannot be derived but simply expresses phenomenology, it is called Fick's first *law* of diffusion.[2]

It is important to realize that the "diffusion coefficient" as introduced here incorporates the effects of the contributions to the diffusion (material flow) in the system due to all (both; see above) mobile components in the system. It is not obvious that these effects can be represented by a single diffusion coefficient in the phenomenological Eq. (8.2). Yet, empirically this was found to be the case. It will be shown theoretically in Sect. 8.7.1 that a single diffusion coefficient indeed suffices to describe the diffusion in the two-component system on the basis of Eq. (8.2).

It is also possible to reconcile Eq. (8.2) with Eq. (8.1): the driving force for diffusion can be taken as a gradient of energy (gradient of chemical potential) by incorporating an additional factor (the so-called thermodynamic factor) into the diffusion coefficient, which will be demonstrated in Sect. 8.7.2. Hence, by using Eq. (8.2), as will be done in the following, one then in fact departs from an equation as Eq. (8.1) and its concept.

The treatment in this section immediately makes clear that the diffusion coefficient in Eq. (8.2) can be concentration (distance/position) dependent. It can also be time dependent (e.g. if microstructural changes occur during the diffusion process in the material considered). Especially also in such cases one often prefers approaches as indicated above to describe diffusion processes, i.e. departing from equations as Eq. (8.2) (or Eq. (8.6); see below), as they facilitate relatively easy computations.

The differential Eq. (8.2) is especially suited to describe diffusion in stationary states (cf. footnote 2 in this chapter). To deal with non-stationary states, i.e. to describe

---

[1] The progress of science is tributary immensely to the process of "thinking in analogies". One of the most striking examples is provided by the (mathematical) similarities in the theories for the conduction of heat in solids (see the book by Carslaw and Jaeger 1959) and for the diffusion of mass in solids (see the book by Crank 1975). Fick was led by such thinking in analogies to his proposal of what we now call Fick's (first and second) laws of diffusion. He remarks that "It was quite natural to suppose that this law for diffusion ... must be identical with that, according to which the diffusion of heat in a conducting body takes place". And he explicitly links his proposals to the earlier "theory of heat" by Fourier and that due to Ohm for the "diffusion of electricity in a conductor" (Fick 1855).

[2] Fick tested successfully the validity of Eq. (8.2) for the case of diffusion in a system composed of dissolved rock salt in water in a stationary (steady) state (cf. Sect. 7.3): Fick kept a bottom layer of water in a vessel saturated with dissolved rock salt and kept the top layer as pure water, leading to a stationary state of constant transport of dissolved salt from bottom to top of the vessel, thus establishing a *time-independent* dissolved salt concentration profile as a function of height in the vessel; the (net; cf. Sect. 8.2) amount of dissolved salt arriving at a certain height, coming from the bottom, equals the (net; cf. Sect. 8.2) amount of dissolved salt leaving this height, moving upward, and this holds at every height in the vessel. The resulting concentration-height profile must be a straight line (integrate Eq. (8.2) for the case J = constant, i.e. does not depend on x (here x = height in the vessel)), as found by Fick (Fick measured the specific mass as function of the height) (cf. Fig. 7.2).

Fig. 8.1 One dimensional
concentration-depth profile
and corresponding
flux-depth profile (with flux
proportional with $\partial c/\partial x$),
showing a linearization of
the flux between x_1 and x_2
for derivation of Fick's
second law

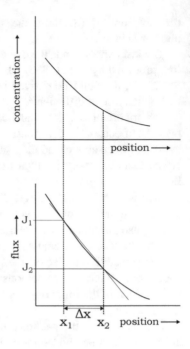

the evolution of concentration profiles as function of time (and temperature), Eq. (8.2) is combined with a material balance, leading to a second differential equation exhibiting an explicit time dependence for the concentration c.

Consider Fig. 8.1. Although the dependence of c on x will not be linear in general, the flux, which is proportional to $\partial c/\partial x$ (Eq. (8.2), can be taken as linearly dependent on x for x values between two locations x_1 and x_2 at infinitesimally small distance from each other. (This type of linearization of dependencies on variables is a "trick" usually applied in finding the differential equations describing complicated processes for which in general analytical formulations cannot easily be derived.) Hence:

$$J_{x_1} - J_{x_2} = -\Delta x \frac{\partial J}{\partial x} \tag{8.3}$$

where $\partial J/\partial x$ is the gradient in J for $x_1 \leq x \leq x_2$. Because $J_{x_1} \neq J_{x_2}$, accumulation of material occurs between the planes $x = x_1$ and $x = x_2$. Per unit area of cross section perpendicular to the diffusion direction this accumulation of material per unit of time is given by $J_{x_1} - J_{x_2}$ which leads to an (infinitesimal) increase of the concentration per unit of time equal to $\partial c/\partial t$ in the volume $1 \cdot \Delta x$, and thus, the material balance reads:

$$J_{x_1} - J_{x_2} = \Delta x \frac{\partial c}{\partial t} \tag{8.4}$$

Combining Eqs. (8.3) ands (8.4), it is obtained:

$$\frac{\partial c}{\partial t} = \frac{-\partial J}{\partial x} \tag{8.5}$$

or, using Eq. (8.2):

$$\frac{\partial c}{\partial t} = \frac{\partial}{\partial x}\left(D\frac{\partial c}{\partial x}\right) \tag{8.6a}$$

which, if D is constant, reduces to:

$$\frac{\partial c}{\partial t} = D\frac{\partial^2 c}{\partial x^2} \tag{8.6b}$$

Equation (8.6) is usually called Fick's second law. However, since Eq. (8.6) can be straightforwardly derived by combining Fick's first law and a material balance, as shown above, the terminology "law" for the resulting Eq. (8.6) in fact is wrong, as this expression is not based on empiricism.[3] Another name for the formula is "continuity equation".

With a view to the possible generalizations of Fick's first law, expressed either as Eq. (8.1) or Eq. (8.2), and as hinted at in the third paragraph below Eq. (8.1), (analogous) generalizations/extensions of Fick's second law are possible as well. A number of such extensions are presented in Philibert (2009). Overlooking the entire natural world (i.e. not only focussing on solid state diffusion as mostly done here), it might even be said that Fick's laws pertain to an only limited class of diffusion phenomena…

8.2 The Atomistic Approach to Diffusion

On a microscopic scale diffusion is due to the jumping, from one site to another, of the basic constituents which build up the piece of matter concerned: atoms or molecular entities. From now on in the discussion we will speak of atoms, which is not a real limitation for what follows.

The jumping process referred to does not necessarily occur in one specific direction. It can be conceived as caused by thermal agitation: energy fluctuations of a thermally vibrating atom occur by collisions with its, also thermally vibrating, neighbours and thereby energy barriers for jumps from one site to another can be (occasionally)

---

[3] In a strict sense use of the notion "law" should be confined to rules, describing the action of forces (as the law of gravity) and the course of processes (as Fick's first law), which have been found (initially) to be valid on the basis of empirical (i.e. relying on experience and observation alone) work. However, in science one is not puristic: for example, one also speaks of Bragg's law (Sects. 4.5 and 6.10), which relation was, also initially, derived theoretically in a straightforward manner and in no way was proposed on the basis of empiricism. Adopting a wide interpretation one could say: a "law" expresses "the regularity of nature", but this approach introduces a broad and ill defined transition region between just a formula/relation and a law.

overcome. Generally, the trajectory followed by an atom has a strongly haphazard nature: a "random walking" occurs. A zig-zag path is observed for an individual atom. Only by considering large numbers of jumping atoms, it becomes possible to observe the *net* diffusional flow along the concentration gradient. This discussion reveals the statistical nature of the diffusion problem.

Simple statistical theory, dealing with the above sketched "random walk problem", indicates that, considering many atoms, initially all at the origin in space, the net, mean square path after time t covered by an atom, $\langle x^2 \rangle$, is given by:

$$\langle x^2 \rangle = \Gamma t a^2 \tag{8.7}$$

where Γ denotes the jump frequency and a is the jump distance. To arrive at this result, it has been assumed that each jump (direction) is independent of (the directions of) all previous jumps that positive (forward) and negative (backward) jumps are equally probable and that the jump distance is constant (think of diffusion in a crystal). Another way for considering this problem is taking Eq. (8.6b) and solving it for the case of all diffusing atoms at the origin in space at $t = 0$. It follows for the net, average mean square distance to the origin of an atom after time t (three-dimensional diffusion):

$$\langle x^2 \rangle = 6Dt \tag{8.8}$$

This equation, already (see further in this chapter), presents a "square root of time dependence" of a "diffusion distance" (here $\langle x^2 \rangle^{1/2}$) (at constant temperature and constant pressure).

From Eqs. (8.7) and (8.8), the following equation, giving the atomistic interpretation of the diffusion, is obtained:

$$D = \frac{1}{6}\Gamma a^2 \tag{8.9}$$

Intermezzo: Brownian Motion

In 1828, R. Brown reported he had observed randomly moving particles (from the pollen of plants) suspended in water. At the time, the origin of these random movements was unclear. Later work showed that the Brownian motion can be observed for suspended particles smaller than, say, 1 μm (the resolution of the light optical microscope is of the order 0.2 μm; see Sect. 6.5.2) and the view emerged that the haphazard movements of the particles are due to collisions with the thermally moving molecules of the liquid (water).

A theoretical analysis was eventually given by Einstein and Smoluchowski in 1905–1906. Each zig or zag as observed under the light optical microscope by making particle-position measurements at times t_1 and t_2, i.e. determination of

x_1 and x_2, is the outcome of very many collisions with the liquid molecules (the number of collisions per second is very great: about 10^{20}/s) and, as a result, from the net distance covered in a time period t (only) a squared average velocity of the particle can be determined which is given by $\langle x^2 \rangle / t^2$, where $x = x_2 - x_1$ and $t = t_2 - t_1$. It should be recognized that the squared instantaneous velocity of each particle is very much larger than this squared average velocity. As follows from Eq. (8.8), a value for the diffusion coefficient can be derived from $\langle x^2 \rangle / 6t$, and thereby, the random walk of a single particle (the microscopic scale) has been related to the diffusion of many particles (the macroscopic scale).

The great contribution of Einstein was to demonstrate that Avogadro's number can be calculated straightforwardly from D, provided the value of the viscosity of the liquid is known. Einstein's method to determine Avogadro's number from Brownian motion and other methods for the determination of Avogadro's number led to similar values in the first decade of the twentieth century, and thereby, the reality of the atoms and molecules was settled once and for all.

An alternative way of arriving at Eq. (8.9) is as follows. Consider a crystalline bar exhibiting a one-dimensional concentration profile of the component considered, A, along the length axis of the bar (see Fig. 8.2). Select two neighbouring (lattice) planes perpendicular to the bar axis at the locations x_1 and x_2. The jump frequency for the atoms is Γ (cf. Eq. (8.7)). Assuming that the jump frequency is the same along the three orthogonal directions and that jumps can occur in positive and negative directions with the same probability, it follows that, if n_1 represents the number of atoms A in the lattice plane at x_1, then per unit of time $1/6\,\Gamma n_1$ jumps of atoms A occur from the lattice plane at x_1 to the lattice plane at x_2. Similarly, if n_2 represents the number of atoms A in the lattice plane at x_2, then per unit of time $1/6\,\Gamma n_2$ jumps of atoms A occur from the lattice plane at x_2 to the lattice plane at x_1. Hence, the flux of A from the lattice plane at x_1 to the lattice plane at x_2 is given by:

Fig. 8.2 The atomistic approach to diffusion. A case of one dimensional diffusion in a crystalline bar with concentration profile (of the component A). The net flux of atoms A from x_1 to x_2 (x_1 and x_2 indicate positions of neighbouring lattice planes) is given by the number of atoms A jumping over the distance a from x_1 to x_2 minus the number of atoms A jumping over the distance a from x_2 to x_1

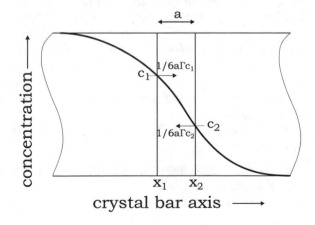

$$J = \frac{1}{6}\Gamma(n_1 - n_2) \tag{8.10}$$

The concentration of A at the lattice plane at x_1 equals $c_1 = n_1/a$, where a is the distance between adjacent lattice planes, and, similarly, the concentration of A at the lattice plane at x_2 equals $c_2 = n_2/a$. Therefore, Eq. (8.10) can be expressed as;

$$J = \frac{1}{6}\Gamma a(c_1 - c_2) \tag{8.11}$$

Again applying an "infinitesimal" consideration and linearizing the concentration profile between x_1 and x_2 (see discussion above Eq. (8.3)), it follows:

$$c_1 - c_2 = -a\frac{\partial c}{\partial x} \tag{8.12}$$

Substituting Eq. (8.12) into Eq. (8.11) it is obtained:

$$J = -\frac{1}{6}\Gamma a^2 \cdot \frac{\partial c}{\partial x} \tag{8.13}$$

Comparing Eq. (8.2) with Eq. (8.13), it follows that the above treatment (Eqs. (8.10)–(8.13)) provides a derivation of Fick's first law. Evidently:

$$D = \frac{1}{6}\Gamma a^2 \tag{8.14}$$

which was also the result of the random walk consideration (Eq. (8.9)).

8.3 Solutions of Fick's Laws

Fick's second law can generally be solved numerically subject to given boundary conditions. Some special, analytical solutions, for binary systems, are possible provided specific constraints are obeyed. Thus, if D is constant at constant temperature (and constant pressure; no function of composition/position) and the diffusion distance, characterized by $\sqrt{(Dt)}$ is small as compared to the size of the specimen in the direction of the diffusion, analytical solutions of Fick's second law at constant temperature (and constant pressure), $c(x, t)$, can be expressed in terms of error functions.

One often considers "semi-infinite" and "infinite"(pair of "semi-infinite") systems:

(i)   For the "semi-infinite" system, the following boundary conditions hold (see Fig. 8.3a):
      $c = c_0$ for $x = 0$ and $t \geq 0$;

Fig. 8.3 Illustration of solutions for Fick's second law:

a semi-infinite system: $c = c_0$ for $x = 0$ and $t \geq 0$; $c = 0$ for $x > 0$ and $t = 0$;

b infinite system (pair of semi-infinite systems): $c = c_0$ for $x < 0$ and $t = 0$; $c = 0$ for $x > 0$ and $t = 0$

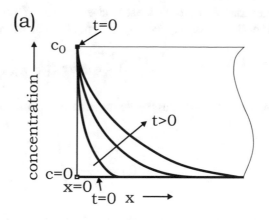

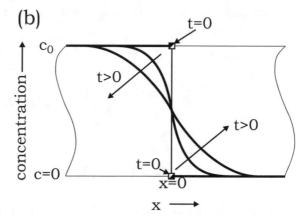

$c = 0$ for $x > 0$ and $t = 0$.

This case can be met for a substrate in contact with a (gas) atmosphere of which a component can dissolve in the substrate.

(ii) For the "infinite" system (pair of "semi-infinite" systems) the boundary conditions are (see Fig. 8.3b):

$c = c_0$ for $x < 0$ and $t = 0$;

$c = 0$ for $x > 0$ and $t = 0$.

This case can be met for an A/B diffusion couple produced by welding together a piece of A and a piece of B (without that appreciable diffusion across the interface has occurred). It is assumed that both components are fully mutually dissolvable and that full homogenization establishes one phase of constant composition (pure A and pure B are amorphous and stay amorphous upon dissolving B or A, respectively, or pure A and pure B have the same crystal structure which is kept upon dissolving B or A, respectively).

The general solution of Fick's second law for cases (i) and (ii) at constant temperature (and constant pressure) and for constant D can be written as:

$$c(x, t) = a + b\,\mathrm{erf}\left\{\frac{x}{2\sqrt{(Dt)}}\right\} \tag{8.15}$$

where the so-called (Gauss) error function, also called "probability integral", is given by

$$\mathrm{erf}(z) = \frac{2}{\sqrt{\pi}} \cdot \int_0^z \exp(-s^2)\mathrm{d}s \tag{8.16}$$

Note that erf $(0) = 0$, erf $(\infty) = 1$ and erf $(-z) = -\mathrm{erf}\,(z)$ (and thus erf $(-\infty) = -1$). The constants a and b in Eq. (8.15) follow by substituting the boundary conditions, and thus, it is obtained:

"semi-infinite" system:

$$c(x, t) = c_0 \cdot \left[1 - \mathrm{erf}\left\{\frac{x}{2\sqrt{(Dt)}}\right\}\right] \tag{8.17}$$

"infinite" system:

$$c(x, t) = \frac{c_0}{2} \cdot \left[1 - \mathrm{erf}\left\{\frac{x}{2\sqrt{(Dt)}}\right\}\right] \tag{8.18}$$

Note that these results can be read as that the plane of concentration c moves along the abscissa with a speed proportional with $\sqrt{(Dt)}$.

Many analytical solutions can be found in the classical, mathematically oriented books by Crank (1956) and by Carslaw and Jaeger (1959), the latter book being devoted to the conduction of heat (in solids) which is governed by differential equations similar to Fick's laws for diffusion. Nowadays the importance of analytical solutions has been reduced considerably in view of the advent of powerful (personal) computers allowing one to solve the governing diffusion differential equations, Eqs. (8.2) and (8.6), numerically. For the development of the corresponding algorithms, see the mentioned book by Crank.

Diffusion in thin film systems has become of great importance in recent days. A number of solutions to Fick's second law for a variety of thin film systems, and of the character discussed above, have been given in the Appendix to this chapter.

As can be seen from Fig. 8.3b, the resulting concentration–distance profile has a point of inversion at $x = 0$. Such symmetry only occurs if the diffusion coefficient is no function of concentration. This is in reality rarely the case, if at all. Already the

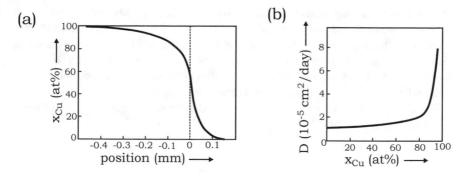

Fig. 8.4 **a** Concentration profile in a Cu-Ni diffusion couple (120 h at 1025 °C; measurements by Grube and Jedele (1932) and **b** corresponding (inter)diffusion coefficients as deduced by Matano (1933) (see Sect. 8.4)

classical, old data obtained by Grube and Jedele (1932) on diffusion in the system Cu-Ni reveal that the diffusion coefficient depends (strongly in the copper-rich region) on concentration (Fig. 8.4, cf. Fig. 8.3b).

8.4 Concentration Dependence of the Diffusion Coefficient; the Boltzmann–Matano Analysis

To characterize and predict diffusion behaviour, knowledge of the diffusion coefficient for the (binary, solid solution) system considered is required. For simple cases of diffusion, where analytical solutions for the concentration profile, at constant temperature (and at constant pressure), are possible, straightforward fitting of these analytical functions, to experimentally acquired data of concentration profiles, suffices in principle to determine diffusion-coefficient values. If the diffusion coefficient is not a constant for each temperature and depends on concentration or on time, this approach becomes generally impossible to apply, with the exception of special cases, which have been considered in the literature and for which specific, also analytical approaches have been proposed yet (Crank 1975).

The most used method for determination of the concentration dependence of the (inter)diffusion coefficient, for a binary solid solution, infinite (cf. Sect. 8.3) system, is the Boltzmann–Matano analysis. For the case that the boundary conditions for solving Fick's second law, as given by Eq. (8.6a) (i.e. recognizing that $D = f(c)$), can be expressed in terms of a single variable, $u = x/\sqrt{t}$, i.e. the so-called *Boltzmann substitution*, Eq. (8.6a) transforms into (using $\partial c/\partial t = dc/du \cdot \partial u/\partial t = -\frac{1}{2}x/t^{3/2} \cdot dc/du$ and $\partial c/\partial x = dc/du \cdot \partial u/\partial x = 1/t^{1/2} \cdot dc/du$):

$$-\frac{u}{2}\frac{dc}{du} = \frac{d}{du}\left(D\frac{dc}{du}\right) \tag{8.19}$$

involving that c is a function of u. This implies, that for cases where this substitution can be made (see above), and at constant temperature, $c = f(u) = f(x/\sqrt{t})$, and thus the plane with concentration c shifts along the x-axis according to $x_c = $ constant. $\sqrt{t}$, with x_c as the instantaneous position of the plane of concentration c. It can be easily verified that the Boltzmann substitution complies with the cases of semi-infinite and infinite diffusion considered in Sect. 8.3 (see below Eq. 8.18).

Upon "multiplication" of both sides of Eq. (8.19) with du and integration from $c = 0$ till $c = c$, it is obtained:

$$-\frac{1}{2}\int_0^c u\,dc = \left[D\frac{dc}{du}\right]_{c=0}^{c=c}$$

The concentration profiles to be analysed have been determined for a specific time of diffusion (at constant temperature). Thus t can then be taken constant, and hence, $u = x/\sqrt{t}$ can be substituted as follows:

$$-\frac{1}{2}\int_0^c x\,dc = Dt\left[\frac{dc}{dx}\right]_{c=0}^{c=c}$$

For the infinite system considered here, it holds $(dc/dx)_{c=0} = 0$. Hence, it is finally obtained:

$$D(c) = -\frac{1}{(2t)}\left[\frac{dx}{dc}\right]_{c=c} \cdot \int_0^c x\,dc \tag{8.20}$$

Before considering this result for $D(c)$, it is noted that, for the infinite system considered, not only it holds that $(dc/dx)_{c=0} = 0$, but also $(dc/dx)_{c=c_0} = 0$. Consequently:

$$\int_0^{c_0} x\,dc = 0 \tag{8.21}$$

The last equation defines the origin of the x-axis: if $c = c_0$ would pertain to pure A and $c = 0$ to pure B, then the amounts of material A and B, that have (net) passed through the plane $x = 0$ since the start of (inter)diffusion, are equal (see the hatched areas in Fig. 8.5). The plane $x = 0$ is identified with the original interface between A and B (see also Sect. 8.7). The plane $x = 0$ is called *Matano interface*.

Now that the origin of the x-axis has been defined, the integral in Eq. (8.20) can be evaluated for every c from the distance–concentration profile, $x(c)$, measured after diffusion time t at fixed temperature (see the cross-hatched area in Fig. 8.5). Also, the derivative $[dx/dc]_{c=c}$ can be determined (even without that the origin of

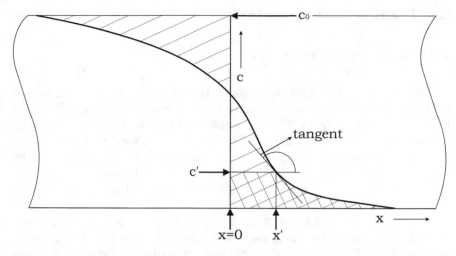

Fig. 8.5 Illustration of the Matano analysis. The concentration (c)—distance (x) profile, for an infinite, binary, solid-solution system. The singly hatched areas comply with Eq. (8.21). The cross-hatched area complies with the integral in Eq. (8.20) with upper bound $c=c'$

the x-axis has been specified) for every c from the tangent of the curve of $x(c)$. This is also shown in Fig. 8.5. In this way, by performing the analysis according to Eq. (8.20), the value of D can be determined as function of the concentration, for the infinite, binary, solid solution system, at the temperature concerned.[4] Obviously the accuracy in determining the tangent at the extremities of the concentration–distance profile can be low, thereby affecting the quality of the diffusion-coefficient data for the concentrations at these locations (i.e. for the case considered here, at $c = 0$ and at $c = c_0$).

Matano (1933) was the first to perform this analysis using the data obtained by Grube and Jedele (1932) for the Cu–Ni system: see Fig. 8.4b. Evidently, in the Cu–Ni system, a strong concentration dependence of D occurs, especially at the copper-rich side, where the value of the diffusion coefficient strongly increases with increasing copper concentration. It will be argued later that this observation can be qualitatively understood; i.e., this result is revealing about the occurring diffusion mechanism (see Sect. 8.6 under (v)).

---

[4] A Matano-like analysis is also possible for a semi-infinite system (Mittemeijer and de Keijser, Scripta Metallurgica, 11(1977), 113–115).

8.5   Diffusion Mechanisms in Crystalline Systems

8.5.1   Exchange Mechanisms

Diffusion is due to the migration of atoms. The perhaps most simple mechanism to be conceived for atoms on a crystal structure is the direct exchange (Fig. 8.6a). During many years (before 1950), this mechanism was thought to prevail, but it can be shown, at least for metals, that the deformation necessary for the two atoms during their passage ("squeezing" together), which acts against their (ion–ion) repulsion, is energetically that unfavourable that this mechanism is very unlikely to contribute significantly to diffusion. An, at first sight, seemingly even less likely exchange mechanism would be the so-called ring mechanism, which is a cooperative process: migration of atoms is realized by rotation of a ring of (four, in the example shown) atoms (Fig. 8.6b). However, as compared to the direct exchange of two neighbouring atoms in metals (Fig. 8.6a), the (ion–ion) repulsion would be reduced, which makes this special mechanism relatively more likely. The concerted exchange mechanism has been proposed to contribute to grain-boundary diffusion (see Sect. 8.8) and in a (very) minor way to the self diffusivity in an elemental semiconductor as silicon. Note that, in contrast with metallic bonding which is undirected, this material is characterized by directional (covalent) bonding (see Chap. 3), involving a tendency to break as few as possible bonds upon exchange.

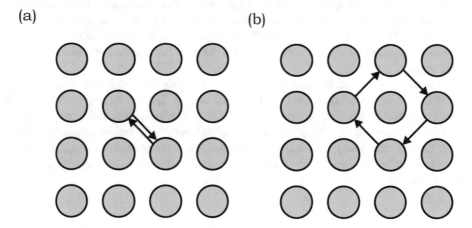

Fig. 8.6 a Direct exchange and **b** ring exchange mechanisms for substitutional diffusion

8.5.2 The Vacancy Mechanism; Substitutional Diffusion

After, say, 1950, the role of crystal defects in diffusion has increasingly been realized. Indeed, the presence of a vacancy as neighbour of the oscillating atom considered would enhance the chance for a jump of the atom considered to the (vacant) neighbouring crystal-structure site. This consideration leads to the vacancy mechanism of diffusion.

The vacancy mechanism can be conceived as the direct exchange of a vacancy with an atom on the same crystal structure (Fig. 8.7). The energy barrier corresponding to this vacancy-atom exchange is only a fraction of that for the direct exchange of two atoms on the same crystal structure. The activation-energy barrier for this process is given by the difference in (potential) energy of the atom before its jump and at its position half-way, implying that the activation energy of this migration process is representative of a distortion energy: the jumping atom has to force its way between adjacent atoms which have to be displaced (see Fig. 8.7, middle of top and bottom parts of the figure). At the half-way position, these displacements are most pronounced and the jumping atom is said to be in the "activated state". The vacancy mechanism has been found to be the dominant diffusion mechanism for the parent atoms (self diffusion) in metals at elevated temperatures and for substitutionally dissolved foreign atoms. Note that, according to this mechanism, a net flow of diffusing substitutionally dissolved atoms in one direction (in case of self

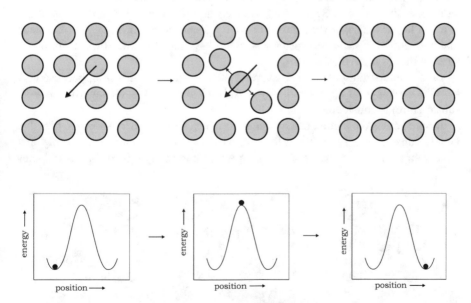

Fig. 8.7 The vacancy mechanism; substitutional diffusion. An atom during a jump from one crystal-structure site to an adjacent one, according to the vacancy mechanism (top part of the figure), and the corresponding change of the energy of the jumping atom (bottom part of the figure)

diffusion there is no net flow of atoms in any direction) is associated with a net flow of vacancies in the opposite direction.

8.5.3 Interstitial Diffusion

Solute atoms as carbon and nitrogen are relatively small and are dissolved at interstitial crystal-structure sites of a parent metal crystal structure, as the octahedral interstitial sites in a metal as iron. Diffusion is realized by the migration of the interstitial atoms. Evidently, for interstitial diffusion, the sites available to the jumping interstitial atoms occur on the substructure of interstitial sites, e.g. the substructure composed of all octahedral interstices of the b.c.c. crystal structure of α-iron (ferrite).

In view of the small solubilities of interstitials as carbon and nitrogen on the interstitial sites of the parent metal crystal structure, the number of vacancies on the substructure of interstitial sites considered is usually enormous, and consequently, the chance that an interstitial atom can jump to a neighbouring vacant substructure site is much larger than for a substitutionally dissolved atom on the parent metal crystal structure (cf. Figs. 8.7 and 8.8). The distortion of the parent crystal structure to let move an interstitial atom to a vacant neighbouring site on the substructure of interstitial sites of the parent crystal structure (see middle part of Fig. 8.8) is of the same order of magnitude as holds for the diffusion of a substitutionally dissolved atom (see discussion in Sect. 8.5.2). Hence, the diffusivity of an interstitial atom is much larger than that for a substitutionally dissolved atom, because the chance that a neighbouring site is a vacancy is much larger for the interstitial atom on its substructure of interstitial sites.

In fact, interstitial diffusion is a variant of substitutional diffusion, because also here the exchange of an (interstitially dissolved) atom with a vacancy (on its own substructure) is considered.

In some cases the picture can be rather complex. Boron can be shown to dissolve largely substitutionally in ferrite (α-iron; b.c.c. crystal structure) at relatively low temperatures. However, at elevated temperatures, the occupation of interstitial sites

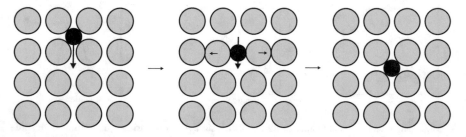

Fig. 8.8 Interstitial diffusion. An interstitial atom during a jump from one interstitial site to an adjacent one

by boron cannot be neglected and the diffusion of boron in ferrite then is governed by an interstitial diffusion mechanism (Fors and Wahnström 2008).

8.6 The Jump Frequency and the Activation Energy of Diffusion

Considering the vacancy mechanism and restricting the treatment for the time being to the diffusion of A in A, which is called "self-diffusion", the jump frequency Γ (see Sect. 8.2) can be given as:

$$\Gamma = z \cdot p_{vac} \cdot p_{mig} \tag{8.22}$$

This equation simply expresses that the jump frequency is, of course, proportional with the number of nearest neighbour crystal-structure sites (the so-called coordination number), z (cf. Sects. 3.5.3 and 4.2.4; $z = 12$ for f.c.c crystals; $z = 8$ for b.c.c. crystals, etc.), the chance that a nearest neighbour site is a vacancy, p_{vac}, and the chance that a jump (a migration) to the vacant nearest neighbour site occurs, p_{mig}.

As discussed in Sect. 5.1, the equilibrium fraction of thermal vacancies is given by Eq. (5.1) and thus:

$$p_{vac} = c_{vac} = \exp\left(\frac{-\Delta G_{vac}}{RT}\right) \tag{8.23}$$

where ΔG_{vac} denotes the Gibbs energy (free enthalpy) of formation of a mole of vacancies, apart from the entropy of mixing (=change in configurational entropy).

The chance that an atom can jump (migrate) into a vacant neighbouring crystal-structure site is proportional with the number of (thermal) oscillations (in the diffusion direction) around the equilibrium position of the crystal-structure site on which the atom resides, given by the frequency υ, and the probability that one of these oscillations (vibrations) is large enough to realize the jump. Adopting a Boltzmann factor for this last probability it follows:

$$p_{mig} = \upsilon \cdot \exp\left(\frac{-\Delta G_{mig}}{RT}\right) \tag{8.24}$$

with ΔG_{mig} as the energy required for the jumping atom to move to the "activated state" (cf. discussion in Sect. 8.5.2). The frequency of the thermal vibrations of the oscillating atoms is often taken to be of the order of the so-called Debije frequency which equals about 10^{13}/s.

Recognizing that $\Delta G = \Delta H - T\Delta S$ (cf. Eq. 7.4), and adopting Eq. (8.14) for the diffusion coefficient, it follows straightforwardly:

$$D = \frac{1}{6}a^2 z \upsilon \cdot \exp\left[\frac{\Delta S_{vac} + \Delta S_{mig}}{R}\right] \cdot \exp -\left[\frac{\Delta H_{vac} + \Delta H_{mig}}{RT}\right]$$

$$= D_0 \cdot \exp\left[-\frac{\Delta H_{vac} + \Delta H_{mig}}{RT}\right] \tag{8.25}$$

In the above derivation actually diffusion in a pure, elemental material (as a pure metal), i.e. "self-diffusion", was considered and D then is a constant. D in solid solutions is generally composition dependent (cf. Sects. 8.4 and 8.7).

The diffusion coefficient for self-diffusion, say of A in A, can be determined by diffusion of the (dilute) radioactive isotope $A*$ in A and subsequently analysing the $A*$ concentration profile: A and $A*$ are chemically identical, and thus, it appears acceptable to take the corresponding diffusion coefficients as equal.

However, the above statement ignores so-called correlation effects. Not all possible jumps of a diffusing single atom $A*$ dissolved in A are equally probable. Adopting the vacancy mechanism for substitutional diffusion, after one jump of the atom $A*$, the next jump of that atom $A*$ is more likely to happen back into the vacancy now on the previous location of the atom $A*$. In other words: the randomicity valid for the jumps of A in A (on that basis Eq. (8.9) was derived) does not hold for the jumps of $A*$ in A. Hence, $D_{A*} = f\, D_A$ with $f < 1$ (of course, for interstitial diffusion, such correlation can be neglected, and hence, $f = 1$ for interstitial diffusion) (for details, see, e.g. Shewmon 1989). The correlation factor f is independent of temperature for cubic structures. The experimental imprecision associated with the determination of concentration profiles usually makes diffusion analyses insensitive for the incorporation of the correlation factor (see Sect. 8.7.2 below Eq. (8.57)).

The diffusion coefficient D_{A*} is called tracer diffusion coefficient. It should be recognized that D_{A*}, as determined for diffusion of $A*$ in a homogeneous binary solid solution $A - B$, depends on the concentration of A in that matrix; further, see Sect. 8.7.2 below Eq. (8.49).

The resulting equation (Eq. (8.25)) is often called "Arrhenius equation" and is normally written as:

$$D = D_0 \cdot \exp\left(-\frac{Q}{RT}\right) \tag{8.26}$$

with

$$Q = \Delta H_{vac} + \Delta H_{mig} \tag{8.27}$$

as the so-called activation energy of diffusion.

A plot of ln D versus $1/T$, *the so-called Arrhenius plot*, should lead to a straight line with a slope given by $-Q/R$, with R as the gas constant. For metals, ΔH_{vac} and ΔH_{mig} are of comparable magnitudes. For f.c.c. metals, typical values for Q range from 1 to 3 eV/atom $\approx$ 100–300 kJ/mol, whereas for b.c.c. metals typical values of Q are larger: of the order 4 eV/atom $\approx$ 400 kJ/mol.

For interstitial diffusion, it can be said that $p_{vac} \approx 1$ (cf. discussion in Sect. 8.5.3), and thus, it follows

$$Q = \Delta H_{mig} \qquad (8.28)$$

which for f.c.c. metals means (see immediately above) that the activation energy for interstitial diffusion is of the order 50% of the value of Q for substitutional diffusion.

An impression of practical results is given by Fig. 8.9, where an Arrhenius presentation of diffusion coefficients for diffusion in various metals is given. These data give rise to the following discussion:

(i)   It is seen that the diffusion coefficient for diffusion of C in α-Fe (ferrite, b.c.c crystal structure for Fe) is much larger than that for diffusion of Fe in α-Fe (self-diffusion) at the same temperature: the activation energy, Q (following from the slope of the straight line in the plot), is distinctly smaller. This illustrates that Fe has to diffuse substitutionally on its own (parent) crystal structure, whereas C diffuses on the substructure of octahedral interstitial sites; i.e., it diffuses interstitially.

Note: The determination of the diffusion coefficient of C in α-Fe has provided the classical example confirming the Arrhenius type of temperature dependence of a diffusion coefficient: the validity of Eq. (8.26), here with $Q = \Delta H_{mig}$, has been confirmed over a very large temperature range (from -40 °C upwards (see Fig. 8.10)); a large part of the data are based on, in particular, internal friction measurements (see also Sect. 12.7 and Fig. 12.14b). However, later compilations of data for the diffusion coefficient of C in α-Fe, acquired by a range of experimental techniques, suggest a deviation from the straight line in an Arrhenius plot at high temperatures: an upward curvature appears to occur (Silva and McLellan 1976). As a possible explanation, it has, for example, been suggested that at such high temperatures some tetrahedral interstices would yet be occupied by the interstitial atom, which normally is considered as unlikely for b.c.c. iron (see the discussion in Sect. 9.5.2.1), and that the activation energy for jumping of carbon from tetrahedral site to tetrahedral site would be smaller than for jumping from octahedral site to octahedral site. Further, it is interesting to remark that in recent years experimental evidence has been obtained demonstrating that below about 100 K the diffusion coefficient of C in α-Fe becomes practically independent of temperature; i.e., the activation energy becomes vanishingly small. Theoretical analysis has suggested that this phenomenon is due to quantum mechanical tunnelling (Dabrowski et al. 2006).[5]

(ii) Evidently, the activation energy for interstitial diffusion of C in γ-Fe (austenite, f.c.c. crystal structure for Fe) is much larger than for C in α-Fe (ferrite, b.c.c.

---

[5] The diffusion coefficient of hydrogen in metals is very large—of the same order of magnitude as found for atomic diffusion in liquids—and, moreover, is characterized by a very low value of the activation energy, except at relatively high temperatures; diffusion of hydrogen in metals is not dominated by the thermally activated atomic jumps over barriers, but rather is governed by quantum mechanical tunneling (Fukai 2005).

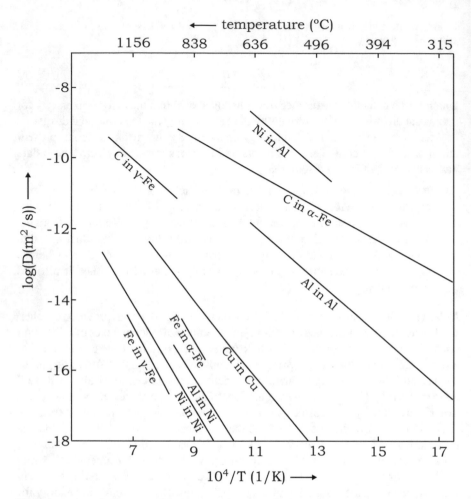

Fig. 8.9 Arrhenius plot of the diffusion coefficient for different (dilute) systems (*A* in *B* where *A* is the diffusing species (solute) and *B* is the matrix (solvent)). Data taken from (i) Landolt-Börnstein, Numerical Data and Functional Relationships in Science and Technology, Vol. 26: Diffusion in Solid Metals and Alloys, Ed. H. Mehrer, Springer Verlag, Berlin, 1990, and (ii) Smithells Metals Reference Book, 7th Edition, Eds. E.A. Brooks and G.B. Brook, Butterworth Heinemann, Oxford, 1992

crystal structure for Fe). This can be discussed as follows. Conceiving the (metal) atoms as rigid spheres, the relative amount of "free space" in the f.c.c. (and h.c.p.) crystal structure (26%) is smaller than in the b.c.c. crystal structure (32%). Carbon occupies the octahedral interstitial sites in both f.c.c.

Fig. 8.10 Diffusivity of carbon in b.c.c. iron as function of $1/T$ (data taken from C. Wert, Physical Review, 79 (1950), 601–605)

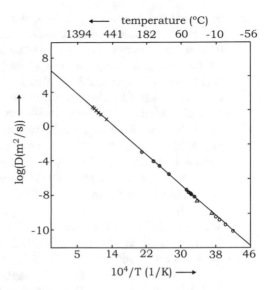

iron and b.c.c. iron (see also Sect. 9.5.2.1).[6] The activation energy for carbon diffusion may be interpreted as associated with the moving apart of a number of iron atoms between two adjacent octahedral interstices in order that the interstitial carbon atom can pass (jump) to the next octahedral interstitial site. The relatively large amount of "free space" in b.c.c. iron would facilitate such moving apart of the iron atoms for the jumping carbon atom and this would explain the relatively small value of the activation energy for carbon diffusion in b.c.c. iron.

(iii) The activation energy for diffusion of C in γ-Fe is reduced considerably if the amount of C dissolved in γ-Fe increases (see Fig. 8.11), which might

---

[6] It is remarkable to observe that the octahedral interstitial sites in the close packed crystal structures (f.c.c. and h.c.p.) are larger than in the less close packed b.c.c. crystal structure (for the same size of the (metal) atom taken as rigid sphere; this approximately holds for iron (see Sect. 9.5.2.1)). Yet, the total relative interstitial site volume is larger in the b.c.c. crystal structure, because per (metal) atom there are three times as much interstices: three octahedral and six tetrahedral interstices per metal atom in b.c.c. and one octahedral and two tetrahedral interstices per metal atom in f.c.c. and h.c.p.

As long as the reasoning is based on the relative amount of "free space" (see the main text), adopting the rigid (hard, solid) sphere model (cf. Sect. 4.2; for the f.c.c. crystal structure the atoms are in touch along the <110> directions (face diagonals of the unit cell); for the b.c.c. crystal structure the atoms are in touch along the <111> directions (body diagonals of the unit cell), a difference in the absolute value for the size of the atom in the one and the other crystal structure is irrelevant. Adopting the atomic volume (i.e. the volume per atom, for example calculated from the unit cell volume divided by the number of atoms in the unit cell) of the element considered as a measure for atom size, it follows, perhaps surprisingly/counter-intuitively, for many metals which can crystallize in f.c.c./h.c.p. and b.c.c. modifications, that the not close packed b.c.c. modification exhibits the smaller atomic volume (Rudman 1965). An exception is iron, where the reverse holds (also see Table 4.4 in Sect. 4.2.5), thereby lending support for the simple argumentation given above, also if one does not depart from the metal atom as a rigid sphere.

be considered as a consequence of the widening of the iron parent crystal structure upon dissolution of carbon: thereby the moving apart of a number of iron atoms, between two adjacent octahedral interstices in order that the interstitial carbon atom can pass (jump) to the next octahedral interstitial site, would become easier. It has also been claimed that the repulsion between carbon atoms at close distance (see discussion in Sect. 9.5.2.1) would enhance the diffusion of the interstitial atoms down along the concentration gradient, which, because of the relatively large interstitial solubility, can be appreciable in austenite (Bhadeshia 2004). A comparable effect for C in α-Fe cannot be observed because the solubility of C in α-Fe is very small and appreciable concentration gradients generally do not develop.[7]

(iv)   Now consider the self-diffusion coefficients of Cu in Cu and Al in Al (both f.c.c.). It follows that at the same temperature the diffusion coefficient for diffusion of Cu in Cu is smaller than that for diffusion of Al in Al. This difference is a direct result from the difference in melting temperature (1083 °C = 1356 K for Cu and 660 °C = 933 K for Al): the thermal equilibrium vacancy concentration is given by the Boltzmann-type expression (see Eq. (5.1) and its discussion in Sect. 5.1.1), and thus, at the same temperature, the vacancy concentration in Cu is much smaller than in Al. Thus, p_{vac} (cf. Eq. (8.23)) is (much) larger for Al than for Cu and this leads to the large difference in diffusion coefficient values assessed accordingly.

(v)   Indeed, the prevailing thermal equilibrium concentration of vacancies of the host crystal structure is an important, determining factor for the substitutional diffusivity. Thus, a substitutionally dissolved foreign element diffuses at relatively high speed through a parent crystal structure if that parent crystal structure is constituted from an element of low melting point (cf. discussion in (iv) above). Compare, for example, in Fig. 8.9 the self-diffusion of Ni with the diffusion of Ni in Al.

Also, this reasoning explains why the interdiffusion coefficient for the Cu–Ni system increases with increasing Cu concentration, as the melting point of Cu (1083 °C = 1356 K) is lower than that of Ni (1455 °C = 1728 K) (see Fig. 8.4b and the remark at the end of Sect. 8.4).

Reversely, the experiments by Grube and Jedele (1932) thus already hint at a vacancy mechanism for interdiffusion in this system, whereas such a mechanism was unknown at the time that the experiments were performed (see begin of Sect. 8.5.2): it is thus understandable that Grube and Jedele in

---

[7] The size of the octahedral interstices in b.c.c. iron is smaller than in f.c.c. iron (see footnote 6 in this chapter and see Sect. 9.5.2.1). This may explain the pronouncedly larger solubility of carbon in f.c.c. iron and the very small solubility of carbon in b.c.c. iron. Thus, the conclusion of the deliberations in the above main text and in this footnote is that (1) the diffusion coefficient of carbon in f.c.c. iron is smaller than in b.c.c. iron, because it is more difficult in f.c.c. to establish the displacement of the iron atoms pertaining to the activated state of the carbon atom jumping from one to the next octahedral interstitial site (less "free space" in f.c.c.), while (2) the solubility of carbon in f.c.c. is larger than in b.c.c. iron, because the size of the octahedral interstitial site in f.c.c. is larger.

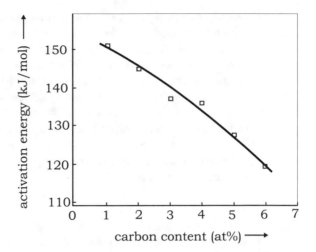

Fig. 8.11 Activation energy versus carbon content for diffusion of carbon in f.c.c. iron (data taken from C. Wells, W. Batz and R.F. Mehl, Transactions AIME, 188 (1950), 553–560)

their interpretation of these results depart from an exchange ("Platzwechsel") mechanism (cf. Sect. 8.5.1) that supposedly occurs faster in pure copper/the copper-rich solid solutions than in pure nickel/the nickel-rich solid solutions (ascribed to (also: as above) the difference in melting point of copper and nickel)… (see p. 805 of the Grube and Jedele (1932) paper).

Recognizing that for substitutional diffusion Arrhenius plots can only provide information on $Q = \Delta H_{vac} + \Delta H_{mig}$, one is tempted to ask if ΔH_{vac} and ΔH_{mig} can be determined separately.

8.6.1 The Determination of ΔH_{vac}

The determination of ΔH_{vac}, the vacancy-formation enthalpy (cf. Eq. (8.25)), can be performed by investigating the temperature dependence of the equilibrium vacancy concentration (cf. Eq. (5.1)). Upon increasing the temperature, the thermal vacancy concentration increases. Because the amount of atoms in the specimen considered remains constant, increase of the number of vacancies implies that the number of crystal-structure sites increases. The increase of temperature also has as consequence that the atomic distances increase (thermal expansion; cf. Sect. 3.1). These processes have different effects on two experimentally accessible specimen parameters upon temperature increase:

(i)   The *specimen length* increases because both the number of crystal-structure sites increases and the lattice parameter increases (see (ii)).

(ii) The *lattice parameter* (a) increases because of the thermal expansion, but (b) decreases because the ratio of the number of atoms and the number of vacancies decreases and the volume of a vacancy is smaller than that of an atom. Effect (a) is larger than effect (b), and thus, the net result is an increase of the lattice parameter.

The above discussion suggests that the relative increase of specimen length is larger than the relative increase of the lattice parameter.

The specimen length and the lattice parameter belong to a select and very small group of material/specimen properties which can be determined with very high accuracy (another such parameter is the specimen mass). The technique to determine the change of specimen length is called *dilatometry* (cf. Sect. 10.13), and the technique to determine the lattice parameter is *X-ray diffraction* (cf. Sects. 4.5 and 6.10). By measuring both the specimen length and the lattice parameter, as a function of temperature, results as shown in Fig. 8.12 can be obtained. Indeed, as indicated by the above discussion, the relative increase in specimen length, $\Delta l/l$, is larger than the relative increase in lattice parameter, $\Delta a/a$.

Consider a specimen (single element) in the form of a cube with edge length l. If the number of crystal-structure sites per unit cell equals α and the unit cell parameter is given by a (here, as implicit in the above discussion, we restrict ourselves to cubic materials with one atom at each crystal-structure site), it holds for the number of crystal-structure sites n:

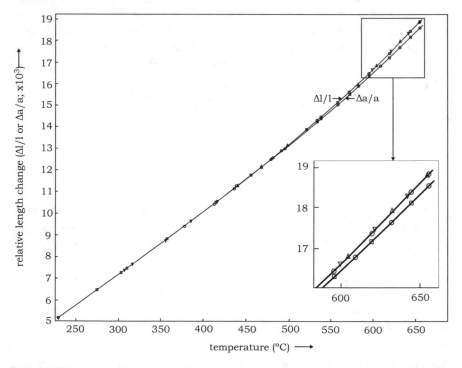

Fig. 8.12 The fractional change in specimen length and lattice parameter of aluminium as function of temperature (with respect to specimen length and lattice parameter at 20 °C), allowing the determination of the change of the vacancy concentration as function of temperature from the difference of the relative length change and the relative lattice-parameter change as function of temperature (data taken from Simmons and Balluffi 1960)

$$n = \frac{\alpha l^3}{a^3} \tag{8.29}$$

Thus, it follows via $\ln n = \ln \alpha + 3 \ln l - 3 \ln a$ and differentiating:

$$\frac{\Delta n}{n} = 3 \left(\frac{\Delta l}{l} - \frac{\Delta a}{a} \right) \tag{8.30}$$

The change of the number of crystal-structure sites is identical to the change of the number of vacancies and thus the fractional change of the vacancy concentration, Δc_{vac}, is given by:

$$\Delta c_{vac} = \frac{\Delta n}{n} \tag{8.31}$$

So, by measuring the difference between the relative length increase and the relative increase of the lattice parameter, Δc_{vac} can be determined as a function of temperature. It is found that Eq. (5.1) is obeyed, and thus, a value for ΔH_{vac} can be obtained (Simmons and Balluffi 1960).

The $\Delta l/l$ and $\Delta a/a$ data, as shown in Fig. 8.12, can in principle also be used to determine the volume of a vacancy. Conceive the crystal as a binary solid solution of atoms A and vacancies V. Suppose the volume of an atom can be given by $a_0{}^3/\alpha$ and the volume of a vacancy can be given by $a_{vac}{}^3/\alpha$, where a_0 and a_{vac} pertain to the hypothetical lattice-parameter values of a fully occupied crystal structure and that of a hypothetical crystal structure fully occupied by vacancies. Suppose the volume of a vacancy is equal to a fraction β of the volume of an atom. Hence: $\frac{a_{vac}^3}{\alpha} = \beta \frac{a_0^3}{\alpha}$.

Then, if the vacancy concentration equals c_{vac}, it follows for the specimen length (using $a_{vac}^3 = \beta \cdot a_0^3$; see above):

$$l^3 = n(1 - c_{vac})a_0^3 + n c_{vac} a_{vac}^3$$
$$= n a_0^3 (1 - c_{vac} + \beta c_{vac}) \tag{8.32}$$

In this, equation two unknowns appear: a_0 and β. Hence, a second equation is needed to solve for a_0 and β. This second equation is obtained considering the lattice parameter for the "atom-vacancy (A-V) solid solution".

Adopting a Végard-like relation (cf. Eq. (4.7) in Sect. 4.1.1) for the lattice parameter, a, of the A-V solid solution, i.e. a linear dependence on the vacancy concentration, it follows (using $a_{vac} = \beta^{1/3} a_0$; see above):

$$a = (1 - c_{vac})a_0 + c_{vac} a_{vac}$$
$$= a_0 + c_{vac} a_0 (\beta^{1/3} - 1) \tag{8.33}$$

Hence, if the length (dilatometry) and the lattice parameter (X-ray diffraction) are known at the same temperature (see, for example, the data in Fig. 8.12) β (and a_0)

can be determined. As a rule of thumb, it can be said that the volume of a vacancy is about ½ that of an atom (i.e. β is about ½).[8]

8.6.2 The Determination of ΔH_{mig}

An experiment to determine ΔH_{mig}, the migration enthalpy (cf. Eq. 8.25), runs as follows. A solid is annealed at elevated temperature for a time long enough that the thermal equilibrium concentration is established. Upon quenching to a low (room) temperature the thermal vacancy concentration of the annealing temperature can be retained. Next annealing experiments can be performed at annealing temperatures much lower than the first annealing temperature. At these annealing temperatures then a very large number of *excess vacancies* is available: the actual, quenched-in vacancy concentration can be orders of magnitude larger than the thermal equilibrium concentration at these temperatures. Then the vacancy concentration operative at the various (lower; cf. above discussion) annealing temperatures can be taken as constant: $p_{vac} = c_{vac}$ is constant, and consequently (cf. Eqs. (8.22) and (8.25)–(8.27)), the activation energy of diffusion reduces to ΔH_{mig}. Hence, "Arrhenius analysis" of the diffusion coefficient (see below Eq. (8.27)), for at least the initial stages of diffusion at these lower annealing temperatures (upon prolonged annealing distinct annihilation of the excess vacancies may occur at sinks as dislocations and grain boundaries), leads to an assessment of $Q = \Delta H_{mig}$.

Against the physical background sketched in the above paragraph, resistometry can be used to determine ΔH_{mig}: the electrical resistance (of a metal) is sensitive to, amongst other microstructural parameters, the vacancy concentration. Thus, starting from a quenched specimen containing a high concentration of excess vacancies, the change of the electrical resistance as a function of time at constant temperature is a measure for the rate of annealing out of excess vacancies. Then, performing annealing experiments at temperatures such that the initial vacancy concentration (dominated by the excess vacancies) is practically constant (see above), the temperature dependence of the rate of change of the electrical resistance allows determination of ΔH_{mig}.

---

[8] Computer simulations have suggested that the vacancy volume in (bcc) metals depends about linearly on an elastic constant (the Poisson constant, ν; see Sect. 12.2): from about $1 \times$ the atomic volume for $\nu = 0.25$ down to about $0.6 \times$ the atomic volume for $\nu = 0.4$ (Kurita and Numakura 2004).

8.7 Intrinsic Diffusion and Material Flow

8.7.1 The Kirkendall Effect

At this stage of the development, and without explicit recognition until now, the treatment of diffusion in a system comprised any material transport as occurring with respect to the "laboratory frame of reference", i.e. as observed by a non-moving person located outside the interdiffusing system and looking at it. Such material transport then can be due to the outcome of individual atomic movements ("jumps"), but also due to movement of a piece of material, as an atomic plane, while the atoms therein remain static (for now this seems an odd remark, but see what follows later). In both cases, material flow in the laboratory frame of reference, i.e. as experienced by the static "observer", is noticed. The diffusion coefficient, D, as defined before, for diffusion in a binary solid solution (see Eq. (8.2) and its discussion in Sect. 8.1), as a matter of fact comprises the outcome of both these types of material transport, which can occur simultaneously, and provides a full description of the homogenization in the system (the coordinate "x" in Eq. (8.2) pertains to the laboratory frame of reference). As will be shown in this section, indeed both types of material transport can occur in interdiffusing systems.

The above text can be made more transparent referring to a well-known example (it appears, in a little different way, in the original paper by Darken (1948), that provides the basis of the treatment in this section, and, again, in Shewmon (1989), p. 133). Consider a boat with a person floating on a flowing river, and, as will be assumed in the hypothetical case dealt with here, with the same speed as the river, and a non-moving observer on a bank of the river looking at the boat. The person in the boat, which flows with the river, lets drop some ink in the flowing river. The person in the boat sees the ink in the water that starts to spread out, as the result of a diffusion process. The observer on the bank sees this too, but additionally the observer on the bank notices that the spreading ink spot moves as a whole with the boat, i.e. with the river. The first type of material flow (spreading out of the ink), as observed by the person in the boat moving with the river, can be conceived as *genuine* diffusion, called *intrinsic* diffusion. This is the diffusion pertaining to the moving (with the river) coordinate system of (attached to) the person in the boat. The second type of material flow, which is (additionally) noticed by the static observer on the bank of the river, is simply caused by the flow of the river and is not caused by intrinsic diffusion of the ink in the water. Both simultaneously occurring types of material flow are described, in the non-moving coordinate system of (attached to) the observer on the bank of the river, by Eq. (8.2) with the single diffusion coefficient D.

The above text makes clear that describing the *intrinsic* diffusion would be more revealing about the diffusion properties of the ink in the water than describing the total material (ink) flow, which is represented by the single diffusion coefficient D mentioned above: the material (ink) flow, associated with the speed of the flowing river, is unrelated to the intrinsic diffusion properties of ink in water.

Now considering (one dimensional) interdiffusion in a binary solid solution diffusion couple, one would intuitively conceive diffusion as the *direct* outcome of atomic jump processes. Then, diffusion described for such a system in the laboratory frame of reference (i.e. the coordinate system attached to the non-moving observer outside the system) would be due to such *intrinsic* diffusion only. However, this interpretation is too limited: material flow in such diffusion couples can also occur by movement (as observed by the outside observer in the laboratory frame of reference) of atomic planes (oriented perpendicularly to the diffusion direction), as a whole, in the diffusion couple. This last process involves material flow that thus is *not directly* caused by the intrinsic diffusion due to jumps of individual atoms. It will be shown below that the vacancy mechanism of interdiffusion in a binary solid solution diffusion couple provides the example exhibiting both types of material transport, as observed in the laboratory frame of reference.

Hence, it is concluded that the diffusion coefficient D, as introduced with Eq. (8.2) in Sect. 8.1, and describing diffusion in the laboratory frame of reference (i.e. as observed by a non-moving observer outside the diffusion couple) not only incorporates the effects of the contributions to the diffusion (material flow) in the system due to *both* mobile components in the system, as remarked in Sect. 8.1, but thereby also represents the outcome of two types of material flow: intrinsic diffusion and material flow due to movement of atomic planes as a whole inside the diffusion couple.

Obviously, the need now emerges to relate characteristics of the intrinsic diffusion processes (of the two components in the system) with the "overall" diffusion coefficient D as present in Eq. (8.2). In the following D will be called "chemical diffusion coefficient" (or "interdiffusion coefficient") and will be given the symbol D_{chem}, in order to allow distinction with the intrinsic diffusion coefficients of the two components to be defined below.

The system considered is an "infinite", one-dimensional, binary (components A and B), solid solution diffusion couple (cf. Sect. 8.3). Consider a lattice plane oriented perpendicular to the length axis (indicating the diffusion direction) of the system. A coordinate system now is fixed to this lattice plane. This coordinate system is called "lattice-fixed frame of reference".[9] Obviously, i.e. within the spirit of the discussion of Eq. (8.2) in Sect. 8.1, the intrinsic diffusion fluxes of components A and B, J'_A and J'_B, respectively, at this lattice plane can be represented in the *lattice-fixed frame of reference* as:

$$J'_A = -D_A \frac{\partial c_A}{\partial x'} \tag{8.34a}$$

---

[9] At this stage of the analysis, instead of defining a *lattice*-fixed frame of reference, it would have been possible just to speak of a frame of reference fixed to a plane oriented perpendicular to the diffusion direction, as in fact done by Darken (1948), thereby avoiding seeming restriction to diffusion in crystalline materials. The lattice-fixed frame of reference is already introduced here to facilitate the interpretation of the material flow with velocity v below Eq. (8.35) as a consequence of the vacancy mechanism of diffusion.

$$J'_B = -D_B \frac{\partial c_B}{\partial x'} \tag{8.34b}$$

where D_A and D_B denote the intrinsic diffusion coefficients of components A and B, respectively, and x' is the coordinate of the lattice-fixed frame of reference parallel to the length axis of the system.

The material flow in the system not caused by the intrinsic diffusion is represented by movement of the lattice plane considered, along x (x is the coordinate of the laboratory frame of reference parallel to the length axis of the system) with velocity v (cf. the material (ink) flow, associated with the speed of the flowing river, in the example discussed above). By definition, the local flux can be written as the product of the average velocity of the moving particles, at the location considered, v, and the number of these particles in a unit volume, at the location considered. Thus, the contributions of the material flow, corresponding to the movement of the lattice plane, to the fluxes of A and B in the *laboratory frame of reference* are equal to vc_A and vc_B, respectively. Hence, it holds for the total fluxes of A and B, J_A and J_B, respectively, in the laboratory frame of reference (using $\partial c_A/\partial x' = \partial c_A/\partial x$ and $\partial c_B/\partial x' = \partial c_B/\partial x$; the x- and x' axis are both parallel to the length axis (diffusion direction) of the diffusion couple with the x'-axis in relative motion with respect to the x-axis)[10]:

$$J_A = -D_A \frac{\partial c_A}{\partial x} + vc_A \tag{8.35a}$$

$$J_B = -D_B \frac{\partial c_B}{\partial x} + vc_A \tag{8.35b}$$

The necessity to introduce material-flow contributions, in the laboratory frame of reference, as expressed by the second terms at the right-hand sides of Eqs. (8.35a and b), was experimentally demonstrated convincingly by Smigelskass and Kirkendall (1947) by the movement, upon one-dimensional interdiffusion, of inert, wire markers, attached to an original interface, with respect to the ends of an "infinite" diffusion couple. As at the ends of the "infinite" diffusion couple no diffusion has occurred, it holds that a coordinate system fixed to the ends of the diffusion couple is nothing else than the laboratory frame of reference defined above. Hence, the marker movement with respect to the ends of the diffusion couple is identical with movement in the laboratory frame of reference. It is such marker movement, which has been called the "Kirkendall effect".

There is no need to introduce a specific mechanism as origin for the marker movement in order to introduce Eqs. (8.35a and b) and, consequently, the final, resulting Eqs. (8.39) and (8.41), given below, can then be considered as a "phenomenological" description of interdiffusion. However, in order to provide a physical mechanism for the material-flow terms vc_A and vc_B in the laboratory frame of reference, the vacancy

---

[10] Note that this approach to the relative movement of one coordinate system with respect to another coordinate system also occurs in Einstein's first, special theory of relativity (1905).

mechanism for substitutional diffusion (cf. Sect. 8.5.2) will now be adopted (As a matter of fact, the experiment performed by Smigelskas and Kirkendall was considered already as an ultimate proof of the vacancy mechanism; see the *"Intermezzo: A revolution in diffusion understanding; "young" versus "old""*).

The vacancy mechanism of diffusion implies that an atom exchanges its position with an adjacent vacancy. Hence, a (net) number of atoms A moving, in the lattice-fixed frame of reference at the location of the lattice plane considered (see above), to one side of the diffusion couple is accompanied with an equally large (net) number of vacancies moving, in the lattice-fixed frame of reference at the location of the lattice plane considered, to the opposite site of the diffusion couple. A similar statement holds for the B atoms. Generally, the intrinsic diffusion coefficients of A and B are unequal, and also, the "number fluxes" of A and B (i.e. fluxes of A and B described as number of atoms) will then normally be unequal. This leads to the conclusion that there is a net number flux of atoms, i.e. the difference of the number fluxes of A and B atoms, through the lattice plane considered to one side of the diffusion couple, accompanied by a similarly large net number flux of vacancies through the same lattice plane to the opposite side of the diffusion couple.

Now, in the following consideration, adopting the vacancy mechanism, some assumptions must be made:

– The molar (atomic) volume is taken constant (concentration independent), implying that the total volume of the system is constant. This is the only assumption explicitly made in the original treatment by Darken (1948), which is *not* based on adoption of any diffusion mechanism.
– The number of lattice sites in the system is fixed.
– The vacancy concentration remains constant, equal to its (desired) equilibrium value, which thus is taken concentration independent.

Then it holds that $c_A + c_B = c_0$, where c_0, the total number of moles (atoms) per unit volume, is a constant. As follows from the above, this holds everywhere in the system and implies that the molar (atomic) volume is constant (see above) and that the vacancy concentration is everywhere the same (it is also negligible as compared to the concentration sum $c_A + c_B$). It follows: $\partial c_A/\partial x = -\partial c_B/\partial x = \partial c_A/\partial x' = -\partial c_B/\partial x'$ (cf. immediately above Eq. (8.35)). Using this last result and Eqs. (8.34a and b), it can then be written for the fluxes in the lattice-fixed frame of reference:

$$J'_A + J'_B + J'_V = 0 \tag{8.36}$$

$$J'_V = -\left(J'_A + J'_B\right) = (D_A - D_B)\frac{\partial c_A}{\partial x} \tag{8.37}$$

The net flux of vacancies is not nil and generally depends on x (x'). If J'_A is larger than J'_B, as shown in Fig. 8.13, for a diffusion couple with the A-rich part at the left-hand side and the B-rich part at the right-hand side, then there is a net flux of

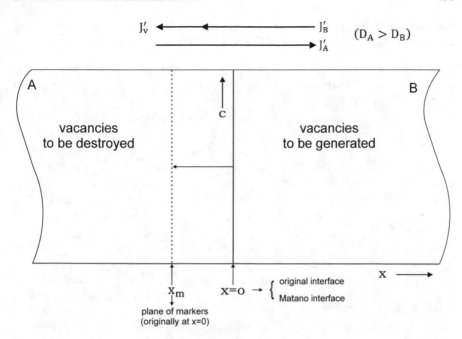

Fig. 8.13 Fluxes of atoms A and B and vacancies through a lattice plane in the lattice-fixed frame of reference, for a one-dimensional, binary (components A and B) diffusion couple subject to substitutional interdiffusion according to the vacancy mechanism (see text)

atoms to the right and a net flux of vacancies to the left in the diffusion couple. The requirement that the vacancy concentration is everywhere the same, and equal to the (concentration independent) equilibrium value (see above), implies that, during the interdiffusion, at the right-hand side of the diffusion couple vacancies have to be generated continuously and that at the left-hand side of the diffusion couple vacancies have to be annihilated/destroyed continuously.

To imagine that such processes are possible, conceive the diffusion couple as a set of lattice planes oriented perpendicular to the diffusion direction. Negative climb of an edge dislocation at the right-hand side of the diffusion couple, i.e. at the right-hand side of the original interface with the inert markers (see below Eq. (8.35)), causes the emission of vacancies and an entire lattice plane can thus be inserted eventually (see Fig. 5.15 in Sect. 5.2.7). Positive climb of an edge dislocation at the left-hand side of the diffusion couple, i.e. at the left-hand side of the original interface with the inert markers, causes the annihilation/destruction of vacancies and eventually an entire lattice plane can thus be removed eventually (again, see Fig. 5.15 in Sect. 5.2.7). Hence, after such insertion of a lattice plane at the right-hand side and such removal of a lattice plane at the left-hand side, it follows that the lattice plane with the markers has moved to the left (over one lattice-plane spacing) with respect to the ends of the diffusion couple, i.e. has moved to the left in the laboratory frame of reference. This discussion implies that the marker movement is opposite to the diffusion direction of the fastest diffusing component (i.e. A in the example discussed).[11] The plane moving with the markers has been called "Kirkendall interface" or "Kirkendall plane".

Intermezzo: A Revolution in Diffusion Understanding;
"Young" versus "Old"
Until late in the forties of the past century, interdiffusion in crystalline (metallic) solids was thought to take place by direct exchange of atoms (also ring mechanisms were proposed), implying equal diffusion rates for the components in the (usually binary) diffusion systems. During preparation of his doctor's thesis Ernest Kirkendall did observe that the original interface of a diffusion couple composed of brass (a Cu–Zn alloy) electroplated with copper had moved from its original location in the couple (an infinite system). However, in the work as it was published in 1939 (Kirkendall was 25 years old), under "pressure" of his thesis supervisor, he could not insist on unequal diffusivities of copper and zinc, as explanation of the interface shift. In 1942, he published a second paper (this time entirely by his own), and, after having repeated his experiments, he now claimed that the exchange mechanism could not explain his observations of migration of the original interface. Further work occurred a few years later, together with a student, Alice Smigelskas. They devised and performed the now famous experiment with the inert, insoluble Mo wires at the location of the original interface to demonstrate unequivocally that the original interface migrates upon interdiffusion and concluded that this is the consequence of unequal diffusivities of Cu and Zn in the Cu–Zn alloy/Cu diffusion couple: Zn diffuses faster in Cu than Cu in the Cu–Zn alloy. The groundbreaking, corresponding paper was published in 1947 (it is taken up in the references at the end of this chapter) and met heavy criticism/unbelief. A main opponent was R. F. Mehl, a leading authority on diffusion at the time, who first refused to accept the paper for publication. After the paper was finally published, the text of the paper was followed by lengthy (printed) discussion, which is still worthwhile to read.

Interestingly, in their celebrated paper Smigelskas and Kirkendall did *not* propose a vacancy mechanism for diffusion (the words vacancy, vacancies or empty lattice sites, and the like, do not even appear in the paper). Their only

---

[11] This conclusion (that the migration of the markers occurs to the side of the diffusion couple where the largest intrinsic diffusion coefficients prevail) is generally accepted (e.g. see Liu et al. 2008), and has been validated also in cases where a concentration dependence of the molar volume has to be accounted for. However, it has been claimed that the direction of the marker movement would be dictated by the sign of the slope of Vegard's relation (Eq. (4.7)) for the component pair considered, in the way that the markers would move in the direction of increasing lattice parameter (marker-movement direction predictions on this basis can (and very often are) but need not be the same as those obtained by the first reasoning above). This was ascribed to the larger tendency of the larger atoms to be surrounded by a vacancy atmosphere, thereby relaxing the local and global misfit, coherency strain and at the same time enhancing the exchange of the larger atoms with vacancies, i.e. increasing their intrinsic diffusion coefficient … (Kirkaldy and Savva 1997).

purpose was "to demonstrate that when diffusion takes place in alpha brass the zinc diffuses much more rapidly than the copper". It is in the discussion contribution by R. Smoluchowski that identification of the vacancies occurs. He remarks that the paper by Smigelskas and Kirkendall "indicates in a most convincing manner that diffusion is not only a question of two kinds of atoms, but that there is also a third "constituent"- the vacancies". And he continues with: "To be sure, I never expected the influence of vacancies to be so large". But others were less convinced at the time (e.g. see further contributions to the discussion of this paper). Moreover, in the discussion of the later Darken (1948) paper, it is stated, in the contribution by C. E. Birchenall, L. C. Corrêa da Silva and R. F. Mehl, that the paper by Darken "is based on one experiment of questionable validity" (meant is the work by Smigelskas and Kirkendall) and that "the Smigelskas and Kirkendall experiment, the only evidence supporting this point of view, contains serious inconsistencies".

It took some time before the "Kirkendall effect" and the reality of the vacancy mechanism for substitutional diffusion were accepted.[12]

The reason to insert this *Intermezzo* in particular derives from the observation (again) how difficult it can be for an (especially young) researcher to convince the establishment in a certain field of science of a revolutionary finding. This phenomenon was also discussed in Chap. 4 (see the "*Intermezzo: A revolution in crystallography; "young" versus "old"*". For this reason, the present *Intermezzo* bears a similar title). In the present case, the main actor, Kirkendall, left the field of science, in also 1947, after his short, scientific career that comprises only three papers. It is unclear if this is primarily due to the resistance to and slow acceptance of his work (at advanced age Kirkendall claimed that financial conditions at the time urged him to accept a position away from science, and that forever), but likely such circumstances make shifting of profession more attractive.

With reference to the discussion in the last paragraph, the concentration of vacancies needed to move with velocity v a unit area of a plane perpendicular to the diffusion direction equals $c_A + c_B = c_0$. Consequently:

$$J_v' = v c_0 \tag{8.38}$$

Thus, it is obtained from Eqs. (8.37) and (8.38)

$$v = (D_A - D_B) \frac{1}{c_0} \frac{\partial c_A}{\partial x} \tag{8.39a}$$

or, similarly derived

---

[12] The history of the discovery of the Kirkendall effect is described by H. Nakajima (JOM, 49 (1997), 15–19), on the basis of personal recollections of Kirkendall.

$$v = (D_B - D_A) \frac{1}{c_0} \frac{\partial c_B}{\partial x} \tag{8.39b}$$

Finally, substitution of Eq. (8.39a) in Eq. (8.35a) gives

$$J_A = -(X_B D_A + X_A D_B) \frac{\partial c_A}{\partial x} \tag{8.40a}$$

or, similarly derived

$$J_B = -(X_B D_A + X_A D_B) \frac{\partial c_B}{\partial x} \tag{8.40b}$$

where $X_A = c_A/c_0$ and $X_B = c_B/c_0$ are the mole (atomic) fractions A and B.

Note that in the laboratory frame of reference $J_A = -J_B$ as $\partial c_A/\partial x = -\partial c_B/\partial x$ (cf. the definition of the Matano interface ($x = 0$) in Sect. 8.4); in the lattice-fixed frame of reference $J'_A \neq -J'_B$.

Comparison of Eq. (8.2) with Eq. (8.40a) provides the following expression for D_{chem}:

$$D_{chem} = X_B D_A + X_A D_B \tag{8.41}$$

Hence, indeed (cf. Sect. 8.1), a single diffusion coefficient, D_{chem}, suffices to fully describe diffusion in the laboratory frame of reference on the basis of Eq. (8.2).

The above analysis provides a physical explanation for the occurrence of mass flow not due to intrinsic diffusion (the vc_A and vc_B terms in Eqs. (8.35a and b). Again it is remarked that the Darken treatment (1948) does *not* present such a physical explanation; in Darken's consideration, the mass flow, as characterized by its velocity v in the laboratory frame of reference, is simply accepted as an empirical fact, and therefore, Darken's treatment can be considered as a phenomenological theory that thereby has a more general meaning than for the case of diffusion by the vacancy mechanism only. Equations (8.39) and (8.41) are the main results of the present analysis which have validity also beyond the vacancy mechanism for substitutional interdiffusion.

It must also be remarked that the example presented in the begin of this Sect. 8.7, to introduce the two types of mass flow in the laboratory frame of reference, intrinsic diffusion and mass flow due to transport of a piece of material as a whole, in an essential part is not in conformity with what happens upon interdiffusion. In the example discussed, the flow of the ink spot as a whole with the flowing river, with velocity v, is fully unrelated to the spreading of the ink spot, i.e. the intrinsic diffusion process. The reverse holds during interdiffusion: the velocity v of the moving (lattice) plane is the indirect result of the intrinsic diffusion fluxes of the two components, as exhibited by Eq. (8.39).

The results obtained, Eqs. (8.39) and (8.41), immediately suggest a route to obtain values for the intrinsic diffusion coefficients D_A and D_B: Eqs. (8.39) and (8.41) can be conceived as two equations with two unknowns, D_A and D_B, provided D_{chem} and v are known. One then can proceed as follows.

The original interface with the markers has been shifted, after a time of interdiffusion t, from the original position, $x = 0$ (to be determined as the Matano interface (see Eq. (8.21) in Sect. 8.4), to a position x_m. For the infinite system considered, the Boltzmann substitution ($u = x/\sqrt{t}$) is possible (see Sect. 8.4). Then, using Eq. (8.39) after multiplication of the left-hand and right-hand sides of this equation with $\sqrt{t}$, the velocity of the plane with the markers, v_m, can be given as

$$v_m \cdot \sqrt{t} = \left[(D_A - D_B)\frac{1}{c_0}\frac{dc_A}{du} \right] \equiv F(u_m) \qquad (8.42)$$

where the term within the square brackets should be taken at $u = u_m$. Since D_A and D_B depend on c_A and c_A is a function of u (see Sect. 8.4), the right-hand side of Eq. (8.42) is function of u_m only, given by $F(u_m)$. Note that at this stage of the consideration u_m is not claimed to be constant. It is observed (at constant temperature and constant pressure) that the plane with the markers, originally at $x = 0$, moves proportionally with $\sqrt{t}$, i.e. $x_m = c_1\sqrt{t}$, and hence:

$$v_m = \frac{dx_m}{dt} = \frac{c_1}{2\sqrt{t}} = \frac{x_m}{2t} \qquad (8.43)$$

with c_1 as a constant. Then it is concluded from a comparison of Eqs. (8.42) and (8.43) that, *for the moving plane with the markers*, it holds that $F(u_m) = c_1/2 = $ constant and thus u_m and, consequently, c_A are constant (It could have been said immediately that the concentration of the plane with the markers, as long as moving proportionally with $\sqrt{t}$, is constant, since any plane moving proportionally with $\sqrt{t}$ is characterized by a constant concentration (see Eq. (8.19) and its discussion in Sect. 8.4). From Eq. (8.42), it is obtained $F(u_m) = \frac{1}{2}\, d(x_m)/d\sqrt{t}$ and because $F(u_m)$ is constant, it follows upon integration that $F(u_m) = \frac{1}{2} u_m$. This last result determines implicitly the value of constant concentration of the plane with the markers (see above definition of $F(u_m)$).

The above treatment can, of course, also be given reversely, i.e. starting with the adoption of a constant concentration of the plane with the markers and then conclude that the plane with the markers must migrate with a velocity proportional with $\sqrt{t}$ (cf. Kirkaldy and Young 1987, Sect. 2.13 vs. Philibert 1991, Sect. II.3). In view of the above discussion on v_m, the first part of the discussion by LeClaire of the Darken (1948) paper is worthwhile to read.

Now, as a result of the above consideration, the velocity of the marker shift (=shift of the Matano interface, i.e. the original interface, of constant concentration) can be determined using $v = dx_m/dt = x_m/(2t)$ (Eq. (8.43)). D_{chem} for the concentration of the plane of the markers (i.e. at the location x_m) can be determined according to the Boltzmann–Matano analysis (Eq. (8.20) in Sect. 8.4). Then, D_A and D_B follow straightforwardly, for the same concentration, by straightforward calculus from Eqs. (8.39) and (8.41).

D_{chem} as function of concentration can be obtained for the whole concentration range from the single concentration–distance curve as measured upon interdiffusion in the initially pure A − pure B infinite diffusion couple, by application of the Boltzmann–Matano analysis. However, after having put inert markers at the original interface in such a couple, the intrinsic diffusion coefficients, D_A and D_B, can only be determined for the concentration pertaining to the moving plane of the markers (see above procedure). To determine D_A and D_B at other concentration values, more experiments with markers in other infinite diffusion couples are required: for example interdiffusion experiments with initially pure A − A_xB_{1-x} diffusion couples, where the moving markers, attached to the original interface, stay during interdiffusion at a concentration different from that during interdiffusion in the initially pure A − pure B diffusion couple, etc. Evidently, this is a laborious procedure (for an other, more often used method, see Sect. 8.7.2).

The validity of the analysis, culminating in Eqs. (8.39) and (8.41), for application in practice, depends on satisfaction in reality of the assumptions made:

(i)   The approximation made for the molar (atomic) volume, by taking it as concentration independent, is not too serious and can be corrected for if deemed necessary.

(ii)   The assumption made for the case of substitutional diffusion, that the vacancy concentration is everywhere at its equilibrium value, is rather difficult to verify. It strikes, that, at the time of writing this book, research on vacancy sources and sinks, as these must be present and active in sufficient quantity at both sides of the diffusion couple, is very rare. Also theoretical work on modelling the role of vacancy sources and sinks on diffusion kinetics is scarce. In the extreme case, that no vacancy sources and sinks are active, the Darken equation (8.41) must be replaced by the so-called Nernst–Planck equation, which was originally derived for ambipolar diffusion of ions (the adjective "ambipolar" points out that two species of opposite charge, here anions and cations, diffuse simultaneously) in ionic crystals subject to maintenance of electrical neutrality (no build-up of space charge; cf. Philibert 1991, Chap. 6). The Nernst–Planck equation (for two diffusing species) describes the concentration dependence of the diffusion chemical (interdiffusion) coefficient according to:

$$D_{chem} = \frac{D_A D_B}{X_A D_A + X_B D_B} \qquad (8.41a)$$

Recent work testing possible deficiencies in the action of sources and sinks of vacancies involved analysis of interdiffusion in *single-crystalline* thin bilayer Ag-Au films and *single-crystalline* thin bilayer Pd–Ag films. Comparing the viability of the Darken formalism and that of the Nernst–Planck formalism, it followed that the Darken approach provided the satisfactory and best description of the interdiffusion in these systems; at least in these cases, sufficient sources and sinks were active, implying that the equilibrium vacancy concentration was (closely) realized in these systems during the interdiffusion (Noah et al. 2016a, b).

(iii)   The above leaves unimpeded that the annihilation/destruction of excess vacancies in the part of the diffusion couple where the fastest diffusion occurs (i.e. the left-hand side of the diffusion couple considered in Fig. 8.13, with A as the fastest diffusor in the lattice-fixed frame of reference), may not occur completely or fast enough. Thereby agglomeration of the excess vacancies in pores may take place, and then obviously the (also geometric) boundary conditions for the case of diffusion assumed are violated. The porosity discussed here is called "Kirkendall porosity". The chance for such pore development obviously is smaller upon selecting initial compositions, for both partners at the original interface, relatively close to each other, as in an $A - A_xB_{1-x}$ system, as compared to an A-B system: the net flux of vacancies then is smaller as both the difference of the operating intrinsic diffusion coefficients, D_A and D_B, and the occurring concentration gradient, $\partial c_A/\partial x$, will be smaller (cf. Eq. (8.39)).

8.7.2   From Thermodynamics to Kinetics

It has already been mentioned that, at a deeper level than as expressed by Eq. (8.2), for D_{chem}, and by Eq. (8.34), for J'_A and J'_B, a local flux can be considered, but still in an approximate way, as being proportional with the (negative of the) local gradient in energy (Eq. (8.1)). Now again considering one-dimensional diffusion in a two component, A-B system, in the presence of a chemical gradient, for describing the flux of one of the components, the "energy" parameter in Eq. (8.1) then can be interpreted as the partial Gibbs energy, i.e. the chemical potential, of the diffusing component considered (cf. Sect. 7.3). This leads for intrinsic diffusion of component A in the lattice-fixed frame of reference to[13]:

$$J'_A = \mathrm{const}_A \cdot - \frac{\partial \mu_A}{\partial x} \tag{8.44}$$

The local flux, by definition, can be written as the product of the average velocity of the diffusing particles, at the location considered, v_A, and the number of these

---

[13] The replacement of the concentration gradient by the chemical potential gradient, as "driving force" for diffusion, indicates the possibility that diffusion can occur in a direction opposite to that of the concentration gradient, i.e. from lower to higher concentration as long as the chemical potential of the component concerned becomes reduced. This can regularly be observed in (but not only) multicomponent systems. Anyway, a (multicomponent) system strives for decreasing its total (Gibbs) energy and this will govern the diffusion directions of the various components and, consequently, considering only one component and its transport by diffusion in a multicomponent system will generally lead to erroneous conclusions.

particles in a unit volume, at the location considered, i.e. the concentration, c_A:

$$J'_A \equiv c_A \cdot v_A \tag{8.45}$$

where v_A is usually written as (cf. Eqs. (8.44) and (8.45)):

$$v_A = M_A \cdot \frac{-\partial \mu_A}{\partial x} = M_A \cdot F_A \tag{8.46}$$

with the M_A as the so-called mobility. $F_A = -\partial \mu_A / \partial x$ is conceived as the "force" acting on the diffusing particles A in the chemical potential field (cf. footnote 1 in Chap. 3). Note that, in the context of Eq. (8.46), F_A is an unusual force, as the velocity v_A is not given by the product of mass and acceleration: the diffusing particle cannot move freely; it continuously changes abruptly its moving direction (cf. Sect. 8.2). The force F thus establishes a constant average velocity for the ensemble of particles subjected to it.

Now, Eqs. (8.44)–(8.46) allow the following reformulation of the flux J'_A:

$$J'_A = c_A M_A \cdot - \frac{\partial \mu_A}{\partial x} = c_A M_A \cdot - \left(\frac{\partial \mu_A}{\partial c_A} \right) \left(\frac{\partial c_A}{\partial x} \right)$$

$$= M_A \cdot - \left(\frac{\partial \mu_A}{\partial \ln c_A} \right) \left(\frac{\partial c_A}{\partial x} \right)$$

On the basis of thermodynamics, a term $\partial \mu / \partial \ln c$ can generally be expressed as[14]

$$\frac{\partial \mu}{\partial \ln c} = RT \left\{ 1 + \frac{\partial \ln \gamma}{\partial \ln c} \right\}$$

Now coming back to the strategy announced in the third paragraph below Eq. (8.2) in Sect. 8.1, it is desired to maintain the form of Eq. (8.2) for describing the diffusional flux, albeit recognizing that the diffusional flux is better described by Eq. (8.44). Then, on the basis of the above treatment, J'_A according to Eq. (8.44) can be written as

---

[14] At this place in this book some thermodynamic knowledge is required that has not been presented before (in Chap. 7). This is no obstruction to understand the main message of the treatment in this section. Yet, as clarification for (re)reading at a later stage, the corresponding short derivation is given in this footnote as follows. The chemical potential of the considered component of concentration c is given by

$\mu = \mu_0 + RT\ln a = \mu_0 + RT\ln \gamma c = \mu_0 + RT\ln \gamma + RT\ln c$, with $a = \gamma c$ as the so-called activity of the component considered of concentration c and with γ as the so-called activity coefficient; μ_0 is a constant (reference chemical potential). Differentiating the left-hand and right-hand sides of this equation with respect to c, and multiplying thereafter both sides with c, leads to the expression for $\partial \mu / \partial \ln c$ given above (Note that μ can also be written as $\mu = \mu'_0 + RT\ln \gamma X$, with the atomic fraction given as $X = c/c_0$ and where-RT$\ln c_0$ has been taken up into μ'_0. This is used in footnote 16, where μ'_0 is yet written without prime).

$$J'_A = M_A RT \cdot \left\{ 1 + \frac{\partial \ln \gamma_A}{\partial \ln c_A} \right\} \cdot - \frac{\partial c_A}{\partial x} \qquad (8.47)$$

implying that the flux J'_A can be represented by

$$J'_A = D_A \cdot - \frac{\partial c_A}{\partial x} \qquad (8.48)$$

with

$$D_A = M_A RT \left\{ 1 + \frac{\partial \ln \gamma_A}{\partial \ln c_A} \right\} = M_A RT \left\{ 1 + \frac{\partial \ln \gamma_A}{\partial \ln X_A} \right\} \qquad (8.49a)$$

where γ_A denotes the activity coefficient of A[14] and use has been made of $c_A/c_0 = X_A$ and consequently $d\ln c_A = d\ln X_A$. It can similarly be derived for D_B

$$D_B = M_B RT \left\{ 1 + \frac{\partial \ln \gamma_B}{\partial \ln c_B} \right\} = M_B RT \left\{ 1 + \frac{\partial \ln \gamma_B}{\partial \ln X_B} \right\} \qquad (8.49b)$$

Thereby, for J'_A and J'_B, a formalism of the mathematical type presented by Eqs. (8.34a and b) has been obtained.

The expressions (8.49a and b) exhibit concentration dependence of the intrinsic diffusion coefficients D_A and D_B both through the factors $\{1 + d\ln\gamma_A/d\ln X_A\}$ and $\{1 + d\ln\gamma_B/d\ln X_B\}$ and through M_A and M_B which generally depend on the concentration of the A-B solid solution. The above result suggests a route to determine D_A and D_B different from the one indicated in the one but last paragraph of Sect. 8.7.1, as follows.

Consider diffusion of the (dilute) radioactive isotope A^* into a *homogeneous* A-B solid solution (see below Eq. (8.25) in Sect. 8.6). The tracer diffusion coefficient D_{A^*}, as determined from this experiment, is generally unequal to the intrinsic diffusion coefficient D_A, as the effect on diffusion of a chemical potential (concentration) gradient, as holds for diffusion in a compositionally *heterogeneous* solid solution, did not play a role in the experiment to determine D_{A^*}. To determine D_A, and assuming that the mobility M_{A^*}, at the concentration of the solid solution concerned, is not influenced significantly by the presence of a superimposed gradient of chemical potential (concentration), D_{A^*} must be multiplied with the factor $\{1 + \partial\ln\gamma_A/\partial\ln X_A\}$ (cf. Eq. 8.49a) for the concentration of the A-B solid solution concerned:

$$D_A = D_{A^*} \left\{ 1 + \frac{\partial \ln \gamma_A}{\partial \ln X_A} \right\} = M_{A^*} RT \left\{ 1 + \frac{\partial \ln \gamma_A}{\partial \ln X_A} \right\} \qquad (8.51a)$$

and, similarly

$$D_B = D_{B^*} \left\{ 1 + \frac{\partial \ln \gamma_B}{\partial \ln X_B} \right\} = M_{B^*} RT \left\{ 1 + \frac{\partial \ln \gamma_B}{\partial \ln X_B} \right\} \qquad (8.51b)$$

where it has been assumed additionally that the mobilities of $A*$ and A and of $B*$ and B are equal because of the chemical identity of A and $A*$ and of B and $B*$ (but see the remark below Eq. 8.57).

Evidently, for the case of diffusion of A in pure A, this is called "self-diffusion";[15] it holds $\{1 + \partial \ln \gamma_A / \partial \ln X_A\} = 1$ (as γ for a pure material equals 1). Hence, the so-called self-diffusion coefficient of A, D_A^s, and similarly that of B, D_B^s, equal

$$D_A^s \equiv D_A(X_A = 1) = D_{A*}(X_A = 1) = M_{A*}(X_A = 1)RT \tag{8.52a}$$

$$D_B^s \equiv D_B(X_B = 1) = D_{B*}(X_B = 1) = M_{B*}(X_B = 1)RT \tag{8.52b}$$

As the result of a small thermodynamic intermezzo, one can deduce[16]:

$$X_A \frac{\partial \ln \gamma_A}{\partial X_A} = X_B \frac{\partial \ln \gamma_B}{\partial X_B} \tag{8.53}$$

and thus (cf. Eq. 8.51)

$$D_A = D_{A*} \cdot \text{TF}; \quad D_B = D_{B*} \cdot \text{TF} \tag{8.54}$$

with TF as the so-called thermodynamic factor given by

$$\text{TF} = \left\{ 1 + \frac{\partial \ln \gamma_A}{\partial \ln X_A} \right\} = \left\{ 1 + \frac{\partial \ln \gamma_B}{\partial \ln X_B} \right\} \tag{8.55}$$

Note that, for diffusion of (a tiny amount of) $A*$ in A ("self-diffusion") and diffusion of (a tiny amount of) $A*$ in B ("impurity diffusion"), it holds that TF $= 1$ and thus $D_A(X_B = 0) = D_{A*}(X_B = 0)$ and $D_A(X_B = 1) = D_{A*}(X_B = 1)$. For all (other) compositions, i.e. $0 < X_B < 1$, the value of the tracer diffusion coefficient $D_{A*}(X_B)$ must be multiplied with the appropriate TF in order to obtain a value of the intrinsic diffusion coefficient $D_A(X_B)$ (cf. Eq. (8.51a) and Eq. (8.54)). Similar remarks pertain to D_B (cf. Eqs. (8.51b) and (8.54)).

---

[15] In the case of self-diffusion there is neither a net flow of vacancies (nor of matter). $D_A^s = D_A(X_A = 1)$ can therefore directly be determined from the diffusion of $A*$ into A by straightforward fitting of a concentration profile for $A*$ as determined on the basis of Eq. (8.6b) with D taken as D_A^s.

[16] At this place, again, some thermodynamic knowledge is required that has not been presented before. Knowledge of and understanding of the background of this footnote (which cannot be expected for a beginning materials scientist) are no prerequisite for understanding the message of this section leading to the final result presented as Eqs. (8.56) and (8.57).
Using:
(i) $\mu_A = \mu_{0,A} + RT \ln \gamma_A X_A$ (cf. footnote 14) and thus $\partial \mu_A = RT[\partial \ln X_A + \partial \ln \gamma_A]$ and, similarly,
   $\partial \mu_B = RT[\partial \ln X_B + \partial \ln \gamma_B]$,
(ii) $X_A + X_B = 1$, and thus $\partial X_A = -\partial X_B$,
(iii) $X_A \partial \mu_A + X_B \partial \mu_B = 0$, i.e. a so-called Gibbs–Duhem equation,
the result presented above as Eq. (8.53) is obtained.

Hence, on the basis of measurements of D_{A*} and D_{B*}, as function of concentration of the binary solid solution, and calculation of TF from thermodynamic data, as function of concentration of the binary solid solution, D_A and D_B can be determined. Finally, using Eq. (8.54), Eqs. (8.39) and (8.41) can be rewritten as

$$v = (D_{A*} - D_{B*}) \cdot \text{TF} \cdot \frac{\partial X_A}{\partial x} \qquad (8.56)$$

$$D_{\text{chem}} = X_B D_A + X_A D_B = (X_B D_{A*} + X_A D_{B*}) \cdot \text{TF} \qquad (8.57)$$

The above procedure provides a possibility to validate the Darken treatment by comparison of D_{chem} values obtained from a Boltzmann–Matano analysis (Sect. 8.4) with those derived from D_{A*} and D_{B*} measurements and application of Eq. (8.57).

In the above the consequence of correlation effects has been ignored (see below Eq. (8.25) in Sect. 8.6), as is usual in tests of the Darken treatment. The reason is the limited impact of such corrections, i.e. involving correlation factors, for experiments performed in practice: the precision (reproducibility) associated with experimentally determined values of D_{chem} is not better than of the order 10% (as apparent upon comparison of corresponding results from different groups of researchers), which is of about the same magnitude as the effect of incorporating correlation effects (Kirkaldy and Young 1987).

For the purpose of illustration, Fig. 8.14a, b demonstrates the courses of D_A, D_B and D_{chem} as function of concentration schematically and for the Ag-Pd system, exhibiting a single solid solution phase over the whole composition range at the temperature (435 °C) (and pressure (1 atm)) concerned. Note that $D_{Ag*}(X_{Pd} = 0)$ is the self-diffusion coefficient of Ag in Ag and $D_{Ag*}(X_{Pd} = 1)$ is the tracer diffusion coefficient of Ag in Pd and is called the "impurity diffusion coefficient" of Ag in Pd (see text below Eq. 8.55). Similarly, $D_{Pd*}(X_{Pd} = 1)$ is the self-diffusion coefficient of Pd in Pd and $D_{Pd*}(X_{Pd} = 0)$ is the tracer diffusion coefficient of Pd in Ag and is called the "impurity diffusion coefficient" of Pd in Ag.

Indeed, as expected from the relatively low melting point of Ag (660 °C) as compared to that of Pd (1555 °C), a higher vacancy concentration occurs in the Ag-rich part of the system than in the Pd-rich part and as consequence the intrinsic and chemical diffusion coefficients are larger in the Ag-rich part (see Fig. 8.14b and cf. the discussion in Sect. 8.6). The self-diffusion coefficients $D_{Ag*}(X_{Pd} = 0)$ and $D_{Pd*}(X_{Pd} = 1)$ differ about six orders of magnitude. This explains why the chemical diffusion coefficient is governed by the intrinsic diffusion coefficient of Ag (see D_{chem} and D_{Ag}, at larger than small values of x_{Pd} in Fig. 8.14b) and compare with Eq. 8.41). Yet, the chemical diffusion coefficient, D_{chem}, varies only about three orders of magnitude, which in this case is a result of the similar shape of the strong dependencies on concentration of the intrinsic diffusion coefficients D_{Ag} and D_{Pd} (Noah et al. 2016b).

Finally:

(i)   Arrhenius equations can be adopted for D_A and D_B, recognizing that the corresponding activation energies will be composition dependent. Perhaps

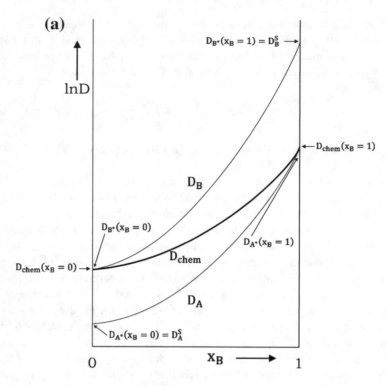

Fig. 8.14 Courses of chemical and intrinsic diffusion coefficients as function of concentration. **a** Schematic illustration of the interrelationships of the chemical diffusion (interdiffusion) coefficient, D_{chem}, with the intrinsic diffusion coefficients D_A and D_B, with the bounding self-diffusion coefficients $D_{A*}(x_B = 0) = D_A^s$ and $D_{B*}(x_B = 1) = D_B^s$ and the bounding impurity diffusion coefficients $D_{A*}(x_B = 1)$ and $D_{B*}(x_B = 0)$, as function of x_B. **b** A real example of chemical and intrinsic diffusion coefficients as function of concentration: the figure presents data for the Ag-Pd system; the self-diffusion coefficients, $D_{Ag*}(X_{Pd} = 0) = D_A^s$ and $D_{Pd*}(X_{Pd} = 1) = D_B^s$ and the impurity diffusion coefficients, $D_{Ag*}(X_{Pd} = 1)$ and $D_{Pd*}(X_{Pd} = 0)$; the intrinsic diffusion coefficients, D_{Ag} and D_{Pd}, and the chemical (interdiffusion) coefficient, D_{chem}, as function of the Pd concentration. All data pertain to the temperature of 435 °C (and a pressure of 1 atm) (redrawn from Noah et al. 2016b). The chemical diffusion coefficient has been determined according to the Darken equation (8.41) but with incorporation of (multiplication of the right-hand side of Eq. (8.41) by) a factor implying correction for the so-called "vacancy wind": Considering one-dimensional substitutional diffusion, the net flux of vacancies in the system upon interdiffusion (denoted as "vacancy wind") causes that atoms are offered more vacancies from one side (the right side in Fig. 8.13) than from the opposite side (the left side in Fig. 8.13). This will enhance the apparent diffusion coefficient of the faster diffusing component (A in Fig. 8.13) and diminish the apparent diffusion coefficient of the slower diffusing component (B in Fig. 8.13). The incorporation of this effect in the calculation of D_{chem} according to Eq. (8.41), as described above, leads to a minor change in the dependence of D_{chem} as function of concentration and is not considered here (further see supplementary material to Noah et al. 2016b)

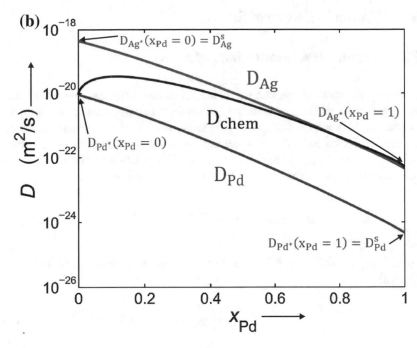

(b)

$$D_{Ag^*}(x_{Pd} = 0) = D^s_{Ag}$$

$$D_{Ag}$$

$$D_{chem}$$

$$D_{Ag^*}(x_{Pd} = 1)$$

$$D_{Pd^*}(x_{Pd} = 0)$$

$$D_{Pd}$$

$$D_{Pd^*}(x_{Pd} = 1) = D^s_{Pd}$$

D (m²/s)

x_{Pd}

Fig. 8.14 (continued)

surprising in view of Eq. (8.57), this appears, on the basis of the experimental evidence, also possible for D_{chem} as a function of concentration, e.g. also for the Ag-Pd system dealt with above (see Fig. 6 in Noah et al. 2016b). The interpretation of the activation energy associated with D_{chem} in terms of the activation energies pertaining to D_A and D_B appears cumbersome. Further, it is remarked that the activation energies of D_A and D_{A^*} generally will be different (similarly for D_B and D_{B^*}). This difference becomes smaller for increasing temperature as TF approaches one; the solid solution becomes more ideal.

(ii)   The identification of a quantifiable link from the thermodynamics to the kinetics of a system, as provided in this section, is a rare event in materials science theory (see the "*Intermezzo: The coupling of thermodynamics to kinetics*" at the end of Sect. 10.16).

8.8   Diffusion in a State of Stress

8.8.1   Pressure/Hydrostatic State of Stress

Considering the vacancy mechanism for substitutional diffusion, one would not be amazed to find out that imposing the system studied to a compressive pressure results in a smaller diffusion coefficient, as it is anticipated that the vacancy concentration may be decreased under compressive pressure. Similarly, a tensile pressure should result in a larger diffusion coefficient. This qualitative discussion can be made quantitative on the basis of Eq. (8.25). D can be written as

$$D = \text{const.} \exp\left[\frac{-\left(\Delta G_{\text{vac}} + \Delta G_{\text{mig}}\right)}{RT} \right] \tag{8.58}$$

As $(\partial G/\partial p)_T = V$ (cf. Sect. 7.3), it follows (pressure dependence of "const" can be neglected):

$$\frac{\partial}{\partial p}\left(\ln\frac{D}{\text{const}} \right)_T = -\left(\frac{1}{RT} \right) \cdot \left(V_{\text{Vac}} + V_{\text{mig}} \right) \tag{8.59}$$

The sum $V_{\text{vac}} + V_{\text{mig}}$ is called the "activation volume". V_{vac} represents the (partial) molar volume of the vacancies. As already argued in Sects. 5.1.1 and 8.6.1, the volume of a vacancy is smaller than that of an atom, as the consequence of relaxation of the surrounding crystal structure. The interpretation of V_{mig} is somewhat less straightforward. It can be expected that V_{mig} is related to the local dilatation of the lattice caused by the jumping atom at the saddle point of the jump ("activated state") (cf. Sect. 8.5.2 and its discussion). It is generally believed that V_{mig} is positive and small as compared to V_{vac}. Most values found for the activation volume per atom for solid metals under compressive pressure indeed fall in a range of 0.5–0.9 times the atomic volume. Obviously, since for interstitial diffusion vacancy formation does not play a role (see discussion in Sect. 8.6), the corresponding activation volume is very small.

A pressure does not have a direction: using the notion pressure indicates that a three-dimensional, spherical state of stress prevails. This is called a "hydrostatic state of stress", which, in the language of Chap. 12, means that the three principal stress components are equal (see Sect. 12.4). If the system considered is subjected to a three-dimensional state of stress with unequal principal stress components, or even only a planar state of stress (this is a three-dimensional state of stress with the third principal stress component equal to zero), the relation of diffusion and stress becomes much more complicated. A few remarks are made in the following.

8.8.2 "Self-Stress"

Quantitative research on the relation between diffusion and stress almost invariably has been restricted to the relation between diffusion and *externally* imposed *pressure*. Investigation of the role of other (more complicated) states of stress has been confined to states of stress invoked by the diffusion process itself. Such stresses can be called "self-stresses".

If, in a one-dimensional *A-B* diffusion couple, atoms *A* diffuse substitutionally into the *B* part, and vice versa atoms of *B* into the *A* part (cf. Figure 8.13), at least the following two origins for self-stress development can be indicated:

(i) *Composition dependence of the molar volume:* If the molar volume in the *A-B* solid solution depends on the composition such that the *A* atoms have a larger molar volume, v_{mol}, than the *B* atoms, $v_{mol,A} > v_{mol,B}$, then, upon interdiffusion, in principle, the *B* part of the diffusion couple can experience a compressive stress parallel to and close to the original interface and the *A* part can experience a tensile stress parallel and close to the original interface (see Fig. 8.15).

(ii) *Difference of the atomic fluxes:* If not sufficient sources and sinks for vacancies are available, and D_A is distinctly smaller than D_B, the *A* part of the couple

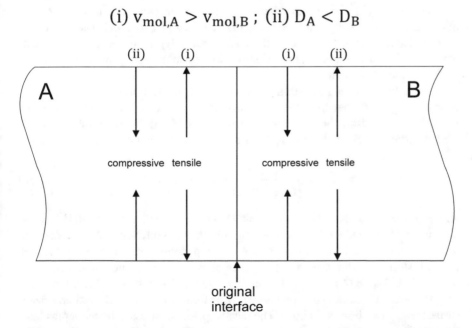

Fig. 8.15 Development of stress upon interdiffusion in a one-dimensional *A-B* diffusion couple, paralled to and close to the original interface, if no mechanisms for stress relaxation operate: (i) in case of a molar volume larger for *A* than for *B*, $v_{mol,A} > v_{mol,B}$ and (ii) in case of an atomic flux of *A* smaller than that of *B*, $J'_A < J'_B$, and no active vacancy sources and sinks

becomes vacancy depleted and the B part of the couple becomes vacancy supersaturated. The volume of a vacancy is a fraction of the atomic volume (see Sects. 5.1.1 and 8.8.1). Consequently, a tensile stress parallel and close to the original interface is expected in the B part and a compressive stress parallel and close to the original interface is expected in the A part (see Fig. 8.15).

However, normally stress-relaxation mechanisms operate too at the temperature of diffusion (as glide and climb of dislocations (cf. Sect. 5.2.7), grain-boundary migration (cf. Sects. 11.2 and 11.3), etc.). This is practically unavoidable, as the necessary mobility of the atoms in order that diffusion can occur, implies that such atomic mobility also suffices to initiate stress-relaxation processes …

As an example, the case of the inward interstitial diffusion of nitrogen or carbon in a piece of iron is considered. The nitrogen and carbon originate from gaseous components which dissociate and adsorb at the surface of the piece of iron and subsequently are absorbed and diffuse inward. The interstitially dissolved nitrogen or carbon tends to expand the iron crystal structure (cf. Sect. 4.4.2 and notes (ii) and (iii) in Sect. 8.6), and thus the surface region as a whole strives for lateral expansion. This desired lateral expansion is counteracted by the core of the specimen where no nitrogen or carbon has arrived. The surface region is tightly connected (cohesively bonded) with the core of the specimen and therefore the surface region cannot expand laterally as desired. As a result, a compressive, cohesive stress parallel to the surface develops in the nitrided or carburized region (also see case (v) and Fig. 12.55 in Sect. 12.18). This is a planar state of stress with two (here equal, because of the rotational symmetry with respect to the surface normal) principal stress components parallel to the surface and a third principal stress component, perpendicular to the surface, which equals nil.

The resulting concentration (nitrogen/carbon)-depth profile, $c(z)$, is schematically shown in Fig. 8.16, together with the stress-depth profile, $\sigma(z)$, as can be calculated straightforwardly from the concentration (nitrogen/carbon)-depth profile, $c(z)$, using the background presented in Chap. 12, if no stress-relaxation mechanisms operate:

$$\sigma(z) = \left[\frac{\beta E}{1 - \nu} \right] \cdot \{ \langle c \rangle - c(z) \} \tag{8.60}$$

with β as the lattice-expansion coefficient (Végard constant; here $\beta > 0$); E and ν as Young's modulus and the Poisson constant, respectively (cf. Sect. 12.2). Because in the case considered the two principal stress components parallel to the surface are equal, it suffices to characterize the state of stress with only one component, σ. It follows from Eq. (8.60) that for $c(z)$ smaller than the depth-averaged concentration, <c>, the stress is compressive (negative), which holds for the (nitrided/carburized) surface region, as discussed above. The tensile (positive) stress at larger depths (also there where no nitrogen or carbon atoms have arrived) arises as a consequence of the requirement of mechanical equilibrium.

Stress-depth profiles as presented above have only partly been observed in reality: in accordance with the above discussion, during the inward diffusion already

Fig. 8.16 Diffusion of a component from the gas phase (e.g. nitrogen/carbon) into a solid (e.g. iron). Schematic presentation of the concentration (c)—depth (z) profile and the corresponding stress (σ)—depth (z) profile. Near the surface pronounced compressive (negative) stresses parallel to the surface prevail; at larger depths modest tensile (positive) stresses occur in order to maintain mechanical equilibrium

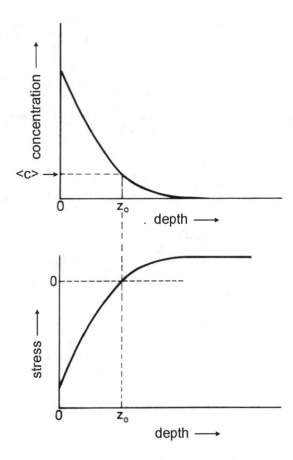

pronounced relaxation of stress occurs in the surface region (where naturally also the largest stresses are induced; cf. Fig. 8.16): processes of dislocation production (Straver et al. 1984) and creep (see Sect. 12.16) can be identified (Kurz et al. 2015).

To describe quantitatively the effect of a three-axial state of stress on diffusion is cumbersome and a satisfactory, complete and rigorous treatment allowing practical application has not been achieved until now. A major problem is associated with the specification of a physically acceptable definition of a "driving force" for diffusion: in the presence of a state of stress, other than the hydrostatic state of stress (in this last case one speaks of a "pressure"; see end of Sect. 8.8.1), adoption and adaptation of the chemical potential of a component are impossible as the notion chemical potential loses its significance in a non-hydrostatic state of stress. This has not always been realized in the existing literature, which has led to erroneous treatments. Larché and Cahn (1985) have recognized this principal difficulty and, considering the effect of compositional, coherency stress (i.e. self-stress) on interdiffusion, replaced the chemical potential for non-hydrostatic states of stress by the so-called diffusion potential. However, their treatment assumes, from the start, so-called network solids,

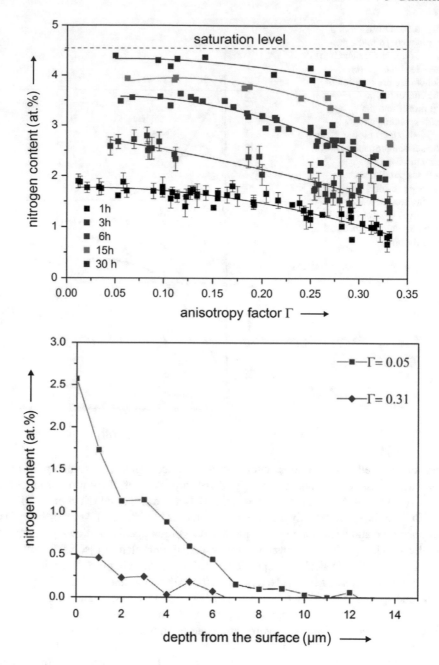

◄**Fig. 8.17** The effect of the dependence of self-stress on the crystallographic orientation of surface adjacent grains: the surface concentrations of the grains with various (*hkl*) planes parallel to the surface will be different, and hence the larger penetration depths occur for the grains with (*hkl*) planes parallel to the surface associated with the larger surface concentrations. **a** Nitrogen content measured on the surface of various grains of Fe-4.5 at.% Cr as a function of the orientation factor, Γ of the corresponding grains, as measured from specimens nitrided for different times at 450 °C. $\Gamma = 0$ (minimum) corresponds with (*hkl*) = (100) and $\Gamma = 1/3$ (maximum) corresponds with (*hkl*) = (111). To exemplify the error of the measured (by EPMA; see Sect. 6.8.3) averaged N content for each grain, error bars for two nitriding times (1 and 3 h) have been shown. The polynomial fits have been given only to guide the eye. Evidently, the surface N content decreases with increasing orientation factor. **b** Nitrogen-depth profiles measured for differently oriented grains (with different orientation factors, Γ, as indicated in the figure) on the cross section of a Fe-4.5 at.% Cr specimen nitrided at 450 °C for 1 h. (taken from Akhlaghi et al. 2015)

implying that crystal-structure sites are neither generated nor annihilated during interdiffusion, which generally does not meet reality (also see above).

A rare case of application of the Larché-Cahn approach to experimental inter-diffusion data, i.e. fitting to concentration-depth profiles, is provided by Noah et al. (2016b), as a side note in their work. It followed, although recognizing that the concept of a "network solid" did not apply, that the effect of the compositional self-stress in this case was in any case marginal regarding the determined values of the diffusion coefficients. It was concluded earlier by Kirkaldy and Savva (1997) that generally the coherency stresses developing upon interdiffusion become fast and completely, locally and globally relaxed. Although, in contrast with this statement, such relaxation may not always be complete (as shown by Noah et al. 2016b; also, see Kurz et al. 2015), a distinct effect of the remaining coherency, compositional state of stress, on the diffusion coefficients as determined from measured concentration-depth profiles, appears, at least in most cases, to be negligible (Noah et al., 2016b). This means that the available databases of diffusion coefficients, determined without any accounting for compositional self-stress, remain in principle usable.

To obtain a manageable theoretical treatment for a more or less crude assessment of the consequences of a three-axial state of stress, one could consider the equivalent hydrostatic stress component, which is given by the average of the three principal stress components.[17] Since the hydrostatic stress is a pressure, it can straightforwardly be incorporated in the expression for the chemical potential. The effect on the diffusion coefficient of the state of stress then can approximatively be given by Eq. (8.59), with p as the equivalent hydrostatic stress component of the state of stress.

The above text focussed on *direct* effects of the diffusion-induced, compositional self-stress on the diffusion parameters, as diffusion coefficients. *Indirect* effects of compositional self-stress on diffusion are important as well. An example has

---

[17] It should be recognized that a planar state of stress is characterized by also *three* principal stress components, albeit one of these equals nil. Hence, the equivalent hydrostatic stress component (given by the sum of the three principal stress components divided by three) for the planar state of stress considered in Fig. 8.16 and pertaining to Eq. (8.60), with two equal principal stress components parallel to the surface, equals $2/3 \cdot \sigma$.

been exposed and (at least semi-)quantitatively explained: again consider the system shown in Fig. 8.16. If the substrate, here a piece of iron, experiencing the inward diffusion of a component, here nitrogen/carbon, is a polycrystal, then the various surface adjacent grains will all experience a (compressive) stress parallel to the surface (see above text connected with Fig. 8.16). Because of the elastic anisotropy of iron (cf. Sect. 12.3), the different grains, with different (hkl) planes parallel to the surface, will experience different values of stress: σ_{hkl}. Next, the equivalent hydrostatic component of the (planar) state of stress (equal to $2/3 \cdot \sigma_{hkl}$; see footnote 17) is determined for the grains of various (hkl). It is straightforwardly possible, on the basis of thermodynamics, to relate the solubility of the inward diffusing component with the prevailing pressure (here taken as the equivalent hydrostatic component: $2/3 \cdot \sigma_{hkl}$). As a consequence, the grains of various (hkl), and thus of different equivalent hydrostatic stress component, will have different surface concentrations (see Fig. 8.17a). Therefore, the diffusional flux from the surface into the grain interiors of the grains of various (hkl) will be different. As a result the concentration–depth profiles for the grains of various (hkl) can differ appreciably, such that the grains with the highest surface concentration, of the inwardly diffusing component, exhibit the largest penetration depth of the inwardly diffusing component (see Fig. 8.17b) (Akhlaghi et al. 2015).[18]

8.9   Microstructure and Diffusion

Diffusion is structure sensitive. Structure defects in crystalline materials pronouncedly influence the diffusion process: for example, think of the vacancy/substitutional diffusion mechanism. There, where the (long range) order of the atomic arrangement in crystalline material is disturbed, one may expect consequences for the diffusion process. Obviously, such symmetry breaks occur at dislocations, grain boundaries and surfaces. The diffusivity at these defects is known to be much faster than through the bulk of a perfect crystal.

   Considering grain-boundary diffusion on the basis of the vacancy mechanism, it can be said that both the enthalpy of formation and the enthalpy of migration (cf. Eqs. (8.25)–(8.27)) may be considerably smaller at the grain boundary than in the bulk. It has also been found that the activation volume (cf. Sect. 8.8.1) for grain-boundary diffusion can be relatively small (Klugkist et al. 2001). This can be interpreted as a consequence of the less perfect state of bonding at the grain boundary. An extreme situation in this sense occurs at the surface, where a much less saturated state of bonding is realized than in the bulk.

---

[18] A higher surface concentration implies the occurrence of a larger concentration gradient and thus (cf. Eq. (8.2)) a larger diffusional flux, which results in a larger penetration depth. Also recognizing that the case considered (cf. Fig.8.16) implies a case of semi-infinite diffusion, Eq. (8.17) (for constant D) in Sect. 8.3 already indicates a larger penetration depth for larger surface concentration, c_0.

The adoption of grain-boundary diffusion as a vacancy-mediated diffusion process appears to be too simple, as indicated by the results of computer simulations. Other mechanisms may contribute to the occurring grain-boundary diffusion. Thus, it has been suggested that (self) interstitials, as defects facilitating mass transport either by jumping from interstitial site to interstitial site or by exchange with "non-interstitial" atoms, may relatively easily form in grain boundaries and can be as important as vacancies for the occurring grain-boundary diffusion (Suzuki and Mishin 2005). The jumps of vacancies and interstitials in/along a grain boundary may involve a group of atomic jumps per defect jump (Suzuki and Mishin 2005).[19] It has also been suggested that ring mechanisms (see Sect. 8.5.1) contribute to grain-boundary diffusion.

Despite the above, complicated picture of grain-boundary diffusion, the experimental results and the computer simulations demonstrate that an Arrhenius-type temperature dependence for grain-boundary diffusion holds (yet), and thus grain-boundary diffusion can be conceived as described by an effective activation energy. The adjective "effective" has been used here because the activation energy of a single atomic jump may not be rate controlling, if a group of atomic jumps per defect jump has to occur (see above and Sect. 10.7; Bos et al. 2007).

As a rule of thumb it holds for the activation energies for diffusion in the bulk (also called volume diffusion), at a grain boundary and at the surface, Q_b (the symbol Q_{vol} is also used), Q_{gb} and Q_s, respectively:

$$Q_b \approx 2Q_{gb} \approx 4Q_s \qquad (8.61)$$

As a consequence of the exponential dependence of the diffusion coefficient on the activation energy (Eq. 8.26), differences in Q as indicated by Eq. (8.61) have an enormous impact on the diffusivity: the corresponding diffusion coefficients differ orders of magnitude. Of course, the impact for a specimen of diffusion along its planar (as grain boundaries and surfaces) and linear (as dislocations) defects also depends on the corresponding defect densities (see Sect. 8.9.1).

Obviously, the precise structure of the grain boundary can have a large influence on the corresponding value of the diffusion coefficient. Moreover, the grain-boundary diffusion can be (highly) anisotropic, i.e. depending on the direction along the grain boundary (see Sect. 8.9.1). However, the higher the temperature, the less pronounced the structural differences between the various grain boundaries, and the differences between the corresponding grain-boundary diffusion coefficients become less outspoken and vanish close to the melting point.[20]

---

[19] This phenomenon parallels the observation for the grain-boundary mobility, where rate control appears to be governed by groups of atomic jumps (in this case not necessarily only along the grain boundary) as well (cf. Sect. 10.7 and Bos et al. 2007).

[20] Upon heating a polycrystalline material, close to the melting point of the bulk material grain boundaries can be "wetted", i.e. be covered by a liquid film. The condition for this process to occur is that the energy of the solid/liquid interface becomes smaller than half the grain-boundary energy. (cf. the discussion on grain-boundary wetting in Sect. 9.4.5). The occurrence of a grain-boundary diffusivity, at temperatures close to the melting temperature of the bulk material, which

Fig. 8.18 Schematic
Arrhenius plot for a
polycrystalline sample,
where at low temperatures,
grain-boundary diffusion
(diffusion coefficient D_{gb})
and at higher temperatures,
volume (bulk) diffusion
(diffusion coefficient D_{vol})
dominates

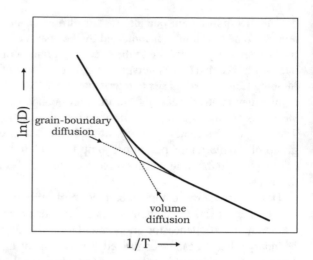

As a concluding remark with respect to the above discussion, it must be said that
the understanding of diffusion along dislocations, grain boundaries and surfaces is
still very incomplete.

Often the term "short-circuit" diffusion is used to characterize the effect of diffusion along planar and linear defects: at a temperature low enough so that bulk diffusion is negligible, the diffusion at the defects can be still considerable and provided
the relative number of atoms in the specimen associated with these defects is large
enough, significant diffusion can still be observed.

As an example, the case of diffusion in a polycrystalline specimen with a significant grain-boundary density is considered; see the Arrhenius diagram shown in
Fig. 8.18. At low temperatures (high values of $1/T$), significant diffusion only occurs
through the grain boundaries as indicated by the straight line at high values for
$1/T$ which has a relatively small slope (i.e. relatively small activation energy; cf.
Eq. (8.61)). At high temperatures (small values of $1/T$), the diffusion is dominated
by bulk (volume) diffusion, and consequently, the slope of the straight line in the
Arrhenius plot is significantly larger. Note that although the contribution of bulk
(volume) diffusion is much larger in this high temperature range, the diffusion at
the grain boundaries is still much faster than the bulk (volume) diffusion: there are,
however, far more atoms in the "bulk" than at the grain boundaries.

---

approaches the diffusivity in the liquid state, is clear indication of grain-boundary wetting in the
system considered (Divinski and Herzig 2008). This is a recent example of the power of diffusion
analysis to reveal the microstructure of a material (see Sect. 8.9.1 for a famous, "old" example). The
occurrence of "superplasticity" at high strain rate (up to 10^2/s) in nanostructured materials upon
plastic deformation at such elevated temperature has been ascribed to such grain-boundary wetting
by a liquid film (see footnote 23 in Sect. 12.16.1).

8.9.1 Diffusion Along the Low-Angle Symmetrical Tilt Boundary

Observations on diffusion can teach us a lot of the defects present in the specimen. A very illustrative example is discussed below.

The atomic arrangement at the low-angle symmetrical tilt grain boundary can be conceived as a (vertical) wall of parallel edge dislocations at a constant distance d equal to $b/2\sin(\theta/2) \approx b/\theta$, where b is the Burgers vector and θ is the angle of rotation around an axis lying in the boundary plane (see Eq. (5.14) and Fig. 5.20). It may be proposed that the diffusion along the dislocation line is relatively fast: the amount of "free space" at the dislocation line/core (beneath the half-plane) is larger than in the bulk of the crystal. Thus, the low-angle symmetrical tilt boundary can be conceived as composed of (*i*) an array/wall of dislocation cores where relatively fast diffusion along the dislocations lines can occur and (*ii*) more or less undisturbed crystalline material in between the dislocation cores where the "normal" bulk diffusion should prevail (Turnbull and Hoffman 1954).

Considering diffusion along the grain boundary in the direction of the dislocation lines, the following reasoning can be applied. If the fraction of the cross-sectional area of the grain boundary (cross section perpendicular to the grain boundary and perpendicular to the dislocation lines) occupied by the dislocation cores/lines equals g, it follows for the effective grain-boundary diffusion coefficient, D_{gb}^{eff}, for diffusion parallel to the dislocation lines (see Fig. 8.19):

$$D_{gb}^{eff} = (1 - g)D_b + gD_{disl} \approx gD_{disl} \qquad (8.61)$$

with D_b as the volume (bulk) diffusion coefficient and assuming that $D_b \ll D_{disl}$.

Taking the thickness of the grain boundary as δ, its length as l and the radius of the dislocation core as $r = \delta/2$, it follows:

$$g = \left(\frac{l}{d}\right)\frac{\pi r^2}{l\delta} = \frac{\pi r^2}{d\delta} \qquad (8.62)$$

Substituting r by b (the length of the Burgers vector; cf. Sect. 5.2.4), implying $\delta = 2b$, and d by $b/2\sin(\theta/2)$, it is obtained from Eqs. (8.61) and (8.62):

$$D_{gb}^{eff}\delta = g\delta D_{disl} = \left\{\pi b2\sin\left(\frac{\theta}{2}\right)\right\}D_{disl} \qquad (8.63)$$

Solving Fick's laws for a case of diffusion along a grain boundary in a bicrystal, it follows that from the measurement of the extent of diffusion only the product $D_{gb}^{eff}\delta$ can be determined. Thus, if $D_b \ll D_{disl}$, the value of $D_{gb}^{eff}\delta$ should increase practically linearly with $2\sin(\theta/2) \approx \theta$. Such results have been obtained (Turnbull and Hoffman 1954; Okkerse 1954a, 1954b). Okkerse in particular showed that diffusion along the symmetrical tilt grain boundary in the direction *perpendicular* to the dislocation

Fig. 8.19 Model for deriving an expression for the effective grain-boundary diffusion coefficient, D_{gb}^{eff}, for a low-angle symmetrical tilt grain boundary. The boundary (see also Fig. 5.20) can be conceived as a vertical arrangement of parallel edge dislocations, with dislocation-core radius $r = \delta/2$, where δ is the grain-boundary width, at vertical distances between the dislocations equal to d

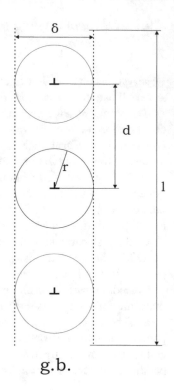

g.b.

lines was much less fast than in the direction *parallel* to the dislocation lines, thereby demonstrating the anisotropic nature of the diffusion in the grain boundary.

It has been the tremendous success of this simple theory and the elegant dedicated diffusion experiments on bicrystals (of lead and of silver) that provided the evidence for the concept of the low-angle symmetrical tilt boundary as a wall of regularly spaced edge dislocations positioned on top of each other (see Figs. 5.20 and 8.19): along the edge dislocation lines ("pipes"; one also speaks of "pipe" diffusion) fast diffusion occurs, whereas through the grain-boundary material in-between these "pipes" relatively slow diffusion (comparable to bulk (volume) diffusion), more or less negligible as compared to the diffusion through the "pipes" (cf. Eq. 8.61), occurs. Hence, a structure model was confirmed indirectly at a time where direct (transmission electron) microscopical evidence did not yet exist[21]: Analysis of diffusion phenomena can lead to profound insight into the (defect) nature of the atomic arrangement of materials.

---

[21] The first observations of (edge) dislocations, made by using a transmission electron microscope, were published in 1956.

8.9.2 Diffusion Along a Moving Grain Boundary

Especially, at relatively low temperatures pronounced diffusional mixing can occur (in thin film systems) that can only be explained as a result of grain-boundary diffusion along *moving* grain boundaries (Mittemeijer and Beers 1980).

Consider Fig. 8.20 showing an *A/B* bicrystal with grain boundaries oriented perpendicular to the *A/B* interface. The temperature is that low that no significant bulk (volume) diffusion across the *A/B* interface can occur. However, grain-boundary diffusion, e.g. inward diffusion of *B* along the grain boundaries oriented perpendicular to the *A/B* interface in the *A* part of the bicrystal, is possible in a significant way (cf. above discussion of Fig. 8.18). If the grain boundary would be immovable/static, a thin deeply penetrating layer along the grain boundary composed of *A/B* mixed

(a)

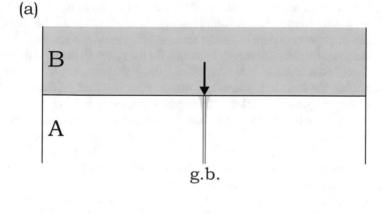

(b)

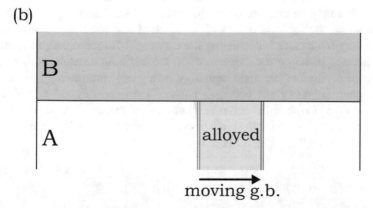

Fig. 8.20 Diffusion along a (migrating) grain boundary in a *B/A* bicrystal. **a** Inward diffusion of *B* into *A* along a (stationary) grain boundary (g.b.). **b** Formation of an alloyed (*AB*) zone adjacent to the *B/A* interface by motion of the grain boundary. The energy reduction upon diffusional mixing is thought to drive the process: "diffusion-induced grain-boundary migration (DIGM)"

material occurs in the A part/sublayer of the system (Fig. 8.20a), but for the entire system then only relatively marginal diffusional mixing is achieved (this is the type of grain-boundary diffusion in thin film systems considered in the last paragraph of the Appendix to this chapter). However, if the grain boundary considered starts to move in a direction more or less parallel to the A/B interface, then in the wake of the moving boundary diffusionally mixed (by grain-boundary diffusion) material is left behind and as a result a layer of relatively strongly mixed material occurs at the original A/B interface (Fig. 8.20b), albeit of a thickness less penetrating than holds for the thin mixed layer along the grain boundary in case the grain boundary remains static. The mobility of the grain boundary is the key factor bringing about that, "globally", appreciable diffusional mixing does occur, whereas grain-boundary diffusion along static grain boundaries would only lead to traces of diffusional mixing along the static grain boundaries. In thin film systems normally a high grain-boundary density occurs. Therefore, if a relatively high degree of intermixing in thin film systems is observed upon annealing at relatively low temperatures (say, at temperatures equal to half the melting temperature in K and below), then this effect can be understood as the result of grain-boundary diffusion along moving grain boundaries.

Diffusional mixing leads to energy reduction for the system. This must provide the "driving force" (see footnote 1 of Chap. 9) for the grain boundary to start to move upon annealing the bicrystal. Therefore, this process is called "diffusion-induced grain-boundary migration (DIGM)". Having said this, the question then is how this driving force can be translated into an atomic mechanism for boundary motion, a topic which has been discussed controversially in the literature (see King (1987) and, for a newer concept, see Klinger and Rabkin 2011). One of the greatest obstacles to further progress in fundamental understanding of DIGM is the lack of knowledge on the atomic structure of *moving* grain boundaries.

Considering the original position of the moving grain boundary in Fig. 8.20b, one may wonder why the grain boundary started to move to the right and not to the left. This may be understood as the result of local kinetic constraints/atomic configurations, the appearance of which can be of statistical nature. Hence, along the boundary the chance for initiating the movement "to the right" alternates with the chance for initiating the movement "to the left". As a result, with respect to the original position of the grain boundary, a "zig-zag" morphology for the entire moving grain boundary can occur, as parts of the same original grain boundary started to move "to the right" and other parts started to move "to the left" upon the begin of interdiffusion along the grain boundary (see the sketch in Fig. 8.21a and the experimental example in Fig. 8.21b).

Intermezzo: Priority and Scientific Decency

Den Broeder (1970, 1972) was possibly the first who extensively described and recognized the occurrence of DIGM in his experiments on interdiffusion in a solid–solid, bulk, Cr-W diffusion couple, by analyzing cross sections taken

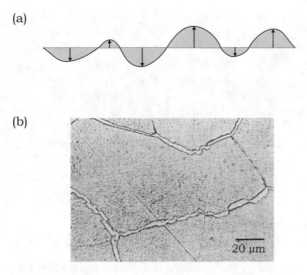

Fig. 8.21 "Zig-zag" shape of a grain boundary experiencing DIGM. **a** Schematic picture indicating the statistical nature of the direction of grain-boundary movement along the grain boundary (here: proceeding along the grain boundary, either to the right or to the left). **b** W-Cr diffusion couple, annealed for 6 h at 1400 C. Light optical micrograph of a cross section prepared at a few μm from the original W/Cr interface in the originally pure tungsten part of the diffusion couple. The cross section is oriented perpendicular to the diffusion direction, i.e. parallel to the original interface (den Broeder 1970, 1972)

parallel to the original Cr/W interface, i.e. *perpendicular* to the diffusion direction in the binary diffusion couple (see Fig. 8.21b). This work remained unnoticed, although published in a high-quality, international journal (1972), and the effect was "rediscovered" later (in 1978 in a study of the inward diffusion of Zn from the vapour phase into solid Fe) and published in the same journal (Hillert and Purdy 1978) without referring to the original work by den Broeder.

These remarks illustrate an aspect of a "priority discussion" also touched upon in the discussion about who "discovered" the Periodic Table (see the *Intermezzo* in Sect. 2.5). Again, upon close inspection, also in this case a number of contributions can be found in the literature where DIGM was "discovered" (According to a review (King 1987) one of these "first" papers is from 1938[22]). As also remarked about the "discovery" of the Periodic Table in the mentioned *Intermezzo*, who was first may not interest us. The point here is: the authors of these "first" papers were apparently unaware of each other's published work. This may be understandable for the time of the discovery of the Periodic Table, but this is unacceptable for scientific work after, say, the Second World War. The "first" papers on DIGM indicated in the mentioned review have *all* been published in well-known, international, first-class scientific journals i.e. journals of great accessibility. The explicit example discussed above concerns two papers even published in the same journal within a time span of six years, and therefore, this particular case of unawareness is remarkable, even recognizing the vast and exponentially growing amount of published literature. The phenomenon of lack of referencing, and thereby acknowledging, reflects an attitude, which appears to become more widespread nowadays, to spend insufficient time on reading and checking the literature.[23] There is no excuse for this flaw. Even the electronic literature-surveillance systems of present day seem not to lead to improvement in this respect. This may have as background human sins as laziness, carelessness or vanity (and then such non-referencing is done deliberately). In all cases it is scientifically indecent behaviour. A prerequisite of scientific research is command of the existing literature. The scientist's moral should then lead to fair, appropriate referencing in papers published as the result of research performed.

It has been recognized that the rapid diffusional mixing along a grain boundary can also lead to the nucleation of new (alloyed) grains, at the moving grain boundary, which grains grow by "diffusion-induced grain-boundary migration". This phenomenon has been called "diffusion-induced recrystallization (DIR)". It may be thought that the vehement nature of the diffusion process in the moving grain boundary, as e.g. near to the interface of a bicrystal where very large gradients in chemical potential along the grain boundary can occur, can lead to large structural rearrangements eventually inducing the nucleation of new grains (see Sect. 11.2.1).

It could be shown that the relatively advanced stage of homogenization in the DIR grains is due to grain-boundary diffusion in the moving grain boundary that is the interface between the new grain and the matrix (Mittemeijer and Beers 1980). Hence, in "large-scale" diffusion couples where recrystallization phenomena in the diffusion zone occur (usually near the original interface), grain-boundary diffusion, in the moving grain boundaries, may contribute considerably to the homogenization.

If the grain-boundary velocity is denoted by v, and the width of the grain boundary by δ, then the time available for homogenization by grain-boundary diffusion, τ, for each piece of material that has been "run over" by the grain boundary, equals δ/v. Realistic values for δ and v are 0.5 nm and 100 nm s^{-1}, respectively, leading to $\tau \approx 5 \times 10^{-3}$ s. This may seem a very small period of time, but this value of τ yet suffices to bring about appreciable homogenization as illustrated by Fig. 8.22.

In contrast with "normal" recrystallization in homogeneous "bulk" materials, where the driving force usually is the reduction in stored cold work (decrease of

---

[22] Upon reading this early note (F.N. Rhines and A.M. Montgomery, Nature, 141 (1938), 413) it is clear that these authors had no idea of the background of their observation of the "disturbed" grain boundary of a Cu bicrystal upon inward diffusion of Zn.

[23] After having written the text of this *Intermezzo* in the first edition of this book, upon preparing the second edition I became aware of a recent, very readable booklet where this disgrace of modern science, and many even more serious ones, have been exposed and illustrated by illuminating examples: see G. Pacchioni, *The Overproduction of Truth,* Oxford University Press, 2018.

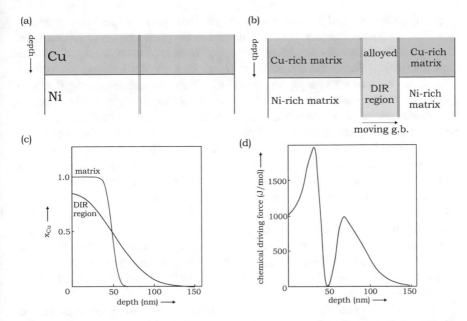

Fig. 8.22 Diffusion induced recrystallization (DIR) in a Cu(48.5 nm)/Ni(100 nm) thin film diffusion couple (Mittemeijer and Beers 1980). **a** Schematic illustration of the bilayer containing a grain boundary traversing the Cu/Ni interface. **b** Schematic illustration of the creation of an alloyed DIR region by movement of the grain boundary. **c** Concentration-depth profiles in the Cu/Ni bilayer for the matrix and for a DIR region (0.5 h at 550 °C). **d** Chemical driving force (=difference of Gibbs energies of matrix and DIR region) for the grain-boundary movement as a function of depth (0.5 h at 550 °C)

dislocation density in the wake of the moving recrystallization front, as compared to the deformed parent material to be recrystallized; cf. Sect. 11.2), here the driving force derives from the advanced stage of homogenization achieved by the diffusional mixing in the moving boundary/recrystallization front, as compared to the less advanced stage of homogenization in the matrix. An example for a Cu–Ni bilayer is shown in Fig. 8.22. Part c of the figure provides a comparison of the concentration-depth profiles in the bilayer as a function of depth for (i) the matrix, where diffusional homogenization according to volume/"bulk" diffusion has occurred, and for (ii) the recrystallized grain, that has achieved its distinctly more advanced state of homogenization by the grain-boundary diffusion occurring in the moving grain boundary (recrystallization front) as it sweeps through the matrix. The corresponding variation in the chemical driving force for the movement of the grain boundary/"recrystallization front", along the moving grain boundary (=difference of Gibbs energies of matrix and DIR region along the moving grain boundary, i.e. as function of depth), is shown in part *d* of the figure. The apparent, pronounced variation in driving force along the grain boundary suggests that the boundary may not move as a whole with the same speed. This can contribute to the irregular nature of the grain-boundary advancement, as observed experimentally.

Finally, it is remarked that the process of "discontinuous (cellular) precipitation" (see Sect. 9.4.3) has features which are related to the phenomenon of DIGM/DIR. The typical discontinuous precipitation reaction involves the formation of a solute depleted (parent) phase and a precipitate phase, as a usually lamellar microstructure, in the wake of a moving grain boundary advancing into (sweeping through) a supersaturated matrix (see Sect. 9.4.3 and Fig. 9.8b). A discontinuous change in both the crystal orientation and the composition (solute concentration) of the parent phase across the moving interface are characteristics of this transformation. As with DIGM/DIR, the change in concentration due to this transformation is realized usually by *diffusion along the grain boundary* moving into the matrix, as holds for DIGM/DIR. However, in the case of discontinuous precipitation, a decomposition is brought about by the diffusion process along the moving boundary, whereas with DIGM homogenization is promoted by the diffusion along the moving boundary. The chance for the reaction front of the "discontinuous precipitation", upon initiation at a grain boundary in the parent crystal, to move "to the right" and not "to the left" can be discussed as above for DIGM and, indeed, a similar "zig-zag" type of morphology, with respect to the original position of the grain boundary, can be observed for the moving grain boundary.

Analysis of the grain-boundary diffusivity along the moving grain boundary in DIGM/DIR has shown that the diffusivity along a moving grain boundary is not distinctly different from the diffusivity along a stationary (i.e. not moving) grain boundary (Mittemeijer and Beers 1980) and a similar result has also been obtained for diffusion along the moving transformation front in a discontinuous transformation (Sect. 9.4.3). Such results oppose speculations that the diffusivities along moving grain boundaries might be orders of magnitudes larger than those along stationary grain boundaries as a consequence of supposed structural differences between moving and stationary grain boundaries.

Evidently, the treatments in Sects. 8.8 and 8.9 and also, to a minor extent, that in Sect. 8.7, show that diffusion, as a phenomenon, is not completely understood and that substantial knowledge for some specific, quantitatively accurate models still lacks. On the one hand, this may come as a surprise, recognizing that diffusion is a classical theme in materials science. On the other hand, it thus is illustrated that important progress in this field of science, essential for understanding material behaviour, is still to be expected (cf. footnote 3 in Chap. 4).

Appendix: Concentration-Depth Profiles in Thin Layer Systems

With a view to the great practical importance of thin film systems, as in the microelectronic industry, a number of solutions to Fick's second law for a variety of thin layer systems is summarized at the end of this chapter in this appendix. Various cases with different initial and boundary conditions can be considered for different diffusion stages. It will be assumed that the diffusion coefficient can be taken as concentration independent.

Case 1 (Fig. 8.23a): *A bilayer (AB) or multilayer (ABABAB ...) for which the thickness of each sublayer is much larger than the diffusion length, $\sqrt{Dt}$, corresponding to a very early diffusion stage for the (multi)layer.*

The solution for the bilayer *AB* (possibly as part of the multilayer *ABABAB...*) with the *A/B* interface at $z = z_i$ and initial conditions $C = C_0$ for $z \leq z_i$ and $C = 0$ for $z > z_i$ is given by (Crank 1975):

$$C(z, t) = \frac{C_0}{2\sqrt{\pi Dt}} \int\limits_{z-z_i}^{\infty} \exp\left(-\frac{\xi^2}{4Dt}\right) d\xi = \frac{1}{2} C_0 \text{erfc} \frac{z - z_i}{2\sqrt{Dt}} \qquad (8.64)$$

where erfc ($=1 - $ erf) denotes the error-function (erf) complement (cf. Eq. 8.16). Note that, evidently, $C = C_0/2$ at $z = z_i$ for all $t > 0$.

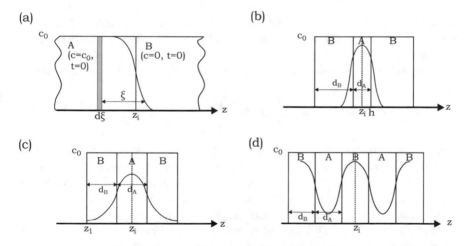

Fig. 8.23 Concentration-depth profiles in thin film systems. **a** Bilayer or multilayer with the thickness of each sublayer much larger than the diffusion length. **b** Trilayer or multilayer with the thickness of each sublayer *A* much smaller than that of sublayers *B* and the diffusion length much smaller than the thickness of sublayers *B*. **c** Trilayer with thickness of each sublayer of the order of the diffusion length. **d** Multilayer with the thickness of each sublayer of the order of the diffusion length

Case 2 (Fig. 8.23b): *A trilayer (BAB) or multilayer (BABABA...) for which the thickness of sublayers A is much smaller than that of the sublayers B, i.e. $d_A \ll d_B$, and $\sqrt{Dt} \ll d_B$.*

For the trilayer *BAB* (possibly as part of the multilayer *BABABA*...) with $z = z_i$ at the centre plane of sublayer *A* and initial conditions $C = C_0$ for $z_i - d_A/2 \leq z \leq z_i + d_A/2$ and $C = 0$ for $z < z_i - d_A/2$ and $z > z_i + d_A/2$, the concentration profile is obtained as (cf. Eq. 8.64)

$$
C(z,t) = \frac{C_0}{2\sqrt{\pi Dt}} \int\limits_{z-(z_i+d_A/2)}^{z-(z_i-d_A/2)} \exp\left(-\frac{\xi^2}{4Dt}\right) d\xi
$$

$$
= \frac{1}{2}C_0\left[\text{erf}\frac{d_A/2 + (z - z_i)}{2\sqrt{Dt}} + \text{erf}\frac{d_A/2 - (z - z_i)}{2\sqrt{Dt}}\right] \tag{8.65}
$$

It is clear that the system can be cut in half by a plane at $z = z_i$ without affecting the distribution, which is symmetrical about $z = z_i$. Therefore, Eq. (8.65) also holds for a bilayer system composed of a sublayer *A* of thickness of $d_A/2$, with the surface or a diffusion barrier at $z = z_i$, on top of the semi-infinite sublayer *B* (substrate, see Fig. 8.23b). Redefining $d_A/2$ as h and taking $z_i = 0$, it follows for the diffusion-induced concentration profile in this case

$$
C(z,t) = \frac{C_0}{2}\left[\text{erf}\frac{h + z}{2\sqrt{Dt}} + \text{erf}\frac{h - z}{2\sqrt{Dt}}\right] \tag{8.66}
$$

Case 3 (Fig. 8.23c): *A trilayer (BAB) for which the thickness of each sublayer is not much larger than the diffusion length $\sqrt{Dt}$, which represents a relatively advanced diffusion stage for the trilayer.*

The outer surfaces of both *B* sublayers are barriers for mass transport. Hence,

$$
\frac{\partial C}{\partial z} = 0 \quad \text{at } z_1 = z_i - \left(d_B + \frac{d_A}{2}\right) \text{ and } z_2 = z_i + \left(d_B + \frac{d_A}{2}\right)
$$

The resulting concentration profile can be constructed by the superposition (reflection) principle as follows (cf. Crank 1975). The concentration profile given by Eq. (8.65) is reflected at the plane at boundary z_2, and this reflected profile is obtained by replacing z_i in Eq. (8.65) by $z_i + (2 d_B + d_A)$. This firstly reflected curve is reflected at the plane at z_i, and the secondly reflected profile is obtained by replacing z_i in Eq. (8.65) by $z_i - (2d_B + d_A)$. Then, the secondly reflected profile is reflected again at the boundary z_2 ($z_i \rightarrow z_i + 2(2d_B + d_A)$) and at z_i($z_i \rightarrow z_i - 2(2d_B + d_A)$) and so on. Therefore, the complete solution as the result of such successive reflections is given by

$$C(z,t) = \frac{1}{2}C_0 \sum_{n=-\infty}^{\infty} \left[\operatorname{erf}\frac{d_A/2 - n(2d_B + d_A) + z - z_i}{2\sqrt{Dt}} \right.$$

$$\left. + \operatorname{erf}\frac{d_A/2 + n(2d_B + d_A) - z + z_i}{2\sqrt{Dt}} \right] \tag{8.67}$$

Case 4 (Fig. 8.23d): *A multilayer (ABABAB…) for which the thickness of each sublayer is not much larger than the diffusion length $\sqrt{Dt}$, which represents a relatively advanced diffusion stage for the multilayer.*

Considering the initial conditions: $C = C_0$ for $z_i - d_A/2 \leq z \leq z_i + d_A/2$ and $C = 0$ for $z_i - (d_A/2 + d_B) < z < z_i - d_A/2$ and $z_i + d_A/2 < z < z_i + (d_A/2 + d_B)$ with z_i denoting the centre plane of sublayer A, and recognizing that for $t \geq 0$ $\partial C/\partial z = 0$ at the centre plane of the sublayers B, the total concentration profile for the trilayer BAB in the multilayer is obtained by the superposition (reflection) principle (see also Case 3) as follows

$$C(z,t)$$

$$= \frac{1}{2}C_0 \sum_{n=-\infty}^{\infty} \left[\operatorname{erf}\frac{d_A/2 - n(d_B + d_A) + z - z_i}{2\sqrt{Dt}} + \operatorname{erf}\frac{d_A/2 + n(d_B + d_A) - z + z_i}{2\sqrt{Dt}} \right] \tag{8.68}$$

The diffusion-induced concentration profiles as given by the Eqs. (8.64)–(8.68) have been derived considering volume (bulk) diffusion. In polycrystalline thin films the role of grain-boundary diffusion is often dominant, recognizing the high grain-boundary density and the usually applied relatively low diffusion annealing temperatures (as compared to the melting point of the components). Often columnar microstructures occur in thin films; i.e. the grain boundaries are oriented more or less perpendicular to the film surface and sublayer interfaces. Then, if the volume diffusion length, $(D_b t)^{1/2}$, is much smaller than the grain-boundary diffusion length, $(D_{gb} t)^{1/2}$, with D_b and D_{gb} as the volume and grain-boundary diffusion coefficients, respectively, it can be shown (see Wang and Mittemeijer 2004) that the laterally averaged concentration profile for Cases 1–4, as induced by grain-boundary diffusion, are also given by Eqs. (8.64)–(8.68), provided C is identified with $\overline{C}$, the laterally averaged concentration, D is identified with the grain-boundary diffusion coefficient D_{gb}, and C_0 is identified with $C_0 \delta \eta$ (δ is grain-boundary width and η is grain-boundary length per unit area for the plane parallel to the surface.

References

General

H.S. Carslaw, J.C. Jaeger, *Conduction of Heat in Solids* (Clarendon Press, Oxford, 1959).
J. Crank, *The Mathematics of Diffusion*, 2nd edn. Oxford University Press (1975).
P.F. Green, *Kinetics, Transport, and Structure in Hard and Soft Materials* (Taylor & Francis, London, 2005).
I. Kaur, Y. Mishin, W. Gust, *Fundamentals of Grain and Interphase Boundary Diffusion, 3rd revised and enlarged edition*. (Wiley & Sons, Chichester, 1995).
J.S. Kirkaldy, D.J. Young, *Diffusion in the Condensed State* (The Institute of Metals, London, 1987).
H. Mehrer, *Diffusion in Solids* (Springer, Heidelberg, 2007).
J. Philibert, *Atom Movements; Diffusion and Mass Transport in Solids* (Les Editions de Physique, Les Ulis, 1991).
P. Shewmon, *Diffusion in Solids*, 2nd edn. (The Minerals, Metals & Materials Society, Warrendale, 1989).

Specific

M. Akhlaghi, M. Jung, S.R. Meka, M. Fonović, A. Leineweber, E.J. Mittemeijer, Dependence of the nitriding rate of ferritic and austenitic substrates on the crystallographic orientation of surface grains; gaseous nitriding of Fe-Cr and Ni-Ti Alloys, Philos. Mag. **95**, 4143–4160 (2015)
H.K.D.H. Bhadeshia, Carbon–Carbon Interactions in Iron. J. Mater. Sci. **39**, 3949–3955 (2004)
C. Bos, F. Sommer, E.J. Mittemeijer, Atomistic study on the activation enthalpies for interface mobility and boundary diffusion in an interface-controlled phase transformation. Phil. Mag. **87**, 2245–2262 (2007)
F.J.A. den Broeder, *Onderzoek naar de Diffusie in het Systeem Chroom-Wolfraam*. Ph.D. Dissertation, Delft University of Technology, pp. 90–95 (in Dutch, 1970)
F.J.A. den Broeder, Interface reaction and a special form of grain boundary diffusion in the Cr-W system. Acta Metall. **20**, 319–332 (1972)
L. Dabrowski, A. Andreev, M. Georgiev, Carbon diffusion in α-iron: Evidence for quantum mechanical tunneling. Metallur. Mater. Trans. A **37A**, 2079–2084 (2006)
L.S. Darken, Diffusion, mobility and their interrelation through free energy in binary metallic systems. Trans. AIME **175**, 184–194 (1948)
S. Divinski, C. Herzig, Radiotracer Investigation of diffusion, segregation and wetting phenomena in grain boundaries. J. Mater. Sci. **43**, 3900–3907 (2008)
A. Fick, *Über Diffusion*. Poggendorff's Annalen, **94**, 59–86 (in German, 1855); published in English, in abstracted form, as A. Fick, On liquid diffusion. Philos. Mag. **10**, 30–39 (1855)
D.H.R. Fors, G. Wahnström, Nature of Boron solution and diffusion in α-Iron. Phys. Rev. B **77**, 132102 (2008)
Y. Fukai, *The Metal-Hydrogen System; Basic Bulk Properties* (Springer, Berlin, 2005)
G. Grube, A. Jedele, Die Diffusion der Metalle im Festen Zustand. Zeitschrift Für Elektrochemie **38**, 799–807 (in German, 1932)
M. Hillert, G.R. Purdy, Chemically induced grain boundary migration. Acta Metall. **26**, 333–340 (1978)
A.H. King, Diffusion induced grain boundary migration. Int. Mater. Rev. **32**, 173–189 (1987)
J.S. Kirkaldy, G. Savva, Correlation between coherency strain effects and the Kirkendall effect in binary infinite diffusion couples. Acta Mater. **45**, 3115–3121 (1997)

L. Klinger, E. Rabkin, Theory of the Kirkendall effect during grain boundary interdiffusion. Acta Mater. **59**, 1389–1399 (2011)

P. Klugkist, A.N. Aleshin, W. Lojkowski, L.S. Shvindlerman, W. Gust, E.J. Mittemeijer, Diffusion of Zn along tilt grain boundaries in Al: pressure and orientation dependence. Acta Mater. **49**, 2941–2949 (2001)

N. Kurita, H. Numakura, Formation volume of atomic vacancies in body-centred cubic metals. Z. Metallkde. **95**, 876–879 (2004)

S.J.B. Kurz, S.R. Meka, N. Schell, W. Ecker, J. Keckes, E.J. Mittemeijer, Residual stress and microstructure depth gradients in nitrided iron-based alloys revealed by dynamical cross-sectional transmission X-ray microdiffraction. Acta Mater. **87**, 100–110 (2015)

F.C. Larché, J.W. Cahn, The interactions of composition and stress in crystalline solids. Acta Metall. **33**, 331–357 (1985)

Y. Liu, L. Zhang, D. Yu, Y. Ge, Study of Diffusion and Marker Movement in fcc Ag-Au Alloys. J. Phase Equilibria and Diffusion **29**, 405–413 (2008)

C. Matano, On the relation between the diffusion coefficients and concentration of solid metals (the nickel-copper system). Jpn. J. Phys. **8**, 109–113 (1933)

E.J. Mittemeijer, A.M. Beers, Recrystallization and interdiffusion in thin bimetallic films. Thin Solid Films **65**, 125–135 (1980)

M.A. Noah, D. Flötotto, Z. Wang, M. Reiner, C. Hugenschmidt, E.J. Mittemeijer, Interdiffusion in epitaxial, single-crystalline Au/Ag films studied by auger electron spectroscopy sputter-depth profiling and positron annihilation. Acta Mater. **107**, 133–143 (2016a)

M.A. Noah, D. Flötotto, Z. Wang, E.J. Mittemeijer, Interdiffusion and stress development in single-crystalline Pd/Ag bilayers. J. Appl. Phys. **119**, 145308 (2016b)

B. Okkerse, *Zelfdiffusie in Lood*, Ph.D. Dissertation, Delft University of Technology, pp. 55–69 (in Dutch, 1954a)

B. Okkerse, Self-diffusion in lead. Acta Metall. **2**, 551–553 (1954b)

J. Philibert, Beyond Fick's equations, an overview. Int. J. Mater. Res. **100**, 744–749 (2009)

P.S. Rudman, The atomic volumes of the metallic elements. Trans. Metallur. Soc. AIME **233**, 864–871 (1965)

J. Silva, R.B. McLellan, Diffusion of carbon and nitrogen in bcc Iron. Mater. Sci. Eng. **26**, 83–87 (1976)

R. Simmons, R.W. Balluffi, Measurements of equilibrium vacancy concentrations in aluminum. Phys. Rev. **117**, 52–61 (1960)

A.C. Smigelskas, E.O. Kirkendall, Zinc diffusion in alpha brass. Trans. AIME **171**, 130–142 (1947)

W.T.M. Straver, H.C.F. Rozendaal, E.J. Mittemeijer, Consequences of the heterogeneous nitriding of α-iron: dislocation production and oriented precipitation. Metallur. Trans. A **15**, 627–637 (1984)

A. Suzuki, Y. Mishin, Atomic mechanisms of grain boundary diffusion: low versus high temperatures. J. Mater. Sci. **40**, 3155–3161 (2005)

D. Turnbull, R.E. Hoffman, The effect of relative crystal and boundary orientations on grain boundary diffusion rates. Acta Metall. **2**, 419–426 (1954)

J.Y. Wang, E.J. Mittemeijer, A new method for the determination of the diffusion-induced concentration profile and the interdiffusion coefficient for thin films by auger electron spectroscopical sputter depth profiling. J. Mater. Res. **19**, 3389–3397 (2004)

Chapter 9
Phase Transformations: Introduction and Typology

The manipulation of the microstructure of materials belongs to the heart of the realm of materials science. Often, but not always, *non-equilibrium* structures/states are produced purposely. The goal of the invoked microstructural changes is to bring about favourable values for the material properties of interest in the application of the material concerned. Mechanical treatments in combination with heat treatments, such as cold rolling followed by annealing to induce recrystallization, provide one example, which is discussed in Chap. 11. Very often the microstructure is changed by deliberately generated phase transformations, which are the focal point of interest in this and the following chapter. A classical example involves (see Fig. 9.1 pertaining to a binary system, and see also Chap. 7):

- annealing at elevated temperature in a one-phase region of a (usually) metallic alloy, so that a homogeneous alloy is established, followed by
- quenching (=very fast cooling, so that the atoms cannot move substantially (no long-range diffusion) during the cooling) down to a relatively low temperature in a two-phase region: i.e. the two-phase thermodynamic equilibrium at this last temperature is not realized; one could say: the atomic arrangement of the high temperature, homogeneous state is "frozen in"; next
- the supersaturated solid solution thus obtained is decomposed by annealing at a moderate temperature in the two-phase region, leading to precipitation of a dispersed, second phase.

The two-phase material produced in this way may, for example, show favourable mechanical properties as high hardness, due to precipitation/dispersion hardening (see Sect. 12.14.4). This is the motivation for the thermal treatments applied to industrially important Al-based alloys. In this sense, also the "quenching and tempering" of steels has to be understood: Here, by quenching from the austenite-phase field, a very hard, but brittle martensite phase is generated, which by annealing ("tempering") at moderate temperature is softened in association with the precipitation

© Springer Nature Switzerland AG 2021
E. J. Mittemeijer, *Fundamentals of Materials Science*,
https://doi.org/10.1007/978-3-030-60056-3_9

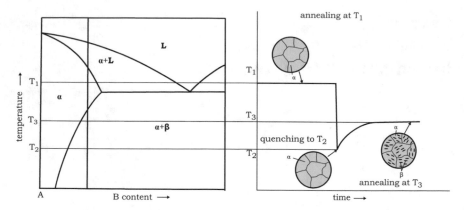

Fig. 9.1 Classical example of phase transformation to change the microstructure of a material to bring about favourable properties: *precipitation/dispersion hardening*. The change of the microstructure upon subjecting the specimen, with a composition indicated by the vertical line in the phase diagram at the left-hand side, to the temperature-time programme indicated at the right-hand side, is shown by the sketches in the figure at the right-hand side:
annealing at elevated temperature, T_1, in a one-phase region of the alloy concerned, so that a homogeneous alloy is established; followed by;
quenching down to a relatively low temperature, T_2, in a two-phase region, such that the atomic arrangement of the high temperature, homogeneous state is "frozen in "; next:
the supersaturated solid solution thus obtained is decomposed by annealing at a moderate temperature, T_3, in the two-phase region, leading to precipitation of a dispersed, second phase

of (metastable) carbides, leading to a lower hardness but a much higher ductility, so that a favourable combination of these mechanical properties results (for further discussion, see the "*Intermezzo: Tempering of iron-based interstitial martensite*" in Sect. 9.5). Understanding phase transformations is crucial for being able to optimize the microstructure of a material for a specific application.

As the last example mentioned above may suggest, the preferred microstructure very often is not the one corresponding to thermodynamic equilibrium: material systems as applied by mankind very often are remote from the equilibrium, "end" state. This implies that an understanding of the kinetics, i.e. the time and temperature dependences, of phase transformations is of overwhelming importance. Therefore appropriate attention is paid to the description of phase-transformation kinetics in this book (Chap. 10).

9.1 Thermodynamics and Kinetics of Phase Transformations; Thermal Activation and the Activation Energy

Two major scientific questions can be formulated which have to be dealt with in order to arrive at a fundamental understanding of phase transformations:

(i) What is the origin of a phase transformation? This question can be reformulated as follows. A phase transformation occurs if it leads to a lowering of the energy of the system: the system strives for thermodynamic equilibrium characterized by a state of minimal energy (cf. Sect. 7.3). The energy difference of the system before the phase transformation and after the completed phase transformation is called "the driving force".[1] So the above question can be converted into: what is the driving force of the phase transformation? This then is asking for the *thermodynamics of the phase transformation*.

(ii) What are the mechanisms and rate of the phase transformation? How does nucleation of the new phase particles occur? What does control their growth? This thus is asking for the *kinetics of the phase transformation*.

One may "feel" that the magnitude of the driving force has something to do with the kinetics of the phase transformation: the larger the driving force, the larger the rate of transformation. It would therefore be extremely useful if we would know how to couple the thermodynamics to the kinetics of a phase transformation. This coupling is by no means obvious. Consider Fig. 9.2. The energy of the system is plotted as function of a so-called "reaction coordinate". During the phase transformation the system proceeds from a level of higher energy to a level of lower energy: this

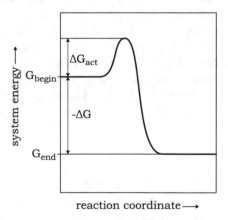

Fig. 9.2 System energy as a function of a reaction coordinate. The "driving force" of the reaction/transformation is given by $-\Delta G$ with $\Delta G = G_{end} - G_{begin}$; the energy barrier to overcome is given by ΔG_{act}

---

[1] The concept "driving force" is somewhat confusing, as this is not a force but an energy difference. The notion "force" could be used justifiably for the derivative of the (thermodynamic) energy function with respect to position (in space) coordinate(s); see Sect. 8.1 and Eq. (8.1).

difference is denoted by $-\Delta G$.[2] On its way to a state of lower energy, the system has to overcome an energy barrier, by thermal activation: ΔG_{act}. Therefore it is immediately clear that the driving force by itself does not provide the full kinetic description: The occurrence of the energy barrier, ΔG_{act}, or a series of energy barriers (!), has to be included in a proposed transformation-mechanism model (Chap. 10 and see Bos et al. 2005). There is no obvious connection between $-\Delta G$ and ΔG_{act}. Yet, mostly erroneously, models have sometimes been proposed in the literature where only $-\Delta G$ appears as a rate-controlling parameter.

The coupling of thermodynamics to kinetics is one of the most exciting areas of activity in research on phase transformations. The coupling of thermodynamics to kinetics may even be considered as the "holy grail" of materials science; no generally valid approach has been formulated until now. In an *Epilogue* to Chap. 10, three important examples of hitherto successful coupling of thermodynamics to kinetics are summarized.

The energy barrier to overcome, ΔG_{act}, cannot be identified fully with what is usually meant with the "activation energy" of a reaction/transformation. Suppose a Boltzmann-type equation holds for the probability that a (thermally activated) process, that costs an amount of energy (here ΔG_{act}), occurs (cf. Eq. (5.1) and its discussion). Then the progress of transformation/reaction can depend on a parameter k, which could be defined as a transformation/reaction rate constant, given by[3]

$$k = k_0 \exp\left(\frac{-\Delta G_{act}}{RT}\right) = k_0 \exp\left(\frac{\Delta S_{act}}{R}\right) \cdot \exp\left(\frac{-\Delta H_{act}}{RT}\right) \qquad (9.1)$$

recognizing that $\Delta G_{act} = \Delta H_{act} - T\Delta S_{act}$ (cf. Eq. (7.4)) and with k_0 as a constant. A plot of $\ln k$ versus $1/T$ results in a straight line with a slope given by $-\Delta H_{act}/R$. The energy term ΔH_{act}, an enthalpy difference, is usually called the activation energy (see the parallel, more explicit discussion presented for the activation energy of diffusion: Eqs. (8.25)–(8.27)). A symbol often used for ΔH_{act} is Q. If the $p\Delta V$ term in Eq. (9.1) is negligible, ΔH_{act} reduces to the internal energy difference ΔU_{act} (see below Eq. (7.3)). Evidently the entropy of activation, ΔS_{act}, does not influence the value of the activation energy of the thermally activated process obeying Eq. (9.1) and therefore ΔG_{act} is not identical with the activation energy. Equations of the type of Eq. (9.1) are called Arrhenius-type equations.

---

[2] ΔG is defined as $\Delta G \equiv G_{end} - G_{begin}$ (cf. Eq. 7.4). Hence, in order that the driving force is positive, if $G_{end} < G_{begin}$, the driving force has to be defined as $-\Delta G$.

[3] The simplest such case corresponds to the progress of transformation being proportional to k: see Eqs. (10.9)–(10.11) in Sect. 10.6.1, where the velocity of a moving transformation front is proportional to k (with k then defined as the "interface mobility", M).

9.2 Energetics of Nucleation; Homogeneous and Heterogeneous Transformations; Homogeneous and Heterogeneous Nucleation

The formation of a particle of a new phase in the matrix of a parent phase leads to a number of different contributions to the total energy change, such as:

- The change in chemical energy, ΔG_{chem}; this is in fact the energy contribution driving the transformation, supposing that the product phase is a stable phase: $\Delta G_{chem} < 0$.
- The change interfacial/surface energy, ΔG_{int}, because product phase/parent phase interface is formed; this energy change opposes the transformation: $\Delta G_{int} > 0$.
- The change in strain energy: ΔG_{strain}, because the assembly of atoms now taken up in the product-phase particle in general will have a volume different from the volume they occupied at the time they still were part of the parent phase; this energy change opposes the transformation: $\Delta G_{strain} > 0$.

The total Gibbs energy change, ΔG_{tot}, upon formation of a product-phase particle equals, if the above list of energy changes is comprehensive:

$$\Delta G_{tot} = \Delta G_{chem} + \Delta G_{int} + \Delta G_{strain} \tag{9.2}$$

Considering the formation of a spherical particle of radius r, the following relations can be given:

- $\Delta G_{chem} = (4/3)\pi r^3 . \Delta G^v_{chem}$, with ΔG^v_{chem} as the change of chemical energy per unit volume ($\Delta G^v_{chem} < 0$, a minimum requirement for the transformation to occur);
- $\Delta G_{int} = 4\pi r^2 \gamma$, with γ as the interfacial energy per unit (interfacial) area ($\gamma > 0$);
- $\Delta G_{strain} = $ const. $(4/3)\pi r^3 \varepsilon^2 = (4/3)\pi r^3 . (\Delta G^v_{strain} > 0)$, with ε as a linear misfit-strain parameter (product particle coherent with the matrix) or as a volume misfit parameter (product particle incoherent with the matrix) and ΔG^v_{strain} as the strain energy per unit volume.

Hence, for the case considered, it follows:

$$\Delta G_{tot} = \frac{4}{3}\pi r^3 \left(\Delta G^v_{chem} + \Delta G^v_{strain}\right) + 4\pi r^2 \gamma \tag{9.3}$$

The dependence of ΔG_{tot} on r is sketched in Fig. 9.3. Evidently, only if r is sufficiently large ΔG_{tot} becomes negative. For small particles the interfacial area is relatively (i.e. per unit volume particle) large and thus the term $4\pi r^2 \gamma$ ($+(4/3)\pi r^3 \Delta G^v_{strain}$), which is in any case positive, dominates.

Straightforward differentiation of ΔG_{tot} with respect to r and equating the result to zero leads to the following results. The radius, r_{crit}, occurring at the maximum increase of Gibbs energy, ΔG^*, is given by:

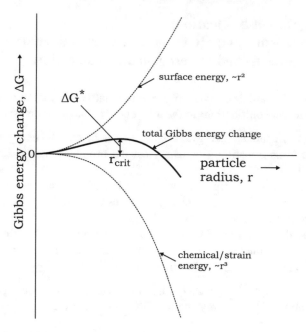

Fig. 9.3 Variation of Gibbs energy of a growing, spherical product-phase particle as function of the particle radius r. Contributions of surface/interface energy (>0), strain energy (>0) and chemical energy (<0), as well as the total Gibbs energy change, have been indicated. Note that the contributions of chemical energy (<0) and strain energy (>0) have been taken together in one curve as both are proportional with r^3 (cf. Eq. (9.3)). The critical radius, r_{crit}, indicates the size a growing particle (called embryo; $r < r_{crit}$) has to overcome to become a nucleus ($r > r_{crit}$) that can grow under release of Gibbs energy; the critical energy of nucleation of nucleus formation is given by ΔG^*

$$r_{crit} = -\frac{2\gamma}{\left(\Delta G^v_{chem} + \Delta G^v_{strain}\right)} \qquad (9.4)$$

The maximum increase of Gibbs energy, ΔG^*, occurring at r_{crit}, is given by:

$$\Delta G^* = \frac{\frac{16}{3}\pi\gamma^3}{\left(\Delta G^v_{chem} + \Delta G^v_{strain}\right)^2} \qquad (9.5)$$

Obviously, at the boundaries of the phase fields in a phase diagram, phases involved in corresponding phase transformations are in equilibrium. In order that a phase transformation runs in one or the opposite direction, a "driving force" (cf. Sect. 7.3, below Eq. (7.4)) has to be provided. This implies that the temperature must be lowered or raised to a value different from that characterizing the phase equilibrium concerned. Then, the chemical Gibbs energies of components of the phases involved are no longer equal for all phases and thereby a mass flux is desired, in order that thermodynamic equilibrium is restored. The difference between the temperature where the phases participating in the transformation are in equilibrium and the temperature where the transformation is actually happening is called "undercooling" or "super-heating", depending on the actual transformation temperature being below or above the equilibrium temperature.

At the temperature where equilibrium between parent and product phases prevails, T_{trans}, it holds (ignoring the role of strain and interface/surface energies): $\Delta G^v_{chem} = 0$, and thus, $\Delta S^v_{chem} = \Delta H^v_{chem}/T_{trans}$ (cf. Eq. (7.10)). Hence, assuming that ΔH^v_{chem} and ΔS^v_{chem} are practically temperature independent in a (restricted) temperature range around T_{trans}, and focussing on a precipitation reaction taking place in a super-saturated, "undercooled" matrix, it follows for ΔG^v_{chem} at a temperature T below T_{trans} (undercooling $\Delta T \equiv T_{trans} - T$):

$$\Delta G^v_{chem}(\Delta T) = \Delta H^v_{chem}\frac{\Delta T}{T_{trans}} \tag{9.6}$$

For the case that ΔG^v_{strain} is negligible, as compared to ΔG^v_{chem}, it thus follows from Eqs. (9.5) and (9.6):

$$\Delta G^*(\Delta T) = \frac{\left(\frac{16}{3}\pi\gamma^3\right)}{\left(\Delta H^v_{chem}\right)^2}\cdot\left(\frac{T_{trans}}{\Delta T}\right)^2 \tag{9.5a}$$

As the result of thermal fluctuation and by chance, a local atomic rearrangement may occur such that effectively a particle is created compatible with the stable phase to be formed under the given conditions. If such a particle occurs with $r < r_{crit}$, the system can lower its Gibbs energy if the (unstable) particle dissolves; it costs Gibbs energy to increase the size of the particle. If, as the result of thermal fluctuation and by chance, a particle with $r > r_{crit}$ occurs, the system can lower its Gibbs energy if the (stable) particle grows. *The formation of particles of supercritical size (from the reservoir of particles of subcritical size) is called "nucleation".* The Gibbs energy barrier for nucleation is given by ΔG^*. In this sense a particle of subcritical size is not a nucleus; it is often called an "embryo". Note that the critical size is *not* given by the value of r where, upon particle growth, ΔG_{tot} for the first time becomes negative.

In the above, with "fluctuation" apparently a change in the arrangement of the atoms, in a small volume of the material susceptible to phase transformation, is meant that realizes a local arrangement of the atoms as in the new phase to develop. One may wonder if the strict separation in energy contributions as expressed by Eq. (9.3) is possible for such small "embryos": for example, do "bulk" atoms occur in small embryos? Yet, this approach is usually adopted as it has led to viable concepts for the kinetics of the transformations in particular (see Chap. 10).

The view on nucleation as above implies that at some intermediate stage of transformation at some locations in the material considered the transformation has progressed and at other locations not. This type of transformation is therefore called "heterogeneous transformation"; it is the type of transformation usually encountered in solids.

If, in an initially homogeneous material, there is no Gibbs energy barrier for nucleation, the transformation can start at all locations in the material considered simultaneously. The material is unstable with respect to any occurring fluctuation as described above. At some intermediate stage of transformation the transformation has progressed at all locations to the same extent. This type of transformation is therefore called "homogeneous transformation". In view of the above discussion this situation is neared if, in the absence of the development of misfit-strain energy, the interfacial energy approaches zero. Thus, in a near homogeneous transformation there can be no sudden, discontinuous change (of properties) at the interface; as a consequence the interface cannot be sharp; i.e. the interface is of diffuse nature. In this case, the fluctuation, as a precursor of the phase transformation, is no longer strongly confined spatially: there is no drastic, local rearrangement of atoms as in the "embryos" considered above.

In the discussion on the occurrence of a fluctuation, the location where the fluctuation has occurred has not been specified. If all locations in the material susceptible to phase transformation are equally possible sites for the occurrence of such fluctuations, then one speaks of "homogeneous nucleation". This phenomenon may hold for transformations occurring in gases. Nucleation in solids is almost always of heterogeneous character and one speaks of "heterogeneous nucleation". The reason for prevalence of heterogeneous nucleation in solids is that the Gibbs energy barrier for nucleation at defects in the microstructure of the material, as grain boundaries, stacking faults, dislocations and (condensed) vacancies, can be smaller than for nucleation in the defect-free remainder of the matrix. This may be obvious: the reduction in the matrix of grain-boundary area, stacking fault area, dislocation-line length, etc., upon occupation of grain boundaries, stacking faults and dislocations, etc., by the developing precipitates, releases grain boundary, stacking fault, dislocation strain, etc., energy, and thereby ΔG^* becomes reduced (see also Sect. 9.4.5 and footnote 13). The Gibbs energy barrier for nucleation may even become zero for heterogeneous nucleation (see the following *Intermezzo*), implying spontaneous, immediate nucleation at defects for which this holds. Note that also in this case the nucleation occurs heterogeneously and the transformation is of heterogeneous character.

Intermezzo: Nucleation of AlN in Fe–Al Alloy

If an iron-based, ferritic Fe–Al alloy is nitrided (nitrogen is taken up from a nitriding atmosphere (e.g. a NH_3/H_2 gas mixture)) AlN should precipitate: Al has a strong affinity for nitrogen. The equilibrium crystal structure of AlN (i.e. the crystal structure to be observed upon formation of AlN from the pure elements at normal pressure and temperature (N_2 gas and Al solid)) is hexagonal. This hexagonal AlN has a very large volume misfit with the α-Fe (ferrite) matrix. An alternative crystal structure of AlN is f.c.c. (NaCl-type crystal structure), which has a much smaller volume misfit with the ferritic matrix, but has a less negative Gibbs energy of formation. Dependent on the presence or absence of (many) dislocations in the initial alloy material, precipitation of one or the other type of AlN precipitate can occur, as is discussed next (Biglari et al. 1995).

If it is assumed that nucleation of AlN occurs in a defect-free material (recrystallized Fe–Al alloy), the corresponding dependence of the Gibbs energy change upon formation of an AlN particle (following the treatment on the basis of Eq. (9.3)) is shown as a function of particle radius in Fig. 9.4a. It follows that the formation of hexagonal AlN is favoured over the formation of cubic AlN in the recrystallized Fe–Al alloy.

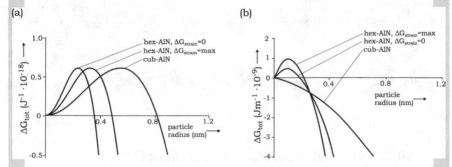

Fig. 9.4 Total Gibbs energy change upon formation of an AlN particle in an Fe-2at%Al alloy upon nitriding, calculated for the cubic (NaCl-type crystal structure) and the hexagonal modifications of AlN (with a maximal and minimal estimate for the strain (mismatch) energy in case of the hexagonal modification) in a dislocation-poor matrix, i.e. AlN precipitation in recrystallized Fe-2at%Al alloy (**a**) and in a matrix containing many dislocations, i.e. AlN precipitation in cold-worked Fe-2at%Al alloy (**b**) (taken from Biglari et al. 1995).

If it is assumed that the nucleation of AlN occurs in a material containing many dislocations (cold-worked Fe–Al alloy), the corresponding dependence of the Gibbs energy change upon formation of an AlN particle is shown as a function of particle radius in Fig. 9.4b. Upon formation of AlN on/around a dislocation, two additional energy effects (additional to the chemical, surface/interface and strain energy terms given in Eq. (9.3)) have to be considered. The formation of a precipitate at a dislocation line can release all or part (also dependent on the incoherent or coherent nature of the precipitate/matrix system) of the elastic energy initially stored in the volume it now occupies, i.e. dislocation-line energy (cf. Sect. 5.2.4), ΔG_{disl}, is released. The interaction energy of the precipitate/matrix and dislocation stress fields, ΔG_{int}, can be made negative by proper positioning of the precipitate: thus, the precipitate should develop in the compressive part (at the side of the half-plane) or the tensile part (below the half-plane) of the strain field of an edge dislocation (cf. Sect. 5.2.1 and Fig. 5.5), depending on the volume misfit of the precipitate and the matrix being negative or positive. The total Gibbs energy change upon formation of an AlN particle on/along a dislocation line is therefore given by (cf. Eq. (9.3)):

$$\Delta G_{tot} = \Delta G_{chem} + \Delta G_{surf} + \Delta G_{strain} - \Delta G_{disl} + \Delta G_{int} \qquad (9.7)$$

It follows from Fig. 9.4b not only that the formation of cubic AlN is favoured over the formation of hexagonal AlN in the cold-worked Fe–Al alloy, but also that the formation of cubic AlN around or along a dislocation line occurs without the occurrence of a Gibbs energy barrier for nucleation. In this case, thermal agitation, as a mechanism to produce by chance by local atomic rearrangement a product-phase particle of supercritical size, is unneeded: there is no nucleation-energy barrier.

These predictions for recrystallized and cold-worked nitrided Fe–Al alloy were confirmed experimentally (Biglari et al. 1995).

9.3   Diffusional and Diffusionless Transformations

Any classification of (heterogeneous) phase transformations is problematic and can be debated controversially. This is partly because intermediate cases of extremes, incorporated as cornerstones in a classification scheme, occur in nature.

Generally one would expect that the formation of a new phase in an originally homogeneous material is accompanied by the realization of compositional changes within the transforming material. This holds for the example discussed at the beginning of this chapter and as illustrated in Fig. 9.1. Unavoidably, development of a compositional change requires the operation of a diffusion process (Chap. 8). Phase transformations for which this holds are called *diffusional phase transformations*. The growth process of the product phase can thus be "diffusion controlled". This does not necessarily imply that the entire phase transformation is "diffusion controlled", as a rate influencing nucleation mechanism can co-determine the transformation kinetics (see Chap. 10)!

If no change in composition occurs upon phase transformation, as in an allotropic phase transformation as, for solid iron, the γ (austenite, f.c.c.) to α (ferrite, b.c.c.) transformation, experienced upon cooling from the austenite-phase field, evidently long-range diffusion is not required for the transformation to occur. The process can be governed by the independent motion of individual atoms at and across the interface between the parent and product phases, as in the mentioned γ to α transformation of iron, and one speaks of *massive transformation*.[4] The growth process of the product

---

[4] Recognizing that the allotropic transformation, as considered here, occurs by breaking the atomic bonds and subsequently rearranging the atoms in a new crystal structure, one also speaks of a *reconstructive* transformation. It should be noted that use of the terminology "reconstructive" for

phase can thus be "interface controlled". This does not necessarily imply that the entire phase transformation is "interface controlled", as a rate determining nucleation mechanism can co-determine the transformation kinetics (see Chap. 10)!

A transformation without occurrence of long-range diffusion can also be governed by the coordinated, simultaneous, dependent movements of thousands of atoms at the interface, as is the case in *martensitic transformations*, which can only occur in solid materials: the degree of coordination of the atoms in the parent phase tends to be preserved in the martensite, product phase, in association with, what could be conceived as, realization of the unit cell of the martensite product phase by straining of the unit cell of the parent phase (see further Sect. 9.5.2.2 and also Sect. 4.2.2). This picture immediately makes clear that macroscopic shape changes can occur upon martensitic transformations.[5]

The terms *civilian transformations* and *military transformations* have also been used for distinguishing transformations where independent atomic movements occur (the first two transformation modes discussed above) from transformations characterized by coordinated, regimented, simultaneous atom movements (the last discussed transformation mode).

The above discussion thus illustrates two kinds of *diffusionless phase transformations*: the massive transformation is a diffusionless, civilian transformation and the martensite transformation is a diffusionless, military transformation.

Nucleation stages in the sense as described in Sect. 9.2 can occur in both diffusional and diffusionless transformations. The growth of the nuclei (defined in Sect. 9.2) can be controlled by diffusion processes in diffusional transformations and by atomic jump processes at the interface between the product and parent phases in diffusionless transformations, which leads to the distinction between diffusion-controlled growth and interface-controlled growth (see further Sect. 10.6). Martensitic transformations can exhibit *athermal* (i.e. not thermally activated) nucleation; further, they are characterized by a usually very fast growth of the (supercritical) particles, which growth can yet be (weakly) thermally activated (see last paragraph of Sect. 9.5.2.4).

Following the above subdivision of types of phase transformations, Sects. 9.4 and 9.5 present a more detailed consideration of a few important phase transformations.

---

a transformation does not necessarily require the absence of the need for long range diffusion, although this is the case here.

[5] Recognizing that the martensitic transformation, as considered here, occurs by a homogeneous deformation of the crystal to be transformed, one also speaks of a *displacive* or *shear* transformation. Evidently, if unconstrained, such homogeneous deformation leads to a macroscopic shape change of the crystal; but see Sect. 9.5.2.2.

9.4 Diffusional Transformations; Examples

9.4.1 Age-Hardening Alloys; "Clusters", Transition and Equilibrium Precipitates

The principle of age hardening has in fact been outlined already at the start of this chapter, restricting ourselves for simplicity and the purpose of illustration to a binary system (cf. discussion of Fig. 9.1). It follows that a distinct solubility of one component should occur at elevated temperature in a one-phase region (at T_1 in Fig. 9.1) and that a much smaller solubility in this phase should occur at lower temperature (at T_2 and at T_3 in Fig. 9.1). The heat treatment to be applied consists of (cf. Fig. 9.1):

(i)　　*solution annealing* at T_1 leading to a homogeneous solid solution (α);
(ii)　　*quenching* to T_2 (often room temperature). The cooling rate should be fast enough to prevent the decomposition in the two-phase ($\alpha + \beta$) region (then one speaks of "quenching") and at T_2 no decomposition should occur during storage of the alloy: the supersaturated solid solution is retained;
(iii)　*precipitation annealing* of the quenched alloy within the two-phase ($\alpha + \beta$) region at a temperature T_3 high enough that the kinetics of the decomposition process allow significant decomposition of the supersaturated solid solution in a reasonable time span. This process is called *ageing*.

The choice of composition and, in particular, of the temperature and time of ageing are important for the size, dispersion, morphology and the coherent/incoherent (cf. Sect. 5.3) nature of the β precipitates developing in the α matrix, which can thereby have a huge effect on the value of the property to be optimized (in the case considered, the mechanical strength, as possibly characterized by the hardness; see Sect. 12.14.4).

In many cases the equilibrium precipitate β does not precipitate in a direct way: intermediate/transition precipitates may occur which may have favourable properties and then the age-hardening process aims at the development of such a stage of precipitation.

A classical example is provided by some Al-based alloys, as Al–Cu. In the case of Al-based Al–Cu alloys the equilibrium precipitate β is known as θ-CuAl$_2$. It has a rather complicated body-centred tetragonal crystal structure and occurs as a fully incoherent precipitate in the α matrix. The associated interfacial energy is (accordingly) that large that the nucleation-energy barrier, ΔG^* (cf. Eq. 9.5), becomes rather large as well. The supersaturated system prone to decomposition then, if possible, chooses a route via metastable, transition precipitates which, upon their formation, release less chemical Gibbs energy but are associated with smaller nucleation-energy barriers. These metastable precipitates exhibit (crystal) structures more closely related to the f.c.c. crystal structure of the α matrix (and therefore the interfacial energy and the misfit-strain energy can be relatively small).

In fact, the precipitation process in an alloy as discussed above may start with the formation of so-called "clusters": very small local enrichments of solute atoms. At this stage, the crystal-structure integrity of the α matrix is maintained. The difference in size of the solute (here Cu) and solvent (here Al) atoms (Cu atoms are smaller than Al atoms) then implies that the development of clusters of solute atoms leads to the development of coherency strains and thereby at this stage already (or in particular; see later) pronounced hardening can occur (cf. Sects. 12.14.3 and 12.14.4). Such "clustering processes" can already occur at room temperature. In the case of the Al–Cu alloy the clustering involves the enrichment of Cu atoms in $\{100\}$ planes of the α matrix. The clusters develop here as discs, one to two atomic planes thick; diameter of the order 10 nm. Diffraction experiments performed independently by Guinier (1938) and Preston (1938) revealed the presence of such clusters (by the occurrence of streaks through matrix reflections in corresponding X-ray diffraction patterns) and therefore these clusters are often named "Guinier–Preston (GP) zones". The technical Al-based alloy "duralumin", not only containing Cu as alloying element, derives its hardness from such metastable transition precipitates.

Also, in case of other alloys (pre)precipitation phenomena as discussed above can occur. The clusters may be of different nature and "structure". In particular, the reader is referred here to the discussion on clustering of carbon interstitials and (pre)precipitation of so-called α''-nitride in iron-based martensites (see the *"Intermezzo: Tempering of iron-based interstitial martensite"* in Sect. 9.5).

After the formation of clusters/GP zones in the Al-based Al–Cu system, subsequent formation of, consecutive, metastable precipitates, called θ'' and θ', is possible; eventually the equilibrium precipitate θ-CuAl$_2$ occurs (Fig. 9.5). The occurrence of a next (transition) precipitate in the precipitation sequence is associated with the disappearance of the previous transition precipitate (cluster); the next transition precipitate may nucleate on particles of the preceding stage, or at heterogeneities in the matrix as dislocations, grain boundaries, etc. The above discussion regarding the emergence of clusters suggests that their nucleation is a homogeneous nucleation

Fig. 9.5 Crystal structures of the matrix, the intermediate/transition θ'' and θ' precipitates and the equilibrium θ precipitate in the system Al–Cu

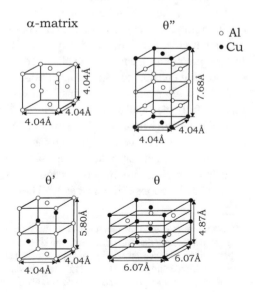

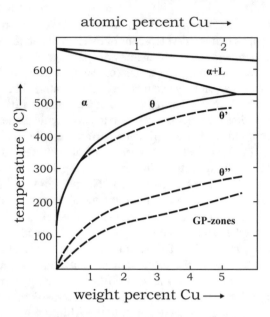

Fig. 9.6 Part of the equilibrium phase diagram for the binary system Al–Cu. Apart from the solvus line denoted with θ, indicating the composition of α in equilibrium with the equilibrium precipitate θ, also such solvus lines, denoted with GP zones, θ'' and θ', indicating the composition of α in metastable equilibrium with the GP zones and the transition/intermediate precipitates θ'' and θ', respectively, have been indicated (after G. Lorimer in *Precipitation Processes in Solids*, (Editors: K.C. Russell and H.I. Aaronson), The Metallurgical Society of AIME, 1978, p. 87).

process, whereas the nucleation of the transition and equilibrium precipitates is (as usual) a heterogeneous nucleation process (cf. Sect. 9.2).

Just as for the equilibrium precipitate, θ-$CuAl_2$, a solvus can be indicated in the equilibrium phase diagram for the GP zones, for the θ' precipitates and for the θ'' precipitates (see Fig. 9.6). Evidently, the full precipitation sequence, from GP zones to the θ precipitates, can only be observed if the supersaturated Al–Cu alloy, for the composition considered, is aged at a temperature below the GP-zone solvus.

If the accommodation of the misfit between a precipitate particle and the matrix is realized fully elastically, the inclusion/precipitate is surrounded by an elastic strain field of long-range nature. If full elastic accommodation of the misfit is not possible, e.g. upon growth of the precipitate particle, or upon transition from one to the next transition precipitate or to the equilibrium precipitate, incoherency at the precipitate-particle/matrix interface may occur and the misfit can be partly or largely accommodated by dislocations at the interface.[6] Then, the remaining elastic strain field can be

---

[6] It should be realized that incoherency at the interface of a particle/inclusion and the matrix by itself does not imply that full elastic accommodation of the volume misfit is impossible. A situation of both full elastic accommodation of volume misfit and incoherency could for example be realized by adding inert, solid particles to a single component melt of a different substance, followed by (i) (isothermal) solidification of the melt upon cooling of the particles/melt mixture and (ii) subsequent

of short-range and less pronounced nature (see also the discussion on dislocation-wall formation in Sect. 5.3 below Eq. (5.14)). Thus it may be understood that the maximum hardening effect is realized at an intermediate stage of the sequential precipitation process, as in the Al-rich Al–Cu alloys discussed above: ageing for maximum strength means ageing until and including the development of the θ'' transition precipitates. Continued ageing, leading to θ' transition precipitates and θ equilibrium precipitates, reduces the mechanical strength; this phenomenon is called "overaging".

Decomposition of a supersaturated solid solution according to precipitation sequences related to the one discussed above for Al–Cu is found for Al–Mg alloys, Al-Zn alloys and ternary variants thereof (like Al–Cu–Mg); this is a non-exhaustive listing.

The precipitation process discussed in this subsection occurs throughout the matrix, irrespective of the occurrence of homogeneous (possibly associated with clustering) or heterogeneous nucleation (on dislocations, grain boundaries, etc.). A characteristic is that the matrix composition at a certain location changes *continuously* with time and temperature. Therefore, this type of precipitation reaction is called *continuous precipitation*. Experimentally, a gradual decrease of solute concentration in the matrix due to the continuous precipitation in an age-hardening alloy, as considered above, can be revealed by a gradually shifting position of the Bragg reflections of the matrix in an X-ray diffraction pattern (cf. Sect. 4.5 and 6.10), because the matrix-lattice parameter depends in a monotonous way on the solute content (see Fig. 9.7; the pendant transformation, the so-called discontinuous transformation, is dealt with in Sect. 9.4.3).

9.4.2 Eutectoid Transformation

Grain-boundary precipitation reactions play an important role in materials science and engineering: the grain boundary acts as the location of nucleation; subsequent growth can occur along the grain boundary or in directions inclined (even perpendicular) to it. A special type of such precipitation processes involves the occurrence of a reaction front that, initiating at the grain boundary, advances into the supersaturated matrix grain leaving behind it a precipitated microstructure consisting of the equilibrium phases arranged in a lamellar structure ("duplex structure"; see Fig. 9.8).

---

cooling of the fully solidified specimen: After process step (i) the interface between particles and solidified melt may be fully incoherent in the absence of any internal stress; in process step (ii) the developing thermal misfit between particles and solidified melt may be accommodated fully elastically if the difference of the thermal expansion/shrinkage coefficients of particles and solidified melt is not too large. Then a situation of full elastic accommodation of volume misfit and incoherency at the particle/matrix interface has been realized. However, precipitation processes in a supersaturated matrix generally involve that occurrence of incoherency at the precipitate/matrix interface is associated with only partly elastic accommodation of the volume misfit.

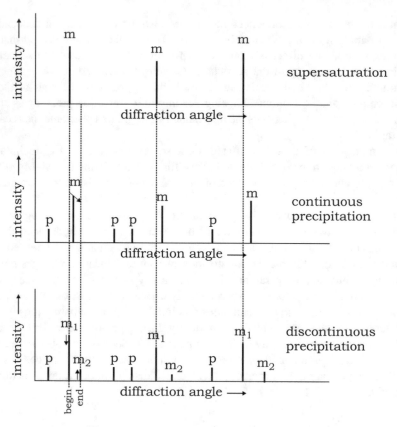

Fig. 9.7 Schematic depiction of the change in (X-ray) diffraction patterns of a supersaturated solid solution upon *continuous* and *discontinuous* precipitation. In both cases, reflections corresponding to the precipitating phase appear (indicated with "*p*" in the figure). In case of *continuous precipitation*, the matrix reflections (indicated with "*m*" in the figure; initial position indicated by m_1) both decrease in intensity and shift in diffraction angle, corresponding to the gradual variation of the lattice parameter of the matrix caused by the gradual change in composition due to solute depletion of the matrix upon precipitation: $m_1 \rightarrow m_2$. In case of *discontinuous precipitation*, the reflections corresponding to the supersaturated matrix decrease in intensity but do not change their position. A second set of matrix reflections at (slightly) different diffraction angles (indicated by m_2) arises, corresponding to the composition of the new matrix phase in equilibrium with the precipitate phase and thus with a different composition (solute depleted) and lattice parameter: $m_1 \downarrow$ and $m_2 \uparrow$

Such a lamellar aggregate, constituted of alternate, parallel crystals of the equilibrium phases (α and β in the figure), is also called "cell" or "colony". A transformed grain can be composed of many colonies. The above description pertains to eutectoid transformations, discussed here, and discontinuous transformations, dealt with in Sect. 9.4.3.

(a)   (b)

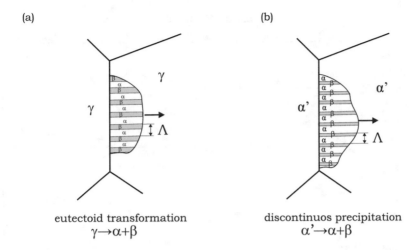

eutectoid transformation   discontinuos precipitation
$\gamma \rightarrow \alpha + \beta$ $\alpha' \rightarrow \alpha + \beta$

Fig. 9.8 Lamellar transformation product microstructures developing by growth from a grain boundary of the parent phase. **a** Schematic depiction of the reaction front of a eutectoid precipitation cell moving from a grain boundary into a grain of the parent phase γ. **b** Schematic depiction of the reaction front of a discontinuous precipitation cell moving from a grain boundary into a grain of the supersaturated matrix α'. Λ is the period of the lamellar product structure

Intermezzo: The Fe-C System; Steels and Cast Irons

In view of the great technological importance of the Fe-C system, a few introductory remarks on the notions "steels" and "cast irons" are in order.

Steels are alloys of basically iron and carbon (up to about 2 wt% C, usually less than about 1 wt% C) and possibly some other elements. Therefore, the iron-rich part of the Fe-C phase diagram is of prime importance (see Fig. 9.9; see also footnote 4 in Sect. 7.5.2.5). In fact two phase diagrams are shown in Fig. 9.9: the equilibrium phase diagram relies on adoption of graphite (hexagonal crystal structure (cf. Sect. 4.2.3.3), usually severely defected) as the stable, solid carbon phase (dashed lines in Fig. 9.9). The metastable diagram based on cementite (Fe_3C, orthorhombic crystal structure), taking the role of graphite, is more important (solid lines in Fig. 9.9)[7] (especially see Fig. 9.24 discussed in the *"Intermezzo: The Fe-C and Fe-N phase diagrams"* at the end

---

[7] In technology the phase boundary between the two-phase field ferrite-cementite and the two-phase fields ferrite–austenite and austenite-cementite is denoted by the symbol A_1; the phase boundary between the two-phase field ferrite–austenite and the single-phase field austenite is indicated by the symbol A_3 and the phase boundary between the single-phase field austenite and the two-phase field austenite-cementite is presented by the symbol A_{cm}. The phase transformations occurring at the boundaries A_1, A_3 and A_{cm} pertain to (metastable) equilibrium conditions, i.e. the transformation temperatures as indicated by A_1, A_3 and A_{cm} can only be observed approximately in practice if very slow heating or cooling rates are applied. Rapid heating or cooling shifts the observed transformation

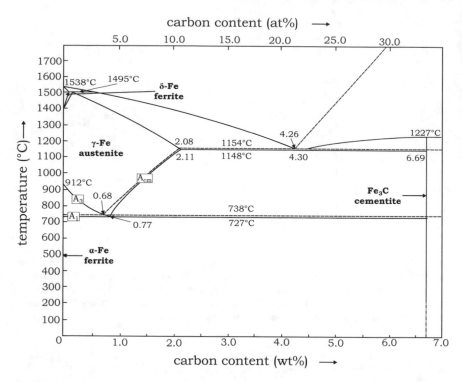

Fig. 9.9 Phase diagram of the binary system Fe-C. The technologically important metastable diagram (including the phase cementite) is given by the solid lines; the stable diagram (including the phase graphite) is given by the dashed lines (taken from *Metallography, Structures and Phase Diagrams*, Vol 8, 8th ed., *Metals Handbook*, American Society of Metals, 1973, pp. 236, 275, 276)

of Sect. 9.5.2.1): Phase transformations induced by heat treatments applied in practice to steels rarely lead to the development of graphite, almost invariably cementite occurs instead. Nevertheless it should be recognized that cementite in iron–carbon alloys/steels (at normal temperatures and pressure) is unstable with respect to its decomposition into ferrite and graphite, and thus annealing at elevated temperature of, for example, a two-phase, ferrite-cementite specimen will eventually lead to the development of graphite.

temperatures to higher or lower values, respectively. These apparent phase boundaries, as observed upon heating or cooling at some fixed rate, are denoted by the symbols Ac_1, Ac_3 and Ac_{cm} or Ar_1, Ar_3 and Ar_{cm}, respectively. (Ac is an abbreviation for "arrêt chauffant"; Ar is an abbreviation for "arrêt refroidissant"). To describe the effect of alloying elements in steel on the temperatures of the phase transformations discussed here, considerable effort has been spent on the development of empirical formulas. Thereby it can safely be assumed that the iron-rich part of the metastable Fe-C phase diagram (Fig. 9.9) is only slightly modified by the presence of (modest amounts of) the alloying elements.

Cast irons are iron-based iron–carbon alloys containing more than 2 wt% carbon; these alloys are not called steels. As follows from Fig. 9.9, these alloys solidify at temperatures below the temperature range for solidification of steels. The eutectic solidification reaction (Eq. (7.19) in Sect. 7.5.2) at about 1423 K (1150 °C) plays an important role: an iron–carbon melt (containing about 4.3 wt% C) solidifies into austenite and either cementite or graphite. The formation of graphite in cast irons is much more likely than in steels because of a larger amount of carbon and as also promoted by the presence of an alloying element as silicon.

A technologically very important example of an eutectoid transformation (Eq. (7.22) in Sect. 7.5.2):

$$\gamma \rightarrow \alpha + \beta \tag{7.22}$$

is the austenite (γ, containing 0.77 wt% C) $\rightarrow$ ferrite (α, containing maximally about 0.02 wt% C (at equilibrium at the eutectoid temperature)) + cementite (Fe_3C, containing 6.67 wt% C) transformation in steels at about 1000 K (727 °C) in the Fe-C system (see Fig. 9.9). The eutectoid reaction parallels the eutectic reaction (see Sect. 7.5.2 and Eqs. (7.19) and (7.22)). During the eutectoid transformation, $\gamma \rightarrow \alpha + Fe_3C$, a lamellar microstructure (as also indicated in Sect. 7.6 for a eutectic reaction), characterized by broad ferrite lamellae and small cementite lamellae, develops.[8] This microstructure is called "pearlite".[9] Applying the lever rule (cf. Eqs. (7.16) and (7.17) in Sect. 7.5.2.1) at a temperature immediately below the eutectoid temperature shows that pearlite consists of about 10 wt% cementite and 90 wt% ferrite. Because the densities of cementite and ferrite are approximately equal (7.9 and 7.7 g/cm^3, respectively) the respective weight and volume percentages of cementite and of ferrite in pearlite are about the same. Note that by applying the lever rule it is assumed that each of the lamellar phases is of uniform, equilibrium composition.

A colony of more or less parallel ferrite and cementite lamellae nucleates at an austenite-grain boundary (upon cooling from a temperature above the eutectoid

---

[8] The lamellar microstructure, discussed here, is only one, albeit an important one, of many possible microstructures which can occur upon eutectoid and eutectic transformations.

[9] The name "pearlite" originates from the "mother-of-pearl" impression made by a polished and etched cross section of a pearlitic microstructure. The cementite lamellae protrude slightly from the surface of the cross section after etching. Because the cementite lamellae of a colony are more or less parallel and equidistant, the cementite lamellae of a colony, protruding from the surface, can more or less act as a diffraction grating (occurrence of such diffraction effects requires that the interlamellar spacing is of the order of the wavelength of the incident light; cf. Eq. (6.7) in Sect. 6.2). Upon illumination (in a light optical microscope), this diffraction effect, for various wavelengths from various colonies of different interlamellar spacing in the cross section, leads to the observation of the so-called "pearl-like lustre".

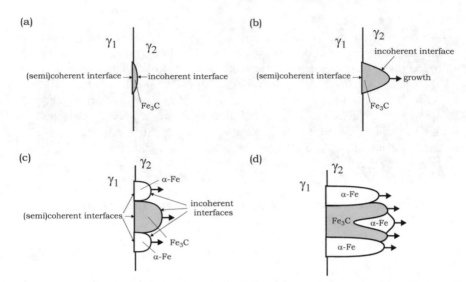

Fig. 9.10 Schematic depiction of the formation of pearlite from austenite. **a** Nucleation of a cementite crystal at the interface of two austenite grains, γ_1 and γ_2, thereby establishing a (semi)coherent interface with γ_1 and an incoherent interface with γ_2. **b** Growth of the cementite grain into γ_2. **c** Nucleation of ferrite at the γ_1/γ_2 interface in the carbon depleted austenite matrix of γ_2 adjacent to the cementite grain. **d** Branching of a cementite lamella

temperature); see Fig. 9.8a. For an alloy of *hypo*-eutectoid[10] composition (i.e. carbon content below about 0.8 wt% C), upon cooling from the austenite-phase field, first a primary development of ferrite particles occurs (formation of *proeutectoid ferrite*), until the eutectoid temperature is reached, whereupon the eutectoid transformation takes place for the remaining, now carbon-enriched austenite phase. For an alloy of *hyper*-eutectoid (see footnote 10) composition (i.e. carbon content above about 0.8 wt% C), upon cooling from the austenite-phase field, first a primary development of cementite particles occurs (formation of *proeutectoid cementite*), until the eutectoid temperature is reached, whereupon the eutectoid transformation takes place for the remaining, now carbon-impoverished austenite phase. The proeutectoid phases, developing before the eutectoid transformation occurs, nucleate and grow at the austenite-grain boundaries.

A pearlite colony starts with the formation of a cementite or ferrite crystal at an austenite-grain boundary. Suppose a cementite crystal nucleates first (Fig. 9.10a). It may be possible for this crystal to realize a good fitting; i.e., a low energy interface is established with one of the neighbouring austenite grains, say γ_1. The cementite/γ_1 interface may thus be (semi)coherent. The associated (energetically) good packing of the atoms at this interface makes transfer of atoms across this interface less easy, and therefore, as generally holds, the (semi)coherent interface is of low mobility.

---

[10] Here the terms "hypo" and "hyper" mean "below" and "above", respectively, a specific value of the composition (here the eutectoid composition).

If the cementite/γ_1 interface is (semi)coherent, then, generally, the cementite/γ_2 interface will be incoherent. The associated (energetically) bad matching of atoms of both phases at this interface makes transfer of atoms across this interface relatively easy, and thus, as generally holds, the incoherent interface is a relatively mobile interface. Consequently, growth of the cementite crystal will take place into the γ_2 grain (Fig. 9.10b). The austenite surrounding the cementite crystal in the γ_2 grain is depleted in carbon, implying an increased driving force for the formation of ferrite. As a result, at the austenite-grain boundary ferrite grains form adjacent to the initial cementite grain. These ferrite grains, by virtue of their orientation relationship with the cementite, will have a (semi)coherent interface with γ_1 and, generally, an incoherent interface with γ_2, as well (Fig. 9.10c). Hence, by this type of sidewise nucleation, a lamellar structure of ferrite and cementite, i.e. a pearlite colony, forms, at the austenite-grain boundary, that grows sideways into the γ_2 grain; the pearlite grows into that austenite grain with which there is, for both the cementite and the ferrite, no specific orientation relationship; see Fig. 9.10. It should be noted that it has also been shown that branching of a cementite lamella (or a ferrite lamella) into parallel lamellae can occur (Fig. 9.10d), which in its extreme form would lead to a pearlite colony composed of a single crystal of cementite and a single crystal of ferrite (all cementite phase and all ferrite phase of a single colony would then be interconnected).

Evidently, the growth of a colony in an eutectoid transformation in an A/B system requires cooperative growth of the two phases at the colony front; one also speaks of "coupled growth": both phases grow simultaneously in a direction more or less perpendicular to the transformation front. Growth of the A-rich α lamella implies that B present in the parent γ crystal at the tip of the α lamella has to diffuse to the tips of the adjacent β lamellae. This substantial atomic redistribution at the colony front can be effectively realized by diffusion along the colony front (more or less perpendicular to the colony-front growth direction) either by volume diffusion in the γ grain ahead of the eutectoid front or via grain-boundary/interface diffusion (i.e. at and along the α/γ and β/γ interfaces). For substitutionally dissolved elements, it may be expected that interface diffusion of the components being redistributed controls the colony growth (grain-boundary diffusion of substitutionally dissolved components is much faster than volume diffusion; cf. Sect. 8.9).

For the eutectoid transformation in the Fe-C system, carbon has to be redistributed. Carbon in austenite is dissolved interstitially, implying fast diffusion (because of the high density of vacancies on the substructure of interstitial sites; cf. Sect. 8.5.3). Then, the occurrence of grain-boundary diffusion as rate-controlling mechanism for colony growth in the eutectoid transformation (cf. previous paragraph) can be less pronounced in the Fe-C system.

Evidently, the rate of growth of the eutectoid colony depends not only on the speed of the redistribution of the components, ahead of the advancing eutectoid transformation front, but also on the interlamellar spacing Λ, i.e. the combined width of an α and β, adjacent lamellae pair (cf. Fig. 9.8); e.g., the larger Λ, the slower the colony growth rate. It would therefore seem advantageous, from a kinetic point of view, to make Λ vanishingly small. Thus, in particular at relatively low temperatures of eutectoid transformation one would expect a small interlamellar spacing to compensate for the low diffusion rate at low temperatures (cf. Eq. (8.26)). On this basis one may also understand why, upon eutectoid transformation, a lamellar product geometry develops, instead of a system of more or less spherical β grains in an α matrix (a β sphere has the smallest surface (i.e. α/β interfacial) area per unit volume of β): diffusion processes leading to that last mentioned transformation product geometry require much more time than those to establish the lamellar product geometry (albeit with the larger α/β interfacial area).[11] However, the increase of energy of the system as a consequence of the increase of α/β interfacial energy upon decreasing Λ prohibits occurrence of vanishingly small values for Λ. On this basis a minimum value for Λ can be derived in dependence on the extent of undercooling (see above Eq. (9.6) for definition of "undercooling"), as follows.

Consider a (geometrically idealized) colony as shown in Fig. 9.8. The α/β interfacial area per unit volume colony is given by $2/\Lambda$ (per Λ there are two α/β interfaces; there are $1/\Lambda$ periods of Λ in a unit volume colony). Hence, the Gibbs energy change experienced by the system upon eutectoid transformation per unit volume, ΔG_{eut}, is given by the (transformation promoting) change in chemical Gibbs energy per unit volume ΔG_{chem}^v (<0; cf. Sect. 9.2) and the (transformation obstructing) formation of interfacial energy per unit volume $(2/\Lambda)\gamma_{\alpha/\beta}$ (>0):

$$\Delta G_{eut} = \Delta G_{chem}^v + \frac{2\gamma_{\alpha/\beta}}{\Lambda} \tag{9.8}$$

According to Eq. (9.6):

$$\Delta G_{chem}^v = \frac{\Delta H_{chem}^v \, \Delta T}{T_{eut}} \tag{9.9}$$

with (here) T_{eut} as the equilibrium eutectoid transformation temperature and the undercooling $\Delta T = T_{eut} - T$. Combining Eqs. (9.8) and (9.9) and setting $\Delta G_{eut} = 0$ leads to the following expression for the minimum interlamellar spacing, Λ_{min} (for even smaller values of Λ, ΔG_{eut} would be larger than zero and the eutectoid transformation would be impossible energetically):

$$\Lambda_{min} = \frac{-2(\gamma_{\alpha/\beta} T_{eut})}{(\Delta H_{chem}^v \, \Delta T)} \tag{9.10}$$

Evidently, the larger the undercooling, the smaller the minimum interlamellar spacing, which is compatible with the qualitative remark made in the preceding paragraph.

If Λ would take its minimum value as prescribed by Eq. (9.10) no driving force for the transformation remains ($\Delta G_{eut} = 0$), and thus the velocity of the transformation

---

[11] This reasoning provides an interesting example of the interplay of thermodynamics (the state of minimal energy a system strives for; cf. Chap. 7) and kinetics (the role of finite mobilities of the atomic species; e.g. see Chap. 8) for understanding microstructures as occurring in reality.

front then is equal to zero. For a progressing transformation ΔG_{eut} must be smaller then nil and consequently the interlamellar spacings observed in practice are larger than Λ_{min}. Various theoretical approaches for predicting the values of Λ occurring in reality have been presented. Some typical results indicate that the occurring values of Λ can be two to three times larger than Λ_{min} (Puls and Kirkaldy 1972).

Much of the above (including Eq. (9.10) and its discussion) could similarly have been said and derived for the eutectic solidification transformation: $L \rightarrow \alpha + \beta$ (cf. Eq. (7.19) in Sect. 7.5.2.2).

It should be remarked that, even after many decades of research, a comprehensive, fundamental understanding of the development of such regular patterns, as the lamellar structures discussed here, has not been reached. So the question, why precisely the interlamellar spacing that is observed in the experiment is selected by nature, at this stage can only be answered incompletely (e.g. see Parisi and Plapp 2008).

9.4.3 Discontinuous Transformation

The simplest type of discontinuous transformation is described by (see Fig. 9.8b):

$$\alpha' \rightarrow \alpha + \beta \tag{9.11}$$

where α' denotes a supersaturated matrix that decomposes into an α phase, of lower or even nil supersaturation, and β, which is the equilibrium precipitate. There is great similarity with the eutectoid transformation: the transformation front, initiating at a grain boundary of the parent, α' microstructure, moves into a supersaturated α' grain. Behind the transformation front a lamellar microstructure develops consisting of α lamellae and β lamellae. The supersaturation of the α lamella is far less then that of the supersaturated matrix, parent α' grain. The α lamella has the same crystal structure as its α' parent, but a crystal orientation different from that of the α' grain it moves into; the α/α' and β/α' parts of the reaction front are incoherent grain boundaries.

The occurrence of a discontinuous change in solute content of the parent phase (across the moving interface that sweeps through the matrix grain) is the origin of the name of this type of precipitation reaction. Note that in a continuous transformation, as defined at the end of Sect. 9.4.1, the transformation does not require moving grain boundaries: the continuous transformation is not confined to a reaction front that sweeps through the material; the continuous transformation occurs throughout the matrix. Experimentally, the abrupt, discontinuous change in matrix composition in a discontinuous transformation (at the transformation front) can be revealed by the emergence in an X-ray diffraction pattern of new Bragg reflections of the matrix, increasing in intensity, at diffraction-angle positions corresponding to the lattice parameters of the product α solid solution phase, while at the same time, the Bragg reflections of the supersaturated α' matrix phase remain at their original diffraction-angle positions (because the composition of the (remaining part of the) α' matrix does not change) but gradually decrease in intensity (see Fig. 9.7).

A competition between the continuous and discontinuous types of transformation modes can occur in a precipitating system. Consequently, ahead of the moving transformation front in a discontinuous transformation, the original, supersaturated matrix may already have decomposed (partly), such that small precipitates of the equilibrium precipitate phase β, called β' particles to distinguish these from the lamellar β phase in the lamellar microstructure due to the discontinuous transformation, have developed by continuous transformation. Then, at the moving discontinuous transformation front, the discontinuous transformation can be described by (see Williams and Butler 1981):

$$\alpha' + \beta' \to \alpha + \beta \tag{9.12}$$

At the discontinuous transformation front the small β' particles, possibly coherent with the α' matrix, are replaced by β lamellae under simultaneous removal (largely) of the supersaturation in the α phase (see Fig. 9.11). In this case positive contributions to the driving force of the transformation are due to not only the decrease of chemical Gibbs energy of the supersaturated matrix ($\alpha' \to \alpha$), but also to the decrease of interfacial ($\alpha'/\beta' \to \alpha/\beta$) energy (area) and the stress relaxation upon the β' (coherent) $\to \beta$ (incoherent) transition. The replacement of the small β' particles by the coarse β lamellae can be considered as a coarsening process for the equilibrium precipitate of the system, and thus the variant of the discontinuous transformation

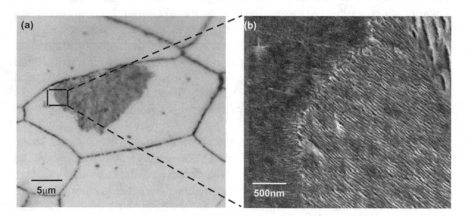

Fig. 9.11 SEM micrographs of the discontinuous transformation/coarsening of VN precipitates in an Fe-4.42at.%V alloy nitrided at 580 °C. **a** Part of a parent matrix grain has experienced the discontinuous coarsening reaction, initiating at a grain boundary of the parent, ferrite matrix: submicroscopical (i.e. invisible in the micrograph), coherent VN precipitates in the ferrite matrix have been replaced by colonies of alternating ferrite and VN lamellae. **b** At higher magnification the lamellar morphology of the discontinuously coarsened region is revealed (taken from S. S. Hosmani, R. E. Schacherl and E. J. Mittemeijer, Acta Materialia, 53 (2005), 2069–2079)

given by Eq. (9.12) can be called *discontinuous coarsening* as well, a name which is particularly fitting if practically all equilibrium precipitate has precipitated as small β' particles in the matrix before the arrival of the reaction front.

About semantics: Discontinuous transformations are also called cellular transformations, with reference to the developing lamellar, cellular microstructure. As follows from the first paragraph of Sect. 9.4.2, eutectoid transformations and discontinuous transformations could then be taken together as subgroups of the group of "cellular transformations", but usually discontinuous transformations are meant specifically if one speaks of cellular transformations. Also, the eutectoid transformation could be defined as a discontinuous transformation, as at the moving transformation front a discontinuous change in composition occurs, whereas in a continuous transformation the composition change occurs continuously throughout the matrix, but, again, this is not done.

Exactly how the cellular transformation is initiated at the moving grain boundary is less clear. What is the nucleus? A grain of one of the two phases of the eventual, duplex structure or a basic unit of the lamellar final microstructure? Experimental analysis (using transmission electron microscopy (TEM) combined with local composition analysis, with typical spatial resolutions of better than 10 nm; cf. Sect. 6.8.7) has shown that, at least in some systems and with reference to discontinuous transformation according to Eq. (9.11), such a transformation may start with the originally precipitate-free transformation front beginning to move, leaving behind a matrix depleted of solute (Fig. 9.12a, b). The corresponding, increasing enrichment of solute in the matrix ahead/at the transformation front after some time, i.e. after some migration of the transformation front, induces a first development of equilibrium precipitate at the transformation front (Fig. 9.12c). The solute atoms caught by the advancing transformation front diffuse along the transformation front to the developing equilibrium precipitates. Lamellar colonies then may grow from these first developing precipitates (Fig. 9.12d–e). Parallel growth of lamellae takes place only after a period of less cooperative evolution of the initial precipitates has been overcome.

A steady state of growth of a lamellar colony is characterized by a constant transformation-front velocity and a constant interlamellar spacing. Assuming that the redistribution of solute and solvent components is realized at the moving transformation front and that volume diffusion in the α' and α phases is negligible, it is clear that the α lamellae cannot be solute depleted down to the equilibrium composition for any transformation-front velocity larger than nil (Cahn 1959). Consequently, in the wake of the moving transformation front and parallel to it, a solute-concentration profile exists in the α lamellae, which, subject to the assumptions indicated above, represents the frozen-in concentration profile at the transformation front. Experimental determination of such concentration profiles (Fig. 9.13) allows determination of the diffusivity along the transformation front, provided the transformation-front velocity and the interlamellar spacing have been determined experimentally

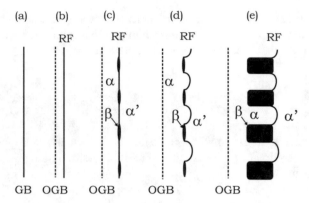

Fig. 9.12 Schematic depiction of the formation of a discontinuous precipitation cell ($\alpha' \rightarrow \alpha +$ β). **a** A precipitate-free grain boundary of the matrix (GB) starts to move (to the right in the figure). **b** Matrix depleted of solute is left behind by the moving reaction front (RF) (OGB = original position of the grain boundary of the matrix). **c** First development of equilibrium precipitates β. **d** Upon continued migration of the RF, the RF bows out between the pinning β precipitates. **e** Elemental redistribution along the RF leads to increase in length of the β precipitates at the extremities of the bowing boundary segments: eventually, a lamellar colony grows from the first developing precipitates

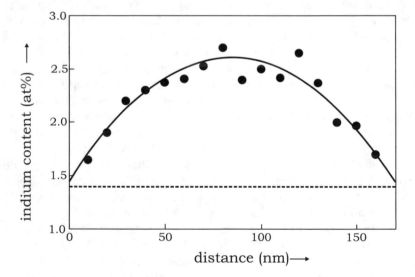

Fig. 9.13 Discontinuous precipitation in a Cu-4.5at.%In alloy according to $\alpha' \rightarrow \alpha + \beta$, with α' as the supersaturated, parent, Cu-4.5at.%In solid solution, and α and β, the lamellar product phases, as the copper-rich Cu-In solid solution containing less indium than the parent phase and β as the Cu$_7$In$_3$ intermetallic phase. The figure shows the indium-concentration profile within an α-lamella parallel to the transformation/reaction front for a transformation temperature of 600K. The equilibrium concentration of indium in the product phase, α, equals 1.4 at.% In and has been indicated by the dashed line in the figure (taken from Lopez et al. 2003)

as well. It has been shown that (i) the diffusivity along the transformation front can vary up to a factor of 10 at constant temperature, which reflects the grain-boundary structure sensitivity of grain-boundary diffusion, and that (ii) the diffusivity along moving grain boundaries (transformation fronts) is not essentially different from the diffusivity along stationary (i.e. not moving) grain boundaries (Lopez et al. 2003; see Sect. 8.9.2). Hence, analysing the kinetics of discontinuous transformations can reveal fundamental data on grain-boundary diffusion in (moving) grain boundaries.

It is important to remark that the interlamellar spacing and the transformation-front velocity are no constants. Even for a single colony/cell changes of the inter-lamellar spacing and the front velocity (i.e. growth rate) occur upon progressing transformation at constant temperature.

9.4.4 The Widmanstätten Morphology

The nucleation and growth of a second phase along a grain boundary is, for example, observed upon the formation of proeutectoid ferrite and proeutectoid cementite (see Sect. 9.4.2). A morphologically strikingly different type of grain-boundary precipitation involves the formation of needles, discs or plates of precipitate phase, upon nucleation at the grain boundary, which during growth penetrate the matrix grain, apparently along specific crystallographic directions, thereby giving rise to the so-called Widmanstätten side plates.

A nucleus of a precipitate phase at a grain boundary can be bounded by coherent or semicoherent and incoherent interfaces[12] with the adjacent matrix grains. In the absence of precipitate-matrix misfit-strain energy, the particle of critical size (see Eq. (9.4) in Sect. 9.2) has a shape that realizes minimal interfacial energy. Coherent and semicoherent interfaces are usually flat (to maintain the good matching, at the interface, of the crystal structures of precipitate particle and adjacent matrix grain) and exhibit themselves as facets, whereas incoherent interfaces (characterized by bad matching, at the interface, of the crystal structures of precipitate particle and adjacent matrix grain) can be curved. As has been made likely in Sect. 9.4.2, while discussing the nucleation of pearlite, (semi)coherent interfaces are generally much less mobile than incoherent interfaces. The shape of the precipitate-phase particle, as it develops during growth, is determined by the relative migration rates of the bounding interfaces. Thus, if kinetics prevails, a precipitate nucleus, at a grain boundary of the matrix phase, with one good matching (semi)coherent interface can become a thin disc or plate upon growth, with the plane of the disc or plate defined by the (semi)coherent interface. On this basis, the occurrence of the Widmanstätten morphology can be understood.

Precipitation along grain boundaries is observed at low undercooling (see next section), whereas Widmanstätten side plates are observed if higher undercooling (i.e.

---

[12] Interfaces between crystals of different phases are also called "interphase boundaries".

higher driving force for the precipitation) is realized. This may be caused by a relatively larger ratio of the migration rates of incoherent and (semi)coherent interfaces at larger undercooling, but may also be explained by the need to overcome larger misfit strains, possibly associated with the Widmanstätten morphology, by a larger chemical driving force contribution, ΔG_{chem}.

9.4.5 Grain-Boundary Wetting

Consider the schematic binary phase diagram shown in Fig. 9.14a. Annealing of a supersaturated α-phase solid solution in the two-phase region $\alpha + L$ must lead to precipitation of the liquid phase L. Annealing of a supersaturated α-phase solid solution in the two-phase region $\alpha + \beta$ must lead to precipitation of the solid-phase β. As discussed before, nucleation of the precipitate phase at grain boundaries of the matrix (α) phase can be favoured, because of the relatively small value of the nucleation-energy barrier, ΔG^* (cf. Sect. 9.2).

Upon nucleation at grain boundaries, various types of precipitation morphologies may occur and some of these have already been touched upon above. The growth of the precipitate-phase grain may be realized by penetration of the matrix grain, starting from the grain boundary (as for the Widmanstätten side plates discussed in the previous section). On the other hand, the growth of the precipitate phase may be restricted to the grain-boundary region and for this case two distinctly different morphologies can occur: (i) individual precipitate particles along the grain boundary (observed as a chain of particles in a cross section of the specimen) or (ii) a layer of the precipitate phase along the grain boundary.

The development of a layer-like morphology of precipitate phase along grain boundaries of the matrix has first been called "*wetting*" for the case that the precipitate phase is a liquid. Later it has been recognized and demonstrated (see, Lopez et al. 2004) that the occurrence of layers of a solid, second phase developing along grain boundaries, as for example holds for the development of (proeutectoid) cementite layers along grain boundaries of the austenite matrix (cf. Sect. 9.4.2), has a similar background (see below) and should be called grain-boundary wetting as well.

In general the presence of precipitate-phase layers along grain boundaries can have either detrimental effects (as enhanced brittleness) or favourable effects (as improved plasticity). An extreme, dramatic example of the influence of this precipitate morphology is the occurrence of so-called superplasticity in case of grain-boundary wetting by a liquid phase (e.g. see Straumal et al. 2003).

The gain in energy obtained upon heterogeneous precipitation at a grain boundary of the matrix, parent phase, as compared to homogeneous precipitation in the bulk of the matrix phase, obviously is due to the resulting reduction in grain-boundary area of the matrix phase in case of grain-boundary precipitation, as has been remarked in Sect. 9.2. In the following the discussion is restricted to the case of precipitation at a grain boundary separating two matrix grains (and thus precipitation at edges where three matrix grains meet, i.e. so-called triple junctions, and precipitation at

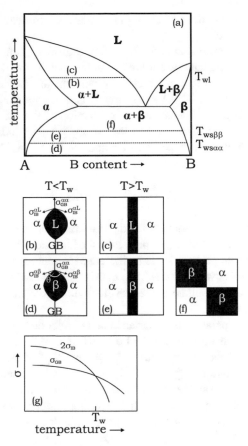

Fig. 9.14 a Schematic phase diagram showing the tie-lines of grain boundary (GB) wetting by a liquid phase and of GB wetting by a solid phase at T_{wl}, and $T_{ws\alpha\alpha}$ and $T_{ws\beta\beta}$, respectively. **b** Liquid-phase L does not wet a GB in the solid-phase α ($T < T_{wl}$). **c** Liquid-phase L wets a GB in the solid-phase α ($T > T_{wl}$). **d** Solid-phase β does not wet a GB in the solid-phase α ($T < T_{ws\alpha\alpha}$). **e** Solid-phase β wets a GB in the solid-phase α ($T > T_{ws\alpha\alpha}$). **f** Only α/β interphase boundaries (IBs) exist in a two-phase, $\alpha + \beta$ polycrystal if wetting occurs at both α/α and β/β GBs ($T > T_{ws\beta\beta}$). **g** Schematic dependencies of σ_{GB} (grain-boundary tension) and $2\sigma_{IB}$ (with σ_{IB} as the interphase-boundary tension) on temperature (for the notions grain-boundary tension and grain-boundary energy, see Sect. 11.3.1). The point of intersection indicates the wetting temperature, T_w (taken from Lopez et al. 2004).

corners where four matrix grains meet (a configuration of more than four grains (edges) at a corner is unstable; cf. Sect. 11.3.1) are not considered). Then, recognizing that upon grain-boundary precipitation a grain boundary (GB) of the matrix is replaced by two interphase boundaries (IBs), a simple criterion based on the change of interfacial tension, σ (see Sect. 11.3.1 for physical background and definition of interfacial/grain-boundary tension and energy), can be given that predicts whether (a chain of) individual grain-boundary (liquid or solid) precipitates or layers of (liquid or solid) precipitate phase develop at grain boundaries (provided a net driving force for precipitation operates):

If $\sigma_{GB}^{\alpha\alpha} < 2\sigma_{IB}^{\alpha p}$, where p stands for L or β, the growing liquid or solid particle tends to reduce its interfacial contact area with the matrix and develops a lens-like shape (here it is supposed that the interfacial tensions are isotropic) characterized by a contact angle θ (see Fig. 9.14b, d) which is given by the static, mechanical equilibrium of the interfacial tensions at the junction of the GB and both IBs (the interfacial tensions must balance in the plane of the grain boundary; for discussion on interfacial energy and interfacial tension and the background of Eq. (9.13), see Sect. 11.3.1)[13]:

$$\sigma_{GB}^{\alpha\alpha} = 2\sigma_{IB}^{\alpha p} \cos(\theta/2) \qquad (9.13)$$

If $\sigma_{GB}^{\alpha\alpha} > 2\sigma_{IB}^{\alpha p}$ the growing liquid or solid particle is not stable mechanically, i.e. an interfacial energy equilibrium cannot be established (cf. Eq. (9.13) and its footnote): the growing precipitate particle tends to increase its contact area with the matrix-grain boundary, it will cover, i.e. "wet", the matrix-grain boundary (the contact angle is nil) and a layer of precipitate phase develops along the matrix-grain boundary (Fig. 9.14c, e).

The role of the temperature is important via its effect on the interfacial energies. In general, with increasing temperature both $\sigma_{GB}^{\alpha\alpha}$ and $\sigma_{IB}^{\alpha p}$ decrease. Consider the schematic dependencies on temperature of $\sigma_{GB}^{\alpha\alpha}$ and $2\sigma_{IB}^{\alpha p}$ shown in Fig. 9.14g. At sufficiently low temperature, $\sigma_{GB}^{\alpha\alpha} < 2\sigma_{IB}^{\alpha p}$. Upon increasing temperature, the temperature dependencies $\sigma_{GB}^{\alpha\alpha}(T)$ and $2\sigma_{IB}^{\alpha p}(T)$ intersect at a temperature T_w: wetting occurs at temperatures equal to or higher than the "wetting temperature T_w". Starting at $T < T_w$, upon increasing temperature the contact angle decreases, becomes nil at T_w and remains nil at $T > T_w$.

One can draw a tie-line (cf. Sect. 7.5.2) in the phase diagram at T_w. At and above this tie-line, the second, liquid or solid phase forms a layer separating the matrix crystals. Such tie-lines have been drawn in Fig. 9.14a for solid wetting and for liquid wetting in the $\alpha + \beta$ and $\alpha + L$ two-phase regions, respectively.

An illustration of the above is provided by Fig. 9.15, dealing with grain-boundary precipitation in the two-phase, Zn-rich phase (solid solution)/Al-rich phase (solid solution) region for a Zn-rich Zn–Al alloy. At sufficiently low annealing temperature,

---

[13] The nucleation-energy barrier for this case of heterogeneous (grain-boundary) nucleation can straightforwardly, and similarly to the treatment given in Sect. 9.2, be derived. It follows (cf. Christian, 1975; Porter and Easterling, 1981) that ΔG^* is given by the formula for homogeneous nucleation (cf. Eq. (9.5)) multiplied by a so-called shape factor which for the case considered (precipitate-particle shape as of a symmetrical doubly-spherical lens (Fig. 9.14b, d)) is given by [1 − (3/2)cos(θ/2) + (1/2)cos³(θ/2)]. Evidently, the shape factor and thus the nucleation-energy barrier reduce upon increasing ratio of $\sigma_{GB}^{\alpha\alpha}/2\sigma_{IB}^{\alpha p}$ (in the range 0 till 1), i.e. upon decreasing contact angle, θ (cf. Eq. (9.13)). If $\sigma_{GB}^{\alpha\alpha}/2\sigma_{IB}^{\alpha p}$ has become equal to one, the contact angle is zero, the shape factor is zero as well and hence the nucleation-energy barrier is nil: development of the precipitate phase is only determined by growth (not by nucleation) and occurs by extension along the matrix-grain boundary.

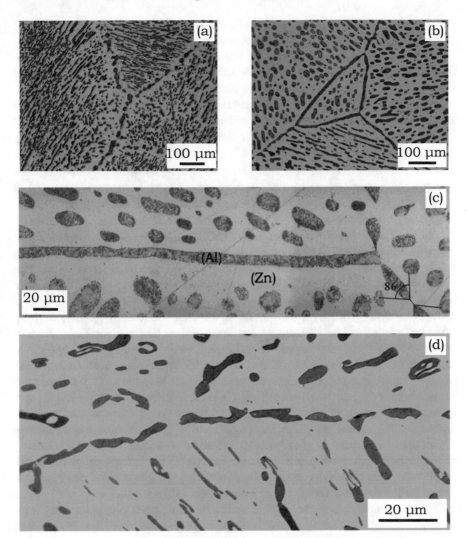

Fig. 9.15 Optical micrographs of the cross sections of Zn-5wt%Al samples annealed for 672 h at 523 K (**a**), 648 K (**b**) and 618 K (**c**), and annealed for 2016 h at 556 K (**d**) (taken from Lopez et al. 2004)

the minority precipitate phase (Al-rich solid solution) develops at the grain boundaries as a collection of individual particles (appearing as a chain of particles along the grain boundary in the specimen cross section considered). At temperatures above about 563 K grain boundaries get wetted by the Al-rich phase developing as a band along grain boundaries of the matrix (cf. Fig. 9.15a, b). Even long time annealing at a temperature a little below 563 K does not lead to wetted grain boundaries (Fig. 9.15d).

In polycrystalline specimens a spectrum of grain boundaries with different energies exists. Therefore, in polycrystals a range of T_w values occurs: from $T_{w,min}$ to

$T_{w,max}$. At $T_{w,min}$ the grain boundaries of highest energy are susceptible to wetting; at $T_{w,max}$ all grain boundaries become wetted; at $T_{w,min} < T < T_{w,max}$, only a fraction of the grain boundaries will be wetted. For example, see the triple junction shown in Fig. 9.15c. The grain boundary positioned horizontally in the micrograph is fully covered by a uniform precipitate (Al-rich phase) layer, whereas the two grain boundaries on the right-hand side of the micrograph are covered by chains of individual Al-rich phase precipitate particles. The contact angle for a precipitate particle at the latter type of grain boundary has been indicated: it is about 86°, implying that $\sigma_{GB}^{\alpha\alpha} = 1.46\,\sigma_{IB}^{\alpha\beta}$ (cf. Eq. (9.13)).

Finally, an important difference between grain-boundary wetting by a liquid phase and grain-boundary wetting by a solid phase is discussed. In the two-phase, $\alpha + \beta$ region of the phase diagram, α/α, β/β and α/β interfaces can in general occur. Depending on the composition of the alloy α/α or β/β grain boundaries may be dominant. The composition of the alloy pertaining to the example considered in Fig. 9.15 is such that α/α grain boundaries were predominant. The alloy composition can be changed such that a significant amount of β/β grain boundaries occurs. The β/β grain boundaries can exhibit wetting phenomena as well. However, the wetting temperatures for first appearance of wetting at α/α grain boundaries and at β/β grain boundaries generally are different, as the energies (energy spectra) of these grain boundaries will be different. Hence two different tie-lines, for $T_{w,min}$, have to be indicated in the phase diagram for solid-phase wetting in the $\alpha + \beta$ two-phase region, whereas only one such tie-line suffices for liquid-phase wetting in a solid phase + liquid, as $\alpha + L$, two-phase region (see Fig. 9.14a). Now, if the temperature for wetting at α/α grain boundaries, $T_{ws\alpha\alpha}$, is lower than that for wetting at β/β grain boundaries, $T_{ws\beta\beta}$, upon increasing the temperature the α/α grain boundaries will be wetted first by β phase, and when the temperature is increased further and becomes higher than $T_{ws\beta\beta}$, the β/β grain boundaries will be wetted by α phase. Eventually, at sufficiently high temperature, only α/β interphase boundaries are stable in the polycrystal. This leads to a configuration of alternating α phase and β phase, similar to that for the light and dark fields on a chess-board (Fig. 9.14f).

The above deliberations serve to demonstrate that simple principles carry a lot of power for understanding a wealth of developing microstructures in materials subjected to phase transformations.

9.5 Diffusionless Transformations; Examples

Diffusionless transformations in solids (see the introductory remarks in Sect. 9.3) can be evoked in many systems provided the cooling rate from a single-(solid)phase field at elevated temperature is high enough. Consider the schematic, partial binary phase diagram in Fig. 9.16. Upon cooling the alloy of composition as indicated in the figure from the high temperature single-phase (γ) field, the alloy passes a two-phase ($\alpha + \gamma$) field and arrives then in the low temperature single-phase (α) field. If slow

Fig. 9.16 Indication of a cooling procedure in a schematic phase diagram for an A-rich, A-B alloy, where depending on cooling rate, a diffusion-controlled transformation or a partitionless transformation can occur, from the solid γ phase to the solid α phase

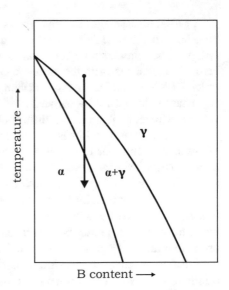

temperature ⟶

γ

α $\alpha+\gamma$

B content ⟶

cooling is imposed, the alloy will start to decompose in the two-phase region: grains of α phase, depleted in solute, nucleate and grow, possibly at the grain boundaries of the parent γ phase, which at the same time becomes enriched in solute. This phase transformation thus is accompanied with long-range diffusion. At a somewhat higher cooling rate, leading to larger undercooling, the developing α phase may exhibit the Widmanstätten morphology (cf. Sect. 9.4.4). At a still higher cooling rate, no pronounced decomposition may occur while passing the two-phase field during cooling. Then, during continued cooling, in the single-phase (α) field, the transformation $\gamma \rightarrow \alpha$ can happen without that a change in composition occurs, i.e. without the necessity of long-range diffusion.[14] In this case of transformations taking place without that a redistribution of solute occurs, one also speaks of *partitionless transformations*. Two types of such (long range) diffusionless transformations can be distinguished: the massive transformation, which occurs at moderately high cooling rate, and the martensitic transformation, which generally requires higher cooling rate (one speaks of quenching).

9.5.1 The Massive Transformation

The diffusionless, massive transformation does occur in many systems. The focus in this section is on the austenite-ferrite transformation in iron-based alloys as the vehicle to discuss the massive transformation.

---

[14] A thermodynamic driving force for this phase transformation without composition change is already available upon cooling in the two-phase ($\alpha + \gamma$) field below the so-called T_0 line (see Sect. 9.5.1 and Fig. 9.19).

A partial phase diagram of the type as shown in Fig. 9.16 holds, for example, for Fe–Mn. Experimental determination of the boundaries of the ferrite (α)—austenite (γ), two-phase field requires very long annealing times, i.e. years in the case considered. Then, starting from the austenite (γ), single-phase field, cooling rates of the order 10 K/min suffice to avoid decomposition in the ferrite (α)—austenite (γ), two-phase field and a $\gamma \rightarrow \alpha$ transformation occurs in the ferrite, single-phase field without Mn redistribution, i.e. diffusionless. The nucleation of the ferrite grains in this transformation initiates at the austenite-grain boundaries. The grown ferrite grains show a more or less equiaxed, massive morphology which gave rise to the name *massive transformation*. Growth occurs by jumping of the individual atoms across the transformation front. Both the nucleation and the growth of the product phase in a massive transformation are thermally activated. The driving force of the transformation can be large (much larger than that involved in grain coarsening, where the decrease of grain-boundary density, with its associated decrease in energy, drives migration of a grain boundary (cf. Sect. 11.3), but still pronouncedly smaller than that involved in martensitic transformations). As a consequence the migration rate of the α/γ transformation front can be relatively large (e.g. a couple of μm/s; see Fig. 9.17), which can lead to a rather irregular shape of the product-grain boundaries (see Fig. 9.18), as a minimum energy grain-boundary shape is kinetically unrealizable.

The thermodynamic, energy condition for the massive transformation to occur, if only the chemical Gibbs energies are considered (cf. Eq. (9.2)), is that the Gibbs energy of the product phase is smaller than that of the parent phase *for the same composition*. The locus of points where $G^{\alpha}_{\text{chem}} = G^{\gamma}_{\text{chem}}$, for the same composition, can be given in the phase diagram as a relation between temperature and composition

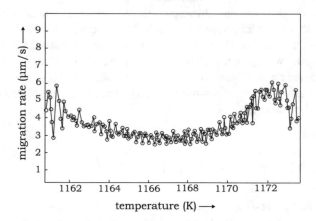

Fig. 9.17 γ(austenite)/α(ferrite) interface-migration velocity for pure iron as a function of temperature; the $\gamma \rightarrow \alpha$ transformation occurs upon cooling from the austenite-phase field. The transformation proceeds in the figure from the right to the left for a cooling rate of 10 K/min (taken from Liu et al. 2004)

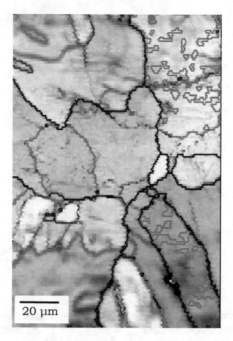

Fig. 9.18 Large-angle grain boundaries (thick black lines) and small-angle grain boundaries between subgrains (thin grey lines) in pure iron after the γ(austenite) $\rightarrow$ α(ferrite) transformation upon cooling from the austenite-phase field (taken from Liu et al. 2004)

(at constant pressure (of 1 atm), of course); this is the so-called T_0 *line*. Such a result for the Fe-C system is shown in Fig. 9.19.

Evidently, the massive transformation could, according to this thermodynamic criterion, already occur during cooling within the austenite-ferrite two-phase region, below the T_0 line. The extent of undercooling (T below T_0) needed for the transformation to run is dependent on the transformation-opposing deformation- and interface-energy contributions associated with the transformation (cf. Eq. (9.2)). Whether the diffusionless, massive $\gamma \rightarrow \alpha$ transformation occurs below the T_0 line and within the ($\alpha + \gamma$) two-phase field, or below the $\alpha/(\alpha + \gamma)$ phase-boundary line, has been a matter of considerable controversy. It has been shown that for substitutional iron-based alloys, as for example Fe-Co and Fe–Mn, where the diffusion of the solute is relatively slow, the massive $\gamma \rightarrow \alpha$ transformation initiates at a temperature below the $\alpha/(\alpha + \gamma)$ phase-boundary line (Liu et al. 2004). However, for iron-carbon alloys a more complicated picture arises:

The diffusivity of an interstitially dissolved solute, as carbon, is relatively high (cf. Sect. 8.5.3), and thus, it can be understood that, provided the carbon concentration is large enough (above about 0.01 at.% C), so that upon cooling enough time is passed in the ($\alpha + \gamma$) two-phase field (cf. Fig. 9.19), decomposition, in solute-depleted α and solute-enriched γ, may already occur upon traversing the ($\alpha + \gamma$) two-phase field. Upon continued cooling, at a certain temperature below the T_0 line, then the

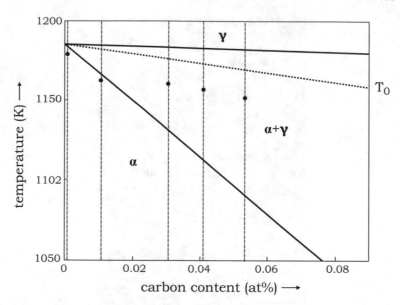

Fig. 9.19 Iron-rich part of the Fe-C phase diagram with indication of the T_0-line, in the α(ferrite) + γ(austenite), two-phase region, where $G^\alpha_{chem} = G^\gamma_{chem}$. If only the chemical Gibbs energies are considered, the massive $\gamma \rightarrow \alpha$ transformation becomes possible if $T < T_0$. The dots indicate the average experimental onset transformation temperatures for the massive part of the $\gamma \rightarrow \alpha$ transformation for ultra-low carbon Fe-C alloys (taken from Liu et al. 2008)

massive, diffusionless $\gamma \rightarrow \alpha$ transformation does set in (i.e. the driving force for the massive transformation has become sufficiently large). The onset temperatures for the massive $\gamma \rightarrow \alpha$ transformation for these iron-carbon alloys, with carbon contents above about 0.01 at.% C, have been indicated in Fig. 9.19; indeed these onset temperatures, for the massive transformation for these iron-carbon alloys, lie below the T_0 temperature but fall within the ($\alpha + \gamma$) two-phase field, in contrast with the substitutional alloys for which the massive $\gamma \rightarrow \alpha$ transformation occurs within the single-phase α phase field (Liu et al. 2008). This leads for the Fe–C alloys considered to the unusual phenomenon of a transition of diffusion control to interface control (cf. Sect. 9.3) for growth of the product ferrite (α) phase of the $\gamma \rightarrow \alpha$ transformation:

Transition of initial diffusionless, naturally interface-controlled transformation to later, diffusion-controlled decomposition (cf. Sect. 9.3) is possible for initially diffusionless transformations at temperatures sufficiently high for distinct solute diffusion to occur: at the end of a transformation it appears likely that diffusion control of the solute can become significant, and thus, transition of interface-controlled growth to diffusion-controlled growth can occur (Sietsma and van der Zwaag 2004). Hence, the above discussed observation of "initial" diffusion-controlled growth in the first part of the $\gamma \rightarrow \alpha$ transformation in the ($\alpha + \gamma$) two-phase region for Fe–C alloys containing more than about 0.01 at.% C implies that even earlier, in the very beginning, for a very small period of transformation, interface-controlled growth must have prevailed. The occurrence of a main, massive part of the transformation, which naturally implies interface-controlled growth, for these alloys upon further cooling, then means that growth of the ferrite (α) phase in these Fe–C alloys, upon cooling from the single-phase, austenite (γ) phase field, proceeds through stages of, subsequently, interface control, diffusion control and (again) interface control (Liu et al. 2006).

Careful analysis of the kinetics of the massive transformation has shown that the transformation-front migration rate can exhibit considerable fluctuation that exceeds the experimental inaccuracy largely. An example is provided by Fig. 9.17. The figure shows the transformation-front, i.e. α/γ-interface, migration velocity as an average for a whole specimen obtained by dilatometric[15] analysis of a transforming pure iron specimen during cooling at 10 K/min (the transformation proceeds in the figure from the right to the left!). The fluctuation of the interface-migration velocity is a consequence of the large deformation energy taken up by the system upon the formation of ferrite. The deformation energy can be of the same order of magnitude as the chemical driving force. These last points are discussed in the following paragraph.

The chemical driving force depends primarily on temperature and not on the degree of transformation, recognizing that the transformation occurs partitionless (i.e. product and parent phases have the same composition). This contrasts with the deformation energy that depends of the degree of transformation and not primarily on temperature. During transformation more and more deformation/strain build-up takes place in the specimen. This has as consequence that, upon progressing transformation, the (further) deformation energy, induced per unit transformed, increases with increasing transformation and can eventually become as large as the chemical driving force per unit transformed. Then the transformation front considered comes to a halt.[16] Next, either the deformation/strain energy relaxes (by recovery processes; see Sect. 11.1) and a net driving force occurs again, and/or, if the transformation takes place during cooling, a net driving force can result upon continued cooling as the chemical driving force increases with decreasing temperature (for the iron-based systems considered here). The transformation then proceeds (again) until the deformation-induced energy (again) is equal to the chemical driving force, etc. This

---

[15] A dilatometer is an instrument that records the length change of a specimen during some (usually thermal) treatment (see also Sect. 10.13). The high resolution instrument used for the data presented in Fig. 9.17 has an absolute length-change accuracy of about 10 nm, which, in view of the length of the specimen of about 10 mm, explains the very high relative accuracy of this technique. It can be shown that the fluctuation of the interface-velocity data in Fig. 9.17 is a factor 100 larger than the experimental inaccuracy and thus must have a physical, transformation-process inherent origin.

[16] In this discussion the role of the austenite (parent phase)/ferrite (product phase) interfacial energy is ignored. The interfacial energy per unit transformed, will, as holds for the deformation energy, depend on the degree of transformation and not primarily on temperature, because the ratio of the created interfacial area and the volume of the produced product phase changes during the transformation. In the present case it appears likely that the contribution of the interfacial energy can be neglected as compared to the chemical and deformation energy contributions.

leads to an irregular nature of the transformation-front velocity. A similar discussion is given for the martensitic transformation in Sect. 9.5.2.4.

As discussed in the preceding paragraph, the growing ferrite grain induces strain and defects in the surrounding austenite. This deformed austenite, immediately in front of the growing ferrite, may allow easier nucleation of ferrite than undeformed austenite (see, in particular, the "*Intermezzo: Nucleation of AlN in Fe–Al alloy*" in Sect. 9.2). Thus, repeated nucleation of ferrite in front of the moving interface may occur (this requires a minimum size of the austenite grain, recognizing that the initial nucleation of ferrite occurs at the grain boundaries of the austenite). The occurrence of repeated nucleation of ferrite in front of the moving interface is called: *autocatalytic nucleation*. This leads to bursts in the nucleation density and corresponding maxima in the transformation rate (Liu et al. 2003, 2004).

The important role of the misfit-induced deformation energy and defects, as discussed above for the massive transformation, returns in the discussion of the martensitic transformation dealt with in the next section.

9.5.2   The Martensitic Transformation

The probably most well-known martensitic transformation (cf. Sect. 9.3) is the one observed upon quenching interstitial iron-carbon and iron-nitrogen alloys from the austenite-phase field. The hardening of carbon steel derives from this transformation. Martensitic transformations do not only occur in metallic, ferrous, crystalline solids but are also met in metallic, non-ferrous, crystalline alloys, as Cu–Zn and Cu-Al, and in particular can be induced as well in crystalline ceramics, as ZrO_2. Because of the paramount technological importance of the hardening of carbon steels, attention is devoted here to the formation of martensite in interstitial iron-based alloys. Although this transformation, named after Adolf Martens, who first identified the martensitic microstructure at the end of the nineteenth century, has been extensively investigated for more than a century, fundamental, deep understanding of this complicated and intriguing transformation has not been completed even today.

9.5.2.1   Interstitials in Iron Crystal Structures
         (cf. Sects. 4.4.2.1 and 4.4.2.2)

The discussion starts with a consideration of the size of the interstices in the f.c.c. (austenite) and b.c.c. (ferrite) iron crystal structures. Two types of interstices can be recognized in both crystal structures: octahedral interstitial sites and tetrahedral interstitial sites; see Fig. 9.20. Adopting a rigid sphere model for the (iron) atoms, it follows, by straightforward calculus, for the maximal radius of an interstitial atom to be incorporated without distortion of the parent f.c.c. crystal structure (the iron atoms are in touch along the <110> directions (face diagonals of the unit cell); cf. Sects. 4.4.2.1, 4.4.2.2 and Table 4.6):

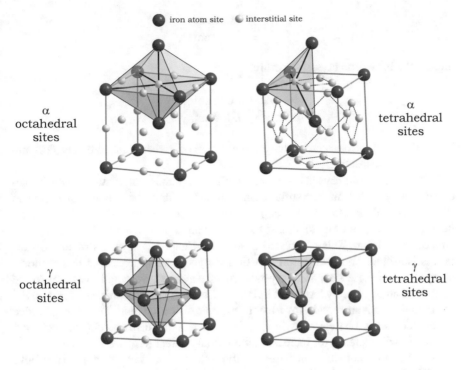

Fig. 9.20 Interstitial octahedral and tetrahedral sites in ferrite (α, b.c.c.) and in austenite (γ, f.c.c.)

$$r_4 = 0.22R \tag{9.14a}$$

$$r_6 = 0.41R \tag{9.14b}$$

with r_4 and r_6 as the atomic radii of the interstitials in the tetrahedral site (fourfold coordination) and the octahedral site (sixfold coordination), respectively, and R as the radius of the atoms of the parent crystal structure. A similar calculation for the b.c.c. crystal structure (the iron atoms are in touch along the <111> directions (body diagonals of the unit cell)) gives:

$$r_4 = 0.29R \tag{9.15a}$$

$$r_6 = 0.15R \tag{9.15b}$$

Adopting the lattice-parameter values for austenite and ferrite as 0.356 nm and 0.286 nm, it follows $R = 0.126$ nm and $R = 0.124$ nm for f.c.c. and b.c.c. iron, respectively. Hence, for the f.c.c. iron crystal structure:

$$r_4 = 0.028 \, \text{nm}$$
$$r_6 = 0.052 \, \text{nm}$$

and for the b.c.c. iron crystal structure:

$$r_4 = 0.036 \, \text{nm}$$
$$r_6 = 0.019 \, \text{nm}$$

The carbon and nitrogen atoms have atomic radii of, rather approximately, 0.08 nm and 0.07 nm, respectively.

For an f.c.c. iron crystal structure, the above, simple calculations then immediately suggest that an interstitial atom, carbon or nitrogen, upon dissolution in the iron crystal structure will occupy an octahedral interstice, rather than a tetrahedral interstice, and that this must be accompanied by considerable local crystal-structure distortion. This associated elastic (cf. Chap. 12) crystal-structure distortion is isotropic. The octahedron constituted by six parent crystal-structure atoms is regular, and these six parent crystal-structure atoms have equal distances to the interstitial atom possibly located at its centre. Consequently, the misfitting, dissolved interstitial atom is surrounded by a local, isotropic distortion field. The average lattice parameter, as can be determined by, for example, X-ray diffraction (see Sects. 4.5 and 6.10), will increase for increasing interstitial content (e.g. see Eq. (4.8)).

A rather different situation arises for the b.c.c. iron crystal structure. The above numerical results would suggest that the interstitial atoms would prefer the tetrahedral interstitial sites over the octahedral interstitial sites. The reverse is true. This can be understood as a consequence of the irregularity of the octahedral interstice in the b.c.c. crystal structure, as illustrated by Fig. 9.21. The six parent crystal-structure atoms constituting the octahedron do not have the same distance to the centre of the octahedron: There are two (the iron atoms labelled 1 and 3 in Fig. 9.21 for the interstitial site labelled z), and not six as holds for the f.c.c. crystal structure, *nearest-neighbour* atoms of the interstitial atom possibly located in the centre of the octahedron (see also Tables 4.5 and 4.6). Upon its insertion, the interstitial will push away these nearest-neighbour iron atoms (in opposite directions along the c-axis, defined by the iron atoms labelled 1 and 3 in Fig. 9.21). The four *next-nearest-neighbour* atoms (along the equivalent a- and b-axes in the figure) have a distance equal to 0.202 nm to the centre of the octahedron, which equals about the sum of the radii of the iron and carbon/nitrogen atoms (0.124 nm + 0.08/0.07 nm), and thus they need not be displaced significantly upon insertion of a carbon or nitrogen atom; in fact, they move a bit towards the carbon/nitrogen atom as a consequence of what could be called Poisson contraction induced by the tensile elongation along the c-axis (see Sect. 12.2). Thus the dissolved, misfitting interstitial atom is surrounded by a local, elastic distortion field of tetragonal symmetry. The insertion of an interstitial into a tetrahedral interstice would require that all four surrounding parent crystal-structure atoms, at equal distances of the interstitial, have to be pushed away and thereby the elastic energy involved in distorting the tetrahedral interstice upon insertion of an

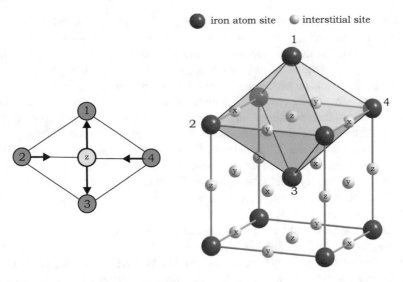

Fig. 9.21 Irregularity of the octahedral interstitial site in ferrite (b.c.c.) (In austenite (f.c.c.) the octahedral interstitial site is regular)

interstitial becomes larger than that required for incorporation of the interstitial on an octahedral interstitial site.

A further consequence of the above deliberations is the suggestion that, because of the much smaller size of the octahedral interstice in b.c.c. iron as compared to f.c.c. iron, the solubilities of carbon and nitrogen in austenite will be much larger than those in ferrite. This is the case indeed (see also footnote 7 in Sect. 8.6): maximally about 0.02 wt% (=0.095 at.%) C and 0.10 wt% (=0.40 at.%) N in ferrite versus 2.1 wt% (=9.1 at.%) C and 2.4 wt% (=9.0 at.%) N in austenite.

There are three octahedral interstitial sites per iron atom in the b.c.c. crystal structure. Hence three sets of octahedral interstitial sites can be recognized in the b.c.c. crystal structure: the x, y, and z sets in Fig. 9.21; see also Fig. 4.43a. For the small amount of interstitials that can maximally be dissolved in ferrite (see above paragraph; then the probability of finding a pair of interstitials close to each other is marginal; but also see next paragraph), interaction (of the surrounding elastic distortion fields) of the dissolved interstitial atoms can be neglected and therefore the numbers of x-, y- and z-type sites occupied by interstitial atoms will be equal. As a result, in spite of the tetragonal nature of the distortion field surrounding each dissolved interstitial atom, the average crystal structure of the ferrite remains body-centred *cubic*, with a lattice parameter that increases with interstitial content. The cubic nature of the average crystal structure cannot be maintained as soon as that a preferred occupation of one (or two) of the three sets of octahedral interstices would occur (see what follows two paragraphs further).

The above geometrical considerations also imply that interstitial atoms in ferrite are highly unlikely to ever occur as a pair of *nearest*-neighbour interstitial atoms at adjacent interstitial sites: the interstitial–interstitial distance would be that small (0.143 nm = half of the lattice parameter of b.c.c. ferrite; see Fig. 9.21) that an enormous repulsion due to *electrostatic and strain* interactions occurs. However, long(er)-range *strain* interaction, recognizing the anisotropic, tetragonal nature of the strain field around an interstitial, may be responsible for the occurrence of (some) clustering of interstitial atoms, albeit with maintaining more remote interstitial inter-atomic distances than corresponding to the geometrically possible nearest distance of two interstitials at octahedral interstices. Such clusters would form as plates along {001} planes of the ferrite (Bhadeshia 2004). In case of nitrogen interstitials in ferrite, the nitrogen atoms in such clusters may even exhibit long-range order and then are described as α''-$Fe_{16}N_2$ precipitates (see van Genderen et al. 1993; cf. the discussion in the "*Intermezzo: Tempering of iron-based interstitial martensitic specimens*" and footnote 22 in Sect. 9.5.2.4).

Consider the Fe-C and Fe-N phase diagrams given in Fig. 9.22a, b (see below the "*Intermezzo: The Fe-C and Fe-N Phase Diagrams*"). Now, upon quenching an iron–carbon or iron–nitrogen alloy from the austenite-phase field (at elevated temperature) to lower, e.g. room temperature, i.e. into the ferrite + cementite two-phase field in case of iron-carbon alloys and into the ferrite + γ'-Fe_4N two-phase field in case of iron–nitrogen alloys (see the phase diagrams; Fig. 9.22a, b) the rate of cooling can be that large that there is no chance for decomposition as prescribed by the phase diagram to occur: diffusion of the interstitial atoms can be made negligible by appropriate quenching. The iron atoms want to adopt a configuration as in ferrite, but the ferrite lattice cannot incorporate the relatively enormous amount of interstitial atoms which were dissolved in the austenite. As a compromise, the interstitial atoms in the resulting

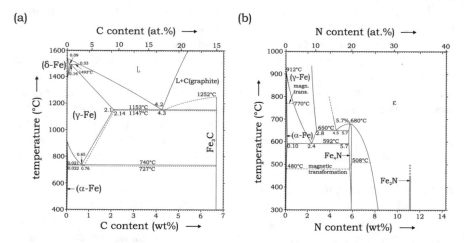

Fig. 9.22 Phase diagrams of **a** Fe-C and **b** Fe-N (redrawn from T.B. Massalski (Editor in Chief), *Binary Alloy Phase Diagrams*, 2nd edition, ASM, Metals Park Ohio 1996.). Considering the Fe-C phase diagram, note, as compared to Fig. 9.9, the slight differences for the numerical data of temperatures and compositions indicated in the diagram. This is of course related to the research that has been performed in between the years of publication of these diagrams

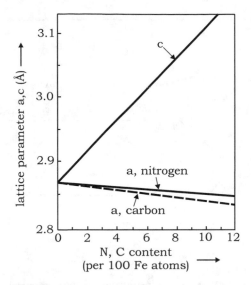

Fig. 9.23 Variation of the lattice parameters c and a of the body-centered tetragonal crystal structure of iron-carbon and iron-nitrogen martensites with interstitial content (taken from Liu Cheng et al. 1990)

"ferrite" now occupy only one, or, more precisely, preferably one, of the three types of octahedral interstices, say the z-type octahedral interstices. This effect can be considered as a consequence of the minimization of elastic energy by alignment of the tetragonal distortion fields of neighbouring interstitial atoms. As a result, the average "ferrite" crystal structure cannot maintain a cubic nature and becomes on average tetragonal: a body-centred tetragonal crystal structure containing a relatively large amount of interstitial atoms (the same amount as in the parent austenite) on preferably one of the three types of octahedral interstices; this is martensite. The symbol used for this b.c.t. martensite is α' (b.c.c. ferrite is denoted by α). In accordance with the discussion in the above paragraphs, the a and c lattice parameters of the average body-centred tetragonal martensite crystal structure decrease modestly and increase pronouncedly, respectively, upon increasing interstitial content (see Fig. 9.23).[17]

On the basis of the atomic radii given above for the carbon and nitrogen atoms, one would expect that the unit cell volume of iron–carbon martensite would be somewhat larger than that of iron–nitrogen martensite. Apparently, this is not the case (see Fig. 9.23). The numerical data given above for the atomic radii pertain to "covalent radii". Indeed the bonds of carbon and nitrogen with the surrounding iron atoms

---

[17] As discussed in Sects. 4.4.1 and 4.4.2 (Eqs. (4.7) and (4.8)), in first order approximation one may generally assume that the lattice parameters of a solid solution are linearly dependent on the number of solute atoms in the unit cell. Then, for *interstitially* dissolved solute atoms, the lattice parameters would be linearly dependent on the number of solute atoms *per* solvent atom, since solvent atoms at their crystal-(sub)structure sites are not replaced by interstitially dissolved solute atoms (in contrast with substitutionally dissolved solute atoms). Therefore the a and c lattice parameters in Fig. 9.23 are presented as function of the number of interstitials per 100 iron atoms, x_C^r or x_N^r (cf. Eq. (4.8)).

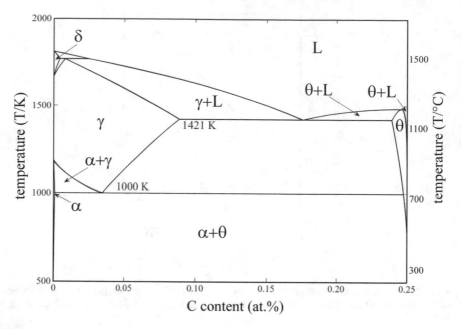

Fig. 9.24 Fe-C phase diagram, evaluated recognizing non-stoichiometry of cementite, Fe_3C_{1-z}. α = ferrite; γ = austenite; θ = cementite = Fe_3C_{1-z}; δ = ferrite; L = liquid. The formation of graphite, diamond and higher carbides has been suppressed. Note the differences with Figs. 9.9 and 9.22a: cementite appears as line compound in Figs. 9.9 and 9.22a, whereas a phase field occurs for cementite in this figure. It must be noted that the presence of the second $\theta + L$ two-phase field (indicated as $(\theta + L)$ in the figure) has no physical meaning; it is an artefact caused by the exclusion in the evaluation of the mentioned phases of higher carbon content (adapted from Göhring et al. 2016a; see the "*Intermezzo: The Fe-C and Fe-N Phase Diagrams*")

can have a strongly covalent nature. However, carbon in iron–carbon martensite may have some positive ionicity (due to transfer of electronic charge to surrounding iron atoms), whereas nitrogen in iron–nitrogen martensite maintains a practically neutral state. This could explain the difference between the lattice-parameter dependencies shown in Fig. 9.23 (Liu Cheng et al. 1990).

This preference of the interstitial atoms for only one of the three types of octahedral interstitial sites can be considered as an ordering phenomenon (cf. Sect. 4.4.2.2). In this special case the interstitials are distributed randomly on a preferred type of interstitial site; one speaks of *Zener ordering*.

Intermezzo: The Fe-C and Fe-N Phase Diagrams

These phase diagrams cannot be considered as presenting genuine thermody-namic equilibrium at 1 atm pressure: the equilibrium solid carbon phase (in equilibrium with ferrite) is graphite and the equilibrium nitrogen phase (in equi-librium with ferrite) is nitrogen gas. The usual Fe-C and Fe-N phase diagrams

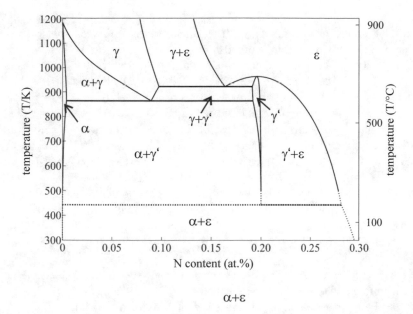

α+ε

Fig. 9.25 Fe-N phase diagram, evaluated incorporating new (from the last decade) data on specific phase-boundary lines. α = ferrite; γ = austenite; γ' = Fe$_4$N$_{1-x}$; ε = Fe$_3$N$_{1+y}$. The formation of N$_2$ (gas) has been suppressed. Note that an $\alpha + \varepsilon$ two-phase field is predicted to occur at temperatures below 443 K (=170 °C). Such predictions are difficult to verify, since at these low temperatures (implying (very) slow kinetics) it can be very demanding to establish genuine equilibrium states (adapted from Göhring et al. 2016b; see the "*Intermezzo: The Fe-C and Fe-N Phase Diagrams*")

thus represent metastable phase equilibria. Iron-based specimens containing cementite, Fe$_3$C$_{1-z}$ (see what follows) or γ' nitride, Fe$_4$N$_{1-x}$, are prone to decomposition of these phases leading to the formation of solid graphite and nitrogen gas (precipitation of N$_2$ gas causing pore formation in the solid matrix), respectively. Such phenomena have special relevance for the carburizing and nitriding of iron-based materials. For detailed discussion, see Mittemeijer and Slycke (1996).

Considering the Fe-C and Fe-N phase diagrams (Fig. 9.22a, b) one cannot avoid noticing their similar appearance. This, of course, is related with other parallels occurring for the elements C and N: of about equal atomic size (as discussed in the main text above), dissolving interstitially in solid Fe and experiencing a largely covalent bonding with Fe (cf. discussion of Fig. 9.23). Then it may surprise that the commonly accepted phase diagrams for Fe-C (see Figs. 9.9 and 9.22a) present cementite as a stoichiometric compound Fe$_3$C (i.e. a line compound; cf. Sect. 4.3), whereas its counterpart in the Fe-N system, γ' iron nitride, Fe$_4$N$_{1-x}$, is recognized to have a certain, significant homogeneity range (see Fig. 9.22b). As may be anticipated, in fact cementite has a certain, distinct homogeneity range as well: vacancies on the C crystal substructure can occur, implying that the composition of cementite has to be indicated as Fe_3C_{1-z} (for conclusive experimental data, see A. Leineweber, S.L. Shang and Z.K. Liu, Acta Materialia, 86 (2015), 374–384). This indeed parallels the occurrence of vacancies on the nitrogen crystal substructure of γ' iron nitride, implying that its composition is indicated as Fe_4N_{1-x} (see Fig. 4.41 discussed in Sect. 4.4.2.1). In view of the now established non-stoichiometry of cementite (θ), a reevaluation of the Fe-C phase diagram was performed (Göhring et al. 2016a). The resulting Fe-C phase diagram is shown in Fig. 9.24, which can be compared with the older versions shown in Figs. 9.22b and 9.9.

As compared to the evaluation of the Fe-N phase diagram as represented by Fig. 9.22b, new data on some important phase-boundary lines in the Fe-N phase diagram have become available. A corresponding reevaluation of the Fe-N phase diagram has been performed (Göhring et al. 2016b). The resulting Fe-N phase diagram is shown in Fig. 9.25 (cf. Fig. 9.22b).

As outcome of the recognition of significant non-stoichiometry of the cementite phase (θ), a phase *field*, instead of a line, for cementite (θ) emerges in the Fe–C phase diagram, which phase field is of morphology similar to that of the phase field for Fe_4N_{1-x} (γ') in the Fe–N phase diagram (compare Figs. 9.24 and 9.25): for example note the occurrence of Fe_3C and Fe_4N as limiting compositions for the phase regions of cementite (θ) and γ' iron nitride.

9.5.2.2 Crystallography of Martensite Formation

Evidently, Fe-C martensite is unstable with respect to decomposition into ferrite and cementite (or, with even larger release of energy, into ferrite and graphite; see the *"Intermezzo: The Fe-C system; steels and cast irons"* in Sect. 9.4.2 and the *"Intermezzo: The Fe-C and Fe-N Phase Diagrams"* immediately above). It develops upon quenching austenite, because long-range atom transport is impossible within the available time at the temperatures during the quench. As a compromise, nature allows the formation of martensite, which also brings about a release of energy (but less than would occur if ferrite plus cementite (graphite) would form). Martensite formation does not require long-range diffusion of the atoms of the components involved. It appears that a coordinated, "military" (cf. Sect. 9.3) transformation/deformation of the iron crystal substructure, in association with shear and possible short-range transport by (single) atomic jumps of interstitial atoms, which would find themselves, after the transformation of the iron (sub)crystal structure, on the "wrong" type (of the three types) of interstitial sites, to the "desired" type of interstitial site (see above discussion), can be the mechanism to bring about this structure.

Martensite formation occurs by a practically simultaneous, cooperative movement of thousands of (iron) atoms and is, compatible with the existing experimental evidence, accompanied by shear parallel to a (macroscopically) undistorted and unrotated plane (the *habit plane*, a plane of the crystalline parent phase on which the crystalline product phase starts to form) common to both the parent austenite and the product martensite. The transformation of the austenite-crystal structure into the martensite crystal structure, if unconstrained (see under (i) below), leads unavoidably to shape change (see Fig. 9.26a, b). If such a shape change is macroscopically impossible, an additional *crystal-structure invariant*[18] deformation of the martensite is required (see under (ii) below), in order to comply with the geometrical constraint of the confining, surrounding austenite, either by slip (dislocation glide) and/or by twinning of the martensite (see Fig. 9.26c, d), while maintaining the martensite crystal-structure symmetry.

These considerations/suppositions comprise a crystallographic theory of martensite formation (see also Wayman 1964) that consists of two main components:

(i) *The lattice correspondence*

The cooperative movement of thousands of iron atoms, to establish the martensite crystal structure from the austenite crystal structure, implies that specific lattice planes and directions in the austenite crystal structure correspond uniquely to specific lattice planes and directions in the martensite crystal structure (i.e. these corresponding lattice planes/directions pertain to the *same* atoms (but as before and as after the transformation)). A simple way to conceive the formation of martensite from austenite, revealing the correspondence of certain lattice planes and directions in austenite and martensite is provided by the so-called Bain lattice correspondence, illustrated in Fig. 9.27 (see also Sect. 4.2.2). Consider the two adjacent unit cells (here we refer to the iron (sub)crystal structure) of the parent austenite phase in Fig. 9.27a. Suppose at the centre of the unit cell at the left-hand side, where an octahedral interstice of the austenite-crystal structure occurs, an interstitial atom resides. At and across the interface of the two adjacent austenite unit cells, a unit cell of b.c.t. type symmetry can be identified. This b.c.t. unit cell can be transformed into a b.c.t. unit cell of the product martensite phase ("Bain deformation") by contraction (of about 17%) along the z direction and (smaller) expansion (of about 12%) along the x and y directions. A few corresponding directions have been explicitly indicated in the figure. The lattice correspondence only implies that the atoms pertaining to certain directions/planes in the parent austenite are the same as those in certain "corresponding" directions/planes in the product martensite; in the specimen frame of reference these corresponding directions/planes of the parent austenite and the product martensite do *not* coincide (with the exception of the habit plane). For example, in terms of Fig. 9.27, the $[10\text{–}1]_\gamma$ direction corresponds with, but as a result of the Bain

---

[18] In the literature, in contrast with the term "crystal-structure invariant deformation", the term "lattice-invariant deformation" is generally used, which, in view of the discussion in footnote 6 in Sect. 4.1.1 of Chap. 4, is unfortunate. To maintain a close link with the existing literature and to avoid unnecessary confusion, for this specific case, in the following (see under (ii)) the term "lattice-invariant deformation" is used yet. Usage of the terms "lattice planes" and "lattice directions" has been discussed in the second paragraph of Sect. 4.1.4 of Chap. 4.

deformation, is not parallel with the $[11-1]\alpha'$ direction. The austenite → martensite transformation according to this "Bain model" involves a minimum of atomic movement, as required for a diffusionless transformation.

(ii)   The lattice-invariant deformation

If unconstrained martensite formation would occur on a planar habit plane in the original austenite-crystal structure, in accordance with the Bain lattice correspondence in association with the Bain deformation, a rotation away from the original habit plane would occur (as indicated by the dashed lines in Fig. 9.26b which represent the habit plane). To assure that the habit plane can be taken as a plane in the austenite, which does not experience, macroscopically, a net distortion and rotation, crystal-structure invariant shears by slip, i.e. by gliding dislocations, or by twinning, should be operative (Fig. 9.26c, d). As suggested by Fig. 9.26d, internally twinned martensite thus is composed of alternate regions in the parent austenite which were subjected to the Bain deformation along different contraction axes such that the distortions in the original habit plane are compensated.[19] Evidently the widths of the martensite crystal parts in twin orientation define the orientation of the habit plane (cf. Fig. 9.26d). If the Bain deformation and the slip or twinning system are given, the specific habit plane, that obeys the requirement of on average (macroscopically) nil distortion and nil rotation, and the austenite-martensite orientation relationship can be calculated. The Bain deformation depends on the interstitial content (note the change of the tetragonality of the martensite crystal structure, as given by the axial ratio c/a, with interstitial content; cf. Fig. 9.23). Hence the habit plane and the orientation relationship depend on interstitial content.

The crystallographic theory of martensite transformation referred to above (for detailed description, see Wayman 1964) has been applied with great success since the 1950s. Some relatively minor discrepancies between prediction and observation have been observed, but its major deficiency is that it does not provide a picture of the product/parent interface on the atomic scale; it is merely imposed that the habit plane is an undistorted and unrotated, i.e. invariant plane of the shape transformation. Against this background dislocation/defect-based model descriptions of the atomic structure at the habit plane have been developed, partially backed by experimental observations; corresponding discussion is beyond the scope of this book (e.g. see Pond et al. 2008).

Experimental results on the orientation relationship (OR)[20] of parent austenite (γ) and product martensite (α') in iron–carbon and iron–nitrogen alloys, as function

---

[19] Because the habit plane is (macroscopically) undistorted and unrotated by the transformation, the Miller indices of the habit plane as defined in the parent austenite crystal structure and the Miller indices of the habit plane as defined in the product-martensite crystal structure are related directly by the Bain lattice correspondence.

[20] An orientation relationship (OR) of the structures of two crystalline phases can be expressed in various ways. One possibility is the indication of three pairs of parallel, independent directions (one such pair involves a direction in the first crystal structure and a direction in the second crystal structure, which are parallel). The usual way is the indication of a plane in the first crystal structure and a plane in the second crystal structure, which are parallel, and a direction in the first crystal structure and a direction in the second crystal structure, which are parallel.

Fig. 9.26 Schematic
depiction of the shape
change of a crystal upon
martensitic transformation in
the unconstrained case,
(**a**) → (**b**), and the slipping
(**c**) and twinning (**d**), which
occur if the martensite
crystal shape is constrained
by a surrounding matrix. γ
can be interpreted as
austenite and α' as
martensite. The dashed lines
in (**c**) and (**d**) represent the
(macroscopically)
undistorted, unrotated habit
plane (see text)

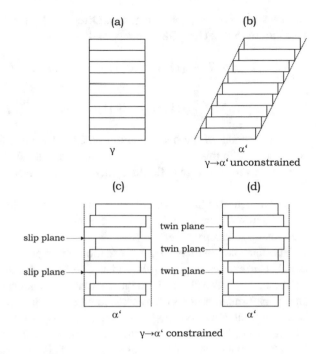

Fig. 9.27 Bain lattice correspondence. **a** A b.c.t. unit cell can be indicated for the pair of adjacent
unit cells of austenite shown. **b** This b.c.t. unit cell can be transformed into a b.c.t. unit cell of the
product martensite phase ("Bain deformation") by contraction (of about 17%) along the z direction
and (smaller) expansion (of about 12%) along the x and y directions. The corresponding $[10\text{-}1]_\gamma$
direction (in **a**) and $[11\text{-}1]_{\alpha'}$ direction (in **b**) have been indicated

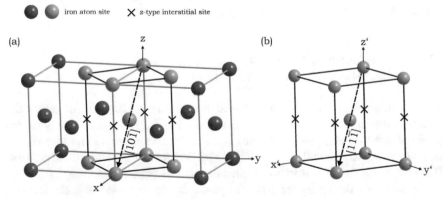

of interstitial content, are ambiguous. Often the Nishiyama-Wasserman (NW) and Kurdjumov-Sachs (KS) ORs are reported:

$$NW : \quad \{111\}_\gamma // \{011\}_{\alpha'} \quad ; \quad < 112 >_\gamma // < 011 >_{\alpha'} \qquad (9.16)$$

$$KS : \quad \{111\}_\gamma // \{101\}_{\alpha'} \quad ; \quad < 110 >_\gamma // < 111 >_{\alpha'} \qquad (9.17)$$

The accuracy of the determination of ORs, by TEM (cf. Sect. 6.8; a technique often applied to this end) is limited (because diffraction spots can be extended and distorted by local stresses); the difference between the NW and KS ORs is only about 5°.

Identification of habit planes requires a microscopic inspection of the microstructure. Therefore attention is paid in the next section to the morphology of the martensitic microstructure.

The austenite–martensite transformation is complex and still incompletely understood. Current research can thus give rise to also controversial discussion. For example, it has been claimed that the martensitic transformation, instead of the single step transformation, $\gamma \rightarrow \alpha'$, as considered above, in fact is a, sequentially running, two-step transformation: $\gamma \rightarrow \varepsilon \rightarrow \alpha'$, with ε as an intermediate h.c.p. martensite. The resulting martensitic microstructure would reveal such ("locked-in" remnants of) intermediate states (Cayron et al. 2010). This proposal was strongly criticized (Bhadeshia 2011; also see the rebuttal by Cayron et al. 2011).

9.5.2.3 Morphology of Martensite; Plate Martensite and Lath Martensite

The morphology of interstitial iron-based martensites as function of interstitial content is shown for iron–nitrogen martensites in Fig. 9.28a–e (light optical micrographs of etched cross sections; for similar microstructures observed for iron-carbon martensites, see Krauss and Marder 1971) In particular for higher interstitial contents (0.9 wt% and above) the martensite grains appear as plates in the shape of a lens (lenticular plate), that spans, for the initial plates, the whole austenite-grain diameter (see Fig. 9.29a). Continued martensite formation induces plates which traverse the distances between the initial plates, etc. As a result a microstructure is obtained that is characterized by product martensite plates of decreasing size within a single-parent austenite grain (see Fig. 9.29b and cf. Fig. 9.28e).

The lattice-invariant deformation (see Sect. 9.5.2.2) for plate (high interstitial) iron-based martensite appears to be realized dominantly by twinning along $\{112\}_{\alpha'}$ planes (a frequently occurring twinning plane for b.c.c. metals is $\{211\}$; cf. Sect. 5.3). The resulting microstructure can be complex. The bright-field transmission electron micrograph (cf. Sect. 6.8.3) shown in Fig. 9.30a shows that the $\{112\}_{\alpha'}$ twinned parts have a width of a few tens nm and less. Because the parent austenite phase could be identified as well in the TEM analysis discussed here, the untwinned part of the

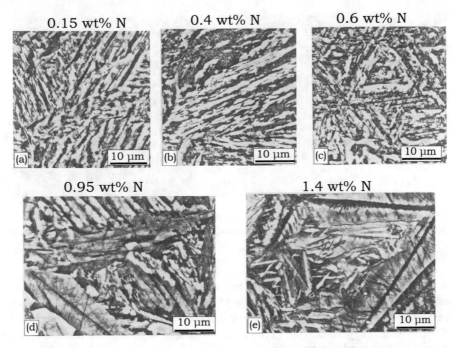

Fig. 9.28 Morphology of iron-nitrogen martensites as a function of nitrogen content: from lath to plate martensite upon increasing nitrogen content (light optical, phase-contrast (cf. Sect. 6.6.3) micrographs of etched cross sections; taken from E. J. Mittemeijer, M. van Rooyen, I. Wierszyllowski, H. C. F. Rozendaal and P. F. Colijn, Zeitschrift für Metallkunde, 74 (1983), 473–483

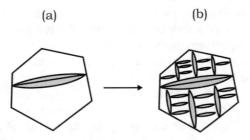

Fig. 9.29 a Schematic depiction of the formation of a lenticular martensite plate within an austenite grain. **b** Continued martensitic transformation leads to formation of plates which traverse the distances between earlier formed plates

martensite (exhibiting an orientation relationship with the parent austenite close to the NW type; cf. Sect. 9.5.2.2) could be distinguished from the martensite in twin orientation. It could be demonstrated that in the twinned part of the martensite planar defects along $\{110\}_{\alpha'_T}$ occur, where α'_T denotes twinned martensite (Fig. 9.30b). These planar defects can be considered as the result of twinning along $\{110\}_{\alpha'}$ planes

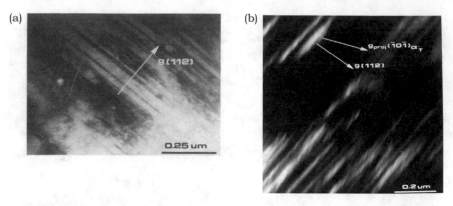

Fig. 9.30 a Transmission electron micrographs of twins on {112} planes in plate martensite in an Fe-1.5wt% N alloy. **b** Within the twinned martensite, also twins on {110} planes occur (taken from van Gent et al. 1985)

of martensite (note that {110} planes are also possible twinning planes in b.c.c. metals). The occurrence of such twinning in the already {112} type twinned part of martensite as well (this type of twinning was also observed for the untwinned part of the martensite) suggests that the {110} type of twinning happens *after* the martensite formation, apparently in order to comply with the state of stress imposed by the surrounding misfitting austenite (van Gent et al. 1985). In contrast with the {112} type twinning *during* the martensite formation, the {110} type twinning implies loss of lattice correspondence (cf. Sect. 9.5.2.2) with the parent austenite.

A characteristic feature of a martensite plate is the often observed "midrib" (see the line drawn between the two lens edges in the cross section of a martensite plate in Fig. 9.29a. A prominent example is shown in Fig. 9.31; see also Fig. 9.28d, e. The "midrib", more or less in the middle of the martensite plate/grain, is usually considered to be the starting plane for the formation of a martensite grain. The midrib region can have a modified fine structure, as compared to the other, later (less "old") parts of the martensite plate/grain: the twin density can be highest near the midrib and also the orientation of the midrib can differ slightly from its surroundings, as revealed by bending of the twin planes in the midrib region (Fig. 9.32). These effects can be the result of the distortion induced by stresses due to the misfit experienced by the martensite plate growing into the austenite matrix in the beginning stage of martensite-plate growth. Etching may then be particularly pronounced in the midrib region: see the broad "ragged" appearance of the midrib in the phase-contrast micrographs shown in Figs. 9.28d, e and 9.31 (phase-contrast microscopy is sensitive to height differences in the etched cross section; cf. Sect. 6.6.3).

The habit plane, i.e. the plane of the crystalline austenite, parent phase on which the crystalline martensite, product phase starts to form, does not necessarily provide the boundary, interfacial plane of the austenite and martensite phases in the system. This is evidently the case if the midrib represents the oldest part of the martensite plate. Then the midrib "plane" can be taken as the habit plane for the austenite → martensite

Fig. 9.31 Midrib of
iron–nitrogen plate
martensite (1.1 wt% N)
revealed by light optical,
phase-contrast (cf.
Sect. 6.6.3) microscopy
(taken from E. J.
Mittemeijer, M. van Rooyen,
I. Wierszyllowski, H. C. F.
Rozendaal and P. F. Colijn:
Zeitschrift für Metallkunde,
74 (1983), 473-483)

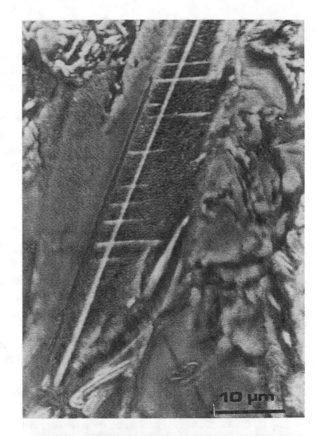

Fig. 9.32 Bending of twin
plates in the midrib region of
iron-nitrogen plate
martensite (1.5 wt% N); the
dashed line indicates the
course of the midrib (taken
from van Gent et al. 1985)

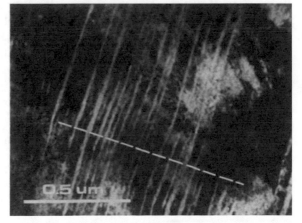

phase transformation. For high interstitial iron-based martensites often $\{259\}_\gamma$-type habit planes occur,[21] in agreement with the theoretical predictions according to the theory touched upon in Sect. 9.5.2.2. However, it has also been observed that the *interfacial plane* between austenite and martensite can be the ($\{259\}_\gamma$-type) habit plane (see van Gent et al. 1985), implying occurrence of a martensite plate without a "midrib".

As compared to the high interstitial content (>0.9 wt%), plate martensite, the low interstitial content (<0.6 wt%), lath martensite has a much finer microstructure: see Fig. 9.28a–c and compare with Fig. 9.28d, e. Packets of the laths occur, in a single, original austenite grain, with the laths of a packet oriented in more or less the same direction. The width of many laths is below the resolution limit of light optical microscopy (widths smaller than 0.2 μm cf. Sect. 6.5). The habit plane of lath martensite is close to $\{111\}_\gamma$; it has been reported that the habit plane would be $\{557\}_\gamma$ (irrational as well, with 12 variants; cf. footnote 21). Because three variants of $\{557\}_\gamma$ occur close to one of the four variants of $\{111\}_\gamma$, the microstructure of lath martensite appears more "regular" than plate martensite (cf. above discussion and compare Fig. 9.28a–c with Fig. 9.28d, e; cf. Marder and Krauss 1969).

The lattice-invariant deformation (see Sect. 9.5.2.2) for lath (low interstitial) iron-based martensite appears to be realized dominantly by slip (cf. Fig. 9.26c). Hence, lath martensite has a very high dislocation density (as high as $10^{16}/m^2$; which can be compared with the dislocation density of a cold-worked metal, which can be as high as $5.10^{15}/m^2$; cf. Sect. 5.2.3). It should be noted that a statement, as that the lattice-invariant deformation of lath martensite is due to slip by dislocation glide, is crude. Twins are observed as well in lath martensite, and, reversely, dislocations are observed too in, dominantly twinned, plate martensite.

Slip by dislocation glide and shear by twinning are deformation modes of metals characteristic of relatively high and relatively low temperatures of deformation. This already suggests that lath martensite forms at relatively high temperature, whereas plate martensite forms at relatively low temperature (see next section).

Not all austenite can usually be transformed in martensite. So plate and lath martensitic microstructures invariably contain some so-called *retained austenite* (see next section).

9.5.2.4 Energetics of Martensite Formation; Retained Austenite

It has been observed upon cooling an austenitic alloy, that, once nucleated at an austenite-grain boundary, martensite grows into an austenite grain, often but not always with a very high rate (see the last paragraph of this section). As a result

---

[21] Because the habit plane is not given by low number Miller indices one also speaks of an "irrational" habit plane. The many orientations of the martensite plates in a single, original austenite grain (see Figs. 9.29b and 9.28d, e), reflect the many possible variants of the irrational habit plane. A $\{hkl\}$ plane for which h, k and l are different has 24 variants. So the many occurring orientations of plate martensite grains in a single, original austenite grain can all comply with the occurrence of a single type of habit plane for the system considered.

the volume fraction of martensite formed at a certain temperature is constant: it is realized virtually at the moment the temperature considered has been reached upon cooling. Only by further cooling, the amount of martensite can be increased.

The amount of energy involved in the transformation austenite $(\gamma) \rightarrow$ martensite (α') can generally be written as (cf. Eq. (9.2)):

$$\Delta G_{tot}^{\gamma \rightarrow \alpha'} = \Delta G_{chem}^{\gamma \rightarrow \alpha'} + \Delta G_{int}^{\gamma \rightarrow \alpha'} + \Delta G_{def}^{\gamma \rightarrow \alpha'} \qquad (9.18)$$

With a view to the discussion in Sect. 9.5.2.2, it may come as no surprise that the transformation-opposing deformation-energy contribution is much larger than the transformation-opposing interfacial energy contribution, in the present case (a similar remark was made with respect to the massive transformation discussed in Sect. 9.5.1). The amount of undercooling, required for the martensite transformation to initiate, is indicated as the temperature difference $T_0 - T_{M_S}$, where T_0 represents the temperature where the chemical Gibbs energies of the austenite phase and the martensite phase, of the same composition, are equal (cf. Sect. 9.5.1)[22] and T_{M_S} denotes the temperature where, upon cooling from the austenite-phase field, the martensite formation starts. At the M_S temperature the chemical Gibbs energy difference driving the reaction has become equal to the transformation-opposing deformation energy induced by the transformation. The chemical Gibbs energy difference driving the reaction at the M_s temperature can be given as (cf. Eq. (9.6)):

$$\Delta G_{chem}^{\gamma \rightarrow \alpha'} (T_{M_S}) = \Delta H_{chem}^{\gamma \rightarrow \alpha'} (T_0) \cdot \frac{(T_0 - T_{M_S})}{T_0} \qquad (9.19)$$

It has been found that the M_S temperature of interstitial iron-based alloys decreases with increasing interstitial content (larger than 1 at.%) in approximately the same way as the T_0 temperature (Fig. 9.33). Then it follows from already the crude approximation provided by Eq. (9.19) (cf. the derivation of Eq. (9.6)) that the chemical Gibbs energy driving the reaction at the start of martensite formation is about constant as function of interstitial content (which is of the order 1 kJ/mol).

The chemical driving force per unit transformed, $\Delta G_{chem}^{\gamma \rightarrow \alpha'}$, primarily depends on temperature and not on the progress of transformation, because the martensitic transformation occurs partitionless (i.e. product and parent phases have the same composition). The deformation energy per unit transformed, $\Delta G_{def}^{\gamma \rightarrow \alpha'}$, however, will in general depend on the degree of transformation and not primarily on the temperature: the amount of strain build-up in the specimen increases during the transformation; per unit of martensite to be formed more deformation energy has to be

---

[22] The chemical Gibbs energy of interstitial iron-based martensite can be described as the sum of (i) the Gibbs energy of the supersaturated solid solution of interstitials in ferrite and (ii) the (strain interaction) energy associated with the alignment of the individual tetragonal strain fields, surrounding the individual interstitial atoms at their octahedral interstices in ferrite, upon occupation of only one set of the three sets of octahedral interstices (the "Zener ordering" energy; cf. Sect. 9.5.2.1), thereby forming martensite.

Fig. 9.33
Martensite-start temperature
(M_S; compiled data taken
from A.R. Marder and G.
Krauss, Transactions ASM,
60 (1967), 651-660) and
martensite-finish
temperature (M_F; taken from
Porter and Easterling 1992)
as function of carbon content
for Fe-C alloys

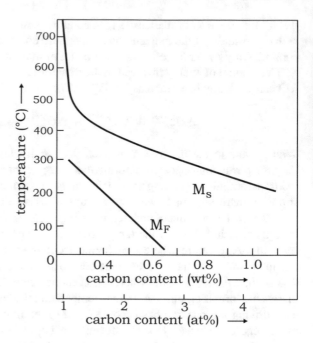

introduced into the specimen for increasing fraction transformed. Thus, at a certain temperature, the extent of martensite formation can be considered as determined by the amount of deformation energy absorbed: at the moment that, per unit martensite formed, the amount of (further) deformation energy induced becomes equal to the chemical driving force acting at the temperature concerned, no further transformation can take place. This discussion parallels the one given for the massive transformation (cf. Sect. 9.5.1). Further cooling enhances the chemical driving force and thus more martensite can be formed, etc. The M_F temperature (the subscript "F" derives from "finish") indicates the stage of completed martensite formation (cf. Fig. 9.33). Even at this stage some remaining austenite may occur; the highly strained, deformed state obstructs further transformation.

The austenite remaining in a partly transformed matrix is denoted "retained austenite". The amount of retained austenite at room temperature increases for increasing interstitial content (Fig. 9.34). Whereas for plate martensite (i.e. above about 0.9 wt% interstitial) retained austenite can be discerned between the (finest) martensite plates in a light optical micrograph of the microstructure (cf. Fig. 9.28e), the small amount of retained austenite in lath martensite (i.e. below about 0.6 wt% interstitial) occurs as films between the laths and can only be observed employing transmission electron microscopy (Thomas 1978).

Also the energetics of the martensitic transformation are still subject of recent research, revealing the complex nature of the martensitic transformation (cf. the end of Sect. 9.5.2.2). Fe-Ni alloys can be subjected to a martensitic transformation applying a relatively low cooling rate (of the order 0.1 K/s, as compared to cooling

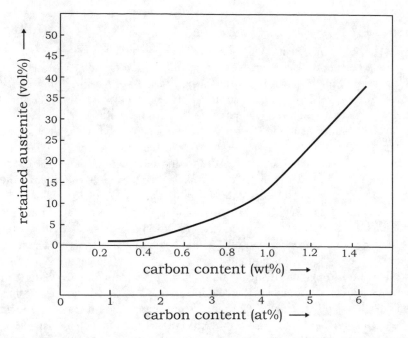

Fig. 9.34 Amount of retained austenite (determined by X-ray diffraction analysis) as a function of carbon content for different Fe-C alloys, water quenched from austenite to room temperature (taken from A. R. Marder and G. Krauss, Transactions ASM, 60 (1967), 651–660)

rates for classical martensitic steels which are of the order 10^4 K/s). This allows to directly measure the rate of formation of martensite during cooling. For the formation of lath martensite in such Fe–Ni alloys it was observed that a *regular, periodic* series of transformation-rate maxima developed upon cooling from the austenite-phase field at constant cooling rate. This behaviour was ascribed to the formation of lath martensite blocks in the austenite grains *in a concerted manner*. After formation and during growth of a first lath martensite block in an austenite grain, elastic and plastic deformation energy develop (in the parent austenite grain), that counteracts continued growth of the block concerned. The reason that subsequent block formation occurs more or less simultaneously in all austenite grains (thereby a transformation rate maximum for the macroscopic specimen occurs) is caused by the operation of a thermally activated relaxation process of deformation energy, that runs fast as compared to the supply of chemical driving force, which is governed by the cooling rate. Thereby, same energetic conditions for the formation of new blocks in all austenite grains are realized (Löwy et al. 2015, 2016).

Intermezzo: Shape-Memory Alloys

Upon heating a martensitic alloy a reverse, martensite $\rightarrow$ austenite transformation can occur, provided no preceding decomposition of the martensite takes place as in the case of iron-based interstitial martensites (see the "*Intermezzo: Tempering of iron-based interstitial martensitic specimens*" below). Then, upon heating of martensite, an A_S temperature (start of austenite formation) can be indicated, analogous to the M_S temperature discussed above and as observed upon quenching from the austenite-phase field. The differences $T_0 - T_{M_S}$ and $T_{A_S} - T_0$ can be conceived as a measure for the deformation energy associated with the austenite $\rightarrow$ martensite transformation and the martensite $\rightarrow$ austenite transformation, respectively (see the discussion below Eq. (9.18)). For certain alloys, exhibiting austenite $\rightarrow$ martensite and martensite $\rightarrow$ austenite transformations, as for example for ordered Ni–Ti, Ni–Al and Cu–Zn–Al alloys, the differences of T_{M_S} and T_0 and of T_{A_S} and T_0 are small (little "hysteresis"), and it can be concluded that in these cases the deformation energy associated with the transformations is small, and thus, the deformation can be entirely of elastic nature; the austenite/martensite interface is "elastically coherent"; thus one speaks of fully reversible, thermo-elastic martensite formation: the critical shear stress (cf. Sect. 12.11) for irreversible, plastic deformation by dislocation movement is relatively very high (as holds for ordered alloys). The mode of transformation thus is based on shear by lattice (crystal-structure; cf. footnote 18 in Sect. 9.5.2.2) invariant twinning (cf. Sect. 9.5.2.2). For such alloys a remarkable deformation behaviour can occur: the shape-memory effect, which implies that a macroscopically deformed specimen returns to its original shape when heated.

If the specimen of an alloy as discussed above is deformed in its (low temperature) martensitic state, then by heating up to a temperature above the A_S temperature, the original shape of the specimen (i.e. before the deformation) is restored (for full shape restoration the deformation in the martensitic state should generally not exceed, say, 5–10%). The effect is understood by recognizing that the deformation in the martensite takes place by the (reversible) movement of twin planes, as pertaining to the preceding martensite formation itself (see above discussion): the deformation can be realized by growth of one of the twin orientations at the cost of the other twin orientation in a twin. By heating above the A_S temperature the need for twinning (originally as required for realization of the martensite upon quenching from the austenite-phase field, as discussed above) is removed and the initial atomic arrangement is reestablished. This could be conceived as that the deformed martensite "remembers" its undeformed condition and restores it by the movement of the twin planes upon heating such that the twins readjust to the thicknesses of the twinned regions as required for the "lattice-invariant" deformation upon martensite formation, and, subsequently, the formation of austenite from the parent martensite takes place under the constraint of shape conservation/retrieval. The ordering in the martensite- and

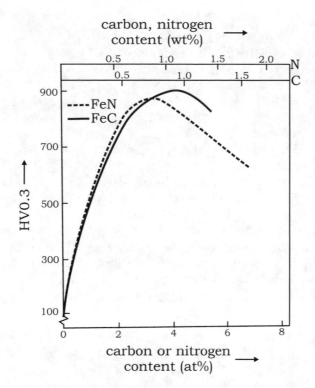

Fig. 9.35 Hardness of martensitic iron-carbon and iron-nitrogen alloys, obtained by quenching the specimens from the austenite-phase field in brine and liquid nitrogen, subsequently, as a function of interstitial content. The hardness was determined according to the Vickers hardness testing technique applying a load of 300g, which explains usage of the symbol HV0.3 along the ordinate (cf. Sect. 12.13) (taken from E. J. Mittemeijer, M. van Rooyen, I. Wierszyllowski, H. C. F. Rozendaal and P. F. Colijn, Zeitschrift für Metallkunde, 74 (1983), 473–483)

the austenite crystal structures reduces the number of possible orientation variants upon transformation, thereby providing support for reverting to the original austenite orientation upon the reverse martensite → austenite transformation (Wayman and Shimizu 1972).

For another, magnetism-induced cause for a shape-memory effect, see Sect. 3.5.3.

Imposing a barrier to the restoration of the original shape, as upon heating from the deformed martensite state for alloys as considered above, a force becomes available that can do work. On that basis, *actuators* can be devised. Many applications of shape-memory alloys are found in the aerospace and medical technologies.

Intermezzo: The Hardness of Iron-Based Interstitial Martensitic Specimens

Iron-based interstitial martensites can be very hard (see Fig. 9.35). This high hardness is due to at least three contributions. The most important effect is the solid-solution strengthening by the dissolved interstitial atoms. Each interstitial atom is surrounded by its tetragonal distortion field (cf. Sect. 9.5.2.1) making unconstrained dislocation glide impossible (cf. Sect. 12.14.3: the effect of solid-solution strengthening).[23] Further the fine grain size, with the fineness of the microstructure being enhanced by the high twin density (cf. Sect. 12.14.2: the effect of grain size on mechanical strength) and the high dislocation density (cf. Sect. 12.14.1: the effect of strain hardening on mechanical strength) hinder dislocation glide in martensite as well.

The decrease of hardness of the martensitic structure observed in Fig. 9.35 for the interstitial concentration increasing beyond about 0.9 wt% (=4.1 at.%) C and about 0.8 wt% (=3.2 at.%) N for Fe-C and Fe-N martensitic specimens, respectively, is ascribed to the increase of the amount of, relatively soft, retained austenite with increasing interstitial content (cf. Fig. 9.34).

Intermezzo: Tempering of Iron-Based Interstitial Martensitic Specimens

The high hardness of virginal martensite makes it also very brittle, rendering it as such to a usually impracticable material. Therefore, the carbon martensitic specimen is annealed at a moderate temperature, leading eventually to the precipitation of carbides, which causes some loss of hardness (but considerable hardness is preserved) but increases the ductility very considerably (In this context, see footnote 12 in Sect. 12.9). This so-called "tempering" has led to the myriad of applications, world-wide, of (carbon) steels as structural materials.[24] In particular by the variation of quenching mode and tempering

---

[23] The substitutional replacement of an iron atom by a solute atom, causes an isotropic distortion and therefore such substitutional solute atoms can only interact with the (relatively minor) hydrostatic component of the strain field of a dislocation. A (strongly) anisotropic, tetragonal distortion is associated with an interstitial atom at an *irregular* octahedral interstice in martensite (b.c.t.) and ferrite (b.c.c.), and therefore a strong interaction with the (dominant) shear components of a dislocation-strain field occurs. This explains why (ferrite and) martensite exhibit strong solid solution strengthening by interstitial solutes. The solid solution strengthening by interstitials in austenite (f.c.c.) is much less outspoken, as the interstitial atom in austenite is positioned at a *regular* octahedral interstice and thus associated with isotropic distortion (see also Sect. 12.9.2).

[24] The combination of "quenching" and "tempering" is described in the German and Dutch languages by a single word: "vergüten" and "veredelen", respectively, which means: "making noble/ennoble", thereby expressing the enormous importance of the profitable material properties for mankind achieved by this combined process.

temperature, an extremely wide property range, corresponding to a similarly wide microstructural range, is accessible to the engineer.

Iron-based interstitial martensites are thermodynamically unstable materials. Then, as the diffusivity of the interstitials is relatively high (cf. Sect. 8.5.3), already at room temperature ageing processes occur: local enrichments of interstitial atoms develop in both carbon and nitrogen martensites. These enrichments, (pre)precipitates (see further), lead to an increase of hardness, as the occurrence of regions of high interstitial content and low interstitial content in the martensite matrix is associated with the build up of coherency strains which can hinder dislocation glide (cf. Sect. 12.14.4 on precipitation/dispersion strengthening). This hardness increase, for ageing at room temperature, provides a simply measurable, direct indication for the occurrence of such processes: significant hardness increase occurs upon ageing at room temperature for times of the order of an hour: see Fig. 9.36: the hardness increases to a maximum during the first day of ageing at room temperature, which maximum hardness is about 50–80 HV larger than the as-quenched hardness (cf. Sect. 12.13 for meaning of the symbol "HV").

The carbon enrichments can be conceived as "clusters" of carbon atoms containing different amounts of carbon interstitials distributed randomly on one type of the three types of octahedral interstices of the bct iron-(sub)crystal structure. The nitrogen enrichments, on the other hand, exhibit an ordered arrangement of nitrogen atoms, on the one type (of three types) of octahedral interstitial sites, leading to the so-called α''-$Fe_{16}N_2$ structure (see van Genderen et al. 1993).[25] Thus, the first stage of ageing carbon and nitrogen martensites involves the development of (carbon) clusters (cf. Sect. 9.4.1) and (coherent α''-nitride) precipitates, respectively. This difference in the behaviour of carbon and nitrogen interstitials is not well understood. It has been suggested that in martensite the size of a nitrogen atom is larger than that of a carbon atom (Liu Cheng et al. 1990; cf. Sect. 9.5.2.1). Then the release of elastic strain energy upon ordering of the interstitials in a local interstitial enrichment, as in an α''-$Fe_{16}N_2$/$Fe_{16}C_2$ structure, would be larger for nitrogen martensite

---

[25] Note that the "Zener ordering", discussed in Sect. 9.5.2.1 as characteristic for interstitial iron-based martensite, involves that the interstitials are located at only one of the three sets of octahedral interstices, but on this one set of octahedral interstitial sites the interstitial atoms are randomly distributed. The α''-$Fe_{16}N_2$ structure implies (only) that (additionally) the interstitial (nitrogen) atoms on this one set of octahedral interstices adopt a long-range order. Nitrogen enrichments developing in the martensite matrix exhibiting such α''-type ordering are nothing else than coherent α''-$Fe_{16}N_2$ nitride particles.

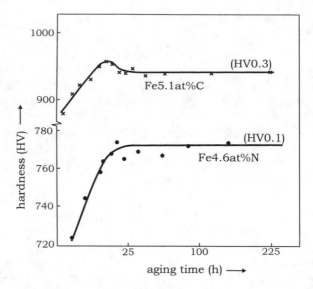

Fig. 9.36 Hardness of a martensitic Fe-5.1at%C alloy and a martensitic Fe-4.6at%N alloy (initially quenched into brine and subsequently into liquid nitrogen) as function of aging time at room temperature (taken from Liu Cheng et al. 1988 and Liu Cheng and E.J. Mittemeijer 1990)

than for carbon martensite. Thus, ordering of the interstitials in an interstitial enrichment in martensite according to the α''-$Fe_{16}N_2$-type structure would be preferred for nitrogen martensites (cf. van Genderen et al. 1997).

The coherency strains developing in the martensite crystal structure as a result of the development of regions of high interstitial content and low interstitial content, by carbon clustering/α''-nitride precipitation, are responsible for the increase of hardness (see above). Continued ageing at room temperature, beyond the time of occurrence of the hardness maximum (see Fig. 9.36), leads to a small decrease of hardness for the carbon martensite (which may reflect coarsening/onset of further transformation of the relatively unstable carbon clusters), whereas the hardness of the nitrogen martensite remains constant (suggesting the relatively large (meta)stability of the α''-nitride regions).

Hardness changes observed as function of annealing temperature are shown for both carbon martensite and nitrogen martensite in Fig. 9.37 (a fresh specimen was annealed for one hour at each temperature). The hardness maximum observed for the carbon martensite at about 100 °C (hardness increase of about 150 HV as compared to the virginal martensite) is due to the precipitation of the so-called transition carbide, which has been designated as ε-$Fe_{2.4}C$ or η-Fe_2C carbide (the precise crystal structure has been discussed controversially in the literature). The hardness maximum observed for the nitrogen martensite at about 75 °C (hardness increase of about 50 HV as compared to the virginal

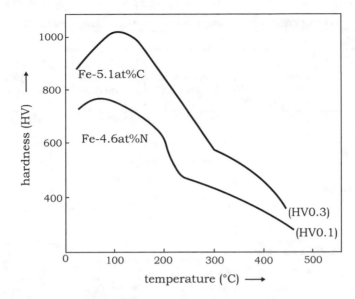

Fig. 9.37 Hardness of a martensitic Fe-5.1 at %C alloy and a martensitic Fe- 4.6 at %N alloy (initially quenched into brine and subsequently into liquid nitrogen) as function of temperature (ageing time at each temperature: 1 h) (taken from Liu Cheng et al. 1988 and Liu Cheng and E.J. Mittemeijer 1990)

martensite) is equal to the hardness maximum observed upon ageing at room temperature (cf. Fig. 9.36) and therefore is ascribed to the development of coherent α'' nitrides. Only upon tempering at temperatures above 75–100 °C does decrease of mechanical strength, as exhibited by decrease of hardness, and increase of ductility occur.

Increase of the temperature beyond the temperature of the hardness maximum up till about 200 °C leads to decrease of hardness because of coarsening and the coherent-incoherent transition of the transition ε/η carbide and the transition α'' nitride.

The precipitation of the transition ε/η carbide in carbon martensite (the associated loss of dissolved carbon in the originally martensite matrix transforms the matrix gradually into ferrite) is traditionally called *the first stage of tempering*.

The next stages in the tempering of carbon martensitic specimens are the decomposition of the retained austenite into ferrite and cementite, in the temperature range 240–320 °C, also called *the second stage of tempering*, which overlaps partly with the subsequent replacement (dissolution) of the ε-$Fe_{2.4}C/\eta$-Fe_2C carbide particles by (precipitation of) cementite-Fe_3C particles in the temperature range 260–350 °C, also called *the third stage of tempering*.

Table 9.1 Tempering stages of iron–carbon martensite and iron–nitrogen martensite

Tempering of carbon martensite

Tempering stage	1	2	3	4
Process	Carbon segregation and clustering	Precipitation of transition carbide	Transformation of retained austenite	Precipitation of cementite
Temperature range (°C)	<100	80–200	240–320	260–350
Volume change	–	– –	+	– –
Enthalpy change	–	– –	– –	– –
Hardness change	+	+ → –		+
Effective activation energy (kJ/mol)	~80	~120	~130 (5.05 at %C)	~200

Tempering of nitrogen martensite

Tempering stage	1	2	3	4
Process	Nitrogen segregation and ordering	Precipitation of α'' nitride	Precipitation of γ' nitride	Decomposition of retained austenite
Temperature range (°C)	<100	100-200	220-300	240–350
Volume change	–	+	– –	+
Enthalpy change	–	– –	+	–
Hardness change	+	–	– –	
Effective activation energy (kJ/mol)	~80	~115	~195	~90 (9.5 at %N)

The temperature ranges, and the corresponding changes in volume, enthalpy and hardness, as well as the effective activation energies, have been indicated. Note that a negative change in enthalpy, H, means that heat, Q, is released; $\Delta H = H_{end} - H_{begin}$ (see footnote 2) and $Q = - \Delta H$ (cf. Liu Cheng et al. 1988 and Liu Cheng and E.J. Mittemeijer 1990)

> Adopting the nomenclature introduced above, the second and third stages of tempering are reversed for nitrogen martensite: The replacement (dissolution) of the α''-$Fe_{16}N_2$ nitride particles by (precipitation) of γ'-Fe_4N nitride particles takes place in the temperature range 220–290 °C, which overlaps partly with the subsequent decomposition of retained austenite into ferrite and γ' nitride in the temperature range 240–350 °C.
>
> The temperature ranges given in the above paragraphs depend on the interstitial content of the martensitic specimens (here 1.13 wt% C = 5.1 at.% C and 1.19 wt% N = 4.6 at.% N). The various stages of tempering for (pure) carbon and (pure) nitrogen martensitic specimens have been summarized in Table 9.1.

As follows from Fig. 9.37, the decrease of hardness upon increasing temperature after completed cementite precipitation in carbon martensites (in Fig. 9.37: above 350 °C) is pronounced. This softening, due to the coarsening of the cementite particles, can be counteracted by adding alloying elements to the steel which have an affinity to carbon: they form fine alloy carbides, at elevated temperatures (say, above 500 °C), which increase the hardness. This process has been called "secondary hardening" and has been designated as *the fourth stage of tempering*.

References

General

J.W. Christian, *The Theory of Transformations in Metals and Alloys, Part I, Equilibrium and General Kinetic Theory*, 2nd edn. (Pergamon Press, Oxford, 1975).

G. Kostorz (ed.), *Phase Transformations in Materials* (Wiley-VCH, Weinheim, Germany, 2001)

G. Krauss, *Steels; Heat Treatment and Processing Principles*, (ASM, Materials Park Ohio, 1995).

I. Manna, S.K. Pabi, W. Gust, Discontinuous reactions in solids. Int. Mater. Rev. **46**, 53–91 (2001)

Z. Nishiyama, *Martensitic Transformation* (Academic Press, New York, 1978).

D.A. Porter, K.E. Easterling, *Phase Transformations in Metals and Alloys*, 2nd edn. (Chapman & Hall, London, 1992).

C.M. Wayman, *Introduction to the Crystallography of Martensitic Transformations* (MacMillan, New York, 1964).

D.B. Williams, E.P. Butler, Grain boundary discontinuous precipitation reactions. Int. Mater. Rev. **26**, 153–183 (1981)

Specific

H.K.D.H. Bhadeshia, Carbon-Carbon interactions in iron. J. Mater. Sci. **39**, 3949–3955 (2004)

H.K.D.H. Bhadeshia, Comments on "The mechanisms of the fcc-bcc martensitic transformation revealed by pole figures." Scripta Mater. **64**, 101–102 (2011)

M.H. Biglari, C.M. Brakman, E.J. Mittemeijer, S. van der Zwaag, The kinetics of the internal nitriding of Fe-2at.% Al alloy. Metall. Mater. Trans. A, **26A**, 765–776 (1995)

C. Bos, F. Sommer, E.J. Mittemeijer, An atomistic analysis of the interface mobility in a massive transformation. Acta Mater. **53**, 5333–5341 (2005)

J.W. Cahn, The kinetics of cellular segregation reactions. Acta Metall. **7**, 18–28 (1959)

C. Cayron, F. Barcelo, Y. de Carlan, The mechanisms of the fcc-bcc martensitic transformation revealed by pole figures. Acta Mater. **58**, 1395–1402 (2010)

C. Cayron, F. Barcelo, Y. de Carlan, Reply to "comments on 'The mechanisms of the fcc-bcc martensitic transformation revealed by pole figures.'" Scripta Mater. **64**, 103–106 (2011)

M.J. van Genderen, A.J. Böttger, R.J. Cernik, E.J. Mittemeijer, Early stages of decomposition in iron-carbon and iron-nitrogen martensites: diffraction analysis using synchrotron radiation. Metall. Trans. A **24A**, 1965–1973 (1993)

M.J. van Genderen, A. Böttger, E.J. Mittemeijer, Formation of α" iron nitride in FeN martensite: nitrogen vacancies, iron-atom displacements, and misfit-strain energy. Metall. Mater. Trans. **28A**, 63–77 (1997)

A. van Gent, F.C. van Doorn, E.J. Mittemeijer, Crystallography and tempering behavior of iron-nitrogen martensite. Metall. Trans. A **16A**, 1371–1384 (1985)

H. Göhring, O. Fabrichnaya, A. Leineweber, E. J. Mittemeijer, Thermodynamics of the Fe-N and Fe-N-C systems: the Fe-N and Fe-N-C phase diagrams revisited. Metall. Mater. Trans. **47A**, 6173–6186 (2016b)

H. Göhring, A. Leineweber, E.J. Mittemeijer, A thermodynamic model for non-stoichiometric cementite; the Fe–C phase diagram. Calphad **52**, 38–46 (2016a)

A. Guinier, Structure of age-hardened aluminium-copper alloys. Nature **142**, 569–570 (1938)

G. Krauss, A.R. Marder, The Morphology of martensite in iron alloys. Metall. Trans. **2**, 2343–2357 (1971)

Y.C. Liu, F. Sommer, E.J. Mittemeijer, Abnormal austenite-ferrite transformation behaviour of pure iron. Phil. Mag. **84**, 1853–1876 (2004)

Y.C. Liu, F. Sommer, E.J. Mittemeijer, The austenite-ferrite transformation of ultralow-carbon Fe-C alloy; transition from diffusion- to interface-controlled growth. Acta Materialia **54**, 3383–3393 (2006)

Y.C. Liu, F. Sommer, E.J. Mittemeijer, Critical temperature for massive transformation in ultra-low-carbon Fe-C alloys. Int. J. Mater. Res. **99**, 925–932 (2008)

S. Loewy, B. Rheingans, S.R. Meka, E.J. Mittemeijer, Modulated martensite-formation behavior in Fe–Ni based alloys; athermal and thermally activated mechanisms. J. Mater. Res. **30**, 2101–2107 (2015)

S. Loewy, B. Rheingans, E.J. Mittemeijer, Transformation-rate maxima during lath martensite formation: plastic versus elastic shape strain accommodation. Philos. Mag. **96**, 1420–1436 (2016)

G.A. Lopez, P. Zieba, W. Gust, E.J. Mittemeijer, Discontinuous precipitation in a Cu-4.5at.%In Alloy. Mater. Sci. Technol. **19**, 1539–1545 (2003)

G.A. López, E.J. Mittemeijer, B.B. Straumal, Grain boundary wetting by a solid phase; microstructural development in a Zn-5 wt% Al Alloy. Acta Mater. **52**, 4537–4545 (2004)

Liu Cheng, C.M. Brakman, B.M. Korevaar, E.J. Mittemeijer, Tempering of iron-carbon martensite; dilatometric and calorimetric analysis. Metall. Trans. A, **19A**, 2415–2426 (1988)

Liu Cheng, E.J. Mittemeijer, Tempering of iron-nitrogen martensite; dilatometric and calorimetric analysis. Metall. Trans. A **21A**, 13–26 (1990)

Liu Cheng, A. Böttger, Th.H. de Keijser, E.J. Mittemeijer, Lattice parameters of iron-carbon and iron-nitrogen martensites and austenites. Scripta Metall. et Mater. **24**, 509–514 (1990)

A.R. Marder, G. Krauss, The formation of low-carbon martensite in Fe-C alloys. Trans. ASM **62**, 957–964 (1969)

E.J. Mittemeijer, J.T. Slycke, Chemical potentials and activities of nitrogen and carbon imposed by gaseous nitriding and carburising atmospheres. Surf. Eng. **12**, 152–162 (1996)

G.D. Preston, The diffraction of X-rays by age-hardening aluminium copper alloys. Proc. R. Soc. Lond. **167**, 526–538 (1938)

A. Parisi, M. Plapp, Stability of lamellar eutectic growth. Acta Mater. **56**, 1348–1357 (2008)

R.C. Pond, X. Ma, J.P. Hirth, Geometrical and physical models of martensitic transformations in ferrous alloys. J. Mater. Sci. **43**, 3881–3888 (2008)

M.P. Puls, J.S. Kirkaldy, The pearlite reaction. Metall. Trans. **3**, 2777–2796 (1972)

R.E. Schacherl, P.C.J. Graat, E.J. Mittemeijer, Gaseous nitriding of iron-chromium alloys. Z. Metallkde. **93**, 468–477 (2002)

J. Sietsma, S. van der Zwaag, A concise model for mixed-mode phase transformations in the solid state. Acta Mater. **52**, 4143–4152 (2004)

B.B. Straumal, G.A. Lopez, E.J. Mittemeijer, W. Gust, A.P. Zhilyaev, Grain boundary phase transitions in the Al-Mg system and their influence on high-strain rate superplasticity. Defect Diffus. Forum **216–217**, 307–312 (2003)

G. Thomas, Retained austenite and tempered martensite embrittlement. Metall. Trans. **9A**, 439–450 (1978)

C.M. Wayman, K. Shimizu, The shape memory ("Marmem") effect in alloys. Metal Sci. J. **6**, 175–183 (1972)

Chapter 10
Phase Transformations: Kinetics

The application of solid-state transformations to tune the microstructure of materials, in order to optimize specific material properties, requires availability of model descriptions of the time–temperature dependencies, i.e. the kinetics, of the phase transformations. The necessary models should not be in particular of atomistic nature, but pertain to mesoscopic and even macroscopic scales (also see the last section of this chapter, Sect. 10.16).

Of course, atomistic simulations can be very useful for the interpretation of the values of the kinetic parameters determined (e.g. regarding the understanding of the activation energy of the mobility of grain boundaries/product-parent interfaces, see Bos et al. (2005). Yet, this last remark leaves unimpeded that a great need exists to have at one's disposal mathematical tools, on a preferably firm, physical basis, that provide an, in any case verifiable, and hopefully reliable representation of the course of a phase transformation. For example, on that basis the austenite–ferrite phase transformation in a steel factory can be and is controlled. On the other hand, fundamental insight into the mechanisms of specific phase transformations can be acquired by the analysis of their characteristic kinetic parameters inherent to the model descriptions applied.

The above remarks serve to indicate the enormous practical and fundamental, scientific interest in the analysis of the kinetics of phase transformations. At some stage of his/her career every materials scientist has to deal with the analysis of phase-transformation kinetics. This is one area where ill-considered statements have been made and analyses have been applied inconsiderately, by many. The treatment presented in Sects. 10.2–10.15, derived largely from the reviews Mittemeijer (1992) and Liu et al. (2007), serves to provide an overview of useful and important approaches with clear indication of their limitations.

However, the first section of this Chapter on the kinetics of phase transformations is devoted to the concepts of time–transformation–temperature (TTT) diagrams and continuous-cooling-transformation (CTT) diagrams, which are used in practice to select alloy compositions and to determine specific heat treatments that lead to specific microstructures. Such diagrams have been determined for technologically

© Springer Nature Switzerland AG 2021
E. J. Mittemeijer, *Fundamentals of Materials Science*,
https://doi.org/10.1007/978-3-030-60056-3_10

applied materials, in particular steels, and are not usually used for a quantitative assessment of phase-transformation kinetics.

10.1 Time–Temperature–Transformation (TTT) Diagrams and Continuous-Cooling-Transformation (CCT) Diagrams

The progress of civilian phase transformations (cf. Sect. 9.3) can be expressed as a function of time at constant temperature (and at constant pressure) in so-called time–temperature–transformation diagrams (TTT diagrams). The TTT diagram, with an ordinate given by the temperature (linear scale) and an abscissa given by the time (logarithmic scale), presents the loci of points of same degree of *isothermal* transformation. An example is shown in Fig. 10.1 for a hypothetical, diffusional transformation, as $\alpha' \rightarrow \alpha + \beta$, $\gamma \rightarrow \alpha + \beta$, etc. The degree of transformation,

Fig. 10.1
Time–Temperature–
Transformation (TTT)
diagram (bottom part of the
figure) for a hypothetical,
diffusional transformation.
Two corresponding degrees
of transformation versus
(isothermal) annealing time
curves have been given as
well for the temperatures T_1
and T_2 ($T_1 > T_2$),
respectively (top part of the
figure); note that at the
higher temperature (T_1) the
transformation progresses
less fast than at the lower
temperature (T_2)

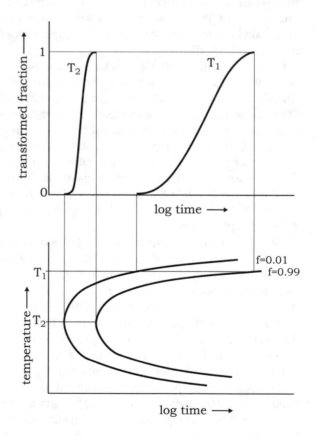

usually indicated by the symbol f, varies from 0 (0%) to 1 (100%) from start to finish of the transformation, respectively.

Such TTT diagrams can be determined by subjecting a supersaturated system to an anneal at a constant temperature for some time and then, e.g. by quenching the system to low (room) temperature, the degree of transformation is frozen in. The degree of transformation characteristic for the anneal at high temperature can be determined at this low (room) temperature by application of some method, e.g. quantitative metallography (i.e. contrast differences in light optical micrographs are used for determination of the relative amounts of phases present) or X-ray diffraction (i.e. the integrated intensities of phase specific reflections are used for determination of the relative amount of phases present), etc. Other methods are possible too, as dilatometry which can be applied at the isothermal annealing temperature: the change of length of a specimen is a measure for the degree of transformation (see also Sects. 10.2 and 10.13).

Suppose the reaction $\alpha' \rightarrow \alpha + \beta$ in the binary (A, B) system represents the diffusional decomposition of a supersaturated solid solution, α', which, according to the phase diagram can occur below a certain temperature T_1. For a very small undercooling, the number of nuclei produced can be small (see also Sect. 10.5) and the transformation rate is relatively slow in spite of a relatively fast diffusion. Increasing the undercooling enhances the driving force for the decomposition (cf. Sect. 9.2) and more nuclei are produced per unit of time promoting the transformation rate. However, upon continued increase of the undercooling the diffusion becomes such slow that the transformation rate decreases. As a result at some intermediate undercooling the optimal combination of nucleation rate and diffusional mobility occurs for yielding the fastest transformation. As a consequence, the locus of points of the same degree of isothermal transformation in the TTT diagram appears as a C-type curve, i.e. with a "nose" pointing in the negative time direction (see Fig. 10.1).[1]

Military transformations (cf. Sect. 9.3) can be represented by horizontal lines in the TTT diagram: the degree of transformation (practically) only depends on the temperature; is constant as function of time at constant temperature. Thus, for martensite formation, each horizontal line between the M_S and M_F temperatures (cf. Sect. 9.5.2.4) corresponds with a certain fraction of produced martensite. The TTT diagram for the Fe-C alloy of eutectoid composition (0.77 wt% C) is shown in Fig. 10.2a. Of course, the transformation from austenite to martensite requires imposition of a cooling (quenching) rate that is sufficiently high (cf. Sect. 9.5.2.1). The cooling (quenching) curve, $T(t)$, can be drawn in the TTT diagram (Fig. 10.2a).

---

[1] In case of a eutectoid reaction, $\gamma \rightarrow \alpha + \beta$, a "nose" in the TTT diagram occurs as well, which can be explained more or less similarly, as follows. The interlamellar, α/β, spacing becomes smaller for increasing undercooling (cf. Sect. 9.4.2), thereby initially more than counteracting the decrease of diffusional mobility upon increasing undercooling (i.e. decreasing temperature): the elemental redistribution at the transformation front requires less distance coverage by diffusion for smaller interlamellar spacing. Continued increase of the undercooling is associated with such pronounced decrease of the diffusional mobility that the transformation rate decreases pronouncedly. Hence, at intermediate undercooling an optimal combination of interlamellar spacing and diffusional mobility occurs that leads to the highest transformation rate.

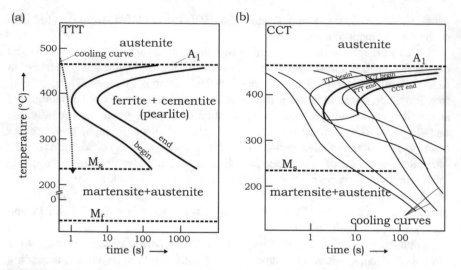

Fig. 10.2 a Time–Temperature–Transformation diagram (TTT) and **b** Continuous-Cooling-Transformation diagram (CCT) of a eutectoid steel. A_1: temperature corresponding to the lower stability limit of austenite (eutectoid composition), M_S: martensite start temperature, M_F: martensite finish temperature. In the TTT diagram, shown in (**a**), the dashed line corresponds to a cooling curve avoiding the nucleation of pearlite and leading to martensite formation. In the CCT diagram, shown in (**b**), thin lines represent experimentally determined cooling curves; the transformation lines of the corresponding TTT diagram shown in (**a**) have also been indicated with thin lines in (**b**) (taken from G. Krauss and J. F. Libsch in *Phase diagrams in Ceramic, Glass and Metal Technology*, A. M. Alper (Editor), Academic Press, New York, 1970; *Atlas of Time–Temperature Diagrams for Irons and Steels*, American Society for Metals, 1991).

Then, one is tempted to conclude that, in order to produce martensite from austenite in an initially completely austenitic specimen, this cooling curve should avoid the "nose" of the eutectoid, pearlite reaction. However, applying this approach it is tacitly assumed that the curves representing the $f(T, t)$ loci, determined for isothermal transformation, also hold for transformation during continuous cooling, which is not necessarily and certainly not generally the case (see what follows and Sects. 10.2 and 10.3).

Against the above background the so-called Continuous-Cooling-Transformation diagrams (CCT diagrams) have been developed. The CCT diagram is characterized by temperature and time axes as hold for the TTT diagram. However, as compared to the TTT diagram, the curves representing the $f(T(t))$ loci are shifted to longer times and lower temperatures in the CCT diagram: see Fig. 10.2b. This can be understood as follows. Upon continuous cooling, the undercooling is initially very small, smaller than during the isothermal annealing at larger undercooling. Therefore, the time needed to reach the same degree of transformation at a certain temperature during continuous cooling will be larger than for the isothermal transformation at that temperature. Similarly, for the same time of transformation, more (eventual) undercooling is required in case of the continuous cooling transformation, as compared to the isothermal transformation, because in the first part of the transformation-time range the undercooling in case of the continuous cooling experiment is smaller than in case of the isothermal transformation experiment. Cooling curves, $T(t)$, have been drawn in Fig. 10.2b. Note, that, whereas in TTT diagrams transformations run from $f = 0$ to $f = 1$ along horizontal lines from the left to the right, transformations in CCT diagrams run from $f = 0$ to $f = 1$ along the cooling curves, i.e. from "top left" to "bottom right" (see Fig. 10.2).

The above discussion implies that the minimal cooling rate, as determined in the TTT diagram such that the "nose" of the eutectoid, pearlite reaction is just "missed" (see above and Fig. 10.2a), in order that martensite is formed in an initially completely austenitic specimen, can involve that this estimate of the minimal cooling rate is too large: according to the CCT diagram for the same system, where the $f(T(t))$ loci are shifted to longer times and lower temperatures (Fig. 10.2b), a less severe cooling rate may suffice.

It should be realized that upon cooling/quenching of a relatively massive, bulky specimen/workpiece the cooling rates for all parts of the specimen/workpiece in general will not be equal: at the surface of the specimen/workpiece the highest cooling rate may occur; transport of heat from the interior of the specimen/workpiece and also the possible production of heat by occurring phase transformations[2] modifies the cooling rates experienced, which generally will depend on the location in the specimen (e.g. the cooling rate is a function of depth beneath the surface).

CCT diagrams can be, and often are, determined utilizing non-isothermal dilatometry, i.e. recording the length change of a specimen subjected to a constant cooling rate. High-precision dilatometers can record length changes with an absolute accuracy of about 10 nm which, in view of a (usually cylindrically shaped, massive or hollow) specimen of length of the order 10 mm, implies that an extremely high relative accuracy is possible in the determination of specific length/volume change (see also footnote 15 in Chap. 9 and Sect. 10.13).

The TTT and CCT transformation diagrams make possible the selection of specific alloy compositions and their specific heat treatments in order to achieve desired microstructures and corresponding properties. This is most dramatically demonstrated for steels and has led to the publication of atlases presenting compilations of TTT and CCT diagrams for a great variety of steels (e.g. Atlas of Time–Temperature Diagrams for Irons and Steels, ASM International, USA, 1991). However, the concept of TTT and CCT diagrams is general and TTT and CCT diagrams have been published for other systems as well (e.g. see Atlas of Time–Temperature Diagrams for Nonferrous Alloys, ASM International, 1991).

After this excursion to a technical/phenomenological description of time and temperature dependencies of phase transformations, a more fundamental approach will be the focus of our attention in the remainder of this chapter.

---

[2] Note that, as a consequence of the cooling rate depending on the location in the specimen/workpiece, a phase transformation is not induced at the same time in all parts of the specimen/workpiece.

10.2   Thermal History and the Stage of Transformation

For the analysis of solid-state transformation kinetics a physical property (e.g. hardness, specific volume/length, electrical resistivity, enthalpy, magnetization) of the material subject to investigation can be traced as a function of time and temperature. Then the degree of transformation (fraction transformed), f, can be defined, for example, as

$$f = \frac{p - p_0}{p_1 - p_0}, \quad 0 \le f \le 1 \tag{10.1}$$

where p is the physical property measured during the course of transformation and p_0 and p_1 correspond with the values of p at the start and finish of the transformation, respectively. In non-isothermal analysis, p_0 and p_1 cannot normally be considered as constants (cf. Fig. 10.3a, b).

For thermally activated transformations, the thermal history of a specimen determines its stage of transformation. Consider the temperature–time, T–t, diagram depicted in Fig. 10.4. A specimen experiencing a thermally activated phase transformation proceeds from "State 1 (t_1, T_1)" to "State 2 (t_2, T_2)" via either path a or path b. Clearly, although the time to proceed from State 1 to State 2 is the same for both paths, the higher temperatures operating along path b cause a stage of transformation in State 2 for path b which is more advanced than that reached along path a. The stage of transformation in State 2 depends, in general, on the path followed: for non-isothermal analysis, *t and T are not state variables* (mathematically speaking: *a state variable is an independent variable*) for the stage of transformation. This

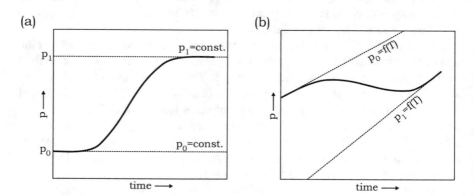

Fig. 10.3 Variation of a physical property p used to trace the degree of transformation **a** as function of time upon isothermal annealing and **b** as function of temperature upon non-isothermal annealing, e.g. with a constant heating rate (one then speaks of isochronal annealing). p_0 corresponds to the initial state and p_1 corresponds to the end state of the transformation. Linear dependencies of p_0 and p_1 on temperature upon non-isothermal annealing (see **b**) may for example hold if p represents the specimen length and the linear coefficients of thermal expansion of parent and product are constant in the temperature range concerned

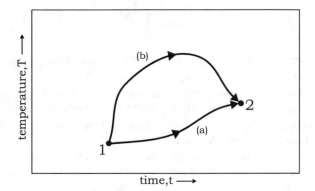

Fig. 10.4 Temperature (T)–time (t) diagram. A specimen experiencing a thermally activated phase transformation proceeds from "State 1 (t_1, T_1)" to "State 2 (t_2, T_2)" via either path a or path b. The stage of transformation in "State 2" depends on the path followed. Hence, in general t and T are not state variables for the stage of transformation

has for example as consequence that the iso-f curves in TTT diagrams, determined by isothermal anneals for various temperatures, do not hold for continuous cooling experiments (see the corresponding discussion in Sect. 10.1).

Thus, it appears appropriate to introduce a "path variable ", β, that is fully determined by the path followed in the temperature–time diagram: $T(t)$ prescribes β. Hence, the fraction transformed is fully settled by the path variable β:

$$f = F(\beta) \tag{10.2}$$

Equation (10.2) does not impose constraints on the type of transformation considered: the relation between f and β, i.e. F, has not been specified; it is only claimed that, given the path followed in the T–t diagram, f is known.

If the transformation mechanism is invariable for the region in the T–t diagram considered, it is tempting to interpret β as proportional to the number of atomic jumps, because T determines the atomic mobility and t defines the duration of the process considered. Against this background the following postulate is given.

for isothermal annealing:

$$\beta = kt \tag{10.3a}$$

for non-isothermal annealing:

$$\beta = \int k(T)\mathrm{d}t \tag{10.3b}$$

with k as the rate constant ; note that $k(T)$ depends on t in Eq. (10.3b). Next, an Arrhenius-type temperature dependence is adopted for the rate constant:

$$k = k_0 \exp\left(\frac{-Q}{RT}\right) \qquad (10.4)$$

implying that the temperature dependence of the transformation, in the region of the T-t diagram considered, can be described by an (effective, cf. Sect. 10.12) activation energy, Q; k_0 and R denote the pre-exponential factor and the gas constant, respectively. Use of an Arrhenius-type equation for rate constants is universally accepted and relies on compatible analyses of experimental data of transformation kinetics, but rigorous theoretical justification for its applicability is lacking. For example, in case of small driving forces (small undercooling or overheating) Arrhenius-type temperature dependences need not hold; see Sects. 10.5 and 10.6 and cf. Sect. 10.10. Moreover, in general it cannot be claimed that, for example, k_0 is independent of the path followed in the T-t diagram, which invalidates the concept of Eq. (10.3); see further Sect. 10.11.

10.3 The Transformation Rate; the Additivity Rule

Accepting the formalism of Eq. (10.3a) in the non-isothermal case but for an infinitesimal lapse of time: $d\beta = k dt$, which leads to Eq. (10.3b). Then it immediately follows that the postulate given by Eqs. (10.3a) and (10.3b) implies that the formulae for the transformation rate in the isothermal and non-isothermal cases are identical:

$$\frac{df}{dt} = \frac{dF(\beta)}{d\beta} \cdot \frac{d\beta}{dt} = k(T) \cdot \frac{dF(\beta)}{d\beta} \qquad (10.5)$$

Hence, β, or f, and T are state variables for the transformation rate. This realization introduces the notion of "*additivity*": after the transformation has progressed at temperatures different from T', having attained a degree of transformation equal to f_0, the course of subsequent transformation at temperature T' is identical to the one followed if the degree of transformation f_0 had been produced by isothermal transformation at T' (Fig. 10.5; see Christian 1975 and Mittemeijer 1992). Again (cf. end of Sect. 10.2), validity of this "additivity rule", as based on Eq. (10.3), is subject to (severe) constraints (see Sect. 10.11).

10.4 Heterogeneous Phase Transformations as a Composite Phenomenon:
Nucleation, Growth and Impingement

An efficacious approach for the quantitative description of phase-transformation kinetics distinguishes the following three, in the course of the heterogeneous (cf. Sect. 9.2) transformation generally overlapping, mechanisms (cf. Fig. 10.6):

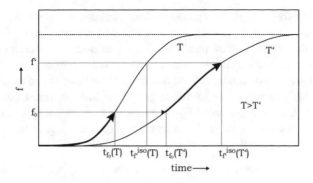

Fig. 10.5 Illustration of the "additivity" rule: annealing for the time $t_{f_0}(T)$ at temperature T and subsequent annealing for the time $t_{f'}^{iso}(T') - t_{f_0}(T')$ at T' leads to the same degree of transformation f' as annealing for $t_{f'}^{iso}(T')$ at T'

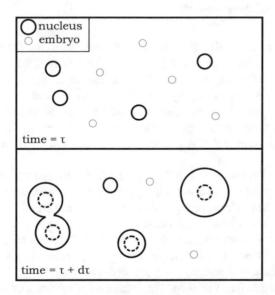

Fig. 10.6 Three, generally overlapping mechanisms controlling heterogeneous phase transformations. *Nucleation*: the formation of nuclei out of the reservoir of embryos (particles of subcritical size); see top part of the figure pertaining to time $= \tau$. *Growth*: The nuclei can grow under the simultaneous formation of new nuclei. *Impingement*: new nuclei cannot occur at locations where growing product-phase particles are present and growth of product-phase particles cannot extend to locations occupied by other product-phase particles and/or can be influenced by depletion of the surrounding matrix by growing neighbouring product-phase particles; see bottom part of the figure pertaining to time $= \tau + d\tau$

1. *nucleation*: the generation of product-phase particles of supercritical size, nuclei, from (the reservoir of) particles of subcritical size, embryos (cf. Sect. 9.2);

2. *growth*: the product-phase particles, starting as nuclei, increase in size by the addition of material from the surrounding matrix of parent phase;

3. *impingement*: new nuclei cannot develop at locations in the system occupied by growing product-phase particles; growth of product-phase particles cannot extend to locations occupied by other product-phase particles and/or can be influenced by depletion of the surrounding matrix by growing neighbouring product-phase particles.

In the following three sections nucleation, growth and impingement are considered separately.

10.5 Modes of Nucleation

The nucleation modes dealt with here, and as presented in the literature, generally pertain to large undercooling or overheating (cf. Sect. 9.2) of the system subject to transformation. Further, at this stage of the development, the equations presented below for the nucleation rate apply to a virtual, infinite volume of untransformed material where the nuclei are not affected by the presence and growth of other nuclei: each product-phase particle is supposed to grow into an infinitely large parent phase, in the absence of other growing particles (see discussion on extended volume in Sect. 10.8).

Upon a phase transformation interfaces develop between the old and the new phases, and (possibly) misfit strain is introduced in the system. Whereas the production of the new phase releases chemical Gibbs energy, the creation of the interfaces and the introduction of misfit strain cost Gibbs energy. According to the classical nucleation theory, a critical particle size of the new phase can be defined such that if the particle is of subcritical size, it costs energy to increase the size of the particle, whereas if the particle (nucleus) is of supercritical size, energy is released if the particle grows further (see Fig. 9.3). The formation of particles of supercritical size from particles of subcritical size is called nucleation.

The nucleation rate, $\dot{N}(T(t)) \equiv dN/dt$, with N as the number of nuclei formed, is determined by the number of particles of critical size and the rate of the jumping of atoms through the interface between the parent phase and the particles of critical size. The frequency of jumping through the interface is given by an Arrhenius term. The number of particles of critical size depends on the critical Gibbs energy of nucleus formation, ΔG^*, which, according to the above description depends on the decrease of the chemical Gibbs energy per unit volume, the interface energy per unit area interface and the misfit-strain energy per unit volume (cf. Sect. 9.2). On this basis the classical theory of nucleation gives the following expression for the nucleation rate per unit volume, $\dot{N}(T(t))$, i.e. the rate of formation of particles of supercritical size (=nuclei):

$$\dot{N}(T(t)) = C\omega \exp\left(-\frac{\Delta G^*(T(t)) + Q_N}{RT(t)}\right) \tag{10.6}$$

with R the gas constant, T the temperature, C the number density of suitable nucleation sites, ω the characteristic frequency factor and Q_N the activation energy for the jumping of atoms through the particle/matrix interface.

10.5.1 Continuous Nucleation

If the undercooling or the overheating is very large, ΔG^* can be considered to be very small as compared to RT. The nucleation rate per unit volume is then only determined by the atomic mobility for transport through the interface, which for isothermally and non-isothermally conducted transformations gives:

$$\dot{N}(T(t)) = N_0 \exp\left(-\frac{Q_N}{RT(t)}\right), \tag{10.7}$$

where $C\omega$ has been combined into N_0, the temperature-independent nucleation rate. Q_N, the activation energy for the jumping of atoms through the interface between the particle of critical size and the matrix, is defined for the remainder of this text as the temperature-independent activation energy for nucleation. This type of nucleation is called continuous nucleation, characterized by a constant nucleation rate at constant temperature; the number of nuclei equals 0 at $t = 0$ (see Fig. 10.7).

It should be noted that for smaller undercooling or overheating, ΔG^* is not very small as compared to RT. In this case the full nucleation-rate equation (Eq. 10.6) must be used. Note that ΔG^* in Eq. (10.6) depends on temperature (cf. Eqs. (9.5) and (9.5a)).

10.5.2 Pre-existing Nuclei

In Sect. 10.5.1 it was assumed that the number of nuclei at the beginning of the transformation is zero. Here we consider the case where there already is a number of pre-existing nuclei (supercritical particles of the new phase) at $t = 0$ and that the further nucleation rate is zero. This implies that the number of product-phase particles is equal to the number of pre-existing nuclei, N^*.

A typical example of such a case can be the preferential nucleation at grain boundaries, edges or corners. Mainly depending on the degree of supercooling (or superheating), saturation of the (grain-boundary) nucleation sites can occur very early in the transformation, effectively leading to a zero nucleation rate for the remainder of the transformation. The term *site saturation* can be used for the general case of pre-existing nuclei at $t = 0$ (cf. Fig. 10.7).

Fig. 10.7 Number of nuclei
a and the corresponding
nucleation rate **b** as a
function of isothermal
annealing time for site
saturation, continuous
nucleation, mixed nucleation
and transient nucleation

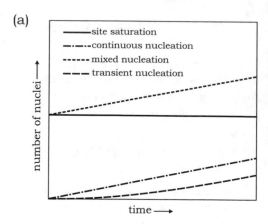

(a)

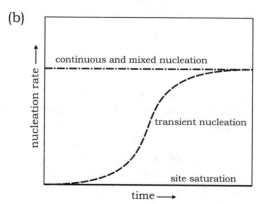

(b)

Another example of site saturation is as follows. Upon rapid cooling/quenching of a phase stable at elevated temperature this phase can become metastable at lower temperatures, e.g. an amorphous alloy or a supersaturated crystalline solid solution may occur, which strives for crystallization or decomposition, respectively. Depending on the precise thermal history of such metastable phases, more or less particles of a new, stable phase, which were generated during cooling in the first, high temperature part of the cooling curve, may have been "frozen in". If a heat treatment is applied subsequently to such a metastable phase with frozen-in particles of the new, stable phase, then those particles larger than the critical size (which are the nuclei of the new, stable phase; see above) can grow, implying occurrence of initial, pre-existing nuclei with no formation of further nuclei: site saturation in the above defined sense. The critical size depends on temperature. Then, given a certain size distribution for the "frozen-in" particles of the new, stable phase, it is evident that the number of nuclei (=supercritical particles) acting in this "pre-existing nuclei nucleation mechanism" is temperature dependent.

10.5.3 Other Modes of Nucleation

To deal with cases intermediate between continuous nucleation and site saturation (at $t = 0$), as described above, the term *mixed nucleation* has been coined representing some weighted sum of the nucleation rates according to the cases of continuous nucleation and pre-existing nuclei; in this way a wide variety of nucleation modes can be described (see Fig. 10.7).

Another, deviating case concerns the occurrence of *transient nucleation*, which implies a nucleation rate which initially may be zero and increases *sigmoidally* up to a steady-state value. The transient may arise because of the time needed to establish a steady-state population of subcritical particles of the product phase (see Fig. 10.7).

10.6 Modes of Growth

Two (extreme) growth models are considered; one for volume-diffusion-controlled growth and one for interface-controlled growth. Volume-diffusion-controlled growth can occur for phase transformations where long-range compositional changes take place. The case of interface-controlled growth can occur if the growth is determined by atomic jump processes in the immediate vicinity of the interface, as holds for the massive austenite $(\gamma) \rightarrow$ ferrite (α) transformation, in substitutional binary Fe-based alloys, and also for some crystallization reactions of amorphous alloys.

The diffusion-controlled and interface-controlled growth modes can be given in a compact form. At time t, the volume Y, of a particle nucleated at time τ is given by:

$$Y(\tau, t) = g \left[\int_\tau^t v dt' \right]^{d/m} \tag{10.8}$$

with g as a particle-geometry factor with the unit $m^3 m^{-d}$, m as growth-mode parameter ($m = 1$ in case of "linear" growth (i.e. for isothermal transformations the transformed volume grows proportional with t; this usually corresponds with interface-controlled growth); $m = 2$ in case of "parabolic", diffusion-controlled growth (i.e. for isothermal transformations the transformed volume grows proportional with $t^{1/2}$), d as the dimensionality of the growth ($d = 1, 2, 3$) and v as the growth velocity (velocity of the product/parent interface) in case of interface-controlled growth ($m = 1$) and as the diffusion coefficient in case of diffusion-controlled growth ($m = 2$).

10.6.1 Interface-Controlled Growth

For the case of interface-controlled growth (then Eq. 10.8 is applied with $m = 1$), the formulation for the interface velocity v can be derived as follows. Consider Fig. 10.8. Upon transfer of an atom from the matrix, parent phase, α, to the product phase, β, a net energy change of the system, $\Delta G = G_\beta - G_\alpha$, occurs (for the case shown in Fig. 10.8, ΔG evidently is negative (i.e. positive driving force; see Sect. 9.1). Along its way, to a state of lower energy, the atom has to overcome an activation-energy barrier, ΔG^a (>0; cf. Sect. 8.6). The number of atoms crossing the interface from α to β per unit of time then equals

$$v' \exp\left(\frac{-\Delta G^a}{RT}\right)$$

with v' as an (atomic) vibration frequency. The formulation for the number of atoms crossing the interface in the opposite direction then reads (now ΔG enters into the expression, as for the reverse direction the total activation energy barrier obviously equals $\Delta G^a - \Delta G$):

$$v' \exp\left(\frac{-(\Delta G^a - \Delta G)}{RT}\right)$$

The net number of atoms crossing the interface from matrix, parent phase (α) to product phase (β) then is given by the difference of both above-indicated material fluxes. This leads to the following formulation for the interface velocity of the moving reaction front between the product phase (β) and the parent, matrix phase (α):

$$v(T(t)) = v_0 \exp\left(-\frac{\Delta G^a}{RT(t)}\right)\left(1 - \exp\left(\frac{\Delta G}{RT(t)}\right)\right) \tag{10.9}$$

Fig. 10.8 Change of system energy upon a jump of an atom from the matrix α to the product phase β. In order that (Gibbs) energy is gained/released by the transformation, ΔG (=$G_\beta - G_\alpha$) must be negative; the "driving force" is defined by $-\Delta G$ (>0)

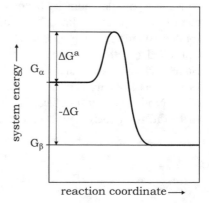

with v_0 as the pre-exponential factor for growth, which incorporates, as compared to v', the atomic jump distance, ΔG^a as the activation energy for the transfer of atoms through the parent phase/new phase interface, and ΔG is the energy difference between the new phase and the parent phase. Note that ΔG can depend on temperature (and thus on time in the case of non-isothermal annealing): $\Delta G(T(t))$. The driving force (see Sect. 9.1) is defined as $-\Delta G = |\Delta G|$ (as $\Delta G < 0$).

For large undercooling or overheating, the driving force $|\Delta G|$ is large compared to RT, and Eq. (10.9) becomes:

$$v(T(t)) = v_0 \exp\left(-\frac{Q_G}{RT(t)}\right) \tag{10.10}$$

with $Q_G \, (=\Delta G^a)$ as the activation energy for growth, and v_0 as the temperature-independent interface velocity. For interpretation of ΔG^a, see Bos et al. (2005).

For small undercooling or overheating, the driving force $|\Delta G|$ is small as compared to RT, and Eq. (10.9) reduces to:

$$v(T(t)) = M(-\Delta G) = M_0 \exp\left(-\frac{Q_G}{RT(t)}\right)(-\Delta G) \tag{10.11}$$

where $Q_G = \Delta G^a$ and M is the temperature-dependent *interface mobility* (the temperature dependence of M_0 can be neglected; see Liu et al. (2007). It is remarked (again) that ΔG in Eq. (10.11) in general depends on temperature. For isothermal transformations Y can still be calculated analytically according to Eq. (10.8), after substitution of Eq. (10.11) if ΔG is constant for the integration. For non-isothermally conducted measurements Y can only be calculated by numerical integration. This has led to limited application of Eq. (10.11), as compared to Eq. (10.10).

10.6.2 Diffusion-Controlled Growth

For the case of diffusion-controlled growth (then Eq. (10.8) is applied with $m = 2$), v in Eq. (10.8) can generally be substituted by v according to Eq. (10.10), where Q_G has to be replaced by the activation energy for diffusion, Q_D, and v_0 has to be replaced by the pre-exponential factor for diffusion D_0 (i.e. $v = D(T(t)) = D_0 \exp(-Q_D/RT)$).

10.6.3 Mixed Growth Mode

In general, growth can exhibit a mixed-mode character: the transformation can start with interface-controlled growth and then a transition to diffusion-controlled growth can occur, as shown by model considerations, e.g. for the isothermal austenite $(\gamma) \rightarrow$

ferrite (α) transformation in Fe-C alloys (Sietsma and van der Zwaag 2004). Such transition from interface-controlled growth to diffusion-controlled growth has been observed experimentally during nano-crystallization of amorphous Al-based alloys (Nitsche et al. 2005). However, the reverse transition, from diffusion-controlled growth to interface-controlled growth, especially in non-isothermally conducted experiments, is possible as well (see Sect. 9.5.1).

10.7 The Activation Energies for Nucleation and Growth

The activation energy introduced for nucleation in Sect. 10.5, Q_N (cf. Eq. 10.6), has been conceived as an activation energy for the transfer (of an atom) from the matrix through the interface between the matrix and the particle of critical size, thereby the particle considered becomes a nucleus. In Sect. 10.6.1, the activation energy introduced for interface-controlled growth (Q_G ($=\Delta G^a$); cf. Eq. (10.9)) has been conceived as an activation energy for transfer from the matrix of an atom through the interface between the matrix and the growing particle (significantly larger than a nucleus). Both activation energies depend on elementary atomic jumps. Yet, they can have considerably different values. This may be due to considerably different structures for the interface with the matrix for the minute embryos (particles smaller than and just equal to the critical size) and for the much larger (up to orders of magnitude) growing particles. For example, in initial stages coherent interfaces may occur, whereas a growing particle may exhibit an incoherent interface. Then, in the initial stage of growth, the activation energy may change due to the occurring changes in the interface structure.

Activation energies for interface mobilities (cf. the factor M in Eq. (10.11)) can be determined by groups of atomic jumps leading to effective activation energies considerably larger than the activation energy for a single atomic jump (Fig. 10.9, see Bos et al. 2005). Such processes may have significantly different net effects for minute embryos and large growing particles, in view of the different interface structures.

Unambiguous results for the activation energies of both the nucleation and growth mechanisms for the same solid–solid state transformation are rare. As demonstrated in Sects. 10.8 and 10.9, the degree of transformation is controlled by an effective activation energy that generally depends on time and temperature and that contains the *constant* activation energies of the operating nucleation and growth mechanisms. The kinetic model fitting discussed in Sect. 10.14 and the methodology presented in Sect. 10.15.6 yet allow the separate determination of the constants Q_N and Q_G in one kinetic analysis. First results with these approaches demonstrate that Q_N can be both larger and smaller than Q_G or Q_D, while of the same order or magnitude (Liu et al. 2004a, b). Clearly, much more experimental data are necessary in order to arrive, possibly with the aid of dedicated computer simulations, at detailed interpretation of values determined for Q_N and Q_G or Q_D.

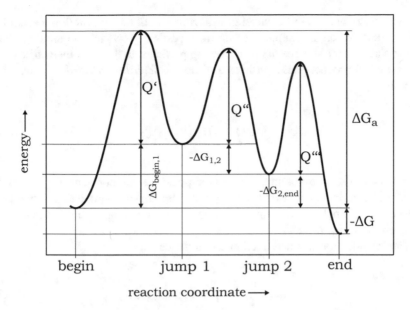

Fig. 10.9 Effective activation energy for interface-controlled growth involving multiple atomic jumps. Many atoms have to take, temporarily, unfavourable positions at the interface, to create space so that the atoms at sites of the parent crystal structure can go to sites of the product crystal structure: atoms at the parent crystal structure at the interface cannot jump generally directly to empty sites of the product crystal structure because empty sites of the product crystal structure are blocked by neighbouring atoms. By series of unfavourable jumps performed by groups of atoms a path is created for the transformation from parent to product to proceed. In the figure the first jump is an unfavourable one, since the system energy is increased by the jump: $\Delta G_{\text{begin},1}$ (>0). The Gibbs energy change driving the transformation is ΔG (<0). For the jump series shown in the figure the resulting effective activation energy is ΔG_a, which is larger than the activation energies for the single atomic jumps (Q', Q'' and Q''')

10.8   Extended Volume and Extended Transformed Fraction

The kinetics of phase transformations are studied either isothermally or non-isothermally, but then often subject to a constant heating or cooling rate $\Phi = \mathrm{d}T/\mathrm{d}t$. Non-isothermal annealing with constant heating/cooling rate is called isochronal annealing. In the following these two cases, isothermal annealing and isochronal annealing will be dealt with in a parallel fashion.

As a first step for calculating the degree of transformation, f Eq. (10.1), on the basis of expressions for the nucleation and growth rates as presented in Sects. 10.5 and 10.6, the so-called extended transformed fraction is determined as follows.

The number of supercritical particles (nuclei) formed in a unit volume, at time τ during a time lapse $\mathrm{d}\tau$, is given by $\dot{N}(T(\tau))\mathrm{d}\tau$, with $\dot{N}(T(\tau))$ according to Eq. (10.6) and derived versions in Sect. 10.5, where it is supposed that each nucleation event takes place in an infinitely large parent phase in the absence of other (growing) nuclei.

The volume of each of these nuclei grows from τ until the current time t according to Eq. (10.8), where it is supposed that every particle grows into an infinitely large parent phase, in the absence of other growing particles. In this hypothetical case, the volume of all product-phase particles at time t, called *the extended volume*, V^e, is given by

$$V^e = \int_0^t V \dot{N}(\tau) Y(\tau, t) d\tau \tag{10.12}$$

with V as the sample volume, which is supposed to be constant throughout the transformation. In order to evaluate Eq. (10.12) for non-isothermal transformation it is necessary to apply explicit time dependences for the temperature T occurring in the expressions for $\dot{N}$ and Y.

The extended transformed fraction, x_e, is defined as

$$x_e \equiv \frac{V^e}{V} = \int_0^t \dot{N}(\tau) Y(\tau, t) d\tau \tag{10.13a}$$

For the precipitation of a second phase in an initially supersaturated matrix of the parent phase, as illustrated in Fig. 9.1, the total volume of product (precipitate) phase after completed transformation, V^p, is only a fraction of the whole specimen volume. Then, for this case of precipitation, the extended precipitate volume has to be normalized with respect to V^p and the extended precipitate-volume fraction is defined as

$$x_e \equiv \frac{V^e}{V^p} = \frac{V}{V^p} \int_0^t \dot{N}(\tau) Y(\tau, t) d\tau \tag{10.13b}$$

The discussion in Sect. 10.6 showed that v in the general expression for Y according to Eq. (10.8) can be substituted by v according to Eq. (10.10) for both diffusion-controlled growth and interface-controlled growth for a high driving force (large undercooling or superheating). Adopting this result for Y, it can be shown for a wide range of nucleation models, including those discussed in Sect. 10.5, by substitution of the appropriate expressions for $\dot{N}$ and Y in Eq. (10.13), that x_e is given by Liu et al. (2007):

for isothermal transformation:

$$x_e = c_{nor} k_0(t)^{n(t)} t^{n(t)} \exp\left(-\frac{n(t)Q(t)}{RT}\right) \tag{10.14}$$

for isochronal annealing[3]:

$$x_e = c_{nor} k_0(T)^{n(T)} \left(\frac{RT^2}{Q\Phi}\right)^{n(T)} \exp\left(-\frac{n(T)Q(T)}{RT}\right) \qquad (10.15)$$

with $c_{nor} = 1$ and $c_{nor} = V/V^p$ for x_e defined according to Eqs. (10.13a) and (10.13b), respectively, and where n, k_0 and Q are functions of the parameters used in the specific nucleation and growth models, as, for example, N_0, d, m and the activation energies Q_N and Q_G. Q is the *effective activation energy* of the phase transformation, which is further discussed in Sect. 10.12; n is the so-called *growth exponent*. The results as given by Eqs. (10.14) and (10.15) can be summarized by

$$x_e = c_{nor} k_0^n (\alpha)^n \exp\left(-\frac{nQ}{RT}\right) \qquad (10.16)$$

with $\alpha = t$ for isothermal annealing and $\alpha = RT^2/(Q\Phi)$ for isochronal annealing.

It is important to realize that the parameters n, k_0 and Q generally vary during the transformation(!): they depend on time (isothermal transformation) and temperature (isochronal transformation). Explicit expressions for n, Q and k_0 in terms of general nucleation and growth mechanisms, for both isothermal and isochronal annealing (heating), have been listed in Tables 1–3 in Liu et al. (2007) (do note that K_0 in these tables for isochronal annealing equals k_0/Q in the current treatment and thus the factor α in Eq. (10.16) for isochronal annealing is given by RT^2/Φ, and not by $RT^2/(Q\Phi)$, in the paper by Liu et al. (2007)). Only for the cases of continuous nucleation and site saturation at $t = 0$ these time dependences (isothermal transformation) and these temperature dependences (isochronal transformation) vanish.

If k_0, n and Q do not depend on time (isothermal transformation) and temperature (isochronal annealing), x_e can be expressed in terms of the path variable β (see Eq. (10.3)): it follows for both cases (isothermal annealing and isochronal heating) that Eqs. (10.14) and (10.15), or thus Eq. (10.16), can then be written as:

$$x_e = \beta^n \qquad (10.17)$$

where c_{nor} has now been incorporated in k_0^n (cf. Eq. 10.4).

10.9 Modes of Impingement

The next step in the calculation of the degree of transformation, f (Eq. 10.1), is the correction for the unrealistic assumption that nucleation and growth of every supercritical particle can occur without taking into account the constraint of already

---

[3] Equation (10.15) is subject to an approximation made for the so-called temperature integral (cf. Eqs. (10.3b) and (10.4)) occurring in the specific nucleation and growth functions. This approximation only holds for the case of isochronal *heating*, not for isochronal cooling (Liu et al. 2007)

present, earlier or later nucleated and thereafter grown product-phase particles. In other words: how to include the effect of "particle interference", i.e. how to derive the function $f = g(x_e)$?

A relation between the actually transformed volume, V^t, and the extended transformed volume, V^e, or between the real transformed fraction, $f = V^t/V$ or $f = V^t/V^p$, and the extended transformed fraction, $x_e = V^e/V$ or $x_e = V^e/V^p$, is required (cf. Eqs. (10.13a and b)). The expressions for the extended transformed volume/fraction do not account for physically impossible nucleation in already transformed volume and the overlap of growing particles. This is called *hard impingement*. Further, in diffusion-controlled transformations, as, for example, can pertain to nano-crystallization of amorphous alloys, the austenite–ferrite transformation in carbon containing alloyed steels and precipitation reactions in general, a solute-depletion zone develops around a growing particle in which zone less likely further nucleation can take place (because of a lesser supersaturation) or even no further nucleation can occur at all (if the supersaturation has become negligible). This is called *soft impingement*.

First the case of hard impingement is considered. Suppose that the nuclei are dispersed randomly throughout the total volume and grow isotropically. If the time is increased by dt, the extended and the actually transformed volumes will increase by dV^e and dV^t, respectively. From the change of the extended volume, dV^e, only a part will contribute to the change of the actually transformed volume, dV^t, namely a part as large as the *un*transformed volume fraction. Hence:

$$dV^t = \left\{ \frac{(V - V^t)}{V} \right\} dV^e \tag{10.18a}$$

and thus:

$$\frac{df}{dx_e} = 1 - f \tag{10.18b}$$

Upon integration of Eq. (10.18b) it is finally obtained:

$$f = g(x_e) = 1 - \exp(-x_e) \tag{10.19}$$

Models for hard impingement in case of anisotropic growth and in case of non-random nucleation have been given in the literature (Liu et al. 2007; see further below).

A rigorous treatment for soft impingement does not exist. However, considering randomly dispersed nuclei and isotropic growth, it can be inferred that a correction for soft impingement in case of growth by solute diffusion in the matrix can be realized by equating the infinitesimal change df with the infinitesimal change dx_e multiplied with the untransformed fraction $(1 - f)$, i.e. the same approach as given above for hard impingement (cf. Eq. 10.18b). Such a treatment of soft impingement, parallel to hard impingement, may be understood as that for the case of soft impingement each precipitate/product particle is supposed to be surrounded effectively by an outer, solute-depleted shell of size such that upon completed precipitation all precipitate particles with their surrounding solute-depleted shells occupy the whole volume of the specimen. According to this picture also treatments of hard impingement for anisotropic growth and for non-random nucleation (see below) may be suitable approaches for soft impingement as well.

Now, for the case of isothermal transformation and k_0, n and Q independent of time (see discussion at the very end of Sect. 10.8), it follows after substitution of x_e in Eq. (10.19) according to Eq. (10.14) (see also Eq. (10.17)):

$$f = 1 - \exp\{-(kt)^n\} = 1 - \exp\{-\beta^n\} \qquad (10.20a)$$

with (cf. Eq. (10.4))

$$k = k_0 \exp\left(\frac{-Q}{RT}\right) \qquad (10.20b)$$

The result for f as given by Eqs. (10.20a) and (10.20b) is known as the Johnson-Mehl-Avrami (JMA) or Johnson-Mehl-Avrami-Kolmogorov (JMAK) equation [see original papers by Kolmogorov (1937), Johnson and Mehl (1939) and Avrami (1939, 1940, 1941)]. This famous and often used, but also abused equation deserves special discussion; see Sect. 10.11.

The result derived above for $g(x_e)$ relies on the assumption of randomly dispersed nuclei and isotropic growth. In case of anisotropically growing particles, the time interval that particles, after their randomly dispersed nucleation, can grow before "blocking" by other particles occurs, is, on average, smaller than for isotropic growth. This blocking effect due to anisotropic growth thereby leads to hard impingement of more pronounced severity than for isotropic growth and can cause strong deviations from the kinetics as described by Eq. (10.19) (and thereby also Eq. (10.20); see Fig. 10.10).

The assumption of randomly dispersed nuclei is untenable if more regular distributions of the product-phase particles occur. This can for example be the case if in a polycrystalline material, with grains of about the same size and morphology (e.g. a case of more or less equi-axed grains with a not too wide grain-size distribution), the grain-boundary junctions serve as preferred nucleation sites. Then, for the case of site saturation at $t = 0$, or, more generally, for cases where nucleation occurs predominantly in the initial stage of transformation, a less severe impingement occurs than that pertaining to a random distribution of nuclei (see below: $\varepsilon \geq 1$; Eq. (10.22)); for the case of transient nucleation (cf. Sect. 10.5), or, more generally, for cases where nucleation occurs appreciably at later stages of transformation, a more severe impingement than pertaining to a random distribution of nuclei may occur (see below: $0 < \varepsilon \leq 1$; Eq. (10.22)). If nucleation takes place not only at grain-boundary junctions but also along the grain boundaries, impingement is generally (much) more severe, as the nuclei now can occur (very) close to each other.

The cases of anisotropic growth and non-random nuclei distribution can be described by (cf. Eq. (10.18b) for random nuclei distribution and isotropic growth):

Fig. 10.10 Transformed fraction, f, as a function of the extended fraction, x_e, for different values of ζ (isotropic ($\zeta = 1$)/anisotropic ($\zeta > 1$) growth) and ε (random ($\varepsilon = 1$)/non-random ($\varepsilon > 1$) distribution of nuclei); see text for details

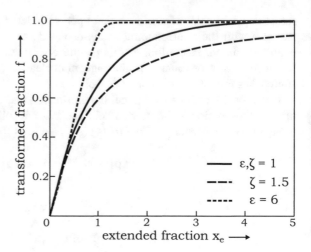

anisotropic growth:

$$\frac{\mathrm{d}f}{\mathrm{d}x_e} = (1 - f)^{\zeta}; \quad \zeta \geq 1 \tag{10.21}$$

non-random distribution:

$$\frac{\mathrm{d}f}{\mathrm{d}x_e} = 1 - f^{\varepsilon}; \quad 0 < \varepsilon \leq 1 \text{ and } \varepsilon \geq 1 \tag{10.22}$$

For $\zeta = \varepsilon = 1$, Eq. (10.18b) results. The effect on f of different cases of impingement is illustrated in Fig. 10.10: anisotropic growth ($\zeta > 1$) induces a stronger impingement, whereas a non-random nuclei distribution due to nucleation at grain-boundary junctions predominantly in the initial stage of transformation causes a less severe impingement ($\varepsilon > 1$), than occurs in the case of a random distribution of nuclei and isotropic growth (for calculation of f from x_e subject to Eqs. (10.21) and (10.22), see Liu et al. (2007)).

10.10   The Transformed Fraction

The general recipe for deriving an explicit *analytical* formulation or calculating *numerically* values for the degree of transformation as function of time and temperature, assuming that hard impingement prevails, is now as follows. The extended transformed fraction x_e is calculated according to Eq. (10.13) using the appropriate nucleation mode and the appropriate growth mode (see Sects. 10.5 and 10.6). The expression for the extended transformed fraction then is substituted into the appropriate impingement correction ($g(x_e)$; see Sect. 10.9) to give the degree of transformation. The procedure has been visualized in Fig. 10.11.

The treatment in this chapter, as presented here in a summarizing way, thus comprises a *modular* constitution of a model for transformation kinetics: nucleation, growth and impingement are separately modelled and then these are combined according to the sketch in Fig. 10.11. Evidently, this in principle allows great modelling flexibility and versatility, far beyond specific transformation-kinetic descriptions as the classical Johnson–Mehl–Avrami (JMA) equation (Eq. (10.20)) or the more general JMA-*like* equations (Eq. (10.23)). This approach yet is constrained: further, see Sect. 10.16.

Analytical descriptions of f provide more direct insight into functional dependences and are often used in practice; analytical descriptions for example allow the more easy identification of the influence of the different nucleation, growth and impingement models. This explains the large, also recent, interest in analytical descriptions of transformation kinetics and their application; an enormous, expanding body of such literature exists.

A few, most important analytical results for the dependence of the degree of the transformation in particular on time and temperature have been given here as well (Eqs. (10.20) and (10.23)). The main limitation in the application of the considered nucleation and growth modes to arrive at *analytical* expressions for f is the requirement that the undercooling or overheating of the transforming system must be relatively large, *in order that Arrhenius-type temperature dependences for the nucleation*

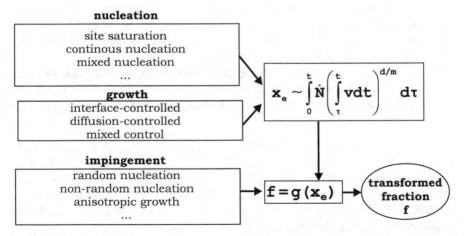

Fig. 10.11 Modular transformation model. Visualization of the procedure of calculating the transformed fraction. The specific models for nucleation, growth and impingement have to be substituted into the expression for the extended transformation fraction, x_e, to calculate, analytically or numerically, the real transformed fraction, f

and growth rates are assured (cf. Sects. 10.5 and 10.6). For small undercooling or overheating, the nucleation and growth modes cannot be described employing an Arrhenius-type temperature dependence with a constant activation energy (cf. for nucleation Eq. (10.6) and for growth Eq. (10.9)). For the special but significant case of site saturation (at $t = 0$) and interface-controlled growth with small undercooling analytical descriptions are possible (Kempen et al. 2002).

10.11 The Classical and Generalized Johnson–Mehl–Avrami Equation; the "Additivity Rule" Revisited

The classical, still very often used, description of transformation kinetics is based on the JMA equation (Eq. (10.20)). As follows from the preceding treatment, application of the JMA equation can be possible if the following list of conditions is satisfied: isothermal transformation, either pure site saturation at $t = 0$ or pure continuous nucleation, high driving force (large undercooling or superheating) and randomly dispersed nuclei which grow isotropically.

In view of these constraints it may come as a surprise that the JMA equation has been used and still is used very often, also in cases where the above-mentioned constraints are not satisfied. Two considerations provide understanding for this phenomenon:

(i)   Inappropriate application of the classical JMA equation is often obscured by fitting to inaccurate experimental data and/or crude or insensitive fitting (e.g. fitting to only one transformation curve, instead of simultaneous fitting to a set of transformation curves measured at various temperatures and/or heating and cooling rates; see Fig. 10.13 in Sect. 10.14. In this context it has also been observed that the classical JMA equation provides a good fit only in the first part of the transformation (Christian 1975).

(ii) Recent developments have shown that the classical JMA equation can be considered as a special case of a family of JMA-like equations which can be applied to a wide(r) range of transformations. The following text provides a corresponding evaluation.

The classical JMA equation as given by Eq. (10.20) cannot be applied to non-isothermal transformations. Only if the JMA equation is given in terms of the path variable, β, the equation can be applied as such to non-isothermal applications as well. Such a variant of the JMA equation, applicable to isothermal and isochronal annealings, is provided by Eq. (10.19) with $x_e = \beta^n$, which holds if k_0, Q and n do not depend on time and temperature (cf. Eqs. (10.17) and (10.20a)):

$$f = 1 - \exp\{-\beta^n\}$$

Equation (10.19), without imposition of the equality $x_e = \beta^n$ (Eq. (10.17)), has a (much) more general validity than the JMA equation, also in its above variant that pertains not only to isothermal but also to isochronal annealing. However, Eq. (10.19) does not contain kinetic information (the time and temperature dependences of a transformation); it only expresses the effect on the degree of transformation of (a special case of) impingement.

Upon substitution of x_e according to Eq. (10.16) into Eq. (10.19) the most general formulation of a JMA-like equation is obtained:

$$f = 1 - \exp\left\{-k_0^n (\alpha)^n \exp\left(\frac{-nQ}{RT}\right)\right\} \tag{10.23}$$

with $\alpha = t$ for isothermal annealing and $\alpha = RT^2/(Q\Phi)$ for isochronal annealing (note that c_{nor} has been incorporated in k_0^n; see below Eq. (10.17)). Here it is implied that k_0, n and Q can be time dependent (isothermal annealing) or temperature dependent (isochronal annealing). Only if the mode of nucleation is either continuous nucleation or site saturation at $t = 0$ these time dependences (isothermal transformation) and these temperature dependences (isochronal transformation) vanish and Eq. (10.23) reduces to the classical JMA equation if, additionally, the transformation occurs isothermally.

Considering isothermal and non-isothermal applications of the generalized JMA equation (Eq. (10.23)) it can be shown, for the only two nucleation modes which comply with the classical JMA equation, that for site saturation (at $t = 0$) the kinetic parameters k_0, Q and n are independent of the path followed in the T-t diagram, whereas this does not hold for continuous nucleation. In the latter case k_0 is different for isothermal annealing and isochronal annealing, but Q and n are the same for both types of annealing (see Table 3 in Liu et al. 2007). Therefore, the values of (constant) kinetic parameters determined by isothermally performed experiments may not be applicable in non-isothermal experiments for the same transformation.

Equations (10.3) and (10.4) can be written explicitly in the following fashion:

$$\beta = \int k(T(t))dt \quad \text{and} \quad k = k_0 \exp\left\{\frac{-Q}{RT(t)}\right\}$$

implying that k does not depend on t other than through T. Then, because of the incompatibility of values of the kinetic parameters for isothermal and non-isothermal annealing as in case of continuous nucleation (see above paragraph), these two equations are incompatible with the case of continuous nucleation. This has as immediate and important consequence that in this case the "additivity rule" cannot be used (cf. Sect. 10.3). Hence, although the JMA equation in the form $f = 1 - \exp\{-\beta^n\}$ (see above discussion) may hold, the "additivity rule" cannot be generally applied.

Consequently, when adopting the JMA equation, applications of the "additivity rule", e.g. to deduce non-isothermal transformation kinetics from isothermal transformation kinetics (by partitioning a non-isothermal T-t path into a series of small isothermal time steps at different temperatures; a usual approach), on the basis of JMA kinetic parameters determined at constant temperature, can be unjustified.

Regarding the concept of "additivity", two further important issues, not always handled sufficiently carefully, deserve attention:

1. "Additivity" requires that the transformation rate is determined by the fraction transformed and the temperature (see Sect. 10.3). This may seem a condition which can be compatible with reality. However, as rarely recognized, for a transformation involving growth, the instantaneous transformation rate, df/dt, is always equal to the product of the total surface area of the growing particles and the instantaneous interface velocity. Hence, such a transformation can only be compatible with the "additivity rule" if the surface area of the growing particles is solely determined by the transformed volume. However, generally the total surface area of the growing particles is determined by the number and the individual sizes (and not average size) of the growing particles.

2. "Additivity" holds if the transformation rate can be expressed as independent of a specific, followed time–temperature path (see Sect. 10.3). Hence, df/dt found by differentiating $f(t)$ given for a specific time–temperature path (e.g. isothermal annealing or isochronal annealing) can *never* be used to prove that a transformation is compatible with the "additivity rule" (examples of such erroneous reasoning abound in the literature).

10.12 The Effective Activation Energy

Experimentally observed variations in the effective activation energy, Q, as derived from the change in f as a function of time and temperature during the course of a transformation, are usually interpreted as the consequence of a change in transformation mechanism. In view of Eqs. (10.14) and (10.15) this reasoning can be flawed:

For a wide range of combinations of nucleation and growth modes, *with Arrhenius temperature dependences* (cf. Sect. 10.10), the effective, overall, activation energy of the transformation, Q, can be given as (Liu et al. 2007):

$$Q = \frac{\frac{d}{m}Q_G + \left(n - \frac{d}{m}\right)Q_N}{n} \tag{10.24}$$

Q (through n) depends on time and temperature, whereas the activation energies for nucleation and growth, Q_N and Q_G, are constants. Therefore, an observation of change of Q with time or temperature, i.e. during the course of a transformation, needs *not* be considered as an experimental artefact or as a consequence of change of transformation mechanism: even if the transformation is isokinetic (i.e. of constant transformation mechanism) Q can vary during the course of the transformation.

The fraction transformed according to the classical JMA equation can be written as

$$f = 1 - \exp\{-(kt)^n\}$$

as done here (Eq. (10.20a)), or, alternatively, as

$$f = 1 - \exp\{-k't^n\}$$

implying $k^n = k'$. Then, if for k' an Arrhenius-type temperature dependence is adopted, $k' = k_0' \exp(-Q'/RT)$, it immediately follows that $nQ = Q'$. Both formulations of the JMA equation are equally valid. However, it is of utmost importance to be aware of the numerical difference of the two corresponding effective activation energies, Q and Q', due to a value of the growth exponent, n, different from one. Because both descriptions for the JMA equation are used frequently, comparing data for the effective activation energy from different sources can easily cause misinterpretations and this has led and still leads to a lot of confusion in the literature.

10.13   Experimental Determination of the Degree of Transformation; Dilatometry and Calorimetry

The degree of transformation can be determined experimentally in various ways (see Sect. 10.2 and Fig. 10.3). Two important methods to determine the degree of transformation are briefly introduced here.

Dilatometry (sometimes denoted by the abbreviation TMA (thermomechanical analysis)) is a technique based on the measurement of the change of length of a (transforming) specimen. Evidently, meaningful application of dilatometry requires that the specific volumes of the untransformed and fully transformed states are different; an additional length-change effect, during isochronal annealing in the temperature range of the transformation, can be due to the thermal expansion coefficients of the untransformed and fully transformed specimens being different (see Fig. 10.3b and its caption). In a dilatometer the actual signal recorded is proportional to the quantity p (specimen length (change)), not f, as a function of time, t, or temperature, T.

Differential scanning calorimetry (DSC) and differential thermal analysis (DTA) are techniques based on the (direct (DSC) or indirect (DTA)) determination of the heat production or absorption by a (transforming) specimen[4]; these methods are

---

[4] A genuine DSC apparatus records directly the *difference of the amounts of heat* absorbed/produced by a sample pan (containing the specimen to be investigated) and a reference pan. A DTA apparatus records the *temperature difference* of a sample pan (containing the specimen to be investigated) and a reference pan. By means of calibration with standard specimens of which heats of transformation (often pertaining to melting) are known, the output signal of a DTA apparatus can be presented as a heat produced/absorbed (by the specimen under investigation) rate, i.e. as holds for a genuine DSC apparatus. Commercial apparatus sold as DSC apparatus often actually are DTA apparatus, in the sense discussed here. Hence, for the discussion in this section DSC and DTA used as DSC are treated in the same way simultaneously (as DSC).

Fig. 10.12 Schematic depiction of **a** length change of a specimen in a dilatometric experiment in the region of a phase transformation upon (isochronal) annealing and **b** rate of enthalpy change in a calorimetric experiment in the region of a phase transformation upon (isochronal) annealing $(d\Delta H/dT = (d\Delta H/dt)/\Phi$ with $\Phi = dT/dt$, the constant heating rate in an isochronal annealing experiment; inset in **b**: enthalpy change due to the phase transformation)

(a)

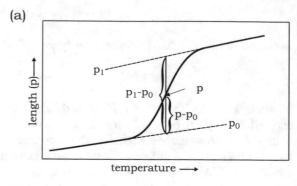

(b)

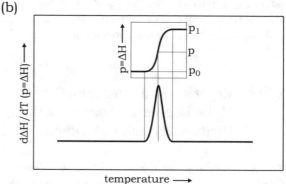

summarized here under the heading *calorimetry*. Clearly, application of calorimetry requires that the change of energy upon phase transformation is associated with the release or absorption of (reaction) heat (enthalpy); an additional heat effect, during isochronal annealing in the temperature range of the transformation, can be due to the specific heat capacities of the untransformed and fully transformed states being different. In an apparatus for calorimetry, the signal recorded is proportional to dp/dt or dp/dT (p represents heat (enthalpy) released or absorbed), not df/dt or df/dT, respectively. The reaction heat released or absorbed is often ascribed, confusingly, to a change of heat capacity of the specimen as well; in the temperature range of the transformation upon isochronal annealing one then speaks of an "apparent heat capacity" as the heat-capacity change in this temperature range is caused by both: the genuine change of heat capacity of the specimen upon phase transformation and the reaction heat released/absorbed by the specimen.

Hence, e.g. upon isochronal annealing, i.e. with constant heating rate, the curves of length change and heat released/absorbed rate could look like as sketched in Fig. 10.12a, b. The calorimetric scan can be considered as a derivative scan of p as function of t or T (dp/dT or dp/dt signal as function of temperature or time[5]),

---

[5] Normally isochronal annealing, i.e. with constant heating rate $\Phi \equiv dT/dt$, is applied, and thus $dp/dT = (dp/dt)/\Phi$.

whereas the dilatometric scan exhibits p as function of t or T (p is proportional to f; cf. Eq. (10.1)). Indeed, upon isochronal annealing, in the temperature range of the transformation, the dilatometric curve shows a sigmoidal shape, whereas a peak is observed in the calorimetric curve (Fig. 10.12a, b). This difference is rather irrelevant; one could simply differentiate the dilatometric curve to obtain a pictorial presentation of the phase-transformation kinetics visually more or less similar to that obtained by the calorimetric scan. Much more important is the recognition that dilatometry and calorimetry can be very differently sensitive to specific phase transformations: upon phase transformation a pronounced change in specific volume (specimen length) may be accompanied with a minor amount of released/absorbed reaction heat of the specimen, and vice versa. Such a consideration can be decisive for the choice of technique to be employed. It should, however, be recognized that the scanning calorimetric techniques discussed here generally are much less accurate for determining relative changes than dilatometry (a high-resolution dilatometer has a length change resolution of 10 nm, implying, for a specimen of length of 10 mm, a relative accuracy of 10^{-4}%; (apparent) heat-capacity changes can normally be determined by DSC techniques not more accurately than with an uncertainty of, say, 5%). This explains a preferred use of dilatometry for analysis of phase-transformation kinetics. In this context it is remarked that most published experimentally determined TTT and CCT diagrams (cf. Sect. 10.1) are derived from dilatometric data.

A value for f is obtained straightforwardly from the dilatometric signal in Fig. 10.12a, recorded upon isochronal annealing, as follows. p_0 and p_1 represent the length changes due to thermal expansion/shrinkage of the untransformed specimen and fully transformed specimen, respectively. By extrapolating $p_0(T)$ and $p_1(T)$ to the temperature range where the transformation occurs, values for $p_0(T')$ and $p_1(T')$ are obtained at the temperatures, T', where f has to be determined. Application of Eq. (10.1) then provides the value of f sought for (see the sketch in Fig. 10.12a). In case of isothermal analysis p_0 and p_1 are constants (do not depend on time: draw horizontal lines in Fig. 10.12a for p_0 and p_1 and replace the temperature coordinate by a time coordinate; see Fig. 10.3a) and the determination of f can be performed in a similar way. Similar procedures hold for other methods where p (and not its derivative with respect to time or temperature) is measured directly (e.g. hardness, magnetization, etc.).

A value for f can be obtained from the DSC signal, recorded upon isochronal annealing as function of temperature, as follows. In the absence of a phase transformation heat is taken up upon temperature rise in accordance with the heat capacity of the initially not transforming specimen. During the phase transformation extra (reaction) heat (enthalpy) is taken up or released. After completed phase transformation heat is taken up upon continued temperature rise in accordance with the heat capacity of the fully transformed specimen. By extrapolation of the heat capacities of the untransformed specimen and the fully transformed specimen to the temperature range where the transformation takes place it is possible in principle to determine separately the heat released/absorbed by the transformation and the degree of transformation.

Fig. 10.13 Rate of enthalpy change divided by the heating rate, $(1/\Phi)\mathrm{d}\Delta H/\mathrm{d}t = \mathrm{d}\Delta H/\mathrm{d}T$, due to isochronal crystallization of amorphous $Pd_{40}Cu_{30}P_{20}Ni_{10}$, at the heating rates indicated, as measured (symbols), and as fitted (lines), by adopting mixed nucleation and volume diffusion controlled growth and randomly dispersed nuclei, after pre-annealing for 600 s at 623 K, 626 K and 629 K (taken from Liu et al. 2004b and 2007). The model parameters were determined by simultaneous fitting to five transformation curves (corresponding to five heating rates) for each pre-anneal; see also Table 10.1

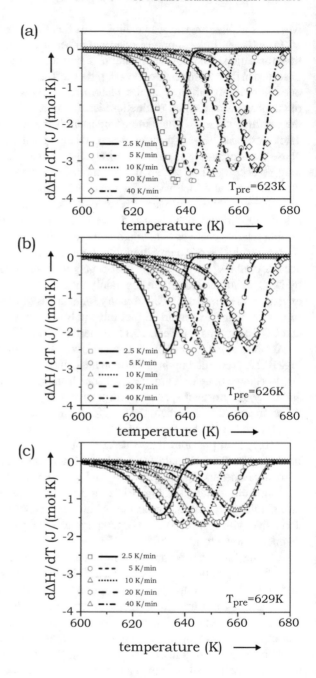

However, it should be recognized that the heat capacity of the transforming specimen, at a certain stage (temperature), is unknown: it is a weighted (by the degree of transformation) mean of the heat capacities of the untransformed specimen and the fully transformed specimen at the temperature considered. Hence, the determination of f is not so straightforward for the DSC scan (dp/dT vs. T scan) as it is for the dilatometric scan (p vs. T scan): for the determination of f by dilatometry the value of the thermal expansion coefficient of the transforming specimen for the stage of transformation f needs not be known (see the procedure described in the one but last paragraph), whereas for the determination by calorimetry the value of the heat capacity of the transforming specimen for the stage of transformation f is needed. At this stage, in view of the discussion further above, one may then suggest to integrate the DSC scan (leading to a curve of heat released/absorbed versus temperature and then apply the procedure as described for the dilatometric scan). This leads to the main problem with the DSC technique: whereas it can be assumed justly that the linear expansion coefficients of the untransformed and the fully transformed specimens are practically constant in the temperature range where the transformation occurs (i.e. the slopes of the $p_0(T)$ and $p_1(T)$ lines are constant; see Fig. 10.12a), this is not generally an acceptable assumption for the heat capacities of the untransformed and fully transformed specimens. A way out of this dilemma for the DSC analysis is an iterative procedure leading simultaneously to values for both the heat of transformation and the degree of transformation as function of temperature, as described by Kempen et al. (2002). In many cases one may proceed less precise: if the difference between the heat capacities of untransformed and fully transformed specimens is modest and if their temperature dependences are not pronounced, both with respect to the amount of reaction heat, some extrapolation of the measured heat-uptake rates, from both sides of the peak in the DSC scan induced by the reaction heat released or absorbed, can be acceptable. After subtraction of such a "base line" from the recorded signal, one obtains the released/absorbed heat (ΔH, with H as enthalpy) rate due to the transformation alone. Such a $d\Delta H/dT$ versus T curve is shown in Fig. 10.12b. The determination of f as function of T then is straightforward. If ΔH_{tot} represents the total heat of transformation (the area under the curve), it holds:

$$\frac{d\Delta H}{dT} = \Delta H_{tot}\frac{df}{dT} \qquad (10.25)$$

and f as function of T follows by stepwise application of Eq. (10.25) (see inset of Fig. 10.12b).

10.14 Fitting of Kinetic Models

Given a certain combination of nucleation, growth and impingement models, the recipe described in Sect. 10.10 leads to explicit calculation of f, either analytically or numerically. Fitting of such calculated results of f as function of t (isothermal annealing) or f as function of T (isochronal annealing) to *single* transformation curves is discouraged: this is an insensitive method in view of the number of fit parameters (about three to six; cf. Liu et al. 2004a, b and 2007; see also the discussion in Sect. 10.11). A number of transformation curves should be determined experimentally (isothermal annealing at various temperatures or isochronal annealing at various heating rates) and the kinetic model should be fitted *simultaneously* to all measured transformation curves. An example of such simultaneous fitting is shown in Fig. 10.13 and discussed next (see also the discussion in Sect. 10.16).

The crystallization of an amorphous alloy is associated with the release of (crystallization) heat. DSC curves, after subtraction of the base line (cf. Sect. 10.13), recorded for five different heating rates, are shown for three different cases of performed pre-anneal in Fig. 10.13a–c. The pre-anneal was intended to produce a certain number of nuclei already at the start of annealing/transformation (i.e. a contribution of site saturation at $t = 0$, as one of the possibly operating nucleation mechanisms, is thereby assured; cf. Sect. 10.5). Fitting was performed employing a wide range of nucleation, growth and impingement models. Good results (values of four fit parameters obtained by simultaneous fitting to five transformation curves, for each pre-anneal) were obtained, in this case adopting mixed nucleation (combination of continuous nucleation and site saturation at $t = 0$ (cf. Sect. 10.5.3)), diffusion-controlled growth and randomly dispersed nuclei (background provided in Sects. 10.5, 10.6 and 10.9). The results obtained (see also Fig. 10.13a–c) for the kinetic model parameters, N_0 (cf. Eq. (10.7), N^* (cf. Sect. 10.5.2), Q_N (cf. Eqs. (10.6) and (10.7)) and Q_D (cf. Eq. (10.8) and Sect. 10.6.2) have been gathered in Table 10.1. A few conclusions can be drawn:

Table 10.1 Kinetic parameters of the modular transformation model, as determined by fitting to (DSC = differential scanning calorimetry) data of the rate of enthalpy change divided by the heating rate, $(1/\Phi)d\Delta H/dt = d\Delta H/dT$, for the case of isochronal crystallization of amorphous $Pd_{40}Cu_{30}P_{20}Ni_{10}$ after different pre-anneals for 600 s at different temperatures T_{pre}

T_{pre} (K)	N^* (m^{-3})	N_0 (s^{-1} m^{-3})	Q_N (kJ/mol)	Q_D (kJ/mol)
620	1.1×10^{19}	4.2×10^{41}	256	330
622	1.3×10^{20}	3.5×10^{41}	255	325
623	2.3×10^{20}	4.1×10^{41}	254	321
625	6.1×10^{20}	5.5×10^{41}	255	315
626	7.4×10^{20}	7.4×10^{40}	255	315
628	2.2×10^{21}	2.2×10^{41}	250	315
629	8.1×10^{21}	2.1×10^{40}	253	320

The model parameters were determined by simultaneous fitting to five transformation curves (corresponding to five heating rates) for each pre-anneal; see also Fig. 10.13. The four fit parameters are: N^* = the number of pre-existing nuclei (Sect. 10.5.2), N_0 = a temperature independent nucleation rate (Sect. 10.5.1), Q_N = the activation energy for nucleation (Sect. 10.5.1) and Q_D = the activation energy for diffusion (Sect. 10.6.2) (data from Liu et al. 2004b and 2007)

Fig. 10.14 Growth exponent n **a** and overall effective activation energy Q, **b** as a function of temperature for isochronal annealing, at the heating rates indicated, for the crystallization of amorphous $Pd_{40}Cu_{30}P_{20}Ni_{10}$ after pre-annealing for 600 s at 626 K (cf. Fig. 10.13b; taken from Liu et al. 2004b and 2007): n and Q depend on temperature

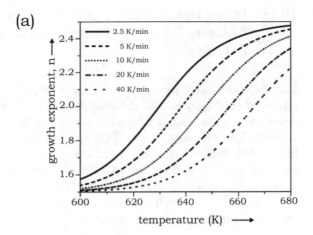

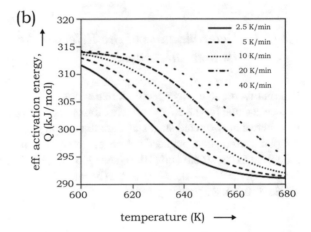

Evidently, the number of nuclei already present at $t = 0$, N^*, increases, and the temperature-independent nucleation rate N_0 of the continuous nucleation mechanism becomes less, with increasing pre-anneal temperature. This is consistent with the expectation: The nucleation mode changes gradually from continuous nucleation to site saturation at $t = 0$ upon increasing pre-anneal temperature.

The values of the activation energies of nucleation and growth (diffusion), Q_N and Q_G, respectively, are practically constant, i.e. do not depend on the pre-anneal performed, as it should be.

In particular for cases of intermediate pre-annealing, the values of the growth exponent, n, and the effective activation energy Q (see Sect. 10.12 and Eq. (10.24)) *do* depend on temperature: see Fig. 10.14a, b, in full accordance with the discussion in Sects. 10.8. and 10.12.

10.15 Direct Determination of the Effective Activation Energy and the Growth Exponent

The discussion in the previous section focussed on full model fitting to the experimental data of the degree of transformation as function of time and temperature. As a result, values of the kinetic parameters controlling the working of the kinetic model were obtained. One may wonder if always such specific model development has to be applied in order to determine the kinetic parameters. It can be shown that, under certain, but fairly relaxed constraints, it is possible to extract from the experimental results, of the degree of transformation as function of time and temperature, a value for in particular the effective activation energy (see Sect. 10.12 and Eq. (10.24)), *without adopting a specific kinetic model.*

10.15.1 Determination of the Effective Activation Energy; Isothermal Anneals

Without recourse to any specific kinetic model, i.e. $F(\beta)$ cf. Eq. (10.2) need not be known, a value for the effective activation energy can be obtained from the lengths of time between two fixed stages of transformation f_1 and f_2, measured at a number of temperatures (f_1 can, but need not be, taken equal to its initial value: 0) (Fig. 10.15a). Adopting Eqs. (10.3a) and (10.4) (see also Eq. (10.17) and its discussion) it follows $k(t_{f_2} - t_{f_1}) = \beta_{f_2} - \beta_{f_1} = $ constant (because $f_2 - f_1 = F(\beta_2) - F(\beta_1) = $ constant; cf. Eq. (10.2)), and consequently

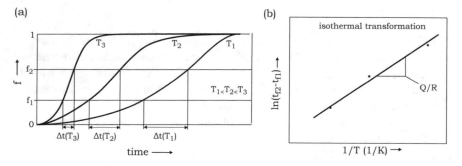

(a)   (b)

Fig. 10.15 a To obtain a value for the effective activation energy from *isothermal* transformations performed at various temperatures (T), the lengths of time, $t_{f_2} - t_{f_1}$, between two fixed stages of transformation, f_1 and f_2, are determined for the various isothermal anneals. **b** Then, in a plot of $\ln(t_{f_2} - t_{f_1})$ versus $1/T$, the effective activation energy can be determined from the slope of the straight line fitted to the data points in this plot. See Eqs. (10.26a, b).

$$\ln\left(t_{f_2} - t_{f_1}\right) = \frac{Q}{RT} - \ln k_0 + \ln\left(\beta_{f_2} - \beta_{f_1}\right) \tag{10.26a}$$

and thus

$$Q = R\,\frac{\mathrm{d}\ln(t_{f_2} - t_{f_1})}{\mathrm{d}(1/T)} \tag{10.26b}$$

Hence, the activation energy can be determined from the slope of the straight line obtained by plotting $\ln(t_{f_2} - t_{f_1})$ versus $1/T$ (Fig. 10.15b). A value for k_0 can only be obtained if β_{f_1} and β_{f_2} are known, implying adoption of a specific kinetic model (i.e. $F(\beta)$ has to be prescribed).

10.15.2 Determination of the Effective Activation Energy; Isochronal Anneals

In accordance with common practice for non-isothermal annealing experiments, only the case of a constant heating rate, $\Phi \equiv \mathrm{d}T/\mathrm{d}t$, is considered (so-called isochronal annealing). Adopting Eqs. (10.3b) and (10.4) β can be approximated by (Mittemeijer (1992); see also footnote 3 in Sect. 10.8):

$$\beta = \frac{T^2}{\Phi} \cdot \frac{R}{Q} \cdot k \tag{10.27}$$

Without recourse to any specific kinetic model, a value for the activation energy can be obtained from the temperatures $T_{f'}$, corresponding to a fixed stage of transformation f' measured for a number of heating rates (Fig. 10.16a). It follows (cf. Eq. (10.27)) that $(T_{f'}^2/\Phi) \cdot (R/Q) \cdot k = \beta_{f'} = \text{constant}$ (because $f' = F(\beta_{f'}) = \text{constant}$; cf. Eq. (10.2)) and consequently

$$\ln\left(\frac{T_{f'}^2}{\Phi}\right) = \frac{Q}{RT_{f'}} + \ln\left\{\frac{Q}{Rk_0}\right\} + \ln \beta_{f'} \tag{10.28a}$$

and thus[6]

$$Q = R\,\frac{\mathrm{d}\ln\left(\frac{T_{f'}^2}{\Phi}\right)}{\mathrm{d}\left(\frac{1}{T_{f'}}\right)} \tag{10.28b}$$

Hence, the activation energy can be determined from the slope of the straight line obtained by plotting $\ln(T_{f'}^2/\Phi)$ versus $1/T_{f'}$ (Fig. 10.16b). A value for k_0 can be

---

[6] Note the typographical errors in the corresponding Eq. (44) in Liu et al. (2007).

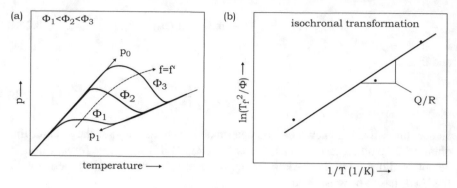

Fig. 10.16   a To obtain a value for the effective activation energy from *isochronal* transformations performed at various heating rates (Φ), the temperatures, $T_{f'}$, corresponding to a certain fixed stage of transformation, f', are determined for the isochronal anneals. **b** Then, in a plot of $\ln(T_{f'}^2/\Phi)$ versus $1/T_{f'}$, the effective activation energy can be determined from the slope of a straight line fitted to the data points in this plot. This is the so-called Kissinger-like analysis (Mittemeijer 1992); see Eqs. (10.28a, b).

obtained if $\beta_{f'}$, is known, implying adoption of a specific model (i.e. $F(\beta)$) has to be prescribed; see Sect. 10.15.3).

Comparing the treatments in Sects. 10.15.1 and 10.15.2 it follows that methods of kinetic analysis for the non-isothermal (isochronal) case on the basis of Eq. (10.28) are full pendants of those derived from Eq. (10.26) for the isothermal case.

10.15.3   *Maximal Transformation Rate and Determination of the Effective Activation Energy*

From Eq. (10.2) (see also Eq. (10.5)) it follows for the maximal transformation rate:

$$\frac{d^2 f}{dt^2} = \frac{d^2 F(\beta)}{d\beta^2} \cdot \left(\frac{d\beta}{dt}\right)^2 + \frac{dF(\beta)}{d\beta} \cdot \frac{d^2 \beta}{dt^2} = 0 \qquad (10.29)$$

In the case of *isothermal* annealing $d^2\beta/dt^2 = 0$ and, consequently, the maximal reaction rate occurs always at *exactly* the same value of β (and thus f) prescribed by $d^2 F(\beta)/d\beta^2 = 0$.

In the case of *isochronal* annealing it follows from Eqs. (10.3a and b):

$$\left(\frac{d\beta}{dt}\right)^2 / \frac{d^2\beta}{dt^2} = \frac{k}{\Phi} \cdot \frac{RT^2}{Q} \qquad (10.30)$$

and thus, for isochronal annealing, using the (approximate) Eq. (10.27):

$$\left(\frac{d\beta}{dt}\right)^2 / \frac{d^2\beta}{dt^2} = \beta \tag{10.31}$$

Therefore it can be concluded that, in case of isochronal annealing, the maximal transformation rate always occurs at *about* (because of the approximate nature of Eq. (10.27)) the same value of β (and thus f) prescribed by (Eqs. (10.29) and (10.31)):

$$\beta\left\{\frac{d^2 F(\beta)}{d\beta^2}\right\} + \frac{dF(\beta)}{d\beta} = 0 \tag{10.32}$$

Hence, the temperature, T_i, where the reaction rate is maximal, i.e. the temperature corresponding to the point of inflection on the curve of f versus t (or T), occurs to a very good approximation at the same value for f *for variable heating rate*.

In the past a family of constant heating-rate procedures has been proposed for the determination of kinetic parameters as activation energies. These methods can all be considered as special cases of the one presented in Sect. 10.15.2: Eq. (10.28) is applied for that stage of transformation where the transformation rate is maximal; i.e. $T_{f'}$, is substituted by T_i in Eq. (10.28) (with "i" indicating the point of inflection on the curve of f (related to p) versus T (e.g. dilatometry; cf. Sect. 10.13), or with i indicating the maximum in the curve of df/dT (related to dp/dT) versus T (e.g. DSC; cf. Sect. 10.13). The best known method of this type is the so-called Kissinger analysis: plotting $\ln(T_i^2/\Phi)$ versus $1/T_i$. The original publication (Kissinger 1957) is one of the most cited papers in the literature on transformation/reaction kinetics.

Kissinger based his analysis on the assumption of homogeneous reactions, whereas most solid-state transformations are heterogeneous (cf. Sect. 9.2), and therefore this analysis cannot be applied apropos of nothing in the latter case, despite the, until and certainly beyond today, enormously large number of applications of this Kissinger-like analysis which were and are performed without giving such proper recognition. It is the treatment leading from Eq. (10.29) to (10.32) that provides a general proof (Mittemeijer 1992) for these Kissinger-like procedures (see conclusion given below Eq. (10.32)). Also note that Eq. (10.28) can be applied for any value of $f = f'$, i.e. the analysis based on Eq. (10.28) not only pertains to the case of f' at T_i as in the Kissinger-like analyses.

For the classical JMA equation, i.e. with time and temperature *in* dependent k_0, Q and n and thus $f = 1 - \exp\{-\beta^n\}$ (Eq. 10.20a), it follows by application of Eq. (10.32) that at maximal transformation rate, i.e. at T_i, $\beta = 1$. Accordingly, the last term at the right-hand side of Eq. (10.28a) vanishes. Then, in the application of Kissinger-like methods, after the activation energy has been determined from the slope of the straight line obtained by plotting $\ln(T_i^2/\Phi)$ versus $1/T_i$, the pre-exponential factor, k_0, can be directly calculated from the intercept of the ordinate at $1/T_i = 0$.

The most severe restriction for all types of analyses considered in Sects. 10.15.1–10.15.3, being based on explicit or implicit assumption of Eqs. (10.3) and (10.4), is the adoption of Arrhenius-type temperature dependences for the nucleation and growth modes (i.e. a high driving force should prevail; cf. Sects. 10.5, 10.6 and 10.10).

10.15.4 Determination of the Growth Exponent; Isothermal Anneals

In contrast with the determination of the effective activation energy Q, as described in Sects. 10.15.1–10.15.3, the determination of the growth exponent n requires, additionally, adoption of, in particular, a specific impingement model. For randomly dispersed nuclei, i.e. impingement according to Eq. (10.19), it straightforwardly follows from the classical JMA equation (Eq. (10.20a)):

$$n = \frac{d\{\ln[-\ln(1-f)]\}}{d\ln t} \tag{10.33}$$

implying that n can be determined from a single isothermal transformation curve as the slope of the straight line obtained by plotting $\{\ln[-\ln(1-f)]\}$ versus $\ln t$.

10.15.5 Determination of the Growth Exponent; Isochronal Anneals

Under the same restrictions as indicated for isothermal annealing (Sect. 10.15.4), it is obtained for randomly dispersed nuclei and more specifically as derived from the classical JMA equation (Eq. 10.20a) using Eq. (10.27) for β[7]:

$$n = -\frac{d\{\ln[-\ln(1-f_T)]\}}{d\ln \Phi} \tag{10.34}$$

with f_T as the degree of transformation at an arbitrary, fixed temperature (e.g. T_i). Hence, provided at least two isochronal anneals of different heating rate have been made, the value of n can be obtained from the slope of the straight line obtained by plotting $\{\ln[-\ln(1-f_T)]\}$ versus $\ln \Phi$.

---

[7] A minus sign as present in Eq. (10.34) is inadvertently missing in the corresponding Eqs. (50–54) in Liu et al. (2007).

10.15.6 Time and Temperature Dependences of the Effective Activation Energy and the Growth Exponent; Determination of the Constant Activation Energies of Nucleation and Growth

Only if extreme boundary conditions are satisfied, it can be assumed that the effective activation energy, Q, and the growth exponent, n, are genuine constants of the phase transformation considered, as has been extensively argued in Sects. 10.8, 10.11 and 10.12; see also the example discussed in Sect. 10.14.

It has been shown that the recipes as described in Sects. 10.15.1 and 10.15.2 for the determination of Q, and as described in Sects. 10.15.4 and 10.15.5 for the determination of n, can also be applied if Q and n vary during the course of the transformation, i.e. if Q and n depend on time and temperature (Liu et al. 2007):

By repeating the analyses based on Eq. (10.26) and (10.28) for a series of (f_1, f_2) and f' values, respectively, the dependence of Q on f is obtained. Here it should be recognized that the dependences of $\ln(t_{f_2} - t_{f_1})$ on $1/T$ (Sect. 10.15.1) and of $\ln(T_f^2/\Phi)$ on $1/T_{f'}$ (Sect. 10.15.2) are no longer given by truly straight lines. However, forced fits of straight lines in these plots are considered as leading, via the slopes of these straight lines, to viable approximations of Q in the concerned (f_1, f_2) range and at the considered value of f', respectively.

For the isothermal determination of n, the dependence of $\{\ln[-\ln(1 - f)]\}$ on $\ln t$ (Sect. 10.15.4) is no longer given by a truly straight line. The local slope at t provides a value for $n(t)$, which can be related to a value of f because the relation between f and t is known experimentally. For the isochronal determination of n, the dependence of $\{\ln[-\ln(1 - f_T)]\}$ on $\ln\Phi$ (Sect. 10.15.5) is also no longer given by a truly straight line. The slope in this plot at a specific value of Φ is considered as leading to a viable approximation of n at the temperature T considered (cf. f_T) for the value of Φ concerned. Repetition of this procedure for various values of T leads to determination of n as function of T, for the value of Φ concerned, which implies that also n as function of f is known, because the relation between f and T for the value of Φ concerned is known experimentally.

Finally, once the values of Q and n have been determined as function of f, the constant values of Q_N and Q_G can be obtained straightforwardly by fitting Eq. (10.24) to the deduced (Q, n) data points. A practical example of such a fit is shown in Fig. 10.17.

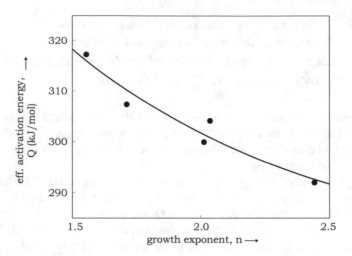

Fig. 10.17 Plot of the effective activation energy, Q, versus the growth exponent, n, obtained by fitting the JMA model (cf. Eq. (10.20)) and the discussion in Sect. 10.11) to data obtained for the transformation kinetics of the isochronal crystallization of $Pd_{40}Cu_{30}P_{20}Ni_{10}$ (cf. Fig. 10.13). The full line drawn in the figure represents a fit of Eq. (10.24) to the data points, involving determination of separate values for the (constant) activation energy of nucleation, Q_N, and the (constant) activation energy of growth, Q_G (taken from Kempen et al. 2002)

10.16   Hierarchy of Models; Some Notes

1. A great many papers have been and are published in which, especially isothermal, transformation-rate data (degree of transformation, f, as function of time at constant temperature) are fitted with the classical JMA-equation (Eq. (10.20)). Such analyses usually yield values for the effective activation energy, Q, and the growth exponent, n (and k_0). Do note that data for Q can also be obtained without implementation of any specific kinetic model (also not the JMA model), but adopting Eqs. (10.26) and (10.28), as described in Sects. 10.15.1 and 10.15.2.

    Then an interpretation of such results may be given. This can be a problematic undertaking, since no direct relation of such a kinetic parameter with a distinct physical process may be apparent (e.g. for Q; for example, see Eq. (10.24)).[8] This can ultimately mean that the kinetic parameters have no physical meaning and that only a phenomenological description of the experiment is presented. Consequently, no knowledge on the operating nucleation and/or growth mechanisms and no microstructural information, as, for example, the density and size of the product particles (see (2) below), can be acquired. It may be that such data, for the parameters (of the JMA equation), still allow application for prediction of transformation kinetics under conditions (modestly) different from

---

[8] This does hold for the classical JMA equation, notwithstanding that specific nucleation and growth processes have been considered separately in its derivation (Sects. 10.5–10.9).

those pertaining to the experiments used in the fitting (which can be useful in technical applications, as in a steel factory), but deep insight in the (nucleation and growth) processes underlying the transformation is not obtained.

2.   The quality of the modelling of transformation kinetics evidently is enhanced if microstructural data are also involved in the fitting. An obvious choice can be the number of product particles at specific times and temperatures. The product-particle density follows from the nucleation rate corrected for impingement (cf. Eq. (10.18a)), as follows:

$$N(t, T) = \int_0^t \dot{N}(\tau) \cdot (1 - f(\tau)) d\tau \qquad (10.35)$$

The product-particle density can be experimentally accessible by, for example, application of transmission electron microscopy. Incorporation of experimental data depending on the kinetic parameters, for nucleation and growth, in a way different from the dependence of the transformation-rate data on the kinetic parameters, allows a more constrained fitting. Indeed, the product-particle density depends only on the nucleation mechanism, not on the growth mechanism. Thereby it becomes possible to arrive at physically plausible values for, for example, Q_N and Q_G, which express themselves, in many cases, in the transformation-rate data largely via an effective activation energy, as described by Eq. (10.24) (Bauer et al. 2011).

As shown in Sect. 10.14, results for Q_N and Q_G can, in principle, also be obtained by fitting to, *simultaneously*, a set of isothermal or isochronal anneals. But this is not always practicable, because of possible lack of sensitivity. Incorporation of microstructural data, which have to be satisfied in the fitting, as here requiring that N as measured equals N as calculated by the model according to Eq. (10.35), as described in the preceding paragraph, then can provide a powerful route to extract genuine values of kinetic parameters characterizing the operating nucleation and growth modes, separately.

3.   The modular approach to modelling transformation kinetics, as summarized in Sect. 10.10, in fact can be considered as a so-called *mean field approach*: the impingement ("particle interference"; cf. Sect. 10.9) is accounted for as the interaction of the developed particles of the product phase with a matrix of the *parent phase of mean degree of transformation* (Eq. (10.18)). The impingement correction, as incorporated in the modular approach, is well suited for the modelling of hard impingement and, to some extent, also of soft impingement (see text below Eq. (10.19)). However, the impingement correction is applied to both the nucleation process and the growth process. In case of precipitation in a matrix, the product-phase particles can grow after nucleation has come to an end and in particular after completed precipitation by a process called coarsening: coarsening is a process where the smaller second phase particles

dissolve, allowing the larger second phase particles to grow (see Sect. 11.3.5). Then an impingement correction as indicated above does not apply. Although the modular approach is and has been used successfully also for precipitation reactions, another mean field approach can in principle be better suited, as discussed in the following.

For modelling phase-transformation kinetics a mean field model generally describes the interaction of a single product particle with a parent phase of to be specified average property. For the models indicated above, this average property has been the average degree of transformation. Another type of interaction of the developed particles of the product phase with the matrix is that with a matrix of the *parent phase of mean composition*. This is an appropriate approach for modelling the kinetics of precipitation of second phase particles in a matrix of a supersaturated parent phase. Hence, considering the matrix of the supersaturated parent phase, during the course of the precipitation (transformation), as being of mean composition, implies that the instantaneous average chemical driving force for precipitation is given by the instantaneous mean composition of the matrix.

Such a specific model for precipitation kinetics has been developed by Kampmann and Wagner (1984). This approach describes nucleation in a classical way (cf. Eq. (10.6)), while growth is considered to be controlled by solute diffusion to product particles of spherical shape. Evidently this transformation model is much more restricted then the modular one. The connection with the thermodynamics of the system is provided along two routes: for nucleation via ΔG^* (cf. Eq. (10.6)) and for product-particle growth via the radius of curvature of the product particle, which, since the product particle is taken spherical, implies that the relation of product-particle growth with the underlying thermodynamics occurs via the size of the particle. The thermodynamic equilibrium compositions of particle and matrix at their common interface are dictated by the radius of curvature of this interface; this is called the Gibbs–Thomson effect or capillary effect (see the discussion in Sect. 11.3.5). On the above basis, the nucleation is composition dependent and the product-particle growth is both composition and size dependent. The model then allows, in each time step, the numerical calculation of the changes of product-particle number densities and product-particle sizes, by numerical integration of the nucleation rate and the growth rate of defined product-particle size classes. Thus, the development of the product-particle size distribution as function of the progress of the precipitation can be calculated and compared with/fitted to the corresponding experimentally determined data.

Whereas the numerical evaluation of the chemical driving force for nucleation (needed for calculation of ΔG^*) is straightforward (e.g. see Hillert 1999), the numerical evaluation of the Gibbs–Thomson effect (to determine the compositions of the parent phase and the growing product-phase particle at their interface) is much more complex (the state of thermodynamic equilibrium, including the energy contribution of the product-phase-particle/parent-phase interface, has to be evaluated). This has led to adoption of simple analytical expressions for the Gibbs–Thomson effect in the Kampmann–Wagner (1984) approach. These expressions are based on solid solution models which generally have no validity for the systems investigated. Severe inconsistency of thermodynamics used in the model thus occurs: incompatible thermodynamic models are used to describe the nucleation and the growth. Until and including the time of writing this text, this has not obstructed application of this approach. Moreover, in published works a contribution of particle/matrix misfit-strain energy has been included in the consideration of nucleation but left out considering the growth, or vice versa. This does not lend trustworthiness to results thus obtained.

It has recently been shown, for a system composed of product-phase particles in a parent-phase matrix, that a single evaluation of thermodynamic equilibrium, common for both nucleation and growth, is possible; the so-called inverse evaluation method. On this basis one can evaluate numerically, in a consistent, physically sound and realistic manner, the nucleation barrier, the critical radius and the Gibbs–Thomson effect (Rheingans and Mittemeijer 2015a, b).

4. Mean field approaches for transformations kinetics, discussed above, have their limitations. As a matter of fact, they depart from the assumption that the driving force for transformation and the nucleation probability and the growth rate are everywhere in the parent phase the same, i.e. it is implied that the initial microstructure is a continuum without structure. Also individual interactions of the product particles are not accounted for. Consequently an exact description of the developing and resulting microstructure cannot be obtained: no spatially resolved microstructural information is accessible in this way. Yet, characteristics of the microstructure can be extracted, e.g. the product-particle number density or, even, the product-particle size distribution, as discussed above.

Atomistic simulations could in principle deliver all microstructural details on a spatially resolved and atomic scale, provided the atomic interactions can be (more or less) exactly described and provided the computational power suffices to consider systems large enough (also to exclude that system boundaries affect the results obtained). Such work, satisfying the conditions mentioned, has not been performed in a rigorous way until now, in fact is considered as impossible at present. This leaves unimpeded that details and useful information are obtained by such simulations, although they are imperfect in the above sense. For example, clustering of solute atoms as a pre-precipitation stage has been studied in this way, etc. Furthermore, simulations on *mesoscopical* scales can be very useful (cf. Raabe 1998). The latter type of simulations can, for example, deal with impingement in an explicit way and apply local rules for growth of product particles. However, in the end it may be concluded that, obtaining a correct description of both the transformation kinetics and the developing and final microstructure, requires such an amount of detailed information on the initial microstructure that this approach can become impracticable (for the power and limitations of this approach, studied for the case of recrystallization (cf. Chap. 11), see Jägle and Mittemeijer 2012).

One may wonder if one must strive for performance of (more or less) perfect, atomistic simulations of the entire course of a transformation. The materials

scientist needs models to describe material behaviour in terms of properties important for practical applications. Such models employ microstructural
parameters allowing prediction of the effect of transformations on material properties, as for example depending on the size distribution of the product particles;
in this sense knowledge of/tracing of the movements of all atoms in the system
does not offer a useful instrument. In this context, the reader is referred to
Sect. 1.4 and the opening paragraphs of this chapter. The mean field models
are much simpler and faster to apply than three-dimensional atomistic simulations or even three-dimensional simulations on larger than atomic scales. If
the mean field models are physically sound and applied correctly (deficiencies
with respect to both these requirements have been indicated in this section), then
they provide the currently best combination of correctness and practicability to
describe phase-transformation kinetics.

Epilogue: The Coupling of Thermodynamics to Kinetics

How to relate the energy landscape of a system (cf. Fig. 9.2) to the kinetics
of a phase transformation for that system is an in general unsolved problem.
Even if the track followed in the energy landscape is known, a universally
valid approach to predict phase-transformation kinetics is not available at
present. Possible treatments restrict themselves usually to the driving force ,
$-\Delta G$, and the activation energy, ΔG_{act}, as the only thermodynamic (i.e. energetic) parameters to be included in a kinetic formalism (see the discussion in
Sect. 9.1). Within this context, this book presents three important examples
of such coupling of thermodynamics to kinetics, which are listed, in a now
retrospective way, below.

(i)   The diffusional flux

The mobile particles in a system, which does not possess an equilibrium
distribution of its component particles (e.g. atoms in a solid solution), will on
average move in a direction to lower their energy. The larger the local energy
gradient, the larger the corresponding local flux of mobile particles of a specific
component will be. This is the background for the proposed diffusional flux
equation in Sect. 8.1:

$$J = -\text{constant.}\frac{d\,(\text{energy})}{dx} \tag{8.1}$$

The "energy" parameter in Eq. (8.1) can be interpreted as the partial Gibbs
energy, i.e. chemical potential, of the diffusing component considered (cf.
Eq. (7.7)). The constant in Eq. (8.1) contains an exponential term of the type
$\exp(-Q/RT)$, where Q is the activation energy, that equals the enthalpy part
of ΔG_{act} (cf. Eq. (8.25)–(8.27)).

An elaboration of this concept is presented in Sect 8.7.2.

(ii)   *The nucleation frequency*

For a phase transformation proceeding by nucleation and growth, nucleation means the generation of product-phase particles of supercritical size. Then the thermodynamically determined energy barrier for nucleus formation, ΔG^*, can be related to the nucleation rate by (cf. Sect. 10.5):

$$\dot{N}(T(t)) = C\omega \exp\left(-\frac{\Delta G^*(T(t)) + Q_N}{RT(t)}\right) \tag{10.6}$$

In the more common nucleation theories, based on the assumption of pronounced undercooling or overheating, this formalism for the nucleation rate is simplified such that the dependence on ΔG^* is neglected (see Sect. 10.5). Thereby a direct coupling of thermodynamics to kinetics is lost.

As a side remark, it is noted that for particle growth a relation between thermodynamics and particle-growth rate has been mentioned in Sect. 10.16 under note 3. The Gibbs–Thomson effect leads to a value for the solute concentration in the matrix at the particle/matrix boundary different from the value of the solute concentration in the bulk of the matrix. This concentration difference can lead to dissolution or growth of the particle according to a diffusion process. Thereby the coupling of thermodynamics and kinetics is of the same type as mentioned under (i) above.

(iii)   *The product/parent interface velocity*

The velocity of the interface between a product phase and the surrounding parent phase can be related to the driving force of the transformation, $-\Delta G$, according to (cf. Sect. 10.6):

$$v(T(t)) = v_0 \exp\left(-\frac{\Delta G^a}{RT(t)}\right)\left(1 - \exp\left(\frac{\Delta G}{RT(t)}\right)\right) \tag{10.9}$$

with ΔG^a as the activation energy for the transfer of atoms through the product/parent interface. Again (cf. above), the coupling of thermodynamics to kinetics gets lost for cases of pronounced undercooling or overheating (cf. Eq. (10.10)). For small undercooling or overheating (as could pertain to recrystallization (cf. Sect. 12.2) the coupling is retained (cf. Eq. (10.11)).

References

General

J.W. Christian, *The Theory of Transformations in Metals and Alloys, Part I, Equilibrium and General Kinetic Theory*, 2nd edn. (Pergamon Press, Oxford, 1975).

G. Kostorz (ed.), *Phase Transformations in Materials* (Wiley-VCH, Weinheim, Germany, 2001)

F. Liu, F. Sommer, C. Bos, E.J. Mittemeijer, Analysis of solid state phase transformation kinetics: models and recipes. Int. Mater. Rev. **52**, 193–212 (2007)

E.J. Mittemeijer, Analysis of the kinetics of phase transformations. J. Mater. Sci. **27**, 3977–3987 (1992)

Specific

M. Avrami, Kinetics of phase change. I. General theory. J. Chem. Phys. **7**, 1103–1112 (1939)

M. Avrami, Kinetics of phase change. II. Transformation-time relations for random distribution of nuclei. J. Chem. Phys. **8**, 212–224 (1940)

M. Avrami, Granulation, phase change, and microstructure. Kinetics of phase change. III. J. Chem. Phys. **9**, 177–184 (1941)

R. Bauer, B. Rheingans, E.J. Mittemeijer, The kinetics of the precipitation of Co from supersaturated Cu-Co alloy. Metall. Mater. Trans. A **42A**, 1750–1759 (2011)

C. Bos, F. Sommer, E.J. Mittemeijer, An atomistic analysis of the interface mobility in a massive transformation. Acta Mater. **53**, 5333–5341 (2005)

M. Hillert, *Phase Equilibria, Phase Diagrams and Phase Transformations; Their Thermodynamic Basis*, 2nd edn. (Cambridge University Press, 2012)

E.A. Jägle, E.J. Mittemeijer, Interplay of kinetics and microstructure in the recrystallization of pure copper: comparing mesoscopic simulations and experiments. Metall. Mater. Trans. A **43A**, 2534–2551 (2012)

W.A. Johnson, R.F. Mehl, Reaction kinetics in processes of nucleation and growth. Trans. Am. Inst. Min. Metall. Eng. **135**, 416–458 (1939)

R. Kampmann, R. Wagner, Kinetics of precipitation in metastable binary alloys—theory and application to Cu1.9at.%Ti and Ni14at.%Al, in *Decomposition of Alloys: the Early Stages* ed. by P. Haasen, V. Gerold, R. Wagner, M.F. Ashby (Pergamon Press, 1984), pp. 91–103

A.T.W. Kempen, F. Sommer, E.J. Mittemeijer, The kinetics of the austenite-ferrite phase transformation of Fe-Mn: differential thermal analysis during cooling. Acta Mater. **50**, 3545–3555 (2002)

H.E. Kissinger, Reaction kinetics in differential thermal analysis. Anal. Chem. **29**, 1702–1706 (1957)

A.N. Kolmogorov, On the statistics of crystallization in metals. Izv. Akad. Nauk SSSR Ser. Mat. **3**, 355–359 (1937) (in Russian with abstract in German)

F. Liu, F. Sommer, E.J. Mittemeijer, Parameter determination of an analytical model for phase transformation kinetics: application to crystallization of amorphous Mg-Ni Alloys. J. Mater. Res. **19**, 2586–2596 (2004a)

F. Liu, F. Sommer, E.J. Mittemeijer, Determination of nucleation and growth mechanisms of the crystallization of amorphous alloys. application to calorimetric data, Acta Mater. **52**, 3207–3216 (2004b)

H. Nitsche, F. Sommer, E.J. Mittemeijer, The Al nano-crystallization process in amorphous $Al_{85}Ni_8Y_5Co_2$. J. Non-Crystall. Solids **351**, 3760–3771 (2005)

D. Raabe, *Computational Materials Science* (Wiley-VCH, 1998)

B. Rheingans, E.J. Mittemeijer, Modelling precipitation kinetics: evaluation of the thermodynamics of nucleation and growth. CALPHAD: Comput. Coupling Phase Diag. Thermochem. **50**, 49–58 (2015a)

B. Rheingans, E.J. Mittemeijer, Analysis of precipitation kinetics on the basis of particle-size distributions. Metall. Mater. Trans. A **46A**, 3423–3439 (2015b)

J. Sietsma, S. van der Zwaag, A concise model for mixed-mode phase transformations in the solid state. Acta Mater. **52**, 4143–4152 (2004)

Chapter 11
Recovery, Recrystallization and Grain Growth

Recrystallization has been identified as a process in metallic solids since the "old days" (last part of the nineteenth century), when it was supposed that cold working of a metallic workpiece destroyed its crystallinity and that subsequent heating restored the crystalline nature by a process then naturally coined with the name "recrystallization". Nowadays, we would define recrystallization as a process that leads to a change of the crystal orientation (distribution) for the whole polycrystalline specimen, in association with a release of the stored strain energy as could have been induced by preceding cold work: a new microstructure results (Fig. 11.1). Recrystallization restores the properties as they were before the cold deformation. Recrystallization (and recovery and grain growth) occurs in all types of crystalline materials, so not only in metals. However, metals are the only important class of materials capable of experiencing pronounced plastic deformation at relatively low temperatures (i.e. low with respect to the melting temperatures), which explains that most of the corresponding research has been and is performed on metallic materials.

The industrial need for understanding the effects of deformation in material-forming production steps and of subsequent annealing processes is obvious. Then it may come as a surprise that even about 150 years of research in this area have not led to comprehensive models describing these processes on the basis of fundamental insight such that reliable application for technological purposes can be guaranteed. One of the main reasons for this deficiency is undoubtedly our still limited understanding of the plastically deformed state (cf. Chap. 12).

Recovery, implying a decrease of the density and a redistribution of defects in the deformed solid, precedes recrystallization. Grain growth can occur in the recrystallized microstructure. Thereby, the sense of a treatment of recovery, recrystallization and grain growth, in this order in this chapter, has been validated. Yet, it is recognized that overlapping of these processes can occur in a significant way.

© Springer Nature Switzerland AG 2021
E. J. Mittemeijer, *Fundamentals of Materials Science*,
https://doi.org/10.1007/978-3-030-60056-3_11

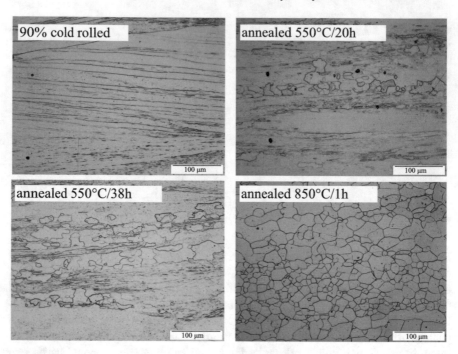

Fig. 11.1 Optical micrographs showing the microstructure of an Fe-4.65at%Al alloy after cold rolling to a degree of deformation of 90% (i.e. a reduction of sheet thickness of 90%), and after subsequent anneals at temperatures and for times as indicated in the micrographs. Upon progressive annealing the elongated grain morphology resulting after cold rolling is gradually replaced by a more or less equiaxed grain morphology as the result of recrystallization, involving the nucleation and growth of new grains in the deformed microstructure (micrographs made by S. Meka, Max Planck Institute for Metals Research)

11.1   Recovery

The defects introduced by plastic deformation processes, as cold rolling, and of importance in subsequent recovery and recrystallization processes, are predominantly dislocations. Point defects, as vacancies, are also introduced upon plastic deformation, but these are usually already annealed out at low temperatures (e.g. in copper at temperatures below room temperature). In particular if the stacking fault energy is low (as holds for silver; copper has a stacking fault energy about three times larger than that of silver and aluminium has a high stacking fault energy about eight times larger than that of silver), dissociation of the dislocations occurs, cross-slip is hindered (cf. Sect. 5.2.8) and twinning becomes a preferred mode of plastic deformation. Also, if not enough slip systems are available, as can occur with hexagonal metals (cf. Sect. 5.2.8), the initial plastic deformation can occur by slip (dislocation glide), but deformation twinning can become important upon progressing plastic deformation.

Recovery, as induced by annealing after plastic deformation, leads to a change of the dislocation microstructure and thereby a *partial* restoration of the material properties as before the plastic deformation is realized. It should be remarked that recovery processes can also operate in materials containing dislocations and non-equilibrium amounts of point defects (as vacancies) and which have *not* been subjected to pronounced plastic deformation by the exertion of external mechanical loads: for example, irradiation (bombardment) by accelerated particles (e.g. ions) induces such a defect structure. In this last case recovery can restore *fully* the original material properties. In this section the discussion is confined to recovery in materials deformed plastically such that distinct permanent shape changes have resulted (as by cold rolling).

During the rearrangement/partial annihilation of the dislocations in the process of recovery, the grain boundaries in the material do not move; the recovery process occurs more or less homogeneously throughout the material, in flagrant contrast with recrystallization, characterized by the sweeping of high-angle grain boundaries through the deformed matrix, which process thus takes place explicitly heterogeneously (see Sect. 11.2 and the discussion on homogeneous and heterogeneous transformations in Sect. 9.2).

The above discussion could be conceived as that recovery is induced, after the plastic deformation (by cold work), by annealing at an appropriate, elevated temperature (say, distinctly below half of the melting temperature in Kelvin). However, if the plastic deformation occurs at elevated temperature (as by hot rolling), recovery processes already run while the material is still deforming; one then speaks of dynamic recovery (similarly, one recognizes dynamic recrystallization).

11.1.1 Dislocation Annihilation and Rearrangement

The driving force for the migration of the dislocations leading to a different dislocation configuration and/or to a partial annihilation of dislocations is a reduction of the strain energy incorporated in the strain fields of the dislocations. This decrease of the stored energy in the material obviously decreases the driving force for the (largely) subsequent recrystallization (recovery and recrystallization may overlap; see later).

The annihilation of dislocations can occur by various mechanisms. Dislocations can migrate by glide along a single slip plane, by cross-slip and by climb (see Sects. 5.2.5, 5.2.6 and 5.2.7).

Evidently (edge) dislocations of opposite sign (cf. Sect. 5.2.3) on the same slip plane can become annihilated by gliding to contact (Fig. 11.2).

If these two initial dislocations of opposite sign are of edge type and on two different glide planes, their possible annihilation requires a combination of climb and glide processes (Fig. 11.3). The climb step is outspokenly thermally activated (cf. Sect. 5.2.7), implying that, according to the mechanism considered here, dislocation annihilation can only occur at elevated temperatures. If the two initial dislocations

Fig. 11.2 Annihilation of
two (edge) dislocations of
opposite sign (cf. Sect. 5.2.3)
by glide

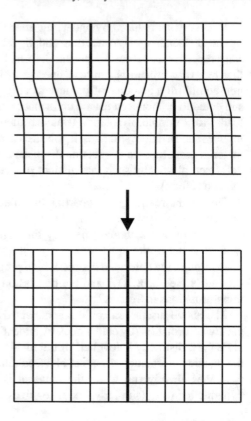

on two different glide planes are of screw type, their annihilation can be established
by cross-slip.

Dislocations may also glide along a slip plane and upon "colliding" with a grain
boundary be incorporated into the grain-boundary structure. Thereby the dislocation
as an isolated defect may loose its identity by local atomic shuffles in the grain
boundary, in association with the loss of strain energy and, in this sense, annihilation
of the dislocation has occurred as well.

Release of strain energy can also be realized by rearrangement of the dislocations
in a single grain of the material. Evidently, if the numbers of dislocations of opposite
sign are unequal, complete dislocation annihilation by any of the first two processes
mentioned above is impossible. The presence of unequal numbers of dislocations
of opposite sign can be the result of bending of a single grain experiencing glide
along a single slip plane: a curved grain results by an excess of edge dislocations
of the same type (cf. the discussion on "geometrically necessary dislocations" in
Sect. 12.14.2); see Fig. 11.4a. Upon annealing, these edge dislocations can strive for
arrangements in "walls" and thus form low-angle tilt boundaries (cf. Sect. 5.3.1).
This rearrangement is realized by climb and short-range glide (see Fig. 11.4b). The
overlapping of "tensile" and "compressive" parts of the long-range strain fields of
neighbouring dislocations in the dislocation wall provides the release of strain energy

Fig. 11.3 Annihilation of
two edge dislocations of
opposite sign (cf. Sect. 5.2.3)
by climb and glide

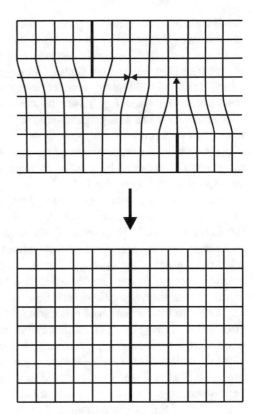

Fig. 11.4 a Bending of a
single grain experiencing
glide along a single slip
plane: a curved grain results
by an excess of edge
dislocations of the same type.
b Upon annealing, these
edge dislocations can strive
for arrangements in "walls",
by climb and short-range
glide, and thus form
low-angle tilt boundaries:
Polygonization of a bended
grain by rearrangement of
edge dislocations

(a)

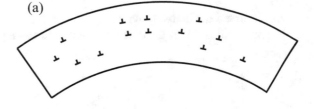

(b)

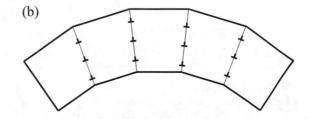

that is the driving force for this process; see the discussion of Eq. (5.15) in Sect. 5.3.1. As a result of the formation of these dislocation walls/low-angle tilt boundaries, the originally (i.e. after the plastic deformation) curved crystal-structure planes of the grain considered become the sides of a polygon: a series of *subgrains* has formed which are slightly differently oriented with respect to each other (with a view to the configuration shown in Fig. 11.4b: the subgrains are slightly rotated with respect to each other around an axis perpendicular to the plane of the drawing). One therefore names this phenomenon: *polygonization*. The process is revealed in X-ray diffraction patterns by the replacement of strongly broadened reflections, observed after the plastic deformation, by a series of neighbouring discrete spots, observed upon subsequent annealing (Cahn 1949).

The simple calculation for the energy per unit area of a low-angle tilt boundary, as given by Eq. (5.15), holds for an infinitely long dislocation wall. In practice, the dislocation walls (segments of low-angle tilt boundaries) in the polygonized microstructure may comprise ten dislocations and less. The process of aligning of edge dislocations of the same sign has also been observed for the misfit dislocations originally present in the interface of an *A/B* bicrystal (see Sect. 5.3.4). In that case diffusion annealing (specific observations were made for a thin Cu/Ni bicrystalline film) leads to the formation of dislocation walls initially comprising even only two edge dislocations (Fig. 11.5). This can be conceived as an extreme case of polygonization, where the driving force also is the release of dislocation-strain energy, albeit the dislocations were not induced by external mechanical action. Although the process is always driven by the release of dislocation-strain energy, the energy gain per dislocation for dislocation walls of (such) limited length cannot be assessed by application of Eq. (5.15); a numerical approach is required (Beers and Mittemeijer 1978).

As discussed qualitatively with respect to Eq. (5.15) already, the energy of the dislocation wall per unit area increases with increasing dislocation density of the

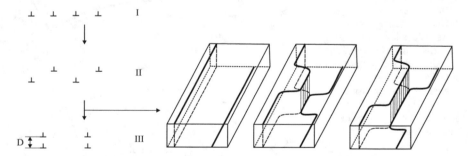

Fig. 11.5 (Edge) Misfit dislocations originally located in the interface of an *A/B* bicrystal (top part of left part of the figure; see "I") upon annealing can move away from the interface by climb (see "II") and, subsequently, by glide can become aligned on top of each other (see "III"). A schematic depiction of this alignment process is shown in the right part of the figure: the dislocations align part by part. Thus dislocation walls can be formed initially comprising only two dislocations. Such observations have been made for Cu/Ni bicrystals (Beers and Mittemeijer 1978)

wall (increase of θ, decrease of D; cf. Eq. (5.14)), but the *energy per dislocation* in the small-angle boundary decreases with increasing dislocation density (increase of θ, decrease of D). Hence, after the polygonization has started, a driving force exists for enhancing the size of the subgrains (cf. Fig. 11.4b) by merging of adjacent dislocation walls/small-angle tilt boundaries.

The obvious mechanism to cause such subgrain coarsening is based on the migration and merging of low-angle boundaries. The migration rate of low-angle boundaries, as symmetrical tilt boundaries, by glide of the edge dislocations, composing the boundary, on their parallel slip planes, is relatively high.

An alternative mechanism leading to subgrain coarsening is the coalescence of adjacent subgrains preceded by subgrain rotation. The driving force for subgrain rotation is understood on the basis of, again, Eq. (5.15). Consider subgrain 1 with its surrounding neighbouring subgrains 2, 3, ... (cf. Fig. 11.6). The decrease, by rotation of subgrain 1 with respect to its surrounding, static neighbours, of the misorientation along the boundary 1/2, separating subgrain 1 from subgrain 2, at the same time, will be associated with decreases or increases of the misorientations along the other boundaries of subgrain 1 with its neighbouring subgrains. Now, for the same change of misorientation, as described by the change of the angle θ, the change of energy according to Eq. (5.15) is the larger the smaller the misorientation, θ. Hence, there is a driving force for making the misorientation of the lowest-angle boundaries (even) smaller, as the cost for making, unavoidably and simultaneously, the misorientation of other, larger-angle boundaries larger, is smaller, because the energy gain (release) for the lowest-angle boundaries is larger per unit area boundary than the energy cost (absorption) for the other, larger-angle boundaries. So, provided the ratio of total amount lowest-angle boundary area and of total amount of larger-angle boundary area is not too small, the subgrain 1 can release energy by rotation such that the lowest-angle boundaries decrease their misorientation and the larger-angle boundaries increase their misorientation (Li 1962). Eventually, the misorientation along the boundary 1/2 vanishes, i.e. coalescence of subgrains 1 and 2 has effectively been realized. This subgrain rotation can be achieved by emittance of dislocations from the lowest-angle boundaries and their migration, as by climb and glide, to the larger-angle boundaries. Additionally, local atomic shuffles in the boundary regions can occur. This intellectually appealing mechanism for subgrain coarsening, by subgrain rotation and coalescence, is a feasible one from an energy point of view. However, conclusive experimental evidence for its importance for the subgrain coarsening occurring in reality lacks, and it has been concluded that subgrain coarsening is dominated by the above first discussed migration of low-angle boundaries (Humphreys and Hatherly 2004).

The simple picture sketched above provides a basis for understanding complex phenomena occurring in complicated dislocation microstructures which result from severe plastic deformation. In a pronounced stage of deformation of a ductile material (as a metal) the dislocations gather in regions of high dislocation density and a dislocation-cell structure develops within the grains, with a high dislocation density in the cell walls and a small dislocation density in-between (cf. Sect. 12.14.1).

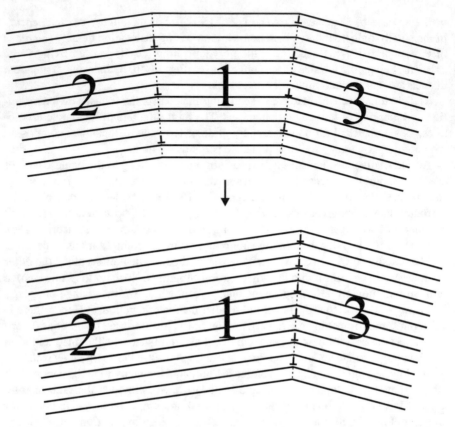

Fig. 11.6 Coalescence of adjacent subgrains by subgrain rotation: By climb and glide of dislocations from the lower-(misorientation) angle grain boundary 1/2 to the higher-(misorientation) angle
grain boundary 1/3, associated with rotation of grain 1, the grain boundary 1/2 is eliminated, while
the misorientation angle of the grain boundary 1/3 is enlarged. This leads to a lowering of the total
dislocation-strain energy of the system (see text)

Annealing-induced recovery in such a microstructure replaces the tangled configuration of the dislocations in the cell walls into more regular arrangements as in
low-angle boundaries and distinctly reduces the dislocation density within the cells.
In the sense discussed above, one can say that the deformation cells have become
subgrains.

The formation of subgrains should not be considered as a recrystallization process:
the orientation (distribution) does not change significantly by the above described
processes of subgrain formation. But the subgrains discussed here can play a role in
the initiation of recrystallization (see Sect. 11.2).

11.1.2   Kinetics of Recovery

The recovery process occurs more or less homogeneously throughout the material. Consequently, the theory of heterogeneous transformations as dealt with in Sects. 10.4−10.15 has no direct relevance for recovery (to a large extent; but see the remark on the determination of the effective, overall activation energy below).

For homogeneous reactions, the probability for the transformation to occur is the same for all locations in the virginal system considered. As a result, the transformation rate decreases monotonically from $t = 0$ onwards. The prescription for the degree of transformation, f, according to Eqs. (10.2) and (10.3), implying dependence only on the "path variable", β, is also fully compatible with the well-known result for homogeneous reactions (cf. Mittemeijer 1992):

$$(1 - f)^{1-m} = 1 - \beta(1 - m) \quad \text{for } m > 1 \tag{11.1a}$$

$$\ln(1 - f) = -\beta \qquad \text{for } m = 1 \tag{11.1b}$$

where m is the so-called order of reaction (known from chemical reaction kinetics). The degree of transformation (here: degree of recovery) can be defined as indicated by Eq. (10.1), where p can be a physical parameter as the electrical resistivity, the hardness, the enthalpy (heat released), the yield limit (cf. Sect. 12.9), etc.

Recipes for the determination of the effective, overall activation energy of the homogeneously occurring recovery, described by Eq. (11.1), are the same as described for heterogeneous transformations in Sects. 10.15.1 and 10.15.2.

Values for kinetic parameters, as the effective, overall activation energy, obtained by fitting expressions as Eq. (11.1) to experimental data, for a parameter p varying upon recovery, may be difficult to interpret. Recovery can be a composite process where various subprocesses may contribute simultaneously (cf. the discussion on and the unravelling of the effects of nucleation, growth (and impingement) modes on the overall kinetics of heterogeneous transformations in Chap. 10). Also, subprocesses may occur consecutively, prohibiting a direct application of Eq. (11.1).

11.2   Recrystallization

The heterogeneous formation of new, strain-free grains growing, by a migrating high-angle grain boundary, into the deformed matrix typifies the recrystallization process. This immediately indicates the driving force for recrystallization: the complete release of the strain energy induced by the preceding process of cold work and as remaining after the subsequent recrystallization-foregoing recovery. Hence, the driving force, $-\Delta G_{recryst}$, is given by (cf. Sect. 5.2.4 and Eq. (5.8)):

$$-\Delta G_{recryst} = E_{elastic} = \text{const} \cdot \rho_d \, G \, b^2 \tag{11.2}$$

with the "const" having a value between 0.5 and 1.0 (see below Eq. (5.8)) and ρ_d as the dislocation density removed by the recrystallization. Strongly deformed, cold rolled metals exhibit dislocation densities as large as 5×10^{15} m^{-2} (cf. Sect. 5.2.3). Taking G and b as for b.c.c. iron (ferrite) and the "const" equal to 1.0 it follows: $-\Delta G_{recryst}$ equals about 2.6×10^7 Pa $= 2.6 \times 10^7$ Nm/m$^3 = 2.6 \times 10^7$ J/m^3, which corresponds to about 0.18 kJ/mol, which should be considered as an upper estimate. This can be compared with the driving force of phase transformations as considered in Chap. 9. Obviously, in principle the driving force for a phase transformation, e.g. the transformation of phase α into phase β, can be very small: at the equilibrium temperature, the driving force, $-\Delta G = G_\alpha - G_\beta$, equals zero. However, many phase transformations are induced remote from the state of equilibrium: for example, the decomposition of a supersaturated solid solution (retained by quenching), α', into the equilibrium phases α and β (see Fig. 9.1 and its discussion in the introduction of Chap. 9), for which the driving force is given by $-\Delta G = -\{(G_\alpha + G_\beta) - G_{\alpha'}\}$. This last driving force can be of the order of 1 kJ/mole. It can thus be concluded that the driving force for recrystallization is rather small as compared to that of the last category of phase transformations.

Recrystallization phenomena have also been observed upon interdiffusion, as in the diffusion zone of A/B diffusion couples (e.g. in thin Cu/Ni bicrystalline films; Mittemeijer and Beers 1980); see the discussion in Sect. 8.9.2. (Note that for the example of the thin Cu/Ni bicrystalline films, also a special variant of polygonization was observed upon annealing; see Fig. 11.5 and its discussion in Sect. 11.1.1).

11.2.1   "Nucleation" of Recrystallization

Recrystallization was formerly conceived as a heterogeneous phase transformation in the sense of the treatment in Sect. 9.2. However, this can be considered a problematic point of view: nucleation as discussed in Sect. 9.2 does not occur in recrystallization. Thermally induced fluctuations in the deformed microstructure do *not* lead to the formation of a strain-free nucleus (particle of supercritical size; cf. Sect. 9.2) separated by a high-angle grain boundary from the matrix.

The above statement can be illustrated by straightforward application of the treatment in Sect. 9.2. Consider Eq. (9.3). Replace ΔG_{chem}^v by $\Delta G_{recryst}$ according to Eq. (11.2), take the interfacial energy, γ, equal to that for a high-angle grain boundary (i.e. of the order 1 J/m^2), and recognize that for recrystallization ΔG_{strain}^v is nil. Then it can be calculated from Eq. (9.3) that the critical Gibbs energy of nucleus formation, ΔG^* (cf. Eq. (9.5)), is very large, in association with a large value of the size for the particle of critical size (cf. Eq. (9.4)). Obviously, this is due to the relatively small driving force (cf. Eq. (11.2) discussed above) and the relatively large value for the interfacial energy. Hence, the nucleation rate, as given by Eq. (10.6), becomes very small. This consideration makes likely that initiation of recrystallization is not a nucleation process according to the theory for heterogeneous phase transformations dealt with in Chap. 9. What then are viable mechanisms for initiating recrystallization?

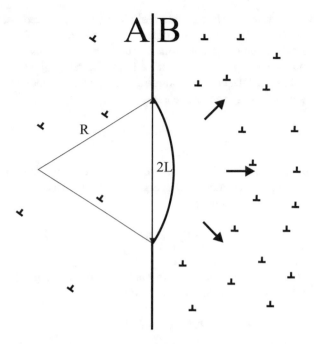

Fig. 11.7 Schematic depiction of strain-induced grain-boundary migration. The inhomogeneity of the deformed microstructure can bring about that a grain A of relatively low stored energy (dislocation poor) is adjacent to a grain B of relatively high stored energy (dislocation rich). Bulging out of the (high-angle) A/B grain boundary into grain B under simultaneous elimination of the surplus stored energy (annihilation of dislocations by the advancing grain boundary) releases stored (deformation) energy and thereby provides a possible mechanism for the initiation of recrystallization. Note: the dislocations as indicated in the figure are identical in grain A (same **b** and **l**; cf. Sect. 5.2.3) and identical in grain B. This has only been done to suggest that grain A and grain B have different crystallographic orientations; of course, in reality, dislocations of varying orientation of **l** and different orientations of **b** can occur in both grains

If genuine nucleation of a strain-free grain, separated by a mobile high-angle grain boundary from the deformed matrix, is impossible, it appears natural to look for regions in the deformed microstructure the growth of which would lead to a reduction of the stored energy in the specimen. In other words, the heterogeneity of the deformed microstructure may provide the key to the initiation of the recrystallization process.

Strain-induced grain-boundary migration is thought to be initiated at a high-angle grain boundary in the deformed microstructure where the dislocation density at both sides of the boundary is significantly different due to the previous (cold) work, which can be a consequence of the dependence on crystal orientation of a grain to applied external loading. The situation can be as sketched in Fig. 11.7, where a (high-angle[1])

---

[1] A *high-angle* grain boundary is required as such a grain boundary has a sufficiently high mobility for bringing about substantial recrystallization, whereas a low-angle grain boundary in this sense has a too low mobility (for the notion "mobility" see Eq. (10.11) in Chap. 10).

grain boundary separates crystals A (relatively low value of stored energy per unit volume, E_A) and B (relatively high value of stored energy per unit volume, E_B). A part of the grain boundary can bulge out, from A into B under simultaneous elimination of a surplus stored energy per unit volume, ΔE_d, in the range given by the possible extremes, $E_B - E_A$, and if the bulging volume approximates a dislocation-free crystal, E_B. Thus, the gain in energy (energy released) is:

$$\Delta E_{strain} = \Delta V \cdot \Delta E_d \tag{11.3}$$

where ΔV is the volume of the "bulge". However, the extension of the grain-boundary area by the "bulging" costs interfacial energy per unit interface area, γ. Thus, the cost in energy (energy absorbed) is:

$$\Delta E_{gb} = \Delta A \cdot \gamma \tag{11.4}$$

where ΔA is the increase in grain-boundary area due to the "bulging". In order that grain-boundary bulging can occur, the condition

$$\Delta E_{strain} > \Delta E_{gb} \tag{11.5}$$

must be fulfilled, and thus

$$\Delta E_d > \gamma \cdot \frac{\Delta A}{\Delta V} \tag{11.6}$$

Now the "bulge" will be approximated as a spherical gap with radius R (see Fig. 11.7). Then, for constant L ($2L$ is the diameter of the initially flat part of the grain boundary that bulges out) and variable R, it follows by straightforward calculus:

$$\frac{\Delta A/\Delta R}{\Delta V/\Delta R} = \frac{2}{R} \tag{11.7}$$

By substitution of the result given by Eq. (11.7) into the condition (11.6) it is finally obtained:

$$R > \frac{2\gamma}{\Delta E_d} \tag{11.8}$$

The smallest possible value of R equals L (cf. Fig. 11.7); then the "bulge" is a hemisphere (cf. the derivation of the largest minimal shear stress for the bowing out of a dislocation pinned at two pinning points, which occurs if the dislocation between the two pinning points is a half-circle; cf. Eq. (5.10) in Sect. 5.2.6). Hence, it follows that strain-induced grain-boundary migration can take place if

$$L > \frac{2\gamma}{\Delta E_{\mathrm{d}}} \tag{11.9}$$

This condition has first been formulated by Bailey (1960).

In the above discussion strain-induced grain-boundary migration was thought to occur along a grain boundary at one or more places, subject to the condition (11.9). Recognizing the microstructural inhomogeneity (even) *within* deformed grains, the above reasoning suggests that a single, large enough subgrain (cell) in a polygonized, dislocation-cell structured (cf. Sect. 11.1) microstructure, located at a *A/B* (high-angle) grain boundary, can act as the region initiating recrystallization (Fig. 11.8). The condition (11.8) can then be formulated as:

$$R_{\mathrm{subgrain}} > \frac{2\gamma}{\Delta E_{\mathrm{d}}} \tag{11.10}$$

where the subgrain shape has been taken (approximated) as a sphere and R_{subgrain} is the subgrain radius. In this case ΔE_{d} is given by the difference of (i) the strain

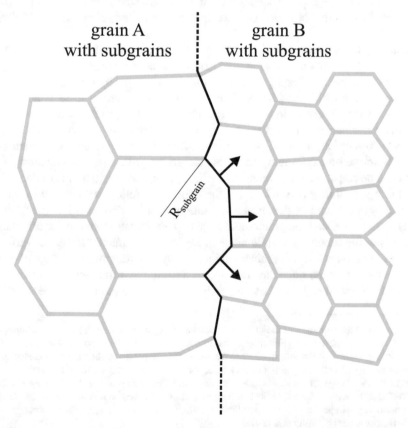

Fig. 11.8 Grains *A* and *B* exhibit a polygonized dislocation-cell (subgrain) structure. If a subgrain, located at the *A/B* (high-angle) grain boundary, is large enough, it can act as a location for the initiation of recrystallization, in accordance with the principle illustrated in Fig. 11.7 (see text)

energy of the polygonized, dislocation-cell structured grain B, into which the large subgrain of the polygonized, dislocation-cell structured grain A grows, and (ii) the strain energy of the large subgrain in grain A, which can be taken as nil (marginal dislocation density *within* the subgrain; cf. Sect. 11.1). Hence, $\Delta E_d = E_B$. For the dislocation-cell structured grain B, the strain energy is governed by the amount of subgrain boundaries. The energy per unit area subgrain boundary in grain B is γ_B. The amount of subgrain boundary per unit volume in grain B is roughly $3/(2 <R_B>)$, with $2 <R_B>$ as the average subgrain diameter of grain B.[2] Consequently, $\Delta E_d = E_B$ $= 3/(2 <R_B>) \cdot \gamma_B$. Substitution of this result into condition (11.10) finally gives:

$$R_{\text{subgrain}} > \left(\frac{4}{3} < R_B >\right) \cdot \left(\frac{\gamma}{\gamma_B}\right) \qquad (11.11)$$

Thereby the condition for recrystallization to be initiated is not expressed as a condition for the difference in strain energy of adjacent grains (cf. conditions (11.8) and (11.9)): merely the size of a subgrain adjacent to the grain boundary is decisive for the mechanism considered here. Even if the stored, strain energies in both grains, A and B, are similar (same average subgrain/dislocation-cell size), the mechanism considered here can operate provided the size distribution of the subgrains is sufficiently wide.

Finally it is remarked that subgrain coarsening, in the bulk of a polygonized/dislocation-cell structured grain, can be a precursor for the initiation of recrystallization. Such subgrain coarsening is dominated by the migration of *low-angle* boundaries (the boundaries of the subgrains), which by itself is no recrystallization (see discussion in Sect. 11.1; note that the subgrain growth discussed in the preceding paragraph involved a subgrain at a *high-angle* grain boundary that grows by migration of this high-angle boundary, thereby changing the orientation of the deformed material into which this subgrains grows: recrystallization). Two cases can be considered: (i) Upon traversing a grain, the orientation variation of the subgrains passed may be random, i.e. the (minor) variation experienced by subsequent subgrain-boundary passages is at random positive and negative. (ii) Alternatively, upon traversing the grain, there may be a systematic trend in the orientation variation of the subgrains[3]: while maintaining the minor magnitude of the variation of the orientation at each subgrain boundary, the systematic (minor) change of orientation can occur in the same direction and as a result the difference in orientation of the "first" subgrain and the "last" subgrain met along the passage can be relatively

---

[2] For a reasonable estimation of the grain-boundary area per unit volume, one cannot assume that the subgrains are spheres (as pertaining to the condition (11.10)), because on that basis a space filling arrangement of subgrains, in order to assure a massive nature of the material considered, is impossible. To estimate the grain-boundary area per unit volume, the shape of the grains can be taken as cubes of edge length $<2R>$. Then it is obvious that per average cube there are six faces of size $(<2R>)^2$, and, as each face is shared by two adjacent cubes, it follows for the estimate of grain-boundary area per unit volume: $3(<2R>)^2/(<2R>)^3 = 3/<2R>$. For spheres of diameter $<2R>$, the surface area per unit volume is $6/<2R>$.

[3] This can be typical for bending as deformation mode, leading to geometrically necessary dislocations, accommodating the orientation variation, which after polygonization cause the aggregate of subgrains to exhibit a systematic trend in the orientation variation (cf. Fig. 11.4 and its discussion in Sect. 11.1).

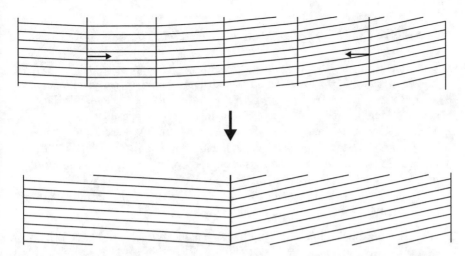

Fig. 11.9 Subgrain coarsening in an orientiation gradient. In the case considered, upon traversing a grain in the polygonized microstructure, there may be a systematic trend in the orientation variation of the subgrains: while maintaining the minor magnitude of the variation of the orientation at each subgrain boundary, the systematic (minor) change of orientation can occur in the same direction, and as a result, the difference in orientation of the "first" subgrain and the "last" subgrain met along the passage can be relatively large. If subgrain coarsening starts at distant locations along the passage considered, the formation of a higher-angle grain boundaries is possible as sketched in the figure

large. Now, for case (ii), suppose that subgrain coarsening starts at distant locations along the passage considered. Evidently, the growing subgrains will meet at some stage, *thereby creating a higher-angle boundary* than found before between adjacent subgrains along the passage (Fig. 11.9). On this basis recrystallization can be initiated as a consequence of *subgrain coarsening in the presence of a gradient in the subgrain orientation*.

The recrystallization mechanisms discussed above all imply that the orientations of the recrystallized material must have been present already in the deformed/recovered material. Yet, observations have been made where the orientations of new, recrystallized grains did not resemble those of the apparent parent grains. It may be speculated that in these cases local, relatively pronounced orientation variations occur in the immediate vicinity of grain boundaries (and grain-boundary junctions!) in the deformed microstructure, as a consequence of the incompatibilities of the intrinsic deformation behaviours of adjacent grains in a massive specimen (cf. the "*Intermezzo: Grain interaction*" at the end of Chap. 6 and see Mishra et al. 2009). If this is so, a mechanism as discussed above could operate, but this can be difficult to observe. Clearly, a similar discussion can be given for the observation of initiation of recrystallization at the interface with second-phase particles.[4]

---

[4] At the same time it should be remarked that, apart from providing a site for initiating recrystallization, second-phase particles may hinder the growth of the recrystallized material by pinning of the migrating recrystallization front.

Intermezzo: The History of an Idea; the Subgrain as Origin
of Recrystallization

Burgers (W.G.; see also the *"Intermezzo: A historical note about the Burgers vector"* in Sect. 5.2.3) wrote the first, extended monograph on recrystallization: W.G. Burgers, *"Rekristallisation, verformter Zustand und Erholung"*, Handbuch der Metallphysik, vol. 3, p. 2, Akademischer Verlaggesellschaft Becker & Erler Kom.-Ges., Leipzig, 1941 (in German). In this book a remarkable discussion about the origin of recrystallization is given (Sects. 106–109 at pp. 233–262). The deformed microstructure is conceived as an assembly of more or less homogeneously strained "blocks" ("Gitterblöcke") separated by highly deformed transition regions/layers. Then two different concepts for the initiation of recrystallization are considered[5]:

(i)   Genuine nucleation of recrystallization nuclei at/in the highly deformed transition regions/layers;
(ii) Growth of "blocks" of, as compared to the surrounding "blocks", relatively low strain energy, *pre-existing* (Burgers speaks of: "präformiert") in the deformed microstructure and which are able to grow upon annealing, driven by the release of energy stored in the deformed surroundings of these "blocks" (cf. Fig. 110 at p. 246 of Burgers' book).

This second hypothesis, as formulated by Burgers, and, by the way, tributary to ideas earlier presented a.o. by Masing in 1920 and Dehlinger in 1933, sounds surprisingly modern: one is immediately tempted to identify the "low-energy block" with the cell/subgrain in a dislocation-cell structured or polygonized grain, presented above as the crucial structural entity to initiate recrystallization. Burgers presented this concept in 1941, which is long before polygonization was first described and its potential importance for the initiation of recrystallization was recognized (Cahn 1949; Beck 1949). Moreover, transmission electron microscopy, capable of revealing the presence of polygonized/dislocation-cell microstructures, emerged as an important technique for microstructural analysis not before the "fifties" of the past century: the first observations by TEM of dislocations were made in 1956.

Whereas Burgers in his evaluation, on the basis of the available experimental information at the time, could eventually not decide between the above extremes for the initialization of recrystallisation (Sect. 110, at pp. 260–262 in his book), research until now has established with certainty that pre-existing, i.e. after deformation/recovery, low-energy "blocks", i.e. the dislocation cell or the subgrain, are the origins of recrystallization (Humphreys and Hatherly 2004).

---

[5] Note that this consideration by Burgers has been wrongly represented in both most recent books on recrystallization (Cotterill and Mould 1976; Humpheys and Hatherly 2004).

11.2.2 Kinetics of Recrystallization

The majority of the kinetic analyses performed of recrystallization adopt an approach as indicated for heterogeneous phase-transformation kinetics; see Chap. 10. "Nucleation", growth and impingement are distinguished as three generally overlapping mechanisms. As shown in Sect. 10.8, this framework can lead to the classical Johnson–Mehl–Avrami equation, describing the degree of transformation (here: fraction recrystallized) as function of time at constant temperature (Eq. (10.20)). To emphasize the restricted validity of the classical JMA equation, the basis assumptions made in its derivation are listed here (again; see Sect. 10.11): isothermal transformation, either pure site saturation at $t = 0$ or pure continuous nucleation, high driving force *in order that Arrhenius-type temperature dependences for the nucleation and growth rates are assured* and randomly dispersed nuclei which grow isotropically.

Particularly problematic with a view to application of the classical JMA equation to recrystallization is the assumption of a large driving force: as indicated at the start of Sect. 11.2, recrystallization is characterized by a small driving force. Further, a random dispersion of "nucleation" sites is unlikely (e.g. strain-induced boundary migration initiating at high-angle grain boundaries, implying a more "regular/periodic" "nucleation"; cf. discussion at the end of Sect. 10.9).

Yet, many applications of classical JMA analysis to recrystallization kinetics have been made. Especially use of inaccurate data and insensitive fitting may have led to seemingly successful fitting of the classical JMA equation (cf. Sect. 10.11).

Also, application of the generalized JMA equation (Eq. (10.23)) cannot be advised, as this equation, although compatible with a range of nucleation and growth modes, is still based on a random distribution of the "nuclei" to describe the effect of impingement. A more promising approach may therefore be adopting the generalized description of the extended volume (Eqs. (10.14)–(10.16)) and combine this with an appropriate impingement mode (e.g. see Eqs. (10.18b), (10.21) and (10.22)) and evaluate the degree of recrystallization (fraction recrystallized) *numerically* on the basis of the recipe described in Sect. 10.10.

However, even then, one still is subject to the assumption of thermally activated "nucleation" and growth according to Arrhenius-type temperature dependencies (large driving force; see above). For example, in case of a small driving force for growth the recrystallization-front velocity, v, can be written as (cf. Sect. 10.6.1):

$$v(T(t)) = M(-\Delta G) = M_0 \exp\left(-\frac{Q_G}{RT(t)}\right)(-\Delta G(T(t))) \qquad (10.11)$$

The driving force, $-\Delta G$, can change with time and temperature, for example, due to ongoing recovery processes, in the not yet recrystallized matrix, while recrystallization runs. Then, an Arrhenius-type temperature dependence for growth generally does not hold. Of course, even in this case a numerical approach remains possible (the volume of the recrystallized particle nucleated at time τ must be calculated now by numerical integration according to Eq. (10.8)).

Finally, it is remarked that interpretation of the value possibly determined for the effective, overall activation energy of recrystallization, is difficult without more ado

The effective activation energy incorporates contributions of "nucleation" and growth (e.g. see Sect. 10.12 and Eq. (10.24)). Procedures for unravelling the activation energies of "nucleation" and growth are possible (e.g. see Sect. 10.15.6).

11.3   Grain Growth

After completion of the recrystallization process as discussed in Sect. 11.2, a coarsening of the microstructure can occur, driven by the release of grain-boundary energy: the larger grains grow at the expense of the smaller grains. As the driving force for this process is (even; cf. discussion at the beginning of Sect. 11.2) distinctly smaller than for recrystallization, the velocity of the migrating grain boundaries is smaller than in the case of recrystallization (cf. (again) Eq. (10.11) given directly above). Two cases of grain growth can be discerned:

- *normal grain growth*, characterized by an approximately uniform velocity for the migrating grain boundaries throughout the specimen, with the consequence that the grain size remains more or less uniform throughout the specimen, but increases during the process;
- *abnormal grain growth*, characterized by mobile grain boundaries for only a few grains, with the result that these few grains become very large as compared to the remaining majority of the grains. This last process has, confusingly, also been called *secondary recrystallization*, as compared to the *primary recrystallization* discussed in Sect. 11.2 where the driving force is the decrease of stored strain energy.

11.3.1   The Grain-Boundary Network; on Grain-Boundary/Interfacial Energy and Tension

Obviously, thermodynamic equilibrium requires elimination of all grain boundaries in the specimen. Normally this ultimate, stable state is not reached. Instead, the arrangement of grain boundaries in a specimen can be such that metastable states occur.

Changes in the arrangement and density of the grain boundaries/interfaces in a material can occur under the constraints of (i) preservation of the massive nature of the specimen (the grains must be space filling), and (ii) establishment of local mechanical equilibrium of grain-boundary/interface tensions at locations where grain boundaries meet, the so-called vertices.

Before proceeding, at this place, some digression on the concepts grain-boundary/interface energy and tension is necessary. The following discussion pertains to interfaces in general, i.e. including surfaces, grain boundaries and interphase boundaries, but only the notion grain boundary will be used, as "pars pro toto".

The atoms at a grain boundary generally possess a higher energy than the atoms in the bulk (of the grain considered), because of their less ideal or incomplete state of chemical bonding. The amount of energy the atoms at the grain boundary have, more than they would have as bulk atoms, is an "excess energy" and, per unit area grain boundary, is called the grain-boundary energy, γ_{GB}. The grain then strives for making the grain-boundary area as small as possible. Hence it costs energy to enlarge the grain-boundary area. Or, in other words, a force has to be applied, in the plane of the grain boundary and acting along a line in the grain-boundary area, in order to extend the grain-boundary area in the direction of the force (cf. Fig. 11.10). This force per unit length, i.e. tension/stress, along the line mentioned is σ_{GB}. On the basis of this reasoning it would follow: σ_{GB} dA (work done) $= \gamma_{GB}$ dA (energy change), with dA as the increase of grain-boundary area per unit length along the line in the grain-boundary area considered. Consequently, the *grain-boundary tension*, σ_{GB}, has the same numerical value as the *grain-boundary energy*, γ_{GB}:

$$\sigma_{GB} = \gamma_{GB} \tag{11.12a}$$

Note that σ_{GB} is expressed in Nm^{-1} and γ_{GB} is expressed in Jm^{-2} (1 J (energy) = 1 Nm (work)).

However, the discussion in the above paragraph has tacitly assumed that γ_{GB} does not depend on (the extension of) A. In order that this is true, it would be necessary that the density and the arrangement (structure) of the atoms in the grain boundary are unchanged upon change of A. This can be true for the surface of liquids, where atoms can rapidly, freely, move from the bulk to the surface, and vice versa, to accommodate imposed shape changes and thereby maintain the overall, equilibrium surface structure. For solids similar phenomena are less likely: a serious straining of the arrangement of grain-boundary atoms may occur (see next paragraph) without relaxation by the transfer of atoms from the bulk, or vice versa: solids are much more viscous than liquids and, in contrast with liquids, can support shear (see Sects. 12.7 and 12.16). Then the numerical values of σ_{GB} and γ_{GB} are not identical. For this

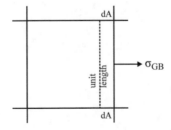

Fig. 11.10 Schematic depiction of the increase in grain-boundary area by moving a grain boundary: a force has to be applied, in the plane of the grain boundary and acting along a line in the grain-boundary area, in order to extend the grain-boundary area in the direction of the force. This is the origin of the notion grain-boundary tension/stress

case, one can proceed as follows. The change in Gibbs energy upon change of grain-boundary area dA is given by d$G =$ d(γ_{GB} A) $= \gamma_{GB}$ d$A + A$ dγ_{GB}. From σ_{GB} dA (work done) $=$ dG (energy change) $= \gamma_{GB}$ d$A + A$ dγ_{GB}, it then follows:

$$\sigma_{GB} = \gamma_{GB} + A \cdot \frac{d\gamma_{GB}}{dA} \qquad (11.12b)$$

which reduces to Eq. (11.12a) if γ_{GB} does not depend on A. It should further be realized that for the results given by Eqs. (11.12a and 11.12b), the grain-boundary tension/stress is taken as isotropic, i.e. σ_{GB} does not depend on direction in the grain-boundary area. In view of the elastic anisotropy (cf. Sect. 12.3), this will generally not be true, but corresponding experimental data are extremely rare.

The above discussion suggests that the differences between σ_{GB} and γ_{GB} are less pronounced for high-angle (more irregular atomic arrangement) than for low-angle (more regular atomic arrangement) grain boundaries.

The origin of the straining in the grain-boundary area of a solid can be discussed as follows. Due to the lack of neighbours or having partly different neighbours, than as for the atoms in the bulk, the atoms in the peripheral grain boundary can have a coordination and bonding different from the bulk atoms, with the result that their strived for atomic volumes (nearest neighbour distances) and strived for arrangement can be different from those of the bulk atoms. However, the atoms at the periphery are constrained to remain in registry with the underlying atomic layers. Hence, the grain boundary experiences a grain-boundary strain/stress with respect to the preferred, strived for atomic positions (cf. Sutton and Balluffi 1995).

The concept of grain-boundary tension now allows defining a local mechanical equilibrium at common grain-boundary edges/"vertices".

Consider Fig. 11.11: three grains, A, B and C, meet at a common edge, perpendicular to the plane of drawing. Given a sufficiently high atomic mobility, the grain boundaries will orient themselves at the edge/triple junction at O such that the grain-boundary tensions $\sigma_{A/B}$, $\sigma_{B/C}$ and $\sigma_{A/C}$ comply with a local mechanical equilibrium

Fig. 11.11 Illustration of local mechanical equilibrium of grain-boundary tensions at a grain-boundary triple junction (edge) of grains A, B and C

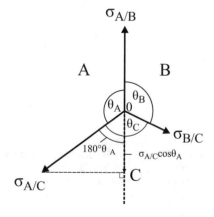

at O given by balance of the three grain-boundary tensions. Thus, vectorial equilibrium of the grain-boundary tension components *along* the A/B boundary plane leads to:

$$\sigma_{A/B} + \sigma_{B/C} \cos \theta_B + \sigma_{A/C} \cos \theta_A = 0 \qquad (11.13a)$$

Equivalent expressions result considering vectorial equilibrium of grain-boundary tension components *along* the B/C and A/C boundary planes. Or, by vectorial equilibrium of the grain-boundary tension components *perpendicular* to the A/B, B/C and A/C boundary planes, a well-known relation is obtained:

$$\frac{\sigma_{A/B}}{\sin \theta_C} = \frac{\sigma_{B/C}}{\sin \theta_A} = \frac{\sigma_{A/C}}{\sin \theta_B} \qquad (11.13b)$$

If the three grain-boundary tensions involved have the same value, it follows that the so-called dihedral angles, θ_A, θ_B and θ_C, are given by 120°. Hence, in case of a single phase material with an isotropic grain-boundary tension, for a two-dimensional, massive arrangement of two-dimensional grains, or, in three dimensions for a massive arrangement of columnar, parallel grains, a microstructure of grains of hexagonal morphology would exhibit metastable (see above and begin of Sect. 11.3.2) equilibrium.

A similar argument as above leads to the conclusion that, for the case of isotropic grain-boundary tension and four grains meeting at a corner (point/vertex), the balancing of the grain-boundary tensions involves that the angles between the grain edges at the corner will be 109° 28', i.e. as pertains to the edges of the regular tetrahedron. Within the present context, it further holds for the three-dimensional grain-boundary network that a configuration of more than four grains (edges) at a corner is unstable, i.e. a balancing of grain-boundary tensions is impossible. The analogous statement for a two-dimensional network is that a configuration of more than three grains (edges) at a corner is unstable. Such an unstable configuration strives for decomposition in metastable configurations in each of which the grain-boundary tensions are balanced.

A special, important case follows if, for the case of three grains meeting at an edge, grains A and B are identical ($A' = A = B$; see Fig. 11.12) and the grain-boundary tensions are isotropic. It follows from the balance of grain-boundary tensions in the plane of the A'/A' boundary (perpendicular to the plane of drawing)[6]:

$$\sigma_{A'/A'} = -2\sigma_{A'/C} \cdot \cos(\theta_{A'}) = 2\sigma_{A'/C} \cdot \cos(\theta_C/2) \qquad (11.14)$$

---

[6] Balancing of the grain-boundary tension components *perpendicular* to the A'/A' boundary is guaranteed by the symmetry of the case considered (the A'/A' plane is a mirror plane). If such symmetry lacks, e.g. if the grain-boundary tensions at both A'/C boundaries are unequal (note that two different A' grains are involved (see Fig. 11.12) and thus dependence of grain-boundary tension on crystal orientation would suffice to cause the effect), then balancing of only grain-boundary tension components in the plane of the A'/A' boundary does not suffice for establishing mechanical equilibrium: local grain-boundary curvatures will be invoked in order that also the balancing of the grain-boundary tension components perpendicular to the plane of the A'/A' boundary is realized.

Fig. 11.12 Illustration of local mechanical equilibrium of grain-boundary tensions at a grain-boundary triple junction (edge) of grains, A', A' and C for the case that the grain-boundary tensions are isotropic (the A'/A' plane is a mirror plane)

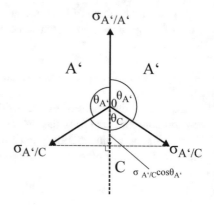

This is the same equation as Eq. (9.13) in Sect. 9.4.5, where the morphology of a second-phase particle developing on a grain boundary of the matrix was discussed in dependence on the strived for contact angle ($\sigma_{A'/A'}$ being smaller or larger than 2 $\sigma_{A'/C}$).

Absolute values for grain-boundary tensions may be difficult to determine; relative determinations, i.e. with reference to a specific grain-boundary tension, are more easily possible by application of Eqs. (11.13) and (11.14). For example, A'/A' may stand for a grain boundary of the specimen, composed of A' grains, intersecting the surface. Then, Eq. (11.14) predicts that, for local mechanical equilibrium of the surface and grain-boundary tensions at the point of intersection, a groove must develop at the point of intersection such that a contact (dihedral) angle θ_V occurs. "Grain C" here then should be interpreted as vacuum or the vapour phase in contact with A' (see Fig. 11.13). On this basis A'/A' grain-boundary tensions can be determined with respect to the same surface tension, supposed to be isotropic, i.e. independent of crystal orientation. To this end precise determination of the contact angle

vacuum/vapour

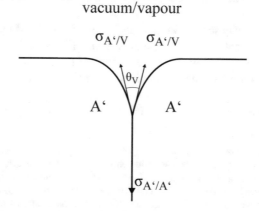

Fig. 11.13 Illustration of the local mechanical equilibrium of grain-boundary and surface tensions at the intersection of a grain boundary with the surface of the specimen, requiring the formation of a groove at the intersection of the grain-boundary with the surface

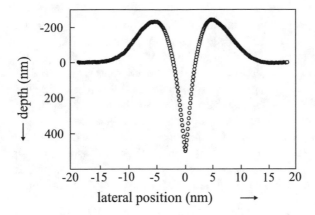

depth (nm)

lateral position (nm) ⟶

Fig. 11.14 Depth profile of a $\Sigma 19a$ grain-boundary groove (for the meaning of the symbol "Σ", see the discussion on the coincidence site lattice (CSL) in Sect. 5.3) at the surface of a Cu-50at.ppmBi bicrystal annealed at 110 h for 1123 K, as measured by atomic force microscopy. The measured contact angle is 140.3°; note the difference in scales along abscissa and ordinate (taken from Schöllhammer et al. 1999)

from the profile of the groove is a prerequisite. This is difficult, because the contact angle is established at the deepest position of the groove where it is very small (see Fig. 11.13). Accurate determination of the contact angle is possible applying atomic force microscopy (see Schöllhammer et al. 1999; cf. the description of scanning probe microscopy in the "*Intermezzo: Combined nanoindentation and scanning probe microscopy*" in Sect. 12.13). An experimental example is shown in Fig. 11.14.

Intermezzo: Interface Stabilized Microstructures

The atoms at an interface of a solid phase with another solid phase, or with a liquid or vapour phase, or with the vacuum, generally possess a higher energy than the atoms in the bulk of that solid phase, because of their less ideal or incomplete state of chemical bonding (cf. the way the concept grain-boundary energy was introduced in the above text). The presence of specific interfaces can cause thermodynamic (energetic) stabilization of phases, which are metastable or unstable according to bulk thermodynamics (energetics). Obviously, corresponding observations can be made especially in thin films and thin film systems, characterized by a high interface density.

An amorphous, solid phase, α', has a higher bulk energy (Gibbs energy; cf. Sect. 7.3) than the corresponding crystalline, solid phase, α. Now consider the situation of this amorphous phase, α', in contact with a crystalline phase, β (both phases of different composition). It can be shown that the energy of the interface between the amorphous phase α' and the crystalline phase β generally is smaller than between the crystalline phase α and the crystalline phase β (Jeurgens et al. 2009). Consequently, considering a layered structure of α' and β, the lower energy of the α'/β interface, as compared to the energy of the α/β interface, can overcompensate the difference in bulk energy of the α' and α phases. Upon increasing thickness of the amorphous layer (phase), the relative contribution of the interface energy (proportional with the interface area), as compared to the contribution of the bulk energy (proportional with the product of interface area and thickness of the layer), decreases. Hence, up to a certain, critical thickness, the layer of the amorphous phase α' is energetically preferred over a layer of the crystalline phase α. In other words: the amorphous phase is the stable phase for a thickness smaller than the critical thickness.

The above reasoning has provided the explanation for the emergence of amorphous phases, instead of the expected, corresponding crystalline phases, at the interface of crystalline *A/B* couples upon diffusion annealing, whereas it was thought before that the presence of such amorphous phases was due to kinetic obstacles for the formation of the crystalline compound (Benedictus et al. 1996; Fig. 11.15a).

Similarly, considering the oxidation of metals, it was shown that the amorphous state for the developing oxide layer can be the energetically stable configuration up to a certain critical thickness of the oxide layer (Reichel et al. 2008). The critical oxide-film thicknesses up to which the amorphous state is preferred energetically, because of its lower sum of interface and surface energies (Fig. 11.15b), as compared to the corresponding crystalline state,

(a)

(b)

< > crystalline modification

{ } amorphous modification

Fig. 11.15 a Formation of a compound phase *AB* at an interface between the two crystalline phases *A* and *B*. Dependent on interface energy values, the compound, product phase can be amorphous up to a certain, critical thickness beyond which the crystalline modification, with the lower "bulk" Gibbs energy, is stable. **b** Formation of an oxide phase at the surface of the crystalline (metal) phase Me upon oxidation in (gaseous) O_2. Dependent on interface- and surface energy values, the oxide phase can be amorphous up to a certain, critical thickness beyond which the crystalline modification, with the lower "bulk" Gibbs energy, is stable (cf. Fig. 11.16)

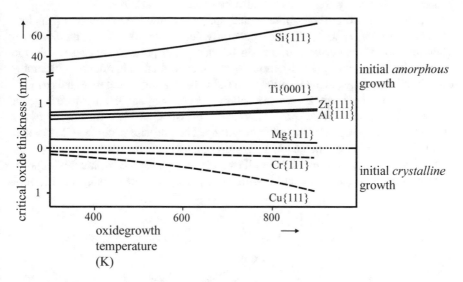

Fig. 11.16 Critical oxide-film thickness, below which oxide films on specific substrates are amorphous (cf. Fig. 11.15b), as a function of oxide-growth temperature, for surfaces of indicated crystallographic orientation of selected materials. A negative value for the critical thickness implies that the oxide film is crystalline from the beginning of oxide-film growth (taken from Reichel et al. 2008)

are shown for various metals as function of temperature in Fig. 11.16. Thus, also the well-known occurrence of an amorphous oxide film on aluminium in ambient at room temperature represents a state of equilibrium and is not the consequence of a kinetically obstructed crystallization, as has often been suggested (Fig. 11.16).

11.3.2 Grain-Boundary Curvature-Driven Growth

The discussion in the above Sect. 11.3.1 described conditions for mechanical equilibrium at locations where grain boundaries meet. Thereby a prescription for a complete state of metastable equilibrium for the entire grain-boundary network has not yet been established.

Obviously, a curved grain-boundary area between two parallel edges has a larger energy than the possible planar grain-boundary area between these two edges. Hence, energy is reduced if the curved grain-boundary area is replaced by the corresponding planar grain-boundary area. This can also be expressed as follows.

A curved surface in three-dimensional space is characterized by two principal radii of curvature, r_1 and r_2 (which generally depend on location at the surface). Therefore, the force per unit area grain boundary, i.e. pressure, acting within the grain at the concave side (see footnote 9) of the curved grain boundary with grain-boundary tension σ_{GB}, at the location with radii of curvature r_1 and r_2, equals $\sigma_{GB}/r_1 + \sigma_{GB}/r_2$. If the boundary is part of a sphere, $r = r_1 = r_2$ and the pressure is given by the well-known result $2\sigma_{GB}/r$.[7] This (extra) pressure enhances the energy of the grain at the concave part of the grain boundary. The pressure becomes nil if r_1 and r_2 (or r) become(s) nil. Then a planar grain-boundary area results.[8]

For isotropic σ_{GB}, it is possible to fill two-dimensional space (a plane) with polygons having *planar* faces compatible with the requirement of mechanical equilibrium

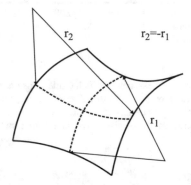

Fig. 11.17 Curved grain-boundary segment with its two principal radii of curvature r_1 and r_2. At the position shown $r_1 = -r_2$, and, consequently, the corresponding, local pressure on the grain interior is nil, a situation which always occurs in case of flat grain boundaries

---

[7] This equation can be derived as follows. Consider a boundary (interface) of spherical shape between two grains: a spherical, small grain 1, with radius r, incorporated in a large grain 2. Because of the occurrence of the grain-boundary energy γ_{GB}, the system could in principle reduce its energy by a decrease of r. A decrease of dr corresponds with a decrease of energy equal to the product of the grain-boundary energy and the decrease of grain-boundary area dA: $\gamma_{GB} \cdot dA = \gamma_{GB} \cdot 8\pi r dr$. To maintain equilibrium (the grain boundary does *not* migrate), an extra pressure ΔP must prevail within grain 1 (i.e. at the concave side (see footnote 9) of the curved (here spherical) grain boundary). This extra pressure performs work against a (n attempted) decrease of r with dr according to $\Delta P \cdot dV$, with dV as the volume change of grain 1 owing to the decrease of its radius dr. Hence, $\Delta P \cdot dV = \Delta P \cdot 4\pi r^2 dr$. For equilibrium it must hold: $\gamma_{GB} \cdot 8\pi r dr = \Delta P \cdot 4\pi r^2 dr$. Hence: $\Delta P = 2\gamma_{GB}/r$. Taking $\sigma_{GB} = \gamma_{GB}$ (cf. Eq. (11.12a) and its discussion in Sect. 11.3.1) the formula $\Delta P = 2\sigma_{GB}/r$ in the main text above is obtained. An equation of this kind can be called a Young–Laplace equation (more well known as explanation for the mechanical stability of a soap bubble, but then with an extra factor 2 in the numerator, recognizing the double interface for (i.e. both sides of) the film of liquid soap).

[8] It is of (at least academic) interest to remark that this is not the only possibility for making the pressure nil. The pressure becomes also nil if $r_1 = -r_2$, i.e. the grain-boundary area at the position considered exhibits radii of curvature of opposite signs (cf. Fig. 11.17).

(a)   (b)

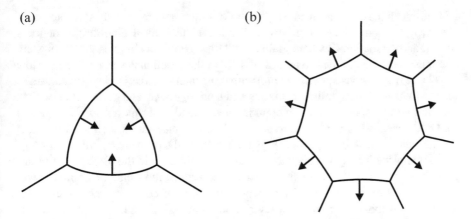

Fig. 11.18 Motions of grain boundaries as driven by grain-boundary tension. Curved grain-boundary segments tend to migrate to their centres of curvatures: **a** concave grain-boundary segments move inwardly, whereas **b** convex grain-boundary segments move outwardly

at the junctions, i.e. the plane is filled with hexagons (see Eq. (11.13)) and thereby a fully (in the sense of the first paragraph of this section) metastable state for the grain-boundary network in two-dimensional space has been realized. However, a similar situation cannot be established in three-dimensional space: no regular polyhedron with planar faces can fill space under the requirement of local mechanical equilibrium of the grain-boundary tensions at the grain-boundary edges. As a consequence (part of) the grain boundaries are curved and a complete metastable equilibrium for the three-dimensional grain-boundary network can never be achieved: *grain growth in the three-dimensional grain-boundary network is unavoidable.*

The above discussion can be summarized by stating that the force due to the grain-boundary tension acting on curved grain boundaries induces grain-boundary migration in order to minimize this force, i.e. the *curved grain boundaries tend to migrate towards their centre of curvature.* Thus, as considered from the point of observation, a concave[9] grain boundary moves inwardly and a convex grain boundary moves outwardly (Fig. 11.18).[10]

---

[9] The observer is at the concave side of a curved surface if neighbouring normals to the surface from this side converge; the observer is at the convex side of a curved surface if neighbouring normals to the surface from this side diverge.

[10] An alternative way to qualitatively understand this phenomenon is as follows. Atoms on the convex side of and adjacent to the grain boundary are more surrounded by the atoms of the grain on the convex side of the grain boundary than atoms on the concave side of and adjacent to the grain boundary (cf. Fig. 11.18). Consequently, the atoms on the convex side of and adjacent to the grain boundary have a lower energy than their counterparts at the concave side of and adjacent to the grain boundary. Hence, a tendency for net transport of atoms from the concave side of the grain boundary to its convex side exists, which can occur in case of sufficient thermal mobility: then the grain boundary migrates to its centre of curvature.

Consider a massive arrangement of parallel, columnar grains with isotropic grain-boundary tension. The system strives for local mechanical equilibrium at locations (edges/"junctions") where grain boundaries meet. This implies that the system attempts to establish dihedral angles of 120° at the junctions/"vertices" (see below Eq. (11.13)). As a consequence, a grain with more than six sides in the planar arrangement will have convex grain boundaries and tend to grow, and a grain with less than six sides will have concave grain boundaries and tend to shrink (cf. Fig. 11.18a and b). Otherwise said: grain boundaries move into the material on their concave side; i.e. the material with the highest energy (subjected to the pressure $\sigma_{GB}/r_1 + \sigma_{GB}/r_2$).

At this place it is appropriate to indicate the difference in the direction of grain-boundary migration between the cases of recrystallization and of grain growth. Recrystallization can proceed by outward migration of concave grain boundaries: e.g. by strain-induced grain boundary migration at a high-angle grain boundary or by subgrain coarsening in the presence of a gradient in the subgrain orientation, as discussed in Sect. 11.2.1. This contrasts with grain growth, where grain-boundary segments of concave nature move inwardly. In the process of recrystallization the (sub)grain on the concave part of the boundary is strain-free, and (yet) this (sub)grain grows into the deformed matrix, i.e. in the direction opposite to that for grain growth, as indicated by the centre of curvature of the boundary. The process is driven by the difference in strain energy of the surrounding, deformed matrix and the growing, recrystallized grain, which suffices to overcompensate the unfavourable extension of grain-boundary length/area by the recrystallization processes indicated. Grain growth, in contrast with recrystallization, occurs in a strain-free matrix and therefore is driven by the decrease of grain-boundary density and thus grain-boundary energy, only.

11.3.3   Kinetics of Grain Growth; Inhibition of Grain Growth

The extra pressure prevailing within the grain at the concave side of a curved grain boundary with grain-boundary tension σ_{GB} equals $\sigma_{GB}(1/r_1 + 1/r_2)$, with r_1 and r_2 as the principal radii of curvature (cf. Sect. 11.3.2). The driving force for grain growth per mole material swept by the moving grain boundary (moving to its centre of curvature; see what follows), $-\Delta G$ (i.e. the release of Gibbs energy upon grain-boundary migration), is given by the product of the extra pressure, as indicated above, and the molar volume, V_m, and thus

$$-\Delta G = \sigma_{GB}\left(\frac{1}{r_1} + \frac{1}{r_2}\right) \cdot V_m \tag{11.15}$$

Now consider a migrating grain-boundary segment with an average radius of curvature r. Then Eq. (11.15) can be written as:

$$-\Delta G = \frac{(cV_{\mathrm{m}})\sigma_{\mathrm{GB}}}{r} \tag{11.16}$$

where the geometrical constant c generally depends on the shape of the moving part of the grain boundary: e.g. c equals 2 for the grain-boundary segment being part of a sphere.

The grain-boundary velocity for small driving forces, as holds for grain growth, is given by Eq. (10.11) and thus, at constant temperature, is proportional with $-\Delta G$. The grain-boundary velocity also equals dr/dt. Hence, from Eqs. (10.11) and (11.16), it is obtained:

$$v = \frac{dr}{dt} = \frac{M(cV_{\mathrm{m}})\sigma_{\mathrm{GB}}}{r} \tag{11.17}$$

with M as the mobility of the grain boundary. Upon integration (at constant temperature) with respect to r with $r = r_0$ at $t = 0$, and assuming that σ_{GB} is isotropic, it follows:

$$r^2 - r_0^2 = 2M(cV_{\mathrm{m}})\sigma_{\mathrm{GB}}\, t \tag{11.18}$$

The above treatment concerns a grain-boundary segment that, taking r_1 and r_2, or r, as positive values, moves to its centre of curvature with a radius of curvature that increases with time (Fig. 11.19). Now, to relate the change of r of an individual grain-boundary segment with a change of grain size, a bold step must be made: the (average) radius of curvature of the moving grain-boundary segment in Eq. (11.17) is equated with the average grain size (equivalent grain radius) of the specimen at each moment of time (at constant temperature) and thereby Eq. (11.18) describes the (average) grain growth occurring in the specimen, if r and r_0 are replaced by $<r>$ and $<r_0>$,[11] respectively:

$$<r>^2 - <r_0>^2 = 2M(cV_{\mathrm{m}})\sigma_{\mathrm{GB}}\, t \tag{11.18a}$$

The basis for a treatment like the above one was given in the middle of the previous century (e.g. see Burke and Turnbull 1952). The result is often written in general form:

$$<r>^{n'} - <r_0>^{n'} = \mathrm{const} \cdot t \tag{11.19}$$

with n' as so-called grain-growth exponent. (cf. the (unrelated) growth exponent introduced in Sect. 10.8 to describe phase-transformation kinetics).

The parabolic relationship indicated by Eq. (11.18a) has often been questioned, as a considerable body of experimental work, after the fifties of the previous century,

---

[11] Note that, according to the above treatment, an initially, truly spherical grain can only shrink: the atoms at the concave side of the spherical grain boundary strive for passage of the grain boundary to get at the convex side; thereby the grain boundary migrates into the material at the concave side, i.e. towards the centre of curvature (see footnote 10). The radius of curvature of grain-boundary segments of this grain must thereby increase, as prescribed by Eq. (11.18). Consequently, the grain cannot maintain its spherical shape in the course of this process.

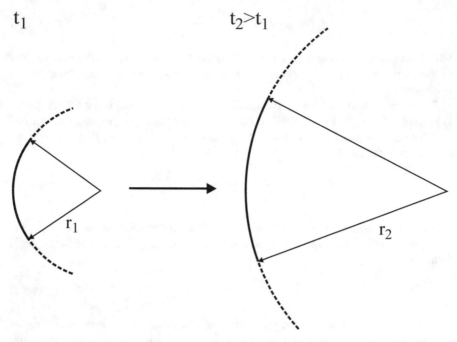

Fig. 11.19 Increase of the radius of curvature of a grain-boundary segment upon its migration. The (average) radius of curvature of the moving grain-boundary segment is equated with the average grain size (equivalent grain radius) of the specimen at each moment of time

has provided values of n' (cf. Eq. (11.19)) larger than 2 (up to 4). Recent theoretical analyses and computer simulations have only confirmed the validity of the parabolic relationship. It appears that much experimental work may have been imprecise (a similar remark was made regarding the application of the JMA equation for phase-transformation kinetics in Sect. 10.11). Further, in particular the ideal situation assumed for the derivation of Eq. (11.18a) can be incompatible with practical situations where small amounts of grain-boundary pinning, second-phase particles are present (see further below).

The driving force for grain growth depends on the value of the grain-boundary tension σ_{GB} (cf. Eqs. (11.15) and (11.16)) and thus depends of the structure of the grain boundary: a low-energy (low-angle) grain boundary experiences a smaller driving force and thus shows a smaller grain-boundary velocity (cf. Eq. (11.17)) than a high energy (high-angle) grain boundary. Hence, the distribution of the type of grain boundaries in the specimen,[12] and thus the crystallographic texture (for the notion texture, see Sect. 4.7) influences the rate of (average) grain growth. The distribution of the type of grain boundaries and the texture can change during grain

_____

[12] The grain boundary between two crystals of the same crystal structure can be defined by the plane of the boundary and the misorientation angle indicating the smallest rotation necessary to realize coincidence of the two crystals; cf. Sect. 5.3.1.

growth, which by itself (i.e. apart from the decrease of grain-boundary density due to grain growth) will lead to a change of the grain-growth rate. The distribution of the type of grain boundaries in the specimen and the texture are a pair of most important parameters characterizing the microstructure, and this recognition has led to a field of activity called "grain-boundary engineering". (Deformation and) Recrystallization and (subsequent) grain-growth procedures are devised in order to arrive at microstructures with optimal properties. Even today, it has to be admitted that most of this work is performed in practice on an empirical basis and that a great need exists for fundamental research in this area. The most significant gap in our knowledge concerns the atomic structure of, in particular moving, (high-angle) grain boundaries (cf. the discussion on "diffusion induced grain-boundary migration (DIGM)" in Sect. 8.9.2).

Obviously, upon continuation of normal grain growth as described above, the driving force for grain growth decreases as the grains become, rather uniformly, larger: see Eq. (11.16). It then becomes conceivable that upon prolonged annealing, the thermal activation, which expresses itself through the grain-boundary mobility, M (cf. Eq. (11.17)), is too small, in view of the strongly decreased value of the driving force, $-\Delta G$, in order to sustain a measurable grain-boundary migration rate, v. Consequently, the process of normal grain growth comes effectively to a halt.

Or, at some prolonged stage of grain growth, the driving force has become that small that a possible grain-boundary pinning force becomes significant in view of the diminished value of the driving force as expressed by Eq. (11.16):

(i)   *Effect of second-phase particles.* The pinning of a grain boundary by a second-phase particle in a matrix may qualitatively be understood as follows. Upon intersection of the particle by the grain boundary, a part of the grain boundary, as large as the area of intersection, has been removed. Thereby grain-boundary energy has been released (see also the discussion at the end of Sect. 9.2): one could say that the grain boundary is attracted to the particle, or otherwise said: it costs energy to remove the grain boundary from the particle. It can be shown that this energy needed to disconnect particle and grain boundary is proportional with the grain-boundary tension and the size of the particle:

Consider a grain boundary intersecting a spherical particle. In order that the grain boundary, experiencing a driving force to move (e.g. see Eq. (11.16)), loses itself from the particle, it bows out (Fig. 11.20), because it thereby exerts a net force on the particle (which is, at the moment of loosening from the particle, equal to the opposite of the drag force exerted by the particle on the grain boundary) due to the grain-boundary tension σ_{GB}: the moving grain boundary exerts a force, per unit length junction grain-boundary/particle surface, in the positive, vertical y-direction, proportional to σ_{GB} (i.e. $\sigma_{GB} \cos\psi$). The total force is obtained by multiplying with the length of the (circular) junction of grain boundary and particle surface (i.e. $2\pi r_p \cos\theta$). It is concluded that the total force to be exerted by the boundary to free itself from the particle is proportional with $\sigma_{GB} r_p$.

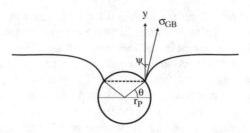

Fig. 11.20 Illustration of grain-boundary pinning by a spherical, second-phase particle. The grain boundary experiences a driving force to move. It looses itself from the particle by bowing out, because it thereby exerts a net force on the particle. The force exerted on the particle equals the product of the component of σ_{GB} acting in the direction of the y-axis, $\sigma_{GB}\cos\psi$, multiplied with the length of the (circular) junction of grain boundary and particle surface, $2\pi r_p\cos\theta$

If the volume fraction of second-phase particles equals φ_p, it follows for the number of (spherical) particles per unit of volume, N_p: $N_p = \varphi_p/((4/3)\pi r_p^3)$. The particles intersecting the (macroscopically planar) grain-boundary are located in a volume defined by planes parallel to the grain boundary and located at distances r_p above and below it. Hence, per unit area grain boundary, there are $2r_pN_p = 2\varphi_p/((4/3)\pi r_p^2)$ intersecting particles.

Thus, it follows from the above treatment that the force (pressure) to be exerted by a grain boundary per unit area, to free itself from the pinning particles, is proportional with $\sigma_{GB}\varphi_p/r_p$. This corresponds with an energy barrier to overcome per mole material swept by the moving boundary, ΔG_{pin}, given by:

$$\Delta G_{pin} = \frac{(c'V_m)\sigma_{GB}\varphi_p}{r_p} \tag{11.20}$$

with c' as a constant (e.g. of value 3/2). A consideration of this type is originally due to Zener; one also speaks of "Zener drag" or "Zener pinning" (cf. Nes et al. 1985).

The net driving force for grain growth now follows from Eqs. (11.16) and (11.20):

$$-\Delta G = \frac{(cV_m)\sigma_{GB}}{\langle r\rangle} - \frac{(c'V_m)\sigma_{GB}\varphi_p}{r_p} \tag{11.21}$$

and (cf. Eq. (11.17)):

$$v = \frac{d\langle r\rangle}{dt} = MV_m\left(\frac{c\,\sigma_{GB}}{\langle r\rangle} - \frac{c'\,\sigma_{GB}\varphi_p}{r_p}\right) \tag{11.22}$$

It follows from this equation that at the start of grain growth, a parabolic growth law is obeyed (cf. Eq. (11.18)), but upon continued growth, the growth rate

diminishes and growth is no longer possible when <r> has become that large
that $-\Delta G$ according to Eq. (11.21) has become nil. The corresponding limiting
value of <r>, <r>$_{final}$, follows from $-\Delta G = 0$ and thus (cf. Eq. (11.21)):

$$<r>_{final} = \left(\frac{c}{c'}\right) \cdot \left(\frac{r_p}{\varphi_p}\right) \tag{11.23}$$

where again (cf. discussion Eq. (11.18)), the (average) radius of curvature r_{final}
is equated with the average grain size (equivalent grain radius). Practical values
of c/c' are in the range 1/3–1/2.

(ii)   *Effect of surfaces.* For a thin layer (or a fibre), the thickness of the layer
       (the diameter of the fibre) can be that small that the grain size becomes of
       the order of the layer thickness (fibre diameter). The tendency to reduce the
       grain-boundary energy in the system provides an explanation for the tendency
       for grain boundaries in thin layers (fibres) to be oriented perpendicular to the
       surface. Moreover, thermal grooving for grain boundaries intersecting the
       surface, as discussed below Eq. (11.14), in order to establish a balancing of
       the surface tension and the grain-boundary tension at the junction of surface
       and grain boundary, is compatible with a perpendicular orienting of the grain
       boundary with respect to the surface (cf. the symmetry of the geometry of
       the case discussed in Eq. (11.14) and see Fig. 11.13). Such surface grooves
       are a barrier for grain-boundary migration, i.e. a force has to be exerted by
       the grain boundary that is about to migrate. This parallels the discussion on
       "Zener drag", i.e. the pinning effect of second-phase particles (see above). As
       a result, a limiting lateral grain size for the grains at the surface occurs, which
       follows from the net driving force being nil (cf. derivation of Eq. (11.23)). In
       practice, the (limiting) lateral grain size in thin films, composed of columnar
       grains traversing the thin film, is about two to three times the layer thickness.

(iii) *Effect of solute atoms.* Solute atoms can influence the mobility of grain bound-
       aries. The energy of a solute atom at the grain boundary is generally different
       from the energy of the solute atom in the bulk of the grain, as a direct conse-
       quence of the difference in the state of bonding (difference in the local atomic
       arrangement). A solute may thus be attracted to the boundary (and thereby
       energy is released) or it may be repelled from the boundary (it costs energy to
       move the solute atom from the bulk to the boundary). If the solute is attracted
       to the boundary, the solute concentration at the boundary is larger than in the
       bulk and one speaks of "solute segregation". In this case, the solute atoms can
       induce a "solute drag" force on the moving boundary. As a result a limiting
       grain size occurs when the net driving force becomes nil (cf. derivation of
       Eq. (11.23)).

11.3.4 Abnormal Grain Growth

Restriction of the mobility of grain boundaries to only a small number of grains causes these grains to grow, by consuming the other, surrounding grains with virtually immobile grain boundaries, and become very large (in case of metals the size of these grains can easily become of the order of a centimetre). The inhomogeneous nature of the process and the (attempted) description of its kinetics in a way analogous to recrystallization (cf. Sect. 11.2.2) has led to the name "secondary recrystallization", but, as indicated at the beginning of Sect. 11.3, the driving force for abnormal grain growth is of different origin and much smaller than the (already modest; see discussion of Eq. (11.2)) driving force of recrystalllization. Other names used, which speak for themselves, are "exaggerated grain growth" and "discontinuous grain growth".

The normal sequence of events upon annealing a deformed material is: recovery, recrystallization, normal grain growth and abnormal grain growth, but overlapping of these processes can occur (cf. the introduction of this Chap. 11).

The kinetic Eqs. (11.18a) and (11.19) pertain to the change of the *average* grain radius (grain size) during normal grain growth, i.e. taking place rather uniformly throughout the specimen. If the treatment is focussed on the growth behaviour of only a *single* grain, in the assembly of grains constituting the specimen, growth of this single grain is governed by the release of energy due to the elimination of the grain boundaries of the surrounding grains, which are consumed, and the counteracting cost of energy due to the increase of grain-boundary area (and thus energy) of the growing grain. As a result, it can be shown that the grain-boundary velocity of the single grain growing into its (static) surroundings is given by:

$$v = \frac{dr}{dt} = MV_m \left(\frac{c'' < \sigma_{GB} >}{<r>} - \frac{c\sigma_{GB}}{r} \right) \tag{11.24}$$

where $<r>$ and $<\sigma_{GB}>$ represent the average grain size and average grain-boundary tension of the static grains and r and and σ_{GB} indicate the grain size and grain-boundary tension of the growing grain, and where for spherical grains[13] $c = 2$ and $c'' = 3/2$. Hence, growth of this single grain can occur if

$$\frac{r}{<r>} > \left(\frac{c}{c''} \right) \cdot \left(\frac{\sigma_{GB}}{<\sigma_{GB}>} \right) \tag{11.25a}$$

which for spherical grains[13] and $\sigma_{GB} = <\sigma_{GB}>$ leads to

$$\frac{r}{<r>} > \frac{4}{3} \tag{11.25b}$$

---

[13] Again, spherical grains cannot constitute a space filling arrangement, i.e. a massive specimen (cf. footnote 2), and therefore, numerical values for c and c'' pertaining to spherical grains to be substituted in Eqs. (11.24) and (11.25a) can only provide crude estimates.

Equation (11.25) provides the criterion to be fulfilled in order that a single grain of (effective) radius r can grow into a static surrounding assembly of grains of average (effective) radius $<r>$. A well-known consequence of the result indicated by Eq. (11.25b) is that *large grains grow at the expense of small grains.*

So far, the occurrence of abnormal grain growth has not been dealt with by the above treatment. After normal grain growth a more or less *uniform* grain size occurs in the specimen. Thus, to explain abnormal grain growth, additional effects have to be considered.

(i) *Effect of second-phase particles.* In the presence of a volume fraction φ_p of (spherical) second-phase particles of radius r_p, the grain-boundary velocity of a single grain growing into its static surrounding grains is given by (cf. Eqs. (11.24) and (11.22)):

$$v = \frac{dr}{dt} = MV_m \left(\frac{c'' <\sigma_{GB}>}{<r>} - \frac{c\sigma_{GB}}{r} - \frac{c'\sigma_{GB}\varphi_p}{r_p} \right) \qquad (11.26)$$

The consequence of Eq. (11.26) for individual grains in the specimen, upon consideration of the inequality $c'' <\sigma_{GB}> / <r> - c\sigma_{GB}/r - c' \sigma_{GB} \varphi_p/r_p > 0$, is that some of them (the larger ones) can grow and others (the smaller ones) cannot. Hence, the grain-size distribution becomes wider during normal grain growth. Normal grain growth in the presence of second-phase particles is inhibited at a final average grain size given by Eq. (11.23). In the end only the largest grains fulfil the inequality indicated, and thus the occurrence of abnormal grain growth in the presence of second-phase particles may be understood.

(ii) *Effect of surfaces.* Evidently, for grains adjacent to the surface, a driving force for (lateral) growth occurs if the surface energy of the grain is lower than those of the surrounding grains at the surface. Clearly, this can be a very important effect for in particular thin films. For this case of abnormal grain growth three contributions to the driving force can be indicated: the decrease of surface energy and the decrease of grain-boundary energy (area) are two contributions driving abnormal growth, whereas the third contribution due to the pinning by surface grooves (see under (ii) at the end of Sect. 11.3.3) opposes the lateral growth of the growing grain. Such surface energy-driven abnormal grain growth obviously is associated with the development of specific crystallographic textures: for example, in case of f.c.c. metals abnormal growth is in particular observed for grains with {111} or {100} planes at the surface. Surface energy-driven abnormal grain growth can lead to grain growth with grain-boundary movement in directions opposite to those which would be expected on the basis of grain-boundary curvature-driven grain growth (cf. Sect. 11.3.2). As a special feature the effect of the outer atmosphere on the surface energy should be mentioned as a means to influence the developing crystallographic texture. Evidently, the occurrence and control of abnormal grain growth in thin films is of great importance to the microelectronic industry.

(iii)   *Effect of texture.* As indicated in Sect. 11.3.3, the distribution of the type
        of grain boundaries in the specimen, and thus the crystallographic texture,
        influences the rate of (average) normal grain growth. Neighbouring grains of
        similar orientation are separated by low-angle grain boundaries of low grain-
        boundary tension (energy) which corresponds with a relatively low driving
        force for (normal) grain growth. Now, consider the presence of a grain of
        crystal orientation strikingly different from that pertaining to the crystallo-
        graphic texture (dominating) component. This grain will generally have high-
        angle, high energy grain boundaries with neighbours compatible with the
        crystallographic texture component. Consequently, this grain may grow into
        its neighbours at a stage where the majority of the grains, belonging to the
        crystallographic texture component, have stopped their (normal) growth (note
        the similarity of the here discussed mechanism for abnormal grain growth
        with the growth of subgrains at a high-angle grain boundary as a mechanism
        for initiating primary recrystallization; cf. Sect. 11.2.1). This mechanism for
        abnormal grain growth will be the more prominent the stronger and sharper
        the crystallographic texture is, since the misorientation of the grains belonging
        to the crystallographic texture component is the smaller the more outspoken
        the crystallographic texture is.

11.3.5   *Particle Coarsening; Ostwald Ripening*

Consider a two-component (A and B) system (A-rich), which, at the temperature,
pressure and composition considered, in equilibrium is constituted of two phases:
the α phase (A-rich), which is the matrix, and the β phase (B-rich), which is finely
dispersed as particles in the matrix, upon precipitation from the supersaturated solid
solution (see Fig. 9.1 and its discussion). Even if the compositions of matrix and
precipitate particles would satisfy the prescription given by the phase diagram for
the "bulk" materials (but see below), genuine equilibrium has not been attained: the
occurrence of many α/β interfaces (interphase boundaries) of variable curvatures
(e.g. the β phase consists of a dispersion of spheres of variable size) provides the
possibility of decrease of energy by letting the larger β phase particles (of larger radii
of curvature) grow at the expense of the smaller β phase particles (of smaller radii
of curvature) which thereby dissolve. This process of particle coarsening is often
denoted as "Ostwald ripening". Because of the similarity in origin of the driving
forces of the particle coarsening process and of the process of grain growth of a
homogenous material, dealt with above in this Sect. 11.3, i.e. decrease of interfacial
area/interface energy, particle coarsening is considered here as well.

The pressure induced on a β phase particle in the α matrix by a curved α/β
interface (of concave nature from the point of view of the β phase particle; cf.
footnote 9) raises the Gibbs energy of the β phase particle, by an amount $2\sigma_{\alpha/\beta}/r$,
with $\sigma_{\alpha/\beta}$ as the interface tension and r as the radius of the β phase particle taken
as a sphere (cf. Eq. (11.16)). This increase of the energy of the β phase particle,

due to the curvature of the α/β interface of specific interfacial tension, is called the "Gibbs–Thomson effect" or "capillary effect".

The Gibbs–Thomson effect has an important consequence: the *local* solubility of B in the surrounding α matrix depends on the radius of curvature of the α/β interface and thus directly on the size of the β phase particle if the β phase particle is a sphere: the local solubility of B in the surrounding α matrix is the larger, the smaller r. It can be shown that this effect becomes important for nanosized β phase particles (i.e. $r < 100$ nm; cf. Sect. 12.14.2). As a consequence, concentration gradients occur in the α matrix containing a dispersion of β phase particles (of spherical shape and) of different sizes. The energy of the system thereby is decreased by a solute (B) flux in the matrix from the small β phase particles to the large β phase particles, i.e. from larger to smaller B concentration in the α matrix (Fig. 11.21). As a result, the small β phase particles become smaller and disappear eventually and the larger β phase particles grow.

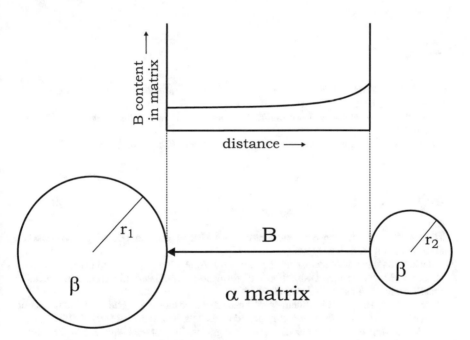

Fig. 11.21 (Precipitate-)Particle coarsening of second-phase, β particles (*B*-rich) in the matrix of parent phase α (*A*-rich) of the two-component system *A-B*. The Gibbs–Thomson effect causes the local solubility of B in the matrix (α) to be larger at the α/β interface for a small β phase particle (small radius of curvature of the particle/matrix interface) than at the α/β interface for a large β phase particle (large radius of curvature of the particle/matrix interface). As a consequence a net flux of B occurs in the matrix from small β phase particles to large β phase particles: the larger β phase particles will grow at the expense of the smaller β phase particles, i.e. coarsening, also called Ostwald ripening, takes place

Adopting volume diffusion of B in the α matrix as rate determining process, the kinetics of β phase particle coarsening is often described by the Lifshitz–Slyozov–Wagner (1961) equation:

$$<r>^3 - <r_0>^3 = \text{const} \cdot t \qquad (11.27)$$

with $<r>$ and $<r_0>$ as the β phase particle radii at t and $t = 0$, respectively (cf. Eqs. (11.18a) and (11.19)). The constant in this equation contains the product of the volume diffusion coefficient of B in the α matrix and the solubility of B in the α matrix for r infinitely large. Therefore the β phase particle coarsening can be strongly temperature dependent, as both the volume diffusion coefficient and the solubility of B in the α matrix generally strongly increase with temperature.

References

General

P. Cotterill, P.R. Mould, *Recrystallization and Grain Growth in Metals*, 2nd edn. (Surrey University Press, London, 1976).
F.J. Humphreys, M. Hatherly, *Recrystallization and Related Annealing Phenomena* (Elsevier, Oxford, 2004).
A.P. Sutton, R.W. Balluffi, *Interfaces in Crystalline Materials* (Clarendon Press, Oxford, 1995).

Specific

J.E. Bailey, Electron microscope observations on the annealing processes occurring in cold-worked silver. Phil. Mag. **5**, 833–842 (1960)
P.A. Beck, The formation of recrystallization nuclei. J. Appl. Phys. **20**, 633–634 (1949)
A.M. Beers, E.J. Mittemeijer, Dislocation wall formation during interdiffusion in thin bimetallic films. Thin Solid Films **48**, 367–376 (1978)
R. Benedictus, A. Böttger, E.J. Mittemeijer, Thermodynamic model for solid-state amorphization in binary systems at interfaces and grain boundaries. Phys. Rev. B **54**, 9109–9125 (1996)
J.E. Burke, D. Turnbull, Recrystallization and grain growth. Progr. Metal Phys. **3**, 220–292 (1952)
R.W. Cahn, Recrystallization of single crystals after plastic bending. J. Inst. Metals **76**, 121–143 (1949)
L.P.H. Jeurgens, Z. Wang, E.J. Mittemeijer, Thermodynamics of reactions and phase transformations at interfaces and surfaces. Int. J. Mater. Res. **100**, 1281–1307 (2009)
J.C.M. Li, Possibility of subgrain rotation during recrystallization. J. Appl. Phys. **33**, 2958–2965 (1962)
S.K. Mishra, P. Pant, K. Narasimhan, A.D. Rollett, I. Samajdar, On the widths of orientation gradient zones adjacent to grain boundaries. Scripta Mater. **61**, 273–276 (2009)
E.J. Mittemeijer, A.M. Beers, Recrystallization and interdiffusion in thin bimetallic films. Thin Solid Films **65**, 125–135 (1980)

E.J. Mittemeijer, Analysis of the kinetics of phase transformations. J. Mater. Sci. **27**, 3977–3987 (1992)

E. Nes, N. Ryum, O. Hunderi, On the Zener drag. Acta Metall. **33**, 11–22 (1985)

F. Reichel, L.P.H. Jeurgens, E.J. Mittemeijer, The thermodynamic stability of amorphous oxide overgrowths on metals. Acta Mater. **56**, 659–674 (2008)

J. Schöllhammer, L.-S. Chang, E. Rabkin, B. Baretzky, W. Gust, E.J. Mittemeijer, Measurement of the profile and the dihedral angle of grain boundary grooves by atomic force microscopy. Zeitschrift für Metallkunde **90**, 687–690 (1999)

Chapter 12
Mechanical Strength of Materials

The response of materials to applied forces concerns a field of material properties which has been of prime interest to human beings since the emergence of mankind. Even as a child, already, one gathers experiences about what we vaguely call the "strength" of a material, by feeling with our fingers how "hard" or "soft" a specific material is.

In fact, a more in-depth consideration, to be introduced in this chapter, makes clear that "strength" may be more explicitly termed, the (ultimate) tensile strength, the hardness, the fatigue resistance, and so on, depending on the type of loading. Strength parameters are often associated with failure of the material: applying loads beyond the limit indicated by the value of such a strength parameter causes some form of disintegration of the material, thereby deteriorating and even making impossible its functioning. Evidently, for application of a workpiece in service, it is imperative to know the limiting load values which can be withstood without inducement of failure.

The challenge for the materials scientist is to describe the material response to loading by a limited number of material-specific constants, as the elastic constants for elastic deformation (see below). To this end a continuum approach (cf. Sect. 8.1) is followed: the material is conceived as a continuous medium; a particulate, atomistic conception is not adopted. This leaves unimpeded that consideration of atomic/molecular mechanisms is required for understanding the origin of a phenomenon as plastic deformation, e.g. for crystalline materials by glide of dislocations (see Sect. 5.2.5), and that, in the case of linear elasticity (see Sect. 12.2; for other forms of elastic behaviour see Sects. 12.6 and 12.7), the elastic constants can be interpreted as a direct consequence of the strength of the chemical bond between the atoms in the material (see Sect. 3.1).

A great insight is the recognition that a material at rest and in the absence of *external* loading can yet be subjected to huge *internal* forces, which can be the result of forced, maintained coherency between different, misfitting parts of a material body: the classical example of the origin of such internal loads is the cooling or heating induced misfit between parts of a heterogeneous body with each part having its own thermal expansion coefficient(s) (see Sect. 12.18). Because these internal

© Springer Nature Switzerland AG 2021
E. J. Mittemeijer, *Fundamentals of Materials Science*,
https://doi.org/10.1007/978-3-030-60056-3_12

forces *result after* some treatment has been applied to the material concerned, these forces are also often called *residual* forces.

As a final note in this introduction to "mechanical strength", it is observed that "mechanical strength" is dependent on size. The important role of defects in the arrangement of the atoms in crystalline workpieces will be outlined below. Very small specimens can be "defect free" (metallic whiskers, carbon nanotubes,…). Such materials can exhibit extraordinarily high levels of intrinsic mechanical strength. One is tempted to envisage practical applications of such materials by scaling up. However, one then ignores the unavoidable introduction of defects in the material workpieces, upon becoming of macroscopic size, which brings about limitations to the mechanical strength that can be utilized in practice. Within this context, it is in particular recalled that defects as thermal vacancies in crystalline materials are equilibrium, and therefore unavoidable defects (cf. Sect. 5.1.1): their number becomes significant for macroscopic specimens. It is already only therefore impossible to simply suppose that mechanical properties observed for nanosized objects can simply be transferred to workpieces of macroscopic dimensions, which is, unfortunately, an erroneous way of thinking followed often (if a new material is presented).

12.1   Elastic versus Plastic Deformation; Ductile and Brittle Materials

Consider a cylindrical body clamped at one end on an unmovable wall. Applying a force along the long axis of the cylinder and acting on the "free" end of the cylinder (see Fig. 12.1) will induce an extension or compression of the body in the direction of the force (extension by "pulling" at the body; compression by "pressing" on the body). If the force is sufficiently small, removal of the force will restore the original shape (and volume) of the body. The deformation of the body subjected to such small forces is nonpermanent and thus one could speak of "conservative deformation"; the usual term is *elastic deformation*. If the force acting on the body is increased, then

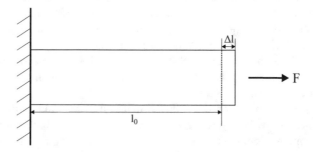

Fig. 12.1 Extension of a cylinder, clamped at the left side on an unmovable wall, upon application of a force F, directed along the long axis of the cylinder, acting on the "free" end of the cylinder. In the case shown, the force "pulls" at the body. l_0 = original length; Δl = length increase by loading

a critical value of the force can be surpassed beyond which, after release of the force, a shape change partly restoring the original shape occurs but a permanent deformation remains and one speaks of *plastic deformation*. This is a "dissipative" form of deformation: the work performed by the load leads to an increase of energy of the deforming body that is not (completely) released upon unloading.

A *ductile* material (e.g. a metal) experiences considerable plastic deformation before fracture occurs, whereas a *brittle* material (e.g. a ceramic) fails before pronounced plastic deformation takes place (see also Fig. 12.16 in Sect. 12.9).

12.2   Basic Modes of Uniaxial Deformation; Concepts of Stress and Strain; Uniaxial Elastic Deformation Laws

The basic deformation modes are tension, compression and shear (see Fig. 12.2a–c).

Consider a body as shown in the figure (a cube). Applying tensile loading forces acting along the length axis of the body, i.e. pulling at the body, leads to an extension of the length of the body: the body experiences a *tensile strain* (see Figs. 12.1 and 12.2a). Increase of the force will lead to an increase of the specimen length. One could plot the applied force as a function of the length of the specimen (see Fig. 12.3a). This plot would characterize the response (length increase) of the body concerned to the applied force. If a body of similar shape but different size would be subjected to the same type of tensile testing, the curve of force versus specimen length would be different from the one of the first specimen of same shape but different size. In order to avoid the trivial effect of size for specimens of similar shape, one desires a replacement of

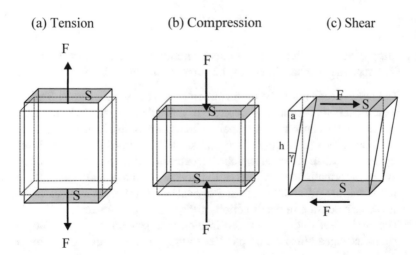

Fig. 12.2 Basic uniaxial deformation modes: **a** tension, **b** compression and **c** shear

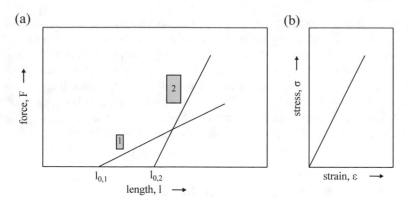

Fig. 12.3 Force-extension diagram **a** and corresponding stress–strain curve, **b** for two differently sized, iso-shaped bodies

the parameters force and change of specimen length by (specimen-size) normalized quantities, as follows.

The force, F, acts in the normal direction on a specimen cross section of size S (cf. Fig. 12.2a). It appears to have sense to normalize the force by defining the parameter stress, σ, as the quotient of force and cross-sectional area onto which F acts:

$$\sigma = \frac{F}{S} \tag{12.1}$$

Similarly, the length change can be normalized by defining the parameter strain, ε, as the quotient of length change, $l_1 - l_0$, and the original specimen length, l_0:

$$\varepsilon = \frac{l_1 - l_0}{l_0} \tag{12.2}$$

By plotting σ versus ε, the resulting curve or straight line does not depend on the size of the specimen for specimens of the same shape (Fig. 12.3b).

(i)   The stress needs not to be uniform across a cross-sectional area S within the specimen. For example, the presence of more than one phase in the specimen would already cause inhomogeneities in the stress distribution across a cross-sectional area S within the specimen, since the various phases need different stresses to realize the same (specimen) extension. Also, the intrinsic anisotropy (of the deformation behaviour) of the crystals (grains) in the specimen brings about that the differently oriented crystals need different stresses to achieve the same extension in the direction of the acting tensile stress.

(ii) The strain needs not to be uniform along the specimen length. The various grains arranged along a line parallel to the length axis of the specimen may experience different strains due to their different crystallographic orientations (the intrinsic anistropy of the deformation behaviour indicated above) or because they belong to different phases.

The discussion under points (i) and (ii) above implies that σ and ε have to be interpreted as *average* stress and *average* strain.

The problem touched upon in the above paragraph (points (i) and (ii)) is closely related to what is called "grain interaction": how do the grains in a massive specimen respond in their deformation behaviour to imposed loads while maintaining the integrity of the specimen? In a massive specimen the separate grains cannot deform as if they were "free standing" because their response is constrained by the surrounding grains (see further the "*Intermezzo: Grain interaction*" in Sect. 6.10).

Because the normalization in the above expressions (12.1) and (12.2) has been realized with respect to the *initial* cross-sectional area and *initial* length, respectively, one also speaks of *engineering* stress and *engineering* strain.

In a tensile testing device the specimen is usually elongated along its length axis at a constant rate while recording the force required. As a result a stress–strain curve is obtained. Often, in the elastic regime, a linear relation between stress and strain occurs (Fig. 12.3b). Such elastic behaviour is describes by Hooke's law given by:

$$\sigma = E\varepsilon \qquad (12.3)$$

where the proportionality constant is called the modulus of elasticity or Young's modulus. Note that E has the same dimension as σ. Very many solid materials deform elastically according to Hooke's law. The value of strain attainable in the regime of linear elasticity is limited: a few tenths of a per cent (for macroscopic specimens). For a validation of Hooke's law on the basis of bond stretching, see Eq. (3.2) and its discussion in Sect. 3.1.

Young's modulus, E, is a so-called elastic constant and is a material property. The higher its value, the higher the value of stress to achieve the same value of strain. For metals the elasticity modulus is of the order 10^5 MPa (e.g. steels with moduli of elasticity of about 2×10^5 MPa). Ceramics (often with an important component of covalent bonding) exhibit on average higher values of E (e.g. silicon carbide with a modulus of elasticity of about 4.5×10^5 MPa). Polymers show an elastic deformation behaviour distinctly different from metals and ceramics: elastic strains are not due to atomic bond stretching, but merely due to stretching of the weak (van der Waals; cf. Sect. 3.6) bonds between neighbouring polymer molecular chains and atomic bond rotation (upon tensile straining the polymer chains become more or less straightened) and therefore much smaller values of E occur for polymers (e.g. Nylon 6,6 and polystyrene with moduli of elasticity of about 3×10^3 MPa). Moreover, Hooke's law may not hold for polymers (cf. rubber, an elastomer; see below).

Compressive uiniaxial loading (Fig. 12.2b) in the elastic regime is also described by Eq. (12.3), if Eq. (12.3) holds upon tensile loading. The only important remark to be made here is that a compressive stress is given as a negative stress, and thus, the strain in compression is also negative.

A strikingly different type of uniaxial loading is required to induce shear. Consider the cube drawn in Fig. 12.2c. The shear forces, F, act tangentially, and in opposite directions, on the top and bottom faces, of area S, of the cube. (See also the discussion on glide of dislocations in Sect. 5.2.5). Similar to the definition of stress acting perpendicular to the cross-sectional area of size S (therefore, this stress is also called *normal* stress), one can now define the *shear* stress, τ, according to:

$$\tau = \frac{F}{S} \tag{12.4}$$

Whereas normal stresses cause changes in the distance between two points in a body, shear stresses induce changes in the angle between two lines in the body. This becomes clear realizing that the shear stresses acting on the cube in Fig. 12.2c cause a rotation of faces perpendicular to the top and bottom faces on which the shear stresses act. The originally perpendicular faces and the horizontal top and bottom faces are no longer at right angles. The shearing has caused a displacement, a, of the top face with respect to the bottom face. By normalization with respect to the height of the cube, h, the shear strain, γ, now is defined as:

$$\gamma = \frac{a}{h} \tag{12.5}$$

The rotation of the faces originally perpendicular to the top and bottom faces has occurred over an angle α for which it holds: $\tan \alpha = a/h = \gamma$. Because α is small, it follows that the shear strain γ can be conceived as the angle of rotation (in radians).

Often, in the elastic regime, a linear relation between shear stress and shear strain occurs, as for the normal stress and normal strain (Eq. 12.3). Hence, Hooke's law then also holds for the shear stress and shear strain:

$$\tau = G\gamma \tag{12.6}$$

where the proportionality constant G is called shear modulus or modulus of rigidity.

The shear modulus is an elastic constant and, thus, a material property. It has the same dimension as the (shear) stress. Its value is usually smaller than the elasticity modulus. For metals the value of G is about 40% of the value of E.

Note that both Eqs. (12.3) and (12.6) are (equally valid) expressions of Hooke's law for uniaxial loading.

Experience teaches us that upon tensile loading, leading to extension of specimen length in the loading direction, contraction of the specimen occurs in directions perpendicular to the loading direction (cf. Fig. 12.4). Suppose tensile loading is imposed in the positive x-direction (the acting (normal) stress is σ_x) for a (cylindrical) specimen oriented with its length axis parallel to the x-axis. The strain in the x-direction obeys (cf. Eq. (12.3)):

$$\varepsilon_x = \frac{1}{E}\sigma_x$$

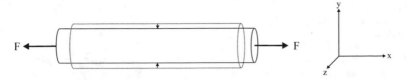

Fig. 12.4 Occurrence of lateral contraction upon applying a tensile force in longitudinal direction on a cylindrical body

The development of this positive strain ε_x in the x-direction is associated with the development of negative strain in the y- and z-directions (adopting a Cartesian frame of reference; see Fig. 12.4). Thus, the so-called Poisson ratio, υ, can be defined:

$$\upsilon = -\frac{\varepsilon_y}{\varepsilon_x} = -\frac{\varepsilon_z}{\varepsilon_x} \tag{12.7a}$$

and the negative strain contributions in the y- and z-directions due to the stress acting in the x-direction obey:

$$\varepsilon_y = \varepsilon_z = -\upsilon\varepsilon_x = -\left(\frac{\upsilon}{E}\right)\sigma_x \tag{12.7b}$$

The Poisson ratio is also an elastic constant, thus also called Poisson constant, and thereby a property of the material investigated. Note that υ is dimensionless. Usual values of υ fall in the range of about 1/4 to maximally 1/2:

metals: values around 1/3: α-Fe (ferrite): 0.29; Al: 0.33; Cu: 0.35
ceramics: TiC: 0.19; Si_3N_4: 0.24; Al_2O_3: 0.27; MgO: 0.36;
network polymers: bakelite: 0.20; ebonite: 0.39;
chain polymers: polystyrene: 0.33; polyethylene:0.40;
rubber (elastomer): 0.49.

On a theoretical basis it can be shown for *elastically isotropic* materials (i.e. having elastic properties independent of direction in the specimen frame of reference; cf. Sect. 12.3) that the Poisson ratio obeys the following limits $-1 \le \upsilon \le \frac{1}{2}$; e.g. see Greaves et al. 2011. This allowable range of values for υ indicates that elastically isotropic materials with a negative Poisson ratio are conceivable. At first sight the occurrence of a negative value for the Poisson ratio appears somewhat odd, as this implies that upon tensile loading in a specific direction expansion (and not contraction) occurs in directions perpendicular to the loading direction. This phenomenon is discussed in the second *Intermezzo* below.

Fig. 12.5 Schematic
depiction of the unit cell of a
cellular, "reentrant"
("directed inwardly")
structure exhibiting a
"negative" Poisson ratio
upon applying the "pulling"
forces F (after Lakes 1987)

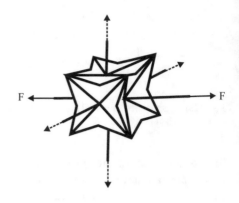

Intermezzo: Short History of the Poisson Constant

In the beginning of the nineteenth century, it was tried to explain the elastic properties of bodies by conceiving a body as a system of "molecules" held together by "molecular" forces acting along the lines connecting the "molecules" (the quotes are used here to relativize the notions "molecule" and "molecular", recognizing the period of time where these calculations were made). On this basis, Poisson showed *for isotropic bodies* (during the years 1820–1830) that the contraction parameter we now call Poisson constant should equal 1/4. This would imply that the number of independent elastic constants needed to describe the elastic deformation of an isotropic body would be one (instead of two; see Sect. 12.3). This result was generally accepted at the time. However, an overwhelming amount of experimental evidence convincingly showed that the Poisson constant for an isotropic body only in rare cases equalled 1/4 and the concept of central elastic forces as described above was thereby shown to be untenable (see, especially regarding metals, also footnote 2 in Sect. 3.1). A theory of elasticity devoid of any hypothesis on the interactions in a supposedly "molecular" structure of elastic bodies was required.

Intermezzo: Negative Poisson Constant

If one stretches a material body in one direction, one "intuitively", i.e. on the basis of experience, expects a contraction in transverse directions (cf. the discussion of Eq. (12.7) above).[1] However, some material bodies expand, i.e. do not contract, in directions perpendicular to the direction of an uniaxially applied *tensile* load, and, similarly, they contract, i.e. do not expand, in directions perpendicular to the direction of an uniaxially applied *compressive* load. In other words: these material bodies exhibit a negative Poisson constant. Such materials are called "auxetic materials". ("auxetic" derives from a Greek word describing increase/growth and thus refers to the expansion in lateral directions upon uniaxial tensile loading; see above).

Obviously, single crystals of intrinsically elastically anisotropic materials (see Sect. 12.3) may in principle reveal a negative Poisson constant in specific directions. This can happen in particular for materials of rather extreme elastic anisotropy (see Fig. 12.6d). But the occurrence of a negative Poisson constant for materials of (macroscopically) elastically isotropic behaviour is of extreme rarity for natural materials. Considerable interest has arisen in man-made material bodies which exhibit negative Poisson constants (e.g. see Crumm and Halloran 2007).[2] To this end, *non-massive* materials, of specific architectures, are made.

Materials constituted of networks of interconnected solid struts and plates, thereby composing large aggregates of cells packed together to fill space, are called "cellular solids" (Gibson and Ashby 1997). In two dimensions one can speak of honeycomb-like structures; in three dimensions one speaks of "foams". Examples of natural materials that are cellular solids are wood and bone.[3] For a specific architecture of the specimen, i.e. a specific distribution of material or, alternatively, of unoccupied internal space (pores, channels) in the specimen, the action of an uniaxially applied load leads to the generation of oriented forces in the network structure, which cause the overall, counter-intuitive displacements in the directions perpendicular to the loading direction. Metal or polymeric foams, of specific architecture (see what follows), can show negative Poisson constants. To this end the foams are compressed in three orthogonal directions: the cell ribs then protrude inwardly rather than outwardly. Such structures are called "reentrant ("directed inwardly") structures": see Fig. 12.5. Stretching by pulling along one pair of the connection bands of the reentrant cell with neighbouring cells causes the cell to unfold, and thereby expansion occurs in the transverse, lateral directions (Lakes 1987)!

Elastic behaviour need not be conform Hooke's law: *nonlinear elasticity* is possible. So-called *elastomers*, i.e. a certain class of polymers, of which rubber provides an example, can experience (still practically time independent; see below) nonlinear elastic behaviour with elastic strains up to thousand percent (think of rubber bands used in households/offices; one may speak of *rubber elasticity*; see further Sect. 12.6).

---

[1] Erroneously, it is sometimes thought that this experience derives from the conservation of volume. However, there is no conservation of volume upon elastic deformation. Only if the Poisson ratio equals ½, volume is conserved during deformation (cf. discussion of Eq. (12.17)).

[2] Apart from the fundamental, scientific interest in materials with negative Poisson constants, their potential applications, e.g. as shock-absorbing material (sound deadening layers) and fasteners raise practical interest.

[3] Cork is a cellular solid that exhibits a Poisson ratio close to zero. This is of obvious importance for its application as stopper of a wine bottle: the stopper must be inserted and removed easily.

Apart from (time independent) nonlinear elastic behaviour, as described above, a further deviation of simple linear elastic behaviour according to Hooke's law involves time dependence of the elastic response. Upon load imposition there will be a part of the strain induced that develops as a function of time after the start of loading, and upon release of the load a part of the strain is not immediately released, but some time is needed to establish complete recovery of the initial, unloaded situation. This type of behaviour is called *viscoelasticity*, as demonstrated by certain polymers, and *anelasticity*, as demonstrated by metals (see further Sect. 12.7). Unless otherwise stated, in the following *linear elasticity* is supposed.

12.3   Elastically Isotropic and Anisotropic Materials

In the discussion until now it has been tacitly assumed that the elastic response of the material does not depend on the direction in the specimen frame of reference along which the loading occurs. This implies that if a single crystal would be loaded uniaxially that then the elastic response would not depend on the (crystallographic) direction along which the load is applied. If this, for single crystals or polycrystalline or amorphous materials, or any material in general, would be true, then the material is said to be elastically isotropic. Hence, the whole treatment is Sect. 12.2 pertains to isotropic bodies only.

Thus the discussion in Sect. 12.2 has led to the introduction of three elastic constants for *isotropic* materials: E, G and v. The false impression could now have been arisen that three elastic constants are needed to describe the elastic behaviour of elastically isotropic materials. Actually, only two of these three elastic constants are independent and consequently needed to predict the elastic behaviour of isotropic materials/bodies. That only two of the three elastic constants mentioned are independent follows from the existence of a relation interrelating these three elastic constants:

$$E = 2G(1 + v) \qquad\qquad (12.8)$$

Now, again considering the elastic response of a single crystal to a load applied in various crystallographic directions, it appears natural to expect that the elastic deformation does depend on the crystallographic direction along which the loading occurs, recognizing that the atomic arrangement is anisotropic in space and that the elastic constants depend on the atomic interaction (chemical bonding; cf. Sect. 3.1). Indeed, truly isotropic elastic behaviour can be expected only for amorphous materials, because the atomic arrangement is (close to) purely random. Crystals of (highest) cubic symmetry already require three elastic constants and (the "worst" case) triclinic crystals need 21 elastic constants to describe the elastic behaviour of these crystals.

As an example, the dependence of Young's modulus, E, on direction $n' = [uvw]$ in a single crystal of cubic crystal symmetry can be expressed as follows:

$$E_{n'}^{-1} = E_{100}^{-1} + 3A \cdot \left(E_{111}^{-1} - E_{100}^{-1}\right) \tag{12.9a}$$

with $E_{n'}$ E_{100} and E_{111} as Young's moduli in the $<n'>$, $<100>$ and $<111>$ directions and A as a geometrical factor:

$$A = \cos^2\left(n', [100]\right)\cos^2\left(n', [010]\right)$$
$$+ \cos^2\left(n', [010]\right)\cos^2\left(n', [001]\right) + \cos^2\left(n', [100]\right)\cos^2\left(n', [001]\right) \tag{12.9b}$$

where $\cos(n', [100])$ denotes a so-called *direction cosine*, here the cosine of the angle between the direction of n' (i.e. $[uvw]$; cf Sect. 4.1.4.2) and the [100] direction in the crystal, etc. Defining $l \equiv \cos(n', [100])$, $m \equiv \cos(n', [010])$ and $n \equiv \cos(n', [001])$, a simpler and usual notation for A is obtained:

$$A = l^2m^2 + m^2n^2 + n^2l^2 \tag{12.9c}$$

The extreme values for E of cubic crystals occur for the $<100>$ and $<111>$ directions. The dependence on crystallographic direction of the elastic modulus, E, is shown for an f.c.c. metal as copper in Fig. 12.6a (note that $1/E$ surfaces are shown in Fig. 12.6). Evidently, E in the $<111>$ directions is *larger* than E in the $<100>$ directions, which is generally true for f.c.c. metals. The metal most closely approaching a truly intrinsically isotropic material is tungsten (W; a b.c.c. metal; see Fig. 12.6b). Alkali halide crystals, and the b.c.c. metals Cr, Nb and Mo, show a reverse behaviour; i.e. E in the $<111>$ directions is *smaller* than in the $<100>$ directions (cf. Fig. 12.6c). A substance of extreme elastic anisotropy is cementite, an important phase in steels (e.g. see the *"Intermezzo" The Fe-C system; steels and cast irons"* in Sect. 9.4.2 and Sect. 9.5; cf. Fig. 12.6d).

Components of metals and ceramics are generally composed of many crystals (grains) and thus are polycrystalline. It can be imagined that, if many crystals (grains) constitute the body concerned and if the distribution (in space) of the orientation of the individual crystals is random, then the *macroscopic* response of the body to applied loads will resemble that of an isotropic body. Such a body is called *quasi-isotropic* (note that the deformation of each separate crystal in the body is in general anisotropic). In accordance with the above discussion, for such a body two macroscopic elastic constants suffice to describe the macroscopic elastic response. This explains the appearance in handbooks of tables of values of E and G (or υ) for (components of) e.g. steels, whereas it has to be realized that individual iron (ferrite, austenite) crystals are essentially elastically anisotropic. As soon as texture (preferred orientation) occurs in such components, for example as the result of deformation (cold work) and heat treatment (recrystallization), such quasi-isotropic elastic behaviour no longer holds.

Polymeric materials can exhibit variation of the orientation of their molecules upon loading. As a result the extent of the anisotropy of polymers can be (much more) pronounced than for crystalline metals and ceramics.

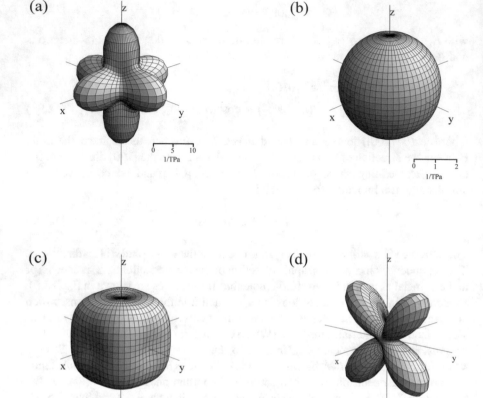

Fig. 12.6 $1/E$ surfaces calculated for **a** copper (f.c.c.), **b** tungsten (b.c.c.), **c** NaCl (rock salt; f.c.c.) and **d** cementite (Fe$_3$C, orthorhombic; nine independent elastic constants). The $1/E$ surfaces were calculated using Wintensor (with permission; *Wintensor*^{TM} developed by W. Kaminsky). The scale bars shown refer to the $[-110]$ direction which lies in the projection plane of the drawings. The data for the elastic constants of copper, tungsten and rock salt were taken from A. G. Every and A. K. McCurdy in: Landolt-Börnstein Numerical Data and Functional Relationships in Science and Technology, Vol. 29a: Second and Higher Order Elastic Constants, Ed. D. F. Nelson, Springer Verlag, Berlin 1992; the elastic constants for cementite were taken from M. Nikolussi, S. L. Shang, T, Gressmann, A. Leineweber, E. J. Mittemeijer, Y. Wang and Z.-K. Liu, Scripta Materialia, 59 (2008), 814–817

12.4   Elastic Deformation Upon Three Axial and Biaxial Loading

Until now cases of uniaxial loading were considered. In this section the state of stress imposed on a body will be generalized to the case of three-dimensional loading.

Fig. 12.7 a Stress acting on a flat area and its decomposition into a normal component σ and a shear component τ and **b** decomposition of a three-dimensional state of stress acting on a point of a body in a three-dimensional Cartesian coordinate system

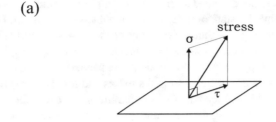

(a)

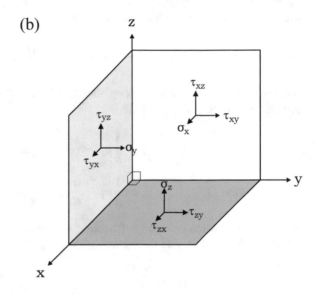

(b)

A stress acting on a certain (flat) area can always be resolved in a component acting perpendicular to that area (the normal component of stress, σ) and a component lying in the plane of that area (the shear component, τ); see Fig. 12.7a. Then, for an arbitrarily oriented Cartesian coordinate system, for describing the state of three-dimensional stress of a point of a body, it follows that three normal stress components are required, σ_x, σ_x and σ_z, and six shear components, τ_{xy}, τ_{xz}, τ_{yx}, τ_{yz}, τ_{zx}, and τ_{zy} (The shear components acting on the planes perpendicular to the x-, y-, z-axes have each been resolved into two components acting parallel to the axes of the Cartesian coordinate system adopted; see Fig. 12.7b). The subscript of σ indicates the direction along which σ operates. The first subscript of τ indicates the normal of the plane upon which the shearing stress component acts (e.g. τ_{xy} operates in a plane perpendicular to the x-axis) and the second subscript of τ indicates the direction in which the shearing stress component acts (e.g. τ_{xy} operates in the direction parallel to the y-axis); see Fig. 12.7b. According to this specification system σ_x should actually be given as σ_{xx} (σ_{xx} acts upon a plane perpendicular to the x-axis and in the direction of the x-axis), but usually one then applies one subscript.

The above consideration would imply that nine stress components are required to define the state of three-dimensional stress in a point. However, mechanical equilibrium requires that the moment of the forces about the three axes of the frame of reference (the x-, y- and z-axes) is nil, and thus: $\tau_{xy} = \tau_{yx}$ and $\tau_{xz} = \tau_{zx}$ and $\tau_{yz} = \tau_{zy}$. Hence, only six stress components are needed to define the state of three-dimensional stress in a point: three normal stresses and three shear stresses.

Consider a cube with its faces parallel to the axes of a Cartesian frame of reference. The strains along the x-, y- and z-axes can be given as follows. The strain along the direction of the x-axis is due to the normal stress σ_x acting in that direction and the (to be subtracted) contributions due to the Poisson contractions caused by the normal stress components acting along the y-axis and the z-axis (cf. Eq. 12.7b):

$$\varepsilon_x = \frac{1}{E}\sigma_x - \frac{\upsilon}{E}\sigma_y - \frac{\upsilon}{E}\sigma_z \tag{12.10a}$$

and similarly for the strains along the directions of the y- and z-axes (cyclic permutation applied to Eq. (12.10a)):

$$\varepsilon_x = \frac{1}{E}\sigma_y - \frac{\upsilon}{E}\sigma_z - \frac{\upsilon}{E}\sigma_x \tag{12.10b}$$

$$\varepsilon_z = \frac{1}{E}\sigma_z - \frac{\upsilon}{E}\sigma_x - \frac{\upsilon}{E}\sigma_y \tag{12.10c}$$

Note that the relation between γ (shearing strain) and τ (shearing stress) is of the same form for uniaxial and three-axial loading (cf. Eq. 12.6):

$$\gamma_{xy} = \frac{1}{G}\tau_{xy}; \gamma_{xz} = \frac{1}{G}\tau_{xz}; \gamma_{yz} = \frac{1}{G}\tau_{yz} \tag{12.11}$$

The elongations described by Eqs. (12.10) and the distortions (shearings) described by Eq. (12.11) are independent of each other. This holds as long as the deformations are small and the actions of the applied forces are not influenced by them. The calculation of a resultant deformation can then be based on the initial shape and size of the body subjected to loading. As a result the principle of superposition holds: a resultant displacement can be described as a linear function of the applied forces (stresses); e.g. see Eq. (12.10). Thus, in the general case of three-dimensional loading, to describe the total state of strain, the three elongations (Eq. 12.10) and the three shearing strains (Eq. 12.11) have to be superimposed.

The normal stress in an arbitrary direction, given by the subscript n ($n = (n_x, n_y, n_z)$, with n_x, n_y and n_z as the components along the x-, y- and z-axes of the Cartesian frame of reference), can be given as:

$$\sigma_n = \sigma_x \cos^2(n, x) + \sigma_y \cos^2(n, y) + \sigma_z \cos^2(n, z)$$
$$+ 2\tau_{xy}\cos(n, x)\cos(n, y) + 2\tau_{xz}\cos(n, x)\cos(n, z) + 2\tau_{yz}\cos(n, y)\cos(n, z) \tag{12.12}$$

where the "direction cosine" $\cos(n, x)$ denotes the cosine of the angle between the direction of n and the x-axis, etc. (If n is interpreted as the normal of the plane perpendicular to σ_n, then $\cos(n, x) = 1/n_x$, recognizing that n has unit length).

For any state of stress, it is always possible to define a Cartesian coordinate system such that only the (three) normal stress components σ_x, σ_y and σ_z occur (these normal stress components then also are of maximal value); i.e. the shear components τ_{xy}, τ_{xz} and τ_{yz} are zero. This coordinate system is called *the principal system*, characterized by the three, mutually perpendicular, *principal axes*, and the corresponding three normal stresses are called *the principal stresses*: σ_x^P, σ_y^P and σ_z^P, which act in the directions perpendicular on *the principal planes*.

In the principal system (frame of reference given by the principal axes) the total stress, σ_{tot}, acting on a plane with an orientation in the principal system characterized by the plane normal, with direction cosines $\cos(n,x)$, $\cos(n,y)$ and $\cos(n,z)$, can be written as[4]:

$$\sigma_{tot} = \left[\left(\sigma_x^P\right)^2 \cos^2(n, x) + \left(\sigma_y^P\right)^2 \cos^2(n, y) + \left(\sigma_z^P\right)^2 \cos^2(n, z) \right]^{1/2} \quad (12.13)$$

Although in the principal system the shearing stress components τ_{xy}, τ_{xz} and τ_{yz} are zero, shearing stresses act on planes which are not principal planes. Whereas the three principal stresses are also the maximal normal stresses, it can be shown that three maximal shearing stresses occur which act upon planes under 45° with the principal normal stresses. These three maximal shearing stresses, also called *the principal shearing stresses*, are given by[5]:

$$\tau_1 = \pm \frac{\left(\sigma_y^P - \sigma_z^P\right)}{2}; \ \tau_2 = \pm \frac{\left(\sigma_x^P - \sigma_z^P\right)}{2}; \ \tau_3 = \pm \frac{\left(\sigma_x^P - \sigma_y^P\right)}{2} \quad (12.14)$$

Thus, τ_1 is that maximal shearing stress that acts on a plane with a normal making (i) angles of 45° with both principal directions along which σ_y^P and σ_z^P act and (ii) an angle of 90° with the principal direction along which σ_x^P acts. (see Fig. 12.8).

---

[4] This result is a direct consequence of the recognition that in the principal system the x, y and z components of the total stress acting on the plane considered are $\sigma_x^P \cos(n, x)$, $\sigma_y^P \cos(n, y)$ and $\sigma_z^P \cos(n, z)$.

[5] These results can be obtained as follows (e.g. see Timoshenko and Goodier 1982). The total stress, σ_{tot}, acting on a plane of arbitrary orientation (not a principal plane), in the frame of reference given by the principal axes, is given by Eq. (12.13). Decompose this total stress into the normal component of the total stress acting on this plane, σ_n (this normal stress is given by the first three terms at the right-hand side of Eq. (12.12), since the shearing stress components in this equation are zero (principal system)) and the shearing stress acting on this plane (cf. Fig. 12.7a). Evidently, this shearing stress is given by $\left(\sigma_{tot}^2 - \sigma_n^2\right)^{1/2}$. Two of the direction cosines defining the orientation of the plane (in the principal system) are independent, because a relation of the type $\cos^2(n, x) + \cos^2(n, y) + \cos^2(n, z) = 1$ holds. Eliminate one of the direction cosines in the expression for the shearing stress by using the relation mentioned. Differentiate the resulting expression for the shearing stress with respect to the two remaining, independent direction cosines. Equating the two resulting differentials to zero leads to values for the maximal shearing stress and the orientation of the corresponding shearing plane.

Fig. 12.8 Illustration of the principal shearing stress τ_1 acting on a plane which makes angles of 45° with the principal directions pertaining to σ_y^P and σ_z^P

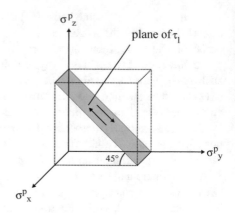

Often one uses the subscripts x, y and z also to indicate the relative magnitudes of σ_x^P, σ_y^P and σ_z^P such that σ_x^P is the largest principal stress component (most tensile or least compressive) and σ_z^P is the smallest principal stress component (most compressive or least tensile). Then τ_2 is the absolute maximum shearing stress.

It should be noted that Eqs. (12.12)–(12.14) hold generally, irrespective of the occurrence of elastic isotropy or anisotropy.

In analogy with the above description for the state of stress, it can be shown that six components of strain suffice to describe the state of strain of a point of a body: three normal strains and three shear strains: ε_x, ε_y, and ε_z, and γ_{xy}, γ_{xz} and γ_{yz}. Also, a Cartesian coordinate system can be defined such that only the three normal strains ε_x, ε_y and ε_z occur; i.e. the shear strains γ_{xy}, γ_{xz} and γ_{yz} are equal to zero. The three coordinate axes then are *the principal strain axes*. The directions in space of the principal stresses and those of the principal strains do not coincide in general for an elastically anisotropic body; for an elastically isotropic body they do.

If the special case of a so-called *hydrostatic (or spherical) state of stress* is considered, it holds for the principal stress components $\sigma_x = \sigma_y = \sigma_z \equiv \sigma$. Then it follows that $\varepsilon_x = \varepsilon_y = \varepsilon_z \equiv \varepsilon$, and from Eqs. (12.10), it is obtained:

$$\varepsilon = \frac{\sigma}{E}(1 - 2\upsilon) \tag{12.15}$$

Suppose the edges of the cube considered have unit length. Then the volume of the cube before imposition of the state of stress equals unit volume, V_0. The volume after elastic deformation due to the state of hydrostatic stress is given by V according to (note that $\varepsilon \ll 1$):

$$V = (1 + \varepsilon)^3 = 1 + 3\varepsilon + 3\varepsilon^2 + \varepsilon^3 \approx 1 + 3\varepsilon \tag{12.16}$$

Combining Eqs. (12.15) and (12.16) and noting that $V_0 = 1$, it is obtained for the volume change by the elastic deformation, $\Delta V = V - V_0$:

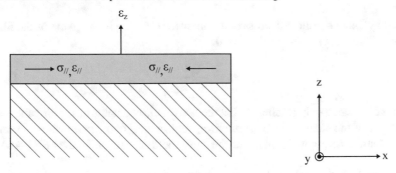

Fig. 12.9 State of planar, biaxial (here compressive) stress (with $\sigma_{//} \equiv \sigma_x = \sigma_y$) acting in a thin film on a substrate

$$\frac{\Delta V}{V_0} = 3\frac{\sigma}{E}(1 - 2\upsilon) \tag{12.17}$$

Because $\upsilon < ½$ (see at the end of Sect. 12.2) and if $\sigma > 0$ (i.e. tensile stress), it follows that $\Delta V > 0$, as expected intuitively. Reversely, if the applied stress is compressive (i.e. <0), $\Delta V < 0$. If $\upsilon = ½$, no volume change would occur; such a case pertains to ideal plastic deformation (see Sect. 12.8).

The above leads to a well-known reformulation of elastic constants for elastically isotropic materials by definition of the so-called bulk modulus, K, which is given by

$$K = \frac{E}{\{3(1 - 2\upsilon)\}} \tag{12.18}$$

and thus the relation between the hydrostatic stress σ and the corresponding relative volume change of the loaded body, $\Delta V/V_0$, becomes (cf. Eq. (12.17)):

$$\sigma = K\frac{\Delta V}{V_0} \tag{12.19}$$

Another case of great practical relevance is a stressed film on a substrate (Fig. 12.9; cf. Sect. 6.10.2). One of the principal axes is oriented perpendicular to the film (say, the z-axis); the other two principal axes (the x- and y-axes) are parallel to the film surface. Because of mechanical equilibrium the normal, principal stress perpendicular to the surface must be equal to zero. Then, if the two principal, normal stress components in the surface are equal, with $\sigma_{//} \equiv \sigma_x = \sigma_y$, it follows from Eq. (12.10) with $\varepsilon_{//} \equiv \varepsilon_x = \varepsilon_y$ and $\sigma_z = 0$:

$$\varepsilon_{//} = \frac{1 - \upsilon}{E} \cdot \sigma_{//} \tag{12.20}$$

which leads to Eq. (6.43) in Sect. 6.10.2. The constant $E/(1 - \upsilon)$ is sometimes called the "biaxial" elastic constant. However, this constant does not suffice to describe the elastic behaviour of the thin film in a biaxial (also called "planar") state of stress: although $\sigma_z = 0$ (see above), ε_z does not equal nil, due to the Poisson contraction caused by the two principal stress components, $\sigma_{//}$, acting in the plane of the film. It follows from Eq. (12.10c) with $\sigma_z = 0$:

$$\varepsilon_z = -2\frac{\upsilon}{E}\sigma_{//}\tag{12.21}$$

So, this discussion only teaches us that the *two* elastic constants describing the elastic behaviour of the *elastically isotropic* film, subjected to two equal principal components of stress acting in the plane of the film, could also be defined as $E/(1 - \upsilon)$ and υ/E.

This all leaves unimpeded that two elastic constants are needed and sufficient to describe the elastic deformation behaviour of elastically isotropic materials.

12.5 Elastic Strain Energy

Upon elastic deformation work is performed by the acting forces. This amount of work is identical with the amount of energy stored in the elastically deformed body.[6] If the forces are released all of the elastic stored strain energy is released. Work done by a force is force times distance covered.

Consider the case of loading of an infinitesimally small cube, with edges dx, dy and dz. A force F acts in the normal direction (taken as the x-direction of a Cartesian coordinate system) on an infinitesimally small face of area dS (Fig. 12.10). Upon elastic deformation the force increases from zero to F and the extension of the cube in the x-direction increases from zero to $\varepsilon_x dx$ (the length of the cube in the x-direction changes from dx to $dx + \varepsilon_x dx$). For a linear relation between force and extension (as given by Hooke's law) it simply follows by integration that the elastic strain energy, U'_{el}, stored in the cube equals half the product of the final value of the force and the final value of the extension.[7] Thus:

$$U'_{el} = \frac{1}{2}F \cdot \varepsilon_x dx = \frac{1}{2}\sigma_x dS \cdot \varepsilon_x dx$$

---

[6] One may ask into which forms of energy the work done is transformed. Obviously, considering the straining of the body concerned upon loading, it appears that the predominant part of the work done is transformed into elastic strain energy. However, consider a gas that is compressed adiabatically (i.e. there is no heat exchange between the system considered and its surroundings). The adiabatic compression induces an increase of temperature of the gas. Similarly, compression of a solid leads to an increase of temperature, albeit a very small one. This minute increase of temperature of a solid upon loading of, say, a couple of tenths Kelvin, corresponds to a very small thermal strain (cf. Sect. 3.1) to be distinguished from the mechanical, elastic strain. This thermal strain, as compared to the elastic strain, is negligible. Would not this not be the case, one should have to distinguish between *adiabatic and isothermal elastic constants*.

[7] If $F = $ constant $\cdot l$, with F and l as force and extension, it follows for the work done = energy stored U:

$U = \int F\, dl = $ constant $\cdot \int l\, dl = $ constant $\cdot \frac{1}{2}(l_{end})^2 = \frac{1}{2}F_{end} \cdot l_{end}$, with F_{end} and l_{end} as the final values of F and l. In the text above, the roles of F_{end} and l_{end} are taken by F and $\varepsilon_x dx$.

Fig. 12.10 a Tensile force, increasing from nil to F, acting on the face $dS = dydz$ of the infinitesimally small cube with edges dx, dy and dz, causing the extension $\varepsilon_x dx$. **b** The corresponding force-length diagram. The grey area below the curve corresponds to the elastic strain energy stored in the volume

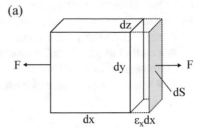

(a)

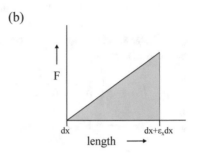

(b)

As $dSdx$ is the (initial) volume of the cube and because the relative increase of volume due to the elastic deformation is marginal, the elastic strain energy per unit volume of the cube, U_{el}, also called *strain energy density*, is given by

$$U_{el} = \frac{1}{2}\sigma_x \varepsilon_x = \frac{1}{2}E\varepsilon_x^2 = \frac{\sigma_x^2}{2E} \qquad (12.22)$$

Evidently, the strains due to the Poisson contraction, here in the directions of the y- and z-axes, do not occur in the above expression of U_{el} because these strains have not been induced by forces acting in the directions of these strains.

Analogously, for uniaxial loading by a shear force acting on a plane perpendicular to the x-axis and in the direction of the y-axis (see Sect. 12.4), it follows for the strain energy per unit volume:

$$U_{el} = \frac{1}{2}\tau_{xy}\gamma_{xy} = \frac{1}{2}G\gamma_{xy}^2 = \frac{\tau_{xy}^2}{2G} \qquad (12.23)$$

If a three-dimensional state of stress prevails, the forces corresponding to the six components of stress, σ_x, σ_y, σ_z, τ_{xy}, τ_{xz} and τ_{yz} (cf. Sect. 12.4), perform work on the volume element $dxdydz$. The total amount of work does not depend on the order in which these six forces are applied.[8] To calculate the total work done, the forces (stresses) can be supposed to increase simultaneously to their final values, while maintaining their relative values; then, the relation between each force and the corresponding displacement it invokes remains linear (e.g. see Eq. (12.10)). Hence, the contributions of the six stress (normal and shear) components to the strain energy density can be simply added (superposition principle; see above):

$$U_{el} = \frac{1}{2}\left(\sigma_x \varepsilon_x + \sigma_y \varepsilon_y + \sigma_z \varepsilon_z + \tau_{xy} \gamma_{xy} + \tau_{xz} \gamma_{xz} + \tau_{yz} \gamma_{yz}\right) \tag{12.24}$$

Note that the formulation given for U_{el} in Eq. (12.22) cannot be substituted into Eq. (12.24), because this formulation has been derived by application of Hooke's law for uniaxial loading. Instead, formulations of Hooke's law of the type given in Eqs. (12.10) and (12.11) have to be applied for the case of three-axial loading. As a result it is obtained:

$$U_{el} = \frac{1}{2E}\left(\sigma_x^2 + \sigma_y^2 + \sigma_z^2\right)$$
$$- \frac{\upsilon}{E}\left(\sigma_x \sigma_y + \sigma_y \sigma_z + \sigma_x \sigma_z\right) + \frac{1}{2G}\left(\tau_{xy}^2 + \tau_{xz}^2 + \tau_{yz}^2\right) \tag{12.25}$$

or, if the strain energy density should be expressed in terms of strains instead of stresses:

$$U_{el} = \frac{1}{2}\lambda\left(\varepsilon_x + \varepsilon_y + \varepsilon_z\right)^2 + G\left(\varepsilon_x^2 + \varepsilon_y^2 + \varepsilon_z^2\right)$$
$$+ \frac{1}{2}G\left(\gamma_{xy}^2 + \gamma_{xz}^2 + \gamma_{yz}^2\right) \tag{12.26}$$

with $\lambda = E\upsilon/\{(1 + \upsilon)(1 - 2\upsilon)\}$.

12.6   Rubber Elasticity; Elastomeric Behaviour

The elastic behaviour of crystalline material is of predominantly linearly elastic nature, in the sense as discussed in Sect. 12.2. This elasticity is, upon tensile loading, due to stretching of atomic bonds and the elastic constants, as the bulk modulus (cf. Eq. 12.18), are related to the curve of potential energy versus interatomic distance (see Sect. 3.1). Elastomers, with rubber as a specific example, are substances which exhibit extremely large values of elastic strains upon tensile loading, say 1000%, as compared to a few tenths of a per cent for macroscopic elastic strains for crystalline materials (see Fig. 12.11a). The associated modulus of elasticity is small: say, in the

---

[8] If this would be the case, a specific cycle of loading and unloading the six forces could be devised that would lead to net production of energy, which would violate the first law of thermodynamics (conservation of energy).

(a)   (b)

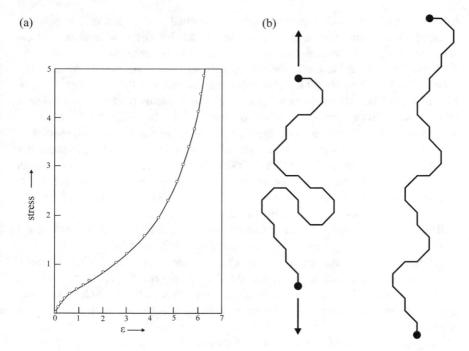

Fig. 12.11 **a** Typical stress–strain curve of a rubber (taken from L. R. G. Treloar, Reports on Progress in Physics, 36 (1973), 755–826). **b** Change of the configuration of a polymeric chain upon extension by an external force (the two black dots could be considered as the location of two neighbouring cross-links on the polymeric chain considered)

range $10-10^3$ MPa (metals: of the order 10^5 MPa; cf. Sect. 12.2) and depends on the value of strain (see the strongly nonlinear nature of the elastic stress–strain curve in Fig. 12.11a).

Elastomers are amorphous, long-chain polymers where the chains of the polymer molecules[9] are distinctly cross-linked. A typical representative of elastomeric materials is rubber and one speaks of rubber elasticity, in contrast with linear elasticity. It should be noted that rubber elasticity can only be observed in a restricted temperature range. At low temperatures (that is beneath the glass transition temperature T_g)[10] elastomers are brittle and exhibit (only) linearly elastic behaviour.

The backbone of the polymer molecule chain must be long and in unloaded state is strongly coiled due to kinking and twisting, which can result from the positioning of carbon–carbon double bonds and side-groups along the chain (see footnote 9). Cross-links between the chains are essential. For example, the so-called vulcanization process of (natural) rubber involves the introduction of sulphur such that each

---

[9] The backbone of the polymeric chain is a string of covalently bonded carbon atoms.

[10] An amorphous polymer beneath the glass transition temperature, T_g, behaves as a "glass", showing linearly elastic deformation and brittle fracture. Above T_g the amorphous polymer behaves as a rubbery solid, until at still higher temperatures a viscous liquid results (see Sect. 12.7).

sulphur atom bonds with a carbon atom in one chain and with a carbon atom of a neighbouring chain. Elastic extension upon tensile loading now involves that the segments of the (neighbouring) chains between the cross-links are straightened and displaced with respect to each other: interchain segmental sliding. As a result, the coiled chains are "unwound"; the chains become elongated in the stress direction (cf. Fig. 12.11b). Upon removal of the load, an almost immediate return to the original coiled configuration occurs, due to the presence of the cross-links; the cross-links are essential for returning to the original dimensions of the elastomeric specimen. Obviously, the higher the cross-link density along the chains, the stiffer (i.e. the larger the value of the modulus of elasticity of) the elastomer. Only after maximal chain stretching by unwinding has been realized does atomic bond stretching become to play a role, which requires a pronounced increase of the tensile load, i.e. the experienced modulus of elasticity increases strongly (see Fig. 12.11a).

The above discussion makes clear that rubber elasticity is not primarily related to a change of potential energy of the system considered, because atomic bond straining, as in linearly elastic behaviour, does not play a role (see the beginning of this section). Elasticity associated with such atomic bond straining can be called *"(potential) energy elasticity"*. It is the change in the configuration of the long-chain polymer molecules that controls the energy change: entropy change (see Sect. 7.3) governs elastomeric behaviour (cf. Fig. 12.11b). Entropy is related to the degree of disorder. The degree of disorder of a certain state is expressed by the number of corresponding distributions/"realizations of the system" (see discussion in Sect. 7.3). The configurational entropy (the degree of disorder) is in equilibrium situations as large as possible. Evidently, stretching of the polymeric chains upon tensile loading reduces the number of corresponding distributions: the entropy is reduced upon tensile loading, which increases the (Gibbs/Helmholtz; cf. Sect. 7.3) energy of the system. The entropy decrease of an elastomer upon tensile loading can thus be directly calculated from the observed increase in (macroscopic) specimen length. Unloading of the specimen allows the long-chain polymer molecules to relax (the entropy becomes maximized) and the original coiled configuration is realized. Elastomeric behaviour thus is also called: *"entropy elasticity"*.

The thermal vibrations of the segments of the cross-linked molecular polymeric chains of tensilely loaded elastomers, related to the kinetic energy of the system, tend to reduce the effective length of the segments, which effect increases with increasing temperature. As a consequence, two remarkable conclusions can be drawn:

(i)   To achieve the same extension of specimen length, a larger tensile load is required at higher temperature. This makes clear that the elasticity modulus of elastomers *increases* with increasing temperature (for crystalline materials a relatively modest *decrease* of the elasticity modulus occurs with increasing temperature).

(ii)   The (linear) coefficient of thermal expansion of elastomers under tensile loading is negative.

As also holds for crystalline materials exhibiting linear elasticity, rubber elasticity involves an immediate and complete return to the original specimen dimensions upon

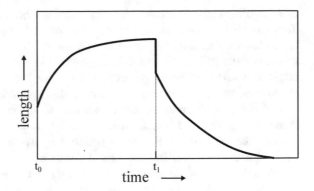

Fig. 12.12 Schematic length-time curve of an anelastic material to which a tensile force is applied between t_0 and t_1

removal of the load, which requires the presence of a sufficient number of cross-links. If the number of cross-links becomes reduced (e.g. by overstretching), an immediate return to the initial specimen dimensions cannot occur: a more or less gradual (and not entirely complete) return occurs to the specimen dimensions as before loading. This time dependence of the elastic behaviour is called viscoelasticity, a phenomenon to be discussed in the next section.

12.7   Viscoelasticity/Anelasticity; Mechanical Hysteresis

Until now in this chapter it was implied that an applied load instantaneously induces the entire corresponding strain and, also, that upon release of the load the strain is completely removed instantaneously as well. However, upon load imposition there can be *a part of* the strain induced that develops as a function of time after the start of loading, and upon release of the load *a part of* the strain may not be immediately released, but some time can be needed to establish complete recovery of the initial, unloaded situation (see Fig. 12.12). For many metallic solids this so-called *anelasticity* is negligible, but it is important for many amorphous, long-chain, not significantly cross-linked polymers at temperatures above the glass transition temperature, T_g, where the material no longer behaves as an amorphous solid but is conceived as a rubbery solid; there one speaks of *viscoelasticity*: an (linearly) elastic (time independent) and a viscous (time dependent) part of the deformation induced upon imposition of a load, can be distinguished. At still higher temperatures the material behaves as a viscous liquid that flows irreversibly upon loading.[11]

The occurrence of viscoelastic behaviour for not significantly cross-linked, long-chain polymers above the glass transition temperature is related to interchain sliding (in case of rubber elasticity only interchain *segmental* (of the segments between the cross-links) sliding occurs), associated with overcoming (with time, upon application

---

[11] The notion "viscosity" of a material is used to indicate the resistance of the material against flow invoked by shear forces.

of a load) the steric hindrance due to side groups/branches attached to the chains. This interchain sliding causes the viscous component of the strain, in addition to the instantaneous, linearly elastic component of the strain (cf. Fig. 12.12). Upon unloading, the strive for maximal entropy provides the driving force to return to the original (before loading) configuration of the polymer molecules, as in the case of elastomers, but this return, in the absence of distinct cross-linking, is time dependent and occurs in addition to the immediate release of the linear elastic component of strain (cf. Fig. 12.12). In the above the case of constant applied load was considered. It should be realized that the rate at which the polymer material stretching occurs, i.e. the "strain rate", can influence the deformation (viscoelastic) properties significantly: a decrease of the strain rate has qualitatively the same effect as an increase of temperature: the material becomes softer/more ductile.

The time delay for the strain observed after the stress has been applied leads to a phenomenon called "damping". This can be illustrated considering the occurrence of *mechanical hysteresis* upon cyclic loading. If the time period of the stress cycling is much larger than the time needed for the material considered to develop the full (i.e. not only the linear elastic component but also the viscoelastic component) strain compatible with the stress applied, a plot of stress applied versus strain observed is a straight line (see Fig. 12.13). The slope of this line gives a value for the so-called *relaxed modulus of elasticity*, which is based on the sum of the linear elastic and viscoelastic strain components. If the time period of the stress cycling is much smaller than the time needed for the viscoelastic component to unfold, a plot of stress applied versus strain observed is a straight line as well (see Fig. 12.13), but with a slope that is representative of the so-called *unrelaxed modulus of elasticity*, which is based on the linear elastic component of strain only. Obviously, the unrelaxed modulus of elasticity is larger than the relaxed modulus of elasticity. Now, if the frequency (=reciprocal of the time period) of stress cycling takes an in-between value, such that the time needed for the viscoelastic component of strain to unfold is of the same order of magnitude as the time needed for one stress cycle, only a certain extent of the maximally possible viscous flow can develop; the viscoelastic strain cannot

Fig. 12.13 Schematic stress–strain curves obtained during cyclic loading of a viscoelastic material: unrelaxed (high frequency) case, relaxed (low frequency) case and intermediate case, when damping occurs (hatched area corresponds to energy loss)

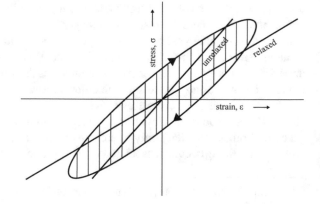

"keep up" (i.e. cannot stay "in phase") with the stress: in the tensile/compressive loading parts of the stress cycle the maximal value of (absolute) strain occurs after the tensile/compressive stress has passed through its maximum (see Fig. 12.13). As a result, the loading and unloading parts of the stress–strain dependencies do not coincide: the area enclosed (hatched area in Fig. 12.13) represents the irreversible energy loss during one stress cycle (in this context, see also footnote 6 in Sect. 12.5). This effect is called elastic mechanical hysteresis. The energy lost is dissipated as heat. Note that this heat is produced by purely elastic deformation! This capacity of a body subjected to a cycling loading stress (the body thereby vibrates) to convert mechanical energy (of vibration) into heat is also called *internal friction* or *damping capacity*.

Internal friction can be measured by putting the (wire) specimen into a cyclic motion, e.g. applying a torsional pendulum device (cf. Fig. 12.14a), and by measuring the amount of energy lost in one cycle during the natural decay (free oscillation) of the system: the decrease of the amplitude of the cyclic movement of the specimen is a measure for the irreversible energy loss during one cycle by mechanical hysteresis.

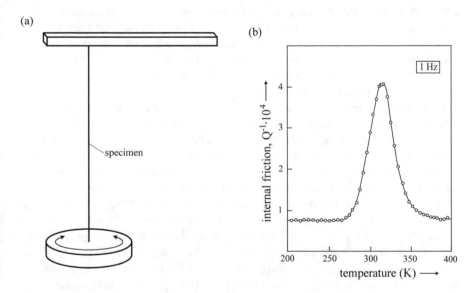

Fig. 12.14 a Schematic representation of a torsional pendulum. **b** The energy loss per cycle, proportional with a dimensionless quantity Q^{-1} usually applied as measure for internal friction and plotted along the ordinate, for a cycle frequency of 1 Hz (i.e. one cycle per second), as function of temperature for b.c.c. iron containing interstitially dissolved carbon (here a ferritic steel): the Snoek effect. The peak height is a measure for the amount of carbon dissolved in the b.c.c. phase and from the result shown here it follows that about 20 at.ppm C is dissolved in ferrite; the detection limit of this method is about 1 at.ppm! From the temperature dependence of the frequency of maximal damping (i.e. the position of the maximum of the Snoek peak as shown in the figure is determined for various frequencies), diffusion data (the activation energy) of interstitially dissolved components can be determined with very high accuracy; cf. Sect. 8.6 (taken from M. Weller, Materials Science and Engineering A, 442 (2006), 21–30)

A plot of the energy loss per cycle as function of the cycle frequency, at constant temperature, can show a number of maxima which can be interpreted as due to specific mechanisms of damping operating in the material investigated. Thus such damping peaks of amorphous polymers can be ascribed to vibrations of the backbone of the polymer molecule or vibrations of the side-groups/branches (see also below).

In metals, the energy lost by mechanical hysteresis is usually only a very small amount of the total elastic strain energy incorporated in the metal upon loading (i.e. the hatched area in Fig. 12.13 is relatively small). In polymers, pronounced mechanical hysteresis can occur. Against this background, in discussions about mechanical hysteresis, for metals one usually applies the term anelasticity and for polymers one generally speaks of viscoelasticity.

Although mechanical hysteresis is a relatively small effect for metals, its atomistic origin is worthwhile considering in view of the large structural differences between crystalline metals and amorphous polymers. Two well-known mechanisms which can give rise to anelasticity of metals are touched upon below.

One may anticipate that especially atoms at the grain boundaries of metallic specimens, which are less strongly bonded than atoms in the bulk of the grains, may be capable of slight relative displacements of elastic nature and thereby a minor amount of viscoelastic strain (grain sliding) can occur; this behaviour can be considered as a pendant of the sliding of the polymer chains discussed above as mechanism for viscoelastic behaviour of amorphous polymers. However, anelasticity for metals is most often associated with a very specific mechanism observed for interstitial atoms in b.c.c. metals, which is discussed next.

Body-centred cubic metals (e.g. W and α-Fe(ferrite)) possess three types of octahedral interstices (see Fig. 9.21; see also Fig. 4.43a): along the a-, b- and c-axes of the crystal structure. Obviously, in the absence of a state of stress, these three types of interstitial sites are equivalent. Upon insertion of an interstitial atom, say a C (or N) atom in α-Fe, into an octahedral interstice (e.g. an octahedral interstitial site along the c-axis; see Fig. 9.21 and Fig. 4.44), the initially irregular octahedron, constituted by six Fe atoms, becomes more regular: the two nearest Fe neighbours of the C atom (i.e. here along the c axis) become displaced outwardly and the four next nearest Fe neighbours of the C atom (i.e. here along the a- and b-axes) are slightly moving inwardly (cf. Poisson contraction; see also Fig. 12.21 discussed in Sect. 12.9.2 and footnote 13 and the extensive discussion in Sect. 9.5.2.1). Hence, insertion of a C atom into the octahedral interstice leads to an elastic deformation field of tetragonal nature. Now consider the case of a (b.c.c.) α-Fe crystal containing a certain amount of C atoms, initially randomly distributed over the three types of octahedral interstitial sites, subjected to a tensile stress acting along the c axis. Obviously, with reference to the above discussion, the C atoms now preferentially occupy the octahedral interstitial sites along the c-axis. If the specimen is subjected to a stress cycle, and if the frequency of the stress cycle and the jump frequency of the interstitial atoms (to move from an unfavourable interstitial site to a preferred interstitial site) are of comparable value, then, in accordance with the discussion above, mechanical hysteresis can be observed. The effect of internal friction by jumping interstitial atoms has become so well known that it is usually named after J.L. Snoek who first explained its origin

(1939): *Snoek effect*. The method has allowed to determine, from the temperature dependence of the frequency of maximal damping, very accurate values of the diffusion coefficients of interstitials, as determined by the jumping frequency, over an unusually large range in temperature and thus has provided a classical example for confirming the Arrhenius type of temperature dependence of a diffusion coefficient (see Fig. 12.14b and see Sect. 8.6 and Fig. 8.10).

12.8 Plastic Deformation Characteristics

Whereas elastic deformation is fully described considering only the initial and final stages of the deformation process, the plastic deformation experienced by a body upon loading beyond the "yield point" (see further) depends on the path in the load-deformation diagram along which the considered final stage of plastic deformation is reached. This makes immediately clear that descriptions of plastic deformation behaviour must be much more complex than those for elastic deformation behaviour.

As a side remark it is noted here that, although elasticity theory has a firm basis, some problems of deep, fundamental nature have not been dealt with definitively: e.g. the elastic grain interaction, i.e. the elastic behaviour of polycrystalline, single- or polyphase materials (cf. Sect. 12.2 and the "*Intermezzo: Grain interaction*" at the end of Sect. 6.10). Such problems are more imminent in case of plastic deformation; e.g. the collective behaviour of a set of dislocations or point defects and the consequences of the presence of inclusions are unsatisfactorily described on the basis of the current state of knowledge.

A major problem is the changing "strength" of the material upon plastic deformation (called "work hardening" or "strain hardening"), which obstructs the identification of truly, genuine material constants describing mechanical strength, as possible for elastic deformation and exemplified by the elastic constants. Upon plastic deformation, for example dislocation production can occur, and, in general, the increase of the dislocation density makes dislocation propagation as a mechanism for glide (cf. Sects. 5.2.5 and 5.2.6) more difficult, implying the application of larger loads to realize the same (additional) extension as at an earlier stage of plastic deformation. Theories for plastic deformation are unavoidably, not only much more complicated, but also much less validated as the theory for elastic deformation.

It is customary to assume that the deformable material, taken as a continuum, is plastically isotropic and that the plastic deformation does not involve volume change, leading to the incompressibility relation that the sum of the principal (cf. Sect. 12.4) strains is zero:

$$\varepsilon_x^P + \varepsilon_y^P + \varepsilon_z^P = 0 \tag{12.27}$$

Thus, for *ideal plastic deformation* Poisson's ratio, υ, equals ½ (see below Eq. 12.17).

Because of Eq. (12.27), the six independent strain components, the three normal strains, ε_x, ε_y, and ε_z, and the three shear strains, γ_{xy}, γ_{xz} and γ_{yz} (cf. Sect. 12.4), reduce to five independent strain components. This immediately leads to the important conclusion that occurrence of compatible plastic deformations of the individual crystals in a polycrystalline, massive specimen, in order to maintain the massive nature and integrity of the loaded specimen, requires that at least five independent slip systems should be available in each crystal (see discussion in Sect. 12.12).

Metals are very ductile materials; i.e. they can be formed by very severe plastic deformation without that the material breaks. Some metals, e.g. gold, can be deformed by cold work even such that very thin foils result, without that the integrity of the piece of metal is lost, as every goldsmith knows. Rolling, forging, (deep) drawing and (hot isostatic) pressing are examples of plastic deformation processes of great industrial importance. Much of plasticity theory therefore has been developed with metals as type of material in mind. In fact the conception of the dislocation, as discussed in Chap. 5, derived largely from the need to explain the plasticity of metals.

The essential difference between elastic deformation and plastic deformation involves that elastic deformation maintains the local atomic arrangements, whereas permanent, plastic deformation (shape change) requires, crudely speaking, the breakage of atomic bonds and the establishment of new atomic bonds. This recognition may make likely that crystalline solids and amorphous solids exhibit essentially different plastic deformation mechanisms: glide of dislocations is a dominant plastic deformation mechanism of crystalline solids (Sect. 5.2.5); viscous flow is the mechanism for plastic deformation of amorphous solids (Sect. 12.7).

12.9   The Tensile Stress–Strain Curve; True Stress and True Strain

The basic, relatively simple test performed to characterize the strength of a material is the measurement of the tensile stress–strain curve: a specimen is subjected to an uniaxially applied tensile load, and it is recorded how the tensile load changes while the specimen length (in the loading direction) increases. Usually the specimen is elongated at a constant (strain) rate at constant temperature. Characteristics obtained from such stress–strain curves are used as essential information for the design of structures: material-acceptance criteria leading to material selection.

A schematic presentation of a stress–strain curve is given in Fig. 12.15. In the sense of the discussion of Sect. 12.2, here the average stress has been plotted versus the average strain, where stress and strain have been defined with respect to the initial cross-sectional area, onto which the applied load acts, and the initial specimen length, respectively, and hence have to be denoted as engineering stress and engineering strain.

The first linear part of the curve obviously represents the (linear) elastic behaviour (Hooke's law); the slope of the straight line in this region is the modulus of elasticity (see Eq. (12.3)). The remainder of this section is devoted to ductile materials

Fig. 12.15 Schematic
stress–strain curve for
uniaxial tensile loading

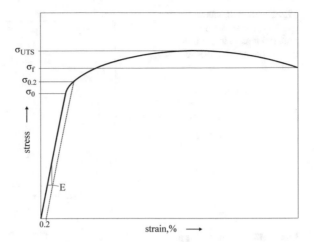

exhibiting pronounced plastic deformation before fracture.[12] Brittle materials, in the
extreme case, do not show plastic deformation at all: failure occurs before the elastic
limit has been crossed (cf. Sect. 12.2). Schematic stress–strain curves for ductile and
brittle materials are shown in Fig. 12.16.

Applying to a ductile material an uniaxial tensile load larger than the one applied
at the end of the linear elastic region causes a permanent deformation remaining after
unloading the specimen. The transition to the plastic deformation region occurs at
the elastic limit called the yield strength or yield stress, σ_0 (corresponding to the
load at and beyond which the material yields).

If the load is released at a stage of modest plastic deformation (see point $\sigma_{0.2}$
in Fig. 12.15 and also see Fig. 12.17) only part of the total strain is recovered as
a reversible elastic strain, the remaining strain is the permanent plastic strain. The
unloading curve is practically a straight line with slope practically equal to the initial
elastic slope. Renewed application of the load causes following the straight line

---

[12] Ductility is the ability of a material to undergo plastic deformation. The term toughness is used
to indicate the (plastic deformation) energy which can be absorbed until fracture occurs (i.e. the
area under the stress–strain curve until fracture): ductile materials are normally tougher than brittle
materials. Evidently, a high toughness requires ductility, but also considerable strength. Usually
ductility and toughness both increase or both decrease upon manipulation of the microstructure.
Ideally, a material for structural application is both strong and tough. However, usually a strong
material has modest toughness and vice versa. Consequently, material treatments and material
developments aim at achieving an *optimal* combination of strength and toughness. For example,
see the discussion on tempering of steels in the *"Intermezzo: Tempering of iron-based interstitial
martensitic specimens"* at the end of Chap. 9. Yet, there are ways to establish the simultaneous
occurrence of high strength and large toughness. Such a route implies the involvement of different
mechanisms to establish plastic behaviour which operate at different length scales (cf. Sect. 1.4),
for example (i) by applying a hard phase with a microstructure that allows local relieve of high
loads by a small amount of plastic deformation, such as provided by sliding along fibres contained
in the hard-phase matrix (of course, dislocations in (metallic) materials in fact have a similar role
in plastic deformation of the body concerned (cf. Sect. 5.2)), and (ii) by subcritical (micro)crack
growth (cf. Sect. 12.15) (Ritchie 2011).

Fig. 12.16 Schematic
stress–strain curves for
uniaxial tensile loading of
a a ductile and **b** a brittle
material

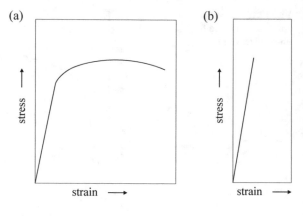

Fig. 12.17 Behaviour of a
plastically deformed material
upon unloading and
subsequent reloading (tensile
testing) at a moderate stage
of plastic deformation

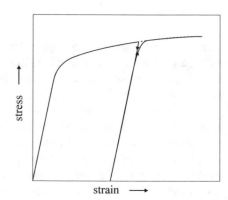

recorded upon unloading, but now in the reverse direction (Fig. 12.17). The moment
of yielding now occurs at a value of stress higher than experienced during the first
loading, which is a consequence of the strain hardening induced by the modest
plastic deformation experienced already before the unloading occurred (for "strain
hardening", see Sects. 12.8 and 12.14.1).

The elastic limit, σ_0, is difficult to establish experimentally from the stress–strain
curve. Therefore, for engineering purposes, the *yield strength* is taken as the stress
that gives a certain small amount of permanent deformation: say, 0.2% permanent
deformation, and the yield strength is then indicated by the symbol $\sigma_{0.2}$. Consider
Fig. 12.15. After the stress has reached the value $\sigma_{0.2}$, upon unloading the specimen
decreases its length according to the dashed straight line indicated in the figure and
the strain of 0.2% remains (the part cut from the abscissa by the dashed line).

The gradual, and thereby difficult to identify (see above) transition from elastic to
plastic behaviour in a tensile stress–strain curve can have an origin in the usual polycrystalline nature of the test specimens. Upon increasing the load beyond a critical
value the specimen does not start to deform plastically homogeneously as a whole:
for example, the grain-interaction effects already discussed before (cf. Sect. 12.2 and

the "*Intermezzo: Grain interaction*" at the end of Sect. 6.10) induce the occurrence of (micro)plastic deformation first at locations within the specimen (e.g. certain grain boundaries and grain junctions) where stress concentrations occur; i.e. on a local, microscopic scale, the state of stress is not uniaxial. Upon increasing the load an increasingly more homogeneous plastic deformation will take place.

Because of the work hardening effect mentioned in the second paragraph of Sect. 12.8 and discussed in Sect. 12.14.1, the load necessary for further plastic deformation increases with continued straining. The curve of engineering stress versus engineering strain shows a maximum. The stress corresponding to this maximum is called the *ultimate tensile strength (U.T.S.)*, σ_{UTS}, which is the maximal stress that the material can bear upon uniaxial tensile loading, but do note that staying at this stress level will induce failure: fracture will occur. Hence, the yield strength and not the ultimate tensile strength is a parameter to be used for material selection for a structure exposed to loading.

The load required to further strain the specimen, beyond the ultimate tensile strength, becomes smaller, because the diameter (cross-sectional area upon which the load acts) of the (cylindrical) specimen, loaded along its length axis, becomes smaller (so-called necking) after having reached the U.T.S. Eventually the specimen fractures at the engineering stress level σ_f, the *fracture strength*.

In the above discussion the notions engineering stress and strain were used with some emphasis on the adjective "engineering". As the dimensions of the specimen change during the loading experiment, it appears appropriate to apply definitions of stress and strain based on the instantaneous dimensions of the specimen, instead of on the original cross-sectional area upon which the load acts, and the original specimen length.

The engineering strain is defined as (cf. Eq. (12.2)):

$$\varepsilon = \frac{\Delta l}{l_0} = \frac{1}{l_0} \cdot \int dl = \frac{l - l_0}{l_0} \tag{12.28}$$

with l_0 and l as the boundaries of integration. The true strain, ε_{true}, should be defined with respect to the instantaneous specimen length:

$$\varepsilon_{true} = \frac{l_1 - l_0}{l_0} + \frac{l_2 - l_1}{l_1} + \frac{l_3 - l_2}{l_2}$$
$$+ \cdots = \sum \frac{l_{i+1} - l_i}{l_i} = \int \frac{1}{l} dl = \ln \frac{l}{l_0} \tag{12.29}$$

using l_0 and l as integration boundaries. From Eqs. (12.28) and (12.29) it follows:

$$\varepsilon_{true} = \ln(\varepsilon + 1) \tag{12.30}$$

Recognizing that no volume change occurs upon (ideal) plastic deformation, and adopting the symbols S and S_0 for the cross-sectional areas, after and before loading, onto which the load acts, it holds:

Fig. 12.18 Schematic *engineering* stress–strain curve for uniaxial tensile loading and the corresponding *true* stress–strain curve

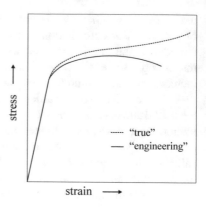

$$S_0 l_0 = S l$$

and thus if follows for the true stress, σ_{true}:

$$\sigma_{\text{true}} = \frac{F}{S} = \frac{F}{S_0} \cdot \frac{l}{l_0} = \sigma(1 + \varepsilon) \tag{12.31}$$

A schematic presentation of the engineering stress - engineering strain curve and the corresponding true stress—true strain curve is provided by Fig. 12.18.

The equations given here for $\varepsilon_{\text{true}}$ and σ_{true} only hold until serious necking occurs, i.e. until the ultimate tensile strength (see below). Beyond the ultimate tensile strength the loading is no longer uniaxial: at the location where necking occurs complicated, no longer uniaxial states of stress prevail.

The true stress—true strain curve, from the onset of yielding until necking begins, is of fundamental importance for plasticity theory. Simple analytical, fundamental and generally valid descriptions of this curve do not exist. An often used, entirely phenomenological, in many cases defective description of this curve reads:

$$\sigma_{\text{true}} = K(\varepsilon_{\text{true}})^n \tag{12.32}$$

K represents the (hypothetical) true stress required to realize a (hypothetical) true strain of 100% (i.e. $\varepsilon_{\text{true}} = 1$). The exponent n is called "strain-hardening coefficient": the larger the value of n (for $0 < n < 1$), the higher the true stress required to attain the same true strain (for $0 < \varepsilon_{\text{true}} < 1$).

The gradual decrease in cross-sectional area upon tensile loading increases the (true) loading stress. In the absence of strain hardening this effect involves that as soon as that the material enters the plastic region; i.e. it starts to yield, the material becomes unstable: it cannot carry the load and at some location of the specimen, where, for example, a local dimensional irregularity or local inhomogeneity exists (stress raiser; cf. Sect. 12.17), an abnormally large reduction of the cross-sectional area occurs: initiation of *necking*, a local breakdown of the uniformity of straining. However, in many cases, the material exhibits strain hardening (work-hardening), i.e. the intrinsic capacity to carry a load increases as plastic deformation progresses. Thus the local initiation of necking can be repaired and the definitive occurrence of necking can be delayed. The strain-hardening (work-hardening) rate decreases with increasing strain: definitive necking occurs as soon as the (increase of) load-carrying capacity of the material can no longer compensate the increase of applied stress due to the decrease of cross-sectional area. The true stress where this happens corresponds with the maximal load (and the maximum in the engineering stress—engineering strain curve). Hence, the instability criterion (see also Hoffman and Sachs 1953), indicating the occurrence of necking, involving that no (further) change (increase) of load can occur, can be given as:

$$dF = d(\sigma_{true} S) = \sigma_{true} dS + S d\sigma_{true} = 0 \qquad (12.33a)$$

and thus

$$\frac{d\sigma_{true}}{\sigma_{true}} = -\frac{dS}{S} \qquad (12.33b)$$

Because of the constancy of volume that holds for ideal plastic deformation (cf. Sect. 12.8) and using Eq. (12.29):

$$\frac{dS}{S} = -\frac{dl}{l} = -d\varepsilon_{true} \qquad (12.34)$$

Combining Eqs. (12.33b) and (12.34), it follows:

$$\frac{d\sigma_{true}}{d\varepsilon_{true}} = \sigma_{true} \qquad (12.35)$$

It is concluded that (definitive) necking occurs at the location in the true stress—true strain curve where the slope of that curve equals the true stress.

In this section the case of uniaxial tensile loading has been considered (recall that when necking occurs, a, nevertheless, triaxial state of stress prevails in the region where necking takes place). The often used material-strength parameters defined above as yield strength and (ultimate) tensile strength have no universal meaning: in practice often other, bi- and triaxial, types of loading govern and corresponding, other strength parameters could be defined with a closer relationship with the "strength" of the material experienced subject to the loading conditions applied (see Sect. 12.10).

12.9.1   Strain and Strain Rate Due to Dislocation Movement

Significant macroscopic plastic deformation in crystalline materials requires the movement of a large number of dislocations. Note that only mobile dislocations can contribute to plastic deformation. A relation is sought for that describes the relation between the mobile dislocation density and the realized plastic strain.

Consider a crystal block of dimensions L_1, L_2 and L_3 containing a set of parallel edge dislocations perpendicular to the plane determined by L_1 and L_2 and with Burgers vector parallel to L_2 (Fig. 12.19). Application of a sufficiently high shear stress, τ, with τ parallel to the slip plane (determined by the dislocation line vector, parallel to L_3, and the Burgers vector, parallel to L_2) and, specifically, parallel to **b,** causes positive edge dislocations and negative edge dislocations (cf. Sect. 5.2.3 and 5.2.5) to move (glide) in opposite directions, thereby inducing a (total) displacement of the top surface with respect to the bottom surface of the crystal block. If each dislocation moves across the entire crystal block, its contribution to the total displacement is the magnitude of the Burgers vector **b**, indicated by b (cf. Sect. 5.2.3). If on average dislocations do not move (glide) across the entire crystal block, but only cover on average the fractional distance $<x>/L_2$, the corresponding contribution of each dislocation to the total displacement is on average $(<x>/L_2) \cdot b$. The density of the mobile dislocations is indicated by ρ_m $(=n.L_3/(L_1.L_2.L_3) = n/(L_1.L_2))$, with n as the number of mobile dislocations in the crystal block). Hence, the total shear strain, γ, is given by (cf. Fig. 12.19)

$$\gamma = n \cdot \frac{[(\langle x \rangle / L_2) \cdot b]}{L_1} = b \cdot \rho_m \cdot \langle x \rangle, \tag{12.36}$$

and thus, the strain rate $d\gamma/dt$ obeys

$$\frac{d\gamma}{dt} = b \cdot \rho_m \cdot \langle v \rangle \tag{12.37}$$

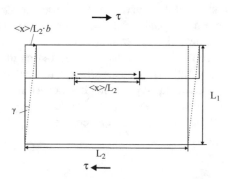

Fig. 12.19 Plastic deformation of a body, crystal block of dimensions L_1, L_2 and L_3, by movement (glide) of an edge dislocation, with its dislocation line perpendicular to the plane determined by L_1 and L_2 and its Burgers vector **b** parallel to L_2, by the distance of $<x>/L_2$ under applied shear stress τ

with $<v> = d<x>/dt$ as the average dislocation velocity. This equation is known as Orowan equation (for Orowan process, see Sect. 5.4).

Such results also hold for screw dislocations and for general, mixed dislocations.

12.9.2 The Yield Drop Phenomenon; Cottrell–Bilby Atmospheres

Especially, but not only, body centred cubic metals may show a "yield drop" after having reached the *upper yield stress* (see Fig. 12.20): after the onset of plastic deformation a, usually 10–20% lower, applied stress is needed for further plastic deformation. Prolonged plastic deformation thereafter can continue at more or less this lower level of applied stress, the *lower yield stress*, for a certain range of plastic deformation; this elongation at constant load is called *yield-point elongation zone*, also called *Lüders extension* (see below). Beyond this range the applied stress to realize further plastic deformation must increase.

The plastic deformation beyond the upper yield point and until the end of the range at the lower yield stress is not uniform in the specimen: the instantaneous dislocation multiplication is restricted to one band (or more bands) of material, the so-called *Lüders band*, that propagates along the whole (length of the) specimen (in this context see the discussion on the occurrence of cross slip and the development of a *glide band* in Sect. 5.2.6), inducing the same plastic strain at every position that it passes. Upon continued deformation, i.e. beyond the yield-point elongation zone, macroscopically homogeneous deformation (macroscopically homogeneous strain/work hardening) occurs; note that strain/work hardening has been happening within the Lüders band from the start of the yield-point elongation zone.

The occurrence of yield drop can be ascribed to the pronounced increase in the density of *mobile* dislocations once the upper yield point is passed. It is recognized that (initially) many dislocations in a material may be immobilized because of their interaction with point defects, as dissolved (interstitial) atoms (e.g. carbon and nitrogen in iron). This can be discussed as follows.

Fig. 12.20 Schematic representation of the yield drop phenomenon and the Lüders strain

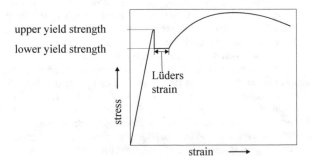

upper yield strength

lower yield strength

Lüders strain

stress

strain

Fig. 12.21 Anisotropy of
the irregular octahedral
interstitial site in ferrite

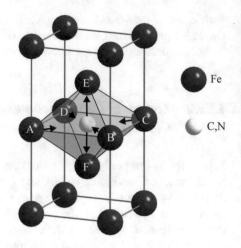

Point defects and dislocations are associated with stress fields. These stress fields will interact; i.e. the configuration strived for by the dislocation and the point defect will be such that the total elastic energy of the material is lowered. This realized decrease in energy, as compared to the presence of the dislocation and the point defect separately, i.e. at infinite distance, is called the *interaction energy*. The more negative the interaction energy, the more energy it costs to separate the dislocation and the point defect.

Dissolved carbon or nitrogen atoms in b.c.c. iron occupy octahedral interstitial sites. These interstitial atoms do not fit ideally at these positions: a tetragonal, elastic deformation occurs: the two nearest neighbours of the interstitial atom in the centre of the *irregular* octahedron of six iron atoms are displaced (i.e. the two iron atoms at *E* and *F* in Fig. 12.21 are moved more apart) under simultaneous slight decrease (cf. Poisson contraction) of the distances between the next nearest neighbours, i.e. the other four iron atoms, at A, B, C and D (see Fig. 12.21).[13] The stress fields of the interstitially dissolved solute atom and of a dislocation can interact. As a result it appears likely, because of the positive misfit pertaining to a carbon or nitrogen atom

---

[13] In ferrite (α-Fe) these tetragonal stress fields around the interstitial atoms are not aligned and, as a result, the average crystal structure of ferrite remains (body centred) cubic, whereas in martensite, because of the higher concentration of interstitials, the interaction, as discussed in the above sense, of the tetragonal stress fields around the individual interstitials, causes an alignment of the individual stress fields such that the *EF*-axes for the interstitials become aligned (in other words: only one of the three types of octahedral interstices becomes occupied) and, as a result, the average crystal structure of martensite is (body centred) tetragonal (see Sect. 9.5.2.1). The octahedral interstice in face-centred cubic iron (γ-Fe) is *regular* and thereby the misfit-stress field introduced upon introduction of an interstitial is spherical (isotropic; here, in first-order approximation, the host matrix (Fe) is assumed to possess elastically isotropic properties), and thus, the (average) crystal structure remains cubic for also large amounts of dissolved interstitial solute (see also the discussion on interstitial diffusion in Sect. 8.5.3 and footnotes 6 and 7 in Sect. 8.6; further note that, although ferrite shows anisotropic elastic behaviour, an overall hydrostatic stress field will lead to an isotropic distortion due to the cubic crystal symmetry of ferrite).

Fig. 12.22 Preferential
occupation by interstitial
atoms of interstitial sites in
tensilely strained regions
near edge dislocations

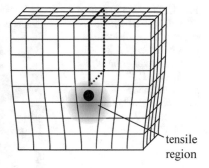

tensile
region

dissolved in ferrite (i.e. the size of the interstitial atom is larger than the "size" of
the interstice), that energy is released (i.e. the interaction energy then is negative)
if these interstitial atoms position themselves adjacent to the dislocation line of,
for example, an edge dislocation in that area of the dislocation-stress field which
is of tensile nature, i.e. below the slip plane and under the half plane; cf. Fig. 5.5
and see Fig. 12.22. Thus concentrations/rows of interstitial atoms develop along
dislocation lines which are called "Cottrell–Bilby atmospheres" or "Cottrell-Bilby
clouds" (Cottrell and Bilby 1949).

The simplest treatment of the here discussed "size effect" in the interaction of a
point defect with a dislocation holds for spherically symmetric misfit-size defects,
with both matrix and point defect elastically isotropic and of equal elastic constants.[14]
In that case the interaction energy of a point defect with a dislocation is given by the
product of the hydrostatic component of the stress field of the dislocation and the
misfit volume (the misfit volume is the difference between the volume of the stress-
free (i.e. before insertion into the "hole" in the matrix) point defect and the "hole" in
the matrix, to be occupied by the point defect). Then, for the case considered, because
the stress field of a screw dislocation has no hydrostatic component (cf. Sect. 5.2.2),
the interaction energy of a point defect with a screw dislocation is nil. The stress field
of an edge dislocation incorporates a hydrostatic component (cf. Sect. 5.2.1), and thus
the point defects are predicted to segregate close to the dislocation line, below the core
of the edge dislocation, as indicated above. As discussed in the previous paragraph,
the misfit-size effect induced by a carbon or nitrogen atom in the irregular octahedral
interstice of ferrite is not spherically symmetric and interaction with not only the
hydrostatic part but also, and in particular, the shear part of a dislocation-stress field
can occur (see the *"Intermezzo: The hardness of iron-based interstitial martensitic
specimens"* at the end of Sect. 9.5.2). Therefore, already the simple theory touched
upon here predicts that these interstitial atoms will enrich at both edge and screw
dislocations.

---

[14] The elastic theory of misfitting inclusions in a matrix has been developed by Eshelby (1956).
His analysis not only has provided fundamental insight into the elastic deformations due to the
inclusion of a misfitting *point defect* in a matrix, but also has been the basis for understanding the
stress fields around misfitting *precipitates* in a matrix (cf. Sect. 12.18).

Fig. 12.23 Precipitation of
α''-Fe$_{16}$N$_2$ iron–nitride
platelets along dislocation
lines in an Fe–N solid
solution (taken from W. T.
M. Straver, H. C. F.
Rozendaal and E. J.
Mittemeijer, Metallurgical
Transactions A, 15A (1984),
627–637)

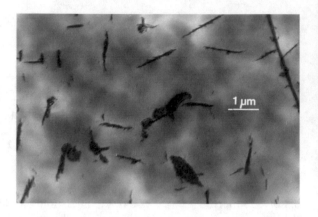

The occurrence of relatively dense Cottrell–Bilby atmospheres in even overall
rather dilute solid solutions can be illustrated by the easy formation of for example
iron-nitride precipitates as platelets along dislocation lines in an Fe-N solid solution
(Fig. 12.23).

The immediate consequence for the mechanical properties of a negative nature
of the interaction energy of point defect and dislocation is clear: as compared to the
absence of such interaction, it will cost more energy to induce movement (glide)
of the dislocation: the dislocation is said to be "locked" and a higher stress, than in
the absence of point defects concentrated at the dislocation line, is required to make
the dislocation mobile. If for the material considered a relatively small number of
dislocations is initially mobile, the relatively few mobile dislocations cannot glide
fast enough to realize sufficient strain (cf. Eq. (12.37); e.g. for a constant strain
rate) and as a consequence the applied stress rises: at the upper yield point initially
immobilized dislocations can become mobile (become "unlocked") and then the
density of the mobile dislocations can thereby rise dramatically. Consequently, the
applied stress necessary to continue yielding can decrease significantly: a yield drop
occurs (Fig. 12.20). In recent years, while maintaining the notion that the yield
drop is due to the sudden increase of the density of mobile dislocations, it has been
argued that, if the unlocking of the immobilized (by Cottrell–Bilby atmospheres or
precipitate particles) dislocations cannot be overcome, the yield drop can be due to
the (abrupt) generation of new, mobile dislocations.

The diffusion of point defects to dislocations is a so-called ageing phenomenon
(cf. Sect. 9.4.1). Upon interrupting the tensile loading experiment, i.e. unloading the
specimen, say at some stage beyond the yield-point elongation zone (see above), and
keeping the specimen for some time at a certain temperature (for an iron–carbon
or iron–nitrogen specimen, this can be room temperature), the point defects (as
interstitial carbon and nitrogen atoms in iron) can diffuse to the unlocked dislocations and lock them (again). Then, upon reloading the specimen, an upper yield
point reappears and at a higher level of applied stress than as observed firstly. This
phenomenon is called *"strain ageing"* (see Fig. 12.24). If the point-defect mobility
is high enough *during* the tensile loading (requiring a sufficiently high temperature

Fig. 12.24 Schematic
stress–strain curve with two
unloading-reloading
interruptions (cf. Fig. 12.17)
in straining. During the
second interruption, strain
ageing takes place

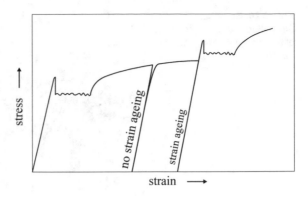

Fig. 12.25 Schematic
representation of the
Portevin-le Chatelier
effect/serrated yielding

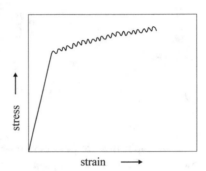

and/or sufficiently low strain rate), it is possible that "unlocked" dislocations become "locked" again *during* the tensile loading, requiring some increase of applied stress to become "unlocked" again, etc. Then, the initial sharp drop from upper yield stress to lower yield stress and the yield-point elongation zone become less pronounced (can disappear at sufficiently high temperature and/or sufficiently low strain rate) and are replaced by a stress–strain curve exhibiting positive and negative variations in the applied stress: the stress–strain curve becomes serrated; one speaks of "dynamic strain ageing" or "serrated yielding", also called the "Portevin-le Chatelier effect" (see Fig. 12.25): the serrations indicate the replacement of the original outspoken upper yield point by many localized yield limits within the component/specimen.

12.9.3 Shear Yielding and Craze Yielding

Plastic deformation as considered until now in this section is thought to occur by shear (possibly concentrated in glide bands; cf. Sect. 5.2.6 and immediately above in Sect. 12.9.2), in isotropic materials, and with no slip plane preference, likely along planes oriented at an angle of 45° with the tensile loading axis because the largest shear stress occurs for such planes (cf. Sect. 12.4 and Eq. (12.14)).

Fig. 12.26 Schematic
illustration of craze yielding
of a thermoplastic polymeric
material

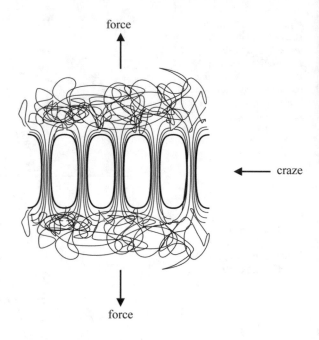

Thermoplastics are polymers characterized by relatively easy relative displacements of adjacent polymeric chains (no extensive cross-linking). They can exhibit
a mechanism for permanent deformation different from shear yielding. In the
case of uniaxial loading, as considered in this section, crazes form upon tensile
loading in directions normal to the loading direction. Crazes, regions in the material where highly localized yielding has occurred, look like cracks. They are open-
ings in the material composed of voids interspersed with fibrils of highly oriented
polymer molecules (in contrast with the more or less random orientation distribution
pertaining to the surrounding matrix) which connect the two opposite surfaces of the
craze (Fig. 12.26). Crazes are microscopic regions of highly localized plastic deformation (as holds for shear bands); the craze thickness is of the order of a micrometre.
Eventual fracture occurs by rupture of the fibrils leading to void enlargement and
crack propagation through the craze. Craze formation does not lead to pronounced
macroscopic plastic deformation: materials that craze are not very ductile: fracture
strain of a few percent.

12.10 Yielding Criteria
    in Cases of Two- and Three-Axial Loading

The yield strength ("yield point") defined for the case of uniaxial loading of an
isotropic material in Sect. 12.9 pertains to the value of principal stress needed to

attain a value of principal strain where plastic deformation is initiated (for principal stress and principal strain, see Sect. 12.4). Against this background, for bi- and triaxial states of loading of isotropic materials it appears plausible to look for definitions of yield criteria which can be expressed in terms of values of the operating principal stresses. This implies that the criterion for yield to occur needs the values of the principal stresses only. Further the yield strength upon two- and three-axial loading is correlated to the yield stress for uniaxial loading. The two perhaps most well-known proposals for yield criteria of this kind bear the names of Tresca and von Mises.

(i)   *The Tresca criterion.* Yielding is predicted to occur if the maximum shearing stress is larger than a critical value. The absolute maximum shearing stress equals the largest of the three maximal shearing stresses given by Eq. (12.14) and thus is of the type (dropping the $\pm$ symbol and see text about relative magnitudes of the principal stress components below Eq. (12.14)):

$$\tau_{max} = \frac{\sigma_x^P - \sigma_z^P}{2}$$

In uniaxial loading, the maximal shearing stress obviously is given by $\tau_2 = \sigma_0/2$. (cf. Eq. (12.14)) for τ_2 with $\sigma_x^P = \sigma_0, \sigma_y^P = \sigma_z^P = 0$), with σ_0 as the yield stress in uniaxial loading (Sect. 12.9). Adopting this critical value as the critical value for τ_{max} in two- and three-axial loading as well, it follows for the Tresca criterion:

$$\tau_{max} = \frac{\sigma_x^P - \sigma_z^P}{2} > \frac{\sigma_0}{2} \qquad (12.38)$$

(ii)   *The von Mises criterion.* Yielding is predicted to occur if the "strain energy of distortion"[15] per unit volume upon two- or three-axial loading exceeds the "strain energy of distortion" per unit volume upon uniaxial loading up till the yield stress σ_0. This leads to the following expression for the von Mises criterion:

$$\left[\left(\sigma_x^P - \sigma_y^P \right)^2 + \left(\sigma_y^P - \sigma_z^P \right)^2 + \left(\sigma_z^P - \sigma_x^P \right)^2 \right]^{1/2} > 2^{1/2}\sigma_0 \qquad (12.39)$$

---

[15] Any state of stress can be subdivided into a hydrostatic state of stress plus a so-called deviatoric state of stress. The total strain energy can be written, for this special case, as a sum of the strain energies of the hydrostatic state of stress component and the deviatoric state of stress component (in general the strain energies of two superimposed states of stress are *not* additive). The first strain energy contribution is called "strain energy of dilatation"; the latter strain energy contribution is called "strain energy of distortion". For the von Mises criterion it then is assumed that the hydrostatic state of stress component $\left(\sigma_x^P = \sigma_y^P = \sigma_z^P \right)$ makes a negligible contribution to the deformation (incompressible body), so that the "strain energy of distortion" is decisive for the occurrence of yielding.

Fig. 12.27 A ball bearing: ball and inner ring in contact upon applying a radial load to the ball. The contact area (black) results from elastic deformation in both steel components (taken from Voskamp and Mittemeijer 1997)

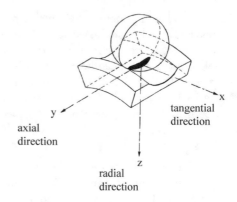

The von Mises criterion is also called "maximum shear energy criterion" (see Eq. (12.23) and τ_{max} above).

The above Eq. (12.39) suggests an alternative formulation. A so-called von Mises equivalent stress, σ_{eq}, can be defined:

$$\sigma_{eq} = 2^{-1/2} \left[\left(\sigma_x^p - \sigma_y^p \right)^2 + \left(\sigma_y^p - \sigma_z^p \right)^2 + \left(\sigma_z^p - \sigma_x^p \right)^2 \right]^{1/2} \qquad (12.40)$$

If σ_{eq} exceeds a certain critical value, say σ_0, yielding will occur. For a certain state of stress imposed on a body, σ_{eq} can be calculated. At locations where σ_{eq} is larger than the critical value, plastic deformation can occur. (Of course, an analogous procedure is possible with the Tresca criterion).

The Tresca criterion is more conservative; i.e. it provides a limit to elastic deformation more severe than the von Mises criterion; the von Mises criterion provides better agreement with experimental reality. It has been argued that application of the von Mises criterion is physically more plausible (R.M. Christensen at https://failurecriteria.com/misescriteriontr.html)

Intermezzo: Application of the von Mises Criterion
to Predict the Location of Failure in Ball Bearings
Rolling bearings are intended to support shafts and other rotating parts, smoothly and safely, in all kinds of machinery. Consider, as an example, ball and inner ring of a radially loaded deep groove ball bearing: the ball rolls in the deep groove of the inner ring while radially loaded (Fig. 12.27). Generally the contact area between ball and ring, which is the result of elastic deformation, can be considered to be of elliptical shape. The state of load-induced stress can be characterized by the three principal stresses σ_x^P, σ_y^P and σ_z^P. The x-axis is parallel with the circumferential (tangential) direction, the y-axis is

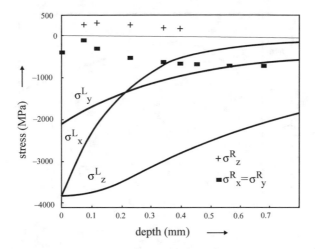

Fig. 12.28 Calculated principal load-induced stresses σ_x^L, σ_y^L and σ_z^L (σ_x^P, σ_y^P and σ_z^P have been indicated with σ_x^L, σ_y^L and σ_z^L, where the superscript "L" denotes "load") as a function of depth under the middle of the contact area shown in Fig. 12.27 for a 6309 type deep groove ball bearing inner ring, i.e. $x = 0$, $y = 0$ and z (depth) is variable, loaded under a radial bearing load of 28 kN causing a maximal Hertzian contact stress of 3.8 GPa between the (highest) loaded ball and the inner ring, and corresponding measured residual stresses σ_x^R, σ_y^R and σ_z^R ($x = 0$, $y = 0$ and z(depth) is variable) for rings endurance tested under the same radial bearing load for 4×10^8 inner ring revolutions at 6000 rpm at a bearing operating temperature of 53 °C (taken from Voskamp and Mittemeijer 1997)

parallel with the axial direction, and the z-axis is parallel to the radial direction (opposite to the surface normal direction).

For this case of (so-called Hertzian) loading, σ_x^P, σ_y^P and σ_z^P within the surface region of the inner ring are all compressive. Their dependence on depth beneath the surface is shown in Fig. 12.28 (here σ_x^P, σ_y^P and σ_z^P have been indicated with σ_x^L, σ_y^L and σ_z^L, where the superscript "L" denotes "load"): σ_z^P has the largest compressive stress value (note that σ_x^P and σ_y^P would be equal if a circular contact area would occur). The von Mises equivalent stress for this applied state of stress, σ_{eq}^L, can now be calculated, using Eq. (12.40), as a function of depth beneath the surface of the inner ring. The result is shown in Fig. 12.29. Evidently, the equivalent stress is largest at some depth beneath the surface. Hence it is suggested that failure is induced not at the surface but underneath it. Indeed, crack initiation, as a final outcome of preceding micro-yielding in the most severely loaded region, occurs beneath the surface.

Subsequent crack growth, parallel to the surface, can be supported by the development of a *tensile* residual stress component in the surface normal direction (see σ_z^R data given in Fig. 12.28): see Fig. 12.30. Rather straight cracks can occur if the {100} planes in the ferrite matrix are preferably parallel to

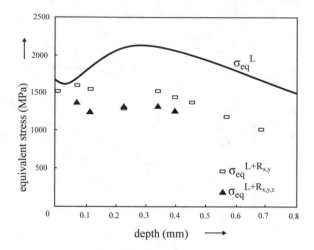

Fig. 12.29 von Mises equivalent stress $\left(\sigma_{eq}^{L}\right)$-depth profile for the 6309 type deep grove ball bearing inner ring, below the centre location of the contact ellipse. The solid line represents the $\left(\sigma_{eq}^{L}\right)$ depth distribution calculated from σ_{x}^{L}, σ_{y}^{L} and σ_{z}^{L} given in Fig. 12.28. The open rectangles represent the equivalent stress values calculated after superposition of σ_{i}^{L} and σ_{i}^{R} (only σ_{x}^{R} and σ_{x}^{R}) using the data of Fig. 12.28. The triangles represent the equivalent stress values calculated after superposition of σ_{i}^{L} and σ_{i}^{R} $\left(\sigma_{x}^{R}, \sigma_{y}^{R}\text{ and }\sigma_{z}^{R}\right)$, using the data of Fig. 12.28 (taken from Voskamp and Mittemeijer 1997)

the surface (Fig. 12.30a); if a different crystallographic texture prevails in the ferrite matrix, the cracks propagate (only) on average parallel to the surface and then can exhibit a zig-zag, facetted appearance (Fig. 12.30b).

In fact, a full description of the state of stress in the inner ring requires knowledge not only of the externally imposed state of loading stress, but also of the development of the internally imposed state of residual stress (the measured residual stress components have been indicated in Figs. 12.28 and 12.29 with the superscript "R"). This modifies the above discussion. See Voskamp and Mittemeijer (1997); see also Sect. 12.18.

12.11   Critical Resolved Shear Stress; the Plastic Deformation of Single Crystals

A total stress σ_{tot} acting along a direction inclined with respect to a plane can always be resolved into a normal stress component σ acting in the normal direction of that plane and a tangential, shearing stress component τ acting along the plane (see top

Fig. 12.30 a Light microscopical micrograph of a section, perpendicular to the surface and parallel to the circumferential (=overrolling=tangential (see Fig. 12.27)) direction, of a fatigue-tested 6309 type deep groove ball bearing inner ring exhibiting a well-developed {100} <110> texture in the ferrite matrix of the subsurface region, with {100} parallel to the surface and <110> parallel to the overrolling direction. The inner ring had experienced 1.6×10^7 rotations under a maximal contact stress of 4.9 GPa at 6000 rpm using a bearing operating temperature of 55 °C. Note the straight path of the crack in the subsurface, parallel to the surface. **b** Light microscopical micrograph of a section, perpendicular to the surface and parallel to the circumferential (=overrolling=tangential (see Fig. 12.27)) direction, of a fatigue-tested 6309 type deep groove ball bearing inner ring exhibiting a well-developed {111} <211> texture in the ferrite matrix of the subsurface region, with {111} parallel to the surface and <211> parallel to the overrolling direction. The inner ring had experienced 1.6×10^7 rotations under a maximal contact stress of 4.9 GPa at 6000 rpm using a bearing operating temperature of 70 °C. The crack is on average parallel to the surface, but has a facetted appearance because the {100} planes of weak coherence in ferrite are not preferably parallel to the surface. The larger facets are parallel to the so-called Low Angle Bands, indicated by LABs in the figure (taken from and for further information see Voskamp and Mittemeijer 1997)

part of Fig. 12.31; see also Fig. 12.7a). The angle made by σ_{tot} with the plane is found by drawing the plane through both σ_{tot} and the surface normal and measuring the angle, θ, between σ_{tot} and the normal. It follows:

$$\sigma = \sigma_{\text{tot}} \cos \theta \qquad (12.41)$$

The shear stress in the plane acts along the intersection of the plane through the normal and σ_{tot} and the plane considered and consequently:

$$\tau = \sigma_{\text{tot}} \sin \theta \qquad (12.42)$$

It holds: $\sigma_{\text{tot}}^2 = \sigma^2 + \tau^2$.

In turn, τ can be resolved further into two components acting along two mutually perpendicular axes lying in the plane considered, and this leads to the specification of the three normal stress components and the six shear stress components as in Sect. 12.4.

As discussed in Sect. 5.2.5, a crystal deforms plastically usually by dislocation glide along slip planes, which commonly are the most densely packed planes, in slip

Fig. 12.31 Geometrical fundamentals of Schmid's law. Note that the slip plane normal, the direction of the applied force F and the slip direction are not within one plane

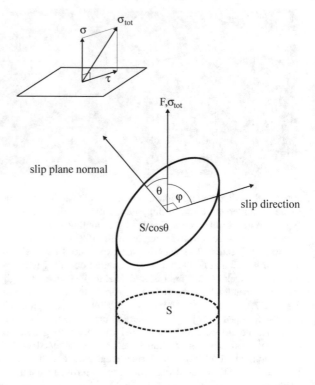

directions which are the most closely packed directions in these slip planes. This leads to the specification of slip systems (= slip plane + slip direction) as illustrated in Table 5.1 for f.c.c., b.c.c. and h.c.p. crystals.

Now consider a single crystal subjected to a load acting in a specific direction with respect to the crystal frame of reference. The tendency to plastic deformation (tendency "to slip") will depend on the orientation of the most favourably oriented slip plane and the most favourable slip direction in that slip plane with respect to the applied load stress. It can be anticipated that slip sets in if the shear stress component acting along the considered specific slip plane and in the considered specific slip direction surpasses a critical value. This critical value is called the *critical resolved shear stress*, τ_{crit}.

To express the critical shear stress in terms of the acting load stress σ_{tot} and the orientation of the crystal, the procedure discussed below Eq. (12.42) could be followed, implying that the angle between the tangential component of σ_{tot} in the slip plane, $\tau = \sigma_{tot} \sin \theta$, and the slip direction has to be defined. Instead one usually proceeds differently, as follows. In addition to the angle θ between slip plane normal and σ_{tot}, the angle between σ_{tot} and the slip direction in the slip plane is defined by φ (see Fig. 12.31). The load F acts in the normal direction on a cross-sectional area of size S, implying $\sigma_{tot} = F/S$. The component of the load F acting along the slip direction is given by $F \cos\varphi$. The (slip plane) area onto which this component acts has the magnitude $S/\cos \theta$. Hence, it follows for τ_{crit}:

$$\tau_{\text{crit}} = \frac{F \cos \varphi}{S / \cos \theta} = \sigma_{\text{tot}} \cos \varphi \cos \theta \qquad (12.43)$$

This expression is known as Schmid's law. Its validity is demonstrated by investigating the onset of yielding of a single crystal (specimen) as function of its orientation: whereas the value of σ_{tot} needed to establish plastic deformation varies greatly as a function of orientation of the crystal, the value of τ_{crit} remains essentially constant. An example is shown in Fig. 12.32, where the critical stress for the occurrence of yielding, i.e. σ_{tot}, has been plotted as function of $\cos\varphi\cos\theta$ for a single crystal of zinc (h.c.p.). The validity of Schmid's law is better demonstrated for a hexagonal metal than for a cubic metal, as the significantly smaller multiplicity of the operating slip system (cf. Table 5.1) allows testing of Schmid's law over a larger range of crystal orientation.

As follows from the above, the tensile (load) stress σ (the subscript "tot" is dropped) applied to a single crystal can be written in terms of the induced resolved shear stress τ acting along the slip plane considered according to:

$$\sigma = M\tau \qquad (12.44)$$

Fig. 12.32 Critical stress for the occurrence of yielding (σ_{tot}) plotted as a function of crystal orientation for a zinc single crystal (taken from D. C. Jillson, Transactions of the American Institute of Mining and Metallurgical Engineers, 188 (1950), 1129–1133)

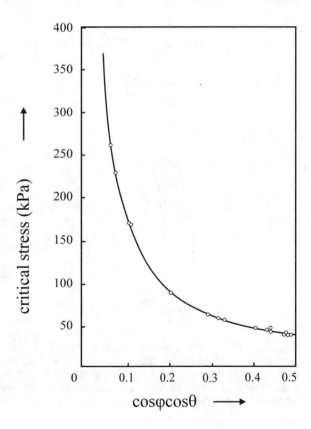

with the orientation factor $M = (\cos\varphi\cos\theta)^{-1}$ (which is the reciprocal "Schmid factor"). Plastic flow of the single crystal (specimen) occurs if the resolved shear stress τ equals the critical resolved shear stress τ_{crit} at the slip plane and in the slip direction in the single crystal where, for the given crystal orientation with respect to the load stress, the resolved shear stress is the highest. This slip system is called the primary slip system.

Upon slip in the slip direction along the slip plane of the loaded single crystal, rotation of the slip direction occurs in the direction of the tensile loading axis (see also beginning of Sect. 12.12 and Fig. 12.33). If, for a case of tensile loading of the single crystal, the shear stress along the slip plane in the slip direction is denoted by τ_{shear} ($>\tau_{crit}$) and the plastic shear strain realized in this direction on the slip plane is denoted by $\Delta\gamma_{shear}$, the corresponding tensile plastic strain in the direction of the load, $\Delta\varepsilon_{tensile}$, obeys approximately (i.e. for small strains, $\Delta\varepsilon_{tensile}$ and $\Delta\gamma_{shear}$, in order that the above mentioned crystal-structure rotation is small):

$$\Delta\varepsilon_{tensile} = \Delta\gamma_{shear} \cos\varphi\cos\theta \qquad (12.45)$$

The above discussion makes clear that the stress–strain curve (case of uniaxial loading) for a single crystal specimen is best presented in terms of shear stress

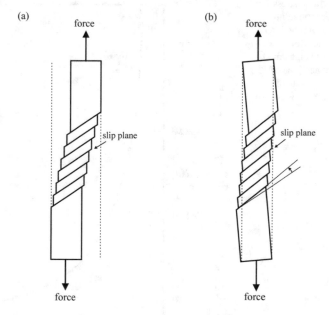

Fig. 12.33 Tensile deformation of a single crystal. **a** If the deformation occurs by glide along a specific set of glide planes and if the deformation is unconstrained, a shape change of the crystal can take place by the same amount of glide along each of these glide planes. **b** In a tensile testing machine uniaxial tensile loading is mediated by grips holding the specimen. These grips must remain in line. As a result rotation of the active glide planes towards the tensile loading axis occurs

versus shear strain, τ versus γ, thereby the differences between the results obtained for different orientations of the single crystal are reduced (not eliminated): the critical resolved shear stress τ_{crit} (i.e. the level of loading where plastic deformation starts) may be the same for the slip system considered irrespective of the orientation of the crystal, but the degree of work hardening, also called strain hardening (cf. third paragraph of Sect. 12.8, the begin of Sect. 12.9 (text above Eq. (12.33)) and Sect. 12.14.1), can be crystal orientation dependent, and thus the resolved shear stress τ ($>\tau_{crit}$) increases for increasing plastic deformation (extension of the crystal) differently for different orientations of the crystal.

12.12 Plastic Deformation of Polycrystals

A single crystal can deform plastically upon uniaxial tensile loading by glide along a certain type of glide plane, which, if the deformation is unconstrained, could lead to an external shape change of the crystal corresponding to the same amount of glide along each glide plane (Fig. 12.33a). However, in a tensile testing machine the specimen is fixed between grips mediating the applied load, which grips must remain in line: the situation sketched in Fig. 12.33a cannot occur. As a result, upon extension of the crystal the glide planes rotate towards the tensile loading axis, as suggested by the sketch in Fig. 12.33b. Such "free" (unconstrained) plastic deformation also cannot be realized for a crystal in a massive, polycrystalline specimen upon tensile loading. In fact the situation resembles the one described by "grain interaction" upon *elastic* loading discussed in the "*Intermezzo: Grain interaction*" at the end of Sect. 6.10. The plastically deforming crystal in the aggregate has to adapt itself to the, possibly also plastically deforming, neighbouring grains. One way to express this problem is the question how to derive the tensile stress–strain curve for a polycrystal from that for the single crystal.

As already pointed out in Sect. 12.8, five independent strain components per crystal are needed in order to realize compatible plastic deformations of the individual crystals in a polycrystalline, massive specimen, in order that the massive nature and integrity of the loaded polycrystalline specimen are maintained. Only in this way any shape change of the loaded body can be realized in principle by plastic deformation; each crystal of the loaded body should undergo the same shape change as the whole body. This means that five independent slip systems should operate in each crystal upon plastic deformation (an independent slip system is a slip system that causes a change in shape that cannot be realized by a combination of other slip systems). For example, in f.c.c. metals 12 equivalent, from a crystallographic point of view (slip plane: {111}; slip direction: <110>), slip systems can be indicated (see Table 5.1), but only five of these are independent.[16] Then it appears natural to suppose that the

---

[16] In rock-salt type crystals six equivalent, from a crystallographic point of view (slip plane: {110}; slip direction: <110>), slip systems can be indicated, but only two of these are independent. Consequently, according to the discussion in the main text, a polycrystal of rock-salt type is brittle. Only

five slip systems required in each crystal of the aggregate, for fulfilment of the above condition, then would be those with the highest resolved shear stresses.

Now, an equation of the type $\sigma = M\,\tau$ (cf. Eqs. (12.43) and (12.44)) may also be adopted for the plastically deforming polycrystalline aggregate, implying that some appropriate averaging for the product $M\,\tau$ can be made, recognizing that five independent slip systems operate in each crystal and that the individual crystals have different orientations. It will be assumed that the critical shear stress is the same for all (crystallographically equivalent) slip systems. Hence:

$$\sigma = \langle M \rangle \tau \tag{12.46}$$

where $\tau = \tau(\gamma)$ represents the shear stress–shear strain curve for the single crystal in the plastic region (cf. end of Sect. 12.11). Assuming that all grains in the specimen, a random aggregate of grains, experience the same amount of plastic deformation ("uniform strain"; cf. the Voigt approach to elastic deformation in a polycrystalline body, discussed in the "*Intermezzo: Grain interaction*" at the end of Sect. 6.10), averaging of the orientation factor M for the case that five slip systems operate (those selected in each grain according the criterion of highest resolved shear stress; cf. above) leads to the result that $<M>$ takes values in the range of about 2 to about 3 for both f.c.c. and b.c.c. materials.

Considering Eq. (12.45), on the same basis it follows from the above for the polycrystalline aggregate that $<M>$ also provides the relation between the normal strain contribution $\Delta \varepsilon_{tensile}$ and the shear strain contribution $\Delta \gamma_{shear}$ for a specific slip system. If the outcome of the averaging over the orientation of the single crystals in the aggregate is not affected by the occurrence of the plastic deformation (i.e. is constant during the plastic deformation), it then holds for the relation between the total normal strain, in the direction of the load, $\varepsilon_{tensile}$ ($=\sum(\Delta \varepsilon_{tensile})_i$, where the summation is carried out over all slip systems) and the total shear strain γ_{shear} ($=\sum(\Delta \gamma_{shear})_i$, where the summation is carried out over all slip systems):

$$\varepsilon_{tensile} = \langle M \rangle^{-1} \gamma_{shear} \tag{12.47}$$

The plastic part of the tensile stress–strain curve of the polycrystalline aggregate can then be constructed by application of Eq. (12.46) starting from the stress–strain relation for the single crystal $\tau = \tau(\gamma)$. The shear strain value γ' corresponds for the polycrystalline specimen with the tensile strain value $<M>^{-1}\gamma'$ (cf. Eq. (12.47)) and to the shear stress value $\tau(\gamma')$ for the single crystal specimen. Thus, combining Eqs. (12.46) and (12.47), a predicted tensile stress—tensile strain, σ versus ε, curve for the polycrystalline specimen is obtained.

---

at higher temperatures, if another slip system becomes operative as well (slip plane: {001}; slip direction <110>), significant plastic deformation of a polycrystal of rock-salt type becomes feasible. This contrasts strongly with many polycrystalline metals, which can experience extensive plastic deformation already at room temperature.

12.13 Hardness Parameters; Macroscopic, Microscopic and Nanoscopic

One way, likely used by our ancestors, to characterize the "hardness" of a material is to assess its sensitivity to scratch it with a (much) harder material.[17] The *indentation hardness* technique, was probably first introduced by Brinell in 1900 and still is the technique of prime importance to characterize the hardness of a material. Hardness measurement and interpretation is a topic of enduring, great scientific and engineering interest. Obviously, hardness testing is often used in industry for quality-control purposes. The recent possibility to measure hardness values on a nanometre scale has led to a focus of research activity on the nanoindentation technique (see further below).

The hardness of a material characterizes, restricting ourselves first to the conventional hardness parameters (see discussion on "contact hardness" further below), the resistance of the material against plastic deformation. The hardness is tested on a local scale, usually by forcing an indenter into the surface of the specimen/component under the action of a specific load for a certain time.

A well-known technique is the Vickers hardness testing, where a diamond indenter of square-base pyramidal geometry is applied. The Vickers hardness value, HV, is given by the ratio of applied load, P (in kg), and the surface area of the indentation as determined from the lengths of the diagonals, L (in mm), of the, ideally square-shaped (see Fig. 12.34a, c), indentation as measured by a light microscope:

$$\mathrm{HV} = \frac{P}{L^2/(2\sin(\theta/2))} = 1.854\frac{P}{L^2} \tag{12.48}$$

where θ represents the angle between opposite faces of the diamond pyramid and is equal to 136°.

The Knoop hardness test resembles the Vickers hardness test, but in this case the diamond pyramid is shaped such that one of the diagonals of the, now lozenge-shaped, indentation, L_1, is considerably larger than the other diagonal, L_2 (see Fig. 12.34b). The Knoop hardness value, HK, follows from:

$$\mathrm{HK} = 14.2\frac{P}{L_1^2} \tag{12.49}$$

with P in kg and L_1 in mm. This hardness measurement technique is especially useful for measuring hardness close to a surface/interface (by aligning L_1 parallel to the surface/interface).

Vickers and Knoop hardness testers are applied especially in research. For routine, technical application more macroscopical, say crude, hardness testing can

---

[17] Although this approach to hardness testing is no longer used in materials science, it is interesting to note that a "scratch test" is (still) often used to investigate the adherence of a thin layer on a substrate.

Fig. 12.34 Schematic illustration of **a** a Vickers and **b** a Knoop hardness indent. **c** Two (Micro)Vickers hardness indents in a grain of nitrided Fe-4.65 at.%Al alloy are shown (SEM image). In the grain interior, a higher hardness than near the grain boundaries prevails (cf. the differently sized indents in the micrograph), because AlN precipitates have (already) developed in the grain interior (dark etching region; taken from and for more details see S. Meka, S. S. Hosmani, A. R. Clauss and E. J. Mittemeijer, International Journal of Materials Research, 99 (2008), 808–814)

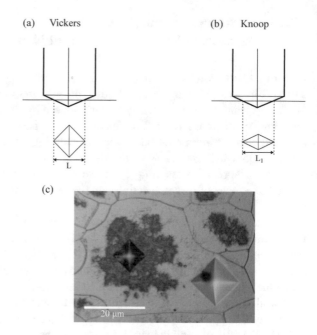

be performed according to the Brinell method, by pressing a spherical indenter of 10 mm diameter, made of steel or tungsten carbide, into the surface of the component, applying loads in the range 500–3000 kg. The Brinell hardness, HB, then follows from the load and the diameter of the indentation. The, also technical, Rockwell hardness test involves, as the only such hardness testing technique discussed here, measuring the *depth* of the indentation produced by a spherical, hardened steel, indenter or a conical, diamond, indenter.

Intermezzo: The Hardest Materials

The hardest natural material is the mineral (i.e. a crystalline substance (element or compound) that is a homogeneous component of the earth crust) diamond. Diamond can be conceived as a ceramic material. This leads to the statement that the hardest known materials are ceramics: diamond, boron carbide, boron nitride, silicon carbide, aluminium oxide, zirconia, quartz.

Zirconia (ZrO_2) is often used as imitation diamond in jewels. Misleadingly, it can be suggested that the ability to cut/scratch glass is a way to distinguish diamond from zirconia. However, zirconia can cut/scratch (is harder than) glass as well. The way to distinguish diamond from zirconia is via their heat conductivities: diamond conducts heat much better than zirconia.

Materials which conduct electricity well generally also conduct heat well. A class of ceramic materials, characterized by simple crystal structures, provides an exception to this rule: they are electrical insulators and thermal conductors at the same time. Diamond belongs to this class, as well as boron nitride (cf. above discussion on distinguishing diamond and zirconia).

Diamond can cut rocks. However, its utility for cutting steels (softer than rocks) is limited: it degrades upon machining by induced reactions under the formation of iron carbides, and diamond is expensive. For cutting steels one can favourably apply the man-made material (cubic) boron nitride (BN), which is almost as hard as diamond. The hardness of diamond and boron nitride derives from the covalent nature of the chemical bonding in both substances: the directionality of the covalent bond obstructs gliding and makes the material rigid (cf. Sect. 3.4). The search for cheap and ultrahard man-made materials continues; ReB_2 is an example.

Diamond is not a stable solid phase of carbon at (normal temperature and) normal pressure (see the discussion on the Fe-C phase diagram in the *"Intermezzo: The Fe-C and Fe-N Phase Diagrams"* at the end of Sect. 9.5.2.1). Graphite is the stable phase at low (normal) pressure, whereas diamond is stable at high pressure (as high as 7 GPa (i.e. 70,000 atm) at 2000 K), which explains that man-made diamond is produced at such high pressures). Interestingly, within the context of this *Intermezzo*, two, for the time being hypothetical, solid phases of carbon, of cubic symmetry, have been proposed which may be of hardness close to but less than diamond (Ribeiro et al. 2006): a b.c.c. phase with 12 C atoms in the unit cell (so a "molecule" C_6 serves as the building unit (motif) of the b.c.c. unit cell, which is constituted of two building units (motifs); cf. Sect. 4.1.2 and the Appendix at the end of Chap. 4) and a primitive cubic phase with a unit cell containing 20 C atoms (so a "molecule" C_{20} serves as the building unit (motif) of the primitive cubic unit cell, which is constituted of one building unit (motif); cf. Sect. 4.1.2 and the Appendix at the end of Chap. 4). Both cubic phases are unstable with respect to graphite and diamond, and thus special, "nonnormal" conditions must prevail in order to allow the development of such phases.

During the last decades the interest in the (variation of) the mechanical properties on a highly localized, small distance (say, of the order of some nm) scale has increased enormously; e.g. in thin films (of cardinal importance to, for example, the microelectronic industry). This implies that the indentation depth (more accurately formulated, the probed volume) should be correspondingly small. Then, the indentation (projected) area induced can no longer be quantitatively recorded by light optical microscopy. SEM, although providing a much larger lateral resolution in principle, is also inappropriate, because of lack of (topological) contrast for very small indentation areas (cf. Sect. 6.9). Therefore, the so-called nanoindentation technique has been developed that avoids recording of the indentation size and shape directly.

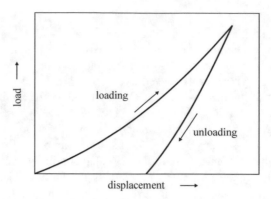

Fig. 12.35 Schematic load–displacment curve recorded by a nanoindenter

The nanoindentation technique involves that a diamond tip[18] is pressed into the surface of the specimen under simultaneously recording of both the load on the indenter, that is continuously increased, and the resulting displacement of the indenter. After that a maximum load has been achieved, the tip is removed by reducing the load until nil. The load–displacement curve is recorded during both the loading and unloading parts of the cycle. A schematic result is shown in Fig. 12.35. Thereby the nanoindentation technique in principle offers the possibility to extract much more information on the elastic–plastic deformation behaviour on a very local scale than provided by a single hardness value.

Intermezzo: Combined Nanoindentation and Scanning Probe Microscopy

The indentations achieved by nanoindentation are too small to be detected by light optical microscopy. The best way to "image" the specimen surface, with the indentation(s) produced, is to combine a scanning probe microscope and a nanoindenter in a single apparatus. To this end the cantilever-laser detection system of a scanning force microscope is replaced by a load–displacement transducer; the same diamond tip that is used for indentation can also be applied as probe tip in imaging mode. Indenting is performed at a fixed position at the surface by moving the diamond tip, up or down with respect to the surface. For the examples shown in Fig. 12.36a, b the diamond tip has been subjected to an electrostatic force (force resolution of the order 100 nN) and the displacement has been measured (depth resolution of the order 0.2 nm) by electrostatic capacity change. Imaging is realized using the same diamond tip, which is brought in contact with the surface and then is moved across the surface in a

---

[18] Four-sided pyramids are use as tips in Vickers and Knoop hardness testers (cf. Fig. 12.34a, b). Three-sided pyramid tips are common in nanoindentation (Berkovich and cube-corner tips), because these are easier to produce with sharp tips than four-sided pyramids for such applications (cf. Figs. 12.34c and 12.36b).

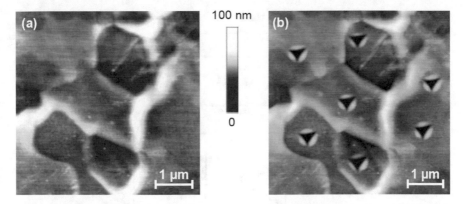

Fig. 12.36 The surface topography of a cross section of a Ti–6Al–4V alloy sample as obtained using the scanning nanoindenter (TriboScope) before (**a**) and after (**b**) nanoindentations have been made. The indentations, made by loading up to a maximum load of 2 mN, were deliberately located in the middle of each grain; only the indentation in the centre and on top of the image was made out of the middle of the grain to avoid the visible tiny step in the surface of the cross section. The height has been indicated by grey scaling (taken from Kunert 2000)

raster pattern. Due to the topography of the surface the tip moves up and down. This movement is detected, as before during the (laterally static) indentation experiment and is used to control a piezoelectric actuator onto which the specimen has been mounted. The piezo, by corresponding contraction or expansion, minimizes the (positive or negative) elevation of the tip and thereby a constant force is realized between the tip and the surface. The resulting up and down movement of the specimen due to contraction/expansion of the piezo upon rastering provides the topographical picture of the surface of the specimen. An example of such a scanning probe image of a surface with nanoindentations is given in Fig. 12.36b.

It immediately appears that the load–displacement curve recorded by the nanoindentation technique bears a relation to the tensile stress–strain curve discussed in Sect. 12.9. The loading curve represents the resistance to elastic and plastic deformation at the location of the indentation; the unloading curve indicates the elastic recovery of the indentation involving a reduction of the displacement. However, even in the case of purely elastic deformation, the stress field under the indenter is complex and in any case not of uniaxial nature as in the macroscopic tensile testing experiment. This already suggests that the direct determination of the elastic modulus, as from the unloading curve, is not trivial. The stress–strain curves, as would be obtained in uniaxial tensile testing, and the corresponding load–displacement curves, as would be obtained by nanoindentaton, are shown in Fig. 12.37 for three cases:

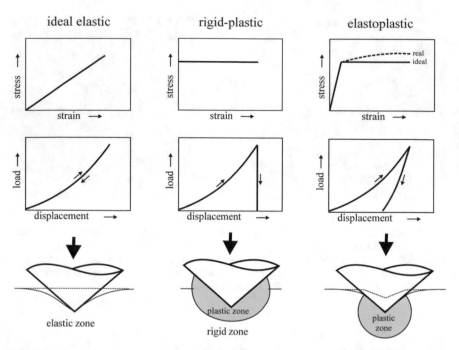

Fig. 12.37 Comparative, schematic presentation of stress–strain and load–displacement curves, and surface profiles at maximum load (full lines) and after complete unloading (dotted lines), for ideal elastic, rigid-plastic, and elastoplastic materials. (Ideal) Plastic deformation conserves the volume of the deforming specimen (cf. Eq. (12.27)). Thus, all of the material displaced by the indenter must be accommodated either by an upward extrusion ("pile-up") around the indent (rigid-plastic materials) and/or by elastic compression (elastoplastic materials) (taken from Kunert 2000)

(i)   if pure, elastic deformation prevails, the loading and unloading curves coincide and no indentation results after unloading;

(ii)   in case of pure plastic deformation of a rigid material no elastic deformation occurs and unloading does not reduce the displacement. Consequently, the displacement (and the indentation as a whole) at maximum load equals the displacement (and the indentation) after unloading;

(iii)   mixed elastoplastic deformation behaviour involves the occurrence of elastic and plastic deformation zones in the loaded material close to the tip. Part of the indentation at maximum load is relaxed upon unloading.

A metal like aluminium provides a good example of a material exhibiting pronounced plastic deformation, and a ceramic material like fused quartz exhibits distinct elastic recovery upon unloading (see the load–displacement curves in Fig. 12.38 and cf. Fig. 12.37).

For quantitative determination of the elastic modulus from the load–displacement curve one usually focusses on the unloading curve, which is dominated by elastic recovery, whereas analysis of the loading curve requires separation of the effects due

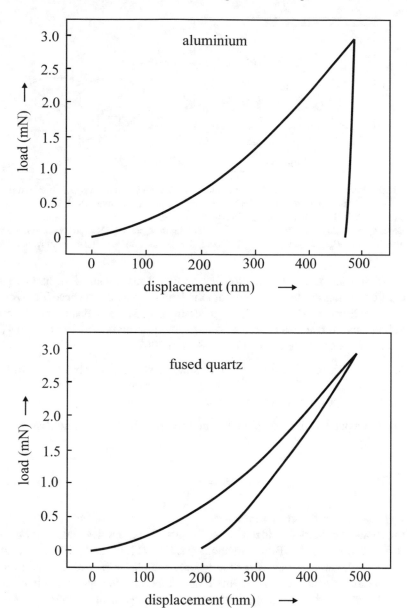

Fig. 12.38 Load–displacement curves for aluminium, as an example of an easily plastically deformable material, and for fused quartz, as an example of a material showing pronounced elastic recovery (taken from Kunert 2000)

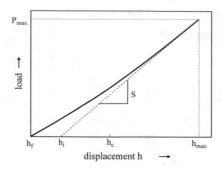
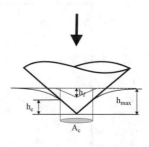

Fig. 12.39 Unloading part of a load–displacement curve and a schematic presentation of a section through the indentation at maximum load (full lines) and after removing the load (dotted lines). Quantities used in the analysis to determine both the hardness and the elastic modulus have been indicated: h_f = residual imprint depth, h_i = intercept depth, h_c = contact depth, h_{max} = maximum indentation depth, P_{max} = maximum applied load, S = contact stiffness (taken from Kunert 2000)

to both elastic and plastic deformation (see above). Thus, values of the hardness and the elastic modulus can be extracted from the load–displacement curve following an analysis by Oliver and Pharr (1992). Two parameters have to be determined from the unloading curve for determination of both the hardness and the elastic modulus (see Fig. 12.39; for a practical discussion, see Kunert 2000):

- the projected contact area (between indenter and specimen) at maximum load, A_c[19], and
- the initial slope of the unloading curve.

The hardness parameter obtained in this analysis is called contact hardness, H_c, and is defined by:

$$H_c \equiv \frac{P_{max}}{A_c} \tag{12.50}$$

where P_{max} denotes the maximum load.

The contact hardness should not be confused with the hardness values as obtained in the Vickers, Knoop and Brinell methods (cf. Eqs. (12.48) and (12.49)): the latter, more classical, hardness values pertain to only the plastic part of the deformation by indentation. Because the size of the indentation left after unloading is nil for a purely elastic material, its hardness according to the classical hardness parameters would be infinitely large. The contact hardness incorporates the effect of elastic deformation as well: the contact area *at maximum load* is also determined by the

---

[19] The projected contact area, A_c, can be written as the product constant . h_c^2, with h_c as the depth of contact between indenter and specimen at maximum load, i.e. the distance along the indenter axis that the specimen is in contact with the indenter (cf. Fig. 12.39). The constant in this expression depends on the shape of the tip of the indenter; for example, for the three-sided pyramid-type Berkovich indenter (cf. footnote 18), often used in nanoindentation experiments, it holds that the constant in case of ideal tip shape equals 24.5; for the ideal cube-corner tip the constant equals 2.6.

occurring elastic deformation (cf. Fig. 12.37, ideal elastic material (case (i) above)). Accordingly, the contact hardness for a purely elastic material has a finite value. Only if the maximal contact area is due to purely plastic deformation (cf. Fig. 12.37, rigid-plastic material (case (ii) above)), the contact hardness value is similar to the hardness values as obtained by the classical methods.

Finally, it should be recognized that the hardness parameter obtained by an indentation technique is not a fundamental material property. Its value can depend on the testing method and the values of the experimental parameters used in its determination. Yet, the importance of the indentation hardness as a material characterizing parameter can hardly be overestimated. Hardness measurements are relatively easy to perform and provide a direct measure for the load-bearing capacity of a material. Hence, they are of great importance in practical applications, but also in fundamental scientific research, e.g. to exhibit the variation in mechanical strength on a highly localized scale (see the *Intermezzo* below).

For those hardness parameters which pertain to plastic deformation only, the hardness and the yield strength (σ_0 or $\sigma_{0.2}$; see Sect. 12.9 and Fig. 12.15) are approximately linearly related:

$$\text{hardness} = \text{constant} \cdot \text{yield strength.} \tag{12.51}$$

For materials where the value of the yield strength is of the order of a percent of Young's modulus (as for metals), it holds that *the hardness is about 2.5 till 3 times the yield strength*; obviously, Eq. (12.51) requires that the hardness and the yield strength are expressed in the same units. The hardness is usually expressed in kg/mm^2 (see above), and the yield strength is usually expressed in MPa (1 MPa = 9.807 kg/mm^2).

In a hardness test applied to a metallic material, under the indenter a region of plastic deformation can be discerned which is surrounded by an, usually much larger, elastically deformed region. This elastically deformed region constrains the plastic deformation in the inner plastic region. A similar constraint does not occur in the uniaxial tensile testing experiment. This explains the numerically larger value for the "strength" according to the hardness test as compared to the tensile loading test (Eq. (12.51)).

Intermezzo: Hardness-Depth Profiling on Nanoscale

The nanoindentation technique allows the characterization of mechanical strength variation over nanoscale distances. Examples are provided by thin film systems and surface regions of surface engineered materials. In these cases the best approach is to measure the hardness perpendicular to the direction of the hardness gradient by making indentations on the specimen cross section along a line parallel to the hardness-depth profile. Measurements in the direction of the hardness gradient, i.e. measurements at the surface of the specimen, suffer from the complicated "averaging" in the load–displacement

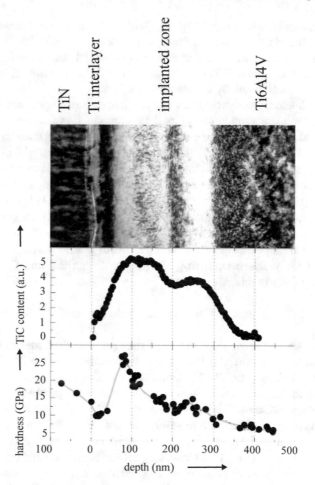

Fig. 12.40 Hardness-depth profiling on nanoscale. Microstructure and contact hardness in the surface adjacent region of a carbon implanted Ti-6.0 wt%Al-4.0 wt%V alloy. From top to bottom: the microstructure (cross-sectional bright field transmission electron micrograph), TiC-content (result of Auger Electron Spectroscopy in combination with sputter-depth profiling), and hardness as measured on the cross section as function of the distance to the specimen surface (taken from Kunert et al. 2001b). Prior to (cross-sectional) sample preparation, a protective TiN layer was sputter deposited on the surface of the specimen after a 5–10 nm thick Ti interlayer was deposited first to improve the adhesion of the TiN layer (see top of the figure). The specimen was doubly implanted with carbon; the second implantation with carbon ions of lower energy and of higher dose. The second implantation caused the highest peak (plateau) in the TiC-content vs. depth profile, indicating that at this depth range a practically continuous TiC layer had formed. At this depth range also the highest hardness occurs. The second TiC-content peak, of lower value and at larger depth, coincides with a second, much less pronounced hardness maximum; at this depth range, the hardness is due to dispersion hardening (by TiC precipitate particles; cf. Sect. 12.14.4 and also see Kunert et al. 2001b)

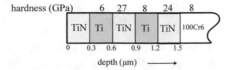

Fig. 12.41 Nanoindentation on a cross section of a five layer, TiN/Ti/TiN/Ti/TiN structure, sputter deposited onto a 100Cr6 tool steel substrate. Each contact hardness value indicated in the figure is the averaged result of three measurements performed in the middle of each sublayer in the cross section. Because of sample-preparation edge effects the top TiN layer was not analysed (taken from Kunert et al. 2001a)

curve of hardness variation along the specimen normal. Only the first, cross-sectional method allows determination of changes in hardness with depth with a depth resolution of a few nanometres, as shown by Kunert et al. (2001a).

An example of hardness-depth profiling on nanoscale is shown for a surface engineered specimen in Fig. 12.40. Carbon implantation improves the wear resistance of a titanium-based commercial alloy as Ti–6Al–4V alloy pronouncedly, in particular by the development of TiC precipitates. The microstructural variation in the surface region, of extent about 350 nm, requires microstructural analysis with a depth resolution of the order 10 nm. The nanoindentation technique applied to the cross section of the specimen meets this requirement for characterization of the mechanical strength variation. The result shown in Fig. 12.40 exhibits the large variation of (contact) hardness in the implanted zone.

The cross-sectional nanoindentation method is in particular useful as the only available method for the determination of the *intrinsic* (contact) hardness of thin layers, as the sublayers in a multilayer structure. An example is shown in Fig. 12.41 for a five layer, TiN/Ti/TiN/Ti/TiN structure.

12.14   Strengthening, Hardening Mechanisms (of Metals in Particular)

Yielding of crystalline materials is possible in a pronounced way for especially metallic specimens on the basis of movement (glide) by dislocations (see Sects. 5.2.5 and 12.9.1). Hence, the resistance against plastic deformation can be enhanced (and thus the hardness can be increased) by hindering this dislocation movement by the presence of obstacles in the microstructure of the material. Such obstacles can be other dislocations (Sect. 12.14.1), grain boundaries (Sect. 12.14.2), solute atoms (Sect. 12.14.3), and particles of another phase (Sect. 12.14.4).

12.14.1   Strain Hardening (Work Hardening)

In the discussion of the uniaxial tensile loading, stress–strain curve in Sect. 12.9, it was mentioned that in the plastic deformation regime, i.e. beyond the elastic limit, the load necessary for further plastic deformation increases with continued straining. This is due to the effect of strain hardening also called work hardening: Upon plastic deformation, dislocation production can occur, and, in general, the increase of the dislocation density makes dislocation propagation as a mechanism for glide (cf. Sects. 5.2.5 and 5.2.6) more difficult, implying the application of larger loads to realize the same extension as at an earlier stage of plastic deformation. The plastic deformation that is not relieved immediately during the loading (e.g. by dynamic recovery and/or dynamic recrystallization; see Sects. 11.1 and 11.2), thus the plastic deformation that remains, is often called "cold work".

A possible mechanism for increase of the dislocation density is an operating Frank-Read source (cf. Sect. 5.2.6). As cross-slip (cf. Sect. 5.2.6) takes place, dislocations intersect, more and more interaction of a moving dislocation with other dislocations occurs, and in general movement of the dislocation becomes increasingly hindered. The dislocation microstructure of a pronouncedly cold-worked metal shows a cell-like structure in a deformed grain: cell walls consisting of tangled dislocations at high density are separated from each other by regions of relatively low dislocation density (see Fig. 12.42).

To push one of two parallel dislocations, with parallel slip planes, past the other one, a shear stress is required that depends reciprocally on the distance between the dislocation lines. Consider a random distribution of the dislocations; random with respect to position, character and sign. Now considering the shear stress needed to move one dislocation past a nearest neighbour in this random distribution, the effect of all other dislocations, except the nearest neighbours of the dislocation considered, is nil (i.e. averages to zero). The average distance between the dislocations in a random distribution can be estimated as the reciprocal of the square root of the dislocation density ($1/\sqrt{\rho}$; see at the end of Sect. 5.2.3). Thus it follows that the increase of the strength (yield strength, shear stress, hardness) upon strain hardening

Fig. 12.42   Schematic depiction of a cell-like structure in a deformed grain: cell walls consisting of tangled dislocations at high density are separated from each other by cell interiors of relatively low dislocation density

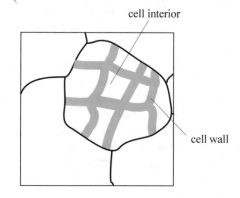

cell interior

cell wall

due to dislocation generation, can be given as:

$$\Delta(\text{strength}) = \text{constant} \cdot \sqrt{\rho} \tag{12.52}$$

This type of dependence of the increase of the (yield) strength on the produced dislocation density upon continued plastic deformation has been often observed. However, deviations occur as well. Many complex theories of strain hardening have been developed. Equation (12.52) can only be considered as a crude attempt to provide a basic understanding.

12.14.2 Grain Size; The Hall–Petch Relation; Nanosized Materials

Glide of dislocations is usually disrupted at grain boundaries, since the slip plane does not continue across the grain boundary. It is found empirically that the strength (yield strength, hardness) is inversely proportional with the square root of the grain size, D, as expressed in a so-called Hall (1951)-Petch (1953) relation:

$$\text{strength} = \text{const}_1 + \text{const}_2 \cdot \frac{1}{\sqrt{D}} \tag{12.53}$$

with const$_1$ and const$_2$ as constants.

The Hall–Petch relation may be directly related to the dependence of the strength on dislocation density resulting from work hardening, as expressed by Eq. (12.52) in Sect. 12.14.1. The hand-waiving reasoning runs as follows. Plastic deformation of a massive polycrystal requires compatible deformations of adjacent crystals: the individual crystals cannot deform "freely". Consequently, to maintain the massive nature of the polycrystalline specimen, strain gradients occur: the plastic strains close to the grain boundaries are different from those far away from the boundary in the bulk of the crystal. Such strain gradients can be realized by dislocations (Example: a curved grain results by an excess of edge dislocations of the same sign; see Fig. 12.43). These dislocations are called *geometrically necessary dislocations*, which provide the

Fig. 12.43 Realization of grain curvature by an excess of edge dislocations of the same sign

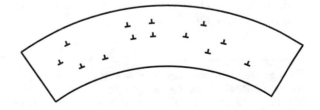

Fig. 12.44 The development of a dislocation pile-up at a grain boundary: a dislocation source produces a series of similar dislocations (of the same Burgers vector) all gliding on the same slip plane that is intersected by a grain boundary

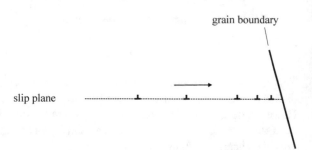

compatability of deformations of neighbouring grains.[20] If the average strain along a side of a cubic grain of size D is ε, without more ado it would overlap (positively or negatively) with its neighbour over a length εD. This overlapping is avoided by introducing strain gradients (as by curvature in Fig. 12.43) invoked by introducing into the grain a number of dislocations (of the same sign) of the order $\varepsilon D/b$ with b as the magnitude of the Burgers vector. Thus the introduced density of geometrically necessary dislocations (number of dislocations per area; cf. Sect. 5.2.3), ρ_{geo}, is of the order $\rho_{geo} = (\varepsilon D/b)/D^2 = \varepsilon/(bD)$. If the total dislocation density is governed by ρ_{geo}, which holds for small strains, then ρ in Eq. (12.52) can be replaced by $\rho_{geo} = \varepsilon/(bD)$ and the above Hall–Petch relation follows immediately.

The classical explanation of the Hall–Petch relation is in terms of the development of pile-ups of dislocations at grain boundaries. Suppose a dislocation source (cf. Sect. 5.2.6) produces a series of similar dislocations (of the same Burgers vector) all gliding on the same slip plane that is intersected by a grain boundary. Dislocations of same sign repel each other and, recognizing this, it becomes conceivable that so-called pile-ups of dislocations are formed at grain boundaries upon plastic deformation (see Fig. 12.44). A large stress concentration occurs ahead of the pile-up, which is proportional to the number of dislocations in the pile-up, which number in turn is proportional to the grain size. Hence, the amount of to be applied stress to activate/nucleate dislocations (i.e. to establish plastic deformation) in the neighbouring grain will be smaller for large grains than for small grains. On this basis the Hall-Petch relation was originally derived. Alternatively, the process can be described as dislocation transmission across grain boundaries: dislocations are absorbed by the grain boundary ("dissociate" into the grain boundary) and are reemitted (into the neighbouring grain). For experimental evidence of such a mechanism, see Javaid et al. (2020). For a review of dislocation models to explain the classical Hall-Petch relation, see Kato (2014).

---

[20] The geometrically necessary dislocations (GNDs) are complemented by the "statistically stored dislocations (SSDs)" such that the total dislocation density is given by the sum of the densities of GNDs and SSDs. Upon plastic deformation the SSDs can be conceived as those dislocations which are introduced in the absence of macroscopic/mesoscopic plastic strain gradients as referred to in the above text. In a strict sense any dislocation is introduced to comply with a geometrical incompatibility....

The favourable effect of grain boundaries on mechanical strength, as discussed here, becomes less outspoken at elevated temperatures, because creep and grain-boundary sliding (cf. Sect. 12.16) can become prominent. Obviously these processes are promoted by a relatively high grain-boundary density. High grain-boundary densities occur in nanosized materials, and thus, grain-boundary sliding has been suggested as a possible mechanism of plastic deformation in these materials at also relatively low temperatures (see further below).

In a pronounced stage of deformation of a ductile material (as a metal) the dislocations gather in regions of high dislocation density and a dislocation-cell structure develops within the grains, with a high dislocation density in the cell walls and a small dislocation density in-between (cf. Sect. 12.14.1). Then, the size parameter in the Hall–Petch relation can no longer be identified with the grain size, but instead the cell size has to be taken for that.

The possible validity of the Hall–Petch relation for *nanosized materials* (grain size smaller than, say, 100 nm) has been the subject of a considerable amount of research (e.g. see reviews by Dao et al. (2007) and by Naik and Walley (2020)). Indeed, very high hardnesses can occur for nanocrystalline materials: for example, the hardness of nanocrystalline copper of grain size 10 nm can be as high as 3000 MPa, implying a yield strength of about 1000 MPa (cf. Eq. (12.51); to be compared with the yield strength of coarse-grained copper, which is about 50 MPa). If the grain size is reduced to values smaller than, say, 100 nm, dislocation-mediated strengthening mechanisms, onto which derivations of the classical Hall–Petch relation have been based (see above), become increasingly more difficult to operate and then grain-boundary mediated processes gain increasing importance.

Twinning can provide a mechanism for strengthening, in particular this could be the case in case of materials with low stacking-fault energy, as silver (and also copper; cf. Sect. 11.1). Twin boundaries are high-angle boundaries (see Sect. 5.3.2) and obstruct the propagation of gliding dislocations and other twins on different twinning planes.[21] It has been shown that twin boundaries give rise to a Hall–Petch relation for the hardness, with the twin-boundary spacing as the (grain-)size parameter, for twin-boundary spacings larger than 150 nm. However, the dependence of the hardness on twin-boundary spacing, D_{twin}, for twin-boundary spacings smaller than 100 nm, as can occur in nanocrystalline materials, is characterized by a $(D_{twin})^{-1}$ dependence, rather than a $(D_{twin})^{-1/2}$ dependence as would be in accordance with Eq. (12.53) (Shaw et al. 2008).

(Further) Strengthening of nanosized material may be achieved by alloying with a very small amount of an element which has practically no solubility in the matrix of the first material. The atoms of this alloying element will then segregate at the defects, as grain boundaries and twin boundaries in the nanosized material. These segregated atoms may pin the grain boundaries and the twin boundaries upon plastic

---

[21] Similarly as discussed above for dislocations: the role of geometrically necessary dislocations (GNDs), to accommodate macroscopic/mesoscopic plastic strain gradients, e.g. as occurring in the vicinity of grain boundaries upon plastic deformation of a massive polycrystalline material, can be taken by twins: "geometrically necessary twins (GNTs)" (Sevillano 2008).

deformation. Especially, if not only a high density of grain boundaries is present, as is intrinsically the case for nanosized materials, but also a high density of twins occurs, these segregated atoms could thus enhance the strength distinctly, as exhibited by increased hardness. It has been shown that upon alloying nanosized silver, with a grain size of about 50 nm and a high density of nanotwins (twin spacings of a few nm), with a tiny amount of copper (silver has negligible solubility for copper), hardness values can be attained which are about 3000 MPa (Ke et al. 2019), i.e. similar to the hardness of nanosized copper with a grain size of 10 nm (see above).

Geometrical effects play a role. Grain boundaries in nanocrystalline materials can possess an amount of so-called "excess free volume" very much larger than for coarse-grained materials (Kuru et al. 2009). The occupation of this excess free volume by segregating alloying element atoms can enhance the mechanical strength (Li et al. 2017).

At this place it is appropriate to remark that segregation of a tiny amount of an alloying element on (grain) boundaries is not generally favourable for the mechanical strength of a material. The cohesion at a grain boundary can be changed as a consequence of (electron) charge transfer between the segregated atoms and the atoms of the matrix material situated at the boundary. Thus it is well known that phosphorous, present as an impurity in tiny amounts in steel as a consequence of the steel-production process, segregates at grain boundaries in the steel and, as a result of subsequent (electron) charge transfer from phosphorous to iron, weakens the steel considerably, which has had disastrous consequences in industrial applications of such steels.

12.14.3   Solid Solution Hardening

In a solid solution heterogeneities occur on the atomic scale: the solute atoms. They can interact with dislocations in a number of different ways.

In Sect. 12.9.2 the focus was on elastic interaction: a point defect experiences a size misfit, characterized by the misfit volume defined as the difference between the volume of the stress-free (i.e. before insertion into the "hole" in the matrix) point defect and the "hole" in the matrix, to be occupied by the point defect. The (elastic) interaction energy could be made negative by proper positioning of the point defect close to the dislocation line. The more negative the interaction energy, the more energy it costs to separate the dislocation and the point defect. This *size effect* is not the only possible form of interaction of a point defect and a dislocation.

Another, usually less important, form of elastic interaction, called *modulus effect*, is due to different elastic constants for the matrix and the point effect. Thus "soft" point defects are attracted to regions of high elastic energy density, leading to a decrease of the elastic energy. Reversely, "hard" point defects are repelled from regions of high elastic energy density.

Further, *electric effects* are possible, e.g. due to interactions induced by solute atoms introducing an excess or deficit of (conduction) electrons.

A special effect, called *Suzuki effect*, occurs when the solubility of a solute is different in the region of a stacking fault (i.e. a local h.c.p. arrangement in f.c.c. or a local f.c.c. arrangement in h.c.p.; see Sect. 5.3.2). For dissociated dislocations, i.e. with a stacking fault in-between (see Sect. 5.2.8), this may have consequences for the mobility of the dislocations.

Notwithstanding the sketched complexity of the possible interaction of a point defect with a dislocation, it appears that the elastic interaction dominates in general.

A solute atom disturbs the ideal crystal-structure regularity. Thus the movement of a dislocation can be hindered by the stress field around a misfitting solute atom: the dislocation can be attracted or repelled, depending on the local signs of the stress fields of the solute atom and the dislocation. The maximum bending (minimum radius of curvature) of a dislocation under the action of a shear stress is given by Eq. (5.9). It follows that it is impossible for the dislocation to take a position of minimal elastic repulsive interaction (with the solute atoms), because the required local curvatures of the dislocation around its neighbouring solute atoms, and the associated increase in dislocation-line length, are much too high. The equilibrium position of the dislocation will thus be determined by the optimum combination of repulsive interaction energy and dislocation-strain energy (dislocation-line length); see also Fig. 12.45. The strength increase by the presence of the solute atoms then follows by the shear stress required to move the dislocation from this equilibrium position.

On the above basis, a number of theories for solid-solution strengthening have been developed through the years. The results display dependencies of the increase of strength (yield strength, critical shear stress, hardness) on the concentration of solute atoms, c, varying from $c^{1/2}$, to $c^{2/3}$ and to c, where these proportionalities may pertain to certain ranges of the solute content (see Hull and Bacon (2001), Haasen (1978)

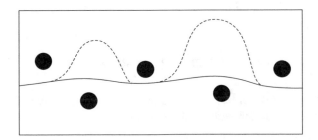

Fig. 12.45 Schematic depiction of a dislocation in a matrix containing solute atoms. The dislocation line will take an equilibrium position determined by the optimum combination of repulsive interaction energy (with the strain fields of the solute atoms) and dislocation-strain energy (dislocation-line length) (see the full line). Upon application of a shear stress, the dislocation bows out (dashed line (plane of drawing as glide plane); the dislocation-line movement for the case shown in the figure and as indicated by the dashed line could occur if the dislocation before application of the shear stress would for example be of (largely) screw character, because for a screw dislocation the dislocation line moves in a direction perpendicular to the Burgers vector under the action of the component of the shear stress in the direction of the Burgers vector (cf. Sect. 5.2.5 (for a parallel discussion, if the dislocation would be of (largely) edge character) and see also Fig. 5.13)

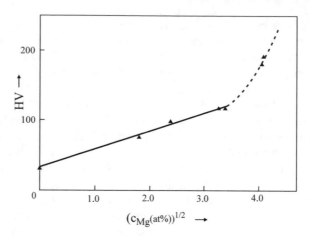

Fig. 12.46 Hardness (Vickers; cf. Sect. 12.13) as function of Mg content for AlMg alloy specimens prepared by melt-spinning. A linear dependence of the hardness on $(c_{Mg})^{1/2}$ is observed (solid solution hardening; full line). The extraordinary increase in hardness for high Mg content (>12 at.% Mg; dashed line) is not due to solid solution hardening, but is attributed to the formation of Guinier–Preston zones (cf. Sect. 9.4.1 and Sect. 12.14.4) (taken from M. van Rooyen, P. F. Colijn, Th. de Keijser and E. J. Mittemeijer, Journal of Materials Science, 21 (1986), 2373–2384)

and de With (2006)). Experimental evidence for these relations has been provided too. The relation most often found and used for the increase of strength by solid solution strengthening may be:

$$\Delta(\text{strength}) = \text{constant} \cdot \sqrt{c} \qquad (12.54)$$

An example is shown in Fig. 12.46.

12.14.4 Precipitation/Dispersion Strengthening

Second phase particles can be introduced in a matrix, in a finely dispersed way, to increase the mechanical strength, by, for example:

– precipitation from a supersaturated matrix phase ("age-hardening"): e.g. the initial precipitation of intermetallic, metastable precipitates (clusters/"Guinier-Preston" zones) and, upon prolonged aging, stable precipitates, as in Al-based Al-Cu and Al–Mg alloys, and the like (e.g. "duralumin" (with Cu as (not the only) alloying element in Al); see Sect. 9.4.1);
– internal oxidation or internal nitriding: e.g. the precipitation of CrN, AlN or $Cr_{1-x}Al_xN$ in steels alloyed with Cr and/or Al upon reaction in an ammonia atmosphere;

– addition of hard particles during material production: e.g. oxide particles during sintering of a metal powder.

The second phase particles are more or less randomly distributed throughout the matrix and intersect glide planes of the matrix. Upon plastic deformation, gliding dislocations in the matrix are obstructed in their movement by these obstacles. They have two options: they either cut the second phase particles, shearing them upon passage (see Fig. 12.47a), or they bow out between the particles, leaving behind dislocation loops around the particles upon passage (Fig. 12.47b). The first mechanism prevails for (tiny) particles coherent with the matrix, as can occur in the beginning stage of a precipitation process. If incoherent (larger) particles occur, the dislocations cannot penetrate the particles and the governing mechanism is the second one.

The second mechanism was already discussed in Sect. 5.4.1: "Orowan process". The critical shear stress needed for a dislocation to pass two adjacent pinning points by bowing out is given by:

$$\tau_0 = \frac{Gb}{d} \tag{5.10}$$

Adopting n_p as the number density of second phase particles, i.e. the number of second phase particles per unit volume, the average distance between particles in a

(a)   (b)

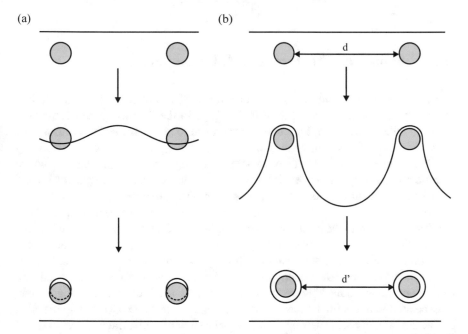

Fig. 12.47 Two mechanisms for dislocation lines to pass obstacles (e.g. precipitates): **a** cutting the obstacles, shearing them upon passage; **b** bowing out between the obstacles and encircling the obstacles (Orowan process). In the latter case, the effective distance between the particles decreases

random distribution of particles with number density n_p can be estimated by $(n_p)^{-1/3}$ (The volume of the matrix "confined to" one particle in the unit volume of the matrix thus is taken as $(n_p)^{-1/3} \cdot (n_p)^{-1/3} \cdot (n_p)^{-1/3} = 1/n_p$; cf. the reasoning to estimate the average distance between dislocations in a random distribution of dislocations as presented at the end of Sect. 5.2.3). Further, taking the particles as spheres of radius r_p, it follows:

$$d = \left(n_p\right)^{-1/3} = \left(\frac{f_p}{\frac{4}{3}\pi r_p^3}\right)^{-1/3} \approx f_p^{-1/3} \cdot r_p \qquad (12.55)$$

where f_p denotes the volume fraction of second phase particles. Thus, by substitution of this result for d in Eq. (5.10), it is obtained for the increase of strength:

$$\Delta(\text{strength}) = \text{constant} \cdot \frac{f_p^{1/3}}{r_p} \qquad (12.56)$$

Evidently, the strength increases with increasing volume fraction of second phase particles and in particular increases with decreasing size of the particles. Further, if d is of the same order as r_p, d in Eq. (5.10) must be replaced by $d - 2r_p$, and, also, once dislocation loops have formed around the particles, the effective d value for particles surrounded by dislocation loops becomes smaller as well (see Fig. 12.47b).

12.15 Failure by Fracture; Crack Propagation

The ultimate feature signifying failure of a material workpiece is its breakage into separate pieces as a consequence of its loading. In the following the treatment is focussed on *tensile fracture*: fracture under tensile loading, as this is the most usual mode of fracture. Tensile fracture involves the separation of atomic bonds (either by tensile fracture of atomic bonds or by atomic shear) across the plane along which fracture proceeds. The design engineer has an obvious interest in knowing the critical value of loading stress at and beyond which such fracture occurs.

Two types of fracture are usually distinguished: *brittle* fracture and *ductile* fracture. As follows from the discussion in the beginning of Sect. 12.9, (i) brittle fracture occurs without significant plastic deformation, whereas a ductile material experiences pronounced plastic deformation before fracture occurs, and (ii) brittle fracture occurs in the absence of reduction of the original cross section of a loaded bar, whereas pronounced necking, reduction of the cross-sectional area of a loaded bar, is associated with ductile fracture. A ductile material has the possibility to absorb a distinct amount of plastic deformation energy before and while fracture occurs (i.e. if sufficient strength is available, so that a considerable toughness exists (see footnote 12 in this chapter)). Brittle fracture implies the very fast propagation of a crack that requires no significant increase of loading stress for its advance. Ductile fracture is characterized by a relatively slow velocity of the crack and its continued propagation requires an increase of the loading stress. Obviously, brittle fracture is the type of fracture to be avoided in engineering applications by all possible means.

Materials may be brittle or ductile in dependence on parameters as, in particular, the temperature and the type of loading (state of stress and stress–strain rate).

- F.c.c. metals show ductile behaviour independent of temperature. This is due to their relatively low intrinsic strength (yield strength) and the large number of possible slip systems (see Sect. 5.2.5).
- B.c.c. metals show brittle fracture at relatively low temperatures, in particular because of the increase of their (yield) strength with decreasing temperature: the fracture stress (see also below for ceramic materials) becomes smaller than the yield (flow) stress.
- H.c.p. metals also show a transition to brittle fracture at relatively low temperatures, in association with the number of active slip systems becoming less for decreasing temperatures.
- Ceramic materials (as the alkali halides (e.g. NaCl, KCl; ionic bonding), the refractory oxides (e.g. MgO, Al_2O_3; mixed ionic and covalent bonding) and the covalent solids (SiC, Si_3N_4; covalent bonding)) are predominantly brittle: their fracture strength is smaller than their yield strength; the fracture strength increases with increase of the contribution of covalent bonding.
- Amorphous and crystalline polymers, in particular thermoplastics characterized by relatively easy relative displacements of adjacent polymeric chains (cf. Sect. 12.9.3), can exhibit a clear ductile–brittle transition as function of temperature, often in association with their glass-transition temperature values. Network polymers, characterized by polymeric chains strongly cross-linked by covalent bonds (see also Sect. 12.6), are of dominantly brittle nature. As compared to metals and ceramics, the fracture strength of a polymer is usually low.

Microstructural changes induced for increasing intrinsic (yield) strength (see Sect. 12.14) usually lead to a decrease of ductility and toughness. The exception with polycrystalline materials is reduction of the grain size, which increases both the strength (Sect. 12.14.2) and the ductility/toughness.

Brittle fracture can lead and has led to sensational cases of catastrophic failure. The sinking of the "Titanic" in 1912 may have been supported by the brittle nature of the steel used for the hull of the ship: the ductile to brittle fracture transition temperature of this steel (b.c.c.; see above) was about 0 °C, so, at the moment of colliding with an iceberg in cold, polar sea water, the hull most certainly was brittle. The example most frequently cited may be the brittle fracture of the so-called Liberty ships used for troop and war material transport across the (Atlantic) ocean in the Second World War: the number of ships lost by brittle failure at sea is larger than the number of ships sunk by torpedo attacks by German submarines. One of the direct reasons for this failure was a ductile to brittle transition temperature of the steel alloy, applied for building the vessels, of about 4 °C, experienced upon cooling in sea water.

Microscopic description of ductile fracture under tensile loading begins with the development of pores/voids in the necking region (e.g. at the interfaces with inclusions/second phase particles, by inclusion/second phase particle fracture and decohesion) assisted by the hydrostatic tensile component of the local state of stress (at the length axis of the specimen, at the centre of necking), which has become triaxial due to the necking (see the discussion around Eqs. (12.33–12.35)). This is followed by the coalescence of the pores/voids leading to an internal crack growing perpendicular to the direction of loading until the remaining rim of material can no longer support the loading stress and fails abruptly by shear along an angle of 45° as this is the angle of maximal shear stress (cf. Sect. 12.4). The ductile fracture surface has a "dimpled" appearance due to the voids/pores formed.

Microscopic description of brittle fracture under tensile loading is usually caught by terms as *cleavage* or grain-boundary, *intergranular* fracture. Cleavage involves fracture along crystallographic planes (e.g. {001} planes for b.c.c. metals, h.c.p. metals and NaCl and {110} planes for ZnS, etc.), where the atomic bonds across the fracture plane are destroyed by tensile separation. Cleavage thus proceeds *transgranular*. Intergranular brittle fracture can occur in particular if the grain boundaries have been weakened mechanically, as due to grain-boundary embrittlement, which can be caused by the segregation of impurity atoms at grain boundaries. The brittle fracture surface usually has a facetted appearance.

The observed values of fracture strength are much lower than the theoretical strength values as calculated for perfect materials. Thus one may envisage the presence of "defects" in real materials which facilitate fracture. To understand the ease of plastic deformation of real, crystalline materials the concept of dislocations (and their glide) was introduced (see Sect. 5.2). Considering brittle fracture, characterized by negligible plastic deformation, one may wonder what kind of defects could be considered to explain (brittle) fracture at stress levels much lower than the theoretical values for fracture of perfect materials.

A first important realization involves the recognition that a nonuniform stress distribution occurs in the vicinity of a crack, or other discontinuity, in a loaded material. Near the discontinuity the stress will be higher than at a location remote from it. Consider a void in a bar subjected to a tensile loading force along the length axis of the bar. The void can be considered as a "stress raiser", leading to a "stress concentration" at the border of the void. This can be considered as a simple consequence of the decrease of material cross section at the depth level where the void resides: the specimen must bear the loading force across all cross sections (see Fig. 12.48). The *stress-concentration factor*, K_σ, is defined as the ratio of the maximum stress at the circumference of the discontinuity, σ_{max}, and the nominal stress acting in the absence of the discontinuity, σ_{nom}:

$$K_\sigma \equiv \frac{\sigma_{max}}{\sigma_{nom}} \tag{12.57}$$

Fig. 12.48 Stress
concentration at a void

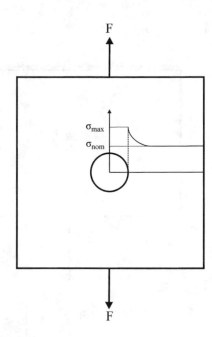

On this basis a first explanation of the discrepancy between real and theoretical fracture strength emerges: fracture can occur if σ_{max} at the border with a discontinuity in the material exceeds the theoretical strength, although σ_{nom} is well below it.

Principally, the condition that σ_{max} exceeds the theoretical strength is only a necessary, not a sufficient condition that crack growth, leading to fracture, can occur. A second recognition involves that a crack can only grow if the total energy change by crack growth is negative; i.e. energy is released. Consider Fig. 12.49a showing an internal crack of length $2l$ in a plate both of thickness t much smaller than $2l$ and of width (measured in the direction of the crack) much larger than $2l$. Two energy contributions controlling brittle crack growth can be discerned:

(i)   The (crack-)surface energy, U_s:

$$U_s = 2(2l)t \cdot \gamma_s \tag{12.58}$$

where γ_s denotes the surface energy per unit of surface.

(ii) The strain energy released upon crack formation, U_{el}, under the action of the tensile loading stress σ_{nom}. For quantitative analysis, in addition to the above indicated geometrical restrictions, it is usually assumed that the crack is of elliptical shape, with main axis $2l$. Then, it can be shown (by nontrivial calculation) that

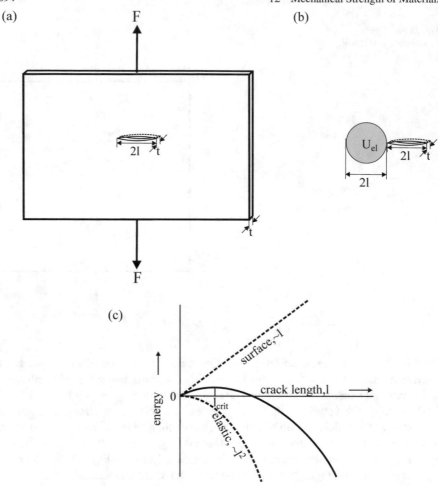

Fig. 12.49 **a** Crack of length $2l$ in a plate of thickness t under applied tensile load F; **b** representation of elastic energy stored (see text) and **c** corresponding surface energy, elastic energy and total energy (full line) as a function of crack length

$$U_{el} = \frac{-\pi l^2 t \cdot \sigma_{nom}^2}{E} \tag{12.59}$$

This result can be conceived crudely as that upon crack growth that part of the strain energy incorporated in the system is released that is more or less concentrated at and largely before the tip of the crack (see discussion with respect to Eq. (12.57)), in a circular region (remember, the plate is very thin as compared to the crack length) of diameter $2l$, with the centre of the circle at some distance in front of the advancing crack (see Fig. 12.49b).

Evidently, crack growth is resisted by the formation of crack surface and promoted by the release of elastic strain energy. Both energy contributions depend differently on the crack length $2l$. Hence, a critical minimum crack length, which is required for a crack to be stable and grow under the action of the loading nominal stress, can be calculated according to

$$\frac{d(U_s + U_{el})}{dl} = \frac{d\left(4lt \cdot \gamma_s - \pi l^2 t \cdot \sigma_{nom}^2/E\right)}{dl} = 0$$

and it follows:

$$l_{crit} = \frac{2E\gamma_s}{\pi \sigma_{nom}^2} \quad \text{or} \quad \sigma_{nom} = \left\{\frac{2E\gamma_s}{\pi l_{crit}}\right\}^{1/2} \tag{12.60}$$

It is simply seen that for l larger than l_{crit} crack growth leads to net reduction of the total energy, and thus crack growth is strived for, whereas for l smaller than l_{crit} reduction of crack length is aimed at (see Fig. 12.49c). Note that if a crack can grow under the action of σ_{nom}, then it can keep growing because the critical value of nominal loading stress becomes smaller for larger crack length (the critical value of σ_{nom} is inversely proportional with the square root of l; cf. Eq. (12.60)). It can be shown that for brittle fracture the stress criterion, on the basis of σ_{max} exceeding the theoretical strength (cf. Eq. (12.57)), leads to practically the same value for critical nominal applied stress as obtained from the energy criterion, given by Eq. (12.60), because the radius of curvature at the tip of the crack in pure brittle fracture is very small (of the order of the atomic spacing).

The result derived and discussed here is not principally different for the case of a crack at the surface (external crack of half-elliptical shape and length l; cf. Fig. 12.49a). The criterion expressed by Eq. (12.60) was first derived by Griffith (1920) and has been named after him.[22]

Now turning to crack growth in ductile materials, it is obvious that the occurrence of plastic deformation in ductile materials, in order to realize crack extension, brings about the incorporation of plastic deformation energy, as a crack-growth counteracting contribution, in the energy balance. Thus, this can be done by replacing in Eq. (12.60) the factor γ_s by $\gamma_s + \gamma_p$, where γ_p is a measure for the plastic deformation energy necessary to increase the crack surface. This approach is justifiable as long as the amount of plastic deformation energy introduced in the specimen is proportional with the crack length, as holds for the crack-surface energy (cf. Eq. (12.58)). For significantly ductile materials γ_s can be neglected as compared to γ_p. Thus, a material-specific (see below) parameter $G_c \equiv 2(\gamma_s + \gamma_p)$ can be defined (cf. Eq. (12.60) for the origin of the factor "2") which is a measure for the crack resistance of a material and is called *fracture toughness*. The more ductile a material, the larger its fracture toughness.

---

[22] The consideration on the basis of an energy balance (Eq. (12.60)), leading to the concept of critical crack length, parallels the treatment leading to the concept of critical size for (second phase) precipitate particle growth in a supersaturated matrix during phase transformation (Sect. 9.2).

An alternative definition of "fracture toughness" has the following background (cf. the two routes for determining a critical value for the applied nominal stress below which a crack of specified length cannot grow (discussion of Eqs. (12.57) and (12.60)). The values of the stress components at the tip of the crack generally are proportional with $\sigma_{\text{nom}} l^{1/2}$. Therefore it is customary to define a *stress-intensity factor*, K, as follows:

$$K \equiv \sigma_{\text{nom}} (\pi l)^{1/2} \tag{12.61}$$

The stress-intensity factor allows specification of the type of stress field surrounding the crack; note the difference between the stress-concentration factor K_σ (Eq. (12.57)) and the stress-intensity factor K. Now, adopting the "stress exceeding a critical value approach" for crack growth to occur (see below Eq. (12.57)), the occurrence of fracture for a specific material can be related to a critical value of K, K_c, which is (also; see above) called *"fracture toughness"*:

$$K_c = \sigma_{\text{nom}} (\pi l_{\text{crit}})^{1/2} \quad \text{or} \quad \sigma_{\text{nom}} = \frac{K_c}{(\pi l_{\text{crit}})^{1/2}} \tag{12.62}$$

Comparing Eqs. (12.60) and (12.62), and replacing γ_s by $\gamma_s + \gamma_p$ (see above), it follows for the relation between the two measures for fracture toughness:

$$K_c = (E G_c)^{1/2} \tag{12.63}$$

In practice a further proportionality factor is introduced into the definition formula of K (Eq. (12.61)) and thus K_c, that allows expressing differences in crack and specimen geometry. With this recognition, K_c can be considered as a material-specific parameter. In designing components such that no fracture in service can occur, material specification is (also) based on a parameter as K_c, rather than G_c.

The above discussion indicates the applied nominal stress, σ_{nom}, the crack length, l, and the fracture toughness, K_c, as the crucial parameters to consider in designing components against failure by fracture. Only two of these can be considered as independent variables: if two are known, the third one follows straightforwardly from, e.g. Eq. (12.62). For example, by special detection methods it may be assured that cracks in a component have a size below the detection limit, l_{max}. Then, for a specific material with the known fracture toughness, K_c, application of Eq. (12.62) results in a maximal applied nominal stress that should not be exceeded to avoid fracture in service. One does not design such that a workpiece is loaded in service at its expected strength limit; usually, additional safety factors are incorporated. Moreover, the applied nominal stress may be kept at such low level that, hopefully, no plastic deformation occurs locally as a result of stress concentration at a stress raiser (i.e. the local stress should not exceed the yield stress; note that the "yield point" depends on the (local) state of stress; see Sect. 12.10).

12.16 Failure by Creep

The discussion in the preceding section suggests, that, if the stress applied to a component is kept at a relatively low value, fracture and even plastic deformation can be avoided and that the time of application of the load plays no role. This holds for metals at temperatures below about 0.4 T_m (T_m denotes the melting temperature in K). However, at relatively high temperatures permanent deformation may occur *over a period of time* at applied stress levels well below the normally accepted yield stress. This phenomenon is called creep. It is usually studied as the time-dependent permanent deformation observed during the application of a constant uniaxially applied tensile load (as for determination of the tensile stress–strain curve (Sect. 12.9)). For metals creep becomes important at relatively high temperatures (above 0.4 T_m), but amorphous polymers can exhibit creep already at room temperature.

A schematic curve showing the dependence of resulting strain as a function of time at constant load, the creep curve, is shown in Fig. 12.50a; an experimental example is given in Fig. 12.50b. Upon application of the load the expected instantaneous elastic strain is realized. Thereafter permanent deformation develops as function of time. Three time ranges are distinguished usually:

- a first, transient stage where the strain rate continuously decreases, which could be interpreted as a consequence of some form of work hardening (cf. discussion of the tensile stress–strain curve in Sect. 12.9);
- a second stage where the strain rate is constant (strain increases linearly with time), which could be interpreted as a steady state where recovery (see Sect. 11.1) and work hardening processes are in balance, implying a steady-state microstructure;
- a third, final stage revealing an increasing strain rate until fracture (in the field of creep often called rupture) occurs. An obvious reason for this phenomenon can be indicated if the creep experiment is performed at constant engineering stress (cf. Sects. 12.2 and 12.9): the occurring reduction of cross-sectional area leads to true stress increase upon progressing time. However, also the advent of

(a)   (b)

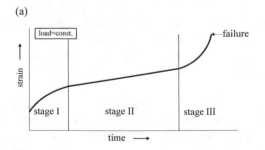
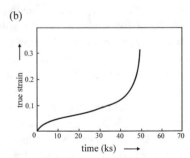

Fig. 12.50 a Schematic creep curve. **b** Experimental creep curve recorded for 9 wt% Cr-1 wt% Cu martensitic steel at 873 K applying a tensile load (uniaxial loading) of 140 MPa (taken from T. Tsuchiyama, Y. Futamura and S. Takaki, Key Engineering Materials, 171–174 (2000), 411–418)

microstructural changes, as due to recrystallization (see Sect. 11.2), second phase particle coarsening (see Sect. 11.3.5) and development of defects as voids and cracks, explain the emergence of the final, tertiary creep stage.

The relative ease of plastic deformation at elevated temperature can bear a strong relation with the mobility of atoms becoming larger with increasing temperature (viscous and diffusional flow), the larger dislocation mobility at higher temperature (climb (cf. Sect. 5.2.7) then becomes more important) and the number of operating slip systems increasing with temperature, etc. Evidently, the creep behaviour, as expressed by the creep curve, depends on applied stress (strain rate) and temperature.

The second, steady-state stage usually is the one of longest duration and the most important one for engineering applications; the constant strain rate of this stage is used as a design parameter characterizing an endurance limit of the component (cf. the discussion on fatigue in Sect. 12.18).

Specific mechanisms, as viscous flow, diffusional flow, and dislocation glide and climb have been considered for steady-state creep:

Viscous flow (i.e. flow invoked by shear forces) is the creep mechanism operating in amorphous materials. The rate of shear strain is proportional with the shear stress:

$$\frac{d\gamma}{dt} = \frac{\tau}{\eta} \tag{12.64a}$$

with η as the so-called viscosity. If η is a parameter that is independent of the applied shear stress (but depends on temperature, see below) one speaks of Newtonian viscosity. The pendant of Eq. (12.64a) for tensile loading is written as:

$$\frac{d\varepsilon}{dt} = \frac{\sigma}{(3\eta)} \tag{12.64b}$$

An applied shear stress, applied to the top and bottom faces of an amorphous body, strives for flow of an upper part with respect to a lower part, of the loaded amorphous material. This can be established by to a certain degree coordinated jumps of the atoms or molecules in the volume loaded. Most of such atomic or molecular movements will occur there where the free volume in the amorphous material is largest. This stress driven motion of the atoms or molecules is associated with energy barriers which must be overcome on the way of the atoms or molecules to energetically more favourable positions in the applied stress field: the viscous flow is thermally activated.

Diffusional mass transport as creep mechanism in polycrystalline solids (metals) can be thought to originate from local (at grain boundaries) differences in vacancy concentration induced by the applied stress field. At a grain boundary perpendicular to an applied tensile stress, a tendency is experienced to separate the atoms, and as a result the local vacancy-formation energy is reduced somewhat. Similarly, at a grain boundary parallel to the applied tensile stress a tendency is experienced to compaction (due to the Poisson contraction effect), and as a result the local vacancy-formation energy is increased somewhat. Thus an increased vacancy concentration occurs at grain boundaries perpendicular to the tensile loading axis and a reduced vacancy concentration occurs at grain boundaries parallel to the tensile loading axis. Consequently, a vacancy flow occurs from the perpendicular to the parallel grain boundaries in conjunction with a counter flow of atoms from parallel to perpendicular grain boundaries (substitutional diffusion; see Sect. 8.5.2). This leads to lengthening of the specimen by grain-size increase in the loading direction (creep; see Fig. 12.51). If the mass flow is realized by volume(bulk) diffusion one speaks of *Nabarro–Herring creep*; if the mass flow is realized by grain-boundary diffusion one speaks of *Coble creep*. Both Nabarro–Herring creep and Coble creep depend on grain size, but obviously Coble creep is much more sensitive to grain size. At relatively low creep temperatures, Coble creep will dominate over Nabarro–Herring creep (cf. the discussion on the contributions of volume and grain-boundary diffusion as function of temperature in Sect. 8.9).

Finally, it should be realized that associated with the grain-size lengthening in the loading direction, there will be a grain-size reduction in the directions perpendicular to the loading direction. If only the diffusional creep mechanisms as discussed here would occur, crack and/or void formation are inevitable as side phenomena. Hence, *grain-boundary sliding* (see Fig. 12.52) is invoked to maintain the structural integrity of the component: the sliding of the grains (a viscous flow process) prevents a grain-boundary separation.

Intermezzo: Whiskering;
Interplay of Internal Stress Gradients and Coble Creep
Since the middle of the past century it has been known that needle-like metal filaments can grow out "spontaneously" from the surface of thin metal films, (already) at room temperature. These metallic "hairs" are usually called "whiskers". Whiskering is observed especially for metals of low melting point, as cadmium, zinc and tin. Whiskers are single crystalline, grow at the bottom by the addition of material from the film, typically have a diameter of 1–10 μm and can reach lengths of several 100 μm; see Fig. 12.53.

Sn whiskers have pronounced, but harmful technological relevance. Sn films on Cu substrates are important parts of microelectronic equipment. Sn whiskers growing from the surface of such Sn(-based) films can bridge the gaps between neighbouring conductors and consequently cause short circuit failure of the component (microchip). Thereby whiskering has led to enormous costs. Hence, next to distinct scientific curiosity about the origin of whisker formation, large technological interest exists in finding routes to avoid whisker formation, preferably on the basis of scientific understanding.

Intuitively one may propose that the presence of an internal, compressive stress (parallel to the surface of the film), as a possible consequence of the preparation process or a subsequent ageing process, is responsible for (is the

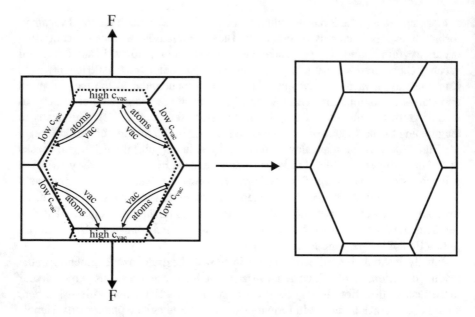

Fig. 12.51 Creep by substitutional diffusional flow in a crystalline solid. Under action of a tensile loading force (uniaxial loading) the equilibrium vacancy concentration, c_{vac}, is enhanced somewhat at grain boundaries perpendicular to the loading axis and reduced somewhat at grain boundaries parallel to the loading axis. The thus induced flows of atoms and vacancies have been indicated within the grain sketched. As a result, elongation of the grain occurs in the direction of the tensile loading. In the sketched example mass flow by volume diffusion has been indicated (Nabarro–Herring creep); the mass flow could also be realized by grain-boundary diffusion (Coble creep)

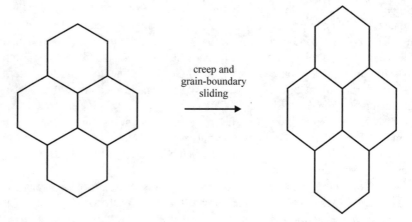

Fig. 12.52 Macroscopic shape change upon uniaxial loading (tensile force acting in the vertical direction of the figure; cf. Fig. 12.51) realized by cooperative creep and grain-boundary sliding

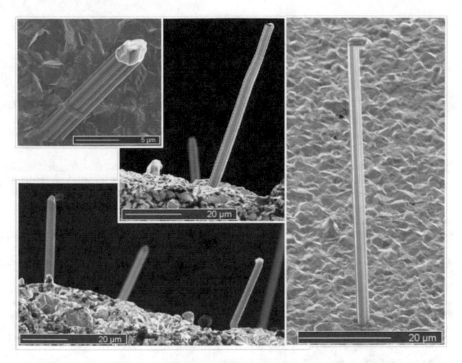

Fig. 12.53 Examples of Sn whiskers grown outwardly from the surfaces of Sn films electro-deposited onto Cu substrates. Scanning electron microscopy (SEM) images of specimens prepared with a focussed ion beam (FIB) workstation (taken from J. Stein, Ph.D. Dissertation, University of Stuttgart 2014)

"driving force for") the development of whiskers. Here one may be led by the experience of tooth-paste protruding from the top of a tooth-paste tube upon compressing the tube... In retrospection it is now easy to state that even the basis of such a concept is naive (see what follows and Sobiech et al. 2011).

The internal stress (also called "residual stress"; cf. Sect. 12.18) arising in the Sn film on a Cu substrate is due to the development of an intermetallic compound (IMC), here Cu_6Sn_5, at the original Sn/Cu interface, within the film and preferably along grain boundaries within the Sn film (see Fig. 12.54). As very common in thin films (see Sect. 11.3.3), the grain boundaries can be oriented largely perpendicular to the surface of the film, leading to columnar grains. As a consequence of the positive volume misfit in the film, associated with the development of IMC within the film (especially along the perpendicular grain boundary), a compressive stress parallel to the Sn/Cu interface (i.e. parallel to the surface of the film) occurs in the depth range of the IMC. Simultaneously a tensile stress (parallel to the surface of the film) is induced near the surface (where no IMC is present), which can be seen as the outcome of

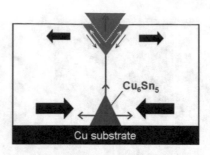

Fig. 12.54 Schematic model of whisker formation in a columnar Sn film on a Cu substrate. Significant penetration of Cu_6Sn_5 into the Sn coating along Sn grain boundaries, directly underneath a surface grain with inclined grain boundaries, leads in particular at these locations to pronounced stress–depth gradients; compressive stress prevails in the bottom part of the Sn coating, whereas in the top part of the Sn coating tensile stress is present. This stress gradient (induces a vacancy-concentration gradient and) tends to realize mass transport of Sn from the bottom to the surface of the Sn layer along the grain boundaries oriented more or less perpendicular to the film surface. At a location where a surface grain with inclined grain boundaries is present, such mass transport can be realized on a more or less permanent basis, because shear stresses acting along inclined grain boundaries (as a consequence of the planar state of stress) can realize shear along these inclined grain boundaries, leading to outward growth of such surface grains (i.e. whiskering). Red arrows denote directions of material transport; filled, black and white arrows denote components of (local) stress (taken from Sobiech et al. 2011)

the realization/maintenance of mechanical equilibrium. As a result, a *negative* stress-depth gradient exists in the film.

This negative stress-depth gradient evokes a vacancy-concentration depth gradient which leads to a mass flux (of Sn atoms) from the bottom part of the film to the top part of the film, likely via the grain boundaries at the low (room) temperature considered. At this stage of the discussion it must have become clear that the mechanism described here complies with the one given for the occurrence of Coble creep (see above in this section). To maintain the massive nature of the film, whisker growth occurs, to accommodate the Sn transported to the film surface along the more or less perpendicular grain boundary.

These whiskers then develop from those (few) surface adjacent grains which have grain boundaries inclined which respect to the surface (and are directly above a perpendicular grain boundary aiding Sn transport from the bottom of the film; see Fig. 12.54): only along the inclined grain boundaries of the surface grain considered, and in view of the prevailing, overall planar state of stress, shear stresses occur (see Fig. 12.54; shear stresses obviously do not occur along the majority, perpendicular grain boundaries; cf. Sect. 12.4). Then, together with the Sn transported from the bottom of the film, shear along these inclined grain boundaries allows outward growth of the surface grain bounded within the film by such inclined grain boundaries, by gliding (i.e. shearing) along these grain boundaries; the surface grain considered thus acts as the "whisker root". The mechanism discussed involves a *localized* process of grain-boundary sliding in the film, whereas in bulk material experiencing Coble creep *globally* occurring grain-boundary sliding was considered (see above in this section). This *local* process of grain-boundary sliding thus allows preservation of the Sn transport to the developing whisker and thereby realizes a *global* (i.e. in the entire film) relaxation of the stress gradient(s) (and consequently the vacancy-concentration gradient(s)).

As a side note it is remarked that, if one (possibly more than one) of the inclined grain boundaries confining the surface grain has a high mobility (as can hold for a high angle boundary and especially in the presence of a specific alloying element at the grain boundary (as Pb)), then extensive lateral (i.e. along the surface) growth of the developing Sn protrusion can occur, as a consequence of the lateral movement of this boundary in combination with the outward transport of Sn and grain-boundary sliding as described above. In that case not a "whisker" but a "hillock" is formed: a hillock has a height to width ratio smaller than 1, whereas the height to width ratio of a whisker obviously is very much larger than 1.

Evidently, it is the *negative nature of the stress gradient* that drives the whiskering. Indeed, as also proven experimentally, (i) even in the presence of a tensile stress in the surface region of the film, whiskering can occur provided the stress gradient is negative (Sobiech et al. 2011), and (ii) even in the presence of pronounced, compressive but homogeneous stress in the film (i.e. in the absence of a stress gradient), no whiskering occurs (Stein et al., 2014).

Finally, on the basis of the above understanding of whiskering, ways to avoid whiskering can be proposed:

- Removal of the internal stress gradient. It was found previously more or less empirically that a so-called post-bake, i.e. annealing of the film with substrate after its preparation, removed the tendency to whisker formation. On the basis of the above insight, it was now possible to conclude and verify that this is due to removal of the stress *gradient*.
- Producing a microstructure of the film constituted of mainly equiaxed grains. This implies the presence of very many more grain boundaries inclined with respect to the surface of the film than occurring in the usual thin film microstructure of largely columnar grains (see above). In the presence of the planar state of stress (see above), all inclined grain boundaries are subjected to shear stresses and are therefore able to accommodate grain-shape changes, as induced by (Coble) creep, by grain-boundary sliding. Then at very many places in the film stress relaxation can occur, as compared to the few locations within in the film with the largely columnar microstructure (with only few surface grains with inclined grain boundaries, i.e. the whisker roots). Alloying the Sn film with Pb was found to suppress whiskering already in the 1960s. The above-described mechanism now provides an explanation for this favourable effect, as it was found

that the microstructure of the Pb-alloyed Sn film is constituted of equiaxed grains. Unfortunately, Pb as alloying element is no longer allowed because of environmental restrictions (cf. discussion of Fig. 7.14 in Sect. 7.5.2.2). Ag as alloying element also enhances the presence of equiaxed grains in the microstructure.
- Obstructing stress (gradient) build-up, as by avoiding IMC formation within the Sn film. This can be done by introducing a barrier (for diffusion) interlayer between the Sn film and the substrate.

Creep by dislocation glide can be important at relatively low temperatures, where the thermal activation originates from the energy barrier for the gliding dislocation to overcome an obstacle in the glide plane on its way to an energetically more favourable position in an applied stress field. Creep by dislocation climb becomes important at high temperatures: the climb of the dislocation makes it possible to side-step an obstacle (and possibly thereafter to continue its movement by glide (and thus passing the obstacle), which in this situation then is not the rate determining step in the creep process). As holds for creep by diffusional flow (see above), grain-boundary sliding, to avoid crack formation as consequence of the changing grain shapes, is a necessary accompanying process of creep by dislocation climb/glide mechanisms.

It has been found, partly on an empirical basis, that the stress and temperature dependences of the constant strain rate, $d\varepsilon/dt$, in the steady-state stage of creep for the processes sketched above, can be summarized, for restricted ranges in temperature and applied stress, as follows:

$$\frac{d\varepsilon}{dt} = \text{const.}\sigma^m . \exp\frac{-Q_{\text{creep}}}{RT} \qquad (12.65)$$

with Q_{creep} as the activation energy of creep and m as a further material and prevailing creep mechanism dependent constant. It is no surprise, in view of the above discussion of creep mechanisms, that Q_{creep} for pure metals is often of the order of the activation energy of self diffusion. Note that equations of the above type allow predicting creep rates at temperatures and loading stresses other than used in the test experiments to determine the constants in the equation. This is scientifically not remarkable (e.g. see Chap. 8 on diffusion), but here it is of special engineering importance, as the (desired) life of components makes creep experiments under the stress and temperature conditions prevailing in practice unfeasible.

An important technological consequence of this section is the realization that, in order to avoid creep of a metal component for applications at elevated temperature, the absolute melting temperature of the alloy used must be high. Because at elevated temperature oxidation can be severe, a second requirement is good oxidation resistance. Both conditions are met by so-called *superalloys*: Co- and Ni-based austenitic (f.c.c. matrix) alloys containing substantial amounts of Cr for oxidation protection. A low grain-boundary density can be favourable for high creep resistance (cf. Coble creep as discussed above). Thus, single crystal turbine blades have been developed. Or turbine blades are made by a *directional solidification* process, which leads to components with grain boundaries more or less in parallel orientation, which is a hindrance to grain-boundary sliding if the loading direction is parallel to the dominant grain-boundary direction.

12.16.1 Superplasticity

For polycrystalline materials (metals and ceramics) constituted of sufficiently small, equiaxed grains (grain size of a few μm) and at temperatures higher than those where creep phenomena normally are first observed (say, $0.5\,T_m$ versus $0.4\,T_m$; see above) and for sufficiently low strain rate (10^{-2}/s–10^{-5}/s), excessively large permanent deformation by creep (tensile extension of the order of 1000%) can occur before failure takes place. This very large tensile elongation is the macroscopic characteristic of superplasticity.[23] The microscopic characteristic is that the grain shape and grain size do not change substantially during the occurrence of superplasticity, whereas, for example, in case of plastic deformation by cold drawing the grain size in the drawing direction is pronouncedly increased, to the same extent as the specimen as a whole, in association with a corresponding decrease in directions perpendicular to the drawing direction.

Grain-boundary sliding (see above) and grain rotation lie at the core of the superplastic behaviour. The resulting misfit within the specimen that would be caused by these primary mechanisms has to be relieved by secondary mechanisms. Several of these have been discussed partially controversially in the literature (e.g. see Courtney 1990) and, departing from explanations offered for steady-state (convential) creep (see above), have been based on local mass flow by diffusion (volume and/or along grain boundaries) and/or dislocation movement.

Superplasticity has distinct technological importance in the forming of metal alloys when (a) very complex shapes, which cannot be realized with methods applying large strain rates, have to be realized, and/or (b) very large total permanent deformations can advantageously be achieved in a single step. Of course, the requirements mentioned at the start of this subsection have to be met. So, (i) in view of the very low strain rate to be applied (in the range 10^{-2}/s–10^{-5}/s), the production time is relatively large and (ii) the microstructure of the alloy should hinder appreciable grain growth at the production temperature, and thus superplasticity is applied in practice to polyphase materials where grain boundaries can be pinned. Eutectic (cf. Sect. 7.5.2.2) and eutectoid (cf. Sect. 7.5.2.5) alloys can provide such fine and grain-growth resistant microstructures in particular.

---

[23] Superplasticity at extraordinarily high strain rates (up to 10^2/s), at temperatures close to the melting point of the bulk material, can be due to grain-boundary wetting by a liquid film (Straumal et al. 2003; see Sect. 9.4.5).

12.17   Failure by Fatigue

The loss of strength of a material, experienced in the course of time when loaded at a stress below the level which would cause instantaneous failure, is called fatigue, ultimately leading to fracture (cf. Sect. 12.16 dealing with creep, another mechanism leading to permanent damage accumulating with time). It has been claimed that fatigue is by far the most prominent failure mechanism occurring in components (e.g. as found in ships, airplanes, turbines and engines in general), and thereby, this topic is of extreme importance for the materials scientist.

Fatigue can occur applying a constant load (static fatigue), cyclic loading (cyclic fatigue) or an arbitrary load–time dependence. The occurrence of failure after some time indicates that permanent damage occurs during the loading. In the following the focus is on cyclic fatigue. After a certain number of cycles failure will occur: some plastic deformation (albeit this can be a very small amount) must occur during each cycle of loading. Fatigue leads to the nucleation of a crack, which propagates slowly during the cyclic stressing, until in a last cycle sudden, complete fracture occurs. The failure surface runs largely perpendicularly to the direction of the applied cyclic stress.

Fatigue leading to failure requires that a part of the stress cycle imposed must be of tensile nature (crack growth is obstructed by a compressive stress normal to the crack faces): fatigue leading to failure does not occur under compression. Often the *dynamic* cyclic stressing occurs superimposed on a state of *static* stress with stress components very much larger than the extremes of the stress cycle. In such a case one may unjustly tend to ignore the effect of the cyclic loading stress. A dramatic example of failure occurring in such a situation, due to the seemingly unsignificant, small cyclic loading stress, as compared to the state of static loading stress, is provided by the spectacular collapse of bridges, after many years of service, under the influence of cyclic fatigue due to the (varying intensity of) traffic crossing the bridge. Rotating machine parts, as crankshafts in motors, provide typical examples of workpieces sensitive to cyclic fatigue where the static loading stress is usually insignificant.

Two often applied types of cyclic loading (of, for example, a cylindrically shaped specimen with constant or varying (along the specimen-length axis) cross section) can be denoted as push–pull (tension–compression (uniaxial)) and rotating bending (flexural). In the first case an entire cross section is homogeneously loaded; in the second case the largest load stress occurs at the surface (with, in case of zero mean stress (see immediately above), zero stress at the centre of the specimen). In both cases: (i) the applied stress can vary symmetrically around a zero mean applied stress level, or (ii) around a specific mean applied stress level (Fig. 12.55a). It is usual to measure the number of cycles to failure, N, as function of, e.g. the maximum tensile stress applied, S. The results are presented as so-called S-N curves—usually S is plotted versus $\log N$—also named Wöhler curves: see Fig. 12.55b. At relatively high values of applied (maximum) stress, failure occurs after a relatively small number of cycles; one speaks of low-cycle fatigue (fatigue life smaller than, say, 10^3 cycles; see Fig. 12.55b). At relatively low values of applied (maximum) stress fatigue progresses

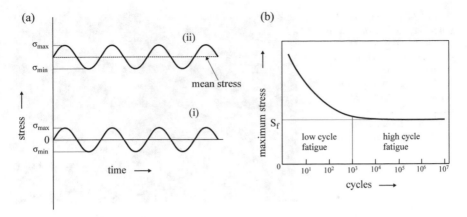

Fig. 12.55 **a** Stress/time curves for cyclic loading: (i) zero-mean stress and (ii) non-zero mean stress. **b** Schematic Wöhler curve (S, the applied maximal tensile stress (in a stress cycle) vs. log N, with N as the number of cycles); S_f denotes the endurance/fatigue limit (the maximal tensile load (in a stress cycle) not causing failure)

very slowly, if at all, and failure may not occur: a fatigue limit, also called fatigue strength or endurance limit, S_f, is observed (horizontal part of the S-log N curve). Such clear fatigue limits are observed for steels. It has been shown that many non-ferrous materials (e.g. aluminium alloys) do *not* exhibit a true fatigue limit (i.e. a horizontal part in the Wöhler curve does not occur), but a maximal applied stress value can be found below which failure occurs beyond 10^7 cycles, and this stress value then yet is called the fatigue strength. Fatigue beyond, say, 10^4 cycles is called high-cycle fatigue; fatigue testing is in practice terminated at 10^7 cycles to keep the time for fatigue testing reasonable.

The onset of fatigue involves the occurrence of localized plastic deformation leading to crack initiation, even while the component is, so to speak, macroscopically only elastically stressed. This can happen at the external surface (most usually, especially if a notch is present) or at the interface with an inclusion. The initiation of the crack occurs during the tensile part of the loading cycle, for the location of the material considered. A second stage corresponds to relatively slow crack growth largely perpendicular to the direction of tensile stress applied. Finally, catastrophic fracture takes place very fast, say during the last cycle: the remaining cross section has become too small, and thus, the local applied tensile stress has become too large, to resist component disintegration, or, alternatively, the stress-intensity factor, typifying the stress field at the crack tip, has exceeded a critical value for tensile fracture to occur (see the discussion of Eq. (12.62)). Inspection of the failure surface by scanning electron microscopy (cf. Sect. 6.9) may reveal numerous striations along that part of the failure surface that was created during the slow crack-propagation stage, where each striation is thought to correspond with an incremental part of crack growth during a single stress cycle. Such an observation is a clear indication of failure by fatigue (see Fig. 12.56). However, in particular with polymers, such striations,

Fig. 12.56 SEM image of the failure surface of a failed practical workpiece of carbon steel. The striations, indicative of the stage of slow fatigue-crack growth, can be clearly observed (micrograph made by S. Kühnemann, Max Planck Institute for Metals Research)

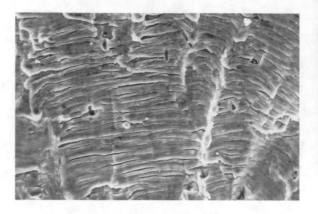

corresponding with the crack-growth contributions by single stress cycles, are often not formed. Whereas for low-cycle fatigue (high applied stress; short fatigue life) most stress cycles correspond to crack propagation, high cycle fatigue (low applied stress; long fatigue life) is characterized by a relatively long crack-initiation stage.

Crack initiation is one of the least understood phenomena in fatigue. The crack-nucleation event usually takes place at the surface (most frequently, e.g. at surface irregularities as scratches and steps)[24] or at an interface (as an inclusion/matrix interface) within the material. Local inhomogeneities, in association with stress concentrations, can give rise to plastic flow, under the action of cyclic stressing, ultimately leading to crack initiation or extension of an initially present small crack. Hence, the first stage of fatigue fracture is governed by plastic flow, rather than by tensile fracture as in the slow crack-propagation stage. Consequently, the initial fracture plane, in correspondence with the process of plastic flow characterized by yielding preferably in the direction of the largest shear stresses, can be inclined with respect to the cyclic stress loading axis (cf. Sects. 12.4 and 12.9.3); the next stage of slow crack growth, in correspondence with the process of tensile fracture, proceeds in directions normal to the cyclic stressing loading axis.

Recognizing that the slow crack-propagation stage is characterized by tensile fracture, it is no surprise to observe that the crack-propagation rate depends on the range experienced for the stress-intensity factor K (cf. Eq. (12.61)) at the tip of the propagating crack:

$$\Delta K = K(\sigma_{\max}) - K(\sigma_{\min})$$

where $\sigma_{\max}$ and $\sigma_{\min}$ take the role of σ_{nom} (cf. Eq. (12.61)) at the moments of maximal and minimal loading stress during a loading cycle (see Fig. 12.55a). Thus:

$$\frac{dl}{dN} = \text{const.}(\Delta K)^m \tag{12.66}$$

---

[24] This needs not hold for surface-hardened (case hardened) metallic components; see the *Epilogue* after Sect. 12.18, as the conclusion of this book.

where const and m are (also material dependent) constants. This equation is known as Paris' law (Paris and Erdogan 1963). Usual values for the exponent m are in the range 2–7. An estimate for fatigue life, in terms of the number of cycles till failure, can now be obtained by integration of Eq. (12.66) from the length l_0, of the initially present crack (the defect present before fatigue, in the sense of the discussion in Sect. 12.15), or of the as nucleated crack (see above), to the critical crack length for fracture l_{crit} (cf. Eqs. (12.60) and (12.62)). Thus, this procedure actually gives an estimate for the number of cycles in the slow crack-propagation stage, before catastrophic final fracture occurs. Note that the estimate for fatigue life obtained in this way presupposes that the crack-propagation stage as described by Eq. (12.66) dominates fatigue life, and thus this estimate could be realistic for low-cycle fatigue in particular.

At this stage it is appropriate to remark that current engineering practice is to accept the presence of some, even propagating cracks in components, but their propagation rate (estimated on the basis of Eq. 12.66) should be that low that they will not reach a length inducing final catastrophic fracture during the desired service life of the component.

In general, notches, surface roughness and inclusions act as local stress raisers and thereby provide sites for crack nucleation. The type of loading most frequently considered is rotating bending, which involves that the cycling load stress is highest at the surface, which leads to crack initiation at or near the surface. Consequently, improving the mechanical strength of the surface region of a component can significantly enhance the fatigue strength of a component. This is the background of a number of "surface engineering" methods as shot peening, induction hardening and thermochemical methods, as carburizing and nitriding, which all lead to improved mechanical properties in the surface region of the component through the effect of induced compressive residual (macro)stress (see next paragraph) and/or microstructural change (e.g. the precipitation of hardness/yield strength increasing precipitates); see the *Epilogue* after Sect. 12.18 (Mittemeijer and Somers 2015).

If a mean applied stress different from zero (see above) occurs, the maximal tensile stress in a loading cycle can be larger (if the mean stress is tensile) or smaller (if the mean stress is compressive). Obviously, fatigue life is increased if the mean stress is compressive (and fatigue life is decreased if the mean stress is tensile). This is the reason to induce compressive residual (macro)stress (see above and see Sect. 6.10.2 for macrostress/strain versus microstress/strain) by microstructural manipulation in those parts of the components which are most severely subjected to fatigue (as the surface regions; again, see the *Epilogue* after Sect. 12.18). The effect of the prevailing residual (macro)stress can be conceived as that of a mean stress as discussed here. Various concepts have been proposed to describe the effect of a mean stress on cyclic fatigue. According to the Goodman approach, the stress necessary to induce final failure after N cycles, i.e. the fatigue limit, changes from S_f to $S_{m,f}$ in the presence of a mean stress σ_m, either externally applied (load stress) or internally applied (residual (macro)stress), according to:

$$S_{m,f} = S_f \cdot \left(1 - \frac{\sigma_m}{\sigma_{UTS}}\right) \qquad (12.67)$$

The occurrence of the ultimate tensile strength (cf. Sect. 12.9) in this equation expresses the dependence of the effect of mean stress on microstructure and material.

Most of the research on fatigue has been done with metals (usually alloys); fatigue is a very often occurring failure mechanism for metallic components. As a rule of thumb it can be said that the fatigue strength of metals equals about one half of the U.T.S. (the hardness is about three times the U.T.S.; cf. Eq. (12.51)).

The description of fatigue, as given above, is not essentially different for other material classes as ceramics and polymers, but specific microstructural differences have to be considered.

Ceramics at low temperatures are brittle: a slow crack-propagation stage cannot occur. "Fatigue" thereby becomes almost immediately fracture, as discussed in Sect. 12.15. However, this picture appears to be too crude (Wachtman 1996). Ceramics, as brittle materials, have a low (fracture) toughness (the ability of the material to absorb energy until fracture; see also Eqs. (12.61)–(12.63) and their discussion in Sect. 12.15). The toughness of ceramic materials can be improved by their polycrystalline nature (a crack proceeding along a cleavage plane in a crystal deflects at the grain boundary with an adjacent crystal), or by the presence of a second phase in the form of fibres or elongated grains which can bridge the (propagating) crack and act as ligaments which carry some load (cf. craze yielding of thermoplastics; see Fig. 12.26 and its discussion in Sect. 12.9.3). Upon cyclic loading (at a stress level below that leading to instantaneous brittle fracture), the ceramic material is damaged by degradation of in particular the toughening elements in the microstructure, as the bridging ligaments, rather than by plastic deformation. Life-time predictions may be based on Paris' law as discussed above (Eq. (12.66)); however, for ceramics values for the exponent m are very high (from 10 up to 40; to be compared with those observed for metals (see below Eq. (12.66)) and for many cases (materials and microstructures) unknown.

Polymers exhibit a fatigue behaviour not basically different from metals. However, for not significantly cross-linked, long-chain polymers above the glass transition temperature, it is essential to recognize the importance of their viscoelastic behaviour related to interchain sliding (discussed in Sect. 12.7). If the frequency (=reciprocal of the time period) of stress cycling is such that the time needed for the viscoelastic component of strain to unfold is of the same order of magnitude as the time needed for one stress cycle, only a certain extent of the maximally possible viscous flow can develop; the viscoelastic strain cannot "keep up" (i.e. cannot stay "in phase") with the stress: "damping" takes place. The resulting irreversible energy loss during one stress cycle is an elastic deformation effect and is called elastic mechanical hysteresis (cf. the discussion in Sect. 12.7). The energy lost is dissipated as heat. Even fatigue of metallic workpieces at room temperature can give rise to energy dissipation in the form of heat and as consequence the component heats up. This effect is much more pronounced for the polymers exhibiting mechanical hysteresis. At relatively elevated temperatures also failure mechanisms controlled by diffusional flow can operate. This requires high temperatures for metals. However, in view of the melting temperatures of polymers, room temperature is already a relatively high temperature for polymers. Hence, for the polymers discussed here the interaction of creep and fatigue is unavoidable at room temperature applications.[25] Regarding the creep–fatigue interaction, it is admitted that fundamental, generally valid model descriptions lack; one mostly relies on empirical relationships. A few remarks can now be made regarding the specific "fatigue" behaviour of polymers:

- If fatigue dominates creep, craze formation (cf. Sect. 12.9.3) can be the mechanism responsible for fatigue-crack initiation in polymers; shear banding gains importance at higher temperatures and in the low-cycle fatigue region (cf. Fig. 12.55b).
- The cyclic loading of polymers differs strongly (much more than for metals) from that observed in tensile testing as discussed in Sect. 12.9. Under cyclic stressing polymers usually soften and never harden, whereas metal alloys can harden (initially "soft" alloy) or soften (initially "hard" alloy). This softening effect for polymers occurs at temperatures and stress-cycle frequencies relevant for practical applications and thereby sets an important constraint for the application of polymers as structural materials.
- At a larger stress-cycle frequency for a polymer, a larger heating can occur due to the mechanical hysteresis and its softening can be more pronounced. Thereby "fatigue" failure becomes a direct consequence of the induced temperature increase: the cyclic stressing parameters at the initial temperature would not have led to failure. Metal fatigue is largely stress-cycle frequency independent.
- The heat produced by mechanical hysteresis in the interior of a polymeric component can be transferred to the surroundings relatively more for smaller workpieces. Thereby the fatigue strength becomes specimen volume/geometry dependent: the smaller the volume/size, for the same geometry, the higher the fatigue strength (for size dependent fatigue strength of case-hardened metallic components, see the *Epilogue* after Sect. 12.18).

12.18 Residual, Internal Stresses

Residual stresses are *internal* stresses which are self-equilibrating [26] stresses existing in materials at uniform temperature and without *external* loading. Such stresses, which can be of macroscopic nature or microscopic nature (see the discussion on macrostrain/stress and microstrain/stress in Sects. 6.10.1 and 6.10.2 and see Fig. 6.31), can be introduced in very many different ways. A few examples are discussed below (see Fig. 12.57):

---

[25] For metallic components subjected to fatigue loading at high temperatures not only the fatigue–creep interaction has to be recognized: the simultaneously occurring oxidation, as a striking example of profound interaction with the environment that occurs as well, can drastically influence service life too.

[26] The sum of the forces acting on a cross section through the body must be nil: force balance. Similarly, a balance of moments is required.

(i)   A growing precipitate in a matrix can be associated with the development
       of a pronounced residual *micro*stress field around the precipitate as the
       consequence of the volume misfit between precipitate and matrix.[27] These
       misfit-stress fields can on average obstruct dislocation movement and thereby
       enhance the yield strength of the material (this effect thus can be incor-
       porated under "precipitation/dispersion strengthening"; cf. Sect. 12.14.4).
       Note that also a temperature change for a system of second phase parti-
       cles/precipitates in a matrix can give rise to volume misfit between second
       phase particles/precipitates and matrix, if the thermal expansion coefficients
       of second phase particles/precipitates and matrix are different.

(ii)   The linear thermal expansion coefficient is anisotropic for noncubic mate-
       rials. Hence, a temperature change experienced by a massive, single phase,
       polycrystalline specimen of noncubic material will lead to a state of residual

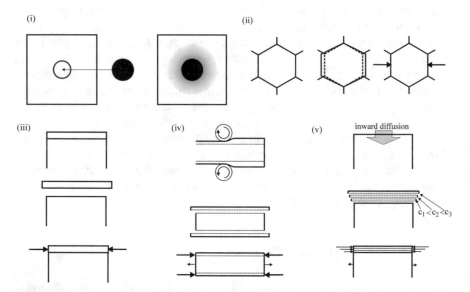

Fig. 12.57 Origins of residual stresses: (i) Misfit-stress-field development around a precipitating,
second phase particle in a supersaturated matrix. (ii) Microstress development due to anisotropic
thermal expansion of grains in a massive specimen upon cooling/heating. (iii) Stress at room temperature in a thin layer on a substrate, e.g. developing upon cooling the layer/substrate system from
elevated temperature, where the layer was grown, in a possibly stress-free state, as a consequence of
layer and substrate having different thermal expansion/shrink coefficients. (iv) Compressive stress
development in the surface region of a specimen upon cold rolling. The specimen tends to elongate
laterally at the surface; the constraint by the core then leads to the development of compressive
stress in the surface adjacent region. (v) Compressive stress-depth profile development caused by
inward diffusion of an alloying element causing depth dependent crystal-structure expansion in the
surface region of a specimen

---

[27] It has not been recognized often that elastic accommodation of the (growth induced or thermally
induced) misfit between a second phase particle/precipitate and the matrix gives (also) rise to a
(hydrostatic) *macro*stress component (see Mittemeijer 2006; Eshelby 1956).

microstress, as the thermal expansions of the (neighbouring) grains in the specimen are incompatible: high, locally strongly varying residual microstresses can occur.

(iii) A thin layer on a thick substrate can have a linear coefficient of thermal expansion distinctly different from the substrate underneath. If layer and substrate were in equilibrium (stress free) at an elevated temperature (e.g. the layer deposition/growth temperature), then either a compressive macrostress or a tensile macrostress develops in the layer upon cooling, depending on the linear coefficient of thermal expansion of the layer being either smaller or larger than that of the substrate.

(iv) Residual stresses can result from inhomogeneous mechanical working, e.g. machining, grinding and cold rolling. Nonuniform plastic deformation leads to a state of residual stress. Suppose the surface adjacent part of a component tends to elongate by tensile deformation and the core of the component is unaffected. Then, since the whole component remains intact, the surface adjacent part and the core of the component must be strained elastically. As a result a compressive stress parallel to the surface occurs in the surface adjacent region, which is compensated by a modest tensile stress in the larger, core part of the component (residual stresses are self-equilibrating; see the first sentence of this section).

(v) Inward diffusion of an alloying element can lead to the development of a residual macrostress-depth profile. This is, for example, the case when pure ferrite (α-Fe; b.c.c.) is nitrided or pure austenite (γ-Fe; f.c.c.) is carburized. At a certain stage of nitriding/carburizing a concentration profile of dissolved alloying element occurs. The dissolved alloying element atoms (nitrogen or carbon atoms at octahedral interstices in both crystal structures (b.c.c. and f.c.c.) cause a volume increase of the material, if unconstrained, which is the larger the larger the concentration of alloying element. Because the surface adjacent material, comprising the concentration-depth profile (of nitrogen/carbon), is part of (cohesively bonded to) the entire specimen, such local expansion cannot occur laterally: the core of the specimen, which does not contain dissolved alloying element atoms, counteracts such lateral expansion. (In fact, this situation parallels the one discussed under (iii) above, where the consequence of different thermal expansion coefficients of layer and substrate upon temperature change is considered). As a result, if full elastic accommodation of the specific volume misfit occurs, a compressive residual macrostress, parallel to the surface, develops in the surface adjacent region of a planar (nitrided/carburized pure iron) specimen according to (see Sect. 8.8.2 where a more extensive discussion of "self-stress", as induced by diffusion, is presented):

$$\sigma(z) = \frac{\beta E}{(1 - v)} \cdot [\langle c \rangle - c(z)] \tag{8.60}$$

where σ denotes the macrostress, z is the depth below the surface, c is the concentration of dissolved alloying element (nitrogen/carbon), $<c>$ is the average concentration of dissolved alloying element in the whole specimen, β is the lattice-expansion coefficient (describing the unconstrained increase of the lattice parameter upon uptake of nitrogen/carbon), E is Young's modulus and v is the Poisson constant. Clearly, at the surface a distinct, compressive residual macrostress parallel to the surface occurs (the surface adjacent region of the specimen tends to expand laterally, which is obstructed by the core of the specimen. The compressive macrostress level decreases with increasing distance to the surface, because the concentration of dissolved alloying element decreases with increasing depth. The condition of mechanical equilibrium (again: residual stresses are self-equilibrating; see the first sentence of this section) causes the presence of a relatively small tensile macrostress parallel to the surface of the specimen in the core of the (planar) specimen, where no dissolved alloying element is present (see Fig. 8.16).

The virtue of residual *micro*stresses, as for example due to locally varying misfitstress fields around precipitates in the matrix (see under (i) above), is an increase of the local intrinsic strength of the specimen, as dislocation movement is on average hindered by the microstress fields. This is illustrated well by the increase of hardness occurring upon increase of the microstress: see Fig. 12.58, where, for a nitrided alloyed steel and a nitrided carbon steel subjected to precipitation of (misfitting) nitride particles as a result of nitriding, the nitriding induced increase of hardness has been plotted versus the nitriding induced increase of microstrain.

The virtue of compressive *macro*stress, as for example invoked by nitriding/carburizing of iron and steels, seems obvious. Crack initiation and growth (e.g. at and perpendicular to the surface) by fatigue will be counteracted by a compressive macrostress parallel to the surface (see also Fig. 6.30): as a result the fatigue strength can be increased very considerably if a compressive macrostress is present in the surface region of the component. This is discussed in more detail in the *Epilogue* after this section.

Residual *macro*stresses can also be detrimental. For example, the removal of thin surface layers causes a redistribution of the residual macrostresses in the remaining body, which can lead to (visible) distortion, which is unacceptable in case of, e.g. precision machine parts. Residual stress can also enhance chemical attack in a specific environment: stress corrosion cracking leading to failure.

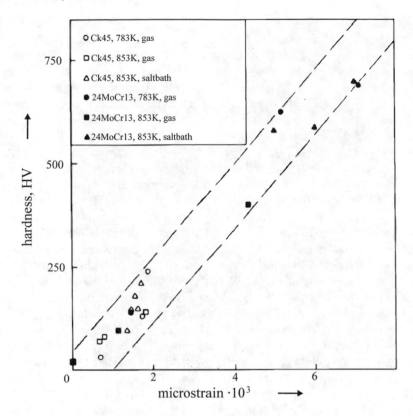

Fig. 12.58 Increase of the hardness (Vickers; cf. Sect. 12.13) as a function of increase of the microstrain (as measured from the X-ray diffraction line broadening; cf. Sect. 6.10.1) for two types of nitrided steels, with respect to the hardness and microstrain, respectively, before nitriding. The increases of hardness and microstrain are due to alloying element (Cr, Mo) nitride precipitates in the ferrite matrix of the alloyed steel (24CrMo13) and dissolved nitrogen and/or precipitated iron nitrides in the ferrite matrix of the carbon steel (Ck45). The nitridings were performed in either a cyanide/cyanate salt bath or in an ammonia-based gas atmosphere (taken from E. J. Mittemeijer, in: "Case-Hardened Steels: Microstructural and Residual Stress Effects" (Ed. D. E. Diesburg), TMS-AIME, Warrendale, Pa., U.S.A., 1984, pp. 161–187)

Epilogue: The Essence of Materials Science;
Optimizing the Fatigue Strength of Ferritic Steels by Nitriding
Nitriding of ferritic steels implies the hardening of the surface adjacent region of the iron-based component (steel) by the precipitation of nitrides in the surface region of the ferritic matrix (the surface adjacent part of the component is called the "case"; one speaks of "case hardening"[28]; see

---

[28] Another classical case-hardening method is carburizing. Carburizing, as nitriding, is a thermo-chemical surface treatment to improve the mechanical properties (wear, fatigue) which depend on

Mittemeijer and Somers 2015). It is important to realize that in the surface adjacent region not only a hardening effect, related tot the development of residual microstresses (see under (i) above), is achieved, but also a residual compressive macrostress is invoked. The last effect is a straightforward consequence of the tendency to volume expansion in the nitrided region due to the precipitation of alloying element nitrides, for example CrN and AlN and/or $Cr_xAl_{1-x}N$, implying a desired lateral expansion of the surface adjacent region of the steel component, inducing, because of maintenance of component integrity, a compressive macrostress parallel to the surface (see under (v) above).

The presence of microstresses in the nitrided surface region, leading to hardness increase in this region, implies that a static strength-depth profile occurs in the specimen. Thus, the intrinsic fatigue strength of the material of the component will be increased in the surface region, as compared to the unnitrided core of the component, due to already only the microstresses present in the surface region. The local fatigue resistance in the surface region is further enhanced by the compressive macrostress parallel to the surface in the surface region, which effect can be described according to the Goodman approach (Eq. (12.67)). As a result the component can be characterized by a fatigue-strength depth profile, $S_f(z)$, as schematically indicated in Fig. 12.59 by the full line.

Now, a case of rotating bending loading, which is of great practical importance (rotating machine parts, as e.g. crankshafts), is considered here for a case-hardened, cylindrical component possessing a fatigue-strength depth profile, $S_f(z)$, as discussed above and indicated in Fig. 12.59a (Mittemeijer 1983). The surface of the component is subjected to the highest applied load that cycles between equal maximal compressive and maximal tensile stress values upon rotation of the component (mean applied stress is zero); the centre (line) of the component experiences zero stress. This is illustrated in Fig. 12.59a by the dashed lines denoted by S_a^0 and S_a, representing the (local) load stress in case of maximal tensile loading before case hardening (S_a^0) and after case hardening (S_a). The situation sketched in Fig. 12.59a, the moment of maximal applied tensile stress at the surface, is the situation of most severe loading: fatigue-crack initiation requires applied *tensile* stress.

---

the quality of the surface adjacent material of the component (see also the "*Intermezzo: Thermochemical surface engineering; nitriding and carburizing of iron and steels*" in Sect. 4.4.2). Carbon and nitrogen are offered by an outward, e.g. gaseous, atmosphere and diffuse into the surface region of the component at elevated temperature. In case of carburizing the treatment is carried out at higher temperatures such that the matrix is austenitic. Upon quenching a hard, martensitic microstructure is induced in the carburized case. The (tendency to) volume expansion associated with the martensite formation contributes to the development of a compressive macrostress parallel to the surface in the surface region. In case of nitriding the treatment is carried out at a lower temperature such that the matrix is ferritic. The precipitation of nitrides in the nitrided surface region leads to the high hardness of the nitrided case. The (tendency to) volume expansion by the precipitation of the nitrides contributes to the development of a compressive macrostress parallel to the surface in the surface region.

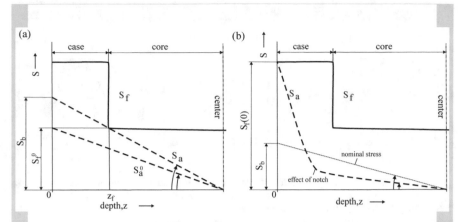

Fig. 12.59 Fatigue strength-depth profile ($S_f(z)$) and applied maximal tensile loading stress-depth profile during rotating bending fatigue (before case hardening: $S_a^0(z)$; after case hardening: $S_a(z)$) for two cylindrical workpieces: **a** unnotched workpiece and **b** notched workpiece. See text for details (taken from Mittemeijer 1983)

The two cases of applied maximal tensile load shown in Fig. 12.59a (before case hardening: S_a^0; after case hardening: S_a) can now be discussed as follows:

(i)   Before case hardening: the applied maximum stress at the surface, $S_a^0(z = 0)$, i.e. the fatigue limit, is equal to S_f^0, which is the depth independent fatigue strength of the component before case hardening. If S_a^0 at the surface becomes larger than S_f^0, crack initiation takes place at the surface of the not case-hardened component.

(ii)   After case hardening: the fatigue limit (indicated by S_b in the figure) is also given by the applied maximum stress at the surface, $S_a(z = 0)$, which is clearly larger than before case hardening. If S_a at the surface becomes larger than S_b, crack initiation in the case-hardened component takes place *not* at the surface but at the depth where the case/core transition occurs (z_f), because there then the local applied load becomes larger than the local fatigue strength.[29] At the surface a surplus amount of S_f ($z = 0$) $- S_b$ in fatigue strength is not utilized.

---

[29] According to the idealized sketch in Fig. 12.59a, the value of the local fatigue strength of the case-hardened component at the case/core transition, $S_f(z_f)$, equals the fatigue strength of the component before case hardening (case (i) above: the fatigue strength of the material before case hardening does not depend on depth), and thus, $S_f(z_f) = S_f^0$. In reality, the fatigue limit for failure initiating at the surface can be smaller than the intrinsic fatigue limit of the material, e.g. due to atmospheric influences. This is one reason why $S_f(z_f)$ can be larger than the value measured for S_f^0 for the not case-hardened component.

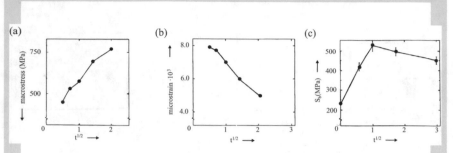

Fig. 12.60 Development of **a** macrostress, **b** microstrain and **c** fatigue limit as function of nitriding time at 580 °C for a notched workpiece of the steel En40B = 24CrMo13 (taken from Mittemeijer 1985)

The component considered in Fig. 12.59a is an unnotched workpiece. If the component contains a notch, the nominally applied stress, at the location of the notch, is not the actual load stress: stress concentration occurs (cf. Eq. (12.57)). The real and nominal applied stress-depth profiles are sketched in Fig. 12.59b (thick and thin, respectively, dashed lines) for the level of applied stress that fatigue-crack initiation occurs: the fatigue limit for the notched component is given by S_b, i.e. the value of the applied nominal stress at the surface leading to crack initiation: if the nominal applied stress at the surface becomes larger than S_b, the actual applied stress at the surface becomes larger than $S_f(z = 0)$ and crack initiation in the notched, case-hardened component does take place at the surface, as holds for the not case-hardened workpiece, but the fatigue limit for the case-hardened component is much larger (increases of more than 100% are possible; see Fig. 12.60c).

The usual, brief explanation of the virtue of case hardening, for spectacular increase of the fatigue limit, relies on the invocation of a residual compressive *macro*stress parallel to the surface in the surface region, which almost visibly (cf. Fig. 6.30) counteracts fatigue-crack initiation and growth. In the above the contribution of the *micro*stresses for the local increase of S_f has been made clear as well. The importance of both macro- and microstresses becomes particularly clear considering the effect of nitriding time on the fatigue limit of nitrided alloyed steels (Mittemeijer 1985). The changes of the compressive macrostress parallel to the surface and the average microstrain (the microstresses are usually represented by some average microstrain as determined from the broadening of X-ray diffraction lines; see Sect. 6.10.1 and Eq. (6.38)) are shown as a function of nitriding time at 580 °C for the steel 24CrMo13 (= En40B) in Fig. 12.60a and b, respectively. Upon nitriding alloying-element nitride precipitates develop in this steel. The measurements shown in Fig. 12.60a and b pertain to the very surface region of the specimens. Hence, because nitrogen saturation of the surface region can be shown to be established in this region for a time shorter than the time of the first data in these figures, the changes observed for macrostress and average microstrain with increasing treatment time must have to do with microstructural development without compositional change: i.e. a so-called ageing phenomenon (cf. Sect. 9.4.1) lies at the core of these phenomena (of course continued nitriding leads to extension to larger depths of the nitrided case by the nitrogen taken up).

The precipitation of alloying element nitride starts with the formation of very small, extremely thin platelets (at most a couple of atomic layers thick) in the nitrided ferrite matrix. At this stage largely elastic accommodation of the volume misfit of tiny nitride precipitate and ferritic matrix occurs; i.e. the platelets are largely coherent with the matrix, severely strained and surrounded by long-range stress fields of strongly varying nature (e.g. see Figs. 6.19a and 6.22a). This already explains that at this stage relatively large values for the average microstrain in the matrix occur (see Fig. 12.60b). Upon continued nitriding, which implies an ageing treatment for the nitrided surface region, the platelets coarsen and become semicoherent/incoherent with the matrix. Then the volume misfit of the precipitates is appreciably/largely accommodated plastically, e.g. by the development of dislocations at the precipitate/matrix interface. Only a part of the volume misfit is still accommodated elastically and the remaining stress fields around the now coarsened precipitates are of (more) short-range nature. Consequently, the average microstrain in the surface region reduces upon continued nitriding, as is observed (Fig. 12.60b). Only after precipitate coarsening, leading to incoherency of precipitate and matrix, the equilibrium structure of the precipitate can be realized and the tendency to volume increase of the nitrided case increases, which effect, because of the counteraction to lateral extension of the nitrided case by the unnitrided core, induces a compressive macrostress in the surface region increasing with time, as is observed (Fig. 12.60a; note that plastic deformation in a surface region of a component can (also) lead to macrostress development by the elastic interaction of case and core as indicated under (iv) above).

The fatigue limit increases with both an increasing compressive macrostress and increasing microstresses (see the discussion at the begin of this *Epilogue*). Considering the results shown in Figs. 12.60a and b, one may then wonder what their net effect on the fatigue limit in the case considered is. For notched components, subjected to rotating bending, fatigue-crack initiation takes place at the surface (see above discussion of Fig. 12.59b). Fatigue-crack initiation at the surface is controlled by the fatigue strength S_f at the surface which depends on both the macrostress at the surface and the microstresses at the surface, for which results as function of treatment time are shown in Fig. 12.60a and b. The measured values for the fatigue limit of notched components of the considered nitrided steel are shown in Fig. 12.60c as function of treatment time. It appears that initially a distinct increase of the fatigue limit occurs, which can be ascribed to an increasing compressive macrostress, but upon continued nitriding the decreasing microstresses cause a significant decrease of the fatigue limit. Hence, as the result of an investigation as discussed here, an optimal nitriding time, to induce the largest fatigue limit possible for the case considered, can be defined.

This example has been presented here at some length, in a way as the conclusion of this book, because it provides a beautiful illustration of the essence of materials science: optimal properties are achieved by microstructural manipulation based on fundamental understanding, which is the contrary of the accumulation of phenomenological knowledge, which lies at the roots of any science.

References

General

T.H. Courtney, *Mechanical Behaviour of Materials* (McGraw-Hill Publishing Company, New York, 1990)

G.E. Dieter, *Mechanical Metallurgy* (McGraw-Hill Book Company, New York, 1961)

P. Haasen, *Physical Metallurgy* (Cambridge University Press, Cambridge, 1978)

O. Hoffman, G. Sachs, *Introduction to the Theory of Plasticity for Engineers* (McGraw-Hill Book Company, New York, 1953)

R.W.K. Honeycombe, *The Plastic Deformation of Metals* (Edward Arnold Publishers Ltd., London, 1968)

W.F. Hosford, *Mechanical Behaviour of Materials* (Cambridge University Press, Cambridge, 2005)

D. Hull, D.J. Bacon, *Introduction to Dislocations*, 4th edn. (Butterworth-Heinemann, Oxford, 2001)

M. Kato, Hall-Petch relationship and dislocation model for deformation of ultrafine-grained and nanocrystalline metals. Mater. Trans. **55**, 19–24 (2014)

S. Suresh, *Fatigue of Materials* (Cambridge University Press, Cambridge, 1991)

S.P. Timoshenko, J.N. Goodier, *Theory of Elasticity*, 3rd edn. (McGraw-Hill Book Company, Singapore, 1982)

P.F. Thomason, *Ductile Fracture of Metals* (Pergamon Press, Oxford, 1990)

S.P. Timoshenko, *History of Strength of Materials* (McGraw-Hill Book Company, New York, 1953)

J.B. Wachtman, *Mechanical Properties of Ceramics* (Wiley, New York, 1996)

G. de With, *Structure, Deformation, and Integrity of Materials, Volumes I and II* (Wiley-VCH Verlag, Weinheim, 2006)

Specific

A.H. Cottrell, B.A. Bilby, Dislocation theory of yielding and strain ageing of iron. Proc. Phys. Soc. A **52**, 49–62 (1949)

A.T. Crumm, J.W. Halloran, Negative Poisson's ratio structures produced from zirconia and nickel using co-extrusion. J. Mater. Sci. **42**, 1336–1342 (2007)

M. Dao, L. Lu, R.J. Asaro, J.T.M. de Hosson, E. Ma, Toward a quantitative understanding of mechanical behavior of nanocrystalline metals. Acta Mater. **55**, 4041–4065 (2007)

J.D. Eshelby, The continuum theory of lattice defects. Solid State Phys. **3**, 79–144 (1956)

L.J. Gibson, M.F. Ashby, *Cellular Solids; Structure and Properties*, 2nd edn. (Cambridge University Press, Cambridge, 1997)

G.N. Greaves, A.L. Greer, R.S. Lakes, T. Rouxel, Poisson's ratio and modern materials. Nat. Mater. **10**, 823–837 (2011)

A.A. Griffith, The phenomena of rupture and flow in solids. Philos. Trans. R. Soc. Lond. **221A**, 163–198 (1920)

E.O. Hall, The deformation and ageing of mild steel. Proc. Phys. Soc. (Lond.) **64B**, 747–753 (1951)

F. Javaid, Y. Xu, K. Durst, Local analysis on dislocation structure and hardening during grain boundary pop-ins in tungsten. J. Mater. Sci. **55**, 9597–9607 (2020)

X. Ke, J. Ye, Z. Pan, J. Geng, M.F. Besser, D. Qu, A. Caro, J. Marian, R.T. Ott, Y.M. Wang, F. Sansoz, Ideal maximum strengths and defect-induced softening in nanocrystalline-nanotwinned metals. Nat. Mater. **18**, 1207–1214 (2019)

M. Kunert, *Mechanical Properties on Nanometer Scale and Their Relations to Composition and Microstructure*, Ph.D. Dissertation, University of Stuttgart (2000)

M. Kunert, B. Baretzky, S.P. Baker, E.J. Mittemeijer, Hardness-depth profiling on nanometer scale. Metall. Mater. Trans. A **32A**, 1201–1209 (2001a)

M. Kunert, O. Kienzle, B. Baretzky, S.P. Baker, E.J. Mittemeijer, Hardness-depth profile of a carbon-implanted Ti-6Al-4 V alloy and its relation to composition and microstructure. J. Mater. Res. **16**, 2321–2335 (2001b)

Y. Kuru, M. Wohlschlögel, U. Welzel, E.J. Mittemeijer, Large excess volume in grain boundaries of stressed, nanocrystalline metallic thin films: its effect on grain-growth kinetics. Appl. Phys. Lett. **95**, 163112 (2009)

R. Lakes, Foam structures with a negative Poisson's ratio. Science **235**, 1038–1040 (1987)

A. Li, I. Szulfarska, Morphology and mechanical properties of nanocrystalline Cu/Ag alloy. J. Mater. Sci. **52**, 4555–4567 (2017)

E.J. Mittemeijer, *X-ray diffraction analysis of the microstructure of precipitating Al-based alloys*, in *Analytical Characterization of Aluminum, Steel, and Superalloys,* ed. by D. Scott MacKenzie and G.E. Totten (Taylor and Francis, London, 2006), pp. 339–354

E.J. Mittemeijer, Fatigue of case-hardened steels; role of residual macro- and microstresses. J. Heat Treat. **3**, 114–119 (1983)

E.J. Mittemeijer, Nitriding response of chromium-alloyed steels. J. Metals **37**, 16–20 (1985)

E.J. Mittemeijer, M.A.J. Somers (eds.), *Thermochemical Surface Engineering of Steels* (Woodhead Publishing, Elsevier, Cambridge, 2015)

S.N. Naik, S.M. Walley, The Hall-Petch and inverse Hall-Petch relations and the hardness of nanocrystalline metals. J. Mater. Sci. **55**, 2661–2681 (2020)

W.C. Oliver, G.M. Pharr, An improved technique for determining hardness and elastic modulus using load and displacement sensing indentation experiments. J. Mater. Res. **7**, 1564–1583 (1992)

P. Paris, F. Erdogan, A critical analysis of crack propagation laws. Trans. ASME J. Basic Eng. D **85**, 528–534 (1963)

N.J. Petch, The cleavage strength of polycrystals. J. Iron Steel Inst. (Lond.) **173**, 25–28 (1953)

F.J. Ribeiro, P. Tangney, S.G. Louie, M.L. Cohen, Hypothetical hard structures of carbon with cubic symmetry. Phys. Rev. B **74**, 172101 (2006)

R.O. Ritchie, The conflicts between strength and toughness. Nat. Mater. **10**, 817–822 (2011)

J.G. Sevillano, Geometrically necessary twins and their associated size effects. Scripta Mater. **59**, 135–138 (2008)

L.L. Shaw, A.L. Ortiz, J.C. Villegas, Hall-Petch relationship in a nanotwinned nickel alloy. Scripta Mater. **58**, 951–954 (2008)

M. Sobiech, J. Teufel, U. Welzel, E.J. Mittemeijer, W. Hügel, Stress relaxation mechanisms of Sn and SnPb coatings electrodeposited on Cu: avoidance of whiskering. J. Electron. Mater. **40**, 2300–2313 (2011)

J. Stein, M. Pascher, U. Welzel, W. Huegel, E.J. Mittemeijer, Imposition of defined states of stress on thin films by a wafer-curvature method; validation and application to aging Sn films. Thin Solid Films **568**, 52–57 (2014)

B.B. Straumal, G.A. Lopez, E.J. Mittemeijer, W. Gust, A.P. Zhilyaev, Grain boundary phase transitions in the Al-Mg system and their influence on high-strain rate superplasticity. Defect Diffus. Forum **216–217**, 307–312 (2003)

A.P. Voskamp, E.J. Mittemeijer, The effect of the changing microstructure on the fatigue behaviour during cyclic rolling contact loading. Zeitschrift für Metallkunde **88**, 310–320 (1997)

P.J. Whithers, Residual stress and its role in failure. Rep. Prog. Phys. **70**, 2211–2264 (2007)

Index

A

Abbe limit, 307
Abnormal grain growth, 598, 614
Absolute maximum shearing stress, 661
Acceptor levels, 105
Activation energy, 417, 470, 576
Activation energy of creep, 704
Activation energy of diffusion, 418, 547
Activation energy of growth, 547, 571
Activation energy of nucleation, 543, 571
Activation energy of recrystallization, 597
Activation volume, 444, 450
Actuators, 525
Additivity rule, 540, 557
A-Fe, 419, 627, 646, 713
A"-Fe$_{16}$N$_2$, 527
Ag-Au, 374
Ag-Cu-Ni, 391
Age hardening, 478, 688
Aging, 478, 658, 719
Ag-Pd, 441, 443
A-iron, 416
Airy disc, 308, 309, 312
Al, 156, 422, 627
Al$_2$O$_3$, 627, 691
Al-Cu, 478, 481, 688
Al-Cu-Mg, 481
Alkali halide crystals, 631
Alkali-metal halides, 58
Allotropes, 172
Allotropy, 166, 172, 367
Alloy, 174
Al-Mg, 481, 688
AlN, 193, 474, 688, 716
α"-Fe16N2, 658
Al-Si, 363, 376
Aluminium, 261, 582, 605, 676

Aluminium oxide, 672
Al-Zn, 481
Amorphous cement theory, 262, 268
Amorphous solids, 122
Amorphous structure, 9
Analytical electron microscopy (AEM), 331, 333, 339
Analyzer, 306
Anelasticity, 630, 643
Angular magnification, 287, 293
Anisotropic growth, 553
Annealing twins, 270
Antibonding band, 66
Antibonding orbital, 65, 85
Anti-cuboctahedron, 155
Antiphase boundaries, 271
Antistructure atom, 234
Aperiodic crystals, 197, 216, 225
Aperture stop, 296
Arrhenius analysis, 426
Arrhenius equation, 418
Arrhenius plot, 418
Arrhenius-type equation for rate constants, 540
Athermal nucleation, 477
Atom fraction, 371
Atomistic approach, 401, 405, 621
Attractive force, 47
Au-Cu, 180
Aufbau Prinzip, 73
Au-Ni, 374, 376
Austenite, 172, 185
Austenite-ferrite transformation, 499
Austenite-martensite orientation relationship, 514
Autocatalytic nucleation, 504
Azimuth effect, 299

Printed in the United States
by Baker & Taylor Publisher Services